The Biochemistry of the Nucleic Acids

ELEVENTH EDITION

Roger L. P. Adams
Department of Biochemistry
University of Glasgow

John T. Knowler
Department of Biological Sciences
Glasgow Polytechnic

David P. Leader
Department of Biochemistry
University of Glasgow

CHAPMAN & HALL

London · Weinheim · New York · Tokyo · Melbourne · Madras

Published by Chapman & Hall, 2-6 Boundary Row, London SE1 8HN, UK

Chapman & Hall, 2-6 Boundary Row, London SE1 8HN, UK

Chapman & Hall GmbH, Pappelallee 3, 69469 Weinheim, Germany

Chapman & Hall USA, 115 Fifth Avenue, New York, NY 10003, USA

Chapman & Hall Japan, ITP-Japan, Kyowa Building, 3F, 2-2-1 Hirakawacho, Chiyoda-ku, Tokyo 102, Japan

Chapman & Hall Australia, 102 Dodds Street, South Melbourne, Victoria 3205, Australia

Chapman & Hall India, R. Seshadri, 32 Second Main Road, CIT East, Madras 600 035, India

First edition 1950 by Methuen and Co. Ltd
Eleventh edition 1992 by Chapman & Hall Ltd
Reprinted 1993, 1996

© 1992 Roger L. P. Adams, John T. Knowler and David P. Leader

Typeset in 10½/12½pt Times by Best-set Typesetter Ltd, Hong Kong
Printed in England by Clays Ltd, St Ives Plc, Bungay, Suffolk

ISBN   0 412 39940 7(PB)   0 412 46030 0(HB)

A Catalogue record for this book is available from the British Library

Library of Congress Cataloging-in-Publication Data available

♾Printed on permanent acid-free text paper, manufactured in accordance with ANSI/NISO Z39.48-1992 and ANSI/NISO Z39.48-1984 (Permanence of Paper).

The Biochemistry of the
Nucleic Acids

Contents

7 Repair, recombination and rearrangement of DNA — 257

Preface

When the first edition of this book was published in 1950, it predated the publication of the double-helical structure of DNA by three years. It is not, therefore, surprizing that nothing of the original book remains in the current edition. Indeed, such is the pace of change in the field of nucleic acids that less than 50% of material incorporated into the 1986 edition has been retained.

The book aims at the advanced undergraduate and at graduates that are undertaking course work or requiring an in-depth background for their research. It also aims to provide the established scientist with a single text that permits updating across the whole field from DNA structure, replication and repair, through gene expression and its control to protein synthesis. Every chapter is accompanied by thorough referencing that enables the reader to evaluate personally the data and methodology that cannot be included in the text. In an attempt to keep this list within bounds, references are limited to about ten per page and, to accommodate the more recent literature, many of the older references have been left out in this latest edition.

The first seven editions emerged from the pen of the late J. N. Davidson and subsequent editions have been produced by his colleagues who have tried to retain something of the character and structure of the earlier editions. This has become increasingly difficult with the introduction of new concepts and the arrangement of the book has changed drastically. One of the important changes introduced with the tenth edition, and retained in this edition, was an appendix describing the rationale behind some of the methods which now form an essential part of the study of nucleic acids. Cross-referencing to this appendix has allowed a better narrative flow in the earlier chapters while enabling the interested student to understand how some of the major advances have been made.

With each new edition the book has grown in size despite every effort to prune away older, less relevant material. In a field in which new developments are occurring so rapidly it is inevitable that new knowledge will accumulate more quickly than it can be integrated into the text, but we have endeavoured to incorporate into this edition material published up to the date of completion of the manuscript in March 1991. This, of course, can lead to difficulties as many of the ideas presented have not proved themselves with time. Nonetheless, we feel it is better to describe the uncertainties present at the forefront of the subject than to hold back new, exciting, yet untempered findings.

It is a pleasure to express our thanks to those who have allowed us to reproduce figures and diagrams, and especially those who have provided original photographs. We would also like to thank the University of Glasgow for the provision of facilities that have made the production of this edition possible.

R.L.P.A.
J.T.K.
D.P.L.
July 1991

Abbreviations and nomenclature

The abbreviations employed in this book are those approved by the Commission on Biochemical Nomenclature (CBN) of the International Union of Pure and Applied Chemistry (IUPAC) and the International Union of Biochemistry (IUB).

Nucleosides

A	adenosine
G	guanosine
C	cytidine
U	uridine
Ψ	5-ribosyluracil (pseudouridine)
I	inosine
X	xanthine
rT	ribosylthymine (ribothymidine)
N	unspecified nucleoside
R	unspecified purine nucleoside
Y	unspecified pyrimidine nucleoside
dA	2'-deoxyribosyladenine
dG	2'-deoxyribosylguanine
dC	2'-deoxyribosylcytosine
dT or T	2'-deoxyribosylthymine (thymidine)

Modified bases or nucleosides when in sequence

m^1A	1-methyladenine
m^6A	N^6-methyladenine
m^6_2A	N^6-dimethyladenine
iA	N^6-isopentenyladenine
m^5C	5-methylcytosine

m^{4}C	N^4-methylcytosine
ac^{4}C	N^4-acetylcytosine
azaC	5-azacytosine
Cm	2'-O-methylcytidine
L	lysidine (2-lysylcytosine)
m^{1}G	1-methylguanine
m^{2}G	N^2-methylguanine
m$_2^2$G	N^2-dimethylguanine
Gm	2'-O-methylguanosine
Um	2'-O-methyluridine
s^{4}U	4-thiouracil
mcm^{5}U	5-(methoxycarbonylmethyl)uracil
mcm^{5}s^{2}U	5-(methoxycarbonylmethyl)-2-thiouracil
mnm^{5}s^{2}U	5-(methylaminomethyl)-2-thiouracil
mo^{5}U	5-methoxyuracil
cmo^{5}U	5-(carboxymethoxy)uracil
mcmo^{5}U	5-(methoxycarboxymethoxy)uracil
cmnm^{5}U	5-(carboxylmethylaminomethyl)uracil
D	5,6-dihydrouracil
m^{1}I	1-methylinosine
Q	queuosine
yW	wybutosine

Nucleotides

AMP	adenosine 5'-monophosphate
GMP	guanosine 5'-monophosphate
CMP	cytidine 5'-monophosphate
UMP	uridine 5'-monophosphate
dAMP	2'-deoxyribosyladenine 5'-monophosphate
dGMP	2'-deoxyribosylguanine 5'-monophosphate
dCMP	2'-deoxyribosylcytosine 5'-monophosphate
dTMP	2'-deoxyribosylthymine 5'-monophosphate
2'-AMP, 3'-AMP, 5'-AMP etc.	2'-, 3'- and 5'-phosphates of adenosine etc.
ADP etc.	5'-(pyro) diphosphates of adenosine etc.
ATP etc.	5'-(pyro) triphosphates of adenosine etc.
ddTTP etc.	2', 3'-dideoxyribosylthymine 5'-triphosphate
araCTP	1-β-D-arabinofuranosylcytosine 5'-triphosphate

Polynucleotides

DNA	deoxyribonucleic acid
cDNA	complementary DNA

mtDNA	mitochondrial DNA
RNA	ribonucleic acid
mRNA	messenger RNA
rRNA	ribosomal RNA
tRNA	transfer RNA
nRNA	nuclear RNA
hnRNA	heterogeneous nuclear RNA
snRNA	small nuclear RNA
Alanine tRNA or tRNA^{Ala} etc.	transfer RNA that normally accepts alanine
Ala-tRNA^{Ala} or Ala-tRNA etc.	transfer RNA that normally accepts alanine with alanine residue covalently linked
poly(N), or $(N)_n$ or $(rN)_n$	polymer of ribonucleotide N
poly(dN) or $(dN)_n$	polymer of deoxyribonucleotide N
poly(N-N'), or $r(N-N')_n$ or $(rN-rN')_n$	copolymer of $-N-N'-N-N'-$ in regular, alternating, *known* sequence
poly(A) · poly(B) or $(A)_n · (B)_n$	two chains, generally or completely associated
poly(A), poly(B) or $(A)_n$, $(B)_n$	two chains, association unspecified or unknown
poly(A) + poly(B) or $(A)_n +$ $(B)_n$	two chains, generally or completely unassociated

Miscellaneous

RNase, DNase	ribonuclease, deoxyribonuclease
P_i, PP_i	inorganic orthophosphate and pyrophosphate
nt	nucleotide
bp	base pair
kb	kilobase or kilobase pair
mt	mitochondrial
M_r	relative molecular mass (formerly 'molecular weight')

Amino acids

Ala or A	alanine
Arg or R	arginine
Asn or N	asparagine
Asp or D	aspartic acid
Cys or C	cysteine
Gln or Q	glutamine
Glu or E	glutamic acid

Gly or G	glycine
His or H	histidine
Ile or I	isoleucine
Leu or L	leucine
Lys or K	lysine
Met or M	methionine
fMet	*N*-formylmethionine
Phe or F	phenylalanine
Pro or P	proline
Ser or S	serine
Thr or T	threonine
Trp or W	tryptophan
Tyr or Y	tyrosine
Val or V	valine

Enzyme nomenclature

In naming enzymes, the recommendations of the Nomenclature Committee of the International Union of Biochemistry (1984) are followed as far as possible. The numbers recommended by the Commission are inserted in the text after the name of each enzyme, e.g. EC 2.7.7.7 for DNA polymerase I from *Escherichia coli*.

1

Introduction

The fundamental investigations which led to the discovery of the nucleic acids were made by Friedrich Miescher [1] (1844–95), who may be regarded as the founder of our knowledge of the chemistry of the cell nucleus. In work carried out in the laboratory of Hoppe-Seyler in Tübingen in 1868, he isolated the nuclei from pus cells obtained from discarded surgical bandages, and showed the presence in them of an unusual phosphorus-containing compound that he called 'nuclein' and which we now know to have been nucleoprotein. It was subsequently shown that nucleic acids were normal constituents of all cells and tissues, and Miescher's investigations of the nucleic acids were continued by Altmann, who in 1889 described a method for the preparation of protein-free nucleic acids from animal tissues and from yeast [2].

On hydrolysis, the nucleic acid from thymus glands was found to yield the purine bases adenine and guanine, the pyrimidine bases cytosine and thymine, a deoxypentose and phosphoric acid. The nucleic acid from yeast on the other hand yielded on hydrolysis adenine, guanine, cytosine, uracil, a pentose sugar and phosphoric acid. Yeast nucleic acid, therefore, differed from thymus nucleic acid in containing uracil in place of thymine and a pentose in place of a deoxypentose. This led to the impression that deoxypentose nucleic acid was characteristic of animal tissues, and pentose nucleic acid was characteristic of plant tissues. It was not long before the validity of this concept was questioned but final proof that ribonucleic acid is a general constituent of animal, plant and bacterial cells was not forthcoming until the early 1940s as a consequence of the ultraviolet spectrophotometric studies of Caspersson [3], the histochemical observations of Brachet [4] and the chemical analysis of Davidson [5, 6].

It took a surprisingly long time also to establish the nature of the sugars in deoxypentose and pentose nucleic acids, but they now form the basis of the names deoxyribonucleic acid (DNA) and ribonucleic acid (RNA). The elucidation of the detailed structure of nucleosides and nucleotides can largely be attributed to Todd and his collaborators (reviewed in [7]), who established the nature of the glycosidic linkage between the sugar residues and the purine or pyrimidine bases and the nature of the phosphate ester bonds. Their work taken together with the studies by Cohn and his colleagues [8] provided final confirmation of the nature of the 3′–5′-internucleotide linkage in both DNA and RNA, and made it possible for clear concepts as to the primary structure of the two types of nucleic acids to be put forward.

These advances established the biology of the nucleic acids on a new foundation. The use of new techniques in cytochemistry and cell fractionation showed that DNA and RNA are normal constituents of all cells, whether plant or animal, DNA being confined mainly to the nucleus while RNA is found also in the cytoplasm [5, 6, 9, 10].

The development of techniques of subcellular fractionation and for the isolation of nuclei made possible chemical measurements of the distributions of DNA and RNA amongst the subcellular fractions of various cell types, and led ultimately to the recognition of RNA in the nuclear, ribosomal and soluble fractions of cells and to the demonstration of the constancy in the average amount of DNA per nucleus in the somatic cells of any given species [11].

The presence of the bases in approximately equimolar proportions led to the development of the tetranucleotide hypothesis for both DNA and RNA, in which both nucleic acids were considered to be polymeric structures containing equivalent amounts of mononucleotides derived from each of the four purine and pyrimidine bases linked together in repeating units. It was only when methods for the quantitative analysis of the nucleic acids had been developed that this hypothesis was finally abandoned as a consequence of the demonstration that the various nucleotides did not necessarily occur in equimolar proportions [12].

In the early 1950s Chargaff [13] drew attention to certain regularities in the composition of DNA, namely that the sum of the purines was equal to the sum of the pyrimidines, that the sum of the amino bases (adenine and cytosine) was equal to the sum of the keto bases (guanine and thymine) and that adenine and thymine,

and guanine and cytosine, were present in equivalent amounts. These observations were to be of crucial importance in the subsequent interpretation of X-ray crystallographic analyses which were performed by Astbury [14], Pauling and Corey [15], Wilkins and colleagues [16] and Franklin and Gosling [17]. The two sets of data were brilliantly combined by Watson and Crick [18] in their now famous double-helical structure made up of specifically hydrogen-bonded base pairs which suggested 'a possible copying mechanism for the genetic material'.

For a while after the elucidation of the structure of the double helix, it was thought that there was one sort of RNA, ribosomal RNA (rRNA), which carried the genetic message from the nucleus to the site of protein synthesis in the cell cytoplasm. By the early 1960s, however, it was realized that rRNA was too stable and too constant in base composition to fulfil this function, and in 1961 Jacob and Monod [19] proposed a short-lived messenger molecule (mRNA). Support for the concept was not long in coming. It had been shown that following infection of bacteria with bacteriophage there was synthesis of an RNA complementary to the DNA, and it was further demonstrated that this RNA associated with bacterial ribosomes and became the template for bacteriophage protein synthesis [20]. Thus was established the concept of the flow of information symbolized as:

$$\text{DNA} \xrightarrow{\text{transcription}} \text{RNA} \xrightarrow{\text{translation}} \text{protein}$$

The third major species of RNA, transfer RNA (tRNA), was first proposed by Crick [21] as an adapter molecule required for the insertion of amino acids against an RNA template. It was also this species which formed the subject of many of the breakthroughs in the analysis of RNA structure, beginning with the elucidation of the primary sequence of a yeast tRNA for alanine [22], and followed nine years later by the elucidation of the three-dimensional structure of the yeast tRNA for phenylalanine [23, 24].

An enzyme catalysing the synthesis of DNA-like polymers from deoxyribonucleoside 5'-triphosphates was first identified in *Escherichia coli* by Kornberg and his collaborators [25]. This enzyme was named DNA polymerase and is now recognized as one of a family of enzymes concerned in the replication and repair of DNA molecules [26].

RNA polymerases, catalysing the synthesis of polyribonucleotides from ribonucleoside 5'-triphosphates, were identified almost simultaneously by several groups around 1960 (reviewed in [27]), so that within the space of a very few years understanding of the whole field of nucleic acid biosynthesis and function underwent a complete transformation.

The first phase of nucleic acid chemistry and biochemistry provided a broad structural and mechanistic understanding of DNA replication and gene expression. In order to obtain a detailed understanding of genetic organization and the regulation of gene expression structural analysis of individual genes was required. The revolution that realized what at one time had seemed an impossible dream arose from the forging together of the enzymology of the restriction endonucleases [28] and the genetics of bacteria and bacteriophages. The cloning of recombinant DNA enabled the isolation of useful quantities of individual genes, the nucleotide sequences of which became accessible through the chemical and enzymic techniques developed by Maxam and Gilbert [29] and Sanger and coworkers [30]. The introduction and automation of phosphoramidite chemistry

for the synthesis of defined oligonucleotides [31] has also made an important contribution in this area, the horizons of which have been further extended by the introduction of the polymerase chain reaction [32].

The effects of this revolution have been felt in all areas of the biological sciences, not only that of nucleic acids, but in the latter area its impact on our understanding of the eukaryotic genome has been especially striking. Examples of this are the discovery of introns, the unravelling of the nucleic acid rearrangements in the field of immunology, and the progress already made to understanding the basic biological problems of differentiation and development.

Not all the excitement in the field of nucleic acids in the last decade has been generated by DNA. The discovery of catalytic RNA [33], at first regarded as something of a curiosity, heralded a radical reassessment of our thinking about the origin of life. It is now widely believed that contemporary 'ribozymes' are molecular fossils of an 'RNA world' that preceded the appearance of DNA and protein. This journey with RNA into the past promises to be just as breathtaking as the headlong rush with DNA into the future.

REFERENCES

1 Miescher, F. (1879) *Die histochemischen und physiologischen Arbeiten*, Leipzig.

2 Altmann, R. (1889) *Arch. Anat. Physiol.*, 524.

3 Caspersson, T. (1950) *Cell Growth and Cell Function*, Norton, New York.

4 Brachet, J. (1950) *Chemical Embryology*, Interscience, New York.

5 Davidson, J. N. and Waymouth, C. (1944) *Biochem. J.*, **38**, 39.

6 Davidson, J. N. and Waymouth, C. (1944–5) *Nut. Abs. Rev.*, **14**, 1.

7 Brown, D. M. and Todd, A. R. (1955) in *The Nucleic Acids* (eds E. Chargaff and J. N.

Davidson), Academic Press, New York, vol. 1, p. 409.

8 Cohn, W. E. (1956) in *Currents in Biochemical Research* (ed. D. E. Green), Interscience, New York, p. 460.

9 Feulgen, R. and Rossenbeck, H. (1924) *Hoppe-Seyler's Zeitschr.*, **135**, 203.

10 Behrens, M. (1938) *Hoppe-Seyler's Zeitschr.*, **253**, 185.

11 Vendrely, R. (1955) *The Nucleic Acids* (eds E. Chargaff and J. N. Davidson), Academic Press, New York, vol. 2, p. 155.

12 Chargaff, E. (1950) *Experientia*, **6**, 201.

13 Chargaff, E. (1955) in *The Nucleic Acids* (eds E. Chargaff and J. N. Davidson), Academic Press, New York, vol. 1, p. 307.

14 Astbury, W. T. (1947) *Symp. Soc. Exp. Biol.*, **1**, 66.

15 Pauling, L. and Corey, R. B. (1953) *Proc. Natl. Acad. Sci. USA*, **39**, 84.

16 Wilkins, M. F. H., Stokes, A. R. and Wilson, H. R. (1953) *Nature (London)*, **171**, 738.

17 Franklin, R. E. and Gosling, R. G. (1953) *Nature (London)*, **171**, 740; **172**, 156.

18 Watson, J. D. and Crick, F. H. C. (1953) *Nature (London)*, **171**, 737.

19 Jacob, F. and Monod, J. (1961) *J. Mol. Biol.*, **3**, 318.

20 Brenner, S., Jacob, F. and Meselson, M. (1961) *Nature (London)*, **190**, 576.

21 Crick, F. (1957) *Biochem. Soc. Symp.*, **14**, 25.

22 Holley, R. W., Apgar, J., Everett, G. A., Madison, J. T., Marguisee, M., *et al.* (1965) *Science*, **147**, 1462.

23 Robertus. J. D., Ladner, J. E., Finch. J. T., Rhodes, D., Brown. R. *et al.* (1974) *Nature (London)*, **250**, 546.

24 Quigley, G. J., Wang, A. H. J., Seeman, N. C., Suddath, F. L., Rich, A. *et al.* (1975) *Proc. Natl. Acad. Sci. USA*, **72**, 4866.

25 Lehmann, I. R., Bessman, M. J., Simms, E. S. and Kornberg, A. (1958) *J. Biol. Chem.*, **233**, 163.

26 Kornberg, A. (1991) *DNA Replication* (2nd edn), Freeman, San Francisco.

27 Smellie, R. M. S. (1963) in *Progress in Nucleic Acid Research* (eds J. N. Davidson and W. E. Cohn), Academic Press, New York, vol. 1, p. 27.

28 Kelly, T. J. and Smith, H. O. (1970) *J. Mol. Biol.*, **51**, 393.

29 Maxam, A. M. and Gilbert, W. (1977) *Proc. Natl. Acad. Sci. USA*, **74**, 560.

30 Sanger, F., Air, G. M., Barrell, B. G., Brown. N. L., Coulson, A. R. *et al.* (1977) *Nature (London)*, **265**, 687.

31 Matteucci, M. D. and Caruthers, M. H. (1981) *J. Am. Chem. Soc.*, **103**, 3185.

32 Saiki, R. K., Scharf, S. J., Faloona, F., Mullis, K. B., Horn, G. T. *et al.* (1985) *Science*, **239**, 1350.

33 Cech, T. R., Zaug, A. J. and Grabowski, P. J. (1981) *Cell*, **27**, 487.

2

The structure of the nucleic acids

2.1 MONOMERIC COMPONENTS

Before any account is given of the structure of the nucleic acids proper it is necessary to describe the structures of their component parts. Nucleic acids are high-molecular-weight polymeric compounds which on complete hydrolysis yield pyrimidine and purine bases, a sugar component and phosphoric acid. Partial hydrolysis yields compounds known as nucleosides and nucleotides. Each of these component parts will be discussed in turn.

2.1.1 Pyrimidine bases

The pyrimidine bases (Fig. 2.1) are derivatives of the parent compound pyrimidine, and the major bases found in the nucleic acids are cytosine found in both RNA and DNA, uracil found in RNA and thymine found in DNA. In certain of the bacterial viruses cytosine is replaced by 5-methylcytosine or 5-hydroxymethylcytosine which may be glucosylated to varying degrees. The pyrimidine bases can undergo keto–enol tautomerism as shown for uracil in Fig. 2.2.

2.1.2 Purine bases

Both types of nucleic acids contain the purine bases, adenine and guanine. It should be noted that the style of numbering of the pyrimidine ring in the purines differs from that used for pyrimidines themselves. Adenine and guanine have the structures shown in Fig. 2.3. As in the pyrimidines, the purine bases can exist in two tautomeric forms.

Other naturally occurring purine derivatives include hypoxanthine, xanthine and uric acid (Fig. 2.4).

Certain 'minor bases' are also found in small amounts in most nucleic acids [1–4]. DNA contains 5-methylcytosine, N^4-methylcytosine or N^6-methyladenine and RNA contains a wide variety of methylated bases, including thymine [3]. These unusual bases comprise less than 5% of the total base content of the RNA and vary in relative amounts between different RNA species. Some of the minor bases in RNA are listed in Table 2.1.

The phenylalanine transfer RNA from yeast contains a most unusual base known as wybutosine, the structure of which is shown in Fig. 2.5 [5–7].

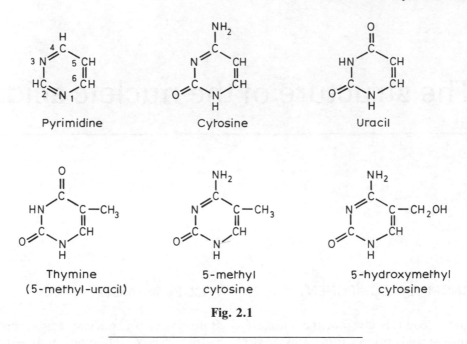

Pyrimidine   Cytosine   Uracil

Thymine   5-methyl   5-hydroxymethyl
(5-methyl-uracil)   cytosine   cytosine

Fig. 2.1

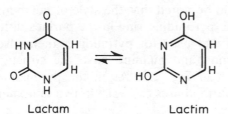

Lactam   Lactim

Fig. 2.2

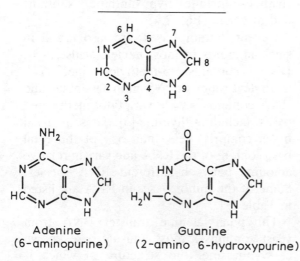

Adenine   Guanine
(6-aminopurine)   (2-amino 6-hydroxypurine)

Fig. 2.3

2.1.3 Pentose and deoxypentose sugars

The sugar component of RNA is D-ribose which in polynucleotides occurs in the furanose form. In DNA this sugar is replaced by 2-deoxyribose also in the D-furanose form (Fig. 2.6). This apparently small difference between the two types of nucleic acid has wide-ranging effects on both their chemistry and structure since the presence of the bulky hydroxyl group on the 2-position of the sugar not only limits the range of possible secondary structures available to the RNA molecule but also makes it more susceptible to chemical and enzymic degradation.

Some RNAs, notably ribosomal RNAs, contain very small amounts of 2-O-methyl ribose.

When the pentose sugars occur in nucleic acids or nucleotides the carbon atoms are numbered as 1′, 2′, 3′ etc. to avoid confusion with the numbering of the ring atoms of the bases.

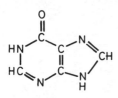

Hypoxanthine
(6-hydroxypurine)

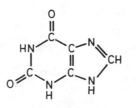

Xanthine
(2,6-dihydroxypurine)

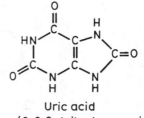

Uric acid
(2,6,8-trihydroxypurine)

Fig. 2.4

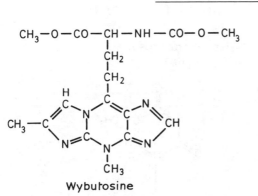

Wybutosine

Fig. 2.5

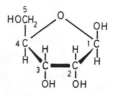

β-D-ribofuranose

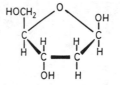

β-D-2-deoxyribofuranose

Fig. 2.6

Table 2.1 Some of the more important minor bases in RNA

1-methyladenine	dihydrouracil
2-methyladenine	5-hydroxyuracil
6-methyladenine	5-carboxymethyluracil
6,6-dimethyladenine	5-methyluracil (thymine)
6-isopentenyladenine	5-hydroxymethyluracil
2-methylthio-6-isopentenyladenine	2-thiouracil
6-hydroxymethylbutenyladenine	3-methyluracil
6-hydroxymethylbutenyl-2-methylthioadenine	5-methylamino-2-thiouracil
1-methylguanine	5-methyl-2-thiouracil
2-methylguanine	5-uracil-5-hydroxyacetic acid
2,2-dimethylguanine	3-methylcytosine
7-methylguanine	4-methylcytosine
2,2,7-trimethylguanine	5-methylcytosine
hypoxanthine	5-hydroxymethylcytosine
1-methylhypoxanthine	2-thiocytosine
xanthine	4-acetylcytosine
6-aminoacyladenine	
7-(4,5-*cis*-dihydroxyl-1-cyclopenten-3-ylaminomethyl)-7-deazaguanosine (Q)	

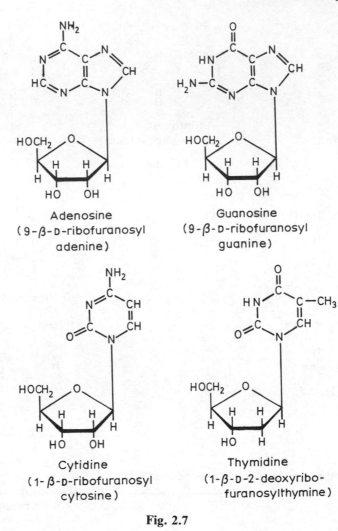

Fig. 2.7

Table 2.2   Minor nucleosides in RNA [8]

1-ribosylthymine	2'-*O*-methyluridine
5-ribosyluracil (pseudouridine)	2'-*O*-methylcytidine
2-ribosylguanine	2'-*O*-methylpseudouridine
2'-*O*-methyladenosine	2'-*O*-methyl-4-methylcytidine
2'-*O*-methylguanosine	

2.1.4 Nucleosides

When a purine or a pyrimidine base is linked to ribose or deoxyribose the resulting compound is known as a nucleoside. Thus, adenine condenses with ribose to form the nucleoside adenosine, guanine forms guanosine, cytosine forms cytidine and uracil forms uridine (Fig. 2.7). The ribonucleoside from hypoxanthine is named inosine. The nucleosides derived from 2-deoxyribose are known as deoxyribonucleosides – deoxyadenosine, deoxyguanosine, deoxycytidine, deoxythymidine and so on.

In addition to the nucleosides listed above several others are found in very small amounts in certain classes of nucleic acids. These are listed in Table 2.2. The nucleoside 5-ribosyluracil has been obtained in small amounts from the digestion products of RNA, particularly tRNA. and has been named pseudouridine (Fig. 2.8).

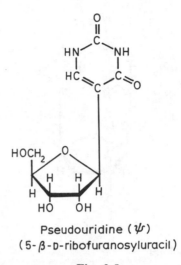

Pseudouridine (*ψ*)
(5-*β*-D-ribofuranosyluracil)

Fig. 2.8

2.1.5 Nucleotides

The structures of nucleotides are dealt with in more specialized terms by Hutchinson [9] and Michelson [10]. They are all phosphoric acid esters of the nucleosides (Fig. 2.9). Those derived from ribonucleosides are usually referred to as ribonucleotides and those from deoxyribonucleosides as deoxyribonucleotides. These terms are sometimes abbreviated to riboside, ribotide, deoxyriboside and deoxyribotide.

Since the ribonucleosides have three free hydroxyl groups on the sugar ring, three possible ribonucleoside monophosphates can be formed. Adenosine, for example, can give rise to three monophosphates (adenylic acids), adenosine 5'-phosphate, adenosine 3'-phosphate and adenosine 2'-phosphate. In the same way guanosine, cytidine and uridine can give rise to three guanosine monophosphates (guanylic acids), three cytidine monophosphates (cytidylic acids) and three uridine monophosphates (uridylic acids), respectively. These are frequently referred to by the abbreviations given at the beginning of the book.

The ribonucleoside 5'-phosphates may be further phosphorylated at position 5' to yield 5'-di- and -tri-phosphates. Thus adenosine 5'-phosphate (AMP) yields adenosine

5'-diphosphate (ADP) and adenosine 5'-triphosphate (ATP) (Fig. 2.10). Adenosine 5'- and guanosine 5'-tetraphosphate have also been described. Similarly the other ribonucleoside 5'-phosphates yield such di- and tri-phosphates as GDP, CDP, UDP, GTP, CTP and UTP. The 5'-monophosphates of adenosine, guanosine, cytidine and uridine together with the corresponding di- and tri-phosphates all occur in the free state in the cell as do the deoxyribonucleoside 5'-phosphates which are referred to as dAMP, dADP, dATP, dTMP, dTDP, dTTP etc. Ribonucleoside 3',5'-bisphosphates and 2',3'-cyclic monophosphates can be formed on hydrolysis of RNA molecules, and ribonucleoside 3',5'-cyclic monophosphates of adenine and guanine occur in many tissues where they play multiple roles in the regulation of metabolic pathways (Fig. 2.11) (reviewed in [11]).

2.2 THE PRIMARY STRUCTURE OF THE NUCLEIC ACIDS

Nucleic acids are polymers made up of hundreds, thousands or millions of nucleo-

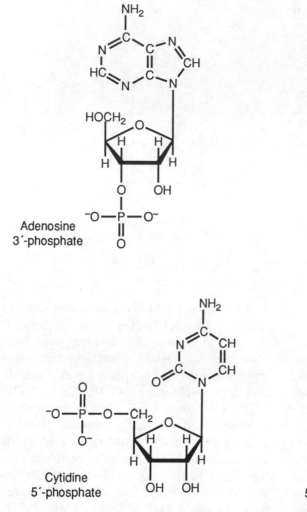

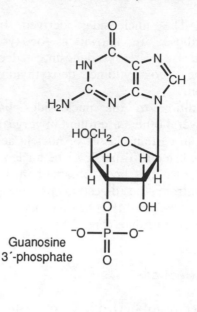

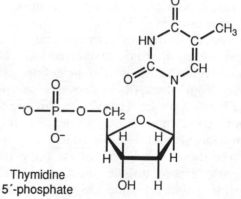

Fig. 2.9

tides coupled together by phosphodiester linkages. In the case of both DNA and RNA the internucleotide linkage involves the 3′- and the 5′-hydroxyls of the sugar, i.e. it is a 3′–5′-phosphodiester bond that joins adjacent nucleotides.

2.2.1 Shorthand notation

The representation of polynucleotide chains by complete formulae is clumsy and it has become customary to use the schematic system illustrated in Fig. 2.12, where the chain, shown in full on the left, is abbreviated as on the right. The vertical line denotes the carbon chain of the sugar with the base attached at the top to C-1′. The diagonal line from the middle of the vertical line indicates the phosphate link at C-3′ whereas that at the bottom of the vertical line denotes the phosphate link at C-5′. This system may be used for either RNA or DNA.

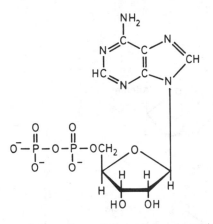

Adenosine diphosphate (ADP)

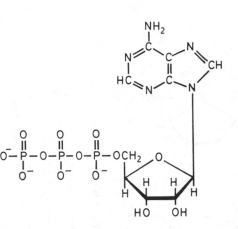

Adenosine triphosphate (ATP)

Fig. 2.10

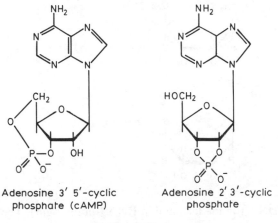

Adenosine 3' 5'-cyclic
phosphate (cAMP)

Adenosine 2' 3'-cyclic
phosphate

Fig. 2.11

An even shorter notation, originally suggested by Heppel *et al.* [12] and now embodied in the rules of the CBN (see Abbreviatons and nomenclature at the beginning of the book) denotes a phosphate group by p and the nucleoside by a capital letter (i.e. A, C, G, T, or U). When the p is placed to the right of the nucleoside symbol (e.g. Cp) the phosphate is attached to the

C-3' of the ribose moiety; when placed to the left of the nucleoside symbol (e.g. pC) the phosphate is attached to the C-5' of the ribose moiety. Thus, UpUp (or U–Up) is a dinucleotide with a phosphodiester bond between the C-3' of the first uridine and the C-5' of a second uridine which also has a second phosphate group attached to its C-3' group. UpU or U–U would be the dinucleoside monophosphate, uridylyl(3'–5')uridine. The letter p *between* nucleoside residues may be replaced by a hyphen, and is often omitted altogether to save space when long nucleotide sequences are being reported. The examples in Fig. 2.13 illustrate the method.

The prefix d (e.g. dA) may be used to indicate a deoxyribonucleoside, but is usually omitted in the case of (deoxy) thymidine. Thymine ribonucleoside is represented as rT to make the distinction.

Cyclic-terminal nucleotides may be represented by using 'cyclic-p' to indicate a 2':3'-phosphoryl group or by means of the symbol >p. Thus, U-cyclic-p or U>p is uridine 2':3'-phosphate and UpU-cyclic-p or UpU>p is the cyclic-terminal dinucleotide.

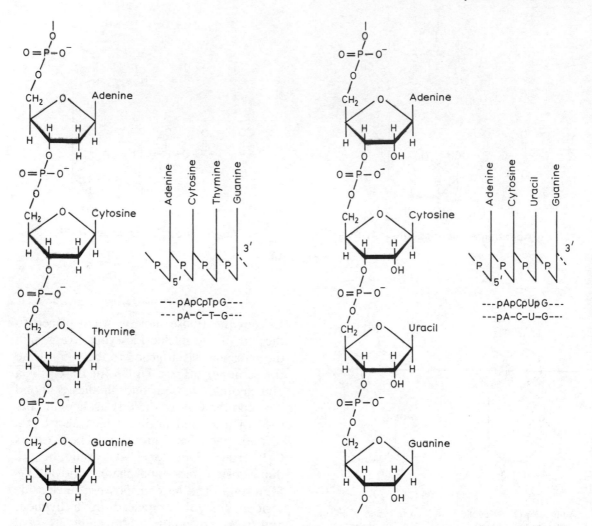

Fig. 2.12   A section of the polynucleotide chain in DNA (on the left) and in RNA (on the right). The shorthand notations are shown alongside.

2.2.2 Base composition analysis of DNA

The common monomeric units of DNA are the four deoxyribonucleotides containing the bases adenine, cytosine, guanine and thymine. Many DNAs, however, contain small amounts of other bases, e.g. 5-methylcytosine, and in a few bacteriophages one of the common pyrimidine bases is completely replaced by a different pyrimidine base, e.g. in the DNA of phages T2, T4 and T6 cytosine is completely replaced with 5-hydroxymethylcytosine, and in PBS1 (a phage which infects *Bacillus subtilis*) uracil replaces thymine.

Methods for determining the molar proportions of bases by hydrolysis and chromatography are discussed in detail by Bendich [13]. More recent methods involving high-performance liquid chro-

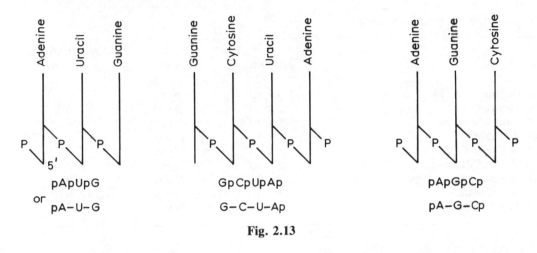

Fig. 2.13

Table 2.3 Molar proportions of bases (as moles of base per 100 moles of phosphate) in DNAs from various sources (data from various authors)

Source of DNA	Adenine	Guanine	Cytosine	Thymine	5-Methylcytosine
Bovine thymus	28.2	21.5	21.2	27.8	1.3
Bovine spleen	27.9	22.7	20.8	27.3	1.3
Bovine sperm	28.7	22.2	20.7	27.2	1.3
Rat bone marrow	28.6	21.4	20.4	28.4	1.1
Herring testes	27.9	19.5	21.5	28.2	2.8
Paracentrotus lividus	32.8	17.7	17.3	32.1	1.1
Wheatgerm	27.3	22.7	16.8	27.1	6.0
Yeast	31.3	18.7	17.1	32.9	–
E. coli	26.0	24.9	25.2	23.9	–
M. tuberculosis	15.1	34.9	35.4	14.6	–
ϕX174	24.3	24.5	18.2	32.3	–

matography are now available [14]. The results of the analysis of a number of DNAs, as shown in Table 2.3, reveal wide variations in the molar proportions of bases in DNAs from different species although the DNAs from the different organs and tissues of any one species are essentially the same. Extensive tables showing the molar proportions of bases have been published [15].

It was Chargaff [16] who first drew attention to certain regularities in the composition of DNA. The sum of the purines is equal to the sum of the pyrimidines; the sum of the amino bases (adenine and cytosine) is equal to the sum of the keto (oxo) bases (guanine and thymine); adenine and thymine are present in equimolar amounts, and guanine and cytosine are also found in equimolar amounts. This equivalence of A and T and of G and C is of the utmost importance in relation to the formation of the DNA helix and is referred to as Chargaff's rule. There are two major deviations from the rule in Table 2.3. (a)

Table 2.4 DNA molecular weights

Source	Mol. wt.	Length	Number of kilobase pairs (kb)
Bacteriophage ϕX174	1.6×10^6	0.6 μm	–
Polyomavirus	3×10^6	1.5 μm	4.5
Mouse mitochondria	9.5×10^6	4.9 μm	14
Bacteriophage λ	33×10^6	17 μm	50
Bacteriophage T2 or T4	1.3×10^6	67 μm	200
Mycoplasma PPLO strain H-39	4×10^8	200 μm	600
H. influenzae chromosome	8×10^8	400 μm	1 200
E. coli chromosome	3×10^9	1.5 mm	4 500
Drosophila melanogaster chromosome	43×10^9	20 mm	70 000

In wheat germ DNA guanine and cytosine are not present in equimolar amounts but this is explained by the scarcity of cytosine being compensated for by the presence of 5-methylcytosine. (b) In ϕX174 DNA and in the DNA of several similar small coliphages the adenine is not equimolar with thymine nor is guanine with cytosine. This is because ϕX174 DNA is single-stranded (see below).

2.2.3 Molecular weight of DNA

The molecular weights of DNA molecules are very difficult to determine accurately by the methods of classical chemistry since they range from 10^6 to more than 10^{10} and the larger molecules are very difficult to obtain intact. The molecular weights of DNAs from a variety of sources are shown in Table 2.4. In more recent years it has become customary to describe the size of DNA molecules in terms of kilobase pairs (kb) rather than daltons or molecular weight. A molecular weight of one million corresponds approximately to 1.5 kilobase pairs. Details of methods used to determine the molecular weight of DNA are given in the Appendix (section A.1.3).

2.3 THE SECONDARY STRUCTURE OF DNA

2.3.1 The basic structures

X-ray diffraction was used extensively in the study of the molecular architecture of DNA by Astbury [17] and later by Franklin and Gosling [18] and by Wilkins and his colleagues [19–21]. Using early information obtained by this technique and chemical observations, Watson and Crick [22–24] in 1953 put forward the view that the DNA molecule is *double stranded* and in the form of a right-handed helix with the two polynucleotide chains wound round the same axis and held together by hydrogen bonds between the bases.

By making scale models they were able to show that the bases could fit in if they were arranged in pairs with a pyrimidine on one strand paired with a purine on the other. The equivalence of adenine and thymine, and guanine and cytosine in most naturally occurring DNA molecules, first observed by Chargaff [16], suggested that the most likely hydrogen-bonding configuration of base pairs be of these two types (Fig. 2.14). These base-pairing arrangements have been confirmed as the only ones possible in the

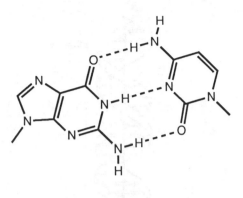

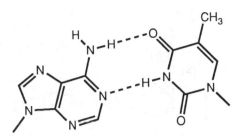

G : C base pair

A : T base pair

Fig. 2.14 The normal base-pairing arrangement found in DNA. (The dashed lines indicate hydrogen-bonds.)

B-form of DNA (see below). Other base-pairing arrangements have been suggested [25] and appear to occur in RNA where they are involved in secondary and tertiary structure stabilization (section 2.6). They are also found in triple-stranded structures of DNA (section 2.5.3) and in some DNA–drug complexes [26].

The pairs of bases are flat and may be stacked one above the other like a pile of plates so that the molecule is readily represented as a spiral staircase with the base pairs forming the treads (Fig. 2.15). The two polynucleotide chains are of opposite polarity in the sense that the terminal nucleotide of one strand has a free 5′-end whereas the complementary strand has a free 3′-end at that same terminus. This means that the two strands are running in opposite directions.

The shorthand method of describing two DNA strands has been to have the strand of 5′ → 3′ polarity on the top line of the sequence with the complementary strand of opposite polarity lying below. For example:

$$5'-AGGTC-3'$$
$$3'-TCCAG-5'$$

It is clear that an important consequence of the base-pairing configuration found in DNA is that the order in which the bases occur in one chain automatically determines the order in which they occur in the other, complementary chain. Apart from this essential condition, there are no restrictions on the sequence of the bases along the chains.

The two helices are right handed (i.e. (a) turn in a clockwise direction as viewed from the near end or (b) turn in the direction of the fingers of the right hand when the thumb indicates the line of sight) and cannot be separated without unwinding. The pitch of the duplex is 3.4 nm and since there are 10 base pairs in each turn of the helix, there is a distance of 0.34 nm between each base pair. The diameter of the helix is 2 nm and examination of the structure illustrated in Fig. 2.15(d) shows there to be two grooves running along the surface of the double helix: these are the major and minor grooves.

The base–sugar linkage is defined by the torsion angle, χ, and is normally *anti* with the C8 of the purine or the C6 of the pyrimidine ring over the sugar (Fig. 2.16). The sugar ring is puckered, with either the

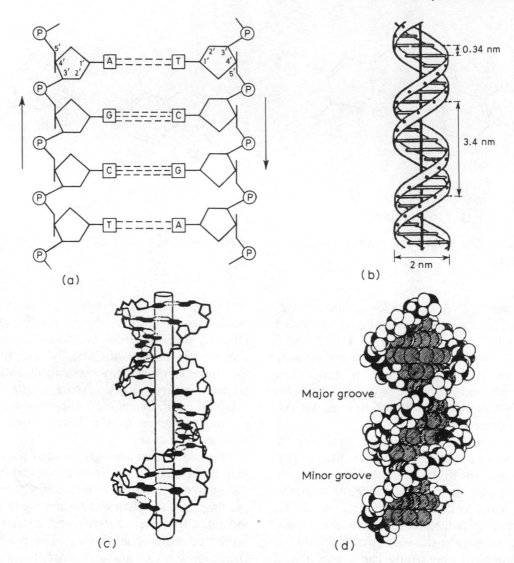

Fig. 2.15 Various diagrammatic ways of representing DNA: (a) showing polarity and base-pairing but no helical twist; (b) showing helical twist and helix parameters but not base pairs; (c) showing helix and base pairs; (d) space-filling representation showing major and minor grooves.

C2′ or the C3′ being displaced from the planar conformation in the direction of C5′ (i.e. the pucker is *endo*). Usually the sugar is in the C2′ *endo* configuration (but see below). The other torsion angles (α, β, γ, ε and ζ) refer to the sugar phosphate chain (Fig. 2.16). These angles are not rigidly fixed, but are free to change slightly to allow the most stable structure to form.

There is a lot of water associated with DNA. Up to 72 molecules of water per 12 base pairs are found largely in the minor groove where they form a spine. The extent of this spine of hydration may be dependent

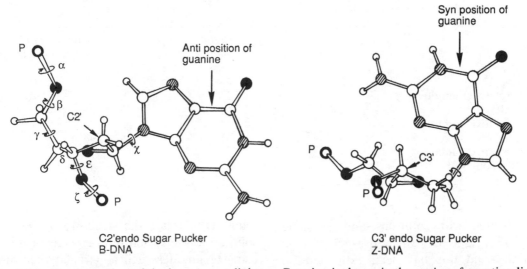

Fig. 2.16 Conformation of the base–sugar linkage. Guanine is shown in the *anti* conformation linked to C2′ *endo* deoxyribose as in B-DNA and in the *syn* conformation linked to C3′ *endo* deoxyribose as in Z-DNA. Reproduced, with permission, from Rich *et al.* [79] – copyright Annual Reviews Inc.

on the base sequence of the DNA [27, 28]. The sugar phosphate backbone points out to give hydrophilic ridges with one negative charge per phosphate.

The stacked bases form a hydrophobic core but the amino and keto groups point into the grooves and allow interaction with the solvent or with protein molecules. In addition the N3 and N7 of the purine rings are potential interaction sites in the minor and major grooves respectively. In the major groove there are two hydrogen bond acceptor sites and a donor site for both an A:T and a G:C pair. In the minor groove both the A:T and G:C pairs have two acceptor sites, whereas the G:C pair has an extra donor site [29].

Although the basic model put forward by Watson and Crick remains close to the accepted structure of the DNA molecule in solution, the more refined X-ray diffraction studies of Wilkins and his colleagues demonstrated that, depending on the conditions chosen to produce the DNA fibres, they can have a variety of possible structures. The major forms studied are called A-, B- and C-form DNA (Table 2.5) with the B-form corresponding closely to the original Watson–Crick model.

The A and C structures are also right-handed double helices but differ in the pitch, the number of bases per turn and the sizes of the major and minor grooves. In both A and C conformations the bases are not flat but are tilted. Because of the presence of the extra 2′-hydroxyl group, RNA appears to be unable to adopt the B conformation, and the A conformation is believed to be very close to the structure adopted by double-stranded RNA and by DNA–RNA hybrids [30]. The sugar in A-form molecules has the C3′ *endo* conformation and this is the consistent difference from B-form DNA in which the sugar has the C2′ *endo* conformation. Various organic solvents [31] and proteins [32] can force DNA into the A form, and the transition from the B conformation to the C conformation appears to occur in

Table 2.5   The different forms of DNA

	Pitch (nm)	Residues per turn	Inclination of base pair from horizontal
A (Na salt) 75% relative humidity	2.8	11	20°
B (Na salt) 92% relative humidity	3.4	10	0°
C (Li salt) 66% relative humidity	3.1	9.3	6°
DNA/RNA hybrid	2.8	11	20°

concentrated salt solutions and ethylene glycol [33]. The B-form is present at high humidity and this has been explained on the basis that it requires two molecules of water to bridge the gap (0.66 nm) between adjacent phosphates in B-DNA. When the relative humidity is reduced, a DNA conformer (the A-form) results that requires only one water molecule to bridge the gap between adjacent phosphates which are now only 0.53 nm apart [34].

X-ray diffraction studies on fibres provide information which relates to the general organization of the sugar phosphate chains but gives no detail at atomic resolution. More recently, as a result of the production of crystals of short oligonucleotides, X-ray crystallography has confirmed the right-handed double-helical structure. In contrast to the fibre diffraction patterns, analysis of crystals of defined DNA fragments can allow the unequivocal positioning of each atom in the helical framework and is a wholly definitive technique. Large oligonucleotides are not readily crystallized so the data currently available relate to small molecules which may well have an atypical structure because the end effects, which do not normally make a substantial contribution to polynucleotide conformation, may play a dominant role.

There have been many experimental attempts to relate the structure of DNA in solution to the structure in crystals or fibres, using techniques such as circular dichroism [33] and low-angle X-ray scattering [35, 36]. That the A-, B- (and Z-) forms of DNA can exist in solution has been confirmed by using NMR (particularly 2D-NOESY) techniques [29, 37, 38]. The fact that DNA is a right-handed double helix in solution has been confirmed by Iwamoto and Hsu [39] in an electron microscopy study of small cyclic molecules, hydrogen-bonded over a 39 base-pair region only.

One of the best experimental systems for the analysis of secondary structure in solution uses circular superhelical DNA (section 2.5.1) and experimental and theoretical analysis of these molecules suggests that the solution structure is close to, but not entirely compatible with, the B structure of DNA. Thus, the molecules are slightly unwound to give 10.4 rather than an integral 10 base pairs per turn [40]. A similar conclusion has been reached by an analysis of the frequency of cleavage, by nucleases or hydroxyl radicals, of DNA adsorbed onto a mica surface [41, 42]. This means that, in solution, the angle between adjacent base pairs (i.e. the helical twist angle) is 34°, rather than 36°.

Because the Watson–Crick structure places the negatively charged phosphates on the outside of the double helix, it was assumed that repulsion between these groups would cause DNA to act as a rigid rod, and studies with the free DNA in solution confirmed this inflexibility for large molecules. However, studies with shorter oligonucleotides [43, 44] indicate that the DNA need not necessarily assume such a straight, rod-like shape, but may be bent in a manner dictated by the nucleotide sequence (see below). This assumes greater significance when considering the interaction of DNA with counter ions and proteins which neutralize the negative charges on the phosphates. It is clear that, in the cell, the DNA can be quite dramatically bent and this is particularly apparent when one considers the packaging of the DNA [45, 46] (chapter 3).

2.3.2 Variations on the B-form of DNA

The structure considered so far is an average structure for DNA of undefined base sequence. With the availability of crystals of oligonucleotides of defined base sequence, X-ray crystallography has enabled the precise structure of DNA of a specified sequence to be determined, and the conclusions have been confirmed by NMR studies in some cases. A particular base pair can vary its orientation, relative to its neighbour, by rotation about its x, y or z axis to give varying degrees of tilt, roll, or twist [47]. In addition it can move relative to adjacent bases by sliding towards or away from the helix axis (Fig. 2.17). (There are a variety of other, less important, movements possible.) When two paired bases roll in opposite directions the base pair assumes a propeller twist. The previously defined A-, B- and C-forms of DNA represent structures with a particular set of values for these continuously variable parameters. For

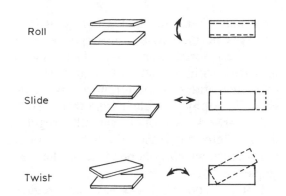

Fig. 2.17 Ways in which adjacent bases may move relative to one another (seen from two different positions).

instance, the bases may slide relative to each other with values of slide (σ) ranging from -0.1 nm to $+0.2$ nm. The A-form DNA has an average value of $\sigma = 0.15$ nm; the B-form has $\sigma = 0$ and in the C-form $\sigma = -0.1$ nm. The value of σ depends on the actual bases involved and so a stretch of DNA may have regions resembling A-, B- and C-form DNA with respect to any particular parameter.

In a similar manner it has become apparent that the helical twist angle is highly dependent on the particular sequence and varies from 27.7° (for ApG) to 40.0° (for GpC). The alteration of the helical twist angle results, in part, from a rolling of adjacent bases over one another along their long axes [48–51]. As the bases of both strands roll to make them more vertical to their helix axis this produces a positive propeller twist. This is favoured because it leads to increased stacking of adjacent bases on the same strand but it also causes interference between the edges of adjacent purines on opposite strands. Calladine [52] proposed various strategies which would relieve this steric clash, the most important of which appears to be the alteration of the helical twist angle. Calladine's rules have been quantified [53] and enable a prediction to be made of the helical twist on going from one specific base to the next specific base, and these predictions have been shown to agree with the structures of particular oligonucleotides as determined by X-ray crystallography. It is clear from these results that B-DNA is not a smooth, regular, double helix but rather the helix shows an irregularity which is dependent on the base sequence.

Not only do the bases have an irregular arrangement, but this imposes a particular deformation to the sugar phosphate backbone which affects the interaction with water molecules and which might be recognized by proteins interacting with DNA.

When the sequence of bases in DNA shows a regular pattern this may have quite a dramatic effect on the structure of the DNA. When there is a region of DNA containing alternating purines and pyrimidines this can lead to a 'wrinkled' form of DNA [54–56] or in more extreme cases to a complete breakdown of the B-form of DNA and conversion to Z-DNA (see below). In wrinkled DNA, although the structure of the purine–pyrimidine dinucleotide (e.g. GpC) is normal, in the next dinucleotide (CpG) the orientation of the linking phosphate is twisted. This has the effect of moving the bases slightly so that the minor groove is made deeper. The wrinkled structure may also affect DNA–protein interactions and may explain why the *lac* repressor (section 10.1.1) binds one thousand times more strongly to poly[d(A-T)].poly[d(A-T)] than to DNA of general sequence. In the normal recognition site (the *lac* operator) three-quarters of the bases are arranged as alternating purines and pyrimidines and they may take up a wrinkled configuration.

If, rather than alternating purines and pyrimidines, DNA has a purine–pyrimidine dinucleotide followed five bases later by a pyrimidine–purine dinucleotide (i.e. R Y n n n Y R n n n R Y n n n Y R) then, rather than being wrinkled, this would lead to the DNA being curved [57, 58]. (Curved DNA is DNA with a non-linear axis arising as a result of intrinsic factors, as opposed to bent DNA where departure from linearity results from the action of external factors. A kink in DNA is a disruption of base-stacking associated with a local structure distortion [59].) Whenever a particular base sequence is repeated in phase with the DNA helical repeat (i.e. every 10–11 bp) this will lead to a summation of small effects which will result in curved DNA. Such intrinsic curvature might be reinforced by the binding of protein resulting in DNA which is strongly bent and this occurs, for example, when the DNA bends around the nucleosome core (chapter 3) [60] or interacts with proteins at the origin of replication (chapter 6) or in promoter regions (chapter 10) [61–63].

Poly(dA).poly(dT) assumes an unusual B-type DNA structure and it has been proposed that the two strands assume different structures. The poly d(A) chain has the C3′ *endo* conformation, typical of A-DNA, whereas the poly d(T) chain has the C2′ *endo* conformation of B-DNA. This requires the bases to be highly tilted. Such DNA has been called heteronomous DNA [64–66]. Although this results in reduced base stacking interactions, it appears to be compensated by increased hydration in the minor groove. When short runs of adenines occur in one strand of a DNA duplex, that DNA adopts an unusual structure which is manifest by its abnormal electrophoretic mobility and its ability to form extremely small circles [67]. This DNA is highly curved and this may result from the in-phase repeating of AAA tracts [68–70], or it may result from a slow change starting at one end of the A-tract leading to the formation of the unusual structure of poly(dA).poly(dT). At the junction where this unusual structure changes back to more normal B-form DNA, there will be a sudden distortion [71–75]. This question has been reviewed recently [76].

2.3.3 Z-DNA

Particularly under conditions of high salt concentration, DNA with an alternating purine–pyrimidine sequence, e.g. poly(dA-dT).poly(dA-dT) tends to form a left-handed double helix known as Z-DNA. This has 12 base pairs per turn in solution. As with wrinkled DNA the repeating unit is the dinucleotide but with the alternating dG.dC polymer the purine sugar linkage is *syn* (Fig. 2.16) placing the N7 and C8 atoms in a very exposed position outside the helix. Because of this, normal base stacking cannot occur but the pyrimidines can stack with pyrimidines on the opposite strands [77–79]. The pucker of the sugar rings is changed so that, although the sugar attached to the pyrimidine is now C2′ *endo*, that attached to the purine is C3′ *endo*. There is one deep groove between the sugar phosphate backbones corresponding to the minor groove in B-DNA. Conversion of B-DNA to Z-DNA brings about a major change in for example the circular dichroism spectrum of DNA (Fig. 2.18).

As well as forming at high salt concentrations, Z-DNA will form in 1 mM MgCl$_2$ if the C5 of cytosine is substituted with methyl, bromo or iodo groups, e.g. in poly(dG-dm^{5}C).poly(dG-dm^{5}C). It can also form in supercoiled DNA molecules where the region of left-handed DNA serves to relax the tension of the supercoiled molecule (section 2.5.1) [81, 82]. Formation of Z-DNA may also allow the accommodation of a G:T mismatch in DNA [83].

Initial evidence for the existence of

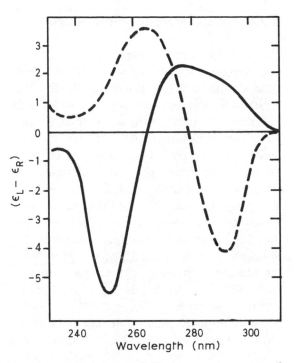

Fig. 2.18 Circular dichroism spectra of poly(dG-dC) in conditions of low salt (0.2 M NaCl—) and high salt (3.5 M NaCl---). (Reproduced, with permission, from [80]; copyright Academic Press Inc.)

Z-DNA in cells came from the use of specific antibodies to Z-DNA. It is not certain, however, to what extent Z-DNA forms during the required fixation processes which involve removal of basic proteins and which may induce supercoiling and hence stabilize regions of Z-DNA [78, 84, 85]. More recently, it has been shown that potential Z-DNA forming sequences interfere with the methylation and restriction of adjacent sequences and this is taken as showing that Z-DNA exists *in vivo*. Moreover, oligonucleotides can be formed which contain short stretches of B-DNA alternating with Z-DNA [86], and similar structures are presumed to be involved in the action of RecA protein during recombination [87] (section 7.4.2).

2.3.4 The dynamic structure of DNA

An additional and separate phenomenon is the dynamic secondary structure of DNA. The double helix is not a totally fixed or rigid molecule but undergoes considerable internal deformation in a continuous manner. This aspect of DNA secondary structure is best shown by ^{3}H-exchange experiments in which the imino protons of thymine and guanine can exchange with ^{3}H from the surrounding water [88]. In the light of recent NMR results, some of the earlier work has been reinterpreted. Exchange requires that the base pair is opened and the probability of finding a particular base pair open is only 10^{-5}; but each base pair will breathe about 100 times a second [89, 90, 170]. Results indicate that small segments of the double helix can swing apart (breathe) and protrude into the external medium in a manner closely dependent on the environment of the DNA molecule and supercoiled molecules show an enhanced tendency to form single-stranded regions [170].

The opening of a base pair is determined by the stability of the base pair and its nearest neighbour and sequences which breathe most readily can be calculated [91–93]. 5′-purine–pyrimidine dinucleotide pairs are much more stable than 5′-pyrimidine–purine dinucleotide pairs as a result of the increased overlap (stacking) of the nucleotide rings. Because of the increased number of hydrogen bonds holding together a G:C nucleotide pair, regions of DNA rich in G + C are more stable than regions rich in A + T. It is clear that, on denaturation (see below) the A + T-rich regions melt first and this is important *in vivo* where A + T-rich regions are involved in the initiation of replication and transcription, processes that require a separation of the two strands of the double helix.

2.4 DENATURATION AND RENATURATION

2.4.1 DNA denaturation: the helix–coil transition

When double-stranded DNA molecules are subjected to extremes of temperature or pH, the hydrogen bonds of the double helix are ruptured and the two strands are no longer held together. The DNA is said to denature and changes from a double helix to a random coil. When heat is used as the denaturant, the DNA is said to melt and the temperature at which the strands separate is the melting or transition temperature, T_m.

The component bases of a polynucleotide absorb light at 260 nm, but in double-stranded DNA this absorption is partially suppressed. This is because the stacking of the bases, one above the other, leads to coupling between the transition of the neighbouring chromophores, i.e. the π-electrons of the bases interact with one another.

When duplex DNA melts the hydrogen bonds break and the bases unstack with the consequence that the absorption at 260 nm rises by about 20–30%. This is the *hyperchromic effect* and is used to monitor the melting of DNA (Fig. 2.19).

As well as a change in absorbance of ultraviolet light the helix coil transition is also accompanied by a change in density of the DNA, the single-stranded molecule being more dense than the corresponding duplex.

The nature of the melting transition is affected by several factors.

1. *The (G + C) content of the DNA*. There are three hydrogen bonds involved in the G:C base pair and only two in the A:T base pair (Fig. 2.14). Because of this the

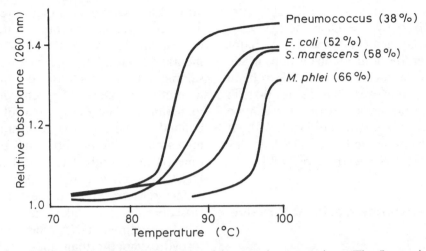

Fig. 2.19 Denaturation by heat of DNAs isolated from various organisms. The figures in parentheses indicate the composition of the DNA in G + C (%) (Reproduced from *Molecular Genetics* by G. S. Stent, W. H. Freeman and Co. Copyright 1971 – after [94]).

higher the G + C content of DNA the more stable will the molecule be and the higher will be the melting temperature. Fig. 2.19 shows the melting of DNAs of different G + C content. The equation:

$$\%GC = (T_m - X)\,2.44$$

[95] expresses the relationship where X is dependent on:

2. *The nature of the solvent*. In low concentrations of counter ion, denaturation occurs at relatively low temperatures and over quite a broad range of temperature. At higher concentrations of counter ion the T_m is raised and the transition is sharp.

3. *The nature of the DNA*. Most DNA molecules are mosaics of regions of varying G + C composition [96, 97] and the A + T-rich regions will melt before the G + C-rich regions. This can result

in the two strands being held together and hence in register by G + C-rich regions. On cooling these double-stranded regions will act as foci to allow rapid reannealing of the two strands of DNA. Only when all the hydrogen bonds are broken does the melting become irreversible on rapid cooling (see below). Short viral DNAs which may be more homogeneous exhibit sharper melting profiles.

On electrophoresis in gradients of increasing alkalinity, AT-rich domains are denatured earlier than GC-rich domains and this leads to a greatly reduced rate of migration through the gel. The patterns produced are characteristic of a particular region and are markedly altered by single base changes and even by methylation of a single base within a sequence [98, 99]. Methylation of cytosine increases the stability of a region of DNA and methylation of adenine has the reverse effect. Adenine methylation in the sequence GATC (i.e. that recognized by the Dam methylase of *E. coli* – section 4.6.1) produces a major clash between the four methyl groups in the major groove and causes the DNA to bend [168]. It is, however, surprising that in some circumstances an extra base can be introduced into one strand of DNA without seriously disrupting the B-form DNA duplex as judged by NMR spectroscopy [169].

2.4.2 The renaturation of DNA: C_0t value analysis

When two DNA strands are returned from the extreme conditions which caused them to melt they may reassociate to re-form a double helix. In order to do so correctly they must align themselves perfectly and this is a process dependent on both the concentration of the DNA molecules and the time allowed for reassociation. Very often imperfect matches may be formed which must again dissociate to allow the strands to align correctly. For the dissociation to happen and to encourage the diffusion of the very large molecules involved, the temperature must be maintained just below the melting temperature for reannealing to occur. If the solution is quenched to 4°C this limits diffusion and prevents the separation of any DNA strands which have become mismatched. Consequently, except for simple DNA molecules which correctly reanneal almost instantaneously, quenching produces solutions of denatured DNA.

If renaturation is allowed to occur under ideal conditions, two DNA samples of identical concentration will take different times to reanneal depending on their complexity. Thus the DNA from a mammal is far more complex and has more, different sequences than does the DNA from a small virus. Each sequence is therefore present in a much lower concentration and will take correspondingly longer to find its complementary strand. The complexity of a DNA sample is thus reflected in the time it takes a solution of DNA of given concentration to reanneal. The $C_0t_{1/2}$ value of a DNA is defined as the initial concentration (C_0) in moles nucleotide per litre multiplied by the time (t) in seconds it takes for 50% of the DNA to reanneal (Fig. 2.20).

The reassociation can be followed spectroscopically or by taking advantage of the fact that a duplex DNA binds more strongly to hydroxyapatite than does single-stranded DNA. This latter method allows the reannealing reactions to be performed over a much broader range of DNA concentration and also allows the isolation of preparations of duplex and single-stranded DNA molecules from a partially reannealed sample. A third method of studying reassociation of DNA molecules makes use of S1 nuclease (chapter 4). This enzyme preferentially digests the single-stranded DNA leaving behind the duplex molecules only. For both the hydroxyapatite and the S1 nuclease method it is convenient to have the DNA radioactively labelled thus allowing rapid and accurate quantification of the proportions of DNA in the reannealed and single-stranded forms.

Further results of renaturation (C_0t) analysis will be considered in chapter 3.

2.4.3 The buoyant density of DNA

The physical properties of DNA are strongly influenced by the percentage of (G + C) in the molecule, and the buoyant density of DNA in concentrated CsCl solutions is no exception. (G + C)-rich DNA has a higher buoyant density than (A + T)-rich DNA [101] and there is a linear relationship

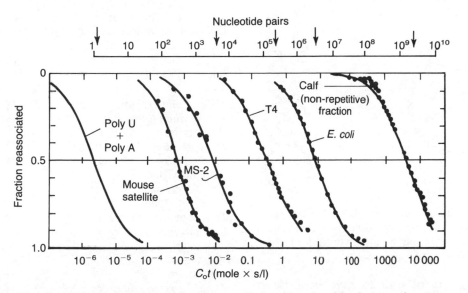

Fig. 2.20 The rates of reassociation of double-stranded polynucleotides from various sources showing how the rate decreases with the complexity of the organism and its genome (from [100]).

between the buoyant densities (ρ) of different DNAs and their (G + C) contents (Fig. 2.21). This can be expressed by the relationship:

$$\rho = 1.660 + 0.098 \text{ (GC)}$$

where GC is the mole fraction of (G + C). The relative (G + C) content can also be determined from the thermal denaturation temperature discussed above [95] and from the ultraviolet spectrum of the DNA [167].

Several dyes and antibiotics have been shown to have strong binding specificity for either A:T or G:C base pairs. Thus distamycin and netropsin are AT-specific and actinomycin is GC-specific [29, 103, 104]. This property can be used to improve the separation of DNA molecules which differ in their GC content in CsCl gradients, since dye binding alters the buoyant density of the molecules. In addition, dyes such as the AT-specific malachite green, or the

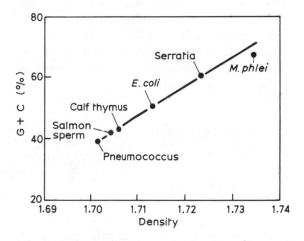

Fig. 2.21 Relationship between bouyant density and content of G + C in DNAs from various sources [102].

GC-specific phenyl neutral red, can be immobilized on polyacrylamide columns and used to fractionate DNA molecules of differing base composition [105].

On the basis of such measurements the relative (G + C) contents of the DNAs from

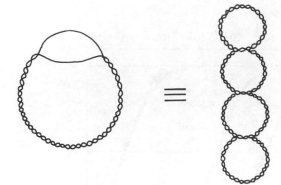

Fig. 2.22 Supercoiling of DNA. The super-coiled molecule on the right is a low-energy equivalent of the partly underwound molecule.

Table 2.6 The relative (G + C) content of DNAs from various sources [109–111]

Source of DNA	Percentage (G + C)
Neocallimastrix spp.	18
Plasmodium falciparum (malarial parasite)	19
Dictyostelium (slime mould)	22
M. pyogenes	34
Vaccinia virus	36
Bacillus cereus	37
B. megaterium	38
Haemophilus influenzae	39
Saccharomyces cerevisiae	39
Calf thymus	40
Rat liver	40
Bull sperm	41
Diplococcus pneumoniae	42
Wheat germ	43
Chicken liver	43
Mouse spleen	44
Salmon sperm	44
B. subtilis	44
T1 phage	46
E. coli	51
T7 phage	51
T3 phage	53
Neurospora crassa	54
Pseudomonas aeruginosa	68
Sarcina lutea	72
Micrococcus luteus	72
Herpes simplex virus	72
Mycobacterium phlei	73

a wide variety of sources have been determined and are shown in Table 2.6. Although mammalian DNAs show a (G + C) content between 40 and 45%, the range of bacterial DNAs is much wider (30–75%). The significance of these variations in base content has been discussed in relation to the taxonomy of bacteria [106] and protozoa [107] and to the evolution of various organisms [108].

2.5 MORE COMPLEX DNA STRUCTURES

2.5.1 Supercoiling

Imagine a covalently closed cyclic DNA molecule (or a linear molecule physically constrained at both ends). Break one of the strands and untwist the double helix by one or two turns and then rejoin the strands. The resulting molecule will try to rewind the two strands back to their normal structure (the most stable form) but will be unable to do so because of the covalent closure. To compensate the duplex will writhe on itself and take on a superhelical configuration (Fig. 2.22). The number and nature of the supercoiled turns will depend on the difference between the secondary structure of the DNA when it was sealed and the secondary structure under the conditions of observation.

Covalently closed, cyclic duplex DNA differs very considerably from linear or open cyclic molecules (i.e. molecules with an interruption in the phosphodiester backbone) of the same size and base composition. The two strands of the DNA are unable to separate and thus a high temperature is required to disrupt the structure of the supercoiled DNA. When these DNA molecules eventually do melt, the two strands cannot separate, but the entire molecule collapses into a compact, fast-sedimenting complex of the two interlocked random coils. The secondary structure of such DNA can differ appreciably from that of the open circular form because of the tertiary restraints on the molecule [112]. Thus the cyclic duplex DNA molecules of the viruses SV40 and polyoma have been shown to have short regions where the two DNA strands are unwound [113, 114, 170]. The intrinsic viscosity is low, and the sedimentation rate and electrophoretic mobility about 20% faster than in the open circular molecules of the same size. The tertiary structure of the superhelix is most commonly represented by a straight interwound superhelix (Fig. 2.22). Other forms such as toroidal and branched structures have been shown to exist in solution, and their formation may depend on the ability of a particular sequence to bend at the end of a supercoiled segment [115, 116].

The linking number (Lk) is the number of Watson–Crick turns present in a DNA molecule. Thus for a molecule of 5000 base pairs Lk is about 500. If the molecule is partially untwisted say by 25 turns there is a change in linking number (ΔLk) equal to -25. The superhelix density (σ) is defined as $\Delta Lk/Lk$ ($= -0.05$ in the above example). In other words the superhelix density of a covalently closed cyclic DNA is the number of superhelical turns per 10 base pairs. Lk is made up from two components: writhe (W) and twist (T) such that

$$Lk = W + T$$

Writhe is a measure of the coiling of the axis of the DNA molecule and twist reflects the winding of one DNA strand around the other. If a molecule is fully relaxed, the writhe is 0 and Lk = T.

In a covalently closed cyclic DNA duplex the linking number is fixed. It can only be changed by processes involving breaking the phosphodiester chain. In our example we changed Lk from 500 to 475 but, by writhing on itself (W $= -25$) the twist can return to 500.

Superhelix density is determined by 'partial denaturation' of the DNA either with alkali [117] or with an intercalating dye [118]. The binding of the dye molecules to the DNA results in an increase in the number of residues per turn in the duplex and thus a decrease in the number of superhelical turns (W). The sedimentation rate of the DNA therefore drops to a minimum and then rises as the binding of the further dye molecules causes the DNA to take on superhelical turns of the opposite sense. Frequently, superhelix density is estimated by gel electrophoresis in the presence of increasing amounts of an enzyme which abolishes superhelical turns [119]. The superhelix density of most covalently closed DNA molecules is -0.06 in neutral caesium chloride [117, 118]. Superhelix density is affected by both temperature and ionic strength [120]. Electron micrographs of relaxed and supercoiled molecules are shown in Fig. 2.23.

There has been doubt as to whether supercoiled DNA exists in eukaryotic cells or whether all the tension introduced by underwinding is removed by association of the DNA with proteins. Although the latter may be true in general it is probable that certain stretches of DNA in the transcriptionally active fraction are in a torsionally stressed state [121–124, 170] (chapter 10).

2.5.2 Cruciforms

Another discontinuity in DNA may be produced by palindromic sequences which

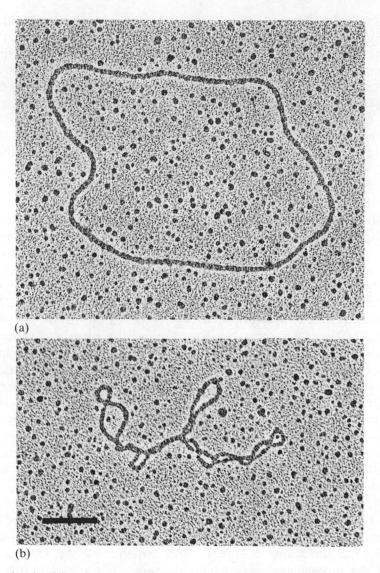

(a)

(b)

Fig. 2.23   Open circular (a) and supercoiled (b) forms of PM2 virus DNA. The bar represents $0.2\,\mu$m. (By courtesy of Dr Lesley Coggins.)

may fold back on themselves to form cruciform structures (Fig. 2.24). Fold-back structures can be formed *in vitro* by *intra-molecular* reassociation of single-stranded DNA but a fold-back structure, even when formed from a perfect palindrome, will never be as stable as the linear duplex DNA as there will always be an unpaired region of DNA in the loop region. The sequence of nucleotides in this loop also has a major effect on the stability of the stem–loop structure [125] and the stem may adopt a B- or a Z-DNA structure [126, 127].

As with Z-DNA, cruciforms may be

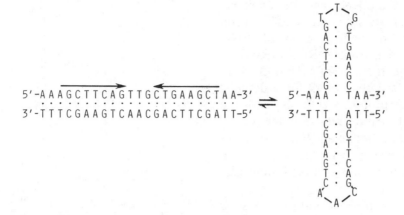

Fig. 2.24 Palindromes and cruciforms. The arrows indicate the palindromic sequence which can fold back on itself to form the cruciform structure.

stabilized by specific binding proteins or by their presence in supercoiled molecules as formation of cruciforms with right handed helical stems is another way in which the molecule can writhe to relieve the supercoil tension. Even in this situation, however, they are unlikely to be present in DNA *in vivo* and form only very slowly *in vitro* [128–130]. Extrusion of a cruciform requires considerable disruption and can occur by one of two mechanisms [131]. Rarely, it involves the denaturation of a long stretch of DNA before its reassociation as a cruciform. For this to occur the flanking DNA must be very AT-rich. More commonly extrusion involves formation of a proto-cruciform which then branch migrates in order to complete the process.

2.5.3 Multi-stranded DNA

Watson–Crick base pairs are not the only ones possible and there is evidence that where regions of DNA contain runs of pyrimidines in one strand these can fold back on themselves to form triple-stranded helices containing one polypurine and two polypyrimidine tracts [132–137]. One of the latter will run parallel with the polypurine tract (i.e. in the same 5′ → 3′ orientation) and form with it Hoogsteen base interactions (Fig. 2.28 and section 2.6).

Tetraplexes could be formed by association of the two arms of a cruciform structure and tetraplex DNA will also form in guanine-rich sequences such as occur in some promoters and also in telomeres (chapter 6). It has been suggested that tetraplexes form at meiosis and are a necessary preliminary structure for genetic recombination [54, 165].

2.6 THE SECONDARY AND TERTIARY STRUCTURE OF RNA

RNA has a variety of functions within the cell and for each function a specific type of RNA is required. The types of RNA differ in chain length and secondary and tertiary structures. Messenger RNA (mRNA) is involved in carrying the genetic message from the DNA to the site of protein synthesis which takes place on small particles known as ribosomes. The ribosomes are made of protein and RNA and the ribosomal RNA (rRNA) itself consists of several different-sized molecules. A small RNA molecule known as transfer RNA (tRNA) is involved in the transfer of amino acids to the ribosome (see chapter 12 for a detailed description of protein synthesis and the structure of the ribosome and tRNA). In addition several other small RNA molecules form parts of enzymes or are involved in enzymic transformation of macromolecules (chapters 6, 9 and 12).

Although RNA molecules do not possess the regular interstrand hydrogen-bonded structure characteristic of DNA, they have the capacity to form double-helical regions. In some RNA molecules in the region of 70% of the bases are involved in secondary structure interactions (chapter 12). Duplexes can be formed between two separate RNA chains, but are more frequently found between two segments of the same chain folded back on itself. The secondary structure is generally similar to the A-form of DNA with tilted bases, since the 2'-OH hinders B structure formation. The helical regions formed in this manner are seldom regular as the segments on the chain brought into opposition do not have entirely complementary sequences. Such molecules frequently show unusual base pairing in addition to the expected A:U and G:C pairs (i.e. G:U pairing is observed) and non-bonded residues may 'loop out' of the structure (Fig. 2.25).

To form a stable duplex requires at least three conventional base pairs and the stability of a duplex region depends on three factors.

1. The various base pairs have differing stability and this is modified by the nature of the adjacent base pairs and the presence of interruptions to the duplex region. The most stable base pairs are

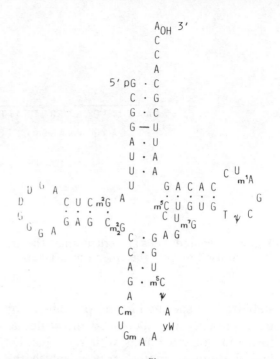

Fig. 2.25 Yeast tRNA^{Phe}. The cloverleaf secondary structure is shown with the hydrogen bonds between standard Watson–Crick base pairs denoted by a dot, and the bond between a G:U base pair denoted by a dash. The identity of the bases can be found by consulting the list of abbreviations at the front of the book.

C:G or G:C base pairs following a G:C base pair. They have a free energy (ΔG) of -20 kJ per mole base pair compared with the very low value of -1 kJ per mole base pair for a U:G base pair following another U:G base pair [138, 139].

2. Along a duplex region there may be occasional unpaired bases forming bulges or short loops. These have a strong destabilizing effect on the duplex region and also induce a kink into the molecule [140].

3. At one end of an intrastrand, base-paired stem there is a hairpin loop (Fig. 2.26). If the nucleotides are in their conventional structure, a loop must have a minimum

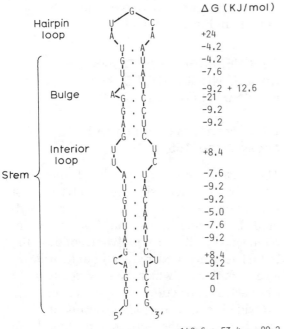

ΔG (KJ/mol)

+24
-4.2
-4.2
-7.6
-9.2 + 12.6
-21
-9.2
-9.2

+8.4

-7.6
-9.2
-9.2
-5.0
-7.6
-9.2

+8.4
-9.2
-21
0

-142.6 + 53.4 = -89.2

Fig. 2.26 A stem–loop structure formed by intrastrand hydrogen bonding of a palindromic region of RNA. The predicted stability of the region is calculated from the data of Tinoco *et al.* [138].

of three unpaired nucleotides. Hairpin loops with six unpaired bases are the most stable but even these reduce the stability of the duplex region by 16–24 kJ per mole oligonucleotide. With shorter loops steric hindrance and base stacking interactions destabilize the loop yet the greater the distance between self-complementary regions the less likely it becomes that duplex regions will be formed. Fig. 2.26 indicates the predicted energy of a hairpin region of palindromic RNA. If the stem–loop region is more stable than $\Delta G = -40$ kJ per mole oligonucleotide there is a possibility that such regions will exist *in vivo* but this is only one of the criteria which must be satisfied before confidence can be placed

in predictions of secondary structure for RNA and methods are continually being refined [141–143, 171].

The sequence UUCG is found frequently in hairpin loops and NMR spectroscopy has shown that only the two central bases are unpaired [144]. This is made possible by formation of a G:U base pair in which the guanine is in the *syn* conformation, and by the central cytosine forming hydrogen bonds with the phosphate joining the two uridines. This and similar 'tetraloops' form particularly stable structures which are not readily dissociated [145].

X-ray crystallographic examination of many small RNA molecules (e.g. tRNA – chapter 12) has shown that extensive folding of the molecule occurs. The structure is stabilized by the formation of base pairs between those bases present in loops and not already involved in secondary structure formation. This frequently aligns short stem regions which can be stabilized by base stacking interactions (Fig. 2.27). The structure is called a pseudoknot as long as the stem regions involved are shorter than a full helical turn [146, 147]. Although the type of hydrogen-bonding involved is usually that of the conventional Watson–Crick base pair, other pairing is possible (e.g. G:U).

Short, triple-stranded regions can occur in RNA in which two of the chains run parallel with one another. Such unusual structures have been studied with model systems where the homopolymer chains poly(A) and poly(U) have been shown to form a triple-stranded structure in which antiparallel poly(A) and poly(U) strands are held together by conventional Watson–Crick base pairs, whereas a second poly(U) strand uses Hoogsteen base pairs to bind in parallel to the poly(A) strand (Fig. 2.28) [133, 148–152, 166].

An unusual structure is found in the

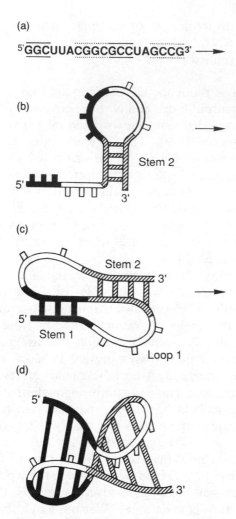

(a)

5'GGCUUACGGCGCCUAGCCG3' ⟶

(b)

Stem 2

5'

3'

(c)

Stem 2

3'

5'

Stem 1

Loop 1

(d)

5'

3'

Fig. 2.27 A pseudoknot in which stems from different regions align to form a continuous duplex molecule. Reproduced from [146] with permission of the authors and publishers.

myxobacterium *Stigmatella aurantiaca* in which a linear, single-stranded DNA molecule is attached by a 2'−5'-phosphodiester bond to a guanosine residue 19 bases from the 5'-end of a 76 nucleotide-long RNA molecule (section 7.7.4(b)). The proposed structure of this molecule involves extensive stem−loop formation and hydrogen bonding between the DNA and RNA [153, 154].

Further details of the secondary and tertiary structures of tRNA and rRNA are to be found in chapter 12.

2.7 CHEMICAL REACTIONS OF BASES, NUCLEOTIDES AND POLYNUCLEOTIDES

2.7.1 Reactions of ribose and deoxyribose [155, 156]

The sugars are readily acylated and alkylated. In the nucleoside the 5'-OH is most susceptible and reaction with triphenylmethyl chloride gives the 5'-*O*-trityl derivative. In the ribopolynucleotides the 2'-OH is acetylated by acetic anhydride to give a molecule which is now resistant to ribonucleases (section 2.7.3 and chapter 4).

Oxidation with periodate occurs at 2',3'-glycols, i.e. at the terminal nucleotide in ribopolynucleotides. The initial reaction gives a dialdehyde which is lost from the polynucleotide by an amine-catalysed cleavage (Fig. 2.29).

Similarly the 2',3'-glycol can react with aldehydes or ketones under acidic conditions or with boric acid to form cross-linked complexes. Depurination reactions (see below) lead to the conversion of the sugars into furfural derivatives which give specific colour reactions with orcinol [157] or diphenylamine [158]. These form the basis of quantitative assays for RNA or DNA respectively.

2.7.2 Reactions of the bases

(a) Nitrous acid reacts with amino groups to convert them into hydroxyls (Fig. 2.30). It thus converts:

cytosine → uracil

adenine → hypoxanthine

guanine → xanthine

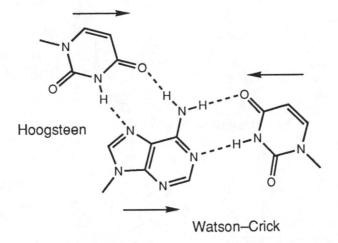

Fig. 2.28 The triple-stranded structure formed by two poly(U) and one poly(A) strand involves a Watson–Crick base paired poly(A).poly(U) with the second poly(U) strand running in the same direction as the poly(A) strand and bonded to it with Hoogsteen hydrogen bonds.

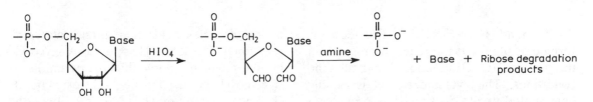

Fig. 2.29 Interaction of the 3′-terminal nucleotide of RNA with periodate.

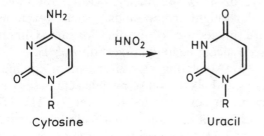

Fig. 2.30 Deamination of cytosine by treatment with nitrous acid.

from treatment have altered base-pairing potential (chapter 7) [159, 160].

(b) Formaldehyde also reacts with free amino groups to produce methylol derivatives but these can then produce cross-links between two bases or between nucleic acid and protein [161].

$$R\text{-}NH_2 \rightarrow R\text{-}NH\text{-}CH_2OH$$

$$\rightarrow R\text{-}NH\text{-}CH_2\text{-}NH\text{-}R'$$

As formaldehyde only reacts with single-stranded regions of DNA it can be used to fix such regions for electron microscopy.

(c) Bisulphite reacts with pyrimidines in single-stranded regions of nucleic acids to form an addition product across the 5–6

It is the free acid which is active and hence reaction must take place at a low pH (about 4.25) and native DNA is not very susceptible to deamination. Nitrous acid is used as a mutagenizing reagent as the bases resulting

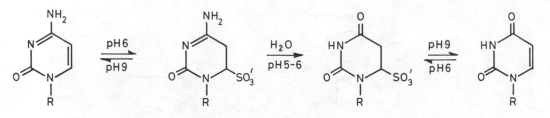

Fig. 2.31 Interaction of cytosine with bisulphite

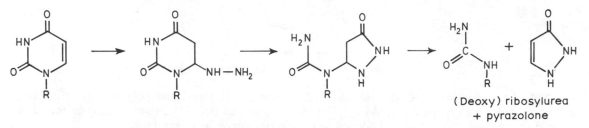

(Deoxy) ribosylurea
+ pyrazolone

Fig. 2.32 The action of hydrazine with pyrimidine bases.

double bond [162] (Fig. 2.31). The adduct is formed at pH 6 but the reaction proceeds in the reverse direction under mild alkaline conditions. The cytosine adduct is rapidly deaminated at pH 5–6 and this provides a mild way of converting cytosine into uracil.

(d) Bromine and iodine will add to pyrimidines to give, for example, 5-bromouracil or 5-bromocytosine and with purines to form the 8-bromo derivative.

(e) Hydrazine produces a nucleophilic addition across the 5–6 double bond of pyrimidine bases (e.g. Fig. 2.32). A slow reaction also occurs with adenine bases. The (deoxy)ribosyl urea breaks down to give apyrimidinic acid with concomitant poly-nucleotide chain cleavage, a reaction which is used in nucleic acid sequencing (section A.5.1).

(f) Hydroxylamine reacts with the amino group of cytosine but it also reacts more slowly with thymine and uracil to bring about pyrimidine ring cleavage, and eventually depyrimidination of nucleic acids.

(g) Strong mineral acids lead to depuri-nation of nucleic acids in an amine-catalysed reaction [155, 158]. The purine-*N*-glycosyl bond is much less stable than the pyrimidine-*N*-glycosyl bond. Purine-*N*-glycosyl bond cleavage occurs much more readily with DNA than with RNA, and concomitant phosphodiester bond cleavage leads to hydrolysis of the nucleic acid to low mole-cular-weight compounds. Such reactions are made use of in the Maxam–Gilbert sequencing technique (section A.5.1).

(h) Alkylating agents are often carcino-gens and alkylation of nucleic acid bases can occur at several sites (chapter 7). The N7 position of guanine is particularly sensitive (Fig. 2.33). The positive charge produced at N7 renders the polynucleotide sensitive to cleavage with piperidine, a reaction which is an important step in the Maxam–Gilbert sequencing technique (section A.5.1).

2.7.3 Phosphodiester bond cleavage

The strong hot acids required to break the

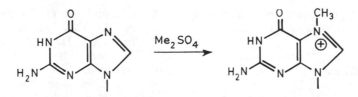

Fig. 2.33 Methylation of guanine with dimethyl sulphate.

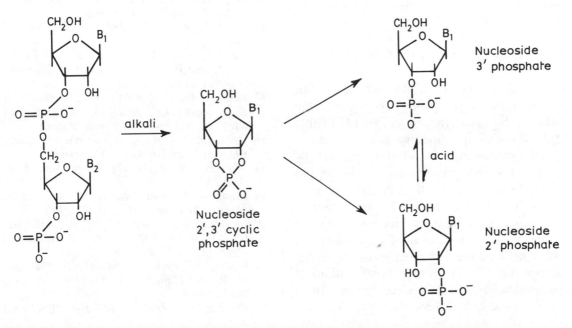

Fig. 2.34 Hydrolysis of a dinucleotide by alkali. The cyclic phosphate is formed as an intermediate and is hydrolysed to give a mixture of 2'- and 3'-phosphates. These can be interconverted (via the cyclic phosphate) under acidic conditions.

phosphodiester backbone of DNA also lead to release of the free bases and breakdown products of deoxyribose. Typical treatments are with 12 M perchloric acid at 100°C or 98% formic acid at 170°C. DNA is not sensitive to mild alkaline hydrolysis.

Alkaline hydrolysis of RNA (0.3 M NaOH at 37°C) yields both the nucleoside 2'- and 3'-phosphates. These isomers are readily interconverted under acidic conditions via the intermediate formation of a nucleoside 2',3'-cyclic phosphate (Fig. 2.34). Hydro-

lysis by pancreatic ribonuclease also proceeds via the formation of cyclic phosphate but treatment with venom phosphodiesterase yields the nucleoside 5'-phosphates (chapter 4).

Alkylation of the 2'-OH group to yield, for instance, 2'-*O*-methylribonucleosides (which are found in ribosomal and transfer RNA) renders the phosphodiester bond resistant to cleavage by alkali or by pancreatic ribonuclease since the 2',3'-cyclic phosphate cannot be formed.

Table 2.7 Wavelength (nm) of maximum absorbance (λ_{max}) of adenine and adenosine at various pHs

pH	Adenine	Adenosine
1	265.5	257.0
7	260.5	260.0
12	269.0	260.0

2.7.4 Photochemistry [133, 163, 164]

The bases absorb ultraviolet light but the spectra of the bases are considerably affected by glycosylation and by pH (Table 2.7). For example adenine has a pK at 4.1 for the protonation of N1 and a second pK at 9.8 for the protonation of the amino group both of which affect the spectrum.

As described in section 2.7 absorption is also affected by base stacking interactions.

Ultraviolet irradiation of pyrimidines in solution catalyses the addition of water across the C5–C6 double bond. Adjacent pyrimidines on a polynucleotide chain form cyclobutane-linked dimers (chapter 7).

REFERENCES

1 Adler, M., Weissman, B. and Gutman, A. B. (1958) *J. Biol. Chem.*, **230**, 717.
2 Littlefield, J. W. and Dunn, D. B. (1958) *Biochem. J.*, **70**, 642.
3 Smith, J. D. and Dunn, D. B. (1959) *Biochem. J.*, **72**, 294.
4 Davis, F. F., Carlucci, A. F. and Roubein, I. F. (1959) *J. Biol. Chem.*, **234**, 1525.
5 Nakanishi, K., Blobstein, S., Funamizu, M., Furutachi, N., Van Lear, G. *et al.* (1971) *Nature (London) New Biol.*, **234**, 107.
6 Thiebe, R., Zachau, H. G., Baczymskyj, L., Biemann, K. and Sonnenbichler, J. (1971) *Biochim. Biophys. Acta*, **240**, 163.
7 Adamiak, R. W. and Gornicki, P. (1985) *Prog. Nucleic Acid Res. Mol. Biol.*, **32**, 27.
8 Nichimura, S. (1972) *Prog. Nucleic Acid Res. Mol. Biol.*, **12**, 49.
9 Hutchinson, D. W. (1964) *Nucleosides and Coenzymes*, Methuen, London.
10 Michelson, A. M. (1963) *The Chemistry of Nucleosides and Nucleotides*, Academic Press, New York.
11 Travers, A. (1980) *Nature (London)*, **283**, 16.
12 Heppel, L. A., Ortiz, P. J. and Ochoa, S. (1967) *J. Biol. Chem.*, **229**, 679.
13 Bendich, A. (1957) *Methods Enzymol.*, **3**, 715.
14 Adams, R. L. P., McKay, E. L., Craig, L. M. and Burdon, R. H. (1979) *Biochim. Biophys. Acta*, **563**, 72.
15 Fasman, G. D. (1976) *CRC Handbook of Biochemistry and Molecular Biology* (3rd edn) vol. 2, CRC Press, Boca Raton.
16 Chargaff, E. (1963) *Essay on Nucleic Acids*, Elsevier/North-Holland, Amsterdam.
17 Astbury, W. T. (1947) *Symp. Soc. Exp. Biol.*, **1**, 66.
18 Franklin, R. and Gosling, R. G. (1953) *Nature (London)*, **171**, 740, **172**, 156.
19 Langridge, R., Wilson, H. R., Hooper, C. W., Wilkins, M. H. F. and Hamilton, L. D. (1960) *J. Mol. Biol.*, **3**, 547.
20 Fuller, W., Wilkins, M. H. F., Wilson, H. R. and Hamilton, L. D. (1965) *J. Mol. Biol.*, **12**, 60.
21 Davies, D. R. (1967) *Annu. Rev. Biochem.*, **36**, 321.
22 Watson, J. D. and Crick, F. H. C. (1953) *Nature (London)*, **171**, 737.
23 Watson, J. D. (1968) *The Double Helix*, Atheneum, New York.
24 Olby, R. (1964) *The Path to the Double Helix*, Macmillan, London.
25 Arnott, S. (1970) *Science*, **167**, 1694.
26 Portugal, J. (1989) *Trends Biochem. Sci.*, **14**, 127.
27 Drew, H. R. and Dickerson, R. E. (1981) *J. Mol. Biol.*, **151**, 535.
28 Subramanian, P. S., Ravishanker, G. and Beveridge, D. L. (1988) *Proc. Natl. Acad. Sci. USA*, **85**, 1836.
29 Neidle, S., Pearl, L. H. and Skelly, J. V. (1987) *Biochem. J.*, **243**, 1.
30 Tunis, M. J. B. and Hearst, J. E. (1958) *Biopolymers*, **6**, 128.
31 Brahms, J. and Mommaerts, W. H. F. M. (1964) *J. Mol. Biol.*, **10**, 73.

32 Shih, T. Y. and Fasman, G. D. (1971) *Biochemistry*, **10**, 1675.

33 Tunis-Schneider, M. J. B. and Maestre, M. F. (1970) *J. Mol. Biol.*, **52**, 521.

34 Saenger, W., Hunter, W. N. and Kennard, O. (1986) *Nature (London)*, **324**, 385.

35 Bram, S. (1971) *J. Mol. Biol.*, **58**, 277.

36 Bram, S. (1973) *Cold Spring Harbor Symp. Quant. Biol.*, **38**, 83.

37 Cohen, J. S. (1987) *Trends Biochem. Sci.*, **12**, 133.

38 Delepierre, M., Langlois D'Estaintot, B. L., Igolen, J. and Roques, B. P. (1986) *Eur. J. Biochem.*, **161**, 571.

39 Iwamoto, S. and Hsu, H-T. (1983) *Nature (London)*, **305**, 70.

40 Wang, J. C. (1979) *Proc. Natl. Acad. Sci. USA*, **76**, 200.

41 Rhodes, D. and Klug, A. (1981) *Nature (London)*, **292**, 378.

42 Tullius, T. D. (1987) *Trends Biochem. Sci.*, **12**, 297.

43 Wu, H-H. and Crothers, D. M. (1984) *Nature (London)*, **308**, 509.

44 Hagerman, P. J. (1984) *Proc. Natl. Acad. Sci. USA*, **81**, 4632.

45 Frederick, C. A., Grable, J., Melia, M., Samudzi, C., Jen-Jacobson, L., *et al.* (1984) *Nature (London)*, **309**, 327.

46 Richmond, T. J., Finch, J. T., Rushton, B., Rhodes, D. and Klug, A. (1984) *Nature (London)*, **311**, 532.

47 Dickerson, R. E., Bansal, M., Calladine, C. R., Diekmann, S., Hunter, W. N. *et al.* (1989) *J. Mol. Biol.*, **205**, 787.

48 Trifonov, E. N. (1982) *Cold Spring Harbor Symp. Quant. Biol.*, **47**, 271.

49 Dickerson, R. E., Kopka, M. L. and Pjura, P. (1983) *Proc. Natl. Acad. Sci. USA*, **80**, 7099.

50 Dickerson, R. E. (1983) *Sci. Am.*, **249** (6), 87.

51 Dickerson, R. E. (1983) in *Nucleic Acids – The Vectors of Life* (eds B. Pullman and J. Jortner), D. Reidel. Dordrecht, p. 1.

52 Calladine, C. R. (1982) *J. Mol. Biol.*, **161**, 343.

53 Dickerson, R. E. (1983) *J. Mol. Biol.*, **166**, 419.

54 Morgan, A. R. (1979) *Trends. Biochem. Sci.*, **4**, 244.

55 Arnott, S., Chandrasekaran, R., Puigjaner, L. C., Walker, J. K., Hall, I. H. *et al.* (1983) *Nucleic Acids Res.*, **11**, 1457.

56 Arnott, S., Chandrasekaran, R., Puigjaner, L. C. and Walker, J. K. (1983) in *Nucleic Acids – The Vectors of Life* (eds B. Pullman and J. Jortner). D. Reidel, Dordrecht, p. 17.

57 Zurkin, V. B., Lysov. Y. P. and Ivanov. V. I. (1979) *Nucleic Acids Res.*, **6**, 1981.

58 Zurkin, V. B. (1983) *FEBS Lett.*, **158**, 293.

59 Diekmann, S. (1987) in *Nucleic Acids and Molecular Biology*, **1**, (eds F. Eckstein and D. J. M. Lilley), Springer-Verlag, New York, p. 138.

60 Calladine, C. R. and Drew, H. R. (1986) *J. Mol. Biol.*, **192**, 907.

61 Snyder, M., Buchman, A. R. and Davis, R. W. (1986) *Nature (London)*, **324**, 87.

62 Zahn, K. and Blattner, F. R. (1985) *Nature (London)*, **317**, 451.

63 Robertson, M. (1987) *Nature (London)*, **327**, 464.

64 Arnott, S., Chandrasekaran, R., Hall, I. M. and Puigjaner, L. C. (1983) *Nucleic Acids Res.*, **11**, 4141.

65 Alexeev, D. G., Lipanov, A. A. and Skuratovskii, I. Ya. (1987) *Nature (London)*, **325**, 821.

66 Nelson, H. C. M., Finch, J. T., Luisi, B. F. and Klug, A. (1987) *Nature (London)*, **330**, 221.

67 Griffith, J., Bleyman, M., Rauch, C. A., Kitchin, P. A. and Englund, P. T. (1986) *Cell*, **46**, 717.

68 Travers, A. and Klug, A. (1987) *Nature (London)*, **327**, 280.

69 Zinkel, S. S. and Crothers, D. M. (1987) *Nature (London)*, **328**, 178.

70 Hagerman, P. (1985) *Biochemistry*, **24**, 7033.

71 Koo, H-S., Wu, H-M. and Crothers, D. M. (1986) *Nature (London)*, **320**, 501.

72 Burkhoff, A. M. and Tullius, T. D. (1987) *Cell*, **48**, 935.

73 Nadeau, J. D. and Crothers, D. M. (1989) *Proc. Natl. Acad. Sci. USA*, **86**, 2622.

74 Koo, H-S. and Crothers, D. M. (1988) *Proc. Natl. Acad. Sci. USA*, **85**, 1763.

75 Burkhoff, A. M. and Tullius, T. D. (1988) *Nature (London)*, **331**, 455.

76 Crothers, D. M., Haran, T. E. and Nadeau, J. G. (1990) *J. Biol. Chem.*, **265**, 7093.

77 Zimmerman, S. B. (1982) *Annu. Rev. Biochem.*, **51**, 359.

78 Rich. A. (1983). *Cold Spring Harbor Symp. Quant. Biol.*, **47**, 1.

79 Rich, A., Nordheim, A. and Wang, A. H. J. (1984) *Annu. Rev. Biochem.*, **53**, 791.
80 Pohl, E. M. and Jovin, T. M. (1972) *J. Mol. Biol.*, **67**, 375.
81 Leng, M. (1985) *Biochim. Biophys. Acta*, **825**, 339.
82 McLean, M. J. and Wells, R. D. (1988) *Biochim. Biophys. Acta*, **950**, 243.
83 Ho, P. S., Frederick, C. A., Quigley, G. J., van der Marel, G. A., van Boom, J. H., Wang, A. H-J. and Rich, A. (1985) *EMBO J.*, **4**, 3617.
84 Hamada, H. and Kakunaga, T. (1982) *Nature (London)*, **298**, 396.
85 Hill, R. J. and Stollar, B. D. (1983) *Nature (London)*, **305**, 338.
86 Bramachari, S. K., Mishra, R. K., Bagga, R. and Ramesh, N. (1989) *Nucleic Acids Res.*, **17**, 7273.
87 Holliday, R. (1989) *Trends Genet.*, **5**, 355.
88 McConnell, B. and Von Hippel, P. H. (1970) *J. Mol. Biol.*, **50**, 297.
89 Guéron, M., Kochoyan, M. and Leroy, J-L. (1987) *Nature (London)*, **328**, 89.
90 Frank-Kamenetskii, M. (1987) *Nature (London)*, **328**, 17.
91 Gotoh, O. and Tagashira, Y. (1981) *Biopolymers*, **20**, 1033.
92 Gotoh, O. and Tagashira, Y. (1981) *Biopolymers*, **20**, 1043.
93 Breslauer, K. J., Frank, R., Blöcker, H. and Marky, L. A. (1986) *Proc. Natl. Acad. Sci. USA*, **83**, 3746.
94 Marmur, J. and Doty, P. (1959) *Nature (London)*, **180**, 1427.
95 Mandel, M. and Marmur, J. (1976) *Methods Enzymol.*, **12**, 195.
96 Adams, R. L. P. and Eason, R. (1984) *Nucleic Acids Res.*, **12**, 5869.
97 Tong, B. Y. and Battersby, S. J. (1979) *Nucleic Acids Res.*, **6**, 1073.
98 Collins, M. and Myers, R. M. (1987) *J. Mol. Biol.*, **198**, 737.
99 Murchie, A. I. H. and Lilley, D. M. J. (1989) *J. Mol. Biol.*, **205**, 593.
100 Britten, R. J. and Kohne, D. E. (1968) *Science*, **161**, 529.
101 Felsenfeld, G. (1968) *Methods Enzymol.*, **12**, 247.
102 Doty, P. (1961) *Harvey Lect.*, **55**, 103.
103 Guttman, T., Votavova, H. and Pivec, C. (1976) *Nucleic Acids Res.*, **3**, 835.
104 Birnsteil, M., Telford, J., Weinberg, G. and Stafford, D. (1974) *Proc. Natl. Acad. Sci.*

105 Bunermann, H. and Muller, W. (1978) *Nucleic Acids Res.*, **5**, 1059.
106 Marmur, J. (1963) *Annu. Rev. Microbiol.*, **17**, 329.
107 Schildkraut, C. L., Mandel, M., Levisohn, S., Smith-Sonnebom, J. E. and Marmur, J. (1962) *Nature (London)*, **196**, 795.
108 Freese, E. (1962) *J. Theor. Biol.*, **3**, 82.
109 Harpst, J. A., Krasna, A. I. and Zimm, B. H. (1968) *Biopolymers*, **6**, 595.
110 Laskowski, M. (1972) *Prog. Nucleic Acid Res. Mol. Biol.*, **12**, 161.
111 Bone, M., Gibson, T., Goman, M., Hyde, J. E., Langsley, G. W. *et al.* (1983) *Molecular Biology of Parasites* (eds J. Guardiola, L. Luzzatto and W. Trager), Raven Press, New York, p. 125.
111a Brownlee, A. G. (1989) *Nucleic Acids Res.*, **17**, 1327.
112 Campbell, A. M. and Lochhead, D. S. (1971) *Biochem. J.*, **123**, 661.
113 Beard, P., Morrow. J. F. and Berg, P. (1973) *J. Virol.*, **12**, 1303.
114 Monjardino, J. and James, A. W. (1975) *Nature (London)*, **225**, 249.
115 Campbell, A. M. (1978) *Trends Biochem. Sci.*, **3**, 104.
116 Laundon, C. H. and Griffith, J. D. (1988) *Cell*, **52**, 545.
117 Pulleyblank, D. E. and Morgan, A. R. (1975) *J. Mol. Biol.*, **91**, 1.
118 Wang, J. C. (1974) *J. Mol. Biol.*, **89**, 783.
119 Keller, W. (1975) *Proc. Natl. Acad. Sci. USA*, **72**, 4876.
120 Wang, J. C. (1976) *J. Mol. Biol.*, **43**, 25.
121 Luchnick, A. N., Bakayev, V. V. and Glaser, V. M. (1983) *Cold Spring Harbor Symp. Quant. Biol.*, **47**, 293.
122 Lilley, D. M. J. (1983) *Nature (London)*, **305**, 276.
123 Shen, C-K. J. and Hu, W-S. (1986) *Proc. Natl. Acad. Sci. USA*, **83**, 1641.
124 Frank-Kamenetskii, M. (1989) *Nature (London)*, **337**, 206.
125 Senior, M. M., Jones, R. A. and Breslauer, K. J. (1988) *Proc. Natl. Acad. Sci. USA*, **85**, 6242.
126 Chattopadhyaya, R., Ikuta, S., Grzeskowiak, K. and Dickerson, R. E. (1988) *Nature (London)*, **334**, 175.
127 Chattopadhyaya, R., Grzeskowiak, K. and Dickerson, R. E. (1990) *J. Mol. Biol.*, **211**, 189.

128 Courey, R. J. and Wang, J. C. (1983) *Cell*, **33**, 817.

129 Sinden, R. R., Carison, J. O. and Pettijohn, D. E. (1980) *Cell*, **21**, 773.

130 Sinden. R. R., Brogles, S. S. and Pettijohn, D. E. (1983) *Proc. Natl. Acad. Sci. USA*, **80**, 1797.

131 Lilley, D. J. M., Sullivan, K. M. and Murchie, K. I. H. (1987) in *Nucleic Acids and Molecular Biology* 1 (eds Eckstein, F. and Lilley, D. J. M.), Springer-Verlag, New York, p. 126.

132 Lee, J. S., Woodsworth, M. L., Latimer, L. J. P. and Morgan, A. R. (1984) *Nucleic Acids Res.*, **12**, 6603.

133 Guschlbauer, W. (1976) *Nucleic Acids Structure, Springer-Verlag*, New York.

134 Mirkin, S. M., Lyamichev, V. I., Drushlyak, K. N., Dobrynin, V. N., Filippov, S. A. *et al.* (1987) *Nature (London)*, **330**, 495.

135 Hanvey, J. C., Shimizu, M. and Wells, R. D. (1988) *Proc. Natl. Acad. Sci. USA*, **85**, 6292.

136 Sklenár, V. and Feigon, J. (1990) *Nature (London)*, **345**, 836.

137 Voloshin, O. N., Mirkin, S. M., Lyamichev, V. I., Belotserkovskii, B. P. and Frank-Kamenetskii, M. D. (1988) *Nature (London)*, **333**, 475.

138 Tinoco, I., Borer, P. N., Dengler, B., Levine, M. D., Uhlenbeck, O. C. *et al.* (1973) *Nature (London) New Biol.*, **246**, 40.

139 Wyatt, J. R., Puglisi, J. D. and Tinoco, I. (1989) *Bioessays*, **11**, 100.

140 Bhattacharyya, A., Murchie, A. I. H. and Lilley, D. M. J. (1990) *Nature (London)*, **343**, 484.

141 Woese, C. R., Magrum, L. J., Gupta, R., Siegel, R. B., Stahl, D. A. *et al.* (1980) *Nucleic Acids Res.*, **8**, 2275.

142 Atmadja, J., Brimacombe, R. and Maden, B. E. H. (1984) *Nucleic Acids Res.*, **12**, 2649.

143 Jaeger, J. A., Turner, D. H. and Zuker, M. (1989) *Proc. Natl. Acad. Sci. USA*, **86**, 7706.

144 Cheong, C., Varani, G. and Tinoco, I. (1990) *Nature (London)*, **346**, 680.

145 Uhlenbeck, O. C. (1990) *Nature (London)*, **346**, 613.

146 Puglisi, J. D., Wyatt, J. R. and Tinoco, I. (1988) *Nature (London)*, **331**, 283.

147 Pleij, C. W. A. (1990) *Trends Biochem. Sci.*, **15**, 143.

148 Massoulie, J. (1968) *Eur. J. Biochem.*, **8**, 423.

149 Massoulie, J. (1968) *Eur. J. Biochem.*, **8**, 439.

150 Thrierr, J. C. and Leng, M. (1972) *Biochim. Biophys. Acta*, **272**, 238.

151 Arnott, S. and Bond, P. J. (1973) *Nature (London) New Biol.*, **244**, 99.

152 Arnott, S., Hukins, D. W. L., Dover, S. D., Fuller, W. and Hodgson, A. R. (1973) *J. Mol. Biol.*, **81**, 107.

153 Furuichi, T., Dhundale, A., Inouye, M and Inouye, S. (1987) *Cell*, **48**, 47.

154 Furuichi, T., Inouye, S. and Inouye, M. (1987) *Cell*, **48**, 55.

155 Kotchetkov, N. K. and Budowsky, E. I. (1969) *Prog. Nucleic Acid Res. Mol. Biol.*, **9**, 403.

156 Brown, D. M. (1974) in *Basic Principles of Nucleic Acid Chemistry* (ed. P. O. P. Ts'O), Academic Press, London, p. 1.

157 Ceriotti, G. (1955) *J. Biol. Chem.*, **214**, 59.

158 Burton, K. (1956) *Biochem. J.*, **62**, 315.

159 Singer, B. and Fraenkal-Conrat, H. (1969). *Prog. Nucleic Acid Res. Mol. Biol.*, **9**, 1.

160 Schuster. H. (1960) *Z. Naturforsch. B*, **15**, 298.

161 Feldman, M. Y. (1973) *Prog. Nucleic Acid Res. Mol. Biol.*, **13**, 1.

162 Shapiro, R., Cohen, B. I. and Servis, R. E. (1970) *Nature (London)*, **227**, 1047.

163 Kotchetkov, N. K. and Budowsky, E. I. (1972) *Organic Chemistry of Nucleic Acids*, Plenum Press, New York.

164 Bush, C. A. (1974) in *Basic Principles of Nucleic Acid Chemistry* (ed. P. O. P. Ts'O), Academic Press, London, p. 91.

165 Sen, D. and Gilbert, W. (1988) *Nature (London)*, **334**, 364.

166 Crick, F. H. C. (1966) *J. Mol. Biol.*, **19**, 548.

167 Sober, H. A. (1968) *Handbook of Biochemistry*, Chemical Rubber Co., Cleveland, Ohio, pp. H-11, H-30.

168 Fazakerley, G. V., Gabarro-Arpa, J., Lebret, M., Guy, A. and Guschlauer, W. (1989) *Nucleic Acids Res.*, **17**, 2541.

169 Joshua-Tor, L., Rabinovich, D., Hope, H., Frolow, F., Appella, E. and Sussman, J. L. (1988) *Nature (London)*, **334**, 82.

170 Lilley, D. J. M. (1988) *Trends Genet.*, **4**, 111.

171 Gouy, M. (1987) in *Nucleic Acid and Protein Sequence Analysis* (ed. M. J. Bishop and C. J. Rawlings) p. 259, IRL Press, Oxford.

3

Genomes of eukaryotes, bacteria and viruses: chromosome organization

3.1 EUKARYOTIC CHROMOSOMES

3.1.1 The nucleus

The function of DNA is to carry the genetic message from generation to generation, and to allow the expression of that message under appropriate conditions. DNA molecules are very large and carry the information for the synthesis of many proteins, not all of which are required at the same time. Problems will obviously arise when it comes to packaging such long lengths of DNA into a cell in such a way as to allow programmed access to the encoded information. These problems are compounded by the fact that the DNA duplex must replicate and the two daughter DNA molecules must segregate to the two daughter cells. In this chapter we shall look at the structures involved in these processes. These structures are called chromosomes and consist not only of the DNA but of a variety of associated proteins. The genome of an organism is the DNA (or RNA) present in a set of chromosomes that carries the genetic information of that organism.

For the study of the larger chromosomes of eukaryotic cells, cytological techniques have been used for many years. Genetical approaches have been most useful for the study of bacterial chromosomes but have been much less successful in the study of the far more complex and slower-growing animal and plant cells. In recent years, the modern techniques of genetic engineering, together with the cloning and sequencing strategies discussed in the Appendix, have enabled rapid advances to be made in our understanding of the organization of the DNA and the genes of all organisms.

In eukaryotes, most of the DNA is found in the nucleus. This is bounded by a double membrane continuous with the endoplasmic reticulum. At intervals in the nuclear membrane are pores through which nucleo-cytoplasmic exchange is believed to occur [1]. Each pore is surrounded by a disc of protein molecules known as the nuclear pore complex which form a channel wide enough to allow the passage of small molecules (M_r less than 5000). Special transport systems exist for the nucleo-cytoplasmic exchange of large RNA and protein molecules [2]. Thus, uptake of protein molecules requires the presence and availability of a nuclear localization signal (consisting of a short run of basic amino acids) and the absence or unavailability of cytoplasmic anchoring signals [3] (section 12.8.6). Uptake is a two-stage process involving the binding of the protein to the nuclear pore complex followed by the ATP-requiring translocation [4, 5]. The pore complex is made of coaxial rings of eight subunits to which is attached a double iris diaphragm [6].

Lying just inside the inner nuclear membrane is the nuclear lamina, the whole complex forming the nuclear envelope [7]. The nuclear lamina is made from three proteins (Lamins A, B and C) of M_r 60 000–70 000 which are polymerized to form a sort of fibrous skeleton, to which the chromosomes are attached, probably by telomeric sequences [8–11]. This attachment of chromosomes to the nuclear lamina may help to align the chromosomes in a manner which is important for their function [12] but as yet little firm evidence is available that this is so. The attachment points do, however, serve as foci for chromosome condensation at mitosis [13–15]. Phosphorylation of lamin proteins causes the disassembly of the lamina during mitosis [16–19]. Reassembly requires the prior formation of the nuclear pore complexes around the chromatin and only then are the lamina and nuclear envelope reformed [1, 20].

Within the body of the nucleus the chromosomes are associated with what is known as the nuclear matrix. During mitosis, the matrix proteins are believed to form the chromosome scaffold. The role of the matrix and scaffold will be considered further in sections 3.3.5 and 10.4.4(e) [8, 21–24].

In prokaryotes the DNA is not housed in a membrane-bound organelle but is attached to the plasma membrane which itself is surrounded by a rigid cell wall to protect the cell from osmotic damage. When the cell wall is eliminated by digestion with the

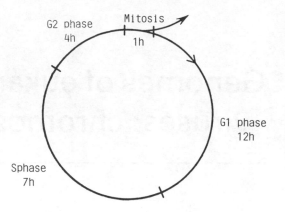

Fig. 3.1 The eukaryotic cell cycle. G1, S and G2 phases are together known as interphase. The actual times involved vary with cell type and growth conditions.

enzyme lysozyme the membrane and its contents are released as the osmotically sensitive protoplast.

In addition to the nuclear DNA of eukaryotes, there is a small amount of DNA present in mitochondria and chloroplasts (section 3.4). Prokaryotes also carry small extra pieces of DNA known as extrachromosomal DNA or plasmids (section 3.6), and these are also found in yeast cells. Viruses are often present in cells where they multiply, eventually leading to the death of the cell. Viruses which grow in bacteria are called bacteriophages or more simply phages.

3.1.2 The cell cycle

In eukaryotic cells, DNA synthesis takes place in a restricted time period during the *cell cycle* (Fig. 3.1). In mitosis the cells divide and form two daughter cells. This is then followed by a gap period designated G1 during which no DNA synthesis occurs. The length of the various gap periods and indeed the length of the entire cell cycle depends very much on the cell type but Fig. 3.1 shows a typical example of a cell in culture with a replication time of 24 hours. G1 is followed by S phase in which DNA synthesis occurs and then by a second, shorter gap period known as G2. The cells then go into mitosis again. In higher eukaryotes, few cells in the organism are actively dividing; most being arrested in the G0 phase shortly after mitosis.

A large number of changes are associated with the onset of DNA synthesis. The replicating DNA becomes associated with the nuclear matrix (section 6.15.5) and many of the enzymes involved in synthesizing DNA are either synthesized or activated at the beginning of S phase [345]. Studies in this area are more advanced with yeasts as these simple eukaryotes are more amenable to genetic analysis than are higher eukaryotes. Crucial to the decision to enter the cycle (G0/G1) and for the onset of S phase and mitosis is a protein kinase (p34) encoded by the *cdc2* gene of *Schizosaccharomyces pombe* (or the *CDC28* gene of *Saccharomyces cerevisiae*). This kinase is activated in a cyclical manner by interaction with cyclins, proteins whose levels rise and fall dramatically with the phases of the cell cycle [27–29, 341, 346, 347] (see also section 6.10.4).

A population of growing cells will normally consist of cells at all stages of the cell cycle but a variety of methods may be used to synchronize the population so that all are dividing at the same time [25]. Alternatively, cells in a particular phase may be selected on the basis of size or adherence to the substrate on which they are growing. The combination of G1, S and G2 phases is known as interphase and under these conditions the cell nucleus is clearly visible using the light microscope [25]. It is only in mitosis that the chromosomes are visible using light microscopy.

3.1.3 Eukaryotic chromosomes

Eukaryotic DNA is contained in a relatively small number of chromosomes, a number which varies according to the species (Table 3.1). No direct correlation can be made between the amount of DNA in the nucleus and the number of chromosomes in which it is contained [30]. Somatic cells of each species have two copies or homologues of each chromosome with the exception of the sex chromosomes for which the female carries two X chromosomes and the male an X and a Y. Germ cells contain only one copy of each chromosome and it is to these cells that the term haploid chromosome content or *haploid DNA content* refers. Whereas most somatic cells are *diploid*, tetraploid cells with four and octaploid cells with eight copies of each chromosome can also be found, particularly in cells in culture. Aneuploid cells have an abnormal chromosome complement which is not necessarily an increase on the diploid condition for each chromosome type so that some may be present in greater numbers than others. The characteristic number and morphology of the chromosomes in any particular cell type is known as the *karyotype* of that cell and is usually determined in metaphase when the chromosomes are highly condensed and readily stained by basic dyes [25, 31]. In interphase, the chromosomes are dispersed within the nucleus and cannot be individually distinguished.

The actual process of mitosis can be subdivided into several discrete stages. At the end of the interphase two poles are formed in the cell by the *centrioles*. At this time each chromosome consists of two identical chromatids produced as a result of DNA replication (chapter 6). Each chromatid is a DNA duplex and the two chromatids are joined at the centromere [32]. In prophase the chromosomes begin

Table 3.1 Haploid DNA content and chromosome number of a variety of organisms

	Haploid DNA content		Haploid chromosome number
	picograms	base pairs	
Simian virus 40	0.000006	5.3×10^3	1
Herpes simplex virus	0.00017	151×10^3	1
E. coli	0.005	4.5×10^6	1
S. cerevisiae (yeast)	0.025	22.5×10^6	17
Arabidopsis thaliana	0.07	64×10^6	5
Drosophila melanogaster	0.17	0.15×10^9	4
Sea urchin	0.45	0.41×10^9	20
Gallus domesticus (chicken)	0.7	0.63×10^9	39
Homo sapiens (man)	2.7	2.4×10^9	23
Mus musculus (mouse)	3.0	2.7×10^9	20
Xenopus laevis (toad)	4.2	3.8×10^9	18
Zea mays (corn)	7.8	7×10^9	10
HeLa cells (human cell culture)	8.5	7.7×10^9	70–164 (mean 82)
Triturus cristatus (newt)	35	31.5×10^9	12
Fritillaria assyriaca (lily)	127	115×10^9	?

to condense and the nuclear envelope disappears (see above). Fine fibres of microtubules form between the two poles and in metaphase the highly condensed chromosomes line up in the centre of the cell at the equatorial region to form the *metaphase plate* which is clearly visible using phase-contrast microscopy. In anaphase the two daughter chromosomes are pulled apart into the two poles and in telophase the lamin proteins are dephosphorylated and re-polymerize causing membrane fragments to associate with individual chromosomes. These fragments then fuse to reform the nuclear envelope [1].

Chromosomes with a centromere towards one end are known as acrocentric (or telocentric if the short arm is too small to see by light microscopy). Those with a centrally located centromere are meta-centric, and those defective chromosomes lacking a centromere are acentric. Additional, secondary constrictions are usually the site of rRNA genes which form the

nucleolus in an interphase nucleus and are known as nucleolar organizers. These are only found on a few chromosomes.

Detailed cytogenetic studies of chromosomes involve the use of fluorescent dyes such as quinacrine mustard or gentle trypsin treatment followed by staining with Giemsa. These treatments give rise to a characteristic band pattern (known as Q-bands or G-bands, respectively) visible using fluorescence or light microscopy [33, 34]. Each chromosome has a banding pattern which allows it to be recognized easily and chromosome deletions or rearrangements may be closely observed. A large variety of banding techniques is now available for cytogenetic studies [7, 25].

In budding yeast (e.g. *Saccharomyces*) chromosome condensation does not occur and the nuclear membrane remains present throughout division. A bud is formed in G1 phase and increases in size during the rest of the cycle. Following nuclear division one of the daughter nuclei migrates into the bud which separates from the parental cell at cell division [35].

Using genetic engineering techniques (see the Appendix) the centromeres of several yeast chromosomes have been isolated and sequenced [36–38]. They contain a 14 bp conserved nucleotide sequence separated from an 11 bp conserved nucleotide sequence by an 82–89 bp AT-rich sequence. The centromere core provides a 220 bp region of the chromosome to which the microtubules can attach at mitosis [39].

Other specific regions of chromosomes which have been isolated are the ends or telomeres which are considered in more detail in section 6.10.3. In addition, for a chromosome to replicate, it requires an origin of replication. Sequences which confer on chromosomes the ability to replicate and segregate have also been cloned and sequenced (section 6.9). By combining these three cloned regions from yeast an artificial chromosome (YAC) has been constructed. On introduction into yeast cells this chromosome appears to act like a natural chromosome in most respects (section A.8.2) [40–42].

3.1.4 The allocation of specific genes to specific chromosomes

The classical genetic approach to study the chromosomal 'linkage' of two gene loci is to determine whether markers co-segregate or segregate independently during genetic crosses between two different inbred strains that show polymorphism for the marker. The more distantly related the parents the greater the polymorphism and the allocation of almost 1000 genetic markers to mouse chromosomes has benefited greatly from the finding that fertile progeny arise on crossing the wild mouse, *Mus spretus*, with common laboratory strains [344]. The genetic ap-

proach is much easier in plants and, by constant back-crossing and selection, almost isogenic strains can be obtained differing only in a small region of the chromosome surrounding the selected marker.

In genetic terms the distance between loci is measured in map units or centimorgans (cM). Two loci are one map unit apart if there is a 0.01 frequency of recombination between them (recombination is discussed in section 7.4). However, as some chromosomal regions are more susceptible to recombination than are others, there is no linear relationship between map units and kilobase pairs (kb). Nevertheless, an average figure can be obtained for a chromosome or for all the chromosomes of a given species and in *Arabidopsis* (a plant with a very small genome) a centimorgan is equivalent to 200 kb whereas in wheat (with a 100-fold larger genome) the corresponding figure is 3500 kb.

In higher organisms externally observable phenotypes associated with polymorphisms are relatively infrequent; however, genetic analysis has been aided by the more numerous cases in which a polymorphism results in differences in electrophoretic mobility of an easily identifiable enzyme. Various approaches are now available for testing the chromosomal location of the genes involved [333].

(a) *Somatic cell genetics*

Cells of two different animal species can be induced to fuse in culture to produce heterokaryons carrying two or more sets of chromosomes [25, 43]. The redundant chromosomes are readily lost from the hybrid cells. The loss occurs at random but in human–mouse hybrids the human chromosomes are lost preferentially and cell clones can be obtained which retain only one or a few human chromosomes. The clones can then be tested for specific enzymes and their presence related to those chromosomes that remain. For this approach to be successful it is important either to be able to select for a particular characteristic (e.g. thymidine kinase activity) or to be able to distinguish the human enzyme from the corresponding mouse enzyme which will be present in all clones. In this way it has been possible to allocate several hundred enzyme functions to specific chromosomes [44].

(b) In situ *hybridization*

The technique of *in situ* hybridization was initially limited to the detection of repetitive gene families (section 3.2.6) but was subsequently refined to allow the assignment of single-copy genes to particular segments of specific chromosomes [45]. The method relies on having a radioactive or fluorescent gene probe (see the Appendix) which can be hybridized to a fixed preparation of metaphase or prometaphase chromosomes [46–48]. Radioautography linked to sensitive chromosome-staining methods leads to the illumination of the gene on the particular chromosome band.

These two methods can be combined by preparing Southern transfers (see the Appendix) of the DNA isolated from the various hybrid clones. These can be treated with the radioactive gene probes to obtain a physical rather than enzymic location of the gene on a particular chromosome [49–51]. This method has to be linked to the use of deletion mutants and the cotransfer of genes on chromosome fragments to obtain more precise assignments to chromosome regions.

(c) *Chromosome isolation*

Rather than relying on hybrid cells as a mechanism of obtaining clones with only one or a few identifiable human chromosomes it is now possible to fractionate a preparation of metaphase chromosomes using a fluorescence-activated cell sorter (FACS). Individual chromosome fractions can then be tested to see which hybridize with the radioactive gene probes [52–54]. Figure 3.2 shows chromosomes from human cells separated in this manner.

Alternatively, DNA molecules of up to 2000 kb can now be fractionated by pulsed field gel electrophoresis (section A.2.1) and this has allowed the production of a molecular karyotype of yeast and trypanosomes [54–57].

(d) *RFLP analysis*

A fourth method, useful for determining whether two genes are close together or distant, illustrates the use of restriction fragment length polymorphisms (RFLPs). This approach also combines recombinant DNA techniques with classical genetics [329]. Advantage is taken of polymorphisms in the nucleotide sequence of sites that affect cleavage by restriction endonucleases and that are found within (or close to) the gene. Such polymorphisms will produce different-sized restriction fragments in the two cases and these can be visualized by hybridization of a radioactive probe to a genomic Southern blot of the DNA (Fig. 3.5 and section A.3). Linkage between the RFLP and the genetic trait is readily observed and has been used in the prenatal diagnosis of human hereditary diseases and the detection of carriers. In addition, however, RFLPs outside the disease locus (which may be unknown) can be diagnostically useful if they show strong genetic linkage to this locus [58].

Long range physical mapping of the genome can be achieved by taking advantage of pulsed field gel electrophoresis (section A.2.1) to separate very large DNA

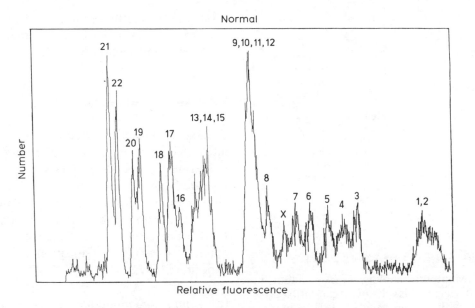

Fig. 3.2 Human chromosome identification by relative fluorescence using flow cytometry (reproduced, with permission, from [52]; copyright Macmillan Journals Ltd).

fragments combined with 'chromosome jumping' to move rapidly from a poorly linked RFLP to one closer to the gene of interest [326, 330–332]. A genetic linkage map of the human genome has been constructed largely using RFLP analysis [325]. Physical maps of the *Drosophila* genome have been constructed by micro-dissection of polytene chromosomes (section 3.3.7) and PCR amplification (section A.7) to obtain probes with which to screen a genomic library [307].

Short, random primers may be used in a set of polymerase chain reactions (PCR) with genomic DNA from two isogenic strains differing in a short region (or a single gene). Any bands appearing in one strain and not in the other must arise from the region where the two strains differ. They, thus, provide a set of probes closely linked to the variant genetic locus and allow more rapid mapping of the chromosome.

3.2 THE EUKARYOTIC GENOME

3.2.1 Haploid DNA content (C value)

One of the most striking features of eukaryotic DNA is the great quantity of it which is present in each cell (Table 3.1). The amount is at least an order of magnitude in excess of that required for the known gene-coding capacities of the cells and the logical conclusion is that the bulk of the DNA is not expressed. Does this mean that it is redundant (junk) or that it has a function which is, as yet, imperfectly understood [59, 60]?

In general terms the minimum size of the genome (i.e. the amount of DNA per cell) increases with the stage of evolutionary development [61] (Fig. 3.3). However, certain amphibia have a C value one hundredfold in excess of man and even more in excess of other amphibia, and the reason for

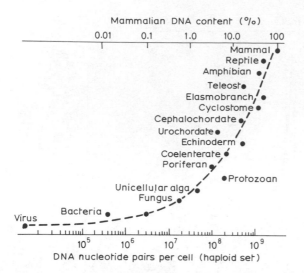

Fig. 3.3 Minimum haploid DNA content in species at various levels of organization (reproduced, with permission, from [61]; copyright the AAS).

this is not clear. The phenomenon is known as the *C value paradox* [62]. Logically, the species with the greater amount of DNA should have the advantage of greater coding potential and the disadvantage of the requirement to replicate very large amounts of DNA before cell division.

In addition to the C value paradox, different cell types within the same species can vary in their haploid DNA content. Amphibian oocytes have a very large amount of cytoplasm to provide with essential components of protein synthesis, such as ribosomes, and consequently have greatly amplified the number of genes coding for ribosomal RNA (section 6.15.6). Eukaryotic cells in culture tend to become aneuploid with the passage of time, apparently adapting to their culture conditions. In addition, they have the capacity to amplify certain genes several hundred-fold in response to external stimuli. Thus the gene coding for the enzyme dihydrofolate reductase is amplified in cells which are

resistant to methotrexate, a drug which interferes with one carbon unit metabolism (chapter 5 and section 6.15.6).

3.2.2 Gene frequency

Eukaryotic DNA is, by convention, divided into three frequency classes of unique DNA, moderately repetitive DNA and highly repeated DNA. There is, in fact, considerable overlap between the three categories which are probably better classified by their $C_0t_{1/2}$ values (section 2.4.2) since this is the manner in which classification is achieved experimentally.

(a) *Unique and moderately repetitive DNA*

In human cells, unique DNA is generally classified as that DNA which has a $C_0t_{1/2}$ value of about 1000 moles of nucleotide seconds litre^{-1}. It comprises about half the total haploid DNA content and is thought to consist of the sequences coding for most enzyme functions for which there is only one or a small number of genes per haploid genome. The genes coding for the various chains of globin (section 9.2.2) or the enzyme glucose 6-phosphatase fall into this category. *Repetitive DNA* includes all the rest of the DNA. Among the repetitive DNA is a fraction which reanneals with a C_0t value of between 100 and 1000 moles of nucleotide seconds litre^{-1}. The sequences represented in this group are generally thought to be those coding for proteins which form major structural components of the cell such as the histones. The genes for rRNA and tRNA also fall into this category.

(b) *Satellite DNA*

At the other extreme DNA with a $C_0t_{1/2}$ value as low as 10^{-3} moles of nucleotide second litre^{-1} consists largely of satellite DNA. This represents highly repeated sequences, of which there may be a million or more copies per haploid genome. These are usually quite short and are arranged in tandem arrays. The origin of the name satellite relates to the method of its isolation on caesium chloride buoyant density gradients of sheared DNA where it will sometimes form a satellite band separate from the main DNA band, due to its differing content of adenine and thymine residues. The simplest known satellite DNA is poly[d(A-T)] which occurs in certain crabs.

Other satellites can have any number, up to several hundred, base pairs which are repeated in tandem fashion along the genome. Human DNA has been shown by density-gradient centrifugation to have four main satellites [63] and by dye-binding [64] and restriction endonuclease cleavage [65] to have two additional satellites.

DNA sequence analysis has shown that the basic repeat unit of satellite DNA is itself made up of subrepeats. For example the major mouse satellite has a repeating structure of 234 base pairs made up of four related 58 and 60 bp segments each in turn made up of 28 and 30 bp sequences [64, 66]. The satellites between and within related species are themselves related in an evolutionary sense by cyclical rounds of multiplication and divergence of an initial short sequence [67]. The nature of the multiplication process is not known for certain but probably involves unequal recombination events (Fig. 3.4 and section 7.4). The divergence involves single base changes (which generate altered restriction enzyme sites – section 4.5.3) and insertions and deletions [68, 69].

Despite our detailed knowledge of the sequence and distribution of satellite DNA our knowledge of the function it has in the cell is rudimentary. The distribution of satellite DNA among chromosomes varies. Some chromosomes have virtually no satellite sequences whereas others (notably the Y chromosome) are largely composed of satellite sequences [70]. In general, satellite DNA appears to be concentrated near the centromere of the chromosomes in the heterochromatin fraction (section 3.3) and be late replicating [48, 71]. In particular, the human α-satellite is centromeric and involved in the sequence-specific binding of one of the centromeric proteins [48].

It was thought that satellite DNA was not transcribed since RNA of corresponding sequence was seldom isolated; but occasional cases of satellite transcription have been reported [72, 73]. On the whole, however, its transcriptional inactivity ties in with its localization in heterochromatin. As satellite DNA is often lost in somatic cells (where it always appears to be heavily methylated – section 4.6.1) it has been proposed that it may have some function in the germ cells [74–77]. This function may relate to the recombination events which occur during gametogenesis and which may be enhanced by the presence of blocks of similar DNA sequences on several chromosomes.

In addition to these satellite DNAs which may make up a considerable fraction of the genome, there are a relatively few copies of a set of different repeats known as *minisatellites* or VNTRs (variable number tandem repeats). These are tandem repeats of short (e.g. 33 bp) sequences of which there are several families. Each member of the first family to be analysed contains a core region of consensus sequence (e.g. 5′-GGGCAGGARG–3′) which was believed to be analogous to the Chi sequence of *E. coli*; i.e. the recognition signal for a 'recombinase' (section 7.4.2). Sequence variation between minisatellite families makes this less likely [78, 79]. Nonetheless,

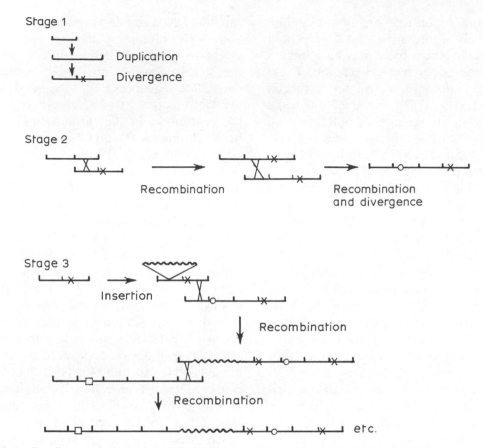

Fig. 3.4 Possible evolution of satellite DNA involving cycles of duplication and divergence.

frequent recombination among members of a particular minisatellite or slippage during replication [80] results in different people having differing numbers of copies of the minisatellite in a repeat and this is the source of a heterozygosity (*DNA fingerprint*) which can be used for pedigree analysis and paternity testing [78, 81, 82] and which has made a notorious entry into forensic science [83] (see below, Fig. 3.5). More recently [84, 85], RFLPs have been detected among the individual repeats of a minisatellite and this has been predicted to make available a capacity to distinguish 10^{70} different allelic states.

(c) *Interspersed repetitive DNA*

This differs from satellite DNA in that it represents sequences which are not clustered together but are dispersed singly throughout the genome [86, 87]. These dispersed repeats fall into two classes: the short interspersed nuclear elements (SINEs) consisting of repeats shorter than 500 bp; and the long interspersed nuclear elements (LINEs). The SINEs are exemplified by the *Alu* family found in human cells. This is a family of related sequences each about 300 bp long which is repeated 300 000 times in the human genome. The name *Alu*

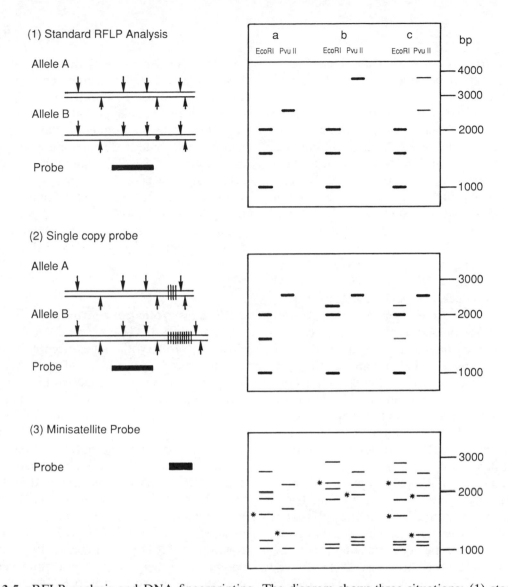

Fig. 3.5 RFLP analysis and DNA fingerprinting. The diagram shows three situations: (1) standard RFLP analysis; (2) analysis near a minisatellite using a single copy probe; and (3) DNA fingerprint analysis using a probe to the repetitive sequence. In each case the left hand side of the diagram shows the physical maps of the alleles in which ↓ indicates *Eco*R I sites and ↑ indicates *Pvu* II sites. The solid bar indicates the location of the probe and ||||||| indicates a variable number of repeats of the mini-satellite sequence. The right hand side of the diagrams indicate the results of Southern blot analysis of DNA from an individual who is (a) homozygous for allele A; (b) homozygous for allele B; and (c) heterozygous. Results are shown for digestions of genomic DNA by the restriction enzymes *Eco*R I and *Pvu* II. (1) Standard RFLP analysis of two alleles (A and B), one of which contains a mutation in a *Pvu* II site. (2) No mutation is present in a restriction enzyme site but the two alleles differ in the number of copies of the minisatellite that is present. The probe is for a unique region of this single-copy gene. (3) DNA from the same individuals tested in (2) is analysed using a probe for the minisatellite sequence. The bands in the *fingerprint* originate from several different genes – those from alleles A and B are indicated by an asterix. No bands are present in the fingerprint of (c) which are not also present in either (a) or (b) indicating that (c) may be the child of (a) and (b). Not all bands are present in (c) indicating that (a) and (b) are not homozygous with respect to all markers.

derives from the presence of a single site for the restriction enzyme *Alu* I in the repeat (section 4.5.3).

The *Alu* I sequence is a head-to-tail dimer of a sequence which closely resembles the B1 family of mouse DNA. The *Alu* sequence is transcribed and it is believed to have become dispersed throughout the genome by a process involving reverse transcription and reintegration (sections 6.4.2(f) and 7.7.4). For this reason it has been termed a retroposon [88].

One of the best-studied LINEs is the L1 family of human DNA [89]. This is a 6–7 kb sequence which is repeated, at least in part, up to 10 000 times per genome [88–90]. These interspersed repeats are also transcribed and may give rise to proteins. It has been suggested that some of the gene products may be involved in the duplication and dispersal of the family which probably also occurs by a reverse transcription mechanism [90]. The mechanism of dispersion of SINEs and LINEs is considered in more detail in section 7.7.4(c).

The interspersion pattern produced in the genome by SINEs was first investigated in detail by renaturation kinetics [91]. The conclusion from these studies was that about half the genome consists of alternating regions of 300 bp of repeated DNA and 1000–2000 bp of unique DNA. This pattern of interspersion (which became known as the *Xenopus* pattern) is typical of most eukaryotes. However, some insect genomes lack this short-period interspersion pattern and consist on average of stretches of about 13 000 bp of unique DNA separated by LINEs [26, 92].

(d) *Fold-back DNA*

A special class of fold-back or palindromic DNA sequences comprising 3–6% of eukaryotic DNA has also been charac-

terized. The size range is from 300 to 1200 base pairs and the molecules have a $C_0t_{1/2}$ value of less than 10^{-5} moles of nucleotide second litre^{-1}. This palindromic DNA is represented in all frequency classes and is widely distributed throughout metaphase chromosomes [93–95]. Some palindromic DNA arises when two copies of the *Alu* sequences are present close to one another in opposite orientations and it was as a result of the ability of such DNA to re-nature instantaneously that the SINEs were discovered.

3.2.3 Eukaryotic gene structure

Whereas prokaryotic genes occupy a single uninterrupted sequence of DNA, the majority of eukaryotic genes so far analysed have been shown to have non-coding sequences inserted into the middle of the gene (section 9.2.1). These intervening sequences, or introns, may be small and single as in the case of the gene for tyrosine suppressor tRNA [96] or large and multiple as is the case for the ovalbumin gene which has seven introns so that the entire gene occupies a length of 7.7 kb with only a small fraction of this DNA being actually used for coding purposes [97]. These intervening regions are transcribed into RNA and then removed by internal processing events in the cell nucleus to produce the final mRNA product (section 9.3.4). Intervening sequences appear to have a wider evolutionary freedom for mutation than the coding sequences of the DNA [98]. Although the majority of structural genes display this phenomenon, it is not universal even in higher eukaryotes and is not found in the majority of tRNA genes or in most of the genes coding for the histones. In general intervening sequences appear to be less frequent in the lower eukaryotes.

In addition to these intervening sequences, eukaryotic genes are preceded by regions of DNA which have a control function. These flanking regions may be over 1000 nucleotide pairs long and their organization is considered in chapters 9 and 10.

It is clear that, at least in part, the extra DNA present in higher eukaryotes can be attributed to repeated sequences and the intervening and flanking sequences of genes.

3.3 CHROMATIN STRUCTURE

Eukaryotic chromosomes in metaphase are generally referred to as chromosomes but in interphase the term 'chromatin' is more generally used to describe the nucleoprotein fibres in the cell nucleus. Originally chromatin was loosely defined and subdivided into only two main classes, euchromatin and heterochromatin. Heterochromatin comprises the dense, readily-stained areas of the nucleus or chromosome and was thought to represent inactive chromatin which was not undergoing transcription, whereas the more loosely packed euchromatin was thought to represent the transcriptionally active material.

Chromatin consists of DNA, RNA and proteins. The actual weight ratio between the three varies greatly with the tissue or cell of origin of the material but, in general, the amount of protein is equal to, or greater than, the amount of DNA, whereas the amount of RNA is comparatively small.

The protein content of chromatin can be further subdivided into the histone and non-histone proteins. The former group consists of a few types of molecule which are present in very large amounts and the latter of a diverse range of protein molecules most of which are present in much smaller quantities. Historically the bulk of the non-histone proteins were referred to as the acidic nuclear proteins but this designation is no longer appropriate.

3.3.1 Histones and non-histone proteins

(a) *Histones*

There are five major types of histone molecule in the eukaryotic cell nucleus. These have been classified as histones H1, H2A, H2B, H3 and H4 since a CIBA Foundation Symposium of 1975 [99]. Previously a variety of differing nomenclatures had been adopted by different laboratories. The histones are basic proteins of low M_r which are readily isolated by salt or acid extraction of chromatin and which can be separately purified on the basis of size or charge [100] (Fig. 3.6). In 2 M NaCl all the histones are dissociated from DNA. However in 0.5 M NaCl histone H1 alone is dissociated so that the functions of the residual or 'core' histones may be investigated [101, 102]. Each molecule consists of a hydrophobic core region with one or two basic arms. Histone H1 is a very lysine-rich protein of about 216 amino acids which shows a high degree of sequence conservation among eukaryotes particularly at the central, apolar region. Histones H2A and H2B are even more highly conserved and are known as the lysine-rich histones. The most conserved of all are the arginine-rich histones H3 and H4. Only three positions in the H4 histone molecule have been found to vary yet quite large regions can be deleted without serious effects being observed [103–105].

Despite the conservation of amino acid sequence there are multiple genes for each histone in all eukaryotes. This reaches extreme proportions in the sea urchin and the fruit fly (*Drosophila*) which contain multiple tandem repeats of a region of DNA containing one gene for each histone

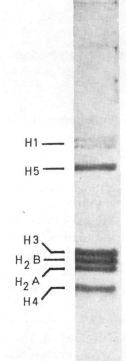

H1 ——

H5 ——

H3

H₂B

H₂A

H4

Fig. 3.6 Polyacrylamide gel electrophoresis pattern of histones from chick erythrocytes (reproduced, with permission, from [115]; Blackwell Scientific Publications Ltd., Oxford).

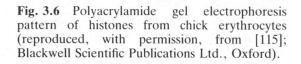

separated from each other by an A + T-rich spacer region (section 9.2.2). Not all these genes produce identical histones. The histone genes present in greatest abundance appear to be active in early development when there is a great demand for new histone synthesis, whereas the minor variants are made at later stages of development and represent the major histones present in somatic cells. The genes for these variant histones may or may not be associated with the tandemly repeated blocks [105–110] (section 9.2).

A special type of histone known as histone H5 is found in the nucleated erythrocyte of

fish, amphibians and birds (Fig. 3.6). It bears many similarities to histone H1 and is thought to maintain the highly repressed state of the chromatin in these non-dividing cell types [111]. In non-dividing cells in mammals histones $H1^0$ and H1e (see below) are present, whereas histones H1a and H1b are present in large amounts only in dividing cells [106, 108].

It is not clear what effect the variations in sequence of the histones has on chromatin structure. Neither is it clear what is the effect of post-translational modification. However, the histones may be methylated, phosphorylated, acetylated or ADP-ribosylated, and some of these modifications, by altering the charge on the molecule, may affect the interactions of histones with each other or with DNA [112–116]. For example there are six subtypes of histone H1 (H1a–e and $H1^0$) giving rise to 14 different phosphorylated forms, the prevalence of which varies during the cell cycle. Acetylation of histone H4 in particular causes unfolding of the nucleosome core particle (see below) and is associated with transcriptionally active regions of chromatin [117, 118]. As much as 20% of histone H2A is covalently linked to a 76-residue polypeptide, called ubiquitin, to form a branched-chain protein known as UH2A or A24 [119]. Marked changes in the level of ubiquitination of histones occur during mitosis [120, 121]. The possible role of modified histones in the control of gene expression is further considered in section 10.6.3.

In sperm cells, histones are replaced by other small basic proteins known as *protamines* [122, 123].

In the absence of DNA, the 'core' histones will associate with each other, the predominant species being a homotypic tetramer in the case of histones H3 and H4 (i.e. a tetramer containing four similar, arginine-rich histones) and a dimer in the case of H2A and H2B. At high ionic strength, all four histones interact together to form complexes which have variously been described as heterotypic tetramers, with one, or octamers, with two of each type of core histone molecule [124, 125].

(b) *Non-histone proteins*

These are present in chromatin in an amount approximately equal to the histones. On SDS/polyacrylamide gels about 100 different proteins can be seen. Some of these are the enzymes involved in replication and transcription (chapters 6 and 8) or form part of the nuclear envelope (section 3.1). Others resemble the histones in being of low M_r (or high mobility on electrophoresis) and are known as the high-mobility group, or HMG proteins. These also resemble histones in being basic proteins and they are present in multiple copies in the chromatin, i.e. they play a structural role. They differ from histones in being only loosely associated with chromatin – they can be extracted with 0.35 M NaCl – and in lacking an apolar centre. They have a basic *N*-terminal region and an acidic *C*-terminal region separated by a short region rich in serine, glycine and proline. The most characterized are HMG1, HMG2, HMG14 and HMG17 [126–128] (section 10.6.3).

3.3.2 The nucleosome

The DNA from a human cell is of the order of 1 m in length and must be condensed into a cell nucleus whose diameter is of the order of 10 μm. The packing must, however, maintain accessibility and prevent tangling during replication. Eukaryotic cells achieve this condensation by a series of packaging mechanisms involving the histones and some other chromosomal proteins, initially to form *nucleosomes* [161].

The elucidation of the structure of the nucleosome is a delightful example of how the results from a large number of experimental approaches can be used to solve a problem. At the same time as the studies on histones were going on, other groups were studying the nucleus using electron microscopy, X-ray and neutron diffraction. Still others employed cross-linking and nuclease studies in which the DNA products were analysed by electrophoresis on agarose or polyacrylamide gels [115, 129, 130].

The initial X-ray diffraction patterns of chromatin indicated the presence of a structure repeating every 10 nm but further interpretation was difficult [131]. Electron microscopy of ruptured nuclei showed the presence of a series of spherical particles joined by thin filaments – the so-called beads on a string picture [132] (Fig. 3.7). The beads have a diameter of 7–10 nm but the length of the filaments is variable. Similar electron micrographs of the SV40 minichromosome present in virus-infected cells indicated the presence of about 21 beads or nucleosomes on each viral DNA molecule [133, 134]. As the contour length of naked SV40 DNA is 1590 nm and that of the minichromosome is only 250 nm it is clear that there has been a six- to seven-fold packing of the DNA into the minichromosome.

Not only is DNA compacted in chromatin, it is also rendered partially resistant to nuclease action (Fig. 3.8). Using micrococcal nuclease which makes double-stranded breaks in DNA it quickly became apparent that at early times of digestion the nuclease was cutting the string (i.e. the linker DNA) which holds the nucleosomes together (Fig. 3.8). An analysis of the size of the DNA showed that the spacing between successive nucleosomes was about 200 bp.

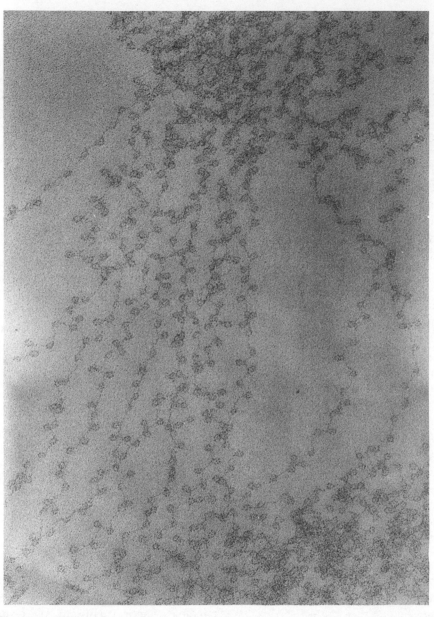

Fig. 3.7 Chromatin fibres streaming out of a chicken erythrocyte nucleus. The bead-like structures (nucleosomes or *v* bodies) are about 7 nm in diameter. The connecting strands are about 14 nm long. The sample was negatively stained with 5 mM uranyl acetate and the magnification is 285 000. (By courtesy of D. E. Olins and A. L. Olins.)

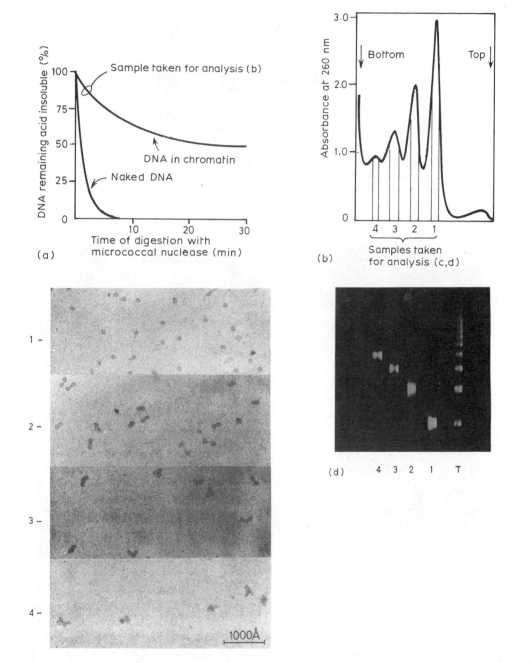

(a)

(b)

(c)

(d)   4 3 2 1   T

Fig. 3.8 The time course of digestion of chromatin with micrococcal nuclease (a). Samples were taken at the position shown and fractionated on a sucrose gradient (b); fractions from positions 1, 2, 3 and 4 of the gradient were shown to contain mono-, di-, tri- and tetra-nucleosomes by electron microscopy (c); the DNA in these fractions was shown by electrophoresis (d) to be of approximate length 200 bp, 400 bp, 600 bp and 800 bp. T shows total DNA before fractionation on the sucrose gradient (reproduced, with permission, from [137]).

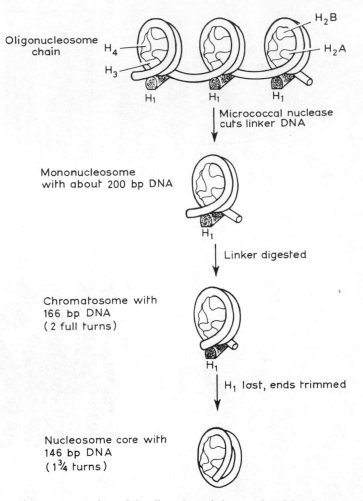

Fig. 3.9 Diagrammatic representation of the digestion of chromatin with micrococcal nuclease showing the relative positions of the various histones in the nucleosome (from [348]).

On further digestion the size of the DNA fragments was reduced first to 166 bp and finally to 146 bp [135–137] (Fig. 3.9).

Meanwhile, cross-linking studies using dimethylsuberimidate [138] had shown that in chromatin there was present an octamer of histones of composition (H2A, H2B, H3, H4)$_2$. Analysis of the stoichiometry of histone and DNA suggested that one octamer was present per 200 bp DNA, i.e. per nucleosome.

Another nuclease (DNase I) is able to make nicks all along the length of DNA in chromatin. The nicks occur at ten base intervals which was interpreted to mean that the DNA was wrapped around a core of histones and was accessible to nuclease on its outer surface; a conclusion which was confirmed by neutron diffraction [115, 139, 140].

Further studies involving X-ray crystallography [140–142] have shown that the nucleosome is a shallow, v-shaped structure around which a 146-bp core of DNA is

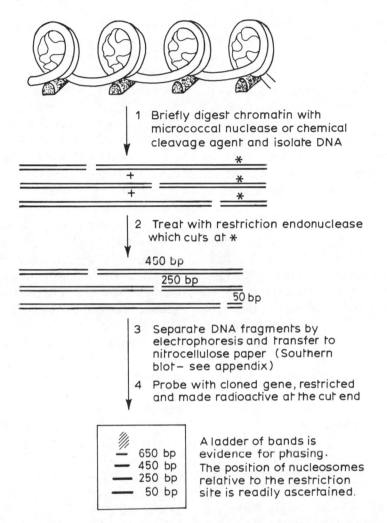

Fig. 3.10 One method for studying nucleosome phasing. Other methods involve more extensive digestion with micrococcal nuclease to produce core particle DNA which can be cloned and sequenced or mapped with restriction enzymes [152, 158, 159].

wrapped making about one-and-three-quarter turns (Fig. 3.9). In order to bend so acutely the DNA double helix makes several sharp bends or kinks [142, 143] which explains how a rigid molecule of persistence length 150 bp [144] can make such tight turns around the histone core. It also explains why certain regions are less sensitive to DNase I than expected.

From DNA–protein cross-linking studies and the X-ray crystallography of nucleosomes it has become clear that an (H3, H4)$_2$ tetramer provides the framework of the nucleosome and an (H2A, H2B) dimer is added to each face of the framework [142, 145]. This is the nucleosome core associated with 146 bp DNA.

Histone H1 holds the two ends of nucleosomal DNA together to form a chromatosome of 166 bp and the remaining DNA forms the linker joining nucleosomes together to form oligonucleosomes (Fig. 3.9).

The length of the linker and hence the repeat frequency of nucleosomal DNA is variable both between species and even within tissues. The repeat frequency ranges from 212 bp for chick erythrocytes (an inactive tissue) to 165 bp for yeast (highly active) [123, 146] and although it would be logical to anticipate that active chromatin might be less closely packed than inactive chromatin this is not the case.

Nucleosomes can be reconstituted *in vitro* from DNA and histones. Reconstitution usually requires the slow dialysis of a mixture of DNA and histones from high- to low-salt conditions [133] but has been achieved under more physiological conditions [20, 147]. *In vivo* proteins are present in the nucleus which are believed to play a part in the assembly of chromatin on newly synthesized DNA (section 6.17).

3.3.3 Nucleosome phasing

The question arises as to whether the nucleosome cores are present in fixed positions on the DNA or are (a) free to move or (b) present in random positions so that in a population of cells all the DNA could be found in the linker region. Such nucleosome positioning or *phasing* may have particular significance with respect to gene expression particularly as certain DNA sequences could be facing into the nucleosome and hence be unavailable for recognition by sequence-recognizing proteins [148–150] (section 10.4.3). Under biological conditions, the nucleosome appears to be remarkably stable and to have little tendency either to travel along a single length of DNA or to move among preformed chromosomes [151]. The approaches to analysing nucleosome phasing have been summarized by Kornberg [152]

but all initially involve the use of micrococcal nuclease (to cleave linker DNA) and a restriction enzyme (section 4.5.3) to make a specific cut in a gene which could be identified by Southern blotting (section A.3). This is illustrated in Fig. 3.10. The limitation with such methods is the preference of micrococcal nuclease for certain regions of DNA rich in sequences such as CATA or CTA [153–155].

This has led to the development of chemical cleavage reagents which show less or no sequence preference [156]. The conclusions reached are that micrococcal nuclease and the chemical reagents will cut naked DNA or chromatin in certain well-defined positions relative to the 5'-ends of genes. This can probably be interpreted as showing the presence of a nucleosome-free region of unusual base composition and secondary structure [157]. When a nucleosome-free region exists, then the nucleosomes present on either side of this region are very likely to take up particular positions as the nucleosome repeat is about 200 bp [152]. Certain positions may be favoured by particular base compositions (see below) and if these regions recur regularly as in satellite or 5S ribosomal DNA then the phasing will be reinforced [158, 159]. As satellite DNAs are built up of a number of subrepeats (section 3.2.2(b)) there may be little to choose between positioning a nucleosome at the beginning of one subrepeat or the next. As these positions may be only 20 or 30 bp apart this results in multiple phases on satellite sequences, each related to the other by the subrepeat sequences of the DNA. Phasing may be strongly reinforced on satellite and possibly on other DNA sequences as a result of proteins which bind to DNA so as to hold the DNA into a looped structure. Such a protein has been found for the α-satellite of African green monkey cells [160].

The favoured positions taken up by nucleosomes referred to above may relate to the requirement for DNA to bend or kink as it wraps around the nucleosome core [142, 161] and on reconstitution of chromatin on DNA of defined sequence the nucleosomes have been shown to form at a specific region [162, 163]. Not only may nucleosomes be positioned relative to genes but there is evidence that they take up a precise position relative to the centromere [39].

3.3.4 Higher orders of chromatin structure

Histone H1 also plays a role in the association of adjacent nucleosomes, and histone H1 molecules can be chemically cross-linked to each other [164, 165]. In the absence of histone H1, nucleosomes and polynucleosomes are soluble over a wide range of ionic strength. However, when histone H1 is present they precipitate above NaCl concentrations of 80 mM and $MgCl_2$ concentrations of 2 mM. This provides a useful mechanism for the separation of nucleosomes with or without histone H1 [166, 167]. In the presence of histone H1, electron microscopy and X-ray diffraction studies show a fibre of 30 nm diameter as well as the 10 nm fibre (which is a chain of single nucleosomes). Although different explanations have been proposed for the structure of the 30 nm fibre it now appears that this is a hollow cylinder or solenoid into which the nucleosomes are coiled [130, 168–170].

There are about six nucleosomes per turn of the solenoid (Fig. 3.11) (although 12 nucleosomes per turn have also been proposed as has a superbead model [334, 335]). The pitch of the solenoid is 11 nm and the faces of the nucleosomes are approximately parallel to the solenoid axis.

3.3.5 Loops, matrix and the chromosome scaffold

When cells, at any stage of the cell cycle, are lysed under certain conditions the DNA can be sedimented intact and displays the dye binding and sedimentation characteristics of supercoiled DNA in loops of 100 kb in size (Fig. 3.12). These loops are stabilized by proteins rather than RNA since proteases but not ribonucleases abolish the supercoiling [171, 172]. Metaphase chromosomes which have been deproteinized are readily visualized in the electron microscope as looped structures with the thread of DNA in each loop entering and leaving the chromosome at the same central point [173]. The loops may be attached to a central core from which they radiate [23].

When the DNA is wrapped around a nucleosome in the cell nucleus its linking number is changed relative to B-form DNA (section 2.5.1) and the DNA is essentially relaxed. However, when the histones are removed the DNA tends to return to the B-form and, as this is prevented by the ends of the loops being fixed at their bases, the DNA is recovered with superhelical turns [161]. We, therefore, have the picture of the chromosome consisting of a series of loops of DNA present in a nucleosomal conformation and condensed at least in part into a solenoid (Fig. 3.11). The degree of solenoid formation may be related to the extent of expression of the DNA (section 10.4.4).

It is possible to digest 99% of the DNA from metaphase chromosome preparations leaving behind a morphologically intact central chromosome 'scaffold'. In a similar manner the material which remains when dehistonated interphase chromatin has most of its DNA removed by nuclease treatment is known as the nuclear 'matrix'. One set of conditions for the removal of chromatin proteins involves the use of non-ionic

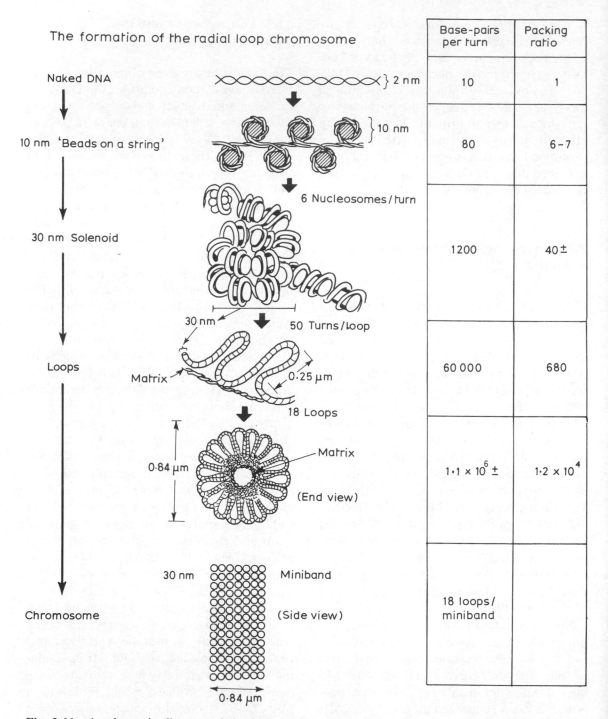

The formation of the radial loop chromosome

Naked DNA

10 nm 'Beads on a string'

30 nm Solenoid

Loops

Chromosome

2 nm

10 nm

6 Nucleosomes/turn

30 nm · 50 Turns/Loop

Matrix · 0.25 µm

18 Loops

0.84 µm

Matrix

(End view)

30 nm · Miniband

(Side view)

0.84 µm

Base-pairs per turn	Packing ratio
10	1
80	6–7
1200	40±
60 000	680
1·1 × 10^6 ±	1·2 × 10^4
18 loops/ miniband	

Fig. 3.11 A schematic diagram of the higher-order organization of a chromatid (reproduced, with permission, from [23]; copyright the Company of Biologists Ltd).

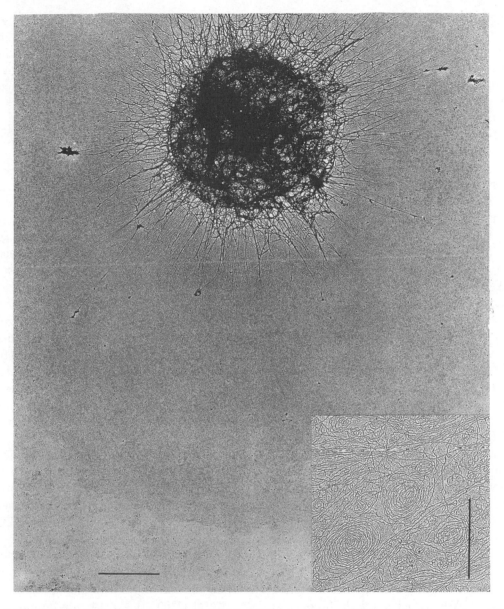

Fig. 3.12 An electron micrograph of some of the DNA in a human (HeLa) cell. The nuclear DNA has been spread throughout most of the field to form a skirt which surrounds the collapsed skeleton of the nucleus. A tangled network of DNA fibres radiates from the nuclear region. The bar represents $5\,\mu$m. *Inset*: Only at the very edge of the skirt can individual duplexes be resolved. Most appear as collapsed toroidal or interwound superhelices, indicating that the linear DNA must be unbroken and looped, probably by attachment to the nuclear skeleton. The bar represents $1\,\mu$m. (By courtesy of Dr P. R. Cook and Dr S. J. McReady).

detergents, which disrupt the membranes, high concentrations of EDTA (ethylene-diaminetetraacetic acid), to chelate the bivalent metal ions which activate nucleases, and high molarities of sodium chloride which dissociate most proteins. However, it is possible that these stringent conditions cause rearrangement of the scaffold/matrix attachment regions (SARs or MARs) and a milder extraction method using lithium diiodosalicylate has been developed [21]. The scaffold proteins include two abundant proteins of M_r 170 000 and 135 000. The larger is DNA topoisomerase II (section 6.5.4) and the smaller binds MARs in a cooperative fashion. This leads to the MARs being gathered together with the intervening DNA looped out [342]. In addition, the matrix contains an ill-defined fibrous protein network and is also associated with the nuclear pore–lamina complex [8, 22, 23] to which the chromosomes are attached (section 3.1). The chromosomes, therefore, appear to take up a precise intranuclear location which is retained throughout the cell cycle (section 3.1) and, when low-salt isolation conditions are used, the matrix-associated DNA may represent sites important in chromosome organization [336].

The loop model of chromosome structure implies that specific DNA sequences may be adjacent to and, therefore, associated with the attachment sites of the loops. Many reports exist showing that active, transcribing DNA is associated with the nuclear matrix (section 10.6.4) and the loop has also been associated with the replicon – the unit of replication (section 6.15.1). Although there has been some controversy over these results [175], it seems clear that MARs occur in particular places and hence particular DNA sequences are associated with the nuclear matrix [21, 177, 337, 338]. Under certain conditions of isolation these se-

quences may be actively transcribing. In addition, both initiation and continued replication of DNA occur in association with matrix proteins and topoisomerase II binding sites are found on matrix associated DNA (section 6.15.5).

3.3.6 Lampbrush chromosomes

The diplotene stage of meiosis may last for several months in developing oocytes and during this time the chromosomes are visible under the phase contrast microscope while still undergoing transcription. Such lampbrush chromosomes have been most extensively studied in the oocytes of *Triturus* (newt) and *Xenopus* (toad).

At this stage the chromosomes are paired, the two homologues being held together by chiasmata. Each member of the pair consists of two chromatids, i.e. two DNA duplexes. The DNA duplexes run the length of the chromosome, being attached to the axial thread in a series of loops. As there are two chromatids, the loops are paired (Fig. 3.13) and the structure has the appearance of those brushes which used to be used to clean lamps but which are now used to clean test-tubes and bottles. The axis consists of protein (the matrix or scaffold) and DNA, most of which is compacted (probably in a solenoidal form) to form beads of about 1 μm in diameter known as chromomeres. These chromomeres contain up to 95% of the total DNA, the rest being largely in the 4000 or so loops in each chromosome.

The loops are the site of transcriptional activity and the DNA in the loops may be only partly condensed into nucleosomes. The loops are made more visible by the accumulated RNA. As the RNA polymerase molecules move around a loop the length of the RNA transcript increases, producing the so-called *Christmas tree effect* (Fig. 3.13).

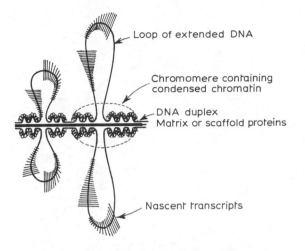

Fig. 3.13 Diagram of part of a lampbrush chromosome showing the paired loops of extended DNA and the Christmas tree effect produced by the nascent transcripts.

In situ hybridization has allowed the localization of histone and ribosomal and transfer RNA gene transcripts to particular loops [115, 178, 179].

3.3.7 Polytene chromosomes

In certain tissues, notably the salivary glands of the fruit fly *Drosophila* and some other species, the DNA molecules do not segregate after replication but remain together during several rounds of the cell cycle. By geometric progression, after ten rounds of replication there are more than one thousand DNA strands lying alongside each other with their specific sequences and specific bound proteins matched. This forms an excellent experimental system for analysis. The centromere remains under-replicated and the four chromosomes of *Drosophila* are readily identified by light microscopy as a highly characteristic, star-like structure with the spokes radiating from the centromere. Unlike metaphase chromosomes, these chromosomes are in the stretched-out, interphase condition and sites can be clearly identified where *puffing* due to transcriptional activity occurs. Puffs can be induced in certain sites by the application of specific insect hormones so that the process of transcription can be effectively visualized with respect to time [180, 181].

Polytene chromosomes have a highly characteristic banding pattern with dark and light bands of differing densities and dispositions at various points in the arms. This makes it possible to correlate the absence of a specific band with the lack of a specific gene function in a *Drosophila* mutant. Although it is not easy to identify unique genes on ordinary metaphase chromosomes because of the low specific activity of the small amount of mRNA or cDNA involved leading to weak radio-autography (section 3.1.4(b)), the large number of binding sites make it possible to perform *in situ* hybridization successfully on polytene chromosomes even in cases where a gene has a low overall frequency with respect to haploid DNA content. As well as their use for the identification of specific RNA hybridization sites, these chromosomes are extensively used in the analysis of the non-histone chromosomal proteins and Z-DNA by immunofluorescence since a specific antibody linked to a fluorescent second antibody can be used to identify the binding site on the DNA of the original antigenic protein or the presence of Z-DNA [182, 183].

3.4 EXTRANUCLEAR DNA

3.4.1 Mitochondrial DNA

Mitochondrial DNA is usually found in cyclic, double-stranded, supercoiled molecules, the exceptions being the linear mitochondrial DNA molecules from *Tetrahymena* and *Paramecium*. Mammalian mitochondrial DNA molecules are not packaged into nucleosomes. They are about 15 kb long and can therefore code for 15 to 20 proteins. However some yeast ones are considerably larger and *Saccharomyces* mitochondrial DNA is 75 kb long. Plant mitochondrial DNA is much longer still. Yeast mitochondrial DNA molecules have been widely studied since they are readily amenable to genetic analysis.

When yeast cells are grown in the presence of mutagens such as ethidium bromide, 'petite' mutants are formed in which large sections of the mitochondrial genome are deleted with the remaining segments being amplified so that the DNA molecules remain the same size. The mutants are readily propagated by the yeast cells but fail to synthesize mitochondrial proteins. Some petites have been isolated whose mitochondrial DNA contains only adenine and thymine [184] yet these DNA molecules are still able to replicate (section 6.11.4).

Clearly, mitochondrial DNA can only code for a small proportion of mitochondrial proteins yet the organization of the mammalian mitochondrial genome has been described as 'a lesson in economy' [185, 186]. Much of the difference in length between mammalian and yeast mitochondrial DNA is a result of the way the DNA is organized, as both appear to contain the same basic set of genes (section 8.5).

Mitochondrial DNA from several mammals, *Xenopus* and *Drosophila*, has been sequenced and the sequence analysed [187–191]. There are genes for two ribosomal RNAs, 22 transfer RNAs and for 13 proteins most or all of which are involved in electron transport (Fig. 3.14) [185, 186, 192, 193]. Some of these protein-coding regions have still only been identified as unassigned reading frames or URFs (section

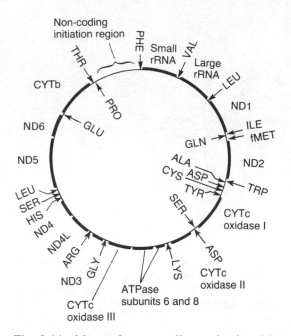

Fig. 3.14 Map of mammalian mitochondrial DNA showing the locations of the genes. Those on the outside are transcribed clockwise from one of the DNA strands; those on the inside anticlockwise from the other. The arrows indicate the positions of the tRNA genes. ND, NADH dehydrogenase complex [192, 193].

9.7). None of the mammalian genes contains introns but these are present in some yeast mitochondrial genes.

The only region of mammalian mitochondrial DNA which is non-coding is the D-loop region involved in the initiation of DNA replication (section 6.11.4). This is also the region at which transcription of both strands is initiated. Transcription continues uninterrupted around the cyclic molecule and the transcripts are then processed to give individual messenger, ribosomal and transfer RNAs. The transfer RNA forms the punctuation between the various protein coding regions and provides the signal for the processing enzymes (section 9.7). One of the proteins coded for by yeast mitochondrial DNA is a maturase which is involved in the splicing of mito-chondrial messenger RNA [194] (section 11.3.3).

Although plant mitochondrial DNA is apparently much more complex than animal mitochondrial DNA it consists of permu-tations of a basic structure related to each other by recombination [195–197].

One theory for the origin of mitochondria suggests that they arose as symbiotic pro-karyotes providing oxidative metabolism to their pre-eukaryotic hosts. Throughout evolution most of the genes were transferred from the symbiont to the nuclear DNA of the host leaving behind only the remnants we find today. The fact that transfer of genes can occur is shown by the finding of one gene (subunit 9 of ATPase) in the mito-chondrial DNA of some organisms but in the nucleus in other organisms [198–201]. Since the rate of production of base pair substitutions in mitochondrial DNA is about 10 times that in nuclear DNA it provides a highly sensitive tool for studying short-time divergences among related species [202].

3.4.2 Chloroplast DNA

Chloroplast DNA is, in general, much larger than mitochondrial DNA, being in the M_r range 100 million (150 kb). In contrast to mitochondria which have from one to ten molecules of DNA per organelle, chloro-plasts tend to have a very large number of copies of the DNA molecule in each organelle, in some cases greater than one hundred. Like mitochondrial DNA, the DNA in chloroplasts carries the coding information for essential membrane com-ponents, tRNA and rRNA [203–205].

Mutants similar to the petite mutants may be isolated but show a tendency to revert, possibly because of the multiplicity of copies of the master gene in the chloroplast. All

known chloroplast DNA molecules are cyclic and supercoiled.

3.4.3 Kinetoplast DNA

The kinetoplast is part of a highly specialized mitochondrion found in certain groups of flagellated protozoa such as the trypano-somes. Its DNA (kDNA) consists of an interlinked series of many thousands of cyclic DNA molecules that vary in size from 0.6 kb to 2.4 kb depending on the type of trypanosome from which they are obtained. These components, which are known as minicircles, are further interlinked with a much smaller number of larger circular DNA molecules of above 30 kb in size, known as maxicircles, in the majority of systems analysed. Whereas maxicircles appear to perform the conventional func-tions of mitochondrial DNA in trypano-somes, minicircles are microheterogeneous in sequence and size and appear to play a role in providing a template for RNA editing [192, 206] (section 11.8).

3.5 BACTERIA

Historically much of the early understanding of gene structure and function has come from the prokaryotes and from the simple bacteriophages which can infect them and grow in them. This has been largely due to the comparative ease with which the pro-karyotes can be grown and infected under laboratory conditions and the fast gener-ation time which has made the isolation of genetic mutants a comparatively simple exercise. In addition, the DNA content of prokaryotic cells is about one hundred-fold less than that of eukaryotic cells so that the analytical task is relatively simple.

By far the majority of genetic experiments

have been performed in varying strains of the bacterium *Escherichia coli* with a much smaller number being in *Bacillus subtilis*, *Micrococcus luteus* and other bacterial species. There is a wide variety of strains of *E. coli* itself, each possessing its own properties with respect to susceptibility of phage infection and to antibiotic resistance. Although bacteria normally divide by binary fission, the transfer of genetic information between two types of bacteria is possible. Strains of bacteria which are able to donate their chromosome are known as high-frequency recombinant or *hfr* strains and have been widely used in the study of bacterial gene function.

Bacteria which possess the full complement of bacterial genes are usually referred to as 'wild type' bacteria or proto-trophs and mutants are classified according to their missing functions. Thus a *thr⁻* mutant is a bacterium which requires thre-onine in its growth medium as it lacks the correct information to make the enzymes involved in threonine synthesis in their fully functional state. Such a nutritional mutant is called an auxotroph. The process of mutant selection is greatly enhanced if the cells are first grown in minimal medium in the presence of an inhibitor which will kill all the growing bacteria when only those mutants survive which are unable to re-plicate in the deficient medium.

Although most bacteria carry all their genetic information on a single, circular chromosome, some possess additional, small, extrachromosomal elements known as plasmids which carry genes coding for functions such as drug resistance. Plasmids are usually comparatively small and cyclic and can be present in one or several copies per cell (section 3.6). They are used widely as tools in genetic engineering. For a more detailed description of bacterial growth and physiology, there are several specialized texts on the subject which can be consulted [207, 208].

3.5.1 The bacterial chromosome

The chromosome of *E. coli* is a single cyclic molecule of 4.5 million base pairs which has an effective circumference of 1 mm but must be contained in a bacterial cell with a diameter in the order of 1 μm. As with eukaryotic chromosomes, the bacterial chromosome is associated with proteins. Some of these resemble the histones but whether they play the same role is not yet certain [340].

A complex packaging mechanism is necessary in order to ensure that all the DNA is folded within the bacterial cell in a manner which will not inhibit transcription, nor allow entanglement of the two daughter strands to occur during replication [176]. This is achieved by two main mechanisms.

First, the DNA is folded into between 40 and 100 loops, and second, each of these quasicircles is itself supercoiled independently of the others [209, 210]. Each loop is maintained as an independent physical area by DNA–RNA interactions since the loops are abolished by ribonuclease but not, in contrast to eukaryotic chromosomes, by proteolytic enzymes. Deoxyribonuclease abolishes the superhelical nature of the loops though limited treatment with this enzyme will attack only a small number of the loops and therefore leads to the abolition of only some of the supercoiling [211, 212]. The intact folded *E. coli* chromosome or nucleoid can be isolated in the presence of non-ionic detergents and high molarities of salt in two possible forms, free and membrane-associated. The free form which sediments at 1600–1700 S consists of about 60% DNA, 30% RNA and 10% protein, the bulk of which is the enzyme RNA

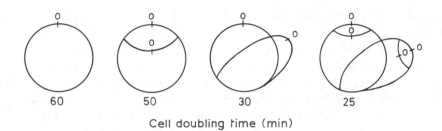

Fig. 3.15 Diagrammatic representation of the structure of one chromosome (of the two present in the cell) immediately before cell division in *E. coli* growing at different rates [214]. O is the origin of chromosomal replication.

polymerase (chapter 8). The membrane-bound form contains an additional 20% of membrane-attached protein. The exact nature of the membrane-attachment sites is not yet known, but they contain the origin of replication (section 6.10.1).

3.5.2 The bacterial division cycle

As in eukaryotes (section 3.1.2), the general rule in bacteria is that chromosome replication (DNA synthesis) and cell division occur alternately. Under slow growth conditions (i.e. doubling times greater than 60 min) there is a gap between completion of a round of DNA synthesis and cell division. Cell division takes place at a fixed time (generally about 60 min) after initiation of a round of DNA synthesis [213].

However, bacteria, under suitable conditions, can grow much more quickly than one division every 60 min and in these situations DNA synthesis becomes continuous, i.e. it no longer occupies a restricted part of the cell cycle [214]. Although cell division still occurs 60 min after initiation of a round of DNA synthesis the cell does not wait until cell division has occurred before initiating a second (or even a third) round of DNA synthesis. This results in some newly divided

cells having just the one copy of part of the DNA (the last to be duplicated) and up to four copies of those regions replicated first (Fig. 3.15).

Protein synthesis is required to initiate a round of chromosome replication and hence an amino acid auxotroph, when starved of essential amino acids, will complete current rounds of replication and come to rest with one unreplicated chromosome per cell. In contrast, a thymine-requiring mutant, when deprived of thymine, cannot make DNA but the chromosome is brought into such a condition that when thymine is restored a second round of replication is initiated.

3.5.3 Bacterial transformation

Pneumococci (*Diplococcus pneumoniae*) may be classified into a number of different types each characterized by the ability to synthesize a specific, serologically and chemically distinct capsular polysaccharide. In 1928, Griffith observed that a particular strain of pneumococci, cultivated *in vitro* under specific conditions, lost the ability to form the appropriate polysaccharide and consequently grew on solid media in so-called 'rough' colonies in contrast to the 'smooth', glistening colonies formed by encapsulated cells. If a living culture of such unencapsulated cells was injected into mice together with killed encapsulated pneumococci of type III, the organisms subsequently recovered from the animals were live, virulent pneumococci of the encapsulated type III. It appeared, therefore, that some material present in the dead type III organisms had endowed the unencapsulated pneumococci with the capacity to synthesize the characteristic type III polysaccharide.

During the next five years it was shown that such pneumococcal transformation could be produced *in vitro* and that a cell-free extract could replace killed cells as the transforming agent and in 1944 Avery, McLeod and McCarty [215] showed that the transforming principle was DNA. This was an experiment of fundamental importance in showing that DNA was the genetic material.

These observations stimulated further research into bacterial transformation, from which it emerged that the reaction was not limited to pneumococci but could be produced in a wide range of bacteria and that these transformations were not limited to changes in serological type since they could also be used to endow bacteria with resistance to specific drugs or antibiotics or the ability to utilize particular nutrients. Transformation can be demonstrated for any characteristic whose acquisition can be measured in the recipient and is made use of in many modern techniques involving genetic manipulation.

The proportion of treated cells which may develop a new characteristic after exposure to appropriate DNA is usually small and some strains of bacteria are much more susceptible than others [339]. *E. coli* requires treatment with $CaCl_2$ in order to act as a recipient at all [216] whereas *Neisseria* are particularly susceptible to transformation (section 7.5.3). In a single population not all cells are competent to take up DNA

and this may be a result of variable cell wall permeability. The most appropriate time for transformation to occur is just after cell division. When pneumococci are cooled to a temperature at which growth is arrested and then rewarmed so that they start to divide synchronously, transformations are exceptionally numerous.

The number of transformation events is proportional to the DNA concentration up to a plateau level and hence the frequency of transformation has been used as a measure of gene frequency (section 6.8.1). Duplex DNA enters the recipient better than does single-stranded DNA and the saturating DNA concentration is reached when about 150 DNA molecules of M_r about 10^7 have been taken up by each bacterium [217].

Just how the transforming DNA enters the cell to be transformed is not fully understood, but it is known that the acquisition of the new characteristic induced by the DNA requires a period up to 1 hour and that for some time after acquiring the new DNA a cell multiplies more slowly than do its unchanged neighbours. The establishment of the mechanism for duplicating the new DNA requires still longer. The mechanism of the transformation reactions has been reviewed [218, 219].

Although duplex DNA is better for transformation, on entry into the recipient cell one strand of the duplex is degraded. The major endonuclease of pneumococcus is associated with the cell membrane and it is responsible for degrading one of the DNA strands following action of an enzyme which introduces single-stranded breaks every 6000 bases [220]. In *E. coli*, the recipient bacterium must be *recBC⁻* otherwise the recBC nuclease degrades the incoming DNA. For expression, the incoming single strand of DNA finds a complement in the recipient cell DNA and displaces one of the original duplex strands in a recombination-type mechanism (chapter 7). This integration event occurs at widely different efficiencies for different markers.

It is possible to transfer more than one inheritable characteristic to susceptible bacteria in a single DNA preparation, e.g. one sample of DNA may carry the three characteristics of resistance to penicillin, resistance to streptomycin, and the ability to form a capsule in pneumococci. Such a sample of DNA might bring about transformation in 5% of the recipient cells. Of the cells transformed, 98% would acquire only one of the three characteristics, 2% would acquire two of the characteristics, and only 0.01% would acquire all three (this is not the case when eukaryotic cells are transformed – see below). Clearly, therefore, the DNA preparation cannot convey a complete set of the donor's characteristics to the recipient, although certain characteristics appear to be linked.

An analogous somatic transformation was first reported for mammalian cells by the Szybalskis [221]. They isolated DNA from cultures of human cells of the strain D98S which contain the enzyme hypoxanthine phosphoribosyltransferase (HPRT) (EC 2.4.2.8) responsible for the reaction:

$$\text{hypoxanthine} + \text{PRPP} = \text{IMP} + \text{PP}_i$$

which is discussed further in chapter 5. The addition of this DNA to cultures of cells of the strain D98/AH-2 which are deficient in the enzyme resulted in the appearance of HPRT-positive, genetically transformed cells detected under highly selective conditions. The transforming activity was abolished by deoxyribonuclease but not by ribonuclease. Modern methods of introducing foreign DNA into animal cells are discussed in section A.9.

3.6 PLASMIDS

Plasmids are duplex, supercoiled DNA molecules which range in size from 2.5×10^6 to 1.5×10^8 daltons. They are stable elements which exist in bacteria and some eukaryotes in an extrachromosomal state. The large ones are present in only one or two copies per cell whereas there may be 20–100 or more copies per cell of the small ones. They enjoy an autonomous, self-replicating status without significantly lowering host viability [219, 222–227].

There is a continuum from the self-replicating plasmid which has no means of transfer from cell to cell, through those plasmids which can be transferred (e.g. the F-factor – see below) and the non-integrative transducing phage (e.g. P1 – section 3.7.8) through to the fully fledged virus which codes for its coat proteins and means of entry into a new host cell (section 3.7).

One type of plasmid is called a sex or F-factor. A bacterium possessing an F-factor (F^+) is male and the F factor induces the bacterium to make a tube or pilus by which the male bacterium becomes attached to a female bacterium, i.e. one not carrying the F-factor. When the F-factor replicates (chapter 6) one linear, single-stranded copy of plasmid DNA may pass through the pilus 5'-end first into the female bacterium. The complementary strand is synthesized in the recipient which thereby becomes male as it now carries the F-factor. Soon all the cells in a population become F^+. The presence of an F-factor in a bacterium excludes the entry of a second, closely related F-factor, but more than one type of plasmid may be stably maintained in a cell if the two are not incompatible.

As well as the F-plasmid, other plasmids and parts of the bacterial chromosome may pass through the pilus and this leads to a low frequency of exchange of genetic markers between bacteria, i.e. a male (F^+) proto-troph bacterium may transfer the gene required for tyrosine synthesis to a female tyr^- auxotroph. This is known as transfer by *bacterial conjugation* [228].

In one in $10\,000\,F^+$ cells the F-factor becomes integrated into the bacterial chromosome by reciprocal crossing over (chapter 7). A plasmid which can exist either autonomously or in an integrated state is called an episome. For insertion the F-factor breaks at a unique site, but this linear piece of DNA integrates into the host DNA at random with regard to position and orientation. Like the λ-prophage (section 3.7), an integrated plasmid loses its ability for independent replication [174]. However, *E. coli* mutants, defective in initiation of chromosomal replication can be rescued by integration of an F-factor which takes over control of initiation. The inte-grated plasmid still causes the formation of pili and can be transferred by conjugation, but now, between the leading 5'-end of the plasmid DNA and the trailing 3'-end, there is the entire bacterial chromosome which gets dragged along the conjugation tube, into the female cell. Since transfer of the whole chromosome takes about 60 min, breakage of the conjugation tube normally occurs before transfer is complete. When this occurs the second half of the F-factor is not transferred and so the recipient remains female. As the fragments transferred cannot replicate autonomously they are either lost, or are integrated into the recipient's chromosome. The culture thereby shows a high frequency of recombination of genetic markers and cells with an integral F-factor are called *hfr*. Because of the mode of integration of an F-factor into the bacterial chromosome, different *hfr* strains transfer the genes of the bacterial chromosome in a different order and with different polarity. However, closer study shows the gene order

to be circularly permuted, providing evi-dence for the linear arrangement of genes on a circular chromosome [229].

Sometimes an *hfr* strain may revert to F^+, and when this occurs a section of the bac-terial chromosome may also be excised and be found in the plasmid (e.g. Flac contains the *lac* operon integrated into an F-factor which is thereby more than doubled in size). Such plasmids are known as F'-factors. At conjugation, the bacterial gene is transferred from the donor to the recipient bacterium (cf. transduction – section 3.7.8) to produce a partially diploid organism and recom-bination readily occurs between the two copies of the gene (chapter 7). Thus in Flac$^+$, *lac* is transferred infectiously but other markers are only transferred at low frequency.

R-factors are plasmids which carry genes conferring resistance to one or a number of drugs. Resistance arises as the result of the production of enzymes which modify or degrade the drug, e.g. chloramphenicol resistance depends on an enzyme which acetylates chloramphenicol; ampicillin resistance by a β-lactamase. Resistance to tetracycline is acquired by virtue of an inability to take up the drug into the cell. R-factors may or may not carry genes en-coding a pilus enabling them to be trans-ferred to other bacteria. For those which can promote their own conjugative transfer it is common to find the antibiotic resistance-determining region is separated from the transfer region. Those R-factors not carrying the transfer (*tra*) genes can be transferred passively from one bacterium to another either by transduction or by relying on mobilization by a conjugative plasmid present in the same cell.

When first discovered, the location of multiple drug-resistance genes or of genes coding for cognate restriction and modi-fication enzymes (chapter 4) on a single plasmid was difficult to explain. However, it transpires that many of these genes are present on transposable genetic elements or transposons. These are discrete segments of DNA which have the ability to move around among the chromosomes and extrachromosomal elements, i.e. plasmids or viruses. Thus the ampicillin-resistance gene referred to above rapidly moves from one plasmid to another and from one location to another on the plasmid in the form of a 4800 bp fragment known as transposon 3 (Tn3). (The structure of transposons and mechanism of transposition are considered in section 7.7.) Transposons also occur in eukaryotes and the first circumstantial evidence for their existence was obtained in studies with maize [230].

A third group of plasmids synthesize chemicals (colicins) which are toxic to bacteria. The larger of these colicinogenic factors can be transferred by conjugation but the smaller ones such as the Col E factors do not code for enzymes to bring about their transmission. Plasmids of *Rhizobium* and *Agrobacterium* carry the genes responsible for nitrogen fixation and crown gall tumour formation in plants [231, 232].

Plasmids, especially the smaller ones existing at high copy numbers, are used as cloning vehicles in genetic engineering (section A.8) and the presence of drug-resistance markers is helpful in the selection regimens used. One of the early vectors used was pSC101 (derived from a fragment of a larger R-factor), but derivatives of Col E1 are now in common use. Thus pBR322 was engineered by removing some regions and inserting others (e.g. the tetracycline-resistance gene from pSC101) to give an autonomous plasmid of M_r 2.7 × 10^6 [233].

Hybrid plasmids can be constructed containing parts of, for example, pBR322 and a small plasmid found in yeast (the 2μ plasmid) or a segment of an animal virus (e.g. SV40). These plasmids are able to grow both in bacteria and in yeast or animal cells and have been much used in the study of eukaryotic gene expression (section 10.4).

3.7 VIRUSES

3.7.1 The structure of viruses

The general structure of a virus is a nucleic acid core surrounded by a protein coat. The nucleic acid may be single- or double-stranded RNA or DNA. The coat may consist of many identical molecules of protein (capsomeres) or may be a complex structure such as one shown in Fig. 3.16(a). The protein coat protects the nucleic acid from damage and also may confer a specific host range on the potential infectivity of the particle.

Viruses infect animals, plants and bacteria and the last are called bacteriophages, or simply phages. As the viruses usually lack any machinery to catalyse metabolism or protein synthesis they can only multiply within a host cell whose metabolism is subverted to serve the requirements of the virus [219, 234, 235].

Bacteriophages contain single-stranded RNA (e.g. MS2), single-stranded DNA (e.g. ϕX174) or double-stranded DNA (e.g. T-phages). The single-stranded DNA phages are either spherical (e.g. ϕX174) or filamentous (e.g. fd or M13) where the circular DNA molecule is wrapped in a protein coat and then formed into a filament of two nucleoprotein strands by bringing the opposite sides of the circle together. All the RNA bacteriophages are spherical. MS2 consists of 180 protein subunits all identical except for one (the maturation protein).

Many of the double-stranded DNA phages have a more intricate structure.

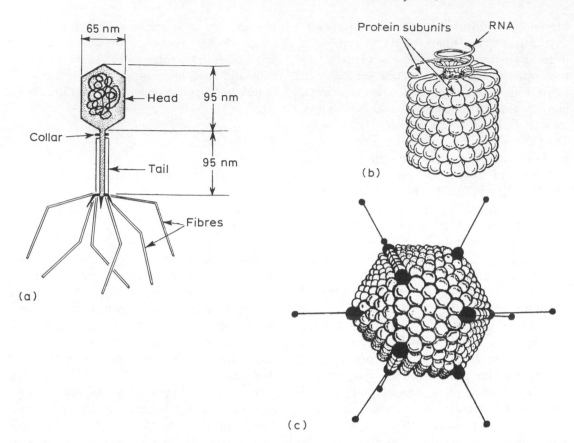

Fig. 3.16 Structure of virus particles. (a) T-even phage particle. (b) A segment of tobacco mosaic virus (TMV) showing the protein subunits forming a helical array [236]; the RNA lies in a helical groove in the protein subunits, some of which have been omitted to show the top two turns of the RNA helix. (c) A model of adenovirus showing the arrangement of the capsomeres and spikes.

The T-even phages (Fig. 3.16) have a head (which contains the DNA), a tail (through which the DNA is injected into the host cell), and a base plate with six tail fibres which recognize, and attach to, sites on the surface of the host cell.

Most plant viruses contain single-stranded RNA as in TMV (tobacco mosaic virus), but a few contain double-stranded RNA as in wound tumour virus; or double-stranded DNA as in cauliflower mosaic virus. The virus particles can be rod-shaped (e.g. TMV) or spherical (e.g. cowpea chlorotic mottle virus).

An extensive study of TMV (for reviews see [236, 237]), which has a particle weight of 4×10^7 and is rod-shaped (measuring $15 \times 300\,\mathrm{nm}$), has led to a detailed picture of its structure (Fig. 3.16). A helical array of about 2100 identical protein subunits of M_r 17 400 surround a single-stranded RNA molecule of M_r 2×10^6. Cowpea chlorotic mottle virus is spherical and is composed of 180 identical subunits of M_r about 20 000 arranged on the surface of an icosahedron in 32 morphological units of 20 hexamers (on the faces) and 12 pentamers (on the vertices) [238].

Table 3.2 Baltimore classification of viruses [239]

Class	Nucleic acid	Replication	Example
I	Duplex DNA	Semiconservative	T4, adeno and herpes viruses
II	Single-stranded DNA (+)	Via duplex replicative form (RF) $(+) \rightarrow (\text{RF}) \rightarrow (+)$	ϕX, minute virus of mouse (MVM)
III	Duplex RNA	Via (+) strand RNA intermediate $\rightarrow (+) \rightarrow$	Reovirus (produces infantile diarrhoea)
IV	(+) strand RNA	Via (−) strand RNA intermediate $(+) \rightarrow (-) \rightarrow (+)$	MS2, polio, foot and mouth disease virus
V	(−) strand RNA	Via (+) strand RNA intermediate $(-) \rightarrow (+) \rightarrow (-)$	Measles, 'flu and rabies viruses
VI	(+) strand RNA	Via DNA duplex $(+) \rightarrow \rightarrow \rightarrow (+)$	Leukaemia virus HIV

A continuous line represents DNA; a dashed line represents RNA.

Of the animal viruses, adenoviruses are icosahedral with 1500 subunits arranged as 240 hexamers on the faces and 12 pentamers at the vertices and a third type of protein forms the fibres which extend from each vertex. Many of the larger animal viruses (e.g. influenza virus) are surrounded by a lipoprotein envelope largely derived from host cell membranes.

3.7.2 Virus classification

Viruses can be classified with respect to their structure or their host cell but the Baltimore classification is based on their mode of gene expression and replication [239]. On transcription of DNA only one strand gives rise to messenger RNA (chapter 8). The mRNA-like strand is given the designation 'plus (+) strand'. The complementary strand is called the 'minus (−) strand'.

There are six classes of virus delineated in Table 3.2. Class I contain duplex DNA which may be cyclic or linear. The DNA is replicated in the standard semiconservative manner as described in chapter 6. Class I viruses infect bacteria, animals and plants and can also bring about cell transformation (section 3.7.9).

Class II viruses carry a single strand of DNA which, upon infection, is converted to a duplex, replicative form (RF). A final stage in the infectious cycle is formation of single-stranded DNA from the RF (section 6.9). The DNA can be linear or cyclic and the viruses infect bacteria or animals. The DNA is plus strand, i.e. it has the same sequence as the mRNA. No minus-strand DNA viruses are known.

Class III viruses include reovirus of man and plant wound tumour virus. The duplex RNA is 'transcribed' to give (+) strand mRNA which later serves as a template to reform duplex RNA. The genomes are segmented, i.e. several different RNA molecules are present in each viral particle (section 9.9).

Class IV viruses have a (+) strand RNA genome. They replicate by synthesis of a minus-strand RNA which in turn serves as a template for viral, (+) strand RNA synthesis (section 8.4).

Class V viruses have a (−) strand RNA genome which is first copied to give (+) strand mRNA which in turn give rise to new molecules of (−) strand RNA (section 9.9).

Class VI viruses are remarkable in that they have an RNA genome ((+) strand) which is converted first to a DNA/RNA hybrid and then to duplex DNA. This is then transcribed to form new (+) strand RNA. This strategy is used by the leukaemia viruses and HIV (AIDS virus) and requires a special enzyme to reverse transcribe the RNA into DNA (section 6.4.8 and section 6.14).

3.7.3 The life cycle of viruses

The replication of a virus can be considered in stages: (1) adsorption of the virus onto the host cell; (2) penetration of the viral nucleic acid into the cell; (3) development of virus-specific functions, alteration of cell functions, replication of the nucleic acid, and synthesis of other virus constituents; (4) assembly of the progeny virus particles;

(5) release of virus particles from the cell [219, 222, 234, 235, 240]. The life cycle of animal viruses is similar to that of phage, the main difference being that a cycle takes 20–60 h, rather than 20–60 min.

(1) *Adsorption.* Viruses will only infect certain specific cells, i.e. they have a limited host range, usually because the coat (or tail) will only recognize and adsorb to specific sites on the appropriate cell walls. The host range of the T-phages is a property of their tail fibres. Some T-even phages are free to adsorb to the bacterial cell wall site only in the presence of tryptophan; in the absence of this amino acid the tail fibres are folded back and attached to the tail sheath [241]. The initial reversible interaction of the tail fibres with the cell wall is followed by the formation of a permanent attachment. The small, male-specific phages (e.g. MS2, R17) attach only to the f-pili of male *E. coli* cells [242] (section 3.6).

(2) *Penetration* of the viral nucleic acid into the cell involves a variety of different mechanisms. After attachment to the cell, the lysozyme present in the base of the bacteriophage T4 tail probably hydrolyses part of the cell wall. This allows the tail core to penetrate into the cell as the contractile tail sheath contracts and the small amount of ATP present in the phage tail is hydrolysed to ADP [243]. How the DNA passes from the head through the tail and into the cell (a process equivalent to passing a 10-metre-long piece of string down a straw) is not understood.

The injection of bacteriophage T5 DNA takes place in two stages: 8% of the DNA (the first step transfer DNA) enters the cell and directs mRNA and protein synthesis. One of the proteins synthesized is required to complete the injection of the remaining DNA [244].

After attachment of bacteriophage MS2

to the f-pilus of male *E. coli* cells, the RNA leaves the phage and is then transported inside the length of the pilus to the cell. This last step requires cellular energy [245]. An alternative mechanism proposed for entry of M13 into male *E. coli* involves retraction of the pilus with the filamentous bacteriophage attached [246]. Replication to the duplex form is necessary for the bacteriophage DNA to be drawn into the host, and when entry is effected a considerable fraction of the capsid protein is deposited in the inner cell membrane [222].

With animal viruses, uptake mechanisms are simpler [240]. Thus polyoma virus attaches to neuraminidase-sensitive sites on mouse cells [247] and is then taken up entirely by pinocytosis and is carried to the nucleus where the DNA is uncoated.

(3) *Development*. Once uncoated, isolated viral DNA will not infect cells as efficiently as does intact virus. Thus, following penetration of viral DNA into a cell, the infective titre drops. With bacteriophage it remains low for perhaps 20 min (the lag phase) and then suddenly rises up to a hundred-fold, indicating the presence of many new mature virus particles. During this lag phase after virus infection, the metabolic processes of the cell are being modified [244]. Viral mRNA directs the synthesis of specific enzymes, and the rates of host cell DNA, RNA and protein synthesis are altered. The viral nucleic acid replicates and virus constituents are synthesized. These different viral functions are divided into two groups, the early and late functions, which appear to be controlled either at the level of transcription (sections 10.1 and 10.2) or translation (section 12.9). Early functions include the biosynthesis of enzymes required for the replication of the nucleic acid, and late functions include the formation of the virus coat and other constituents.

T-even bacteriophages turn off the synthesis of host cell DNA and redirect the synthesis of DNA precursors to fit their particular requirements (e.g. hydroxymethylcytosine triphosphate is produced – chapter 6). On the other hand, the small tumour viruses, SV40 and polyoma, stimulate the synthesis of host cell DNA [248, 249] particularly if the cells are in a resting state before infection [250] (section 6.10.4).

The replication of virus nucleic acids is described in chapter 6.

(4) *Assembly* of virus particles may be either spontaneous (self-assembly) or may involve a series of virus-directed steps [251, 252]. More complex viruses are pieced together step by step. This process has been partially characterized for T4 [235, 253]. Wood and Edgar [251] have shown that about 45 viral genes are involved in T4 assembly, and many of the steps have been characterized by *in vitro* complementation using the partially finished pieces (e.g. heads, tails, fibres) found in cells infected with different mutants under non-permissive conditions. Eight genes have been assigned to the component parts of the head, and a further eight implicated in the assembly of these components. The head, which then appears to be morphologically complete, requires the action of two more gene products before it can interact spontaneously with assembled tails. The base plate components (12 genes) are assembled in two steps, then the core (three genes) and the sheath (one gene) are added by progressive polymerization from the base plate. Two gene products are required to finish the tails before they can be joined to the heads. The assembled heads and tails are modified (one gene) before the tail fibres are added. This last step requires complete tail fibres (two genes for components and three for assembly) and a labile enzyme (L). Phage particles assembled *in vitro* are active and

possess characteristics which vary with the source of the parts (e.g. the genotype of the head and the host range of the tail fibres).

In vitro systems for the reconstitution of phage lambda particles (packaging) have proved very important for the reintroduction of cloned DNA into cells following its *in vitro* modification [255] (section A.8.2).

The reconstitution of the rod-shaped TMV has been studied extensively [235, 256, 257]. When the coat protein and viral RNA are mixed in the correct ionic environment, virus particles are formed which possess up to 80% of the original infectivity.

Attempts to reconstitute small spherical viruses have proved more difficult. MS2 RNA and coat protein form morphologically complete particles which are not infective, possibly because they lack the maturation protein [258]. Cowpea chlorotic mottle virus can be partially degraded and reassembled to form particles indistinguishable from the original virus, and separated protein and nucleic acid have been mixed under conditions such that infectious particles are formed which have the same appearance, serological properties and sedimentation coefficient as intact virus [259]. Such experiments are consistent with the suggestion that the nature of the virus coat subunits alone directs the size and shape (spherical or rod) of the completed virus particles [260].

(5) *Release*. Interference with the normal metabolic processes may lead to the eventual death of the infected cell, followed by natural lysis but, in some instances, it has been shown that the virus actively causes cell lysis. Bacteriophage T4 codes for a lysozyme [261] which digests the host cell wall, causing the release of the progeny virus particles. Bacteriophage M13 and related filamentous viruses do not kill the host bacterium but pass out through the cell

membrane picking up coat protein on the way [222].

In contrast to production of obvious cytopathic effects, some viruses disturb cellular metabolism yet do not cause cell death. Thus lymphocytic choriomeningitis virus infections lead to changes in growth and glucose regulation and the infected pituitary glands do not make growth hormone, yet no pathological injury is apparent [262]. The tumour viruses cause a loss of growth regulation of the infected cells (section 3.7.9).

3.7.4 The Hershey–Chase experiment

In their classic experiment Hershey and Chase [254] showed that on infection of *E. coli* with T2 bacteriophage only the DNA enters the bacterium. Thus only the DNA replicates and it must, therefore, be DNA which carries the information to specify new virus. To demonstrate this they grew T2 phage in the presence of ^{32}P-labelled phosphate and ^{35}S-labelled amino acids. This resulted in phage containing DNA labelled with ^{32}P (there is no sulphur in DNA) and protein labelled with ^{35}S.

The labelled virus was then allowed to infect unlabelled bacteria and the progeny isolated. It was found that much of the ^{32}P (i.e. the DNA) was present in the progeny virus, but none of the ^{35}S-labelled protein. Vigorous shaking of the culture within a few minutes of infection was found to dislodge the empty ^{35}S-labelled protein coat of the virus from the outside of the bacteria. On centrifugation this coat remained in the supernatant fraction when the bacteria sedimented. The bacteria carried the ^{32}P-labelled viral DNA which replicated and produced new phage particles. This shows that the viral DNA carries the genetic material and that the coat protein is not essential once the DNA is inside the bacterium.

3.7.5 Virus mutants

The viruses provide a unique opportunity to characterize completely the information content of a functional genome.

A mutation in a viral genome can cause the inactivation of a gene function. If the missing function can be characterized (e.g. a missing enzyme) then the nature of the gene, and the mutant, is defined. However, if the function is essential, the mutation is lethal and the mutant cannot be propagated and studied. The discovery of *conditional lethal mutants* [263], which lack a gene function under one set of conditions but regain it under another set of conditions, revolution-ized the study of virus, and in particular, phage genetics. Stocks of conditional lethal mutants can be grown under permissive conditions and then the mutant gene func-tion studied and identified under non-permissive conditions.

Two classes of conditional lethal muta-tion are particularly useful. Temperature-sensitive mutants [263] are not viable at the non-permissive temperature (e.g. 42°C) but grow at the permissive temperature (e.g. 30°C). This is due to a single base change in the DNA causing the introduction of the wrong amino acid into the protein, reducing the stability of its configuration at the higher temperature.

Amber mutants [264] of a bacteriophage will only grow in a permissive host which contains a suppressor. These mutants result from a single base change altering an amino acid-coding triplet to UAG (section 12.2), which is read as stop under the normal, non-permissive conditions. The protein is terminated at that point. Permissive host bacteria, which suppress the mutation, have a species of tRNA which translates UAG as an amino acid and allows the protein to be completed. Different amino acids are added by different classes of permissive host.

When two virus mutants with mutations in different genes infect the same cell under non-permissive conditions the missing function of each can be supplied by the other, i.e. complementation takes place, and progeny viruses are formed. If the two mutants have mutations in the same gene, complementation cannot occur and no progeny are formed. Thus complementation between mutants is used to determine the number of complementation groups, i.e. the number of different genes in which the mutations occur. If sufficient mutants have been isolated for each gene of the virus to be represented, then the total number of essential viral genes can be estimated.

The joint growth of two phages in the same cell can also result in the formation of a few progeny phage, recombinants, which carry genetic characters of both parental bacteriophages (chapter 7). A study of the frequency of formation of recombinants from parental phages carrying known mutations allows the construction of a genetic map; the frequency of recombinants with both genetic characters is related to the distance between the two mutation points on the genome. Such a map shows the relative positions of the mutations and therefore of the genes in which these mutations occur, and it can be linear or circular.

3.7.6 Virus nucleic acids (Table 3.3)

Viral DNA varies in M_r from a little over 10^6 to more than 10^8 (cf. the value of 2.2×10^9 for *E. coli* DNA) and, unlike the DNA of bacteria and eukaryotes, it can often be

Table 3.3 Properties of some viral nucleic acids

	Host cell	Base pairs (× 10³)	Single- or double-stranded	Shape
DNA phages				
T2	E. coli	200	Double	Linear
T5	E. coli	130	Double	Linear
T7	E. coli	38	Double	Linear
λ	E. coli	48	Double	Linear
φX174	E. coli	5.4	Single	Cyclic
P22	Salmonella	39	Double	Linear
RNA phage				
MS2	E. coli (male)	3.3	Single	Linear
DNA animal viruses				
Polyoma	Mammals	5.3	Double	Cyclic
Herpes	Man	151	Double	Linear
RNA animal viruses				
Poliovirus	Man	6.7	Single	Linear
Reovirus	Mammals	18.2	Double (segmented)	Linear
Plant virus				
TMV	Tobacco	6.1	Single	Linear

extracted from the virus without degradation. Such intact DNA molecules have revealed a striking variety of tertiary structure and some of the known arrangements are as follows.

(a) *Cohesive (sticky) ends*

When DNA extracted from phage λ (M_r 30 × 10⁶) is heated to 65°C and cooled slowly, its sedimentation coefficient is 37S, but when it is rapidly cooled its sedimentation coefficient is only 32S [264]. Electron microscopy has shown the 37S form to be cyclic and the 32S form to be linear. This behaviour is a result of the 5'-ends of the DNA projecting for 12 nucleotides beyond the 3'-ends, the two single-stranded regions being complementary. These cohesive ends can base-pair and convert the DNA into a cyclic molecule which is disrupted at 65°C. The hydrogen-bonded, cyclic form can be converted with polynucleotide ligase

(section 6.5.1) into a cyclic form with both strands continuous. *E. coli* DNA polymerase, which adds on nucleotides to the 3'-ends, abolishes the ability to form cyclic molecules. Some other lysogenic phages (e.g. φ80) contain DNA with a similar structure.

(b) *Terminal repetition*

The DNA of phage T7 (M_r 25 × 10⁶) is double-stranded and linear, and the sequence (about 0.7% of the total) at the beginning is repeated at the end of each molecule [265]. Treatment of such molecules with *E. coli* exonuclease III results in the formation of cohesive ends which can form circles under suitable conditions. Terminal repetition has been detected in the DNA from several phages (e.g. T2, T4, T3, P22) and is believed to play a role in replication by enabling multiple-length concatamers to be formed (section 6.16.3).

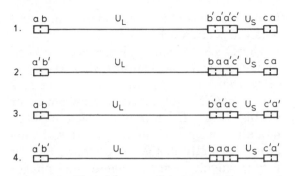

Fig. 3.17 The four possible arrangements of herpes simplex DNA. U_L and U_S are the long and short unique regions, both of which are bounded by repeated sequences.

The genome of adenoviruses (M_r 20–25 × 10^6) has a sequence of about 100 nucleotide pairs repeated in an inverted fashion at the two ends [266] and the 5'-end of both strands is covalently attached to a protein of M_r 55 000 (section 6.12).

(c) *Circular permutation*

If the linear double-stranded DNA of phage T2 (M_r 130 × 10^6) is denatured and allowed to reanneal slowly, circular molecules are formed which can be detected with the electron microscope [267]. These are formed because the phage DNA molecules do not have a unique sequence, but the population is a collection of molecules with sequences which are circular permutations of each other. The DNAs from phages T4 and P22 also exhibit circular permutation and terminal repetition.

The genome of herpes simplex virus (HSV) (M_r c. 10^8) is remarkable in being made up of two parts joined in tandem. Each part has a region which is repeated at the two ends in an inverted orientation. As the two parts can be joined in either orientation there are four possible sequences present in the HSV virus particles [268, 269] (Fig. 3.17). Unlike the small animal viruses,

the HSV genome has the expected amount of the dinucleotide CpG which is deficient in the genome of the host and the smaller viruses [259, 260, 270–272].

(d) *Nicks*

Three specific breaks have been found by electron microscopy in one strand of the double-stranded linear DNA extracted from bacteriophage T5 [273]. Accordingly, when the DNA is denatured, five single-stranded pieces are produced (instead of two) from each T5 DNA molecule. Bacteriophages SP8 and SP50 may also contain specific breaks as does the DNA from cauliflower mosaic virus and hepatitis B virus (section 6.14).

(e) *Hairpins*

The smallest animal viruses (the parvoviruses) have a linear single-stranded DNA genome of M_r 1.2–2.2 × 10^6. Some (e.g. minute virus of mouse – MVM) have a hairpin duplex at both ends of the DNA whereas others (e.g. adenoassociated virus – AAV) have a hairpin at one end only [274].

(f) *Segmented genomes*

Each reovirus particle contains about 12 × 10^6 daltons of RNA which, when extracted, is in pieces of three sizes (M_r 2.3 × 10^6, 1.3 × 10^6, and 8 × 10^5), plus 50–100 single-stranded oligonucleotides rich in adenine [275].

The virion of the RNA tumour viruses (retroviruses) contains two identical copies of the 35S viral RNA together with several copies of transfer RNA derived from the host cell (section 6.14).

(g) *Modification of viral DNA*

The DNAs of the T-even phages contain 5-hydroxymethylcytosine in place of cytosine

[276] and the hydroxyl group of this base can be glucosylated (Table 3.4) [277]. Growth of the phages in a UDP-glucose-deficient host produces phage DNA which is not glucosylated, and such DNA is degraded when the phages infect the normal host.

Table 3.4 Percentage glucosylation of hydroxy-methylcytosine in the DNA of T-even phage [277]

	T2	T4	T6
Unglucosylated	25	0	25
α-glucosyl	70	70	3
β-glucosyl	0	30	0
Diglucosyl	5	0	72

(This modification is considered in more detail in section 6.11.2.)

The DNAs of many phages are modified by host-specific mechanisms when grown in certain strains of *E. coli* (e.g. *E. coli* K12 and *E. coli* B), but not when grown in other strains (e.g. *E. coli* C). Bacteriophage modified by growth in K12 or B will only grow efficiently in the same strain, K12 or B respectively or in C. Unmodified phages will only grow efficiently in C.

When the infecting phages fail to grow (i.e. growth is restricted), the phage DNA is degraded. The restriction/modification system is considered in detail in section 4.5.3.

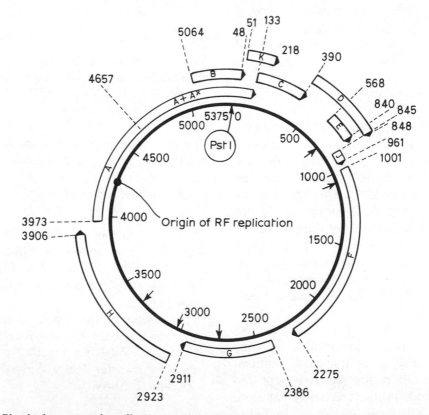

Fig. 3.18 Physical map and coding potential of φX174 DNA. The numbers refer to nucleotides relative to the single *Pst* I site. The other arrows in the inner circle show the sites of action of *Hpa* II. The open bars indicate the regions coding for the different proteins (A–K) and bars of different radii demonstrate the use of all three coding frames. The thick arrows indicate the *C*-termini of the proteins.

(h) *Chromatin structure of animal virus DNA*

The small animal tumour viruses SV40 and polyoma (papovaviruses) exist in the cell in the form of nucleoprotein particles where the cyclic duplex DNA molecule (5224 and 5292 nucleotide pairs respectively) is wrapped around histone octamers to form a nucleosomal structure – the minichromosome (section 3.3.2). There are 21 nucleosomes associated with the cyclic chromosome which, because of the presence of topoisomerases (section 6.5.4), is present in a form of minimum energy. In the cell, or when isolated in low-salt conditions, the minichromosome is highly compact. When the histones are removed, the isolated DNA is present in a supercoiled form (form I) with about 26 superturns per molecule (i.e. there is an average of 1.25 superhelical turns per nucleosome). A single nick in one of the DNA strands will convert the duplex into a relaxed open circular form (form II) and cleavage of both strands (e.g. with a restriction endonuclease which cuts only once in the genome) gives rise to linear (form III) DNA.

Three forms of double-stranded DNA can be isolated from purified polyoma virus [278, 279]: the supercoiled form (21S, component I), the open cyclic form (16S, component II) and a linear form (14.5S, component III). However, viral DNA forms a minor fraction of component III which, in the virion, is mainly cellular DNA that has been encapsulated in virus particles [280]. Electron microscopy (Fig. 2.23) has revealed structures which can be clearly identified with the supercoiled, open cyclic and linear forms and the three forms are readily separated by electrophoresis on agarose gels.

The DNA of the larger animal viruses (e.g. adenovirus or herpes simplex virus) although not naked in the cell does not appear to have a nucleosomal structure similar to that of the host cell DNA. It is for this reason that the smaller SV40 and polyoma virus have been used as models for animal cell chromatin structure.

3.7.7 The information content of viral nucleic acids

(a) *DNA phages*

The large phages have many genes. Thus T4 is known to have 135 genes of which 53 are concerned with phage assembly; 34 of these code for structural proteins. About 60 genes in T4 are not essential but enable the virus to cope with suboptimal conditions and allow it to extend the host range [222].

The small single-stranded phages possess very few genes. ϕX174 and the related phage G4 have eleven genes, six of which are concerned with coat-protein synthesis or assembly, one with cell lysis and the remainder with DNA synthesis (section 6.9). The combined M_r of these eleven proteins adds up to 262 000 which is greater than can be coded for by the small genome if traditional concepts apply. However, in ϕX174 the nucleotide sequence [281] shows that gene B is encoded within gene A but in a different reading frame (section 12.9.6) and gene E is encoded within gene D. Gene K is read in a second reading frame and overlaps the gene A/C junction (Fig. 3.18). In fact gene C starts five nucleotides before the end of gene A so, in this region, all three reading frames are used. Gene A' produces a protein identical to the carboxy terminus of the gene A protein: in this case there are two sites for initiation of translation.

It is clear that the small bacteriophages make full use of their DNA and detailed comparisons of the nucleotide sequence of the genomes of ϕX174 and G4 suggest the possible existence of several other genes [282, 283].

(b) RNA phages

The small RNA phages (e.g. MS2) contain all their genetic information in a single strand of RNA about 3300 nucleotides long (which can code for about 1100 amino acids or about 3–4 proteins of average size). Four viral-coded proteins have been identified including an RNA-dependent RNA polymerase or replicase of M_r about 50 000. In the infected cell, the replicase is an early protein made about 10 minutes after infection, whereas the coat and maturation proteins are late proteins made (in appropriate amounts) about 20 minutes later. Aspects of the mechanism of this control are considered in section 12.9.

(c) *Animal DNA viruses*

The large animal viruses such as herpes simplex virus (HSV) can code for more than 100 proteins, and, as with the large phages (e.g. T4) many of these proteins duplicate the functions of host proteins.

In general, virus diseases cannot be treated with antibiotics because the virus uses the cell's machinery to replicate itself. Thus any interference with virus metabolism automatically affects the infected animal. However, because the larger viruses substitute their own enzymes (e.g. DNA polymerase and thymidine kinase) for those of the cell it is possible to find some chemicals to which only the viral enzymes are susceptible. Acyclovir (9-(2-hydroxymethyl)-guanine) is one such chemical which can be phosphorylated by the virally induced thymidine kinase (in contrast to the host enzyme). The triphosphate inhibits the virally coded DNA polymerase [284, 285].

Acyclovir is used to treat herpes infections but cannot bring about a complete cure as the virus exhibits a latent phase in which the thymidine kinase is inactive. In a similar manner infections with HIV can be treated with azidothymidine which is phosphorylated and incorporated by the virally coded reverse transcriptase leading to production of defective viral DNA.

The other approach to viral therapy is immunological and many vaccines are now available, mainly to proteins on the virion coat or envelope. Recent advances in this field involve the use of small chemically synthesized viral peptides (fragments of viral coat proteins) as immunization with these minimizes the risk of side effects [286, 287].

The papovavirus genome codes for two (SV40) or three (polyoma) early proteins which are known as tumour antigens as they are expressed in transformed cells (see below). Later in infection three virion coat proteins (VP1, VP2 and VP3) are synthesized. The two (or three) tumour antigens are translated from messenger RNA molecules derived by processing a single transcript from the 'early half' of the genome starting near the origin of replication. This region of the DNA is 2700 nucleotide pairs long and in polyoma codes for large T (M_r about 100 000), middle T (M_r 55 000) and small t (M_r 17 000) antigens. In order to achieve this, more than one reading frame is used (cf. ϕX174) although all three proteins have the same N-terminus, showing they are coded for by the same reading frame to start with. This frame has a termination signal after about 600 nucleotides leading to production of small t antigen. RNA splicing (sections 11.3, 11.9) changes the reading frame in some of the transcripts to frame 2 (large T antigen) or frame 3 (middle T antigen).

The late region of polyoma virus (coding

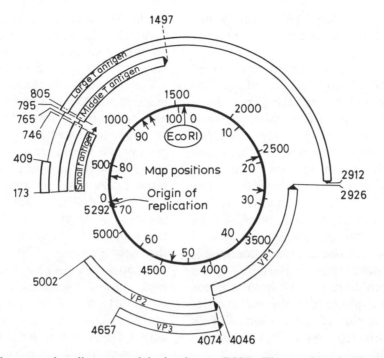

Fig. 3.19 Physical map and coding potential of polyoma DNA. The numbers inside the inner circle show the polyoma DNA divided into 100 map units starting at the single *Eco*R I site. Other arrows show the sites of action of *Hpa* II. All other numbers refer to nucleotides measured from the *Hpa* II 3/5 junction. The open bars indicate the regions coding for the various proteins – the thick arrows indicate the *C*-termini. The single line joining the open bars for the antigens indicates that these proteins are coded for by non-contiguous regions of DNA (section 11.3). N.B. Bars of different radii do not imply different reading frames.

for the virion proteins) also extends from near the replication origin but in the opposite direction to the early region. It is about 2100 nucleotide pairs long and codes for proteins requiring two reading frames. All three messenger RNAs have the same leader but splicing leads to synthesis of VP2 and VP3 in one reading frame (VP3 is identical to the *C*-terminus of VP2) and VP1 in a different, partly overlapping frame (Fig. 3.19) [230, 271, 272].

Thus, like φX174, polyoma virus and SV40 have more coding information than a co-linear relationship between DNA, RNA and protein would suggest but this is

achieved in a somewhat different manner from that with the small phages.

3.7.8 Lysogeny and transduction [288, 289]

When a virulent phage infects a cell, the virus replicates and the cell is killed. With temperate phage an alternative outcome is possible. The cell may be lysogenized with the viral DNA integrated into the cell genome as a prophage. Prophages are either integrated at a specific site on the bacterial chromosome (e.g. λ which normally attaches near the *gal* locus of *E. coli* [290], chapter

7) or at random (e.g. phage mu which causes inactivation of a gene into which it integrates).

The genes in a prophage are transcribed (in a controlled way), replicated and inherited along with the cellular genes. Most of the phage functions (e.g. those involved in the lytic development) are repressed, but others (e.g. those involved in the maintenance of lysogeny) are not. One gene function which is not repressed in the lysogenic cell confers immunity against further infection of the cell specifically by other copies of the same phage. The survival of virulent phages depends on a continuous supply of susceptible bacteria (e.g. in sewage, a rich source of phages) whereas the temperate phage can survive and replicate with a limited population of cells which are protected from further infection.

The prophages in a lysogenic cell can be induced to replicate independently of the host by various agents, most of which damage DNA. This results in the phage DNA being cut out of the cell genome by a recombination event [291] (section 7.5.1). The DNA can then replicate and function in a virulent manner, resulting in cell lysis and liberation of a burst of progeny phage. Phage lambda is the most studied example of a lysogenic phage. In the integrated state only one prophage gene (cI) is expressed. The cI gene directs the synthesis of the lambda repressor which turns off all other viral genes while stimulating its own production. Induction inactivates the repressor and switches on the expression of the *cro* gene. The Cro protein switches off cI transcription and switches on transcription of the other genes involved in lytic growth. Lysogenic phage are induced as a result of treatments (e.g. uv radiation) which damage DNA. As described in section 7.3.4, damage to DNA causes activation of RecA protein leading to the proteolytic cleavage of lambda repressor. Hence repressor synthesis is

no longer stimulated and Cro begins to accumulate in the cell. For more information on the regulation of transcription in prokaryotes, refer to sections 10.1 and 10.2.

Occasionally, on induction, mistakes are made in the excision of the prophage DNA and small pieces may be left behind or sections of bacterial DNA, which can encompass whole genes, may be excised along with the prophage DNA. This bacterial DNA may be replicated along with the phage DNA and packaged into the virus particle. On reinfection of a new bacterium these genes may become integrated into the new genome. This process is called *transduction*. Thus phage λ, which integrates next to the *gal* gene, may pick up this gene on induction to form the composite phage λgal. The *gal* gene is then replicated along with λ DNA and is introduced into a new bacterium. If this recipient is *gal*$^-$, it could thereby acquire the ability to metabolize galactose.

Some prophages (e.g. P1 [292]) are not integrated but remain in a stable, extrachromosomal state like plasmids (section 3.6). Only very occasionally are they packaged and virions released. This packaging is very non-specific, in that any DNA of the right length, including fragments of host chromosome, can be packaged (i.e. a 'headful' mechanism). On infecting a new bacterium, genes of the original host are introduced at random and P1 is known as a generalized transducing phage.

Along with transformation and conjugation, transduction is one of the three ways in which DNA can be transferred between cells. Transfection is the rather inefficient process whereby cells may be infected with naked DNA rather than with intact virus. It involves uptake of fragments of DNA or of the entire viral genome and is described further in sections A.7 and A.8. However, transfection is also used to describe infections with recombinant phage.

3.7.9 Tumour viruses and animal cell transformation

Some animal viruses (tumour viruses) can alter (transform) the infected cell without killing it, so that the cell is endowed with new properties which are typically neoplastic [293–295]. For example, uninfected hamster cells do not form tumours when injected into new-born hamsters, and do not grow when suspended in nutrient agar, but hamster cells transformed by polyoma virus form tumours and grow in agar [293].

Several DNA-containing tumour viruses have been identified, including polyoma, SV40, rabbit papilloma, human papilloma, adenoviruses 7, 12 and 18 and some herpes viruses. Several lines of evidence show that the viral genome is present in the transformed cells in a stable, inheritable, integrated state (cf. the lysogenic state of some bacteria) [296] and SV40-transformed mouse cells can be fused with green monkey cells when the hybrid cells liberate active SV40 particles [295, 297]. Although some SV40-transformed cells appear to contain only one copy of the viral genome per cell [298], others contain up to nine copies [299]. Although the whole of the viral genome may be present in the transformed cells, all but the early viral functions are suppressed.

Some of these viruses can interact with different cells in different ways. Polyoma virus will transform hamster cells but will replicate in, and kill, mouse cells. SV40, on the other hand, transforms mouse cells and kills green monkey cells. As described in section 6.10.4, T antigen makes a specific interaction with DNA polymerase α and the polyoma T-antigen reacts with murine polymerase but fails to interact with the hamster enzyme whereas the SV40 T antigen reacts with primate polymerase α only. This indicates that this interaction is required for virus replication and that failure to interact may lead to cellular transformation. However, a temperature-sensitive mutant of polyoma [300] will transform mouse cells at the non-permissive temperature, 38.5°C, and then replicate when the temperature of the transformed cell is reduced to 31°C [301]. This suggests that at least one virus function necessary for replication is not required for transformation.

The large T antigen binds to the origin of replication of the virus and middle T antigen is found on the cell surface where it may mediate cell/cell interactions. The middle T antigen may be the (tumour specific) transplantation antigen (TSTA) which causes rejection of transplantable tumours by animals previously immunized with the virus. The possible roles of T antigen in transformed cells have been reviewed [302, 303] and it is clear that SV40 large T antigen interacts with and stabilizes a cellular protein known as p53. This interaction may be enhanced when the T antigens cannot interact with DNA polymerase α and this could lead to a depletion of free p53 in the cell. In a similar way, the EIA protein of adenoviruses can bind and inactivate the product of the retinoblastoma gene. These two cellular proteins are antioncogenes which act to regulate cell proliferation and so their inactivation leads to the uncontrolled growth typical of tumour cells [304–306].

The leukaemia viruses and the sarcomaviruses are closely related RNA tumour viruses or retroviruses. The leukaemia viruses are a natural cause of leukaemia and the sarcomaviruses transform cells in tissue culture. The structure of the genome of these retroviruses, together with the mechanism of their conversion into the DNA duplex proviral form, is shown in Fig. 6.35. They code for three enzymes only: *gag*, *pol* and *env*; but integration into the host genome is essential for their replication. Integration (section 7.7) is often accompanied by the enhanced expression of adjacent genes which, if these are cellular oncogenes, will lead to cell transformation [308–311].

Some retroviruses (e.g. Rous sarcoma virus of chickens) have incorporated into their RNA genome the transcript of a cellular oncogene (c-*onc*) which is transferred from cell to cell in an infectious manner. This gene is now known as a viral oncogene (v-*onc*) and may be identical to the cellular homologue or it may be a shortened or mutated version of it. Retroviruses carrying a v-*onc* gene always produce tumours on infection, although they are usually themselves defective as the v-*onc* gene may replace part of the *env* gene. It is necessary to coinfect with a non-defective virus to maintain these viruses in the laboratory – they do not persist in nature.

It is by hybridizing v-*onc* sequences to cellular DNA that a series of more than 20 c-*onc* genes have been isolated [312] and this is contributing greatly to our increasing understanding of cellular growth control and cancer [313].

3.7.10 Viroids and pathogenic RNAs [314, 315]

Viroids are much smaller than viruses and consist of a short strand of circular RNA, with considerable secondary structure, and nothing more. They infect higher plants causing diseases such as potato spindle tuber disease (PSTV) and cadang-cadang, a disease of the coconut palm. Variants exist (known as virusoids and satellite RNAs) in which the circular form, or a linear version of it, are encapsidated with the intervention of a helper virus. Hepatitis delta virus is a similar satellite of human hepatitis B virus.

The RNA is in the form of a cyclic mole-

cule which may be as small as 246 nucleotides up to the 1700 nucleotides of the hepatitis delta virus. The RNA does not code for any protein (hepatitis delta viral RNA is the exception) and shows a high degree of intrastrand base-pairing, forming a hammerhead structure with self-cleaving properties. Replication involves the host RNA polymerase acting by a rolling circle mechanism (section 6.9) to produce first (−) strand concatamers (which may or may not be circularized) followed by (+) strand concatamers which are self-cleaved and ligated to regenerate the infectious RNA [316]. It is possible that the pathogenic effect of these RNA molecules lies in their ability to catalyse cleavage of other RNAs (e.g. mRNA).

The hepatitis delta virus is slightly different in that, although it replicates by a rolling circle mechanism, this requires a protein encoded by part of the antigenomic RNA. This RNA appears to be a short 'transcript', also synthesized in the nucleus by the host RNA polymerase [317].

3.7.11 Prions [343]

These are slow 'viruses' producing diseases such as scrapie in sheep and Creutzfeld–Jacob disease and Kuru in man and bovine spongiform encephalopathy (BSE) in cattle [318–320]. They may also be the cause of Alzheimer's disease [321]. They are of particular interest as they have not been shown to contain any nucleic acid [322]. Rather they may contain nothing but a single glycoprotein of M_r about 30 000 which aggregates into rods present as amyloid plaques in the brains of infected animals.

The protein, PrP, is encoded by the host and the time of onset of the disease differs in animals containing different variants of PrP [323]. The protein is expressed in both normal and infected brains, but not in identical form in that amino acid polymorphisms are present which affect the structure of the protein.

The nature of the infectious agent is in doubt. The evidence for the lack of a nucleic acid component in prions has been criticized on several grounds [327, 328] but it now seems clear that PrP from infected brain contains no nucleic acid and is capable of causing the disease. Various intriguing ways in which a protein could be a self-replicating agent are reviewed by Prusiner [322] and Brunori and Talbot [324].

REFERENCES

1 Alberts, B., Bray, D., Lewis, J., Raff, M., Roberts, K. *et al.* (1989) *The Molecular Biology of the Cell* (2nd edn) Garland Publishing, New York.
2 Starr, C. M. and Hanover, J. A. (1991) *BioEssays*, **12**, 323.
3 Hunt, T. (1989) *Cell*, **59**, 949.
4 Newmeyer, D. D. and Forbes, D. J. (1988) *Cell*, **52**, 641.
5 Richardson, W. D., Mills, A. D., Dilworth, S. M., Laskey, R. A. and Dingwall, C. (1988) *Cell*, **52**, 655.
6 Dingwall, C. (1990) *Nature (London)*, **346**, 512.
7 Dutrillaux, B. (1977) in *The Molecular Structure of Human Chromosomes* (ed. J. Yunis), Academic Press, New York, p. 233.
8 Capco, D. G., Wan, K. M. and Penman, S. (1982) *Cell*, **29**, 847.
9 Geraci, L., Comeau, C. and Benson, M. (1984) *J. Cell Sci.*, *Suppl.*, **1**, 137.
10 Benavente, R., Krohne, G., Schmidt-Zachmann, M. S., Hugle, B. and Franke, W. W. (1984) *J. Cell Sci.*, *Suppl.*, **1**, 161.
11 Chung, H. M. H., Shea, C., Fields, S., Taub, R. N. and Van der Ploeg, L. H. T. (1990) *EMBO J.*, **9**, 2611.
12 Agard, D. A. and Sedat, J. W. (1983) *Nature (London)*, **302**, 676.
13 Hiraoka, Y., Minden, J. S., Swedlow, J. S., Sedat, J. W. and Agard, D. A. (1989) *Nature (London)*, **342**, 293.
14 Heslop-Harrison, J. S. and Bennett, M. D. (1990) *Trends Genet.*, **6**, 401.
15 Henikoff, S. (1990) *Trends Genet.*, **6**, 422.
16 Gerace, L. and Blobel, G. (1980) *Cell*, **19**, 277.
17 Peter, M., Nakagawa, J., Dorée, M., Labbé, J. C. and Nigg, E. A. (1990) *Cell*, **61**, 591.
18 Ward, G. E. and Kirschner, M. C. (1990) *Cell*, **61**, 561.
19 Moreno, S. and Nurse, P. (1990) *Cell*, **61**, 549.
20 Laskey, R. A. and Leno, G. H. (1990) *Trends Genet.*, **6**, 406.
21 Mirkovitch, J., Mirault, M-E. and Laemmli, U. K. (1984) *Cell*, **39**, 223.
22 Lewis, C. D., Lebkowski, J. S., Daly, A. K. and Laemmli, U. K. (1984) *J. Cell Sci.*, *Suppl.*, **1**, 103.
23 Pienta, K. J. and Coffey, D. S. (1984) *J. Cell Sci.*, *Suppl.*, **1**, 123.
24 Jackson, D. A., McCready, S. J. and Cook, P. R. (1984) *J. Cell Sci.*, *Suppl.*, **1**, 59.
25 Adams, R. L. P. (1990) *Cell Culture for Biochemists* (2nd edn), Elsevier/North-Holland, Amsterdam.
26 Wells, R., Royer, H-D. and Hollenberg, C. P. (1976) *Mol. Gen. Genet.*, **147**, 45.
27 Nurse, P. (1990) *Nature (London)*, **344**, 503.
28 Lewin, B. (1990) *Cell*, **61**, 743.
29 Draetta, G. (1990) *Trends Biochem. Sci.*, **15**, 378.
30 Yunis, J. J. (1976) *Science*, **191**, 1268.
31 Yunis, J. J. (1977) *Molecular Structure of Human Chromosomes*, Academic Press, New York.
32 Pluta, A. F., Cooke, C. A. and Earnshaw, W. C. (1990) *Trends Biochem. Sci.*, **15**, 181.
33 Caspersson, T., Hulten, M., Linsten, J. and Zech, L. (1971) *Hereditas*, **67**, 147.
34 Wang, H. C. and Federoff, S. (1973) in *Tissue Culture, Methods and Applications* (eds P. F. Kruse and M. K. Patterson), Academic Press, London, p. 782.
35 Nurse, P. (1985) *Trends Genet.*, **1**, 51.
36 Carbon, J. (1984) *Cell*, **37**, 351.
37 Blackburn, E. H. and Szostak, J. W. (1984) *Annu. Rev. Biochem.*, **53**, 163.
38 Carbon, J. and Clarke, L. (1984) *J. Cell Sci.*, *Suppl.*, **1**, 43.
39 Bloom, K. S., Fitzgerald-Hayes, M. and Carbon, J. (1982) *Cold Spring Harbor Symp. Quant. Biol.*, **47**, 1175.
40 Blackburn, E. H. (1984) *Cell*, **37**, 7.

41 Murray, A. W. (1985) *Trends Biochem. Sci.*, **10**, 112.

42 Schlessinger, D. (1990) *Trends Genet.*, **6**, 248.

43 Ruddle, F. H. (1973) *Nature (London)*, **242**, 165.

44 Ringertz, N. R. and Savage, R. E. (1976) *Cell Hybrids*, Academic Press, London.

45 Gall, J. G. and Pardue, M. L. (1971) *Methods Enzymol.*, **21**, 470.

46 Zabel, B. U., Naylor, S. L., Sakaguchi, A. Y., Bell, G. I. and Shows, T. B. (1983) *Proc. Natl. Acad. Sci. USA*, **80**, 6932.

47 Lawrence, J. B., Villnave, C. A. and Singer, R. H. (1988) *Cell*, **52**, 51.

48 Willard, H. F. (1990) *Trends Genet.*, **6**, 410.

49 Camerino, G., Grzeschik, K. H., Jaye, M., De La Salle, H., Tolstoshev, P. *et al.* (1984) *Proc. Natl. Acad. Sci. USA*, **81**, 498.

50 Cheung, P., Kao, F-T., Law, M. L., Jones, C., Puck, T. T. and Chan, L. (1984) *Proc. Natl. Acad. Sci. USA*, **81**, 508.

51 Koutides, I. A., Barker, P. E., Gurr, J. A., Pravtcheva, D. D. and Ruddle, F. H. (1984) *Proc. Natl. Acad. Sci. USA*, **81**, 517.

52 Davies, K. E., Young, B. D., Elles, R. G., Hill, M. E. and Williamson, R. (1981) *Nature (London)*, **293**, 374.

53 Lebo, R. V., Carrano, A. V., Burkhart-Schulz, K. J., Dozy, A. M., Yu, L-C. *et al.* (1979) *Proc. Natl. Acad. Sci. USA*, **76**, 5804.

54 Watson, J. V. (1984) in *Molecular Biology and Human Disease* (eds A. Macleod and K. Sikora), Blackwell Scientific, Oxford. p. 66.

55 Kemp, D. J., Corcoran, L. M., Coppel, R. L., Stahl, H. D., Bianco, A. E. *et al.* (1985) *Nature (London)*, **315**, 347.

56 Carle, G. F, and Olson, M. V. (1984) *Nucleic Acids Res.*, **12**, 5647.

57 Cox, F. E. G. (1985) *Nature (London)*, **315**, 280.

58 Botstein, D., White, R. L., Skolnick, M. and Davis, R. W. (1980) *Am. J. Hum. Genet.*, **32**, 314.

59 Doolittle, W. F. and Sapienza. C. (1980) *Nature (London)*, **284**, 601.

60 Cavalier-Smith, T. (1980) *Nature (London)*, **285**, 617.

61 Britten, R. J. and Davidson, E. H. (1968) *Science*, **165**, 349.

62 Ohno, S. (1971) *Nature (London)*, **234**, 134.

63 Evans, H. J., Gosden, J. R., Mitchell, A. R. and Buckland, R. A. (1974) *Nature (London)*, **251**, 346.

64 Manuelidis, L. (1978) *Chromosoma*, **66**, 1.

65 Maio, J. J., Brown, F. L. and Musich, P. R. (1977) *J. Mol. Biol.*, **117**, 637.

66 Horz, W. and Altenberger, W. (1981) *Nucleic Acids Res.*, **9**, 683.

67 Southern, E. M. (1975) *J. Mol. Biol.*, **94**, 51.

68 Streeck, R. E. (1981) *Science*, **213**, 443.

69 Taparowsky, E. J. and Gerbi, S. A. (1982) *Nucleic Acids Res.*, **10**, 5503.

70 Miklos, G. L. G. and John, B. (1979) *Am. J. Hum. Genet.*, **31**, 264.

71 Selig, S., Ariel, M., Goitein, R., Marcus, M. and Cedar, H. (1988) *EMBO J.*, **7**, 419.

72 Varley, J. M., Macgregor, H. C. and Erba, H. P. (1980) *Nature (London)*, **283**, 686.

73 Jamrich, M., Warrior, R., Steele, R. and Gall, J. G. (1983) *Proc. Natl. Acad. Sci. USA*, **80**, 3364.

74 Gautier, F., Bunemann, H. and Grotjahn, L. (1977) *Eur. J. Biochem.*, **80**, 175.

75 Adams, R. L. P., Burdon. R. H. and Fulton, J. (1983) *Biochem. Biophys. Res. Commun.*, **113**, 695.

76 Sanford, J., Forrester, L., Chapman, V., Chandley, A. and Hastie, N. (1984) *Nucleic Acids Res.*, **12**, 2823.

77 Bostock, C. (1980) *Trends Biochem. Sci.*, **5**, 117.

78 Jarman, A. P. and Wells, R. A. (1989) *Trends Genet.*, **5**, 367.

79 Dover, G. A. (1989) *Nature (London)*, **342**, 347.

80 Jeffreys, A. J., Royle, N. J., Wilson, V. and Wong, Z. (1988) *Nature (London)*, **332**, 278.

81 Jeffreys, A. J., Brookfield, J. F. Y. and Semeonoff, R. (1985) *Nature (London)*, **317**, 818.

82 Hill, W. G. (1987) *Nature (London)*, **327**, 98.

83 Lander, E. S. (1989) *Nature (London)*, **339**, 501.

84 Dover, G. A. (1990) *Nature (London)*, **344**, 812.

85 Jeffreys, A. J., Neumann, R. and Wilson, V. (1990) *Cell*, **60**, 473.

86 Georgiev, G. P., Kramerov, D. A. Ryskov, A. P., Skryabin, K. G. and Lukanidin, E. M. (1983) *Cold Spring Harbor Symp. Quant. Biol.*, **47**, 1109.

87 Jelinek, W. R. and Haynes, S. R. (1983) *Cold Spring Harbor Symp. Quant. Biol.*, **47**, 1123.

88 Hastie, N. (1985) *Trends Genet.*, **1**, 37.

89 Fanning, T. G. and Singer, M. F. (1987) *Biochim. Biophys. Acta*, **910**, 203.

90 Singer, M. F. and Skowronski, J. (1985) *Trends Biochem. Sci.*, **10**, 119.

91 Graham, D. E., Neufeld, B. R., Davidson, E. H. and Britten, R. J. (1974) *Cell*, **1**, 127.

92 Crain, W. R., Davidson, E. H. and Britten, R. J. (1976) *Chromosoma*, **59**, 1.

93 Jelinek. W. R. (1978) *Proc. Natl. Acad. Sci. USA*, **75**, 2679.

94 Hardman, N., Bell, A. J. and McLachlan, A. (1979) *Biochim. Biophys. Acta*, **564**, 372.

95 Schmid, C. W. and Deininger. P. L. (1975) *Cell*, **6**, 345.

96 Goodman, H. M., Olson. M. V. and Hall, B. D. (1977) *Proc. Natl. Acad. Sci. USA*, **74**, 5454.

97 Dugaiczyk, A., Woo, S. L., Lai, E. C., Mace, M. L., McReynolds, C. *et al.* (1978) *Nature (London)*, **274**, 328.

98 Lomedico, P., Rosenthal, N., Efstratiadis, A., Gilbert, W., Kolodner, R. *et al.* (1979). *Cell*, **18**, 545.

99 Fitzsimons, D. W. and Wolstenholme, G. E. W. (1975). *Ciba Found. Symp. Struct. Funct. Chromatin*, **28**.

100 Von Holt, C. and Brandt, W. F. (1977) *Methods Cell Biol.*, **16**, 205.

101 Frederiq, E. (1971) in *Histones and Nucleohistones* (ed. D. M. P. Philips), Plenum Press New York, p. 136.

102 Christiansen, G. and Griffith, J. (1977) *Nucleic Acids Res.*, **4**, 1837.

103 Kedes, L. H. (1979) *Annu. Rev. Biochem.*, **48**, 837.

104 Isenberg, I. (1979) *Annu. Rev. Biochem.*, **48**, 159.

105 Behe, M. J. (1990) *Trends Biochem. Sci.*, **15**, 374.

106 Lennox, R. W. and Cohen, L. W. (1983) *J. Biol. Chem.*, **258**, 262.

107 Dashkevich, V. K., Nikolaev, L. G., Zlatanova, J. S., Glotov, B. O. and Severin, E. S. (1983) *FEBS Lett.*, **158**, 276.

108 Sittman, D. B., Graves, R. A. and Marzluff, W. F. (1983) *Proc. Natl. Acad. Sci. USA*, **80**, 1849.

109 Woodland, H. R., Warmington, J. R., Ballantine, J. E. M. and Turner, P. C. (1984) *Nucleic Acids Res.*, **12**, 4939.

110 Wu, R. S. and Bonner, W. M. (1984) in *Eukaryotic Gene Expression* (ed. A. Kumar), George Washington University Medical Centre Spring Symposium, Plenum Press, New York, p. 37.

111 Schiffman, S. and Lee, P. (1974) *Br. J. Haematol.*, **27**, 101.

112 Waterborg, J. H. and Matthews, H. R. (1983) *Biochemistry*, **22**, 1489.

113 Loidl, P., Loidl, A., Puschendorf, B. and Grobner, P. (1983) *Nature (London)*, **305**, 446.

114 Perry, M. and Chalkley, R. (1982) *J. Biol. Chem.*, **257**, 7336.

115 Bradbury, E. M., Maclean, N. and Matthews, H. R. (1981) *DNA, Chromatin and Chromosomes*, Blackwell, Oxford.

116 Smulson, M. (1979) *Trends Biochem. Sci.*, **4**, 225.

117 Oliva, R., Bazett-Jones, D. P., Locklear, L. and Dixon, G. H. (1990) *Nucleic Acids Res.*, **18**, 2739.

118 Tazi, J. and Bird, A. (1990) *Cell*, **60**, 909.

119 Findlay, D., Cicchanover, A. and Varshavsky, S. (1984) *Cell*, **37**, 43.

120 Mueller, R. D., Yasuda, H., Hatch, C. L., Bonner, W. M. and Bradbury, E. M. (1985) *J. Biol. Chem.*, **260**, 5147.

121 Jentsch, S., Seufert, W., Sommer, T. and Reins, H-A. (1990) *Trends Biochem. Sci.*, **15**, 195.

122 Somer, J. B. and Castaldi, P. A. (1970) *Br. J. Haematol.*, **18**, 147.

123 Swart, A. C. W. and Hemker, H. C. (1970) *Biochim. Biophys. Acta*, **222**, 692.

124 Campbell, A. M. and Cotter, R. (1976). *FEBS Lett.*, **70**, 209.

125 Thomas, J. O. and Butler, P. J. G. (1977) *J. Mol. Biol.*, **116**, 769.

126 John, E. W. (ed.) (1982) *The HMG Chromosomal Proteins*, Academic Press, London.

127 Cartwright, I. L., Abinger, S. H., Fleischman, G., Lovenhaupt, K., Elgin, S. *et al.* (1982) *Crit. Rev. Biochem.*, **13**, 1.

128 Seale, R. L., Annunziato, A. T. and Smith, R. D. (1983) *Biochemistry*, **22**, 5008.

129 Thomas, J. O. (1979) in *Companion to Biochemistry*, **2** (eds R. T. Bull, J. R. Lagnado, J. O. Thomas and K. F. Tipton) Longman, London, p. 79.

130 Thomas, J. O. (1984) *J. Cell Sci., Suppl.*, **1**, 1.

131 Clark, R. J. and Felsenfeld, G. (1971) *Nature (London), New Biol.*, **229**, 101.

132 Olins, A. L. and Olins, D. E. (1974) *Science*, **183**, 330.

133 Germond, J. E., Hirt, B., Oudet, P., Gross-

Bellard, M. and Charnbon, P. (1975) *Proc. Natl. Acad. Sci. USA*, **72**, 1843.

134 Shure, M., Pulleyblank, D. E. and Vinograd, J. (1977) *Nucleic Acids Res.*, **4**, 1183.

135 Hewish, D. R. and Burgoyne, L. A. (1973) *Biochem. Biophys. Res. Commun.*, **52**, 504.

136 Noll, M. and Kornberg, R. D. (1977) *J. Mol. Biol.;* **109**, 393.

137 Finch, J, T., Noll, M. and Kornberg, R. D. (1975) *Proc. Natl. Acad. Sci. USA*, **72**, 3320.

138 Thomas, J. O. and Kornberg. R. D. (1975) *Proc. Natl. Acad. Sci. USA*, **72**, 2626.

139 Noll, M. (1977) *J. Mol. Biol.*, **116**, 49.

140 Pardon, J. F., Worcester, D. L., Wooley, J. C., Cotter, R. I., Lilley, D. M. J. *et al.* (1977) *Nucleic Acids Res.*, **4**, 3199.

141 Finch, J. T., Lutter, L. C., Rhodes, D., Brown, R. S., Rushton, B. *et al.* (1977) *Nature (London)*, **269**, 29.

142 Richmond, T. J., Finch, J. T., Rushton, B., Rhodes, D. and Klug, A. (1984) *Nature (London)*, **311**, 532.

143 Hogan, M. E., Rooney, T. F. and Austin, R. H. G. (1987) *Nature (London)*, **328**, 554.

144 Hagerman, P. J. (1981) *Biopolymers*, **20**, 1503.

145 Mirzabekov, A. D., Bavykin, S. G., Karpov, V. L., Preobrazhenskayer, O. V., Ebralidze, K. K. *et al.* (1982) *Cold Spring Harbor Symp. Quant. Biol.*, **47**, 503.

146 Morris, N. R. (1976) *Cell*, **9**, 627.

147 Ellison, M. J. and Pulleyblank, D. E. (1983) *J. Biol. Chem.*, **258**, 13307, 13314 and 13321.

148 Weintraub, H. (1980) *Nucleic Acids Res.*, **8**, 4745.

149 Piña, B., Brüggemeier, U. and Beato, M. (1990) *Cell*, **60**, 719.

150 Simpson, R. T. (1990) *Nature (London)*, **343**, 387.

151 Beard, P. (1978) *Cell*, **15**, 955.

152 Kornberg, R. D. (1981) *Nature (London)*, **292**, 579.

153 Horz, W. and Altenberger, W. (1981) *Nucleic Acids Res.*, **9**, 2643.

154 Dingwall, C., Lomonossoff, F. R. and Laskey, R. A. (1981) *Nucleic Acids Res.*, **9**, 2659.

155 Keene, M. H. and Elgin, S. C. R. (1981) *Cell*, **27**, 57.

156 Cartwright, I. L., Hertzberg, R. P., Dervan, P. B. and Elgin, S. C. R. (1983) *Proc. Natl. Acad. Sci. USA*, **80**, 3213.

157 Samal, B., Worcel, A., Louis, C. and Schedl. P. (1981) *Cell*, **23**, 401.

158 Zhang, X-Y. and Horz, W. (1984) *J. Mol. Biol.*, **176**, 105.

159 Bock, H., Abler, S., Zhang, X-Y., Fritton, H. and Igo-Kemenes, T (1984) *J. Mol. Biol.*, **176**, 131.

160 Strauss, F. and Varshavsky, A. (1984) *Cell*, **37**, 889.

161 Morse, R. H. and Simpson, R. T. (1988) *Cell*, **54**, 285.

162 Simpson, R. T. and Stafford, D. W. (1983) *Proc. Natl. Acad. Sci. USA*, **80**, 51.

163 Thoma, F. and Simpson, R. T. (1985) *Nature (London)*, **315**, 250.

164 Thomas, J. O. and Khabaza, A. J. A. (1980) *Eur. J. Biochem.*, **112**, 501.

165 Ring, D. and Cole, R. D. (1983) *J. Biol. Chem.*, **258**, 15361.

166 Campbell, A. M. and Cotter, R. I. (1977) *Nucleic Acids Res.*, **4**, 3877.

167 Goodwin, G. H., Mathew, C. G. P., Wright, C. A., Venkov, C. D. and Johns, E. W. (1979) *Nucleic Acids Res.*, **7**, 1815.

168 Finch, J. T. and Klug, A. (1976) *Proc. Natl. Acad. Sci. USA*, **73**, 1897.

169 McGhee, J. D., Nickol, J. M., Felsenfeld, G. and Rau, D. C. (1983) *Cell*, **33**, 831.

170 Labhart, P., Koller, T. and Wenderli, H. (1982) *Cell*, **30**, 115.

171 Cook, P. R. and Brazell, I. A. (1978) *Eur. J. Biochem.*, **84**, 465.

172 Benyajati, C. and Worcel, A. (1976) *Cell*, **9**, 393.

173 Adolph, K. W., Cheng, S. M. and Laemmli, U. K. (1977) *Cell*, **12**, 805.

174 Nishmura, Y., Caro, L., Berg, C. M. and Hirota, Y. (1971) *J. Mol. Biol.*, **55**, 441.

175 Zakian, V. A. (1985) *Nature (London)*, **314**, 223.

176 Schmid, M. B. (1988) *Trends Biochem. Sci.*, **13**, 131.

177 Jackson, D. A., McCready, S. J. and Cook, P. R. (1984) *J. Cell Sci., Suppl.*, **1**, 59.

178 Callan, H. G., Gross, K. W. and Old, R. W. (1977) *J. Cell Sci.*, **27**, 57.

179 Sommerville, J. (1979) *J. Cell Sci.*, **40**, 1.

180 Lewis, M., Helmsing, P. J. and Ashburner, M. (1975) *Proc. Natl. Acad. Sci. USA*, **72**, 3604.

181 Lonn, U. (1982) *Trends Biochem. Sci.*, **7**, 24.

182 Silver, L. M. and Elgin, S. C. R. (1977) *Cell*, **11**, 971.

183 Zarling, D. A., Arndt-Jovin, D. J., Robert-

Nicoud, M., McIntosh, L. P., Thomae, R. and Jovin, T. M. (1984) *J. Mol. Biol.*, **176**, 369.

184 Fangman, W. L. and Dujon, B. (1984) *Proc. Natl. Acad. Sci. USA*, **81**, 7156.

185 Attardi, G. (1981) *Trends Biochem. Sci.*, **6**, 86.

186 Attardi, G. (1981) *Trends Biochem. Sci.*, **6**, 100.

187 Van Etten, R. A., Michael, N. L., Bibb, M. J., Brennicke, A. and Clayton, D. A. (1982) in *Mitochondrial Genes* (eds P. Slonimski, P. Borst and G. Attardi) Cold Spring Harbor, New York, p. 73.

188 Anderson, S., Bankier, A. T., Barrell, B. G., de Bruijn, M. H. L., Coulson, A. *et al.* (1982) in *Mitochondrial Genes* (eds P. Slonimski, P. Borst and G. Attardi), Cold Spring Harbor, New York, p. 5.

189 Anderson, S., Bankier, A. T., Barrell, B. G., de Bruijn, M. H. L., Coulson, A. R. *et al.* (1981) *Nature (London)*, **290**, 457.

190 Bibb, M. J., Van Etten, R. A., Wright, C. T., Walberg, M. W. and Clayton, D. A. (1981) *Cell*, **26**, 167.

191 Clary, D. O., Wahleithner, J. A. and Wolstenholme, D. R. (1984) *Nucleic Acids Res.*, **12**, 3747.

192 Grivell, L. A. (1983) *Sci. Am.*, **248** (3), 60.

193 Chomyn. A., Cleeter, M. W. J., Ragan, C. I., Riley, M., Doolittle, R. F. *et al.* (1986) *Science*, **234**, 614.

194 Lazowska, J., Jacq, C. and Slonimski, P. P. (1980) *Cell*, **22**, 333.

195 Palmer, J. D. and Shields, C. R. (1984) *Nature (London)*, **307**, 437.

196 Fox, T. D. (1984) *Nature (London)*, **307**, 415.

197 Levings, C. S. (1983) *Cell*, **32**, 659.

198 Borst, P. and Grivell, L. S. (1978) *Cell*, **15**, 705.

199 Bandlow, W., Schweyen, R. J., Wolf, K. and Kaudewitz, F. (1977) *Genetics and Biogenesis of Mitochondria*, De Gruyter, Berlin.

200 Slonimski, P., Borst, P. and Attardi, G. (1982) *Mitochondrial Genes*, Cold Spring Harbor Symposium, New York.

201 Yaffe, M. and Schatz, G. (1984) *Trends Biochem. Sci.*, **9**, 179.

202 Barton, N. and Jones, J. S. (1983) *Nature (London)*, **306**, 317.

203 Klein, A. and Bonhoeffer, E. (1972) *Annu. Rev. Biochem.*, **41**, 301.

204 Smith, H. (1975) *Nature (London)*, **254**, 13.

205 Ohta, N., Sager, R. and Inouye, M. (1975) *J. Biol. Chem.*, **250**, 3655.

206 Borst, P. (1991) *Trends Genet.*, **7**, 139.

207 Mandelstam, J. and McQuillen, K. (1973) *Biochemistry of Bacterial Growth*, Blackwell Scientific, Oxford.

208 Dawes, I. W. and Sutherland, I. W. (1976) *Microbiological Physiology*, Blackwell Scientific, Oxford.

209 Stonington, O. and Pettijohn, D. (1971) *Proc. Natl. Acad. Sci. USA*, **68**, 6.

210 Worcel, A. and Burgi, L. (1972) *J. Mol. Biol.*, **71**, 127.

211 Pettijohn, D. and Hecht, R. (1973) *Cold Spring Harbor Symp. Quant. Biol.*, **38**, 31.

212 Worcel, A., Burgi, E., Robinton, J. and Carlson, C. L. (1973) *Cold Spring Harbor Symp. Quant. Biol.*, **38**, 43.

213 Cooper, S. (1979) *Nature (London)*, **280**, 17.

214 Cooper, S. and Helmstetter, C. E. (1968) *J. Mol. Biol.*, **31**, 519.

215 Avery, O. T., McLeod, C. M. and McCarty, M. (1944) *J. Exp. Med.*, **79**, 137.

216 Mandel, M. and Higa, A. (1970) *J. Mol. Biol.*, **53**, 159.

217 Hayes, W. (1968) *The Genetics of Bacteria and their Viruses* (2nd edn), Blackwell Scientific, Oxford.

218 Notani, N. K. and Setlow, J. K. (1974) *Prog. Nucleic Acid Res. Mol. Biol.*, **14**, 39.

219 Lewin, B. (1977) *Gene Expression-3*, John Wiley and Sons, New York.

220 Lacks, S. and Neuberger, M. (1975) *J. Bacteriol.*, **154**, 1321.

221 Szybalska, E. H. and Szybalski, W. (1962) *Proc. Natl. Acad. Sci. USA*, **48**, 2026.

222 Kornberg, A. (1980) *DNA Replication*, W. & H. Freeman & Co., San Francisco.

223 Bukhari A. I. Shapiro, J. A. and Adhya, S. L. (1977) *DNA Insertion Elements, Plasmids and Episomes*, Cold Spring Harbor, New York.

224 Meynell G. G. (1972) *Bacterial Plasmids*, Macmillan, London.

225 Sherratt, D. J. (1974) *Cell*, **3**, 189.

226 Campbell, A. (1969) *Episomes*, Harper and Row, New York.

227 Novick, R. P. (1980) *Sci. Am.*, **243** (6), 76.

228 Willetts, N. and Wilkins, B. (1984) *Microbiol. Rev.*, **48**, 24.

229 Jacob, F. and Wollman, E. L. (1961) *Sexuality and the Genetics of Bacteria*, Academic Press, New York.

230 McClintock, B. (1965) *Brookhaven Symp. Biol.*, **18**, 162.

231 Nuti, M. P., Lepidi, A. A., Prakash, R. K., Hoogkaas, P. J. J. and Schilperoort, R. A. (1982) in *Molecular Biology of Plant Tumours* (eds G. Khan and J. Schell), Academic Press, New York, p. 561.

232 Braun, A. C. (1982) in *Molecular Biology of Plant Tumours* (eds G. Khan and J. Schell), Academic Press, New York, p. 155.

233 Bolivar, F. (1979) *Life Sci.*, **25**, 807.

234 Luria, S. E., Darnell, J. E., Baltimore, D. and Campbell, A. (1978) *General Virology* (3rd edn), John Wiley and Sons, New York.

235 Primrose, S. B. and Dimmock, N. J. (1980) *Introduction to Modern Virology*, Blackwell Scientific, Oxford.

236 Klug, A. and Caspar, D. L. D. (1960) *Adv. Virus Res.*, 7, 225.

237 Markham, R. (1963) *Prog. Nucleic Acid Res.*, **2**, 61.

238 Bancroft, J., Hills, G. and Markham, R. (1967) *Virology*, **31**, 354.

239 Baltimore, D. (1971) *Bacteriol. Rev.*, **35**, 235.

240 Simons, K., Garoff, H. and Helenius, A. (1982) *Sci. Am.*, **246** (2), 46.

241 Stent, G. S. and Wollman, E. L. (1950) *Biochim. Biophys. Acta*, **6**, 307.

242 Crawford, E. M. and Gesteland, R. F. (1964) *Virology*, **22**, 165.

243 Kozloff, L. M. and Lute, M. (1959) *J. Biol. Chem.*, **234**, 534.

244 McCorquodale, D. J., Oleson, A. E. and Buchanan, J. M. (1967) *The Molecular Biology of Viruses* (eds J. S. Colter and W. Paranchych), Academic Press, New York, p. 31.

245 Brinton, C. C. and Beer, H. (1967) *The Molecular Biology of Viruses* (eds J. S. Colter and W. Paranchych), Academic Press, New York, p. 251.

246 Marvin, D. A. and Hohn, B. (1969) *Bacteriol. Rev.*, **33**, 172.

247 Crawford, L. V. (1962) *Virology*, **18**, 177.

248 Dulbecco, R., Hartwell, L. H. and Vogt, M. (1965) *Proc. Natl. Acad. Sci. USA*, **53**, 403.

249 Winocour, E., Kaye, A. N. and Stollar, V. (1965) *Virology*, **27**, 156.

250 Fried, M. and Pitts, J. D. (1968) *Virology*, **34**, 761.

251 Wood, W. B. and Edgar, R. S. (1967) *Sci. Am.*, **217** (1), 60.

252 Casjens, S. and King, J. (1975) *Annu. Rev. Biochem.*, **44**, 555.

253 Hendrix, R. W., Roberts, J. W., Stahl, F. W. and Weisberg, R. A. (1983) *Lambda II*, Cold Spring Harbor, New York.

254 Hershey, A. D. and Chase, M. (1952) *J. Gen. Physiol.*, **26**, 36.

255 Kaiser, D. and Masuda, T. (1973) *Proc. Natl. Acad. Sci. USA*, **70**, 260.

256 Klug, A. (1980) *The Harvey Lectures*, Academic Press, New York.

257 Butler, P. J. G. and Klug, A. (1978) *Sci. Am.*, **239** (5), 52.

258 Hohn, T. (1967) *Eur. J. Biochem.*, **2**, 152.

259 Bancroft, J. B. and Hiebert, E. (1967) *Virology*, **32**, 354.

260 Crick, F. H. C. and Watson, J. D. (1956) *Nature (London)*, **177**, 473.

261 Streisinger, G., Mukai, F., Dreyer, W. J., Miller, B. and Horiuchi, S. (1961) *Cold Spring Harbor Symp. Quant. Biol.*, **26**, 25.

262 Oldstone, M. B. A., Rodriguez, M., Daughaday, W. H. and Lampert, P. W. (1984) *Nature (London)*, **306**, 278.

263 Edgar, R. S. and Lielausis, I. (1964) *Genetics*, **49**, 649.

264 Hershey, A. D., Burgi, E. and Ingraham, L. (1963) *Proc. Natl. Acad. Sci. USA*, **49**, 748.

265 Ritchie, D. A., Thomas, C. A., MacHattie, L. A. and Wensinv, P. C. (1967) *J. Mol. Biol.*, **23**, 365.

266 Tolun, A., Alestrom, P. and Pettersson, V. (1979) *Cell*, **17**, 705.

267 Thomas, C. A. and MacHattie. L. A. (1964) *Proc. Natl. Acad. Sci. USA*, **52**, 1297.

268 Subak-Sharpe, J. H. and Timbury, M. C. (1977) *Comp. Virol.*, **9**, 89.

269 Hayward, G. S., Jacob, R. J., Wadsworth, S. C. and Roizman, B. (1975) *Proc. Natl. Acad. Sci. USA*, **72**, 4243.

270 Reddy, V. B., Thimmappaya, B., Dhar, R., Subramanian, K N., Zain, B. S., Pan, J. *et al.* (1978) *Science*, **200**, 494.

271 Fiers, W., Contreras, R., Haegeman, G., Rogiers, R., Van de Vorde, A. *et al.* (1978) *Nature (London)*, **273**, 113.

272 Soeda, E., Arrand, J. R., Smoler, N., Walsh. J. E. and Griffin, B. E. (1980) *Nature (London)*, **283**, 445.

273 Bujard, H. (1969) *Proc. Natl. Acad. Sci. USA*, **62**, 1167.

274 Astell. G. R., Smith, M., Chow, M. B. and Ward, D. C. (1979) *Cell*, **17**, 691.

275 Bellamy, A. R. and Joklik, W. K. (1967) *Proc. Natl. Acad. Sci. USA*, **58**, 1389.

276 Wyatt, G. R. and Cohen, S. S. (1953) *Biochem. J.*, **55**, 774.

277 Lehman, I. R. and Pratt, E. A. (1960) *J. Biol. Chem.*, **235**, 3254.

278 Dulbecco, R. and Vogt, M. (1963) *Proc. Natl. Acad. Sci. USA*, **50**, 236.

279 Weil, R. and Vinograd, J. (1963) *Proc. Natl. Acad. Sci. USA*, **50**, 730.

280 Winocour, E. (1967) *The Molecular Biology of Viruses* (eds J. S. Colter and W. Paranchych), Academic Press, New York, p. 577.

281 Sanger, F., Air, G. M., Barrell, B. G., Brown, N. L., Coulson, A. R. *et al.* (1977) *Nature (London)*, **265**, 687.

282 Godson, G. N., Barrell, B. G., Staden, R. and Fiddes, J. C. (1978) *Nature (London)*, **276**, 236.

283 Fiddes, J. C. and Godson, C. W. (1979) *J. Mol. Biol.*, **133**, 19.

284 Fyfe, J. A., Keller, P. M., Furman, P. A., Miller, R. L. and Elion, G. B. (1978) *J. Biol. Chem.*, **253**, 8721.

285 Derse, D., Bastow, K. F. and Cheng, Y-C. (1984) *J. Biol. Chem.*, **257**, 10 251.

286 Lerner, R. A. (1982) *Nature (London)*, **299**, 592.

287 Dreasman, G. R., Sanchez, Y., Ionescu-Matiu, I., Sparrow. J. T., Six, H. R. *et al.* (1982) *Nature (London)*, **295**, 158.

288 Ptashne, M., Johnson, A. D. and Pabo, C. O. (1982) *Sci. Am.*, **267** (5), 106.

289 Ptashne, M. (1986) *A Genetic Switch*, Blackwell Scientific Publications and Cell Press, Cambridge, Mass.

290 Lederberg, E. M. and Lederberg, J. (1953) *Genetics*, **38**, 51.

291 Campbell, A. (1962) *Adv. Genet.*, **11**, 101.

292 Jacob, F. and Wollman, E. L. (1957) *The Chemical Basis of Heredity* (eds W. D. McElroy and B. Glass), Johns Hopkins, Baltimore, p. 468.

293 Tooze, J. (ed.) (1981) *DNA Tumor Viruses* (2nd edn), Cold Spring Harbor Laboratory Monograph **10B**.

294 Weiss, R. A., Teich, N. M., Varmus, H. E. and Coffin, J. M. (1982) *RNA Tumor Viruses* (2nd edn), Cold Spring Harbor Laboratory Monograph **10C**.

295 Dulbecco, R. (1967) *Sci. Am.*, **216** (4), 28.

296 Benjamin, T. L. (1966) *J. Mol. Biol.*, **16**, 359.

297 Watkins, J. F. and Dulbecco, R. (1967) *Proc. Natl. Acad. Sci. USA*, **58**, 1396.

298 Gelb. L. D., Kohne, D. E. and Martin, M. A. (1971) *J. Mol. Biol.*, **57**, 129.

299 Ozanne, B., Vogel, A., Sharp, P., Keller, W. and Sambrook, J. (1973) *Lepetit Colloq. Biol. Med.*, **4**.

300 Fried, M. (1965) *Proc. Natl. Acad. Sci. USA*, **53**, 486.

301 Cuzin. F., Vogt, M., Dieckmann, M. and Berg, P. (1970) *J. Mol. Biol.*, **47**, 317.

302 Crawford, L. V. (1980) *Trends Biochem. Sci.*, **5**, 39.

303 Rigby, P. (1979) *Nature (London)*, **282**, 781.

304 Whyte, P., Buchkovich, K. J., Horowittz, J. M., Friend, S. H., Raybuck, M. *et al.* (1988) *Nature (London)*, **334**, 12.

305 Weinberg, R. A. (1988) *Sci. Am.*, **259** (3), 34.

306 Finlay, C. A., Hinds, P. W. and Levine, A. J. (1989) *Cell*, **57**, 1083.

307 Garza, D., Ajioka, J. W., Carulli, J. P., Jones, R. W., Johnson, D. H. *et al.* (1989) *Nature (London)*, **340**, 577.

308 Bishop, J. M. (1982) *Sci. Am.*, **246** (3), 68.

309 Weinberg, R. A. (1983) *Sci. Am.*, **249** (5), 102.

310 Waterfield, M. D., Sarace, G. T., Whittle, N., Stroobaut, P., Johnsson, A. *et al.* (1983) *Nature (London)*, **304**, 35.

311 Lane, D. P. (1984) *Nature (London)*, **312**, 596.

312 Verma, I. M. (1984) *Nature (London)*, **308**, 317.

313 Steele, R. E. (1990) *Trends Biochem. Sci.*, **15**, 124.

314 Diener, T. O. (1984) *Trends Biochem. Sci.*, **9**, 133.

315 Diener, T. O. (1981) *Sci. Am.*, **244** (1), 58.

316 Symonds, R. H. (1989) *Trends Biochem. Sci.*, **14**, 445.

317 Taylor, J. M. (1990) *Cell*, **61**, 371.

318 Weissman, C. (1990) *Nature (London)*, **338**, 298.

319 Kimberlin, R. H. (1990) *Nature (London)*, **345**, 763.

320 Diener, T. O. (1987) *Cell*, **49**, 719.

321 Wurtman, R. J. (1985) *Sci. Am.*, **252** (1), 48.

322 Prusiner, S. B. (1984) *Sci. Am.*, **251** (4), 48.

323 Lowenstein, D. H., Butler, D. A., Westaway, D., McKinley, M. P.,

Dearmond, S. J. *et al.* (1990) *Mol. Cell. Biol.*, **10**, 1153.

324 Brunori, M. and Talbot, B. (1985) *Nature (London)*, **314**, 676.

325 Donis-Keller, H., Green, P., Helms, C., Castinhous, S., Weiffenbach, B. *et al.* (1987) *Cell*, **51**, 319.

326 Wicking, C. and Williamson, R. (1991) *Trends Genet.*, **7**, 288.

327 Masters, C. (1985) *Nature (London)*, **314**, 15.

328 Robertson, H. D., Branch, A. D. and Dahlberg, J. E. (1985) *Cell*, **40**, 725.

329 White, R. (1985) *Trends Genet.*, **1**, 177.

330 Gemmill, R. M., Coyle-Morris, J. F., McPeek, F. D., Ware-Uribe, L. F. and Hecht, F. (1987) *Gene Anal. Techn.*, **4**, 119.

331 Nguyen, C., Pontarotti, P., Birnbaum, D., Chimini, G., Rey, J. A. *et al.* (1987) *EMBO J.*, **6**, 3285.

332 Jordan, B. R. (1988) *BioEssays*, **8**, 140.

333 Patterson, D. (1987) *Sci. Am.*, **257** (2), 42.

334 Walker, P. R. and Sikorska, M. (1987) *J. Biol. Chem.*, **262**, 12 218 and 12 223.

335 Felsenfeld, G. and McGhee, J. D. (1986) *Cell*, **44**, 375.

336 Stief, A., Winter, D. M., Strätling, W. H. and Sippel, A. E. (1989) *Nature (London)*, **341**, 343.

337 Mirkovitch, J., Gasser, S. M. and Laemmli, U. K. (1988) *J. Mol. Biol.*, **200**, 101.

338 Phi-Van, L. and Strätling, W. H. (1988) *EMBO J.*, **7**, 655.

339 Smith, J. M., Dowson, C. G. and Spratt, B. G. (1991) *Nature (London)*, **349**, 29.

340 Schmid, M. B. (1990) *Cell*, **63**, 451.

341 Reed, S. I. (1991) *Trends Genet.*, **7**, 95.

342 von Kreis, J. P., Buhrmester, H. and Strätling, W. H. (1991) *Cell*, **64**, 123.

343 Carlson, G. A., Hsiao, K., Oesch, B., Westaway, D. and Prusiner, S. B. (1991) *Trends Genet.*, **7**, 61.

344 Copeland, N. G. and Jenkins, N. A. (1991) *Trends Genet.*, **7**, 113.

345 Lowndes, N. F., Johnson, A. L. and Johnston, L. H. (1991) *Nature (London)*, **350**, 247.

346 Nasmyth, K. A. (1990) *Cell*, **63**, 1117.

347 Hunt, T. (1991) *Nature (London)*, **350**, 462.

348 Richmond, T. J., Finch, J. T. and Klug, A. (1982) *Cold Spring Harbor Symp. Quant. Biol.*, **47**, 503.

4

Degradation and modification of nucleic acids

4.1 INTRODUCTION AND CLASSIFICATION OF NUCLEASES

Enzymes which catalyse the breakdown of nucleic acids by hydrolysis of phosphodiester bonds have been found in almost all biological systems [1, 2]. Some, the *ribonucleases*, are quite specific for RNA, others, the *deoxyribonucleases*, act only on DNA, whereas a third group of non-specific nucleases is active against either nucleic acid. The *phosphorylases*, polynucleotide phosphorylase and pyrophosphorylase, are also capable of depolymerizing RNA, but their degradative role *in vivo* is uncertain. The *phosphomonoesterases* act on polynucleotides or oligonucleotides with a terminal phosphate group or on a mononucleotide to liberate inorganic phosphate. Their substrates will often be the products of nuclease action.

Classification schemes for the nucleases have been discussed by Laskowski [1] and by Barnard [2]. Three main features of nuclease action can be used as a basis for classification. The first of these is substrate specificity, i.e. action on RNA, DNA or both, as discussed above. The second is mode of attack; polynucleotides can be attacked at

points *within* the polymer chain *endolytically* or stepwise from one end of the chain *exolytically*. Thus we may have endonucleases which produce oligonucleotides and cause rapid changes in physical properties (e.g. in viscosity of DNA), and exonucleases which produce mononucleotides but initially with rather less drastic effects on the physical properties of the nucleic acid. A few enzymes appear to act as both endo- and exo-nuclease, e.g. micrococcal nuclease. The third feature is mode of phosphodiester bond cleavage. Most biological polymers, like proteins and carbohydrates, can be split in only one way; but, as illustrated in Fig. 4.1, polynucleotides can be cleaved at the phosphodiester bond on either side of the phosphate.

Of the ribonucleases, those enzymes that produce 3'-phosphates and 5'-hydroxyl termini are concerned with the degradation of RNA whereas those enzymes that are involved in the processing of RNA precursor molecules generate 5'-phosphate and 3'-hydroxyl termini.

Additional criteria may be used to define further the action of a nuclease. These include specificity towards secondary structure of substrate, direction of attack by

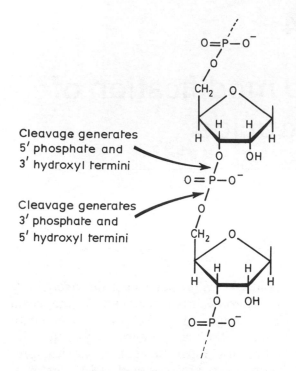

Cleavage generates
5′ phosphate and
3′ hydroxyl termini

Cleavage generates
3′ phosphate and
5′ hydroxyl termini

Fig. 4.1   The cleavage of phosphodiester bonds.

exonuclease (3′ → 5′ or 5′ → 3′), and pre-ferential endonucleolytic bond cleavage, e.g. GpN → Gp, by ribonuclease T1. However, in only a few cases (e.g. the restriction endo-nucleases) are base specificities absolute, and relative differences in reaction rates with different bases are more common. Experimental details for the preparation and handling of several of these enzymes are to be found in *Methods in Enzymology*, volume 65 [3].

4.2 NON-SPECIFIC NUCLEASES

4.2.1 Non-specific endonucleases

(a) *Micrococcal nuclease (EC 3.1.31.3)*

This enzyme is found in cultures of *Staphylococcus* and degrades DNA to a mixture of nucleoside 3′-monophosphates and oligo-nucleotides with 3′-phosphate termini [4]. It attacks RNA and heat-denatured DNA preferentially, although it also actively degrades native DNA when it acts initially as an endonuclease producing double-stranded breaks, and subsequently as an exonuclease. It shows a certain amount of sequence specificity; the primary cutting sites are adenine- and thymine-rich regions of DNA [5, 6]. It requires Ca^{2+} for maximum activity. Its structure and chemical properties have been reviewed [7, 8].

(b) Neurospora crassa *nuclease [9]*

This enzyme has been considerably purified from conidia of *Neurospora* and attacks DNA or RNA to give oligonucleotides with a 5′-phosphate terminus. It exhibits a pre-ference for guanosine or deoxyguanosine residues, but its most interesting property is an absolute requirement for denatured poly-nucleotides, particularly in the presence of high concentrations of NaCl [10]. It is active under a wide range of conditions and requires Ca^{2+} or Mg^{2+}.

(c) *Nuclease S1 from* Aspergillus oryzae *(EC 3.1.30.1)*

This enzyme is very similar to the *N. crassa* nuclease in that it hydrolyses phosphodiester bonds in single-stranded DNA or RNA [10, 11]. It has an acid pH optimum and shows a requirement for Zn^{2+}.

(d) *Nuclease P1 from* Penicillium citrinum

This splits 3′,5′-phosphodiester bonds in RNA and single-stranded DNA as well as 3′-phosphomonoester bonds in mononucleo-tides and oligonucleotides with limited speci-ficity. It also is a Zn^{2+}-requiring enzyme and works well at 70°C [10, 12].

(e) *Mung bean nuclease I*

This is another Zn^{2+}-dependent single-stranded nuclease which acts on both RNA and heat-denatured DNA. It also has an acid pH optimum but is strongly inhibited by salt. It can dephosphorylate 3'-mononucleotides [10, 13].

(f) *BAL 31 nuclease*

An extremely stable enzyme isolated from the marine bacterium *Alteromonas espejiano* (BAL 31). It will degrade both DNA and RNA acting endonucleolytically on single-stranded molecules and exonucleolytically on duplex DNA when it degrades both the 3'- and 5'-termini [10, 14, 17]. It is active at high temperature (60°C), high salt (up to 7 M CsCl) and in the presence of 5% sodium dodecyl sulphate provided that Ca^{2+} and Mg^{2+} ions are present. It has an optimum pH of about 8.0.

The *in vivo* function of these, and many other endonucleases present in a variety of mammalian cells, invertebrates, plants and bacteria is largely unknown. However, they have proved of immense use to molecular biologists as tools for the study of nucleic acids and chromatin [10].

4.2.2 Non-specific exonucleases

(a) *Venom phosphodiesterase [15]*

The venom of several species of snake contains a phosphodiesterase which is commonly employed in the preparation of nucleoside 5'-phosphates. The enzyme occurs naturally in association with a high concentration of phosphomonoesterase from which it can be freed by chromatography and acetone fractionation. Venom diesterase hydrolyses RNA to nucleoside 5'-mono-

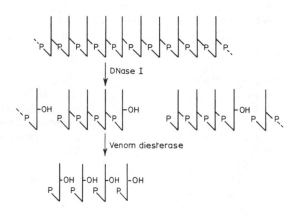

Fig. 4.2 The digestion of DNA by DNase I followed by venom diesterase to yield deoxyribonucleoside 5'-monophosphates.

phosphates starting at the 3'-hydroxyl end of the chain, and is also active in hydrolysing the oligonucleotides produced by the action of deoxyribonuclease I on DNA (Fig. 4.2). The presence of a 3'-phosphoryl terminal group confers resistance on the substrate.

(b) *Spleen phosphodiesterase (EC 3.1.16.1) [16]*

This hydrolyses RNA to nucleoside 3'-monophosphates starting at the 5'-hydroxyl end and also acts on the mixture of oligo-nucleotides produced from DNA by spleen deoxyribonuclease II (Fig. 4.3). It is inactive with oligonucleotides carrying a 5'-phos-phomonoester end group.

4.3 RIBONUCLEASES (RNases)

The following discussion of ribonucleases reviews some of the enzymes which have been purified and studied in detail. They tend to be nucleases associated with RNA degradation and are also often those enzymes most extensively used in the laboratory as tools to cleave or digest RNA. Many of the ribonucleases involved in the specific processing of RNA precursor species are poorly

Table 4.1   Selected ribonucleases*

Enzyme	Source*	Type	Product
Pancreatic RNase A	Mammalian pancreas	Endonuclease	$3'$-PO_4, $5'$-OH
RNase T1	*Aspergillus oryzae*	Endonuclease	$3'$-PO_4, $5'$-OH
RNase T2	*Aspergillus oryzae*	Endonuclease	$3'$-PO_4, $5'$-OH
RNase U1	*Ustilago sphaerogena*	Endonuclease	$3'$-PO_4, $5'$-OH
RNase U2	*Ustilago sphaerogena*	Endonuclease	$3'$-PO_4, $5'$-OH
RNase L	Mammalian cells	Endonuclease	$3'$-PO_4, $5'$-OH
RNase I	*E. coli*	Endonuclease	$3'$-PO_4, $5'$-OH
RNase P1	*Penicillium citrinum*	Endonuclease	$5'$-PO_4, $3'$-OH
Mung bean nuclease 1	Mung bean	Endonuclease	$5'$-PO_4, $3'$-OH
Micrococcal nuclease	*Staphylococcus*	Endonuclease	$3'$-PO_4, $5'$-OH
S1 nuclease	*Aspergillus*	Endonuclease	$5'$-PO_4, $3'$-OH
Endoribonuclease H	Many sources	Endonuclease	$5'$-PO_4, $3'$-OH
Exoribonuclease H	Retroviruses	Exonuclease $5' \rightarrow 3'$ and $3' \rightarrow 5'$	$5'$-PO_4, $3'$-OH
RNase II	*E. coli*	Exonuclease $3' \rightarrow 5'$	$5'$-NMP
Spleen phosphodiesterase	Bovine spleen	Exonuclease $5' \rightarrow 3'$	$3'$-NMP
Venom phosphodiesterase	Snake venom	Exonuclease $3' \rightarrow 5'$	$5'$-NMP
Polynucleotide phosphorylase	Micro-organisms	Exonuclease $3' \rightarrow 5'$	$5'$-NDP
RNase III	*E. coli*	Endonuclease	$5'$-PO_4, $3'$-OH
RNase IV	*E. coli*	Endonuclease	$3'$-PO_4, $5'$-OH
RNase P	*E. coli*	Endonuclease	$5'$-PO_4, $3'$-OH
RNase E	*E. coli*	Endonuclease	$5'$-PO_4, $3'$-OH
RNase M5, M16, M23	*E. coli*	Endonuclease	Presumed to be $5'$-PO_4, $3'$-OH
RNase, F.Q.Y.D.T., PH and BN	*E. coli*	Endo and exonucleases	Presumed to be $5'$-PO_4, $3'$-OH

* Note that ribonuclease activities related to those described above have been detected and in some cases isolated from many other species. Some of these are described in the text or in the given references. ss and ds refer to single stranded and double stranded respectively.

characterized. As indicated in Table 4.1, those that have been identified are discussed in those sections of the book that deal with the processing events that they catalyse.

4.3.1 Endonucleases which form 3'-phosphate groups

(a) *Pancreatic ribonuclease (Ribonuclease A) (EC 3.1.27.5) (reviewed in [18, 19])*

In 1920 Jones [20] described a heat-stable enzyme present in the pancreas which was capable of digesting yeast RNA. The enzyme was purified by Dubos and Thompson [22] and was crystallized in 1940 by Kunitz [23] who named it ribonuclease. The crystallization of pancreatic RNase free from proteolytic contaminants was described by McDonald [24] but more recent purification methods have been much simplified by the use of affinity chromatography [19].

Pancreatic RNase is a very small protein, M_r 13 700, is stable over a wide pH range, and is remarkably resistant to heat in slightly

Table 4.1   *(continued)*

Specificity	Function	Reference
Cuts 3′ to pyrimidine residues	Degradative	Section 4.3.1a
Cuts 3′ to guanosine residues	Degradative	Section 4.3.1b
Partial specificity to 3′ side of adenosine	Degradative	Section 4.3.1b
Cuts 3′ to guanosine residues	Degradative	Section 4.3.1b
Cuts 3′ to purine residues	Degradative	Section 4.3.1d
Cuts 3′ to uridine residues	Degradative	Section 4.3.1e
Non-specific	Assumed to be degradative	Section 4.3.1c
Single-stranded RNA or DNA	Not known	Section 4.2.1d
Single-stranded RNA or DNA	Not known	Section 4.2.1e
Double- or single-stranded RNA or DNA	Degradative	Section 4.2.1a
Single-stranded RNA or DNA	Degradative	Section 4.2.1c
RNA of RNA:DNA duplexes	Degradative	Section 4.3.4, Chapter 6
RNA of RNA:DNA duplexes	Degradative	Section 4.3.4
Non-specific but prefers ss RNA	Assumed to be degradative	Section 4.3.3
RNA and DNA, prefers ss substrate	Degradative	Section 4.2.2b
RNA and DNA, prefers ss substrate	Degradative	Section 4.2.2a
Reversible reaction, non-specific	Not known	Section 4.4
ds RNA, unknown method of recognition	Processing	Section 11.6.1a
Specific sites but recognition method unknown	Not known	Section 4.3.2
Specific cuts to pre-tRNA	Processing	Section 11.7.1a, 11.3.4
Specific cuts to pre-5SRNA	Processing	Section 11.6.1b
Maturation of 5S, 23S and 16S precursor	Processing	Section 11.6.1a
Maturation of pre-tRNA	Processing	Section 11.7.1a

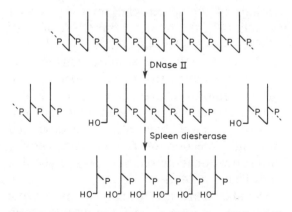

Fig. 4.3 The digestion of DNA by DNase II followed by spleen diesterase to yield deoxyribonucleoside 3′-monophosphates.

acid solution; although it is readily in-activated by alkali. It has no action on DNA and is strongly antigenic. Its maximum activity is in the range pH 7.0–8.2, with the optimum at pH 7.7. Its optimum temperature is 65°C. The enzyme is commonly referred to as ribonuclease A but glycosylated variants occur which are given different notations. Thus, bovine pancreatic ribonucleases B, C and D contain different carbohydrate side chains attached to the asparagine-34 side chain (the latter derivative should not be confused with *E. coli* RNase D – Table 4.1 and section 11.7.1(a)). A man-made derivative of RNase A, which has featured in many studies of the catalytic activity of the enzyme, is known as RNase S. It is produced by cleavage between residue 20 and 21 by the protease, subtilisin.

Since the initial sequence determination of bovine ribonuclease by Moore and collaborators [25], pancreatic ribonucleases have been isolated and sequenced from many sources. Beintema and colleagues have been particularly active in this respect and their many sequences [26] together with those of others have been tabulated by Blackburn and Moore [19]. The three-dimensional structure of pancreatic ribonuclease has been resolved by X-ray diffraction both by crystallographic studies [27–29] and in solution [30] and the nature and mechanism of action of the active site has been determined (reviewed in [18] and [19]). The amino acids involved in catalysis and in maintaining the secondary and tertiary structure are those that are highly conserved. The enzyme has been chemically synthesized by both solid phase and liquid phase methods [31–33].

Pancreatic ribonuclease cleaves the ester bond between the phosphate and the 5'-carbon of the ribose of a pyrimidine nucleotide. It does this in two steps (Fig. 4.4). The ester bond is first transferred to the 2'-hydroxyl group so forming a cyclic ester. This is then specifically hydrolysed to the 3'-monoester. The kinetics and geometry of the two stages have been thoroughly investigated [34–38] but their detailed consideration is outside the scope of this book. Pancreatic RNase, and other RNases whose mode of action is similar, have the theoretical possibility of forming both the 2'- and 3'-monoester by hydrolysis of the cyclic phosphate (Fig. 4.4); most, if not all, form the 3'-ester exclusively.

The action of pancreatic RNase may be illustrated as follows. The pentanucleotide shown in Fig. 4.5(a), in which R and Y represent purine and pyrimidine residues, respectively, will be hydrolysed at the points shown by the broken lines, whereas the ribopolynucleotide chain shown in Fig. 4.5(b), which may also be expressed as pApCpUpGpAp, will be broken at positions 5 and 7 to yield pApCp + Up + GpAp.

To take a slightly more elaborate case, the polynucleotide

$$\text{ACCCCAGGGUUUAGUCp}$$

would be split by RNase thus:

$$\text{AC/C/C/C/AGGGU/U/U/AGU/Cp}$$

to yield

$$\text{ACp + AGGGUp + AGUp + 4Cp + 2Up}$$

Pancreatic RNase also digests certain homopolyribonucleotides. Thus poly(A) and poly(I) are not split by the enzyme whereas poly(C) and poly(U) yield the 3'-mononucleotides. However, the specificity of pancreatic RNase for pyrimidines is not absolute, since Ap diester bonds in a polynucleotide are also attacked, albeit considerably less readily [39].

Modern methods of assaying the activity of ribonuclease continue to employ modifications of the spectrophotometric assay of Kunitz [40], the release from RNA of acid-soluble nucleotides [41] and the hydrolysis of 2',3'-cyclic CMP [42]. Modifications include the use of radioactive or chromogenic substrates, improved precipitations and fluorimetric assays and have been reviewed [19]. Several methods have also been described for the detection of ribonuclease activity after fractionation on SDS/polyacrylamide gels (see e.g. [43]).

Similar activities to that of pancreatic ribonuclease have been described from many other extracellular and intracellular sources [43, 44]. An activity in bovine semen

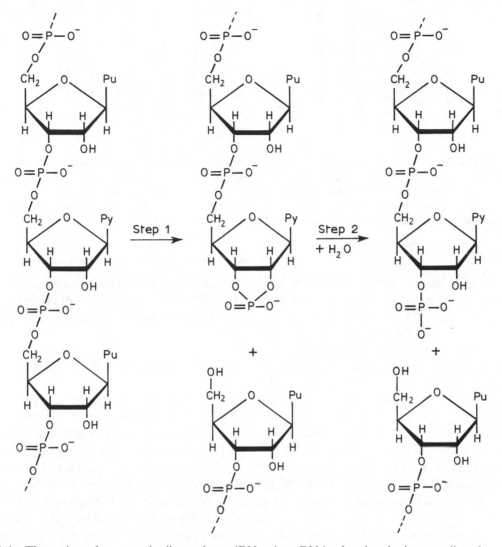

Fig. 4.4 The action of pancreatic ribonuclease (RNase) on RNA, showing the intermediate formation of cyclic phosphates.

represents more than 2% of its total protein. It shares considerable sequence homology with the pancreatic enzyme and also preferentially cuts RNA 3′ to pyrimidine residues. However, it differs in charge, in normally existing as a dimer and in its ability to hydrolyse double-stranded RNA. Furthermore, it appears to have non-catalytic functions that include antispermatic, antitumour

and immonosuppressive activities (reviewed in [45]).

Many, probably most, mammalian tissues contain cytoplasmic enzymes that in many respects resemble pancreatic ribonuclease ([46–48] and references therein). They are, however, present in very small quantities and are hence difficult to study. They have been called type I ribonucleases and char-

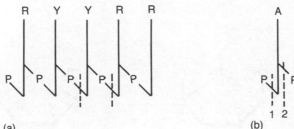
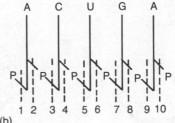

(a)   (b)

Fig. 4.5 (a) The pentanucleotide containing three purine and two pyrimidine nucleotides is split by pancreatic ribonuclease at the broken lines. (b) A pentanucleotide containing two adenines and one residue each of cytosine, uracil and guanine with monoesterified phosphates at each end is split by pancreatic ribonuclease at positions 5 and 7; by ribonuclease T1 at position 9; by 5'-monoesterase at 1; by 3'-monoesterase at 10; by venom diesterase at 2, 4, 6 and 8; and by spleen diesterase at 3, 5, 7 and 9.

acteristically cleave RNA at pyrimidine residues in two steps via a 2',3'-cyclic phosphate (Fig. 4.4). Barton *et al.* [49] subdivided type I enzymes into secreted and non-secreted forms. The secreted enzymes of the pancreas, serum and urine tend to be heat and acid-stable and have alkaline pH optima. RNase A typifies this group. The non-secretory RNases typically have pH optima around 7.0 and are relatively heat-labile. A further member of the super family is angiogenin, a protein that induces growth of blood vessels but is 35% identical to human RNase and has some RNase activity [50].

Type I enzymes, including angiogenin, occur as latent complexes with a proteinaceous inhibitor known as RNasin [51–53]. The inhibitor is an *N*-acetylated polypeptide of 456 amino acids and can be subdivided into 15 homologous, leucine-rich repeats [54, 55]. Cellular ribonucleases have the potential to do a great deal of harm to all cytoplasmic species of RNA and they must clearly be controlled in their activity. It would appear that type I enzymes are normally controlled by RNasin whereas the other main group of degradative enzymes, type II RNases, are compartmentalized in

lysozomes. These latter enzymes are non-specific RNases with acidic pH optima. They are thermolabile and insensitive to RNasin [56].

(b) *Ribonuclease T1 (EC 3.1.27.3)*

This has been the subject of intensive study, and our knowledge of its chemistry approaches that of pancreatic RNase (reviewed in [57]). It is obtained from *Aspergillus oryzae* and at neutral pH specifically hydrolyses the internucleotide bonds of RNA between 3'-GMP and the 5'-OH groups of adjacent nucleotides (Fig. 4.5(b)). Like pancreatic ribonuclease, it does this in a two-stage reaction producing 3'-phosphate termini via 2',3'-cyclic phosphate intermediates.

Ribonuclease T1 is a small (M_r 11085) heat-stable and acid-stable extracellular endonuclease. A second enzyme from the same source, *ribonuclease T2*, preferentially attacks Ap residues and will digest tRNA almost totally to 3'-monophosphates [58]. Guanine-specific enzymes similar to ribonuclease T1 are common in fungi and bacteria. These include ribonuclease U1 from *Ustilago sphaerogena* [59], and ribonuclease N1 from *Neurospora crassa* [60].

Such enzymes may play a role in processing RNA molecules as well as in their degradation [283].

(c) *Ribonuclease I from* E. coli *(reviewed in [61])*

This endonuclease normally occurs in the periplasmic space of the bacterial cell but on conversion of the cells into spheroplasts it is released into the medium. In broken cell preparations it is found artifactually associated with 30S ribosomal subunits [62] and is the only prokaryotic RNase known to be able to attack the intact ribosome [63]. Its intracellular role is unclear since strains lacking the enzyme are perfectly viable [64]. It has been considered to have a scavenging role as have the latent ribonucleases of mammalian cells. Again, however, rRNA turnover is unchanged in mutants lacking the enzyme [64]. The enzyme is a low-Mr basic protein with a single subunit and a pH optimum of 8.1. It digests single-stranded RNA with no absolute base specificity although it hydrolyses poly(U) faster than other homopolymers. Like the previous two enzymes, it produces 3'-phosphate termini via 2',3'-cyclic phosphate intermediates [65]. The enzyme has recently been purified [66].

(d) *Ribonuclease U2 from* Ustilago sphaerogena *(EC 3.1.27.4)*

This endonuclease, like those previously considered, hydrolyses phosphodiester bonds in two separable steps; transphosphorylation and hydrolysis [67]. The enzyme is however unique in its specificity in that it cleaves the polynucleotide to the 3'-side of a purine residue [68]. Under certain conditions, it is highly specific for adenyl residues [69]. It consists of a single polypeptide of M_r 12 490 and exhibits considerable sequence and catalytic similarity to RNase T1 [67, 70].

(e) *2–5A-dependent endoribonuclease (RNase L)*

Exposure of cells to interferon inhibits viral replication by arresting the synthesis of viral mRNA and protein (reviewed in [71]). Induction of the antiviral state involves the synthesis of a number of new proteins. One of these, produced in response to double-stranded RNA of viral origin, is 2–5A synthetase which synthesizes the unusual oligonucleotide $ppp(A2'p)_n$ where $n = 2-4$. Known as 2–5A, this nucleotide in turn activates a latent endoribonuclease (RNase L) [72] which can degrade mRNA [73] and rRNA [74] (section 12.9.3(c)). The catalytic mechanism of the enzyme has been extensively studied [75, 76] and the mouse enzyme has been purified [77].

(f) *Other endonucleases of value in RNA sequence and structural analysis*

Most of the above enzymes have played an important role in the base-specific cleavage of RNA for sequence and structural analysis [78] and they continue to be important experimental tools despite the fact that DNA sequencing has largely replaced RNA sequencing. Other enzymes that have been useful in this respect include the pyrimidine-specific endonuclease of *Bacillus cereus* which cleaves RNA after all pyrimidines [79, 80]. Pancreatic nuclease is less reliable in this respect and cleaves between two adjacent pyrimidines with difficulty. Two other useful enzymes PhyI and PhyII have been isolated from *Physarum polycephalum* [81]. PhyI is a 25 000-M_r protein with a pH optimum of 4.3 and cleaves all phosphodiester bonds with an order of susceptibility UpN>ApN>GpN>CpN. It has been useful in distinguishing between C and U in RNA sequences [78]. A relatively unfractionated preparation known as PhyM

[82] is commercially available, and under denaturing conditions cleaves specifically at U and A residues. RNase PhyII shows outstanding preference for GpN leaving ApN and YpY doublets practically unsplit [81]. Another enzyme which has proved useful in distinguishing between C and U in RNA sequence and in structural investigations is RNase CL3 from chicken liver [83]. It is 61 times more active on cytidylic bonds than uridylic bonds [83].

Other endonucleases that have been detected in the cells and tissues of a wide range of species are not considered here because they are too inadequately characterized for their mode of action, function or their relationship to the above enzymes to be ascertained.

4.3.2 Endonucleases which form 5′-phosphate groups

The mechanism by which these nucleases cleave the phosphodiester bond is noncyclizing. A direct attack of water on a 3′,5′-phosphodiester is catalysed and thus a 2′-OH group is not required. For this reason, many 5′-monophosphate-forming endonucleases will attack RNA and DNA. Many of the enzymes which catalyse RNA cleavage in this way are involved in the specific processing of RNA precursor species. They include the *E. coli* ribonucleases III, P, E, M5 etc. which, as indicated in Table 4.1, are discussed in the relevant sections of text on RNA processing. It is certain that many other activities still await discovery, particularly in eukaryotic cells. At least one of these activities, *E. coli* RNase P, is a ribonucleoprotein of which the RNA is the catalytic component of the enzyme. This and other catalytic RNAs are discussed in sections 11.3.2., 11.3.3 and 11.3.4.

One further, well-characterized, *E. coli*

enzyme which generates 5′-phosphates from single-stranded RNA is considered here because its function is unknown.

(a) *Ribonuclease IV from* E. coli *(reviewed in [61])*

RNase IV cleaves a number of RNAs at a limited number of specific sites by an unknown mechanism of recognition. The 26S RNA of the bacteriophage R17 is cleaved at five sites in a short section of the RNA so generating 15S and 22S fragments. It seems likely that the enzyme is recognizing some feature of the secondary structure of the substrate and Adams *et al.* [84] have related the cleavage sites to the proposed secondary structure of the phage RNA. The cuts occur in, or adjacent to, proposed single-stranded regions. Similarly, the single cut made by the enzyme in 5S RNA is in a region which is thought to be single stranded [85]. The specificity of the enzyme indicates a function as a structure-related, site-specific nuclease.

4.3.3 RNA exonucleases

(a) *Ribonuclease II from* E. coli

RNase II is a processive exonuclease; that is, it remains bound to its substrate until it has hydrolysed all phosphodiester bonds. This distinguishes it from distributive exonucleases which dissociate after every or many catalytic events, e.g. spleen exonuclease [86]. It degrades RNA in a $3′ \rightarrow 5′$ direction producing 5′-nucleotide monophosphates but cannot degrade DNA or RNAs with secondary structure [87]. It is one of the most active enzymes in lysates of *E. coli* and consists of a single subunit with a M_r of 72 000–75 000 [87, 88]. Donovan and

Kushner [89] have presented evidence that RNase II works in conjunction with poly-nucleotide phosphorylase (section 4.4) in the degradation of bacterial mRNA. They suggest that the initial steps are endo-nucleolytic cleavage by polynucleotide phosphorylase followed by exonucleolytic degradation by RNase II. Mutants lacking both enzymes do not survive but accumulate fragments of mRNA.

(b) *Spleen phosphodiesterase (spleen exonuclease) and venom phosphodiesterase (venom exonuclease)*

See section 4.2.2.

4.3.4 Ribonucleases which act on RNA:DNA hybrids (RNase H)

An RNase which specifically digested the RNA of an RNA:DNA hybrid was first described by Stein and Hausen [90]. Since then, RNase H activity has been demon-strated in a wide range of species and cell types ranging from prokaryotes, lower eukaryotes, higher animals and plants (reviewed in [91]). In addition, the reverse transcriptases of retroviruses (section 6.4.8) exhibit RNase H activity [91].

E. coli RNase H has an M_r of 17 580 and its structure is known at 2 Å resolution [92]. It is involved in bacterial DNA replication in which it can apparently serve multiple functions. It inhibits initiation at sites other than the *oriC* locus [93], and it cleaves the RNA preprimer of Col E1 plasmid thus defining the origin of plasmid replication [94]. It also shares with the $5' \rightarrow 3'$ exo-nuclease activity of DNA polymerase I the ability to remove RNA primers from Okazaki fragments [95]. Despite these multiple activities, mutations lacking the

enzyme activity are not lethal. Besides requiring a RNA:DNA hybrid as a sub-strate, the enzyme requires Mg^{2+} ions and shows little base specificity. It acts as an endonuclease and releases 5'-phosphate and 3'-hydroxyl products.

The RNase H activity of reverse trans-criptase resides in much larger enzymes, some of which act as dimers. Early evidence indicated that they acted as exonucleases, however, more recent data have established that they are endonucleases [96]. Further-more, they are much more similar to the *E. coli* RNase H than was at first apparent as the RNase activity is in the *C*-terminal third of the reverse transcriptase poly-peptide ([97] and references therein) and this exhibits considerable homology with the bacterial enzyme [98]. The viral RNase H activity of reverse transcriptase degrades the genomic viral RNA after it has served as a template for the synthesis of an RNA:DNA hybrid. Once the RNA is degraded, the minus-strand DNA is itself used as a template to make plus-strand DNA and the duplex is then integrated into the host cell genome.

4.3.5 Double-stranded RNA-specific ribonucleases

Ribonuclease VI isolated from the venom of the cobra *Naja naja oxiana* [99] has proved useful in mapping the secondary structure of RNA because its cleavage is primarily restricted to double-stranded regions of RNA [99–101]. Investigations with various tRNA species [101] reveal that the enzyme does not recognize any specific nucleotide or nucleotide sequence but appears to re-cognize the stacked nucleotides of RNA duplex. It has been suggested [101] that it interacts with the minor groove of the duplex.

4.3.6 Ribonuclease inhibitors

Many nucleotide substrate analogues have been studied with respect to their effect on the catalytic mechanism of ribonucleases [18, 19].

White and co-workers have investigated oligonucleotides as competitive inhibitors of RNA hydrolysis. ApUp ($K_i = 0.5$ mM) was the best of those examined at inhibiting pancreatic ribonuclease [102] whereas G2′–5′G ($K_i = 0.165$ mM) was the best at inhibiting RNase T1 [103].

Neither these nor other derivatives, such as those in which ribose is replaced by arabinose [104], have found much application in the protection of RNA during its isolation and manipulation. The key to such protection is to inhibit endogenous RNase, especially during the early stages of extraction and to avoid exogenous contamination by traces of RNase from glassware, solutions and the hands of investigators.

Older methods relied on non-specific adsorbants, such as the inert acidic clay derivatives, bentonite and macaloid [105, 106] and on polyanions which protect RNA from attack as a result of their electrostatic interaction with the cationic enzyme [107]. The latter included polyvinyl sulphate [108] and heparin [109]. Although some of these inhibitors are still in use, various investigators have found them less than totally reliable and most modern methods of preparing and working with RNA employ guanidinium thiocyanate [110] and sodium dodecyl sulphate [111] for the disruption of cells and simultaneous inactivation of nuclease. Vanadyl–ribonucleoside complexes [112] and the proteinaceous RNase inhibitor, RNasin (section 4.3.1(a)), are used in many other situations where RNA must be protected. The latter is the inhibitor of choice where RNA must be protected during enzymic reactions [113, 114]. RNase con-tamination of solutions is normally excluded by the inclusion of 0.1% diethylpyrocarbonate. This inhibitor destroys ribonuclease [115, 116] but must then itself be destroyed by autoclaving or it can inactivate RNA by carboxymethylation [117].

4.4 POLYNUCLEOTIDE PHOSPHORYLASE (PNPase, EC 2.7.7.8)

Polynucleotide phosphorylase was first discovered in *Azotobacter vinelandii* [118] but is now known to be widely distributed in bacteria [119] (reviewed in [119, 120]). It catalyses the reversible reaction

$$n\text{NDP (NMP)}_n + n\text{P}_i$$

and was the first enzyme to be discovered that can catalyse the formation of an RNA. In doing this, however, it shows little resemblance to RNA polymerase as it uses no template. Rather it catalyses the polymerization of a mixture of nucleoside diphosphates into a random polymer. With a single species of diphosphate it will make a homopolymer. It is also able to elongate an oligonucleotide primer which it does without the lag which otherwise characterizes the formation of the first internucleotide bond. The *E. coli* enzyme is composed of three identical subunits of M_r, 84 000–95 000 [121] with a pI of 6.1 and a requirement for Mg^{2+} which can be partially replaced by Mn^{2+}.

Polynucleotide phosphorylase has proved a useful polymerizing tool for the biochemist, and has been used to make homopolymers and heteropolymers [119, 122, 123] as well as RNAs containing modified or radioactive nucleotides [124]. Polymers made in this way were of particular importance in the elucidation of the genetic code (section 12.2). Nevertheless, the function of the enzyme *in vivo* is far from clear and it appears likely that it normally catalyses the reverse reaction, i.e. the phosphorolytic degradation of RNA. As already discussed in section 4.3.3, evidence has been presented that RNase II and polynucleotide phosphorylase may act co-operatively to degrade bacterial mRNA [86]. Neither enzyme is essential to the bacteria but mutants in which both activities are lost appear not to be viable.

The phosphorolytic activity of the enzyme is also used as an analytical tool, particular use being made of the fact that at 0°C it will degrade poly(A) tails without digesting the rest of a mRNA [125].

4.5 DEOXYRIBONUCLEASES (DNases)

4.5.1 Endonucleases

The two deoxyribonucleases which were the first to be purified and characterized are both endonucleases. The first type, exemplified by pancreatic deoxyribonuclease (DNase I), hydrolyses DNA to yield 5'-phosphomonoesters. The second type (DNase II) which is found in spleen and thymus hydrolyses DNA to give 3'-phosphomonoesters (Figs 4.2 and 4.3). As a result of the characterization of these two enzymes it was suggested that there were two classes of endodeoxyribonuclease (Class I and Class II), but subsequent discoveries have shown that the product of action is only one of the characteristics that distinguish one nuclease from another [1, 126, 153].

The activity of DNA endonucleases is generally measured by estimating the release of acid-soluble products from DNA, whether as ultraviolet-absorbing material or as radioactive label. These methods are, however, useful only in the presence of extensive endonuclease action. Where extreme sensitivity has been required, supercoiled circular viral or plasmid DNA is used as a substrate. Only one phosphodiester bond cleavage is required to alter the physical properties of such molecules and allow separation of intact and cleaved molecules.

(a) *Pancreatic deoxyribonuclease (DNase I) (EC 3.1.21.1) (reviewed in [4, 154])*

This enzyme breaks down DNA into oligonucleotides of average chain length four units with a free hydroxyl group on position 3' and a phosphate group on position 5' (Fig. 4.2). It is produced in the pancreas and used as a digestive enzyme after secretion into the small intestine.

The enzyme has an M_r of 31 000, an optimum pH in the range of 6.8–8.2 and an isoelectric point of 4.7. It has two disulphide bonds which are very readily reduced by 2-mercaptoethanol with concomitant inactivation. Ca^{2+} ions prevent this inactivation. The enzyme is activated by magnesium ions (optimum concentration 4 mM) or manganese ions, and the nature of the divalent cation qualitatively affects specificity [127]. Citrate, borate and fluoride inhibit the enzyme by removing the activating magnesium ions.

Pancreatic DNase hydrolyses native DNA more rapidly than denatured DNA. As the early products of the reaction are poorer substrates than the initial DNA, the enzyme is autoretarding. In the early stages of the reaction, single-stranded nicks are produced towards the centre of the DNA molecule [128] but later on the RpY bond is preferentially cleaved leading to a final product of di- and oligo-nucleotides. The biosynthetic polymers poly(dA).poly(dT), poly(dI).poly(dC), poly(dG).poly(dC) are degraded in part by pancreatic DNase. In the latter two copolymers, the resistance of the poly(dC) chain to hydrolysis by the enzyme is overcome by adding Ca^{2+} to the Mg^{2+} or by replacing Mg^{2+} by Mn^{2+} [127]. A complex

Table 4.2 The properties of DNase I and DNase II

	DNase I	*DNase II*
Substrate	DNA	DNA
pH optimum	7–8	4–5
Activators	Mg^{2+}, Mn^{2+}, Ca^{2+}	0.3 M Na^+
Inhibitors	Citrate, EDTA	Mg^{2+}
Initial action	Nicks	Double-strand breaks
Product	5′-Phosphoryl-terminated oligonucleotides	3′-Phosphoryl-terminated oligonucleotides

of DNase I with self-complementary oligo-nucleotides has been analysed by X-ray crystallography and the specificity of the enzyme seems to reside, in part, in the stiff-ness and width of the DNA minor groove into which an exposed loop of the enzyme binds [269].

DNases I have been isolated from a variety of animal (and plant) tissues and exist in multiple forms. In bovine pancreas forms A and B differ from forms C and D by a single amino acid substitution and forms A and C have a high-mannose oligosaccharide linked to them whereas forms B and D are linked to more complex oligosaccharides [132, 133].

Actin, in the monomeric, globular form associates to form a 1:1 complex with DNase I, thereby inhibiting its action [4, 154].

(b) *Deoxyribonuclease II (DNase II) (EC 3.1.22.1) (reviewed in [129])*

A deoxyribonuclease of M_r about 40 000 with a pH optimum in the range 4.5–5.5, and no requirement for magnesium ions, has been isolated from spleen and thymus. DNase II from porcine spleen is a dimeric protein with subunits of 35 000 and 10 000 M_r [268]. Along with other hydrolases with acid pH optima it is found in lysosomes and is believed to be involved in intracellular breakdown of DNA.

Double-stranded DNA is degraded by splenic DNase II, in part by a 'one-hit' pro-cess that hydrolyses both strands of the double helix at the same point [128, 130]. This initial phase of the reaction is followed by the slower release of oligonucleotides of chain length from 14 to 100 nucleotides. The final stage produces oligonucleotides of average chain length six units, which have a free 5′-hydroxyl group and a phosphate residue on position 3′ (Fig. 4.3).

The properties of the two main animal DNases are summarized in Table 4.2 but many more nucleases have been character-ized. These have been tabulated by Linn [131]. In addition, many viruses induce new DNase activities [136, 137].

(c) *Endodeoxyribonucleases from* E. coli

The plethora of nucleases is nowhere better illustrated than in *E. coli*. Within this single organism there have been reported nine endonucleases and exonucleases (Tables 4.3 and 4.4) [134, 135] not to mention topoisomerases (section 6.5.4) and plasmid- and phage-coded enzymes.

(i) *Endonuclease I* [138]. This enzyme is the 12 000-M_r product of the *endA* gene. It is an endonuclease which attacks DNA producing scissions at many points along the DNA chain. At each scission an exonucleolytic

Table 4.3 Endodeoxyribonucleases of *E. coli*

Name	Gene	M$_r$	Substrate	Product
Endonuclease I	*endA*	12 000	Duplex DNA	Oligonucleotides with 5'-phosphoryl group
Endonuclease II		No enzyme known		
Endonuclease III	–	27 000	AP sites and u/v-irradiated duplex	Nicks with 3'-deoxyribose and 5'-phosphoryl termini
Endonuclease IV	–	33 000	AP sites	Nicks with 3'-OH and 5'-deoxyribose termini
Endonuclease V		27 000	ss DNA, damaged duplex	Short oligonucleotides
Endonuclease VI	*xthA*	Same as exonuclease III		
uvrABC endonuclease	*uvrA*	114 000	Damaged DNA	Excises short damaged regions
	uvrB	84 000		
	uvrC	68 000		
Restriction endonucleases	*hsd R*	135 000	Duplex DNA with unmodified recognition sequence	Long duplex fragments
	hsd S	60 000		
	hsd M	55 000		

activity removes about 400 nucleotides [139] which are released as a mixture of oligonucleotides of average chain length seven units terminated by a 5'-phosphoryl group. It is highly specific for DNA, attacking native DNA seven times more readily than denatured DNA to give random double-stranded breaks [140]. It is found in the periplasmic space of logarithmic phase cultures of *E. coli*, often in association with double-stranded RNA which inhibits its action. For this reason it is activated by ribonuclease treatment [134].

(ii) *Endonuclease II*. Some confusion is associated with this enzyme as the activity originally given this name has turned out to be a mixture of enzymes (section 7.3.2). The name was given to the activity which produces single-strand breaks in double-stranded DNA which has been alkylated with monofunctional alkylating agents (e.g. methyl methanesulphonate) [141]. The name is still reserved for such an enzyme but none is known to exist. Rather the activity is the result of the action of an *N*-glycosylase (which removes the alkylated base) and an *AP endonuclease* (an enzyme which cleaves DNA lacking a base, i.e. apurinic or apyrimidinic DNA). The original enzyme preparation contained an AP endonuclease which was given the name endonuclease VI (see below) which has since been shown to be identical to exonuclease III (section 4.5.2) [142, 143].

(iii) *Endonuclease III* [144]. This is a single 27 000-M$_r$ enzyme which shows the combined activities of glycosylase (specific for DNA irradiated with ultraviolet light) and an AP endonuclease (section 7.3.2). The glycosylase produces AP-DNA and the endonuclease nicks the duplex DNA 3' to the AP site to give a 3'-terminal sugar (Fig. 4.6). For this reason it is classified as a Class II AP endonuclease [134]. Similar enzymes have been found in *M. luteus* and cells infected with phage T4 [145, 146].

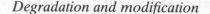

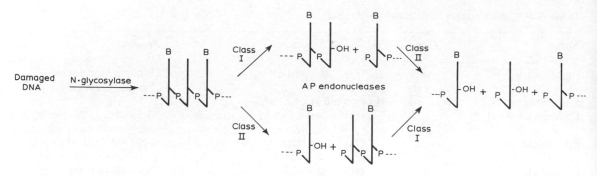

Fig. 4.6 AP endonucleases. The consecutive action of class I and class II AP endonuclease (as exemplified by *E. coli* endonucleases II and IV) removes the baseless sugar phosphate from the DNA.

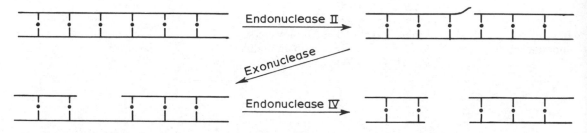

Fig. 4.7 Proposed scheme for the degradation of cell DNA after infection by bacteriophage T4.

(iv) *Endonuclease IV* [147, 148]. This is a Class II AP endonuclease of about 33 000 M_r. It represents about 10% of the AP endo-nuclease activity of the cell, the rest being due mainly to exonuclease III (endonuclease VI).

(v) *Endonuclease V* [149]. This is a 27 000-M_r nuclease which shows greatest activity with single-stranded DNA. It also acts on duplex DNA damaged by ultraviolet ir-radiation and duplex DNA containing uracil which it nicks to give 3'-OH and 5'-phosphoryl termini. Eventually the DNA is degraded to short oligonucleotides.

(vi) *Endonuclease VI* [135]. This is the product of the *xthA* gene and has been shown to be identical to exonuclease III (section 4.5.2) [142, 143]. It is a Class II AP endonuclease which also possesses $3' \rightarrow 5'$

exonuclease and 3'-phosphatase activities which may combine to increase the gap size during the repair of damaged DNA (section 7.3.2).

(vii) *uvrABC endonuclease*. This acts on damaged DNA especially at pyrimidine dimers to produce gaps in DNA. It is con-sidered in detail in section 7.3.2.

(viii) *Restriction endonucleases*. These cleave duplex DNA at specific sequences. At least one such enzyme (e.g. *Eco*B) is present in uninfected *E. coli* cells and many others are carried on plasmids or prophages. These enzymes are considered in section 4.5.3.

(ix) *ATP-dependent endonucleases*. A number of enzymes are known which can hydrolyse or unwind duplex DNA while hydrolysing ATP. Their roles in DNA re-

Table 4.4 Exodeoxyribonucleases of *E. coli*

Name	Gene	M_r	Substrate	End group attacked	Product
Exonuclease I	sbcB	72 000	ss DNA	3'-OH	Mono-and di-nucleoside 5'-phosphates
Exonuclease II	polA	109 000	ss DNA or nicked duplex	3'-OH	Nucleoside 5'-phosphates
	polB	120 000	ss DNA	3'-OH	Nucleoside 5'-phosphates
	polC	140 000	ss DNA or nicked duplex	3'-OH	Nucleoside 5'-phosphates
Exonuclease III	xthA	28 000	Duplex (AP sites)	3'-OH or 3'-OP	Nucleoside 5'-phosphates and ss DNA
Exonuclease IV	–	–	Oligonucleotides	3'-OH	Nucleoside 5'-phosphates
Exonuclease V	recB	140 000	Duplex or	3'-OH or	Oligonucleotides
	recC	130 000	ss DNA	5'-OH	(ATP-dependent)
Exonuclease VI	polA	109 000	Nicked duplex, RNA/DNA	5'-OH or	Short
	polC	140 000	hybrid and ss DNA	5'-OP	oligonucleotides
Exonuclease VII	xseA	88 000	ss DNA	3'-OH or 5'-OH	Oligonucleotides
Exonuclease VIII	recF	(140 000)₂	Duplex	5'-OH or 5'-OP	Nucleoside 5'-phosphate

plication and recombination are considered in chapters 6 and 7.

(d) *Phage-induced endonucleases*

Nucleases are induced following infection of bacteria with a variety of bacteriophages. Endonuclease II and endonuclease IV are induced after infection of *E. coli* with phage T4 [150, 151] (section 6.11.2). T4 endonuclease II makes single-stranded breaks in double-stranded DNA other than that of T4 to give products (at least from phage λ) of about 10^3 nucleotides. These have 5'-phosphoryl and 3'-OH termini. It differs from the host cell endonucleases in its inability to attack T4 DNA (glucosylated or not). T4 endonuclease IV hydrolyses single-stranded DNA to give considerably smaller products than endonuclease II, again with 5'-phosphoryl termini, but with dCMP exclusively in that position. DNA containing

hydroxymethylcytosine (i.e. T4 DNA) is inert as a substrate.

A mechanism has been suggested [151] whereby these enzymes, with the help of a bacteriophage-induced exonuclease, may be involved in the degradation of host DNA after T4 infection (Fig. 4.7).

Other endonucleases play specific roles in the replication of viral DNA (chapter 6), for example that coded for by gene A of ϕX174 breaks a single phosphodiester bond to initiate phage DNA synthesis [152].

4.5.2 Exonucleases

As with the endonucleases a great number of exonucleases have been described from both prokaryotes and eukaryotes [131, 153, 156, 169]. We shall here show the scope of their action by referring to those present in *E. coli* (Table 4.4) [134, 135, 155].

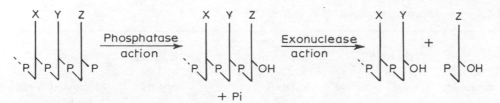

Fig. 4.8 Sequential action of *E. coli* exonuclease III (DNA phosphatase-exonuclease) on a DNA chain terminated by a nucleotide carrying a 3′-phosphate group.

(a) E. coli *exonuclease I (EC 3.1.11.1)*

This enzyme hydrolyses single-stranded DNA and has hardly any effect on native, double-stranded DNA. It is an exonuclease hydrolysing the DNA chain stepwise beginning at the 3′-hydroxyl end, and releasing deoxyribonucleoside 5′-monophosphates until only a dinucleotide is left. The enzyme does not cleave free dinucleotides or the 5′-terminal dinucleotide portion of a polydeoxyribonucleotide chain, but it can degrade bacteriophage DNAs containing glucosylated hydroxymethylcytosine. It has no effect on polyribonucleotides. It is the 72 000-M_r product of the *sbcB* gene but, as mutations in this gene have no deleterious effects on replication, recombination or repair, its function *in vivo* is obviously not essential.

(b) E. coli *exonuclease II*

This is the 3′ → 5′ exonuclease activity associated with DNA polymerase I (see below and section 6.4.3(d)).

(c) E. coli *exonuclease III (DNA phosphatase-exonuclease) [154, 157] (EC 3.1.11.2)*

This enzyme is found in close association with the DNA polymerase of *E. coli* but can be separated from it by chromatography. Its exonuclease action is very similar to the 3′ →

5′ exonuclease activity of DNA polymerase I but, in addition, it acts as a phosphatase highly specific for a phosphate residue esterified to the 3′-hydroxyl terminus of a DNA chain (Fig. 4.8). It does not release inorganic phosphate from monophosphates, from short oligodeoxyribonucleotides or from 3′-phosphoryl-terminated RNA, but it does attack DNA with a phosphoribonucleotide terminus.

As an exonuclease, it carries out a stepwise attack on the 3′-hydroxyl end of the DNA chain, releasing mononucleotides. It acts only on double-stranded DNA, degrading it until 35–45% has been digested. If the enzyme begins its attack from both 3′-hydroxyl ends of the double-stranded molecule (Fig. 4.9), then action ceases when nearly half the DNA has been degraded. The residual, acid-insoluble DNA will be single-stranded and resistant to further attack although it is still susceptible to the action of exonuclease I.

Exonuclease III will also degrade the RNA strand of a DNA:RNA hybrid (RNase H activity) at more than 10 000 times the rate at which the DNA strand is degraded. More than 85% of the AP endonuclease activity of an *E. coli* cell is present as exonuclease III (also called endonuclease VI – section 4.5.1(c)). These multitudinous activities can be reconciled in a common-site model [135] which assumes the enzyme has three regions. One region recognizes a deoxyribose on strand A and a second region cleaves 3′-

phosphates on strand B of a duplex molecule. However, cleavage only occurs if the third region fails to find a base-paired deoxyribonucleotide, a 3'-terminal phosphate or a 3'-deoxyribonucleotide which because of its terminal position is transiently unpaired.

Exonuclease III is the 28000-M_r product of the *xthA* gene, mutants in which show little phenotypic change other than an increased level of genetic recombination. Double mutants with *dut* (defective in dUTPase – section 6.3.2) are, however, lethal pointing to a role for the enzyme in the repair of AP sites which accumulate in *dut*⁻ cells as a result of removal by uracil *N*-glycosylase of the occasional uracils which arise in DNA during replication or following cytosine deamination (section 7.2).

Similar enzymes are found in *Diplococcus pneumonia* [158] and *B. subtilis* [159].

(d) E. coli *exonuclease IV*

This exonuclease shows little activity towards single- or double-stranded DNA, and exhibits a considerable preference (20-fold) for DNA predigested with pancreatic deoxyribonuclease. In this sense it could be termed oligonucleotide diesterase. It can be separated by DEAE-cellulose chromatography into two fractions (IVA and IVB) [160]. A similar enzyme is induced following infection of *E. coli* with phage T4. This enzyme which liberates 5'-monophosphates from a 3'-terminus may be involved in the degradation of cellular DNA (Fig. 4.7) [167, 168].

(e) E. coli *exonuclease V (recBCD nuclease)*

The role of this enzyme in repair and recombination is discussed in section 7.4.2 and by Muskavitch and Linn [161] who also consider similar enzymes from other sources. The enzyme is made of three subunits which

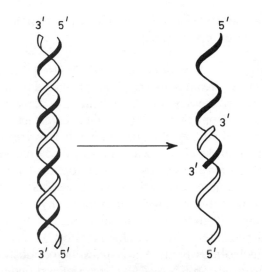

Fig. 4.9 Mechanism of action of stepwise attack of *E. coli* exonuclease III on native DNA beginning at the 3'-hydroxyl termini.

are the products of the *recB*, *recC* and *recD* genes. It acts exonucleolytically on both double- and single-stranded DNA to produce oligonucleotides in a reaction coupled with the hydrolysis of ATP (i.e. it is a DNA-dependent ATPase and an ATP-dependent DNase). Magnesium ions are required for nuclease action but in their absence, or in the presence of calcium ions, the enzyme unwinds duplex DNA. This unwinding action also requires ATP and is stimulated by single-stranded DNA-binding proteins (section 6.5.2). The enzyme also shows endonuclease activity on single-stranded DNA. It is the unwinding action to produce single-stranded DNA which is believed to be the important function of this enzyme *in vivo*.

(f) *Exonuclease associated with* E. coli *DNA polymerase I*

The terms exonuclease II and VI have been used to define the 3' → 5' and the 5' → 3' DNA exonuclease activities which form part of the protein of *E. coli* DNA polymerase I (section 6.4.3) [162].

(i) *3′ → 5′ exonuclease activity* [163]. This activity resides in the same protein domain that possesses DNA polymerase activity. Like exonuclease I it commences attack at the 3′-hydroxyl terminus of a polydeoxyribonucleotide chain, with the stepwise release of deoxyribonucleoside 5′-monophosphates, but unlike exonuclease I it also attacks dinucleotides. It will, for example, hydrolyse single-stranded DNA in preference to native duplex DNA [164] but has no effect on oligonucleotides bearing a 3′-phosphomonoester group or on RNA. Most bacterial and phage DNA polymerases so far investigated (including *E. coli* DNA polymerase II and III holoenzyme) have an associated 3′ → 5′ exonuclease involved in proofreading [162, 164].

Evidence that the enzyme is in fact an exonuclease with these properties comes from several sources. Exhaustive digestion of ^{32}P-labelled d(A-T) copolymer results in the conversion of 99% of the ^{32}P into an acid-soluble form which can be accounted for in terms of 5′-monophosphates. Partial digestion results in the release of a proportion of radioactivity which is the same as the proportion of monophosphates formed. When d(A-T) copolymer specifically labelled with [^{32}P]dTMP at the 3′-hydroxyl end is used as substrate, 90% of the ^{32}P-labelled material is made acid-soluble when less than 1% of the unlabelled nucleotides from the interior of the chain have been released. This indicates attack from the 3′-hydroxyl end of the chain.

(ii) *5′ → 3′ exonuclease activity*. This activity is specific towards native DNA and will function in the presence or absence of a 5′-phosphate group on the substrate. The products are mostly mononucleotides, with 20–25% dinucleotides or longer [165]. The enzyme can act to remove pyrimidine dimers and other regions of damaged DNA and will also remove ribonucleotides from an RNA: DNA hybrid. These properties are important in excision repair (section 7.3.2) and in the removal of RNA primers from Okazaki pieces during DNA replication (section 6.3). This enzyme is utilized to radiolabel DNA in the process known as nick translation (section 6.4.3 (e) and Fig. A.8).

The DNA polymerase I molecule is susceptible to protease action such that it is specifically split into two fragments. One of these (76 000 M_r) has DNA polymerase and 3′ → 5′ exonuclease activity; the other (34 000 M_r) shows 5′ → 3′ exonuclease activity but only if the cleavage is carried out in the presence of DNA [166]. *B. subtilis* DNA polymerase I contains no 5′ → 3′ exonuclease activity [159] but a separate enzyme exists with this function.

A 5′ → 3′ exonuclease activity is absent from *E. coli* DNA polymerase II but a modified activity is present in *E. coli* DNA polymerase III. In this case the exonuclease will only act on single-stranded DNA or on duplex DNA with a 5′-single-stranded tail to which the enzyme can initially attach.

(g) *Exonuclease VII (EC 3.1.11.6)*

This degrades single-stranded DNA from either end, initially releasing large oligonucleotides which are further degraded to smaller oligonucleotides. Despite this finding the enzyme acts exonucleolytically but cleaves only occasionally as it processes along the DNA strand [135, 155, 170–172]. Exonuclease VII is the 88 000-M_r product of the *xseA* gene, mutants of which show increased frequency of recombination. As it can also act at nicks in duplex DNA and excises thymine dimers it may partly substitute for other enzymes involved in excision repair (section 7.3.2). It does not require ATP.

(h) *Exonuclease VIII*

The gene for this enzyme (*recE*) is closely linked to that for exonuclease I (*sbcB* or *xonA*) but it is only expressed in *sbcA* cells when exonuclease VIII can substitute for exonuclease I (section 7.4.2). It can also substitute for exonuclease V in recBCD$^-$ cells. It is a dimer of 140000-M_r subunits which degrades linear duplex DNA from the 5'-ends, whether phosphorylated or not, to yield 5'-mononucleotides and residual single stranded DNA [173, 174], i.e. it is similar but of opposite polarity to exonuclease III. A similar enzyme is found in bacterial cells infected with phage λ and exonuclease VIII can substitute for λ exonuclease in lambda recombination [137].

4.5.3 Restriction endonucleases

When a bacteriophage is transferred from growth on one host to a different host, its efficiency is frequently impaired several thousand-fold. However, when those phage which do survive and multiply are used to reinfect the second host they now grow normally.

The initial poor growth is caused by the action on the phage DNA of a highly sequence-specific bacterial endonuclease (known as a restriction endonuclease because it restricts phage growth). Following the initial cleavage, the invading phage DNA is rapidly degraded to nucleotides by exonuclease action, although certain regions may be rescued by recombination [175].

Host (i.e. bacterial) DNA is not degraded because the nucleotide sequence which is recognized by the restriction endonucleases has been modified (usually by methylation –

section 4.6.1). The few molecules of phage DNA which survive the initial infection do so because they are themselves modified by the host methylase before the restriction enzyme has time to act. This methylation of progeny DNA renders it resistant in a second infection [176].

An inverse phenomenon is the degradation of cytosine-containing DNA in *E. coli* infected with T-even bacteriophages whose own DNA contains hydroxymethylcytosine (section 6.11.2).

The *E. coli* chromosome encodes a type I restriction endonuclease known as *Eco* B (in *E. coli* strain B) or *Eco*K (in *E. coli* strain K12). Other restriction enzymes are coded for by plasmids and these are mostly type II enzymes, but a few phage or plasmid-coded type III enzymes are known which are intermediate in properties between type I and type II enzymes [177–183]. Type I and type III enzymes catalyse both endonucleolytic cleavage and methylation of DNA, but these two functions are carried out by separate type II enzymes.

(a) *Type I*

These are complex multifunctional proteins which cleave unmodified DNA in the presence of *S*-adenosyl-L-methionine (AdoMet), ATP and Mg^{2+}. They are multi-subunit enzymes, which are also DNA methylases (section 4.6.1) and ATPases. The subunits are present in different proportions in the *Eco*B and *Eco*K enzymes [184–186]. They are coded for by three contiguous genes known as *hsdR* (host specificity for DNA restriction), *hsdM* (modification) and *hsdS* (specificity). The three subunits are respectively responsible for the nuclease action, the methylation and the sequence specificity of the enzyme. Before interacting with DNA the *E. coli* K restriction enzyme binds to AdoMet [187, 188]. Binding to

```
EcoK 5′ – A A C N N N N N N N G T G C – 3′
           T T G N N N N N N N C A C G
                *
EcoB 5′ T G A N N N N N N N N N T G C T – 3′
         A C T N N N N N N N N N A C G A
                                 *
```

Fig 4.10 The recognition sequences for the *Eco*K and *Eco*B restriction modification enzymes.

AdoMet takes place rapidly and is followed by a slower allosteric modification of the enzyme to an activated form which then interacts with DNA at a non-specific site. The enzyme moves along the DNA to the recognition site and its subsequent reaction depends on the state of this recognition site [204].

The recognition sites for the *Eco*K and *Eco*B type I restriction enzymes are shown in Fig. 4.10. The adenines with an asterisk in the *Eco*B sequence are methylated by the *Eco*B enzyme and the corresponding adenines in the *Eco*K sequences are the site of methylation by the *Eco*K restriction methylase [189–191]. The recognition sequences both consist of a group of three bases and a group of four bases separated by six (*Eco*K) or eight (*Eco*B) unspecified bases. Four of the seven specified bases are conserved and the methylated adenines in both cases are separated by eight bases. The specificity subunit consists of a number of domains which can be interchanged using protein engineering techniques. This has led to the conclusion that separate domains dictate the specificity for the first group of bases, the second group of bases and the spacing between them [270, 271].

If the recognition site is methylated in both strands of the DNA the enzyme does not recognize it. If the site is methylated on one strand only (as would be the case with DNA immediately following synthesis –

section 4.6.1), the enzyme binds to this site and methylates the second strand in a reaction stimulated by ATP [192]. Methylation can also occur at completely unmethylated sites but only at a rate about 0.2% of that at half-methylated sites. Normally, if the site is unmodified in both strands, the enzyme is triggered into its nuclease (restriction) mode.

Restriction does not occur at the binding site. Rather, in the presence of ATP, the DNA is made to loop past the enzyme which remains bound to the recognition site. This translocation leads to formation of supercoiled loops which become relaxed on subsequent cleavage [185, 193–195]. Cleavage of the looped-out DNA occurs at non-random sites and in two stages (first one strand, followed some time later by a break some distance away on the opposite strand). At this point the enzyme ceases to be a nuclease (i.e. it performs only one restriction event) and becomes a vigorous ATPase. About 10^5 ATP molecules are hydrolysed for each restriction event [175, 205].

(b) *Type II*

These (e.g. *Eco*R I, *Hap* II) are simpler enzymes which require only Mg^{2+} for activity. Most have values of M_r in the range from 30 000 to 40 000 and have two identical subunits. These enzymes, which are neither methylases nor ATPases, appear to translocate along the DNA until they recognize a specific site and, if this is unmodified, cleavage occurs at this site [203]. (Type IIS restriction enzymes cleave a certain number of nucleotides away from the recognition site – see below.) DNA modified on one strand is not a substrate for restriction, but is a substrate for a complementary methylase which recognizes the same specific nucleotide sequence (section 4.6.1).

The vast majority of sites are four to six

Table 4.5 Class II DNA restriction and modification enzymes

Enzyme	Restriction and modification site	Bacterial strain
*Eco*RI	G ↓ A*ATTC	*E. coli* RY13
*Eco*RII	↓ C*C$_T^A$GG	*E. coli* R245
*Ava*I	C ↓ Y^{0}CGRG	*Anabaena variabilis*
*Bam*HI	G ↓ GATCC	*Bacillus amyloloquifaciens* H
*Bgl*II	A ↓ GATCT	*Bacillus globiggi*
*Bsu*RI	GG ↓ *CC	*Bacillus subtilis* R
*Hha*I	G*CG ↓ C	*Haemophilus haemolyticus*
*Hae*III	GG ↓ *CC	*Haemophilus aegyptius*
*Hind*II	GT$_C^T$ ↓ $_G^A$*AC	*Haemophilus influenzae* Rd
*Hind*III	*A ↓ AGCTT	*Haemophilus influenzae* Rd
*Hinf*I	G ↓ ANTC	*Haemophilus influenzae* Rf
*Hpa*II	C ↓ *CGG	*Haemophilus parainfluenzae*
*Hga*I	GACGCNNNNN ↓	*Haemophilus gallinarum*
*Sma*I	CC*C ↓ GGG	*Serratia marcescens* Sb
*Dpn*I	G*A ↓ TC	*Diplococcus pneumoniae*

↓ shows the site of cleavage and * shows the site of methylation of the corresponding methylase where known. ^{0} shows the site of action of a presumed methylase, i.e. methylation at this site blocks restriction but the methylase has not yet been isolated. More complete lists are found in references [200–202, 273].

nucleotide pairs long and have twofold rotational symmetry (Table 4.5) which suggests that the two enzyme subunits may be arranged with twofold symmetry [176, 196]. Cleavage at the site may be staggered by up to five nucleotides (e.g. *Eco*R I) when identical self-complementary, cohesive termini are produced. *Bsu*R I, *Sma* I and *Hind* II produce even breaks without single-stranded termini (Table 4.5). The recognition site for the *Hinf* I restriction enzyme has one unspecified base [197] and frequently bases are only partially specified, i.e. they must be a purine or a pyrimidine. With *Eco*R II the cleavage site is a sequence of five nucleotide pairs and hence not all the ends are mutually cohesive.

Type IIS enzymes do not cleave within the recognition sequence; thus enzyme *Hph* I cleaves DNA eight bases before the sequence TCACC [198] and neither the recognition nor the cleavage sequence is symmetrical.

At least one restriction endonuclease (*Dpn* I) is specific for methylated DNA and cleaves the sequence GmATC. It is found in the bacterium, *Diplococcus* (or *Streptococcus*) *pneumoniae*, a separate strain of which contains an enzyme of opposite specificity (*Dpn* II) together with the associated methylase [199, 272]. The genes for *Dpn* I and *Dpn* II plus M*Dpn* I are dissimilar, but are surrounded by regions of homology that allow interchange of cassettes between strains in a manner similar to the interchange of yeast mating type genes (section 7.5.4). Most cell populations will contain cells of both strains and interchange by transformation-mediated recombination is thought to be a common occurrence. Should one strain be wiped out by a viral infection, cells of the other strain will be resistant.

Hundreds of type II restriction enzymes are now known covering a very broad range of sequence specificities and comprehensive lists can be found in several reviews [200–202, 273].

A restriction enzyme is found in crown gall tumours [206] and an enzyme which cuts a monkey satellite DNA into discrete size fragments has been isolated from the testes of the African Green monkey, but neither this nor a similar site-specific endonuclease from yeast is a restriction endonuclease as defined by the properties of the type II enzymes herein described [207–209].

(c) *Type III*

These (e.g. *Eco* P1, *Eco* P15 and *Hin*f III) have two different subunits; a nuclease and a combined methylase and specificity factor. They also differ from type I enzymes in not catalysing extensive ATP hydrolysis following nuclease action [182, 183, 216, 217]. The holoenzyme is essential for restriction and the two activities (endonuclease cleavage and methylation) compete with one another. The recognition sites are not symmetrical and only one strand of the DNA is modified. The endonuclease makes staggered cuts 20 to 30 bases from the 3′-end of the recognition site.

(d) *Nomenclature*

The restriction endonucleases are named according to the system described by Smith and Nathans [210]. The first three letters give the name of the bacterium (e.g. *Eco* for *E. coli*) and the fourth indicates the strain (e.g. *Eco*R and *Hin*d for *E. coli* strain R and *Haemophilus influenzae* strain d, respectively). Where more than one restriction enzyme is found in a particular strain these are indicated by Roman numerals (e.g. *Hin*a I, *Hin*a II). Enzymes recognizing identical nucleotide sequences are called isoschizomers. However, isoschizomers do not necessarily cleave in the same position

and may respond to methyl groups on different bases in the recognition sequence.

(e) *Applications*

Since their discovery, the class II restriction enzymes have become increasingly used as tools for the biochemist. Because of their ability to make relatively few specific cuts they can be used as a first step in the sequencing of DNA (section A.5) or in the isolation of specific genes.

The use of a series of restriction enzymes allows a physical map to be made of genes or small chromosomes. Thus the small chromosome of the tumour virus SV40 (simian virus 40) is cleaved by each restriction enzyme into a small number of discrete pieces which are readily separated by electrophoresis in gels of polyacrylamide or agarose [179, 211, 212]. The size of the fragments can be determined from their speed of migration in the gel or by direct length measurements using electron microscopy (section A.1.3). The order of the fragments in the genome can be determined by a variety of techniques such as analysis of partial digests or successive cleavage by multiple restriction endonucleases [179, 213].

As cleavage by many of these restriction enzymes is staggered, this leads to production of fragments with 'sticky' ends (i.e. termini with overlapping self-complementary sequences) which can be rejoined by DNA ligase (section 6.5.1). Moreover, fragments from different genomes can be joined together to form hybrid genomes, and this is the basis for genetic engineering considered in detail in the Appendix.

The use of certain pairs of restriction enzymes has enabled the extent of methylation of sites in individual eukaryotic genes to be investigated. Thus *Hpa* II will only cleave the sequence CCGG if it lacks a methyl group on the internal cytosines of both strands of the duplex DNA whereas *Msp* I will cleave the same sequence irrespective of whether the internal cytosine is methylated [214, 215]. Most bacteria contain a methylase (Dam) which methylates the sequence GATC and so *Dpn* I has been used to follow the replication of DNA which becomes resistant to the nuclease as the replication fork passes [282].

4.6 NUCLEIC ACID METHYLATION

4.6.1 DNA methylation

Methylated bases occurring in DNA fall into two classes. The first class contains thymine and hydroxymethylcytosine (in T4 DNA) which are incorporated into DNA from dTTP or hydroxymethyl-dCTP by DNA polymerase. The second class contains bases arising from methylation of a pre-formed polydeoxynucleotide, i.e. methylation occurs after DNA synthesis. This second class of methylated bases contains only 5-methylcytosine, first discovered by Hotchkiss in calf thymus DNA [218] and 6-methyladenine and ^4N-methylcytosine which are largely restricted to the DNA from lower organisms.

The extent of DNA methylation is very variable, ranging from almost nothing in some insect and yeast DNA and in chloroplast DNA, through the bacteria (which may methylate one adenine and/or cytosine in 200) and animals (with 3–5% of cytosines methylated) to the higher plants where the nuclear DNA can have a third of the cytosines present as 5-methylcytosine [219–222]. Usually animal virus DNA is not methylated but exceptions are known [223, 224]. Within the genome of eukaryotes there is an uneven distribution of methylcytosine. This is seen to an extreme degree in the sea urchin whose DNA can be separated into methylated and unmethylated compartments on the basis of its sensitivity to the methyl-sensitive restriction enzyme *Hpa* II [225].

Methylation of DNA takes place at the polynucleotide level [226], and the reactions are catalysed by specific enzymes, the DNA methylases (methyltransferases). The source of methyl groups is methionine, in the form of an intermediate with a high free energy of hydrolysis, *S*-adenosyl-L-methionine [221]. The mechanism is illustrated for cytosine methylation in Fig. 4.11.

A number of DNA methylases have been purified from prokaryotes. These enzymes are either part of the type I or type III restriction enzymes or complement the type II restriction enzymes by recognizing and methylating a common sequence and thus protecting it from the action of the cognate nuclease (section 4.5.3). In addition, *E. coli* (and most other prokaryotes) has two other methylases (the products of the *dam* and *dcm* genes) which respectively methylate adenines in the sequence GATC and cytosines in the sequence $CC^A/_TGG$. In this case the methylase modifies two cytosines in somewhat different sequences.

Sequence analysis of type II methylases shows them to have highly homologous structures with very similar *N*-terminal and *C*-terminal ends of about 240 amino acids and 110 amino acids, respectively. In between there is a cassette of about 50 amino acids which determines specificity. Using protein engineering techniques, these specificity cassettes have been interchanged and enzymes with multiple specificity generated having more than one cassette [274, 275]. The type IIS methylases which act at a non-complementary site appear to solve the specificity problem in a number of different ways. Thus M*Fok* I methylates two adenines in different sequences and resembles two enzymes fused together. M*Hga* I is actually

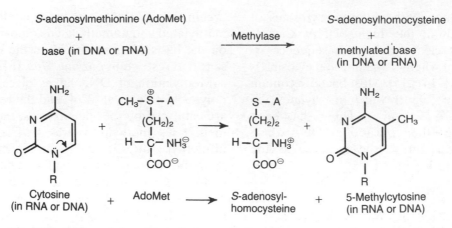

Fig 4.11 The nucleic acid methylase reaction (A represents the adenosyl radical).

two separate enzymes; and M*Ear* I is both a cytosine and an adenine methylase.

Although most methylases are present to protect the host DNA from the complementary restriction endonuclease, it is not certain what are the functions of the Dam and Dcm methylases though it is known they play a role in mismatch repair (section 7.3.3), recombination (section 7.4) and the initiation of DNA replication (section 6.10).

In eukaryotes the DNA sequence recognized and methylated by the methyltransferases, in animals, is

$$5'-CG-3'$$
$$3'-GC-5'$$

and also, in plants, is

$$5'-CNG-3'$$
$$3'-GNC-5'$$

As with the prokaryotes these recognition sequences are symmetrical and are methylated on both strands. However, in eukaryotes not all the recognition sites are modified and the pattern of methylation is tissue-specific [219–222]. The recognition of the increased (epi)genetic information with which eukaryotic DNA is endowed by the presence of a fifth base has led to much speculation as to the function of DNA methylation in eukaryotes. A role in the control of gene expression is considered in section 10.6.2, but it is pertinent to point out here that a base change from, say, cytosine to methylcytosine will have a marked effect on DNA–protein interactions as is exemplified by the restriction endonucleases (section 4.5.3) [276].

As in prokaryotes, methylation of DNA in eukaryotes may have more than one function and the finding that marked changes in the pattern of methylation occur during gametogenesis and in the very early vertebrate embryo indicates that some methylation may be important in determining the potential of a gene for future activity [221, 227, 232]. Such speculation arises from the finding that in somatic cells the DNA methyltransferase serves only to maintain the pattern of methylation in newly synthesized DNA (Fig. 4.12). Thus a pattern of methylation established in, say, the presumptive liver cell DNA in the early embryo can be maintained through the action of the maintenance methylase alone and can be present to affect gene expression in the adult liver.

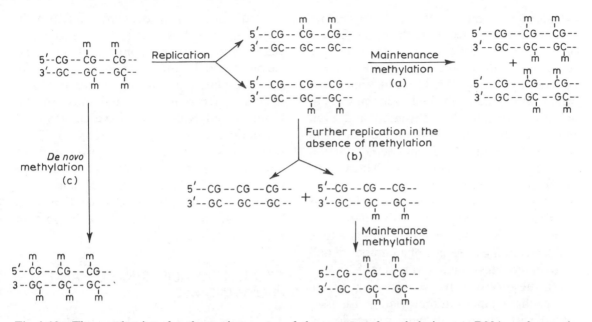

Fig 4.12 The mechanism for the maintenance of the pattern of methylation on DNA and ways in which the pattern may be changed. The top line shows how the preferential action on hemimethylated DNA of the vertebrate DNA methyltransferase reproduces, on newly replicated DNA, the pattern of methylation found in parental DNA (a). If, during two rounds of replication, methylation is blocked in some region of the DNA (e.g. by bound proteins), that region of DNA will thereafter lack methylcytosine in at least half the progeny cells (b). *de novo* methylation of unmethylated, paired CG dinucleotides takes place to a marked extent only in early embryos (c).

The pattern of methylation can be changed in two ways (Fig. 4.12). Methyl groups can be gained by *de novo* methylation of unmethylated –CG– sequences, a process thought to occur to a significant extent only in the early embryo [277, 278]. Very little *de novo* methylation occurs on DNA transfected into animal cells but viral DNA can integrate into the host cell chromosome to bring about transformation and such a process is accompanied by the inactivation and concomitant *de novo* methylation of much of the viral genome [233, 234]. The packaging of foreign DNA into inactive chromatin and its subsequent methylation may be an effective defence against viral infection. Such an association between DNA methylation and gene and chromosome inactivation was postulated by Sager and Kitchin [235] as a prelude to the permanent loss of DNA in lower eukaryotes and is consistent with the models for the permanent inactivation of one of the X-chromosomes in female mammals [236–238].

Such inactivation can be partly reversed by treatment with 5-azacytidine, an analogue of cytidine which, when incorporated into DNA, leads to inactivation of cellular DNA methyltransferase, an inhibition of methylation and a transient overall demethylation of the genome [238–240]. 5-Azacytidine has also been used to bring about phenotypic changes in cultured cells, and in an attempt to induce expression of the foetal globin gene in patients suffering from thalassaemia (a disease caused by a defect in an adult globin gene) [241, 242].

Methyl groups can be lost from DNA

cytosines in a more natural way by replication in the absence of methylation. Such a block in methylation may arise when the recognition site of the methyltransferase is covered by protein. Such passive demethylation will take two cell generations to achieve complete demethylation and even then the DNA in only half the cells will be demethylated. More rapid demethylation has been observed [279, 280] which must occur by an active, site-directed process, possibly involving very short patch repair (section 7.3.3) [281]. This active demethylation may take place on one strand at a time and leads to the presence of a number of half methylated sites in eukaryotic DNA.

All these observations add weight to the theory that changes in methylation are associated with changes in gene expression in eukaryotes.

In *E. coli* certain spontaneous base substitutions occur as a result of deamination of 5-methylcytosine to thymine [266]. Although this reaction was postulated many years ago as a mechanism underlying eukaryotic cell differentiation [267] there is little evidence for such a function, although a role in cell ageing remains possible. Nevertheless, such a deamination occurring over many, many generations is believed to have resulted in the depletion of the −CG− dinucleotide from vertebrate DNA in which it is found at only about one-fifth the expected frequency [228–231].

4.6.2 RNA methylation and other RNA nucleotide modifications

Methylation and numerous other forms of nucleotide modification are known for all four of the bases of RNA. They occur most commonly, and in by far the greatest diversity, on tRNA as a result of post-transcriptional modification and are con-

sidered under the processing of tRNA in section 11.7. Ribosomal RNA is also methylated and contains some pseudouridine (section 11.6.3); mRNA and its precursors are methylated particularly in their 5′-cap structures (section 11.4) and most of the U-series snRNAs also have methylated nucleotides at their 5′-ends (section 11.2). In most cases, the enzymes involved in these modifications are ill-characterized but they have been reviewed [243, 244] and some of the tRNA modifying activities that have been examined are tabulated in section 11.7.2 (Table 11.2).

4.7 NUCLEIC ACID KINASES AND PHOSPHATASES

4.7.1 Bacteriophage polynucleotide kinase

Polynucleotide kinases produced by bacteria phage infected with T2-, T4- and T6 are enzymes with a broad specificity which transfer the γ-phosphate of a nucleoside triphosphate to the 5′-OH of DNA, RNA, oligonucleotides or nucleoside 3′-phosphates (reviewed in [245, 246]). The enzyme has been purified from T2-infected [247] and T4-infected cells and the latter, in particular, has been extensively characterized [248] and employed in a wide variety of research applications. The phosphorylation reaction (Fig. 4.13) requires magnesium ions and the presence of compounds with reduced −SH groups. Phosphoryl transfer does not involve an enzyme−phosphoryl intermediate [249] and the reaction is reversible [250]. Dephosphorylation is not, however, as efficient as phosphorylation and has a pH optimum of 6.2 compared with 7.6 for phosphorylation.

The enzyme also exhibits a totally separate 3′-phosphatase activity which catalyses the hydrolysis of 3′-phosphoryl groups of nucleic acids (Fig. 4.14). Richardson [246] has

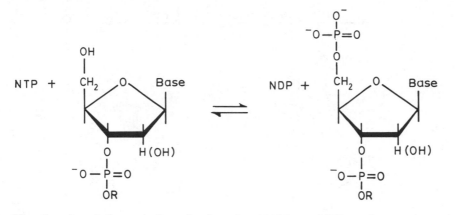

Fig. 4.13 The phosphorylation reaction of polynucleotide kinase. NTP can be a variety of nucleoside 5'-triphosphates. R can be H, nucleoside, oligonucleotide or polynucleotide.

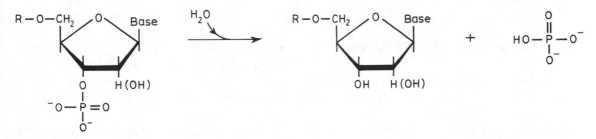

Fig. 4.14 The phosphatase reaction of polynucleotide kinase. R can be H, a nucleotide or a polynucleotide.

reviewed the evidence that this phosphatase activity is probably identical to the previously purified 3'-phosphatase isolated from T4-infected bacteria despite the fact that early evidence indicated that this latter enzyme was specific for DNA [251]. The phosphatase and kinase activities of T4 polynucleotide kinase appear to involve separate active sites [252] and there is no evidence for a concerted reaction involving transfer of phosphate from the 3'- to the 5'-side of a nick so that it could then be a substrate for DNA ligase [253]. (DNA ligase is considered in section 6.5.1.) Indeed the role of polynucleotide kinase in bacteriophage infection is not understood.

Polynucleotide kinase has multiple research uses. The reversibility of the phosphorylation reaction in the presence of ADP allows its use to catalyse the exchange of ^{32}P between the γ-phosphate of ATP and the 5'-phosphate group of a polynucleotide [250]. Such end-labelled polynucleotides [254], and those end-labelled by the forward reaction after the removal of the 5'-phosphate by alkaline phosphatase [255], are used in the analysis of nucleotide 'finger prints', end-group analysis, the mapping of restriction endonuclease-generated fragments and most of all in the sequencing of RNA and DNA (sections A.4.2 and A.5.1, and reviewed in [245, 246]).

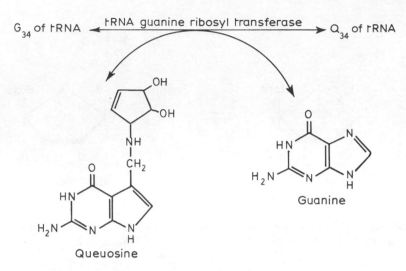

G$_{34}$ of tRNA ⟵ tRNA guanine ribosyl transferase ⟶ Q$_{34}$ of tRNA

Queuosine

Guanine

Fig. 4.15 Insertion of queuosine into the wobble position of tRNA.

4.7.2 Eukaryotic DNA and RNA kinases

DNA kinase activities have been purified from the nuclei of rat liver [256] and bovine thymus [257]. They differ from polynucleotide kinase principally in their inactivity with RNA as a substrate and their lower pH optimum of 5.5 (reviewed in [258]). Again their function is unknown but they are thought to play a role in DNA repair and are used experimentally to introduce labelled 5'-phosphate groups at single-strand breaks in duplex DNA [259]. The resultant labelled DNA can be used to assay DNA ligase activity [259].

An RNA kinase activity which transfers phosphate to the 5'-OH of RNA much more efficiently than DNA has been isolated from HeLa cell nuclei [260].

4.8 BASE EXCHANGE IN RNA AND DNA

The generation of two modified purines, queuosine and inosine, which occur in the anticodon wobble position of a number of tRNAs (section 11.2), occurs by direct base replacement. An enzyme known as tRNA guanine ribosyl transferase or tRNA guanine transglycosylase (EC 2.4.2.29) catalyses the exchange of guanine in the wobble position of the relevant pre-tRNAs for the highly modified base, queuosine, in the mature tRNA [261]. The reaction involves breaking and re-formation of a C–N glycoside bond rather than a 3'–5' phosphodiester bond and it requires no energy (Fig. 4.15). For reviews of queuosine synthesis and insertion into tRNA see [262, 263].

It is now known that inosine biosynthesis in the wobble position of the anticodon occurs by a similar mechanism in which the base at position 34 of the relevant tRNA precursors (probably adenine) is replaced by hypoxanthine (inosine is the nucleoside of hypoxanthine) [264]. The hypoxanthine is first derived from adenosine which is converted into inosine by adenosine deaminase and thence into hypoxanthine by purine nucleoside phosphorylase.

Nucleoside   deoxyribosyl   transferases,

which catalyse the exchange of bases in DNA, occur in bacteria [265].

REFERENCES

1 Laskowski, M. (1967) *Adv. Enzymol.*, **29**, 165.

2 Barnard, E. A. (1965) *Annu. Rev. Biochem.*, **38**, 677.

3 Grossman, L. and Moldave, K. (1980) *Methods Enzymol.*, **65**.

4 Laskowski, M. (1971) *The Enzymes* (3rd edn) (ed. P. D. Boyer), Academic Press, New York, vol. 4, p. 289.

5 Dingwall, C., Lomonossoff, G. P. and Laskey, R. A. (1981) *Nucleic Acids Res.*, **9**, 2659.

6 Horz, W. and Altenberger, W. (1981) *Nucleic Acids Res.*, **9**, 2643.

7 Cotton, F. A. and Hazen, E. E. (1971) *The Enzymes* (3rd edn) (ed. P. D. Boyer), Academic Press, New York, vol. 4, p. 153.

8 Anfinsen, C. B., Cuatrecasas, P. and Taniuchi, H. (1971) *The Enzymes* (3rd edn) (ed. P. D. Boyer), Academic Press, New York, vol. 4, p. 177.

9 Linn, S. and Lehman, I. R. (1965) *J. Biol. Chem.*, **240**, 1287, 1294.

10 Shishido, K. and Ando, T. (1982) in *Nucleases*, Cold Spring Harbor Monograph, **14**, p. 155.

11 Shishido, K. and Ando, T. (1972) *Biochim. Biophys. Acta*, **287**, 477.

12 Fujimoto, M., Kuninaka, A. and Yoshina, H. (1974) *Agric. Biol. Chem.*, **38**, 785.

13 Sung, S-C. and Laskowski, M. (1962) *J. Biol. Chem.*, **237**, 506.

14 Gray, H. B., Winston, T. P., Hodnett, J. L., Legevski, R. J., Nees, D. W., Wel, C-F. and Robberson, D. L. (1981) in *Gene Amplification and Analysis*, Elsevier/North-Holland, New York, vol. 2, p. 169.

15 Laskowski, M. (1971) in *The Enzymes* (3rd edn) (ed. P. D. Boyer), Academic Press, New York, vol. 4, p. 313.

16 Bernardi, R. and Bernardi, G. (1971) in *The Enzymes* (3rd edn) (ed. P. D. Boyer), Academic Press, New York, vol. 4, p. 319.

17 Wei, C-E., Alianell, G. A., Beneen, G. H. and Gray, H. B. (1983) *J. Biol. Chem.*, **258**, 13 506.

18 Richards, F. M. and Wyckoff, H. W. (1971) in *The Enzymes* (3rd edn) (ed. P. D. Boyer), Academic Press, New York, vol. 4, p. 647.

19 Blackburn, P. and Moore, S. (1982) in *The Enzymes* (3rd edn) (ed. P. D. Boyer), Academic Press, New York, vol. 15, p. 317.

20 Jones, W. (1920) *Am. J. Physiol.*, **52**, 203.

21 *Enzyme Nomenclature* (1984) Recommendations of the International Union of Biochemistry, Academic Press, London.

22 Dubos, R. J. and Thompson, R. H. S. (1938) *J. Biol. Chem.*, **154**, 501.

23 Kunitz, M. (1940) *J. Gen. Physiol.*, **24**, 15.

24 McDonald, M. R. (1955) *Methods Enzymol.*, **2**, 427.

25 Smyth, D. G., Stein, W. H. and Moore, S. (1963) *J. Biol. Chem.*, **238**, 227.

26 Beintema, J-J. and Lenstra, J. A. (1982) in *Macromolecular Sequences in Systematics and Evolutionary Biology* (ed. M. Goodman), Plenum Press, New York.

27 Kartha, G., Bello, J. and Harker, D. (1967) *Nature (London)*, **213**, 862.

28 Wyckoff, H. W., Hardman, K. D., Allewell, N. M., Inagami, T., Johnson, L. N. and Richards, F. M. (1967) *J. Biol. Chem.*, **242**, 3984.

29 Carlisle, C. H., Palmer, R. A., Mazumdar, S. K., Gorinsky, B. A. and Yeates, D. G. R. (1974) *J. Mol. Biol.*, **85**, 1.

30 Timchenko, A. A., Ptitsyn, O. B., Dolgikh, D. A. and Eedorov, B. A. (1978) *FEBS Lett.*, **88**, 105.

31 Gutte, B. and Merrifield, R. B. (1969) *J. Am. Chem. Soc.*, **91**, 501.

32 Hirschmann, R., Nutt, R. F., Veber, D. F., Vitali, R. A., Varga, S. L. *et al.* (1969) *J. Am. Chem. Soc.*, **91**, 507.

33 Bernd, G. and Merrifield, R. B. (1971) *J. Biol. Chem.*, **246**, 1922.

34 Rubsamen, H., Khandker, R. and Witzel, H. (1974) *Hoppe Seyler's Z. Physiol. Chem.*, **355**, 687.

35 Walker, E. J., Raiston, G. B. and Darvey, I. G. (1975) *Biochem. J.*, **147**, 425.

36 Walker, E. J., Raiston, G. B. and Darvey, I. G. (1978) *Biochem. J.*, **173**, 1.

37 Usher D. A., Richardson D. I. and Eckstein, F. (1970) *Nature (London)*, **228**, 663.

38 Usher D. A., Erenrich, E. S. and Eckstein, F. (1972) *Proc. Natl. Acad. Sci. USA*, **69**, 115.

39 Beers, R. F. (1960) *J. Biol. Chem.*, **235**, 2393.

40 Kunitz, M. (1946) *J. Biol. Chem.*, **164**, 563.

41 Anfinsen, C. B., Redfield, R. R., Choate, W. L., Page, J. and Carroll, W. R. (1954) *J. Biol. Chem.*, **207**, 201.

42 Crook, E. M., Mathias, A. P. and Rabin, B. R. (1960) *Biochem. J.*, **74**, 234.

43 Blank, A. and Dekker, C. A. (1981) *Biochemistry*, **20**, 2261.

44 Maor, D. and Mardiney, M. R. (1979). *CRC Crit. Rev. Clin. Lab. Sci.*, **10**, 89.

45 D'Alessio, G., Di Donato, A., Parente, A. and Piccoli, R. (1991) *Trends Biochem. Sci.*, **16**, 104.

46 Niwata, Y., Ohgi, K., Sanda, A., Takizawa, Y. and Irie, M. (1985) *J. Biochem. (Tokyo)*, **97**, 923.

47 Brockdorff, N. A. and Knowler, J. T. (1987) *Eur. J. Biochem.*, **163**, 89.

48 Hofsteenge, J., Matthies, R. and Stone, S. R. (1989) *Biochemistry*, **28**, 9806.

49 Barton, A., Sierakowska, H. and Shugar, D. (1976) *Clin. Chim. Acta*, **67**, 231.

50 Shapiro, R., Riordan, J. F. and Vallee, B. L. (1986) *Biochemistry*, **25**, 3527.

51 Roth, J. S. (1967) *Methods Cancer Res.*, **3**, 151.

52 Blackburn, P. (1979) *J. Biol. Chem.*, **254**, 12 484.

54 Lee, F. S., Fox, E. A., Zhou, H-M., Strydom, D. J. and Vallee, B. L. (1988) *Biochemistry*, **27**, 8545.

53 McGregor, C. W., Adams, A. and Knowler, J. T. (1981) *J. Steroid Biochem.*, **14**, 415.

55 Hofsteenge, J., Kieffer, B., Matthies, R., Hemmings, B. A. and Stone, S. R. (1988) *Biochemistry*, **27**, 8537.

56 Blackburn, P. and Moore, S. (1982) in *The Enzymes* (3rd edn) (ed. P. D. Boyer), Academic Press, New York, vol. 15, p. 317.

57 Arni, R., Heinemann, U., Tokuoka, R. and Saenger, W. (1988) *J. Biol. Chem.*, **263**, 15 358.

58 Uchida, T. and Egami, F. (1967) *J. Biochem.* (Tokyo), **61**, 44.

59 Kenney, W. C. and Dekker, C. A. (1971) *Biochemistry*, **10**, 4962.

60 Kasai, K., Uchida, T., Egami, F., Yoshida, K. and Nomoto, M. (1969) *J. Biochem. (Tokyo)*, **66**, 389.

61 Shen, V. and Schlessinger, D. (1982) in *The Enzymes* (ed. P. D. Boyer), Academic Press, New York, vol. 15, p. 501.

62 Neu, H. C. and Heppel, L. A. (1964) *J. Biol. Chem.*, **239**, 3893.

63 King, T. C., Sirdeskmukh, R. and Schlessinger, D. (1986) *Microbiol. Rev.*, **50**, 428.

64 Gesteland, R. (1966) *J. Mol. Biol.*, **16**, 67.

65 Cohen, L. and Kaplan, R. (1977) *J. Bacteriol.*, **129**, 651.

66 Meador, J., Cannon, B., Cannistraro, V. J. and Kennell, D. (1990) *Eur. J. Biochem.*, **187**, 549,

67 Minato, S. and Hirai, A. (1979) *J. Biochem. (Tokyo)*, **85**, 327.

68 Rushizky, G. W., Mozejko, J. H., Rogerson, D. and Sober, H. A. (1970) *Biochemistry*, **9**, 4966.

69 Randerath, K., Gupta, R. C. and Randerath, E. (1980) *Methods Enzymol.*, **65**, Part 1, 638.

70 Yasuda, T. and Inoue, Y. (1982) *Biochemistry*, **21**, 364.

71 Clemens, M. J. and McNurlan, M. A. (1985) *Biochem. J.*, **226**, 345.

72 Jacobsen, H., Krause, D., Friedman, R. M. and Silverman, R. H. (1983) *Proc. Natl. Acad. Sci. USA*, **80**, 4954.

73 Clemens, M. J. and Williams, B. R. G. (1978) *Cell*, **13**, 565.

74 Wrescher, D. H., James, T. C., Silverman, R. H. and Kerr, I. H. (1981) *Nucleic Acids Res.*, **9**, 1571.

75 Suhadolnik, R. J., Lee, C., Kariko, K. and Li, S. W. (1987) *Biochemistry*, **26**, 7143.

76 Torrence, P. F., Brozda, D., Alster, D., Charubala, R. and Pfleiderer, W. (1988) *J. Biol. Chem.*, **263**, 1131.

77 Silverman, R. H., Jung, D. D., Nolan-Sorden, N. L., Dieffenbach, C. W., SenGupta, D. N. *et al.* (1988) *J. Biol. Chem.*, **263**, 7336.

78 RajBhandary, U. L., Lockard, R. E. and Wurst Reiliy, R. M. (1982) in Nucleases (eds. S. M. Linn and R. J. Roberts), Cold Spring Harbor Laboratory, New York, p. 275.

79 Tabor, M. W., Leake, B. H. and MacGee, J. (1976) *Fed. Proc. Abs.*, **35**, 298.

80 Lockard, R. E., Alzner-Deweerd, B., Heckman, J. E., MacGee, J., Tabor, M. W. *et al.* (1978) *Nucleic Acids Res.*, **5**, 37.

81 Pilly, D., Niemeyer, A., Schmidt, M. and Bargetzi, P. (1978) *J. Biol. Chem.*, **253**, 437.

82 Donis-Keller, H. (1980) *Nucleic Acids Res.*, **8**, 3133.

83 Levy, C. C. and Karpetsky, T. P. (1980) *J. Biol. Chem.*, **255**, 2153.

84 Adams, J. M., Cory, S. and Spahr, P. F. (1972) *Eur. J. Biochem.*, **29**, 469.

85 Bellemore, G., Jordan, B. R. and Monier, R. (1972) *J. Mol. Biol.*, **71**, 307.

86 Thomas, K. R. and Olivera, B. M. (1978) *J. Biol. Chem.*, **253**, 424.

87 Gupta, R. S., Kasai, T. and Schlessinger, D. (1977) *J. Biol. Chem.*, **252**, 8945.

88 Cudny, H. and Deutscher, M. P. (1980) *Proc. Natl. Acad. Sci. USA*, **77**, 83.

89 Donovan, W. P. and Kushner, S. R. (1986) *Proc. Natl. Acad. Sci. USA*, **82**, 6427.

90 Stein, H. and Hausen, P. (1970) *Cold Spring Harbor Symp. Quant. Biol.*, **35**, 709.

91 Crouch, R. J. and Dirksen, M.-L. (1982) in *Nucleases* (eds S. M. Linn and R. J. Roberts), Cold Spring Harbor Laboratory, New York, p. 211.

92 Yang, W., Hendrickson, W. A., Crouch, R. J. and Satow, Y. (1990) *Science*, **249**, 1398.

93 deMassey, B., Fayet, O. and Kogoma, T. (1984) *J. Mol. Biol.*, **178**, 227.

94 Itoh, T. and Tomizawa, J. (1980) *Proc. Natl. Acad. Sci. USA*, **77**, 2450.

95 Kitani, T., Yoda, K., Ogawa, T. and Okazaki, J. (1985) *J. Mol. Biol.*, **184**, 45.

96 Krug, M. S. and Berger, S. L. (1989) *Proc. Natl. Acad. Sci. USA*, **86**, 3539.

97 Prasad, V. R. and Goff, S. P. (1989) *Proc. Natl. Acad. Sci. USA*, **86**, 3104.

98 Doolittle, R. F., Feng, D-F., Johnson, M. S. and McClure, M. A. (1989) *Rev. Biol.*, **64**, 1.

99 Lockard, R. and Kumar, A. (1981) *Nucleic Acids Res.*, **9**, 5125.

100 Grinnell, B. W. and Wagner, R. R. (1984) *Cell*, **36**, 533.

101 Auron, P. E., Weber, L. D. and Rich, A. (1982) *Biochemistry*, **21**, 4700.

102 White, M. D., Bauer, S. and Lapidot, Y. (1977) *Nucleic Acids Res.*, **4**, 3029.

103 White, M. D., Rapoport, S. and Lapidot, Y. (1977) *Biochem. Biophys. Res. Commun.*, **77**, 1084.

104 Pollard, D. R. and Nagyvary, J. (1973) *Biochemistry*, **12**, 1063.

105 Brownhill, T. J., Jones, A. S. and Stacey, M. (1961) *Biochem. J.*, **73**, 434.

106 Stanley, W. M. and Bock, R. M. (1965) *Biochemistry*, **4**, 1302.

107 Mora P. T. (1962) *J. Biol. Chem.*, **237**, 3210.

108 Cheng T., Polmar S. K. and Kazazian, H. H. (1974) *J. Biol. Chem.*, **249**, 1781.

109 McKnight G-S. and Schimke R. T. (1974) *Proc. Natl. Acad. Sci. USA*, **71**, 4327.

110 Chirgwin, J. M., Przybyla, A. E., MacDonald, R. J. and Rutter, W. J. (1979) *Biochemistry*, **18**, 5294.

111 Girard, M. (1967) *Methods Enzymol.*, **12a**, 581.

112 Berger, S. L. and Birkenmeier, C. S. (1979) *Biochemistry*, **81**, 5143.

113 Scheel, G. and Blackburn, P. (1979) *Proc. Natl. Acad. Sci. USA*, **76**, 4898.

114 de Martynoff, G., Pays, E. and Vassart, G. (1980) *Biochem. Biophys. Res. Commun.*, **93**, 645.

115 Solymosy, F., Fedorcsák, I., Gulyás, A., Farkas, G. L. and Ehrenberg, L. (1968) *Eur. J. Biochem.*, **5**, 520.

116 Wiener, S. L., Wiener, R., Urivetzky, M. and Meilman, E. (1972) *Biochim. Biophys. Acta*, **259**, 378.

117 Ehrenberg, L., Fedorcsák, I. and Solyrnosy, F. (1976) *Prog. Nucleic Acid Res. Mol. Biol.*, **16**, 189.

118 Grunberg-Manago, M., Ortiz, P. J. and Ochoa, S. (1956) *Biochim. Biophys. Acta*, **20**, 269.

119 Grunberg-Manago, M. (1963) *Prog. Nucleic Acid Res.*, **1**, 93.

120 Littauer, U. Z. and Soreq, H. (1982) in *The Enzymes* (ed. P. D. Boyer), Academic Press, New York, vol. 15, p. 517.

121 Portier, C. (1975) *Eur. J. Biochem.*, **55**, 573.

122 Brenneman, F. N. and Singer, H. F. (1964) *J. Biol. Chem.*, **239**, 893.

123 Leder, P., Singer, M. F. and Brimacombe, R. L. C. (1965) *Biochemistry*, **4**, 1561.

124 Trip, E. M. and Smith, M. (1978) *Nucleic Acids Res.*, **5**, 1539.

125 Soreq, H., Nudel, U., Salomon, R., Revel, M. and Littauer, U. Z. (1974) *J. Mol. Biol.*, **88**, 233.

126 Laskowski, M. (1982) in *Nucleases* (eds S. M. Linn and R. J. Roberts), Cold Spring Harbor Laboratory, New York, p. 1.

127 Bollum, F. J. (1965) *J. Biol. Chem.*, **240**, 2599.

128 Young, E. T. and Sinsheimer. R. L. (1965) *J. Biol. Chem.*, **240**, 1274.

129 Bernardi, G. (1971) in *The Enzymes* (3rd edn) (ed. P. D. Boyer), Academic Press,

New York. vol. 4, p. 271.

130 Bernardi, G. (1965) *J. Mol. Biol.*, **13**, 603.

131 Linn, S. (1982) in *Nucleases* (eds S. M. Linn and R. J. Roberts), Cold Spring Harbor Laboratory, New York, p. 341.

132 Salnikow, J. and Murphy, D. (1973) *J. Biol. Chem.*, **248**, 1499.

133 Abe, A. and Liao, T-H. (1983) *J. Biol. Chem.*, **258**, 10283.

134 Linn, S. (1982) in *Nucleases* (eds S. M. Linn and R. J. Roberts), Cold Spring Harbor Laboratory, New York, p. 291.

135 Weiss, B. (1981) in *The Enzymes* (ed. P. D. Boyer), Academic Press, New York, vol. 14, p. 203.

136 Burlingham, B. T., Doerfler, W., Pettersson, U. and Philipson, L. (1971) *J. Mol. Biol.*, **60**, 45.

137 Pogo, B. G. T. and Dales, S. (1969) *Proc. Natl. Acad. Sci. USA*, **63**, 820.

138 Lehman, I. R. (1971) in *The Enzymes* (3rd edn) (ed. P. D. Boyer), Academic Press, New York, vol. 4, p. 251.

139 Radloff, R., Bauer, W. and Vinograd, J. (1967) *Proc. Natl. Acad. Sci. USA*, **57**, 1514.

140 Studier, F. W. (1965) *J. Mol. Biol.*, **11**, 373.

141 Friedberg, E. C. and Goldthwaite, D. A. (1969) *Proc. Natl. Acad. Sci. USA*, **62**, 934.

142 Yajko, D. M. and Weiss, B. (1975) *Proc. Natl. Acad. Sci. USA*, **72**, 688.

143 Ljungquist, S., Nyberg, B. and Lindahl, T. (1975) *FEBS Lett.*, **57**, 169.

144 Radman, M. (1976) *J. Biol. Chem.*, **251**, 1438.

145 Grossman, L., Riazuddin, S., Haseltine, W. and Lindan, K. (1978) *Cold Spring Harbor Symp. Quant. Biol.*, **43**, 947.

146 Demple, B. and Linn, S. (1980) *Nature (London)*, **287**, 203.

147 Mosbaugh, D. W. and Linn, S. (1980) *J. Biol. Chem.*, **255**, 11743.

148 Lindahl, T. (1979) *Prog. Nucleic Acid Res.*, **22**, 135.

149 Demple, B. and Linn, S. (1982) *J. Biol. Chem.*, **257**, 2848.

150 Hurwitz, J., Becker, A., Gefter, M. and Gold, M. (1967) *J. Cell Comp. Physiol.*, *Suppl. 1*, **70**, 181.

151 Sadowski, P. D. and Hurwitz, J. (1969) *J. Biol. Chem.*, **244**, 6182, 6192.

152 Henry, J. J. and Knippers, R. (1974) *Proc. Natl. Acad. Sci. USA*, **71**, 1549.

153 Linn, S. (1981) in *The Enzymes* (ed. P. D.

Boyer), Academic Press, New York, vol. 14, p. 121.

154 Moore, S. (1981) in *The Enzymes* (ed. P. D. Boyer), Academic Press, New York, vol. 14, p. 281.

155 Chase, J. W. and Richardson, C. C. (1974) *J. Biol. Chem.*, **249**, 4545.

156 Healy J. W., Stollar D. and Levine, L. (1963) *Methods Enzymol.*, **6**, 49.

157 Richardson, C. C., Lehman, I. R. and Kornberg, A. (1964) *J. Biol. Chem.*, **239**, 251.

158 Lacks, S. and Greenberg, B. (1967) *J. Biol. Chem.*, **242**, 3108.

159 Okazaki, T. and Kornberg, A. (1964) *J. Biol. Chem.*, **239**, 259.

160 Jorgensen, S. E. and Koerner, J. F. (1966) *J. Biol. Chem.*, **241**, 3090.

161 Muskavitch, T. K. M. and Linn, S. (1981) in *The Enzymes* (ed. P. D. Boyer), Academic Press, New York, vol. 14, p. 233.

162 Kornberg, A. (1974) *DNA Synthesis*, Freeman, San Francisco.

163 Lehman, I. R. and Richardson, C. C. (1964) *J. Biol. Chem.*, **239**, 233.

164 Cozzarelli, N. R., Kelly, R. B. and Kornberg, A. (1969) *J. Mol. Biol.*, **45**, 513.

165 Kelly, R. B., Atkinson, M. R., Huberman, J. A. and Kornberg, A. (1969) *Nature (London)*, **224**, 495.

166 Klenow, H. and Overgaard-Hansen, K. (1970) *FEBS Lett.*, **6**, 25.

167 Oleson, A. E. and Koerner, J. F. (1964) *J. Biol. Chem.*, **239**, 2935.

168 Short, E. C. Jr and Koerner, J. F. (1969) *J. Biol. Chem.*, **244**, 1487.

169 McAuslan, B. R. (1971) in *Strategy of the Viral Genome*, Ciba Symposium Volume, Churchill Livingstone, Edinburgh and London, p. 25.

170 Chase, J. W. and Richardson, C. C. (1974) *J. Biol. Chem.*, **249**, 4553.

171 Chase, J. W. and Richardson, C. C. (1977) *J. Bacteriol.*, **129**, 934.

172 Vales, L. D., Chase, J. W. and Richardson, C. C. (1979) *J. Bacteriol.*, **139**, 320.

173 Joseph, J. W. and Kolodner, R. (1983) *J. Biol. Chem.*, **258**, 10411.

174 Joseph, J. W. and Kolodner, R. (1983) *J. Biol. Chem.*, **258**, 10418.

175 Arber, W. (1974) *Prog. Nucleic Acid Res. Mol. Biol.*, **14**, 1.

176 Murray, K. and Old, R. W. (1974) *Prog.*

Nucleic Acid Res. Mol. Biol., **14**, 117.

177 Boyer, H. (1971) *Annu. Rev. Microbiol.*, **25**, 153.

178 Meselson, M., Yuan, R. and Heywood, J. (1972) *Annu. Rev. Biochem.*, **41**, 447.

179 Nathans, D. and Smith, H. O. (1975) *Annu. Rev. Biochem.*, **44**, 273.

180 Meselson M. and Yuan R. (1968) *Nature (London)*, **217**, 1110.

181 Bickle, T. A. (1982) p. 85 and Modrich, P. and Roberts, R. J. (1982) p. 109 in *Nucleases* (eds S. M. Linn and R. J. Roberts), Cold Spring Harbor Laboratory, New York.

182 Iida, S., Meyer, J., Bachi, B., Stahlhammer Carlamalin, M., Schrikel, S., Bickle, T. A. and Arber, W. (1983) *J. Mol. Biol.*, **165**, 1.

183 Kauc, L. and Piekarowicz, A. (1978) *Eur. J. Biochem.*, **92**, 417.

184 Lautenberger, J. A. and Linn, S. (1972) *J. Biol. Chem.*, **247**, 6176.

185 Suri, B., Nagaraja, V. and Bickle, T. A. (1984) *Curr. Top. Microbiol. Immunol.*, **108**, 1.

186 Eskin, B. and Linn, S. (1972) *J. Biol. Chem.*, **247**, 6183.

187 Hadi, S. M., Bickle, T. A. and Yuan, R. (1975) *J. Biol. Chem.*, **250**, 4159.

188 Yuan, R., Bickle, T. A., Ebbers, W. and Brack, C. (1975) *Nature (London)*, **256**, 556.

189 Ravetch, J. V., Horiuchi, K. and Zinder, N. D. (1978) *Proc. Natl. Acad. Sci. USA*, **75**, 226.

190 Lautenberger, J. A., Kan, N. C., Lackey, D., Linn, S., Edgell, M. H. and Hutchison, C. A. (1978) *Proc. Natl. Acad. Sci. USA*, **75**, 2271.

191 Kan, N. C., Lautenberger, J. A., Edgell, M. H. and Hutchison, C. A. (1979) *J. Mol. Biol.*, **130**, 191.

192 Vovis, G. F., Horiuchi, K. and Zinder, N. D. (1974) *Proc. Natl. Acad. Sci. USA*, **71**, 3810.

193 Bickle, T. A., Brack, C. and Yuan, R. (1978) *Proc. Natl. Acad. Sci. USA*, **75**, 3099.

194 Yuan, R., Hamilton, D. L. and Burckhardt, J. (1980) *Cell*, **20**, 237.

195 Yuan, R. (1981) *Annu. Rev. Biochem.*, **50**, 285.

196 Frederick, C. A., Grable, J., Melia, M., Samudzi, C., Jen-Jacobsen, L. *et al.* (1984) *Nature (London)*, **309**, 327.

197 Subramanian, K. M., Weissman, S. M., Zain, B. S. and Roberts, R. J. (1977) *J. Mol. Biol.*, **110**, 297.

198 Kleid, D., Humayun, Z., Jeffrey, A. and Ptashne, M. (1976) *Proc. Natl. Acad. Sci. USA*, **73**, 293.

199 Lacks, S. and Greenberg, B. (1975) *J. Biol. Chem.*, **250**, 4090.

200 Roberts, R. J. (1976) *CRC Crit. Rev. Biochem.*, **4**, 123.

201 Roberts, R. J. (1982) *Nucleic Acids Res.*, **10**, r117.

202 Roberts, R. J. (1984) *Nucleic Acids Res.*, **12** (Suppl) r167.

203 Ehbrecht, H-J., Pingoud, A., Urbanke, C., Maass, G. and Gualerzi, C. (1985) *J. Biol. Chem.*, **260**, 6160.

204 Endlich, B. and Linn, S. (1985) *J. Biol. Chem.*, **260**, 5720.

205 Endlich, B. and Linn, S. (1985) *J. Biol. Chem.*, **260**, 5729.

206 Le Bon, J. M., Kado, C., Rosenthall, L. J. and Chirikjian, J. G. (1978) *Proc. Natl. Acad. Sci. USA*, **75**, 4097.

207 Brown, E. L., Musich, P. R. and Maio, J. J. (1978) *Nucleic Acids Res.*, **5**, 1093.

208 Watabe, H., Iino, T., Kaneko, T., Shibata, T. and Ando, T. (1983) *J. Biol. Chem.*, **258**, 4663.

209 Shibata, T., Watabe, H., Kaneko, T., Iino, T. and Ando, T. (1984) *J. Biol. Chem.*, **259**, 10499.

210 Smith, H. O. and Nathans, D. (1973) *J. Mol. Biol.*, **81**, 419.

211 Sharp, P. A., Sugden, B. and Sambrook, J. (1973) *Biochemistry*, **12**, 3055.

212 Sugisaki, H. and Takanarni, M. (1973) *Nature (London) New Biol.*, **246**, 138.

213 Danna, K. J., Sack, G. H. Jr and Nathans, D. (1973) *J. Mol. Biol.*, **78**, 363.

214 Singer, J., Roberts-Ems, J. and Riggs, A. D. (1979) *Science*, **203**, 1019.

215 Waalwijk, C. and Flaveli, R. A. (1978) *Nucleic Acids Res.*, **5**, 3231.

216 Hadi, S. M., Bachi, B., Iida, S. and Bickle, T. A. (1983) *J. Mol. Biol.*, **165**, 19.

217 Reiser, J. and Yuan, R. (1977) *J. Biol. Chem.*, **252**, 451.

218 Hotchkiss, R. D. (1948) *J. Biol. Chem.*, **175**, 315.

219 Burdon, R. H. and Adams, R. L. P. (1980) *Trends Biochem Sci.*, **5**, 294.

220 Adams, R. L. P. and Burdon, R. H. (1982) *CRC Crit. Rev. Biochem.*, **13**, 349.

221 Adams, R. L. P. and Burdon, R. H. (1985)

Molecular Biology of DNA Methylation, Springer-Verlag, New York.

222 Razin, A., Cedar, H. and Riggs, A. D. (1984) _DNA Methylation_, Springer-Verlag, New York.

223 Willis, D. B. and Granoff, A. (1980) _Virology_, **107**, 250.

224 Diala, E. S. and Hoffman, R. M. (1983) _J. Virol._, **45**, 482.

225 Bird, A. P., Taggart, M. H. and Smith, B. A. (1979) _Cell_, **17**, 889.

226 Borek, E. and Srinivasan, P. R. (1966) _Annu. Rev. Biochem._, **35**, 275.

227 Groudine, J. and Conkin, K. F. (1985) _Science_, **228**, 1061.

228 Salzer, W. (1977) _Cold Spring Harbor Symp. Quant. Biol._, **42**, 985.

229 Bird, A. P. (1980) _Nucleic Acids Res._, **5**, 1499.

230 Adams, R. L. P. and Eason, R. (1984) _Nucleic Acids Res._, **12**, 5869.

231 Max, E. E. (1984) _Nature (London)_, **310**, 100.

232 Jaenisch, R. and Jahner, D. (1984) _Biochim. Biophys. Acta_, **782**, 1.

233 Pollack, Y., Stein, L., Razin, A. and Cedar, H. (1980) _Proc. Natl. Acad. Sci. USA_, **77**, 6463.

234 Doerfler, W. (1984) _Curr. Top. Microbiol. Immunol._, **108**, 79.

235 Sager, R. and Kitchin, R. (1975) _Science_, **189**, 426.

236 Wolf, S. F., Jolly, D. J., Lunnen, K. D., Friedmann, T. and Migeon, B. R. (1984) _Proc. Natl. Acad. Sci. USA_, **81**, 2806.

237 Kratzer, P. G., Chapman, V. M., Lambert, H., Evans, R. E. and Liskay, R. M. (1983) _Cell_, **33**, 37.

238 Shapiro, L. J. and Mohandas, T. (1983) _Cold Spring Harbor Symp. Quant. Biol._, **47**, 631.

239 Adams, R. L. P., Fulton, J. and Kirk, D. (1982) _Biochim. Biophys. Acta_, **697**, 286.

240 Tanaka, M., Hibasami, H., Nagai, J. and Ikeda, T. (1980) _Aust. J. Exp. Biol. Med. Sci._, **58**, 391.

241 Taylor, S. M. and Jones, P. A. (1979) _Cell_, **17**, 771.

242 Ley, T. J., Chiang, Y. L., Hiadaris, D., Anagnou, N. P., Wilson, V. L. _et al._ (1984) _Proc. Natl. Acad. Sci. USA_, **81**, 6618.

243 Sol, D. and Kline, L. K. (1982) in _The Enzymes_ (ed. P. D. Boyer), Academic Press, New York, vol. 15, p. 557.

244 Kline, L. and Sol, D. (1982) in _The Enzymes_ (ed. P. D. Boyer), Academic Press, New York, vol. 15, p. 567.

245 Hindley, J. (1983) _DNA Sequencing_ (eds T. S. Work and R. H. Burdon), Elsevier Biomedical Press, Amsterdam.

246 Richardson, C. C. (1981) in _The Enzymes_ (ed. P. D. Boyer), Academic Press, New York, vol. 14, p. 299.

247 Novogrodsky, A., Tal, M., Traub, A. and Hurwitz, J. (1966) _J. Biol. Chem._, **241**, 2933.

248 Lillehaug, J. R. (1977) _Eur. J. Biochem._, **73**, 499.

249 Jarvest, R. L. and Lowe, G. (1981) _Biochem. J._, **199**, 273.

250 Van de Sande, J. H., Weppe, K. and Khorana, H. G. (1973) _Biochemistry_, **12**, 5050.

251 Becker, A. and Hurwitz, J. (1967) _J. Biol. Chem._, **242**, 936.

252 Soltis, D. A. and Uhlenbeck, O. C. (1982) _J. Biol. Chem._, **257**, 11340.

253 Soltis, D. A. and Uhlenbeck, D. C. (1982) _J. Biol. Chem._, **257**, 11332.

254 Berkner, K. L. and Folk, W. R. (1977) _J. Biol. Chem._, **252**, 3176.

255 Maxam, A. M. and Gilbert, W. (1980) _Methods Enzymol._, **65**, 499.

256 Levin, C. L. and Zimmerman, S. B. (1976) _J. Biol. Chem._, **251**, 1767.

257 Tamura, S., Teraoka, H. and Tsukada, K. (1981) _Eur. J. Biochem._, **115**, 449.

258 Zimmerman, S. B. and Pheiffer, B. H. (1981) in _The Enzymes_ (ed. P. D. Boyer), Academic Press, New York, vol. 14, p. 315.

259 Tsukada, K. and Ichimura, M. (1971) _Biochem. Biophys. Res. Commun._, **42**, 1156.

260 Shuman, S. and Hurwitz, J. (1979) _J. Biol. Chem._, **254**, 10396.

261 Okada, N., Noguchi, S., Kasai, H., Shindo-Okada, N., Ohgi, T., Goto, T. and Nishimura, S. (1979) _J. Biol. Chem._, **254**, 3067.

262 Nishimura, S. (1983) _Prog. Nucleic Acids Res. Mol. Biol._, **28**, 49.

263 Singhal, R. P. (1983) _Prog. Nucleic Acids Res. Mol. Biol._, **28**, 75.

264 Elliott, M. S. and Trewyn, R. W. (1984) _J. Biol. Chem._, **259**, 2407.

265 McNutt, W. S. (1955) _Methods Enzymol._, **2**, 464.

266 Coulondre, C., Miller, J. H., Farabrough, P. J. and Gilbert, W. (1978) _Nature (London)_, **274**, 775.

267 Scarano, E., Iaccarino, M., Grippo, P. and

Parisi, E. (1967) *Proc. Natl. Acad. Sci. USA*, **57**, 1394.

268 Liao, T-H. (1985) *J. Biol. Chem.*, **260**, 10 708.

269 Suck, D., Lahm, A. and Oefner, C. (1988) *Nature (London)*, **332**, 464.

270 Cowan, G. M., Gann, A. A. F. and Murray, N. E. (1989) *Cell*, **56**, 103.

271 Price, C., Lingner, J., Bickle, T. A., Firman, K. and Glover, S. W. (1989) *J. Mol. Biol.*, **205**, 115.

272 Lacks, S. A., Mannarelli, B. M., Springhorn, S. S. and Greenberg, B. (1986) *Cell*, **46**, 993.

273 Roberts, R. J. (1990) *Nucleic Acids Res.*, **18** Suppl, 2331.

274 Trautner, T. A., Balganesh, T., Wilke, K., Nayer-Weidner, M., Rauhut, E. *et al.* (1988) *Gene*, **74**, 267.

275 Wilke, K., Rauhut, E., Nayer-Weidner, Lauster, R., Pawlek, B. *et al.* (1988) *EMBO J.*, **7**, 2601.

276 Adams, R. L. P. (1990) *Biochem. J.*, **265**, 309.

277 Monk, M., Boubelik, M. and Lehnert, S. (1987) *Development*, **99**, 371.

278 Silva, A. J. and White, R. (1988) *Cell*, **54**, 145.

279 Wilks, A., Seldran, M. and Jost, J.-P. (1984) *Nucleic Acids Res.*, **12**, 1163.

280 Razin, A., Szyf, M., Kafri, T., Roll, M., Giloh, H. *et al.* (1986) *Proc. Natl. Acad. Sci. USA*, **83**, 2827.

281 Wiebauer, K. and Jiricny, J. (1989) *Nature (London)*, **339**, 234.

282 Ogden, G. B., Pratt, M. J. and Schaechter, M. (1988) *Cell*, **54**, 127.

283 Tsagris, M., Tabler, M. and Sänger, H. L. (1991) *Nucleic Acids Res.*, **19**, 1605.

5

The metabolism of nucleotides

5.1 ANABOLIC PATHWAYS

A supply of ribonucleoside and deoxyribo-nucleoside 5′-triphosphates is required for the biosynthesis of nucleic acids (chapters 6 and 8). The synthesis of these compounds occurs in two main stages: (1) the formation of the purine and pyrimidine ring systems and their conversion into the parent ribonucleoside monophosphate, inosine 5′-monophosphate (IMP) and uridine 5′-monophosphate (UMP); (2) a series of interconversions involving the reduction of ribonucleotides to deoxyribonucleotides and the phosphorylation of these nucleotides to form a balanced set of 5′-triphosphates (Fig. 5.8).

5.2 THE BIOSYNTHESIS OF THE PURINES

The pathway of biosynthesis of the purines leads directly to the production of their nucleoside 5′-monophosphates and this subject has been so extensively reviewed [1–4] that only an outline need be given here. It is known from experiments with isotopes that the sources of the atoms in the purine ring are as shown in Fig. 5.1. The first step in the biosynthetic pathway is the

phosphorylation of ribose 5′-phosphate by transfer of the pyrophosphoryl residue of ATP to carbon-1 of the ribose moiety to form 5-phosphoribosyl-1-pyrophosphate (PRPP) (Fig. 5.2) which is involved in reactions other than those of nucleotide biosynthesis. Synthesis of PRPP is catalysed by *PRPP synthetase* an enzyme subject to control particularly by ADP which acts allosterically as well as competing with ATP [5, 6].

Assembly of the purine ring itself begins with the transfer of the amido-NH_2 group of glutamine to PRPP forming 5-phosphoribosylamine (PRA) (Fig. 5.3) under the influence of the enzyme *phosphoribosyl pyrophosphate amidotransferase* (*amido-phosphoribosyl transferase*, EC 2.4.2.14).

Glycine then reacts with PRA to give glycinamide ribonucleotide (GAR) a nucleotide-like compound in which the amide of glycine takes the place of the usual purine or pyrimidine base. The reaction sequence continues by formylation from N^5N^{10}-methenyltetrahydrofolic acid to give formylglycinamide ribonucleotide (formyl GAR), then amination from glutamine to give formylglycinamidine ribonucleotide (formyl GAM). Ring closure ensues, producing the imidazole ring compound 5-aminoimidazole ribonucleotide

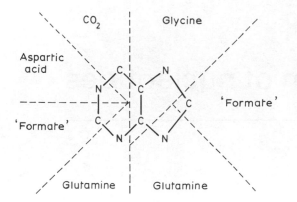

Fig. 5.1 Sources of the atoms in the purine ring.

(AIR). Carboxylation of this compound gives 5-aminoimidazole-4-carboxylic acid ribonucleotide (carboxyl-AIR). The corresponding amide, 5-aminoimidazole-4-carboxamide ribonucleotide (AICAR) is produced in the subsequent two reactions via an intermediate compound, 5-aminoimidazole-4-succinocarboxamide ribonucleotide (succino-AICAR). The purine ring system is completed (Fig. 5.3) when *N*-formyltetrahydrofolic acid donates its formyl group to the 5-amino group of the imidazole carboxamide ribonucleotide. The complete parent ribonucleotide is inosinic acid (inosine 5′-monophosphate, IMP). By virtue of the role of folic acid derivatives as donors of 1-carbon units at two different stages (B and D) in this sequence of reactions, analogues of folic acid such as aminopterin and amethopterin (methotrexate) are powerful inhibitors of purine biosynthesis. Because of this and because,

as will be seen later, they also inhibit the formation of thymine nucleotides, these compounds interfere with the biosynthesis of nucleic acids and are widely used in the treatment of certain forms of cancer and related diseases [7, 8]. In cell biology they are employed in studies on cell hybridization and transformation (section 5.7).

Two other important inhibitors of purine nucleotide biosynthesis are azaserine:

$$N^- = N^+ = CH-CO-OCH_2CH(NH_2)-COOH$$

and 6-diazo-5-oxonorleucine (DON):

$$N^- = N^+ = CH-CO-CH_2-CH_2-CH(NH_2)-COOH$$

both of which are analogues of glutamine:

$$H_2N-CO-CH_2-CH_2-CH(NH_2)-COOH$$

and inhibit the sequence at the amination steps (A and C) from PRPP to PRA and from formyl-GAR to formyl-GAM [9, 10].

Inosine 5′-monophosphate is the common precursor of both adenosine and guanosine 5′-monophosphates. Amination of IMP to AMP proceeds in two stages with the intermediate formation of adenylosuccinic acid (Fig. 5.4). This reaction, in which the amino group of aspartate is transferred to C6 of IMP to give AMP, resembles the reaction above in which 5-aminoimidazole-4-carboxamide ribonucleotide is formed from 5-aminoimidazolecarboxylic acid ribonucleotide (Fig. 5.3). One difference, however, is the requirement for GTP as coenzyme in the reaction forming adenylosuccinic acid from IMP.

The formation of GMP from IMP is also

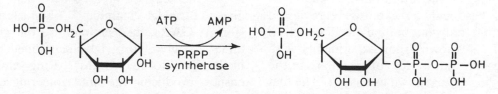

Fig. 5.2 Formation of 5-phosphoribosyl-1-pyrophosphate (PRPP).

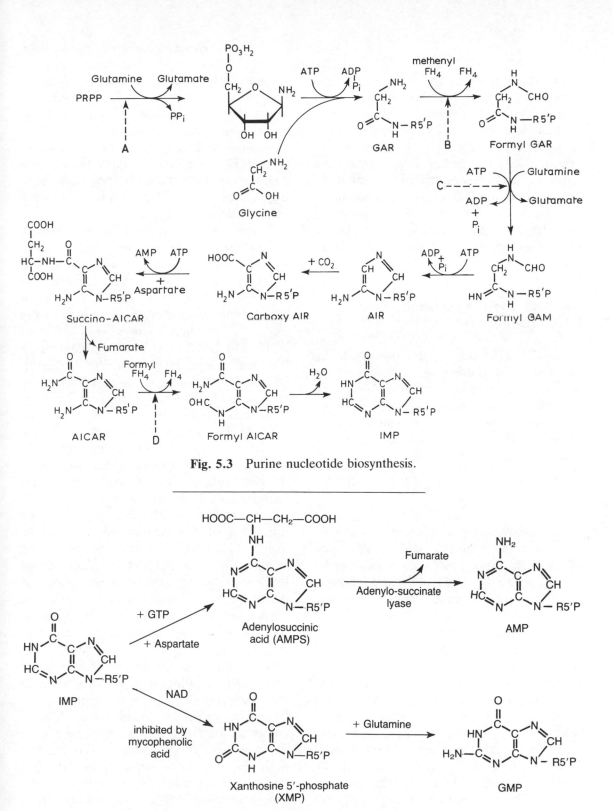

Fig. 5.3 Purine nucleotide biosynthesis.

Fig. 5.4 Formation of AMP and GMP from IMP.

a two-stage reaction in which xanthosine 5′-monophosphate (XMP) is initially formed in a reaction sensitive to inhibition by mycophenolic acid [115]. This is then aminated to give GMP (Fig. 5.4). This amination reaction, like the earlier reactions in the sequence that utilize glutamine, is inhibited by azaserine and DON [11]. The two purine mononucleotides AMP and GMP are phosphorylated by kinases through the diphosphate stage to give ATP and GTP (section 5.8).

5.3 PREFORMED PURINES AS PRECURSORS

In 1947, Kalckar [12] demonstrated the interaction of purine bases and ribose 1-phosphate to yield nucleosides and inorganic phosphate. Such reactions, which are forms of transglycosidation, are reversible and are catalysed by enzymes termed *nucleoside phosphorylases* (EC 2.4.2.1):

Hypoxanthine + ribose 1-phosphate

$\rightleftharpoons$ Inosine + P_i

5-Aminoimidazole-4-carboxamide can take the place of a purine base in this type of reaction. Since most nucleosides can be phosphorylated by ATP under the influence of appropriate phosphokinases, a route exists for the biosynthesis of ribonucleotides from preformed purines.

A much more important mechanism for the conversion of bases into nucleotides involves PRPP which is also concerned in the *de novo* pathway described above. Under the influence of enzymes originally termed nucleotide pyrophosphorylases, but more correctly called *phosphoribosyltransferases*, bases can react with PRPP to form nucleotides and pyrophosphate [13]. The reactions catalysed by phosphoribosyltransferases are illustrated in Fig. 5.5.

One such enzyme, *adenine phosphoribosyltransferase* (EC 2.4.2.7) (APRT), forms AMP from adenine and PRPP. A second enzyme, *hypoxanthine–guanine phosphoribosyltransferase* (EC 2.4.2.8) (HGPRT) converts hypoxanthine and guanine into IMP and GMP, respectively, in the presence of PRPP. *E. coli* has a *xanthine phosphoribosyltransferase*.

Sophisticated control mechanisms govern the pathway of purine nucleotide biosynthesis although there are major variations in these mechanisms from organism to organism [14]. A generalized and consolidated summary of these controls is shown in Fig. 5.5 from which it can be seen that both AMP and GMP exert feedback control on the formation of PRA from PRPP [15, 16]. In some systems, AMP also controls the formation of AMP from IMP while in others GMP controls the formation of XMP from IMP. GTP is required as a cofactor in the synthesis of AMP from IMP. In addition to feedback inhibition most of the enzymes of the pathway are repressed by purine bases, nucleosides and nucleotides.

The enzyme HGPRT also catalyses the formation of the corresponding thio-IMP from the drugs 6-mercaptopurine and azathioprine (imuran). This analogue of IMP inhibits the formation of PRA and AMP and so prevents biosynthesis of the normal purine nucleotides [17]. Because such drugs inhibit purine (and therefore nucleic acid) biosynthesis they have been used as immunosuppressive or cancerostatic agents [18]. Azathioprine is also of value in the treatment of gout by inhibiting purine formation. In the rare condition in children known as the *Lesch–Nyhan syndrome* there is a deficiency of the enzyme hypoxathine–guanine phosphoribosyltransferase [19, 20]. Because of this there is an increase in the concentration of PRPP which in turn leads to increased synthesis of PRA and con-

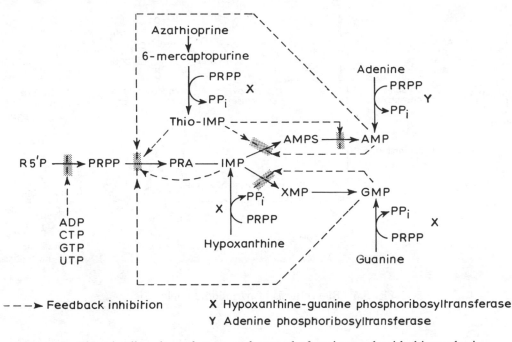

Fig. 5.5 Phosphoribosyltransferases and control of purine nucleotide biosynthesis.

sequently to overproduction of purine nucleotides and ultimately of uric acid. The condition is associated with excessive uric acid synthesis and is resistant to the action of azathioprine presumably because this compound is only active after conversion into thio-IMP (Fig. 5.5), a reaction that requires the missing enzyme HGPRT.

5.4 THE BIOSYNTHESIS OF THE PYRIMIDINES

Whereas purine nucleotide biosynthesis proceeds by growth of the purine ring on PRPP, the formation of the pyrimidine nucleotides involves assembly of a pyrimidine derivative, orotic acid, and the subsequent combination of this moiety with PRPP. The complete series of enzymic reactions giving rise to the parent pyrimidine mononucleotide (uridine

5′-monophosphate, UMP) is shown in Fig. 5.6 [3, 21].

The starting compounds are aspartic acid and carbamoylphosphate which combine under the influence of *aspartate carbamoyltransferase* to form carbamoylaspartate. Formation of the pyrimidine ring is then effected by the action of *dihydro-orotase* giving dihydro-orotic acid, dehydrogenation of which produces the important pyrimidine intermediate orotic acid.

A phosphoribosyltransferase reaction then follows in which orotic acid accepts a ribose 5-phosphate group from PRPP. The resulting product is orotidine 5′-monophosphate (OMP) and inorganic pyrophosphate is eliminated. Decarboxylation of orotidine 5′-monophosphate gives uridine 5′-monophosphate (UMP) which is then converted by kinases through uridine 5′-diphosphate (UDP) into uridine 5′-triphos-

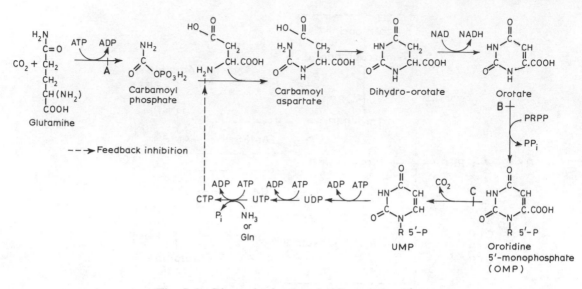

Fig. 5.6 Biosynthesis of pyrimidine nucleotides.

phate (UTP) and it is at this level of phosphorylation that conversion of uracil into cytosine takes place [22] under the influence of the enzyme *CTP synthetase* (EC 6.3.4.2):

$$UTP + NH_3 + ATP \rightarrow CTP + ADP + P_i$$

In eukaryotes, the first three enzymes in the pyrimidine biosynthetic pathway (carbamoylphosphate synthetase, aspartate carbamoyltransferase and dihydro-orotase) are all part of a single multifunctional protein (CAD) of M_r 200 000 [23–26]. The carbamoyltransferase is allosterically controlled, being activated by ATP and inhibited by CTP [27, 28]. UMP exerts feedback inhibition on the synthesis of carbamoylphosphate [14, 27] and inhibits the decarboxylation of OMP to UMP [16]. The effects of these allosteric modulators have been studied by X-ray crystallography [28].

Inhibitors of pyrimidine biosynthesis include azaserine and DON (Fig. 5.6, A) (section 5.2) which inhibit carbamoylphosphate synthesis in those systems utilizing glutamine as the amino group donor, 5-azaorotate, which inhibits the formation of OMP from orotic acid (B) and 6-azauridine which blocks the decarboxylation of OMP to UMP (C) [29].

Exposure of cultured cells to *N*-phosphonoacetyl-L-aspartate (a transition state analogue of aspartate carbamoyltransferase) leads to the selection of mutant cells which have amplified the gene coding for the multifunctional enzyme [23].

Preformed pyrimidines can be taken up by cells and converted into nucleotides by *salvage pathways*. This may proceed by direct reaction with PRPP or via a phosphorylase and kinase as occurs with purines (section 5.3) [30]. In bacteria uridine and deoxyuridine are frequently converted into uracil before incorporation, whereas cytidine is rapidly deaminated to uridine [30–32]. The presence of thymidine induces a thymidine phosphorylase in *E. coli* which converts thymidine into thymine [33]. As radioactive thymine and thymidine are often used to label DNA, the presence of this inducible enzyme means that thymine should always be used for longer labelling times.

5.5 THE BIOSYNTHESIS OF DEOXYRIBONUCLEOTIDES AND ITS CONTROL

The conversion of ribose into deoxyribose takes place at the nucleotide level without breakage of the glycosidic linkage [34] since compounds such as uniformly labelled [^{14}C]cytidine are incorporated into the dCMP residues of DNA without change in the relative specific radioactivities of sugar and base. Four related but readily distinguishable systems have been identified for the reduction of the ribosyl moiety of ribonucleotides to the corresponding deoxyribosyl derivative and two have been extensively investigated [34–37].

In *Lactobacillus leichmannii* the reductase uses ribonucleoside triphosphates and requires a cobamide coenzyme. The enzyme is monomeric and has a M_r of 76 000. There is a single catalytic site at which all four ribonucleotides are reduced. There is also a single allosteric site which can bind any of the four deoxyribonucleoside triphosphates. Binding of an effector increases the affinity of the substrate site for a particular ribonucleotide and is also essential for the binding of the cobamide coenzyme [35, 38]. For example, reduction of ATP is favoured by the binding of dGTP to the allosteric site. In association with a trans-*N*-deoxyribosylase [39] this enables a balanced supply of deoxyribonucleoside triphosphates to be produced. The initial step in the reaction is the homolytic cleavage of the cobalt–carbon bond to yield two free radicals one of which removes a dithiol hydrogen to produce HS-E-S˙ [36]. This cyclizes on abstraction of the 2′-OH from the ribonucleotide. With retention of the stereochemistry the second hydrogen is added to form the deoxyribose.

In *E. coli*, mammalian cells and most other systems investigated, the reductase uses ribonucleoside diphosphates and requires no coenzyme. The *E. coli* enzyme has two atoms of iron and two pairs of non-identical subunits, i.e. it has the constitution $\alpha\alpha'\beta_2$. α (the product of the *nrdA* gene), has a M_r of 80 000 and has sulphydryl groups at the active site. α differs from α' in the *N*-terminal amino acid. β (the product of the *nrdB* gene) has a M_r of 39 000 [35]. The *nrdA* and *nrdB* genes are adjacent to each other on the *E. coli* chromosome and are transcribed to form a polycistronic mRNA (chapter 8). (The phage T4 *nrdB* gene is one of the few prokaryotic genes to contain an intron [40].) Control of transcription is dependent on growth conditions [41]. The complete nucleotide sequence of both genes has been determined [42].

The iron-dependent reductases are more complex than the cobamide-dependent enzyme. The two α chains are tightly associated to form the B1 subunit (which binds the substrates and the allosteric effectors) and the two β chains form the B2 subunit which is associated with the active site of the enzyme. This contains a binuclear iron centre and a single tyrosyl free radical per pair of β chains [35, 37, 38]. The tyrosyl radical is buried within the protein and can only be accessed by long-range electron transfer processes.

The eukaryotic enzyme has a similar constitution but the two subunits are called M1 and M2 [43] or R1 and R2 [44] and the enzyme shows tight cell-cycle regulation with maximum levels of transcript being present in early S-phase [44]. As the free radical can be detected quantitatively in frozen packed cells by EPR spectroscopy it has been possible to follow enzymic activity during the cell cycle. Moreover, as the EPR signal from deuterated enzyme differs from that of the normal enzyme the synthesis and degradation of the enzyme can be readily followed [45]. The genes for the mouse M1 subunit, *E. coli* B1 subunit and the herpes virus-encoded ribonucleotide reductases show considerable homology [46].

Although NADPH is the ultimate hydrogen donor, a series of hydrogen carriers are involved in the reduction of ribonucleotides. The mechanism in *E. coli* is known as the thioredoxin system. Thioredoxin is a sulphur-containing protein with some 108 amino acid residues [47]. It is reduced by NADPH under the influence of *thioredoxin reductase*, a flavoprotein containing FAD. The reduced thioredoxin in turn reduces ribonucleotides to the corresponding deoxy derivatives, becoming itself reoxidized to thioredoxin. Recent evidence has suggested that other reducing agents, e.g. glutathione, may be able to substitute for thioredoxin [35]. Reduction involves the temporary transfer of the 3'-H to a radical in the enzyme (unlikely to be the tyrosyl radical) thereby transferring the radical to the substrate which is reduced by thiols within the active site by one-electron transfers [36]. The same H is ultimately returned to the 3'-position after reduction.

Ribonucleotide reductase reduces the four nucleotides, ADP, GDP, CDP and UDP to the corresponding deoxyribonucleotides. These are then phosphorylated by kinases to the triphosphates. dATP, dGTP and dCTP are used directly for DNA synthesis but dUTP is rapidly hydrolysed to dUMP by an active dUTPase. This prevents incorporation of dUTP into DNA (section 6.3.2). dUMP is converted by a series of steps into dTTP (see below).

As with the cobamide-dependent enzyme the diphosphate to be reduced is determined by the conformation of the enzyme [35, 48]. Enzyme conformation is altered by nucleotides bound at the allosteric sites. Although most studies have been performed using the enzyme from *E. coli*, the mammalian enzyme responds similarly.

One of the allosteric sites (the l or low-affinity site) binds ATP or dATP. When ATP is bound the enzyme is active but dATP inhibits all activity, possibly by causing aggregation [27, 43]. Deoxyadenosine is a potent inhibitor of DNA synthesis as it is rapidly converted into dATP which inhibits ribonucleotide reductase [49]. Binding to the other allosteric site (the h or high-affinity site) is more complex. When ATP (also dATP in *E. coli*) is bound to the h site, reduction of CDP and UDP is enhanced. dGTP in the h site inhibits reduction of GDP, UDP and CDP but stimulates reduction of ADP. dTTP bound to the h site stimulates reduction of GDP but inhibits (possibly via dGTP) reduction of CDP and UDP (Fig. 5.7).

The fine control exerted over the activity of ribonucleotide reductase is in line with its being a key enzyme in the programmed production of a balanced supply of deoxyribonucleoside triphosphates for synthesis of DNA [35, 50]. In support of this is the finding that the tissue level of ribonucleotide reductase correlates with the growth rate of the tissue and shows variations in activity during the cell cycle [43, 51–53]. This variation in activity has been attributed both to changes resulting from allosteric effects and to variation in the absolute amount of enzyme present. However, direct measurement of the deoxyribonucleoside triphosphate pool sizes in cells [53–57] shows them to be very low in non-dividing cells and to rise during S phase in parallel with increased ribonucleotide reductase activity. Pool sizes reach a maximum in G2 phase and then fall after mitosis. As the amounts of the four deoxyribonucleotides are not the same it is possible that these pools do not represent precursors for DNA synthesis but either excess production or the allosteric pool

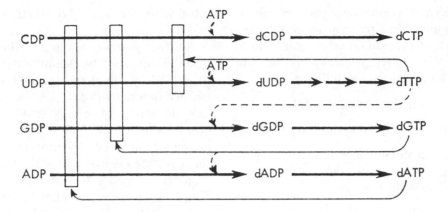

Fig. 5.7 Allosteric control of ribonucleotide reductase; open bars show inhibition, dashed lines stimulation (based on Thelander and Reichard [35]).

which acts to regulate production. The concentration of the components of this regulatory pool would be regulated by balanced synthesis and degradation. When DNA polymerase is inhibited the excess dNTPs are rapidly degraded but degradation ceases when ribonucleotide reductase is inhibited by hydroxyurea [55, 58].

Ribonucleotide reductase is inhibited by hydroxyurea [35, 59] which acts as a free radical scavenger [60–62]. One of the few differences between the *E. coli* enzyme and the mammalian enzyme is that the action of hydroxyurea is irreversible in *E. coli* but is readily reversed in mammalian systems allowing its use as a chemotherapeutic agent and in cell synchronization [59, 63]. 2-Acetylpyridine thiosemicarbazone selectively inhibits the ribonucleotide reductase encoded by herpes simplex virus while having no effect on the mammalian enzyme and so has potential as an antiherpetic agent [64].

Addition to cells of high concentrations of thymidine leads to the formation of a large intracellular pool of dTTP [55, 63, 65, 66]. This feeds back (either directly or through stimulating reduction of GDP) to inhibit further reduction of UDP and CDP bringing about a deficiency of dCTP and hence an inhibition of DNA synthesis. Addition of deoxycytidine, which is rapidly phosphorylated to dCTP, bypasses the block once again allowing DNA synthesis to occur [63, 65–67].

By treating cultured cells with normally lethal concentrations of deoxyribonucleosides an amplification of the gene coding for ribonucleotide reductase can be induced [49] (section 6.15.6).

A second source of deoxyribonucleotides in animal cells is by salvage from circulating deoxyribonucleosides. Indeed in cells such as thymocytes which have active salvage enzymes this may be the major source of dCTP and dTTP [68].

5.6 THE BIOSYNTHESIS OF THYMINE DERIVATIVES

The essential step in the formation of thymine nucleotides is the methylation of deoxyuridine monophosphate (dUMP) to produce thymidine monophosphate (dTMP) (dUMP → dTMP) under the influence of *thymidylate synthetase*. The process is elaborate and takes place in several stages. The source of the additional carbon atom at C5 is N^5N^{10}-methylenetetrahydrofolic acid [3, 69]. The reaction is as follows:

N^5N^{10}-methylenetetrahydrofolate +

dUMP → dihydrofolate + dTMP

The dihydrofolate is reduced again to tetrahydrofolate under the influence of dihydrofolate reductase:

dihydrofolate + NADPH + H$^+$ →

tetrahydrofolate + NADP$^+$

This reaction is powerfully inhibited by the folic acid analogues aminopterin and amethopterin (methotrexate). This leads to the depletion of tetrahydrofolate and accumulation of dihydrofolate, thereby inhibiting the synthesis of thymine derivatives.

The two enzymes, thymidylate synthetase and dihydrofolate reductase (DHFR), are encoded by overlapping DNA sequences in phage T4 [70] (where the thymidylate synthase gene was the first intron-containing, prokaryotic gene to be discovered [71, 72]) and are co-ordinately regulated in *B. subtilis* [73]. This association is taken to extremes in protozoa where the two activities are found in the same polypeptide chain [74–76]. The mammalian thymidylate synthetase shows a high level of amino acid sequence homology with prokaryotic enzymes [77].

Thymidylate synthetase activity is low in non-growing cells but increases in activity in growing cells [78]. It is inhibited by 5-fluoro-dUMP. For this reason 5-fluorodeoxyuridine is an inhibitor of DNA synthesis.

Extracts of *E. coli* infected with a T-even phage contain the enzyme deoxycytidylate hydroxymethylase which brings about the formation of 5-hydroxymethyldeoxycytidylic acid from formaldehyde and deoxycytidylic acid in the presence of N^5N^{10}-methylene-

tetrahydrofolic acid [79] (sections 3.7.6(g) and 6.11.2).

Another pathway leading to the synthesis of dTTP involves the deamination of dCMP to dUMP by the enzyme *dCMP deaminase* (*dCMP aminohydrolase* EC 3.5.4.5). This enzyme is present in only small amounts in non-dividing tissues but is active in rapidly growing tissues such as embryos, spleen and regenerating rat liver [80–82]. It thus shows a similar response to growth as thymidine kinase, thymidylate synthetase, ribonucleotide reductase and DNA polymerase; enzymes present in the replitase complex (section 6.15.5). Although substrate for the enzyme may arise by phosphatase action on dCDP and dCTP the levels of circulating deoxycytidine probably are the major source of cellular dCMP. dCMP deaminase is under allosteric control [83]. The enzyme is inhibited by dTTP and stimulated by dCTP and hence cellular dCMP is channelled towards the pyrimidine deoxynucleoside triphosphate present in limiting amounts.

5.7 AMINOPTERIN IN SELECTIVE MEDIA

Because of the part played by folic acid in purine and thymidylate biosynthesis, analogues of folic acid have become important in cell biology as ingredients in HAT medium (medium containing hypoxanthine, aminopterin and thymidine) [63, 84]. This growth medium is used for selection of animal cells able to bypass the aminopterin-imposed blocks in purine and thymidylate metabolism by incorporating exogenous hypoxanthine and thymidine. Mutant cells lacking the enzymes hypoxanthine phosphoribosyltransferase (HGPRT) or thymidine kinase (TK) cannot grow in HAT medium. However, fusion of a TK$^-$ cell with an HGPRT$^-$ cell leads to a hybrid cell which can grow in HAT medium [85]. Similarly metabolic co-operation is shown by mixed cultures of TK⁻ and HGPRT⁻ cells which exchange thymidylate and purine nucleotides across gap junctions and hence grow in HAT medium [86].

HAT medium is used extensively to select for transformants of TK⁻ cells treated with DNA containing the thymidine kinase gene from herpes simplex virus, linked by genetic engineering techniques to any other DNA [87] (section A.9.3) and to select for monoclonal-antibody-producing cells arising as a result of fusion of non-growing spleen cells with HGPRT⁻ myeloma cells [88].

5.8 FORMATION OF NUCLEOSIDE TRIPHOSPHATES

Kinases convert nucleosides into nucleoside monophosphates, diphosphates and triphosphates by transferring the γ-phosphate from ATP to the substrate. The kinases which act on the pyrimidine nucleosides and deoxyribonucleosides enable the preformed materials to be utilized for nucleic acid synthesis by the so-called 'salvage' pathway, e.g.

thymidine → dTMP → dTDP → dTTP [89]

The thymidine kinases are unique in that they are active in dividing tissues but low in non-growing tissues [90, 91] and their activity shows regular changes during the cell cycle [91, 92]. Some animal viruses code for an additional thymidine (or deoxypyrimidine nucleoside) kinase [93] (section 3.7).

Thymidine kinase is also able to phosphorylate 5-fluorodeoxyuridine but the 5-FdUMP produced is an inhibitor of thymidylate synthetase and because of this fluorodeoxyuridine is an inhibitor of DNA synthesis.

Thymidine kinase is inhibited by dTTP and by dCTP [94–96] but there is also an element of forward promotion [97] which means that on incubating cells with thymidine the size of the dTTP pool produced is proportional to the external thymidine concentration [63, 97].

Nucleotide interconversions are summarized in Fig. 5.8.

5.9 GENERAL ASPECTS OF CATABOLISM

It is recognized that cellular DNA tends to be strongly conserved and that the amount of degradation of DNA in normal circumstances is small. Some species of RNA have a relatively short life span and are therefore degraded quite rapidly. There is also evidence that nucleic acids may not undergo total degradation but that some of the intermediate products such as nucleotides and nucleosides may be reutilized by so-called 'salvage' pathways.

RNA and DNA are hydrolysed by nucleases and diesterases first to oligonucleotides and eventually to mononucleotides and nucleosides, and the glycosidic linkages between the purine or pyrimidine bases and the sugar moieties are cleaved either hydrolytically or phosphorolytically to yield free purine and pyrimidine bases [4, 98]. The nature and function of nucleases and phosphodiesterases have been considered in chapter 4, and the pyrophosphatases and phosphatases that attack nucleotides have been reviewed extensively [99–101].

5.10 PURINE CATABOLISM

The breakdown of purine nucleotides has been studied over many years. Adenine and its nucleosides and nucleotides can be deaminated hydrolytically under the in-

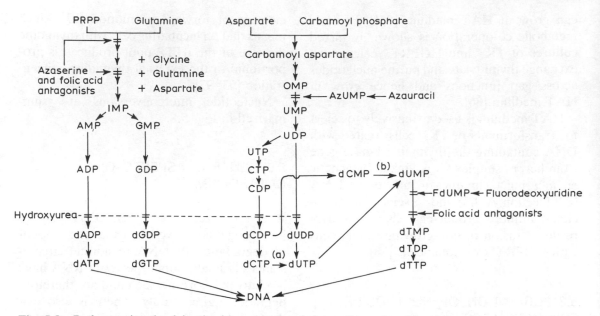

Fig. 5.8 Pathways involved in the biosynthesis of deoxyribonucleoside triphosphates: (a) bacterial systems only; (b) eukaryotes. The blocking action of some antimetabolites is indicated.

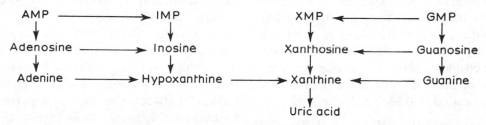

Fig. 5.9 The degradation of purines at the levels of nucleotides, nucleosides and bases.

fluence of the enzymes *adenine deaminase* (present only in some micro-organisms [102, 103]), *adenosine deaminase*, and *adenylate deaminase* to yield hypoxanthine, inosine or inosine monophosphate, respectively (Fig. 5.9), and guanine nucleotides are similarly attacked by guanine, guanosine or guanylate deaminases to yield xanthine or its ribose derivatives (Fig. 5.9). The balance between dephosphorylation and deamination at any stage is quite variable and in T-lymphoblasts although about half the GMP is degraded

by deamination, little if any deamination of dGMP occurs [104]. This can result in the accumulation of cytotoxic levels of deoxy-guanosine in those immunodeficient patients lacking purine nucleoside phosphorylase.

Hypoxanthine and xanthine are oxidized under the influence of xanthine oxidase to yield uric acid (Fig. 5.10).

Although the distribution of the enzymes involved is far from uniform in different species, this scheme of purine degradation appears to be of fairly general application,

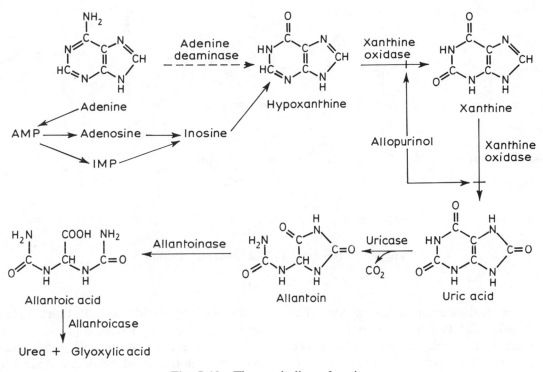

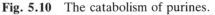

Fig. 5.10 The catabolism of purines.

and experiments with ^{15}N have shown that, as might be expected, the administration of labelled purines to animals is followed by the appearance of the isotope in the excreted uric acid or in its further degradation products.

Uric acid itself is excreted by only a few mammals, since most non-uricotelic animals are provided with the enzyme *uricase*, which oxidizes uric acid to the much more soluble allantoin, and under certain conditions to other end products as well [105]. The conversion of uric acid into allantoin appears to involve a number of intermediate compounds including the symmetrical compound hydroxyacetylenediureinecarboxylic acid [4, 106] (Fig. 5.11).

Man and certain higher apes, however, are unable to bring about this step owing to absence of uricase from their tissues and,

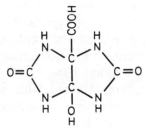

Fig. 5.11 Hydroxyacetylenediureincarboxylic acid.

in these species, the end product of purine metabolism is uric acid which is excreted in the urine along with very much smaller amounts of xanthine and hypoxanthine [107]. The Dalmatian coach-hound is peculiar in that it excretes uric acid in preference to allantoin, owing to lack of tubular reabsorption of uric acid in the kidney [4].

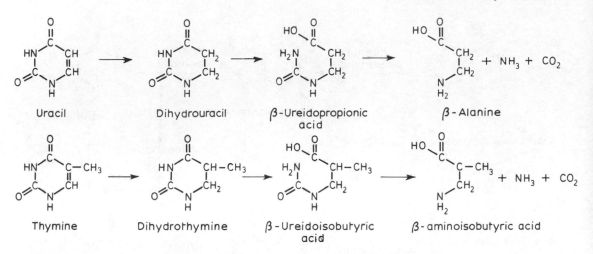

Fig. 5.12   The catabolism of uracil and thymine.

In the pathological condition known as gout, uric acid is deposited in the joints, particularly in the big toe, and under the skin as nodules called tophi. In this disease the miscible pool of uric acid in the human body is increased to as much as 15 times the normal value of about 1 g [108, 109]. The substance allopurinol, which has a structure very similar to that of hypoxanthine, acts as a competitive inhibitor of xanthine oxidase, and so prevents uric acid formation. It is therefore sometimes used in the treatment of gout. Patients treated with allopurinol excrete xanthine and hypoxanthine in place of uric acid [107]. Mention has already been made of the use of azathioprine in the treatment of gout by inhibiting the synthesis of PRA from PRPP, IMP from hypoxanthine and AMP from IMP [29].

In fishes, amphibia and more primitive organisms allantoin is broken down by allantoinase to allantoic acid, and this in turn may be degraded by allantoicase to urea and glyoxylic acid. The main nitrogenous excretory product in the spider is not uric acid but guanine. These aspects of comparative biochemistry are discussed in detail in the books by Baldwin [110], Florkin [111] and Henderson and Paterson [4].

In the birds and the uricotelic reptiles uric acid formation is most pronounced since in them uric acid rather than urea is the main nitrogenous excretory product.

5.11 PYRIMIDINE CATABOLISM

The catabolism of pyrimidine nucleotides, like that of purine nucleotides, involves dephosphorylation, deamination and cleavage of glycosidic bonds, and many of the phosphatases that act on purine nucleotides act also on the corresponding pyrimidine derivatives. As with purine nucleosides, so pyrimidine nucleosides may be hydrolysed to form pyrimidine bases and sugar, or they may be involved in phosphorolytic cleavage [4].

Cytosine can be deaminated by *cytosine deaminase*; this has been demonstrated in yeasts and other micro-organisms [31], and cytosine nucleosides are broken down to uridine nucleosides by *cytidine deaminase* which is widespread in animal tissues [32] as well as in bacteria.

The catabolic pathways of uracil [112] and thymine [113, 114] in mammalian tissues involve reduction of the pyrimidines to the dihydro derivatives, ring opening to give the appropriate ureido-acid, and the removal of ammonia and CO_2 to give β-alanine or its methylated derivative (Fig. 5.12). In some bacteria, uracil and thymine undergo oxidative degradation via barbituric acid and methylbarbituric acid to urea and malonic or methylmalonic acids [4].

REFERENCES

1 Colowick, S. P. and Kaplan, N. O. (1978) *Methods Enzymol.*, **51**.
2 Muller, M. M., Kaiser, E. and Seegmiller, J. E. (1977) *Advances in Experimental Medicine and Biology. Purine Metabolism in Man – II*, Plenum Press, New York, vol. 76A and 76B.
3 Kornberg, A. (1980) *DNA Replication*. Freeman, San Francisco.
4 Henderson, J. F. and Paterson, A. R. P. (1973) *Nucleotlde Metabolism*, Academic Press, New York.
5 Jensen, K. F. (1983) in *Metabolism of Nucleotides, Nucleosides and Nucleobases in Microorganisms* (ed. A. Munch-Petersen), Academic Press, London, p. 1.
6 Becker, M. A. and Seegmiller, J. E. (1979) *Adv. Enzymol.*, **49**, 281.
7 Rhoads, C. P. (ed.) (1955) *Antimetabolites and Cancer*, American Association for the Advancement of Science, Washington, DC.
8 Bresnick, E. (1974) in *The Molecular Biology of Cancer* (ed. H. Busch), Academic Press, New York, p. 278.
9 Levenberg, B., Melnick, I. and Buchanan, J. M. (1957) *J. Biol. Chem.*, **225**, 163.
10 Hartman, S. C. (1963) *J. Biol. Chem.*, **248**, 3036.
11 Abrams, R. and Bentley, M. (1959) *Arch. Biochem.*, **79**, 91.
12 Kalckar, H. (1947) *Symp. Soc. Exp. Biol.*, **1**, 38.
13 Murray, A. W. (1971) *Annu. Rev. Biochem.*, **40**, 811.
14 Mandelstam, J. and McQuillen, K. (1973) *Biochemistry of Bacterial Growth* (2nd edn), Blackwell Scientific, Oxford, p. 236.
15 Bojarski, T. B. and Hiatt, H. H. (1960) *Nature (London)*, **188**, 1112.
16 Creasey, W. A. and Handschumacher, R. E. (1961) *J. Biol. Chem.*, **236**, 2058.
17 Kelley, W. N., Rosenbloom, F. M., Henderson, J. F. and Seegmiller, J. E. (1967) *Proc. Natl. Acad. Sci. USA*, **57**, 1735.
18 Roitt, I. (1974) *Essential Immunology* (2nd edn), Blackwell Scientific, Oxford.
19 Fujimoto, W. Y., Subak-Sharpe, J. H. and Seegmiller, J. E. (1971) *Proc. Natl. Acad. Sci. USA*, **68**, 1516.
20 Rubin, C. S., Dancis, J., Yip, L. C., Bowinski, R. C. and Balis, M. E. (1971) *Proc. Natl. Acad. Sci. USA*, **68**, 1461.
21 Reichard, P. (1959) *Adv. Enzymol.*, **21**, 263.
22 Lieberman, I. (1956) *J. Biol. Chem.*, **222**, 765.
23 Stark, G. R. (1977) *Trends Biochem. Sci.*, **2**, 64.
24 Coleman, P. F., Suttle, D. P. and Stark, G. R. (1977) *J. Biol. Chem.*, **252**, 6379.
25 Makoff, A. J., Buxton, F. P. and Radford, A. (1978) *Mol. Gen. Genet.*, **161**, 297.
26 Grayson, D. R., Lee, L. and Evans, D. R. (1985) *J. Biol. Chem.*, **260**, 15 840.
27 Kantrowitz, E. R., Pastra-Landis, S. C. and Lipscomb, W. N. (1980) *Trends Biochem. Sci.*, **5**, 124.
28 Kantrowitz, E. R. and Lipscomb, W. N. (1990) *Trends Biochem. Sci.*, **15**, 53.
29 Roy-Burman, P. (1970) *Analogues of Nucleic Acid Components*, Springer-Verlag, New York, p. 16.
30 Neuhard, J. (1983) in *Metabolism of Nucleotides, Nucleosides and Nucleobases in Microorganisms* (ed. A. Munch-Petersen), Academic Press, London, p. 95.
31 O'Donovan, G. A. and Neuhard, J. (1970) *Bacteriol. Rev.*, **34**, 278.
32 Wisdom, G. B. and Orsi, B. A. (1969) *Eur. J. Biochem.*, **7**, 223.
33 Mollgaard, H. and Neuhard, J. (1983) in *Metabolism of Nucleotides, Nucleosides and Nucleobases in Microorganisms* (ed. A. Munch-Petersen), Academic Press, London, p. 149.
34 Larsson, A. and Reichard, P. (1967) *Prog. Nucleic Acid Res. Mol. Biol.*, **7**, 303.
35 Thelander, L. and Reichard, P. (1979) *Annu. Rev. Biochem.*, **48**, 133.

36 Stubbe, J. (1990) *J. Biol. Chem.*, **265**, 5329.

37 Nordlund, P., Söjberg, B-M. and Eklund, H. (1990) *Nature (London)*, **345**, 593.

38 Hunting, D. and Henderson, J. F. (1982) *CRC Crit. Rev. Biochem.*, **13**, 325.

39 Witt, L., Yap, T. and Blakley, R. (1978) *Adv. Enzyme Regul.*, **17**, 157.

40 Sjöberg, B-M., Hahue, S., Mathews, C. Z., Mathews, C. K., Rand, K. N. and Gait, M. J. (1986) *EMBO J.*, **15**, 2031.

41 Hanke, P. D. and Fuchs, J. A. (1983) *J. Bacteriol.*, **154**, 1140.

42 Carlson, J., Fuchs, J. A. and Messing, J. (1984) *Proc. Natl. Acad. Sci. USA*, **81**, 4294.

43 Engstrom, Y., Eriksson, S., Thelander, L. and Akerman, M. (1979) *Biochemistry*, **18**, 2941.

44 Fernandez-Sarabia, M. J. and Fantes, P. A. (1990) *Trends Genet.*, **6**, 275.

45 Eriksson, S., Graslund, A., Skoog, S., Thelander, L. and Tribukait, B. (1984) *J. Biol. Chem.*, **259**, 11695.

46 Caras, I. W., Levinson, B. B., Fabry, M., Williams, S. R. and Martin, D. W. (1985) *J. Biol. Chem.*, **260**, 7015.

47 Stryer, L., Holmgren, A. and Reichard, P. (1967) *Biochemistry*, **6**, 1016.

48 Eriksson, S., Thelander, L. and Akerman, M. (1979) *Biochemistry*, **18**, 2948.

49 Meuth, M. and Green, H. (1974) *Cell*, **3**, 367.

50 Elford, H. L., Freese, M., Passamani, E. and Morris, H. P. (1970) *J. Biol. Chem.*, **245**, 5228.

51 Turner, M. K., Abrams, R. and Lieberman, I. (1968) *J. Biol. Chem.*, **243**, 3725.

52 Murphree, S., Stubblefield, E. and Moore, C. E. (1969) *Exp. Cell Res.*, **58**, 118.

53 Albert, D. A. and Gudes, L. J. (1985). *J. Biol. Chem.*, **260**, 679.

54 Walters, R. A. and Ratliff, R. L. (1975) *Biochim. Biophys. Acta*, **414**, 221.

55 Adams, R. L. P., Berryman, S. and Thomson, A. (1971) *Biochim. Biophys. Acta*, **240**, 455.

56 Skoog, K. L., Nordenskjold, B. A. and Bjursell K. G. (1973) *Eur. J. Biochem.*, **33**, 428.

57 Skoog, K. L. and Bjursell, G. (1974) *J. Biol. Chem.*, **249**, 6434.

58 Nicander, B. and Reichard, P. (1985) *J. Biol. Chem.*, **260**, 5376.

59 Adams, R. L. P. and Lindsay, J. G. (1966)

J. Biol. Chem., **242**, 1314.

60 Thelander, L., Larsson, B., Hobbs. J. and Eckstein, F. (1976) *J. Biol. Chem.*, **251**, 1398.

61 Atkin, C. L., Thelander, L., Reichard, P. and Lang, G. (1973) *J. Biol. Chem.*, **248**, 7464.

62 Thelander, M., Gräslund, A. and Thelander, L. (1985) *J. Biol. Chem.*, **260**, 2737.

63 Adams, R. L. P. (1990) *Cell Culture for Biochemists* (2nd edn) (eds R. H. Burdon and P. H. van Knippenberg), Elsevier, Amsterdam.

64 Spector, T., Avarett, D. R., Nelson, D. J., Lambe, C. U., Morrison, R. W. *et al.* (1985) *Proc. Natl. Acad. Sci. USA*, **82**, 4254.

65 Bjursell, G. and Reichard, P. (1973) *J. Biol. Chem.*, **248**, 3904.

66 Morris, N. R. and Fischer, G. A. (1960) *Biochim. Biophys. Acta*, **42**, 183.

67 Reynolds, E. C., Harris, A. W. and Finch, L. R. (1979) *Biochim. Biophys. Acta*, **561**, 110.

68 Cohen, A., Barankiewicz, J., Lederman, H. M. and Gelfand, E. W. (1983) *J. Biol. Chem.*, **258**, 12334.

69 Friedkin, M. (1963) *Annu. Rev. Biochem.*, **32**, 185.

70 Purohit, S. and Mathews, C. K. (1984) *J. Biol Chem.*, **259**, 6261.

71 Belfort, M., Pedersen-Lane, J., West, D., Ehrenman, K., Maley, G., Chu, F. and Maley, F. (1985) *Cell*, **41**, 375.

72 Chu, F. K., Maley, G. F., West, D. K., Belfort, M. and Maley, F. (1986) *Cell*, **45**, 457.

73 Myoda, T. T. and Funanage, V. L. (1985) *Biochim. Biophys. Acta*, **824**, 99.

74 Coderre, J. A., Beverley, S. H., Schimke, R. T. and Santi, D. V. (1983) *Proc. Natl. Acad. Sci. USA*, **80**, 2132.

75 Beverley, S. M., Coderre, J. A., Santi, D. V. and Schimke, R. T. (1984) *Cell*, **38**, 431.

76 Meek, T. D., Garvey, E. P. and Santi, D. V. (1985) *Biochemistry*, **24**, 678.

77 Takeishi. K., Kaneda, S., Ayusawa, D., Shimizu, K., Gotoh, O. *et al.* (1985) *Nucleic Acids Res.*, **13**, 2035.

78 Conrad, R. H. and Ruddle, F. H. (1972) *J. Cell Sci.*, **10**, 471.

79 Flaks, J. G. and Cohen, S. S. (1957) *Biochim. Biophys. Acta*, **25**, 667.

80 Scarano, E., Talarico, M., Bonaduce, L. and de Petrocellis, B. (1960) *Nature (London)*, **196**, 237.

81 Maley, G. F. and Maley, F. (1964) *J. Biol. Chem.*, **239**, 1168.

82 Potter, V. R. (1964) *Cancer Res.*, **24**, 1085.

83 Rossi, M., Dosseva, I., Pierro, M., Cacace, M. G. and Scarano, E. (1971) *Biochemistry*, **10**, 3090.

84 Littlefield, J. W. (1964) *Science*, **145**, 709.

85 Littlefield, J. W. and Goldstein, S. (1970) *In vitro*, **6**, 21.

86 Pitts, J. D. (1971). in *Growth Control in Cell Cultures*. Ciba Found. Symp. (eds G. E. W. Wolstenholme and J. Knight), p. 89.

87 Mantei, N., Boll, W. and Weissmann, C. (1979) *Nature (London)*, **281**, 35.

88 Campbell, A. M. (1991) *Monoclonal Anti body Technology* (2nd edn) (eds R. H. Burdon and P. H. van-Knippenberg), Elsevier, Amsterdam.

89 Grav, H. J. and Smellie, R. M. S. (1965) *Biochem. J.*, **94**, 518.

90 Weissman, S. M., Smellie, R. M. S. and Paul, J. (1960) *Biochim. Biophys. Acta*, **45**, 101.

91 Bello, L. J. (1974) *Exp. Cell Res.*, **89**, 263.

92 Stubblefield, E. and Dennis, C. M. (1976) *J. Theor. Biol.*, **61**, 171.

93 Keir, H. M. (1968) *Soc. Gen. Microbiol. Symp.*, **18**, 67.

94 Potter, V. R. (1964) *Metabolic Control Mechanisms in Animal Cells*, National Cancer Institute, Monograph No. 13, p. 111.

95 Bresnick, E. and Karjala, R. J. (1964) *Cancer Res.*, **24**, 841.

96 Okazaki, R. and Kornberg, A. (1964) *J. Biol. Chem.*, **239**, 275.

97 Ives, D. H., Morse, P. A., Jr. and Potter, V. R. (1963) *J. Biol. Chem.*, **248**, 1467.

98 Smellie, R. M. S. (1955) in *The Nucleic Acids* (eds E. Chargaff and J. N. Davidson), Academic Press, New York, vol. II, p. 393.

99 Kielley, W. W. (1961) *The Enzymes* (2nd edn) (eds P. D. Boyer, H. Hardy and K. Myreback) Academic Press, New York, vol. 5, p. 149.

100 Morton, R. K. (1965) *Compr. Biochem.*, **16**, 55.

101 Bodansky, O. and Schwartz, M. K. (1968) *Adv. Clin. Chem.*, **11**, 277.

102 Nygaard, P. (1983) in *Metabolism of Nucleotldes, Nucleosides and Nucleobases in Microorganisms* (ed. A. Munch-Petersen), Academic Press, London, p. 27.

103 Zielke, C. L. and Suelter, C. H. (1971) in *The Enzymes* (3rd edn) (ed. P. D. Boyer), Academic Press, New York, vol. 4, p. 47.

104 Barankiewicz, J. and Cohen, A. (1985) *J. Biol. Chem.*, **260**, 4565.

105 Canellakis, E. S. and Cohen, P. O. (1955) *J. Biol. Chem.*, **213**, 385.

106 Dalgliesh, C. E. and Neuberger, A. (1954) *J. Chem. Soc.*, 3407.

107 Balis, E. W. (1968) *Fed. Proc.*, **27**, 1067.

108 Wyngaarden, J. B. (1966) *Adv. Metab. Disorders*, **2**, 1.

109 Bishop, C., Garner, W. and Talbott, J. H. (1951) *J. Clin. Invest.*, **30**, 879.

110 Baldwin, E. (1949) *An Introduction to Comparative Biochemistry*. Cambridge University Press, London.

111 Florkin, M. (1949) *Biochemical Evolution*, Academic Press, New York.

112 Schulman M. P. (1954) *Chemical Pathways of Metabolism* (ed. D. M. Greenberg), Academic Press, New York, vol. 2, p. 223.

113 Canellakis, E. S. (1957) *J. Biol. Chem.*, **227**, 701.

114 Fink, K., Cline, R. E., Henderson, R. B. and Fink, R. M. (1956) *J. Biol. Chem.*, **221**, 425.

115 Franklin, T. J. and Cook, J. M. (1969) *Biochem. J.*, **113**, 515.

6

Replication of DNA

6.1 INTRODUCTION

Each daughter cell produced on cell division contains an identical copy of the genetic material. Since DNA carries the genetic blueprint of the cell encoded in the sequence of nucleotides, the question as to how DNA is reproduced in the cell has attracted a great deal of attention.

Studies with intact cells have shown that replication is semiconservative (section 6.2) and that it occurs in a discontinuous manner (section 6.3), and work with purified enzymes has indicated the detailed mechanism of the polymerization reaction (section 6.4). However, it was the realization of the complexity of the reaction which is required to replicate DNA faithfully and rapidly, that led to the development of several *in vitro* systems and finally to the reconstitution from purified components of complexes capable of limited replication (section 6.7).

Modern techniques of gene cloning, *in vitro* mutagenesis and DNA sequencing have facilitated the dissection of several origins of replication (sections 6.9–6.15) and we are now beginning to gain some insight into the controls exerted over the event of initiation of replication which is an important, and possibly the major, site at which cell multiplication is controlled.

6.2 SEMICONSERVATIVE REPLICATION

At the end of their paper [1] suggesting the double-stranded structure for DNA with its complementary base pairs, Watson and Crick wrote: 'It has not escaped our notice that the specific pairing we have postulated suggests a possible copying mechanism for the genetic material'.

Let us suppose that a short length of a DNA double helix has the nucleotide sequence shown in Fig. 6.1 (top), and further that the two strands can be untwisted and separated from one another to form two single chains, as in Fig. 6.1 (middle). Let each base in the single strands attach to itself the complementary deoxyribonucleotide. This would involve the same hydrogen-bonding which exists in the intact DNA double helix. Finally, if these attached mononucleotides are polymerized to form a polynucleotide chain as in Fig. 6.1 (bottom), the end result will be the formation of two complete DNA double helices identical with each other and with the original molecule. One strand of each daughter molecule will be derived from the original DNA molecule; the other will be the product of the new synthesis.

This mechanism is described as *semiconservative* to distinguish it from the other

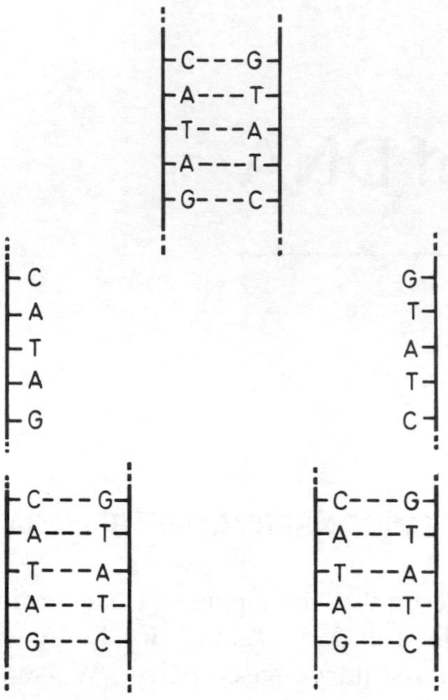

```
        -C---G-
        -A---T-
        -T---A-
        -A---T-
        -G---C-

 -C           G-
 -A           T-
 -T           A-
 -A           T-
 -G           C-

-C---G-      -C---G-
-A---T-      -A---T-
-T---A-      -T---A-
-A---T-      -A---T-
-G---C-      -G---C-
```

Fig. 6.1 Schematic illustration of the separation of the two strands of a portion of DNA with the formation of a new strand on each.

possible mechanisms. In the *conservative* mechanism the two strands do not come apart but act together as a template to form a completely new double-helical molecule; in this case one daughter molecule would be

wholly new and the other totally derived from the parent. In the *dispersive* mechanism the parental molecule is partly degraded and the fragments are incorporated into two new daughter double helices. These possibilities are illustrated in Fig. 6.2.

Convincing evidence for semiconservative replication of DNA was obtained by Meselson and Stahl [2] who grew *E. coli* in a medium containing $^{15}NH_4Cl$ of 96.5% isotopic purity for fourteen generations so as to label the DNA very heavily with ^{15}N. The cells were then transferred to a medium containing $^{14}NH_4Cl$ and samples of bacteria were withdrawn at intervals for several generations. Each sample was lysed by means of sodium dodecyl sulphate and was centrifuged in a concentrated solution of caesium chloride at $140\,000\,g$ for 20 hours to enable the DNA to attain sedimentation equilibrium. The bands of DNA were found in the CsCl gradient in the region of density $1.71\,g\,cm^{-3}$ and were well isolated from all other macromolecular components of the bacterial lysate. Ultraviolet absorption photographs taken during the course of the run revealed the position of the DNA bands.

At the start of the experiment the DNA appeared as one single band corresponding to the heavy ^{15}N-labelled nucleic acid (Fig.

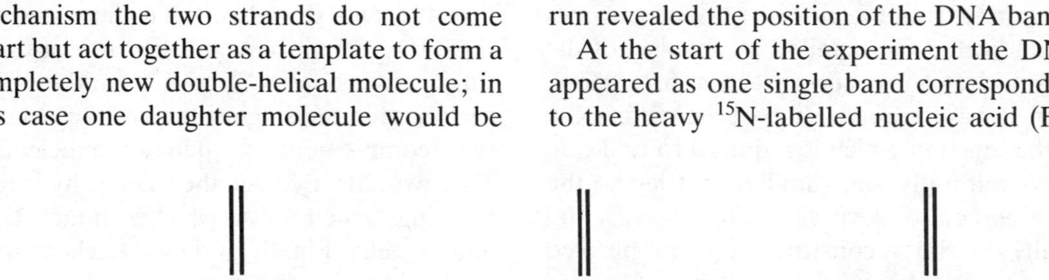

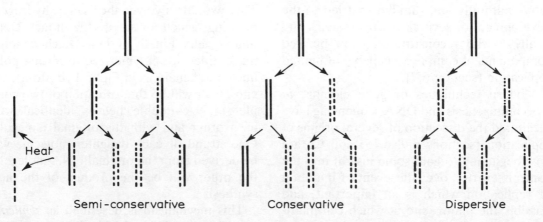

Heat

Semi-conservative　　　　　Conservative　　　　　Dispersive

Fig. 6.2 Possible mechanisms of replication. The dotted lines represent the daughter strands.

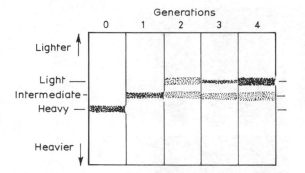

Fig. 6.3 The pattern of results from the Meselson and Stahl experiment.

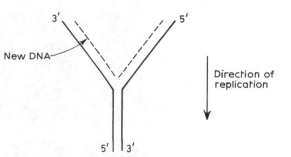

Fig. 6.4 A replication fork.

6.3). Macromolecules containing half this quantity of ^{15}N then began to appear, and one generation time after the addition of ^{14}N these hybrid (light–heavy, ^{14}N–^{15}N) molecules alone were present. Subsequently a mixture of light–heavy (^{14}N–^{15}N) DNA and unlabelled (light–light, ^{14}N only) DNA was found. When two generation times had elapsed after the addition of ^{14}N, the half-labelled and unlabelled DNA molecules were present in equal amounts (Fig. 6.3). During subsequent generations the un-labelled DNA accumulated. Moreover, when the hybrid ^{14}N–^{15}N molecules were heated they separated to give a ^{14}N strand and a ^{15}N strand.

Such experiments indicate that in DNA synthesis each existing DNA molecule is split into two subunits (Fig. 6.2), each subunit going to a different daughter molecule. The other subunit of each daughter molecule is the product of new synthesis. These subunits do not undergo any fragmentation but remain intact for many generations.

Experiments at the chromosomal level also support the view that DNA replication is semiconservative. When the replication of the *E. coli* chromosome is followed by radioautography after labelling of the DNA with ^{3}H, the amount of ^{3}H per unit length of

newly synthesized DNA is consistent with the presence of only one newly synthesized strand in a daughter chromosome [3].

In plants to which [^{3}H]thymidine has been given as a specific precursor of DNA during the period of DNA synthesis, both the daughter chromatids are found to be labelled at the time of cell division. However, at the next round of duplication after withdrawal of the [^{3}H]thymidine, these two chromosomes each produce one labelled and one unlabelled chromatid as would be expected [4].

6.3 THE REPLICATION FORK

6.3.1 Discontinuous synthesis

In the above section we have shown the two parental DNA strands separating from one another before the daughter strands are synthesized. However, in a long DNA molecule replication only occurs over one short stretch at a time and the two parental strands separate only at the point of replication to produce a Y-shaped molecule as the replication fork passes along the DNA (Fig. 6.4).

Replication forks can be seen in electron micrographs when they appear as bubbles in replicating DNA molecules (Fig. 6.5).

As the two DNA chains are antiparallel (i.e. one is running 5′ → 3′ and other 3′ →

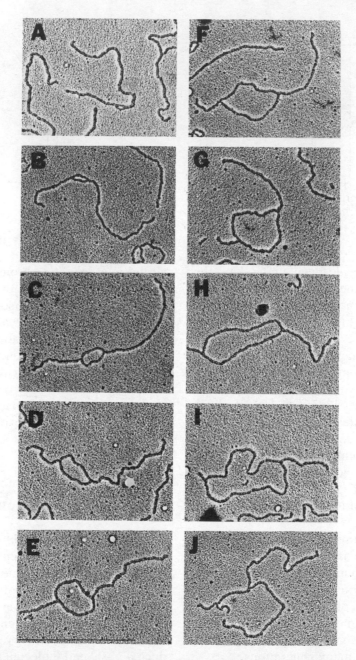

Fig. 6.5 Replicating SV40 DNA molecules. The molecules have been cut at a unique site with the restriction endonuclease *Eco*R I and have been arranged in increasing degree of replication (A to J) and oriented with the short branch at the left. (Reproduced with permission from Fareed, Garon and Salzman [5].)

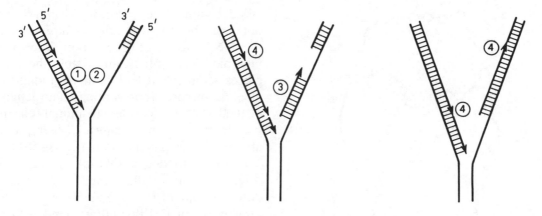

Fig. 6.6 A working model of DNA synthesis. Synthesis occurs in the $5' \rightarrow 3'$ direction down the left-hand strand of the parental DNA molecule (1). This synthesis is continuous and it exposes the right-hand strand of the parental DNA molecule. This single-stranded region (2) is stabilized by association with DNA-binding proteins. At some point, synthesis starts on the right-hand strand (3) which is copied backwards (i.e. $5' \rightarrow 3'$) until the growing strand has filled the gap. DNA ligase then joins the newly synthesized pieces (4).

$5'$) this poses the problem as to whether the two daughter strands are synthesized by two different mechanisms. One daughter DNA chain grows in the $5' \rightarrow 3'$ direction and this occurs by the addition of an incoming deoxyribonucleotide to the 3'-hydroxyl at the end of the growing *primer* chain. This is the reaction catalysed by DNA polymerase and considered in more detail in section 6.4.2. However, the other strand would appear to grow by addition to its 5'-end, which is not the same reaction. It was suggested that this may involve the use of deoxyribonucleoside 3'-triphosphates. Alternatively, some DNA molecules may have terminal 5'-triphosphates which could attack the 3'-OH of an incoming nucleotide. Such termini have not been found and, moreover, addition of nucleotides to a 5'-triphosphate-terminated primer is incompatible with the proofreading function of DNA polymerase (sections 6.4.3(f) and 6.6). This function requires the removal of a mismatched terminal residue. If this residue held the energy for the poly-merization, i.e. if it were a triphosphate, then this energy would be dissipated on proofreading and polymerization would halt.

Other evidence argues against the extension of DNA chains at the 5'-ends.

1. Many DNA polymerases have been investigated and they all add deoxyribonucleoside 5'-phosphates onto a 3'-OH group on the growing DNA chain, (i.e. the primer).
2. No enzymes have been found which will generate or polymerize deoxyribonucleoside 3'-triphosphates, and neither have these putative substrates been found in cells.
3. Both daughter strands have been shown to grow in the $5' \rightarrow 3'$ direction.

The alternative explanation put forward by Okazaki *et al.* [6] is that both chains are synthesized by the same mechanism but that one is made 'backwards' in short pieces which are subsequently joined together by DNA ligase (Fig. 6.6). Careful investigation

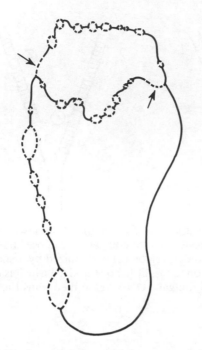

Fig. 6.7 A partial denaturation map of the replicating chromosome of phage lambda. Broken lines represent single-stranded regions of DNA and the arrows indicate such regions occurring at the replication fork. (Drawn from the electron micrographs of Schnös and Inman [7].)

of electron micrographs (which give a resolution of about 100 nucleotides) shows that on one side of a replication fork the DNA is single-stranded (Fig. 6.7) and this supports the '*discontinuous*' mechanism. In this model the strand of DNA which is made continuously in the direction of fork movement is called the *leading strand* whereas that made discontinuously in the reverse mode is called the *lagging strand*.

6.3.2 Okazaki pieces

The major support for discontinuous synthesis stems from the work of Okazaki with *E. coli* and bacteriophage T4. He showed that the most recently synthesized DNA, i.e. that

labelled with a brief pulse of tritiated thymidine, can be isolated, *after denaturation* of the DNA, as short pieces now known as Okazaki pieces [6]. In this system the short pieces sediment at 8–10S on a gradient of alkaline sucrose, representing chain lengths of 1000–2000 nucleotides. In animal cells the fragments are much shorter, being only about 100–200 residues long (i.e. 4–5S) [8].

In order to detect Okazaki pieces it is essential to lower the temperature to slow down the rate of reaction and to give a *very brief* pulse of [^{3}H]thymidine. Under these conditions, considerably more than half the radioactivity is found in small pieces which led to the proposal that Okazaki pieces may be made on both sides of the replication fork. This proposal was given some support when a mutant was isolated that produced short Okazaki fragments (*sof*) only 100–200 nucleotides long [9]. It was later shown, however, that this mutant (now known at *dut*) has a defective dUTPase [10]. *dut* cells have elevated concentrations of dUTP and this competes with and is incorporated into DNA in place of dTTP. A uracil *N*-glycosylase removes the uracil and the damaged DNA is repaired by an excision repair mechanism which involves cutting the DNA chain on the 5′-side of the damage (chapter 7).

If short fragments arise in the *dut* mutant as a result of DNA repair mechanisms, is it possible that in wild-type cells some uracil is incorporated into DNA and that discontinuities are introduced into DNA during repair of DNA? Such discontinuities could be mistaken for replication intermediates. It has been calculated that in *E. coli* during DNA replication one uracil may be accidentally incorporated into DNA for every 300 thymines [11]. As the uracil is excised, this could lead to a cut in the DNA every 1200 bases; a size not too different from that of the Okazaki pieces. However, the frequency of uracil incorporation in *B. subtilis* is very low and in those *E. coli* mutants (*ung⁻*) which fail to excise it, uracil is present less than once per 5000 bases [12]. Furthermore, in double mutants lacking the uracil *N*-glycosylase (*ung⁻*) as well as the dUTPase the DNA contains significant amounts of uracil which is not excised [13] and yet normal-sized Okazaki fragments are formed showing that the majority of these fragments are not normally produced as a result of repair but are true intermediates in DNA replication [14]. Moreover Okazaki pieces accumulate in mutant *E. coli* cells deficient in DNA ligase or DNA polymerase I whether or not uracil excision occurs [15, 16] and this again implies that they are true intermediates in DNA synthesis and that these two enzymes are involved in their subsequent incorporation into high-molecular-weight DNA. (See section 6.3.4 for further discussion of this topic.)

6.3.3 Direction of chain growth

When bacteriophage T4, growing at 8°C to reduce the rate of DNA synthesis, is incubated with [¹⁴C]thymidine for 150 seconds, and for the final 6 seconds with [³H]thymidine, Okazaki pieces can be isolated which apparently contain ³H at only the 3'-end [17, 18]. This was shown by degrading the isolated Okazaki pieces with exonuclease I of *E. coli* (which degrades single-stranded DNA from the 3'-end – chapter 4) when the ³H label is released before the ¹⁴C label. The complementary experiment using a nuclease from *B. subtilis* which acts from the 5'-end causes release of much of the ¹⁴C before the ³H is rendered acid-soluble.

These experiments demonstrate that Okazaki pieces are made in the 5' → 3' direction and support the discontinuous mechanism as outlined in Fig. 6.6.

6.3.4 Initiation of Okazaki pieces

The proposal that the lagging strand is synthesized in the form of short pieces may solve one problem but it creates another. DNA polymerase requires a primer, so how are these Okazaki pieces initiated?

In wild-type *E. coli*, rifampicin (an inhibitor of *E. coli* RNA polymerase) inhibits replication of phage M13 and certain plasmids. Replication is normal in mutants with a rifampicin-resistant RNA polymerase [19]. Moreover, some RNA synthesis has been shown to be essential for the conversion of M13 single strands into the duplex form *in vitro* [20]. Such evidence, considered with the fact that all DNA polymerases require a 3'-hydroxyl priming end, suggested that this end may be provided by an olig*oribo*nucleotide. That rifampicin does not inhibit continued replication of *E. coli* DNA (it does block the initiation of new rounds [21] – section 6.10.1) or the replication of other phages such as φX174 was interpreted to mean that in these cases a second RNA polymerase (resistant to the drug) was involved in the synthesis of the priming oligoribonucleotides. Indeed, the product of the *dnaG* gene of *E. coli* is believed to be such an enzyme and this and similar enzymes are called primases (section 6.5.5).

Okazaki fragments containing oligoribonucleotides at their 5'-end have now been isolated from both prokaryotic and eukaryotic systems. The experimental proof of their existence was initially subject to much controversy because non-covalent RNA–DNA interactions gave some spurious results [22]. However, methods to detect RNA covalently linked to the 5'-end of DNA molecules [23] have shown that, although not all Okazaki pieces have ribonucleotides at their 5'-end, the shortest ones do (see below).

A review of the methods used to detect primers was published in 1976 [24] and some

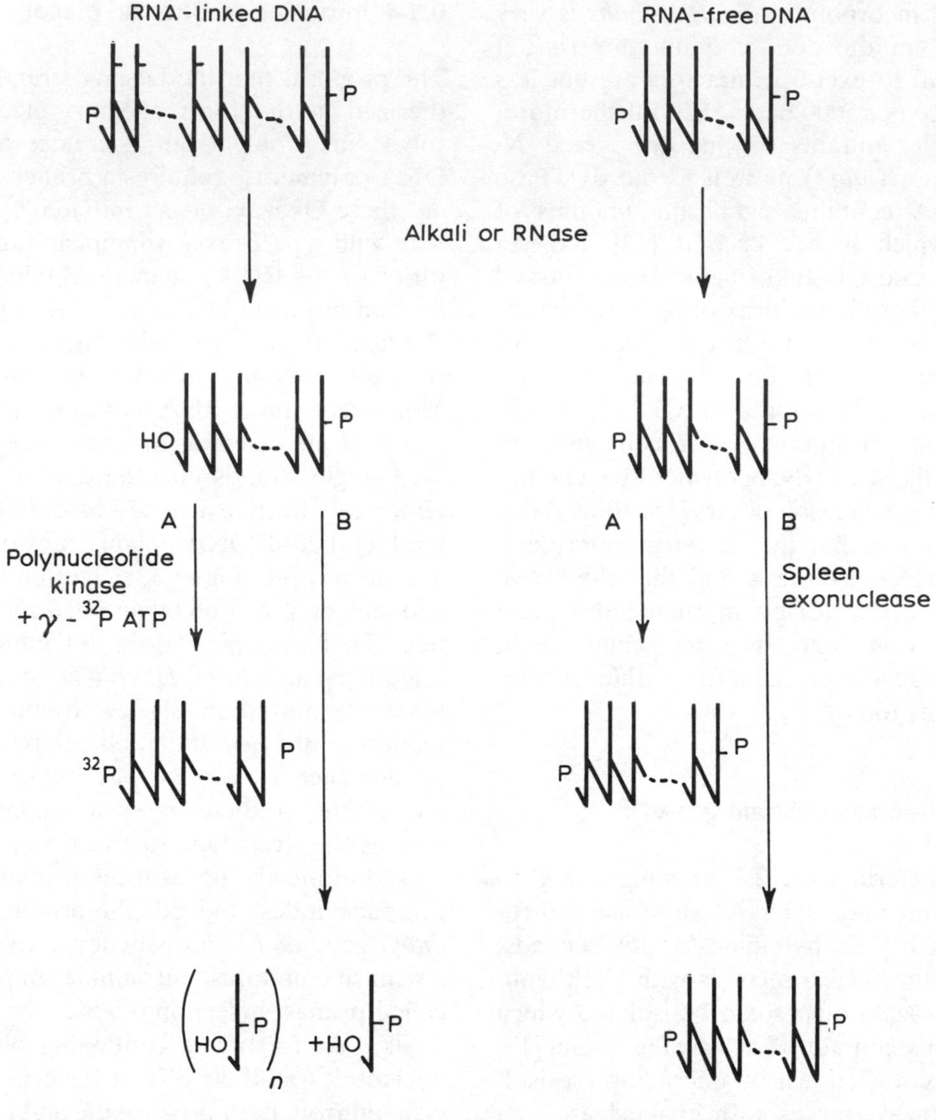

Fig. 6.8 Methods to detect RNA-linked Okazaki pieces produced *in vivo* [22, 23]. The isolated Okazaki pieces are first treated with polynucleotide kinase and non-radioactive ATP to ensure that all 5′-ends are phosphorylated. Subsequent treatment with alkali or ribonuclease removes any RNA from the 5′-end and leaves a 5′-hydroxyl group. This group can be detected using polynucleotide kinase and $[\gamma\text{-}^{32}P]$ATP (Method A). However, this method is not selective for nascent DNA. Method B overcomes this disadvantage by starting with ^{3}H-labelled Okazaki pieces resulting from a pulse labelling experiment. Spleen exonuclease is used under conditions where only the 5′-OH terminated DNA (i.e. that initially linked to RNA) is degraded. These methods have been criticized on the grounds that treatment with alkali of DNA cleaved during repair (chapter 7) may also yield 5′-OH groups [25, 26].

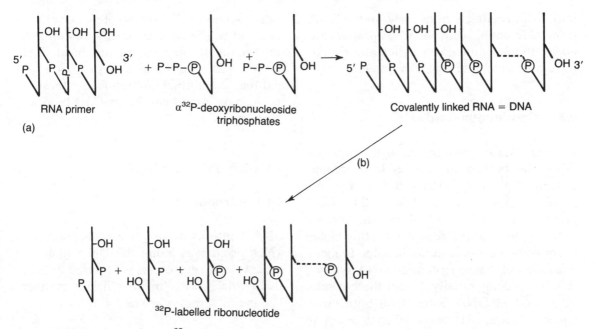

Fig. 6.9 The formation of ^{32}P-labelled ribonucleotide is indicative of the occurrence of a covalently linked RNA–DNA molecule where the radioactive phosphate forms the bridge between the two nucleic acids. (a) Extension of RNA primer; (b) Alkaline hydrolysis of product; Ⓟ = radioactive phosphate.

of these are described in Figs 6.8 and 6.9. The use of toluenized *E. coli* enables [α-^{32}P]deoxyribonucleoside triphosphates to penetrate the cell membrane, when some transfer of radioactivity to a ribonucleotide occurs [27]. This is definite evidence for a covalent attachment of RNA to DNA (Fig. 6.9). When [α-^{32}P]deoxyribonucleoside triphosphates are injected into the slime mould *Physarum polycephalum* [28] there is similar transfer of the ^{32}P to ribonucleotides.

RNA-linked Okazaki pieces accumulate in mutant *E. coli* cells deficient in either the polymerase or the $5' \rightarrow 3'$ exonuclease of DNA polymerase I (section 6.4.3). However, even in such mutant cells, RNA primers were detected on only about 30% of the Okazaki pieces [29] and in a wild-type *E. coli* cell only 3–6 of the total of 20–40 chains in the size range 3000–9000 bases chase into high-molecular-weight DNA [30]. In animal

cells about 40% of Okazaki pieces have RNA primers [31, 32]. These results can be partly explained if removal of primers is a very efficient mechanism. However, a full explanation requires that some of the small fragments arise by repair mechanisms.

Although RNA is not normally found in mature DNA, an exception is the DNA of the plasmid Col E1 grown in the presence of chloramphenicol. In this case it is inferred that the drug prevents the normal excision of the RNA primers [33]. Mitochondrial DNA also frequently has a number of ribonucleotides left in the mature duplex [34].

When precautions are taken to eliminate all but covalent attachment of RNA to Okazaki pieces the size of the RNA segment obtained is very short; only one to three nucleotides long, and the RNA does not have a particular sequence [35]. This contrasts strongly with the 50–100 nucleotides initially attributed (erroneously) to the RNA primer. In animal cells the RNA primers are about 10 nucleotides long. They also do not have a specific base sequence [8, 36].

6.3.5 Continuous synthesis

Discontinuous synthesis of one strand of DNA was postulated in order to overcome certain problems (section 6.3.1), but no problems appear to exist with the chain growing in the $5' \rightarrow 3'$ direction, and so there is no *a priori* reason why this chain should not be made continuously. Indeed, this may well be the case when DNA synthesis is proceeding rapidly. Under these conditions nascent DNA is found in both large and small pieces. However, when deoxyribonucleotides are limiting and the rate of chain elongation slows, both daughter strands appear to be made discontinuously. One possible explanation is that there is competition between the propagation of the growing chain and the initiation of new chains [37].

Any initiation of an Okazaki piece on the leading strand is unlikely to be far in advance of the growing strand and so ligation will occur very rapidly, most probably before the Okazaki piece has grown to a detectable length. This is in contrast to the synthesis of an Okazaki piece on the lagging strand which (a) cannot be ligated until its synthesis is completed and (b) cannot be initiated until a region of single-stranded DNA of comparable length has been exposed on the lagging strand. Under conditions where primer removal and ligation is restricted (i.e. in the absence of a functional DNA polymerase I) Okazaki pieces are formed on both lagging and leading strands of the phage P2 [38]. However, under non-restrictive conditions discontinuities are not detected on the leading strand [39].

Studies *in vitro* using the cellophane disc assay (section 6.7), or isolated nuclei [40] have led to the conclusion that discontinuities on the leading strand are most likely produced as a result of uracil excision [41] and that the normal continuous synthesis of the leading strand can be masked by uracil excision.

6.4 DNA POLYMERASE

6.4.1 Introduction

DNA synthesis occurs at a replication fork which progresses along the DNA molecule and a variety of proteins is required to bring about this process in an efficient manner. The major proteins are as follows.

1. A DNA polymerase is required to add dNTPs to the growing, leading strand and to growing Okazaki pieces on the lagging strand. In general DNA polymerases act alongside other proteins which increase their fidelity, and processivity.
2. A primase is required to help initiate Okazaki pieces.
3. An exonuclease is required to remove the primers.
4. A ligase is required to join the Okazaki pieces together.
5. A helicase is required to help unwind the duplex DNA at the replication fork.
6. A topoisomerase is required to help the helicase unwind the DNA and also to relax the tension engendered by unwinding the duplex DNA.
7. A binding protein is required to stabilize single-stranded DNA exposed by progression of the leading strand before initiation of Okazaki pieces.

The action of these proteins is summarized in Fig. 6.10. The remainder of this section is devoted to the biochemistry of DNA poly-

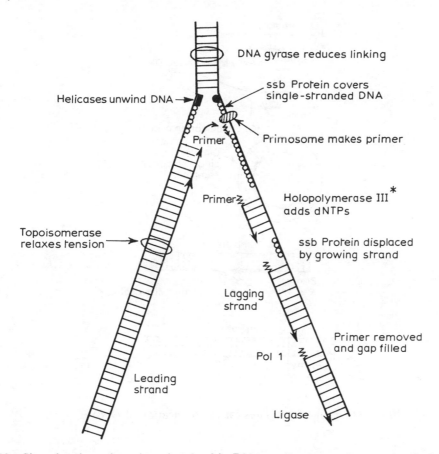

Fig. 6.10 Site of action of enzymes involved in DNA replication at the growing fork in *E. coli*.

merases, and the other proteins are considered in section 6.5.

6.4.2 Mechanism of action of DNA polymerase

DNA polymerase is the name given to an enzyme which catalyses the synthesis of DNA from its deoxyribonucleotide precursors. Such enzymes have been purified from various sources (bacterial, plant and animal) but, apart from some details discussed below, the mechanism of the polymerization reaction is common to all preparations.

DNA polymerase catalyses the formation of a phosphodiester bond between the 3'-hydroxyl group at the growing end of a DNA chain (the *primer*) and the 5'-phosphate group of the incoming deoxyribonucleoside triphosphate. Growth is in the $5' \rightarrow 3'$ direction, and the order in which the deoxyribonucleotides are added is dictated by base-pairing to a *template* DNA chain. Thus, as well as the four triphosphates and Mg^{2+} ions, the enzyme requires both primer and template DNA. No DNA polymerase has been found which is able to initiate DNA chains.

The simplest case to consider is that in which a single-stranded template has bound to it a growing strand of primer terminating

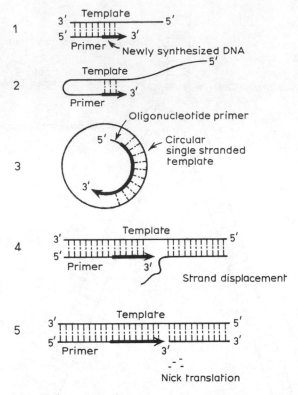

Fig. 6.11 Primer/templates for the replication of DNA.

at the growing point in a 3'-hydroxyl group (Fig. 6.11(1)).

The polymerase binds to the single-stranded template in the region of the 3'-hydroxyl end of the primer (Fig. 6.12). An incoming deoxyribonucleoside triphosphate containing a base which can pair with the corresponding base on the template becomes attached to the triphosphate binding site. The polymerase then catalyses a nucleophilic attack by the 3'-hydroxyl group of the primer on the α-phosphate of the deoxyribonucleoside 5'-triphosphate (Fig. 6.13). Inorganic pyrophosphate is released, a phosphodiester bond is formed, and the chain is lengthened by one unit. The enzyme moves along the template by the distance of one unit and the newly added nucleotide with its 3'-hydroxyl

group now occupies the primer terminus site. The process is then repeated until the enzyme reaches the end of the template strand. *In vivo* the pyrophosphate is hydrolysed by a pyrophosphatase which drives the equilibrium in favour of DNA synthesis.

In the replication of linear, single-stranded DNA the 3'-hydroxyl terminus may loop back on itself and serve as a priming strand as shown in Fig. 6.11(2), or short lengths of oligonucleotide may act as primers by becoming hydrogen-bonded to the template. When a cyclic, single-stranded DNA is used as template, replication can occur only in the presence of short oligonucleotides which become attached by base-pairing to the template and serve as primers (Fig. 6.11(3)).

DNA polymerase cannot use double-

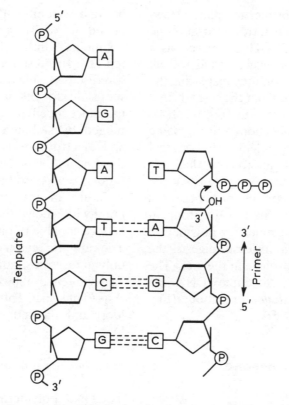

Fig. 6.12 Greater detail of the mechanism shown in Fig. 6.11 (1).

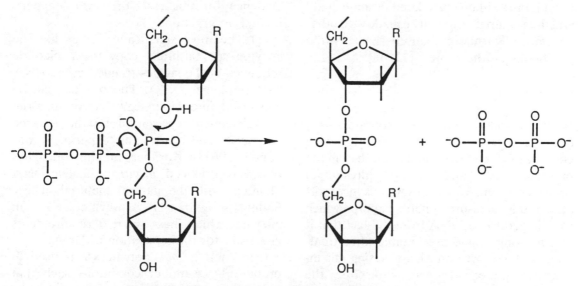

Fig. 6.13 The mechanism of action of DNA polymerase.

stranded DNA as a primer/template. However, if a nick or gap is introduced into one strand, the exposed 3'-OH group acts as a primer and the new growing strand will fill the gap and then may continue displacing the complementary strand from the 5'-end. In a strand displacement reaction (Fig. 6.11(4)) this can lead to complications since at some point the enzyme may leave the original template strand and begin to copy the complementary strand so that *in vitro* a branched structure is formed. In the presence of a 5' → 3' exonuclease, the 3'-OH group at the nick can act as a primer for polymerase action while the exonuclease digests the DNA from the 5'-phosphate at the nick. The nick thus moves along the duplex DNA in a so-called *nick translation* reaction (Fig. 6.11(5), section 6.4.3(e)).

6.4.3 *E. coli* DNA polymerase I (EC 3.7.7.7)

Since a great amount of effort has been put into the study of DNA polymerase I from *E. coli* (the Kornberg enzyme) this archetypal polymerase will be considered in some detail and then comparisons will be made with other enzymes. Kornberg himself has written two books on the subject [42, 43].

(a) *Historical*

In 1956 Kornberg and his collaborators used cell-free extracts from exponentially growing cultures of *E. coli* to demonstrate that, in the presence of ATP and Mg^{2+}, [^{14}C]thymidine, a specific precursor for DNA, was incorporated into an acid-precipitable material which was judged to be DNA from the fact that it was no longer acid-precipitable after treatment with deoxyribonuclease. Using this incorporation method for assay purposes, the bacterial extract was fractionated and it soon

became evident that crude extracts contained a kinase system which converted thymidine into thymidine triphosphate (dTTP). In subsequent work, labelled dTTP was used as substrate and the enzyme responsible for its incorporation into DNA has since been purified to homogeneity as judged by sedimentation, chromatography and electrophoresis. The gene coding for the enzyme has been sequenced and the sequence was used to predict the secondary structure of the protein. X-ray crystallography has shown the enzyme to have a deep crevice capable of binding duplex DNA [44]. The catalytic region of many DNA and RNA polymerases contains a conserved amino acid sequence, the core of which is Tyr-Gly-Asp-(Thr)-Asp. This is present in an exposed loop and flanked by hydrophobic regions [45].

(b) *Evidence for the copying of the template*

The DNA polymerase is an enzyme with the unusual property of taking directions from a template and faithfully reproducing the sequence of nucleotides in the product. The evidence for this rests on several experimental observations.

(1) The most significant fact is that the enzyme can faithfully copy the nucleotide sequence of the single-stranded cyclic DNA bacteriophage ϕX174. The product can be converted into biologically active material by subsequent cyclization by the enzyme polynucleotide ligase. This experiment was done in 1967 by Kornberg and his colleagues [46] who produced, *in vitro*, infective DNA identical with the natural material. These findings show that DNA polymerase I is an enzyme which has all the requirements necessary for the replication of DNA.

(2) When *E. coli* polymerase is used to prepare DNA under conditions such that only 5% of the sample produced comes from

the template, the product has many of the same physical and chemical properties as DNA isolated from natural sources. The product appears to have a hydrogen-bonded structure similar to that of natural DNA and undergoes molecular melting (chapter 2) in the same way. It shows the same equivalence of adenine to thymine and guanine to cytosine that characterizes natural DNA. Moreover the characteristic ratio of A:T pairs to G:C pairs of a given DNA primer is imposed on the product whether the net DNA increase is 1% or 1000%. The base ratios in the product are not distorted when widely differing molar proportions of substrate are used. This is best illustrated by the use of the synthetic template poly(dA-dT) when, from a mixture of all four deoxyribonucleoside triphosphates, only dAMP and dTMP are incorporated into the product.

(3) The nucleotide sequence in the template DNA is reproduced in the product. Although this can now be established by direct sequencing methods, it was established by the Kornberg group using the technique of *nearest-neighbour sequence analysis* [47]. The partially purified *E. coli* enzyme is incubated with a particular template DNA and all four deoxyribonucleoside triphosphates, one of which, say dATP, is labelled with ^{32}P in the innermost (α) phosphate. During the synthetic reaction this ^{32}P becomes the bridge between the nucleoside of that labelled triphosphate (A) and the nearest-neighbour nucleotide containing the base Z at the growing end of the polynucleotide chain (Fig. 6.14). After the synthetic reaction is complete the DNA is isolated and degraded with micrococcal nuclease and spleen phosphodiesterase (chapter 4) to yield the deoxyribonucleoside 3'-monophosphates. The ^{32}P is thus transferred to the 3'-carbon of the neighbouring nucleotide in the chain (Z in Fig. 6.14), i.e. the one with which the labelled triphosphate has reacted

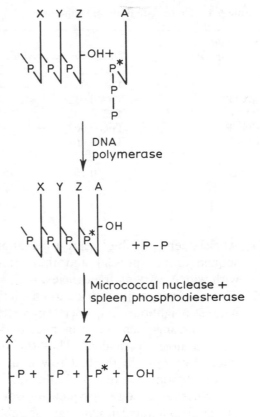

Fig. 6.14 Illustration of the method of nearest-neighbour sequence analysis.

(Z might be any one of the four bases). The four deoxyribonucleoside 3'-monophosphates are isolated by paper electrophoresis or HPLC, and their radioactivities measured to give the relative frequency with which the originally labelled nucleotide locates itself next to other nucleotides in the new chain. This procedure is carried out four times with a different labelled triphosphate each time so as to determine the relative frequencies of all sixteen possible nearest-neighbour (or dinucleotide) sequences [47]. The results of such an experiment with DNA from *Mycobacterium phlei* as primer are shown in Table 6.1.

They illustrate several points.

Table 6.1 Nearest-neighbour frequencies of *Mycobacterium phlei* DNA [48]

Labelled triphosphate	*Deoxyribonucleoside 3'-phosphate isolated*			
	Tp	Ap	Cp	Gp
dATP	TpA 0.012	ApA 0.024	CpA 0.063	GpA 0.065
dTTP	TpT 0.026	ApT 0.031	CpT 0.045	GpT 0.060
dGTP	TpG 0.063	ApG 0.045	CpG 0.139	GpG 0.090
dCTP	TpC 0.061	ApC 0.064	CpC 0.090	GpC 0.122
Sums	0.162	0.164	0.337	0.337

1. All sixteen possible nearest-neighbour sequences are present and they occur with widely varying frequencies.

2. There is a very striking deviation from the nearest-neighbour frequencies predicted if the arrangement of mononucleotides were completely random. Thus the frequency of TpA in the first row is quite different from that of ApT in the second row whereas these two frequencies would have to be identical in a random assembly. The nucleotides have therefore been assembled in accordance with a definite pattern.

3. The sums of the four columns show the equivalence of A to T and of G to C in the product and indicate both the validity of the analytical method and the replication of the overall composition of the substrate DNA.

4. The results indicate that base-pairing occurs in the newly synthesized DNA and that its two strands are of opposite polarity. According to the Watson–Crick model the two strands of the double helix are of opposite polarity, and it is presumed that each can act as a template for the formation of a new chain so as to give precise replication with the formation of two daughter helices identical with each other and with the parent helix. The re-

sults of nearest-neighbour sequence analysis support this mechanism. Naturally, the frequencies of ApA and TpT sequences are equivalent, and so are the frequencies of GpG and CpC. However, the matching of the other sequences depends on whether the strands of the double helix are of similar or opposite polarity.

If the strands are of opposite polarity the following matching sequences can be predicted:

CpA and TpG

GpA and TpC

CpT and ApG

GpT and ApC

whereas, if the strands are of the same polarity, the matching sequences would be:

TpA and ApT

CpA and GpT

GpA and CpT

TpG and ApC

ApG and TpC

CpG and GpC

The results in Table 6.1 favour the helix with strands of opposite polarity. (The ApA and TpT, and the CpC and GpG sequences match in both models.)

5. The nearest-neighbour frequencies measured by the method described above are those of the newly synthesized DNA. To verify that they are an accurate reflection of those in the original DNA template, an enzymically synthesized sample of calf thymus DNA in which only 5% of the total DNA consisted of the original template was itself used as template in a sequence analysis. The results showed good agreement between the sequence frequencies of the products primed by native DNA and by enzymically produced DNA whereas DNAs from other sources gave quite different results.

It can be concluded therefore that the polymerase yields a DNA product with strands of opposite polarity and that the sequence of bases is faithfully reproduced.

(c) *The structure and action of the enzyme*

The purified enzyme is a protein of M_r 109 000 in the form of a single polypeptide chain [49]. This chain can be unfolded in guanidine–HCl–mercaptoethanol so as to denature the protein. When the reagent is diluted out, renaturation occurs with restoration of activity. The protein migrates as a single band on SDS/polyacrylamide gel electrophoresis; it contains only one sulphydryl group and one disulphide group; the residue at the *N*-terminus is methionine.

As well as DNA polymerase activity, the purified enzyme shows several other activities each of which is associated with the same enzyme molecule. Thus exonuclease activity (both $3' \rightarrow 5'$ and $5' \rightarrow 3'$) is purified along with the polymerase activity in a ratio which is not altered by fractionation.

On the basis of binding experiments Kornberg has concluded that the enzyme is able to catalyse the following operations [43].

1. Extension of a DNA chain in the $5' \rightarrow 3'$ direction by the addition to the 3'-hydroxyl terminus of mononucleotides from deoxyribonucleoside triphosphates at the rate of 1000 nucleotides per minute.
2. Hydrolysis of a DNA chain from the 3'-hydroxyl end in the $3' \rightarrow 5'$ direction to yield 5'-monophosphates (the exonuclease II action referred to in chapter 4).
3. Hydrolysis of DNA chain from the 5'-phosphate (or 5'-hydroxyl) terminus in the $5' \rightarrow 3'$ direction to yield mainly 5'-monophosphates.
4. Pyrophosphorolysis of a DNA chain from the 3'-end; this is essentially the reversal of the polymerization reaction.
5. Exchange of inorganic pyrophosphate with the terminal pyrophosphate group of a deoxyribonucleoside triphosphate as a result of alternating polymerization and pyrophosphorolysis.

Reactions 4 and 5 are of doubtful significance *in vivo* since they require a high concentration of inorganic pyrophosphate.

Treatment of the enzyme with the protease *subtilisin* from *B. subtilis* breaks it into two fragments, a larger fragment, known as the Klenow fragment, of M_r 76 000 which retains polymerase activity and $3' \rightarrow 5'$ nuclease activity but not the $5' \rightarrow 3'$ nuclease activity, and a smaller fragment of M_r 34 000 which retains nuclease $5' \rightarrow 3'$ activity in the presence of DNA.

The active centre of the enzyme must comprise at least five major sites.

1. A site for the binding of the single-stranded template chain to which a primer is base paired.
2. A site for the special recognition of the terminal 3'-hydroxyl group of the primer, at which incoming nucleotides are added.

3. A second site for the terminal 3′-hydroxyl group of the primer when it is mismatched (see below). This point is the start of the 3′ → 5′ hydrolytic cleavage.
4. A triphosphate binding site for which all four triphosphates compete.
5. A distant site which allows for the 5′ → 3′ cleavage of a 5′-phosphoryl-terminated chain. It is presumably this area that is broken off by subtilisin.

These sites determine the nature of the DNA which can bind to the enzyme. For example, linear single-stranded DNA binds readily on site (1) whereas an intact linear duplex such as the DNA of bacteriophage T7 does not bind if it has been prepared with great care so as to avoid internal breaks. An intact circular duplex such as plasmid DNA or ϕX174 replicative form DNA does not bind to the enzyme until a 'nick' has been introduced in one of the strands by an appropriate nuclease yielding 3′-hydroxyl and 5′-phosphate termini. Such nicks are active points for replication whereas nicks introduced by micrococcal nuclease with 5′-hydroxyl and 3′-phosphoryl termini are not replication points although they bind the enzyme. One molecule of enzyme is bound at each nick in either case.

Normally DNA polymerase I is highly processive. For each DNA/enzyme association event, hundreds of nucleotides are polymerized before dissociation occurs [53]. However, a mutant of DNA polymerase I (PolA5) shows all the normal properties of the wild-type enzyme but has a reduced rate of polymerization caused by a lowered affinity of the enzyme for the DNA primer/template [54].

(d) The 3′ → 5′ exonuclease and proofreading

Before it was realized that this was an integral function of the polymerase, the 3′ → 5′ exonuclease activity was ascribed to an exonuclease II. It appears that this exonuclease functions to recognize and eliminate a non-base-paired terminus on the primer DNA [55]. When *E. coli* DNA polymerase I is provided with four triphosphates and a primer/template with a mismatched end, the non-matching terminus is removed by the 3′ → 5′ exonuclease before polymerization begins. (In the absence of the triphosphates the exonuclease continues to remove nucleotides from the frayed ends of the DNA molecules [56].)

Evidence has been obtained that there are two distinct sites, separated by 3 nm, which recognize the 3′-OH group of the primer [50–52]. When the terminal nucleotide is correctly matched to the template the 3′-OH group resides in the polymerase site. However, if the terminal nucleotide is not correctly matched to the template the 3′-OH group is displaced into a site on the same or another enzyme molecule, where it is susceptible to the 3′ → 5′ exonuclease.

Such a nuclease, by correcting errors occurring on polymerization, may be expected to increase dramatically the fidelity of the base-pairing mechanism [42, 57] (section 6.6). Certain *E. coli* mutants show an increased or decreased rate of mutation which can be correlated with a change in the relative activities of the polymerase and 3′ → 5′ activities of DNA polymerase I. In *E. coli* infected with phage T4 (which codes for its own DNA polymerase/exonuclease; the latter residing towards the *N*-terminus of the molecule [58]) there are *mutator* and *antimutator* polymerases which show (a) decreased or increased rates of the pyrophosphate exchange reaction and (b) increased or decreased rates of incorporation of non-complementary nucleotides, both of which have been correlated with changes in 3′ → 5′ exonuclease activities [59, 60]. This is considered in more detail in section 6.6.

This proofreading mechanism provides a justification for involving 5'-deoxyribo-nucleoside triphosphates in the 5' → 3' extension of a 3'-OH on the growing DNA chain. Addition of nucleotides to a growing chain terminating in a triphosphate would result, during proofreading, in the removal of the energy required for further extension.

(e) The 5' → 3' exonuclease

This activity, which is present in the smaller fragment released from DNA polymerase I by subtilisin treatment, cleaves base-paired regions of DNA, releasing oligonucleotides from 5'-ends. Because of its ability to jump several bases at a time, this nuclease can act on DNA molecules containing mismatched bases or distortions which render them unsuitable as substrates for polymerase [61]. It may thus serve a function, for instance, in the elimination of thymine dimers from DNA exposed to ultraviolet radiation (chapter 7).

When *E. coli* DNA polymerase I binds to a nick on double-stranded DNA two reactions occur simultaneously. Polymerization extends the 3'-OH end and 5' → 3' exonuclease degrades the 5'-phosphate terminus. This results in *nick translation*. If DNA polymerase I is incubated with a nicked duplex DNA molecule and radioactive deoxyribonucleoside triphosphates then the nick translation reaction results in replacement of long sections of the DNA molecule with a radioactive stretch of identical sequence, a reaction of importance in recombinant DNA work (section A.4.1). Nick translation may end only when the enzyme reaches the end of the DNA molecule [62]. Alternatively nick translation may end if the 5'-terminus is released from the nick thereby rendering it no longer susceptible to the exonuclease. The branched product typical of

the *in vitro* reaction with a native DNA primer/template would then be formed.

(f) *The role of the Kornberg enzyme* in vivo

Although the DNA polymerase I is exceedingly effective in the copying of a single-stranded DNA template when provided with a primer, it is much less effective with double-stranded DNA. This observation and other considerations led to doubts as to the role of the Kornberg enzyme (DNA polymerase I) in the replication of DNA *in vivo* and to the suggestion that it is concerned merely in maintenance and repair of DNA (chapter 7). These considerations may be summarized as follows.

1. The purified enzyme cannot replicate intact double-stranded DNA semiconservatively to yield a biologically active product.
2. The purified enzyme catalyses the incorporation of 1000 nucleotides per minute per molecule of enzyme whereas the estimated rate of incorporation *in vivo* is 100 times faster.
3. Mutants of *E. coli* have been isolated which contain apparently normal Kornberg enzyme but are defective in DNA duplication. This demonstrates that other enzymes (which *may* include other polymerases) are required for DNA synthesis *in vivo*, and this may help to explain the deficiencies enumerated under (1) and (2) where only the purified Kornberg polymerase was present.

The evidence against the Kornberg enzyme being essential for replication *in vivo* is based on the properties of a mutant of *E. coli* (PolA1 or PolA$^-$) isolated by de Lucia and Cairns [63]. This mutant, and several others discovered later, multiplies normally but contains 1% or less of the Kornberg enzyme activity present in wild-type cells.

Such mutants, however, show a reduced ability to join Okazaki fragments [64] and an increased sensitivity to ultraviolet [63] and ionizing radiation [65] and to alkylating reagents [63]. This implies a role for the enzyme in 'gap' filling, and also in excision of RNA primers and mismatched base pairs. However, these functions are not completely lacking in mutants lacking DNA polymerase I and it is suggested that they may also be performed to a limited extent by other enzymes, e.g. polymerases II and III and the RecBC nuclease (chapter 7). Double mutants of PolA and RecB are not viable [66], suggesting an essential function which can be carried out by more than one enzyme.

(g) *Poly d(A-T) and poly (dG).poly (dC)*

When DNA polymerase I is incubated without DNA in the presence of dATP, dTTP and Mg^{2+} an interesting polymer is formed containing adenine and thymine nucleotides [43]. Polymer formation occurs only after a lag period of several hours and then takes place rapidly until 60–80% of the triphosphates have been utilized.

The product of the reaction contains equal amounts of adenine and thymine. Nearest-neighbour frequency analysis shows that the frequencies of ApT and TpA are each 0.5, whereas the sequences ApA and TpT are undetectable. The polymer therefore contains alternating residues of A and T.

The molecular weight calculated from the sedimentation value and reduced viscosity is between 2×10^6 and 8×10^6. The polymer melts sharply at 71°C with an increase of 37% in the absorbance at 260 nm. The process is completely reversible on cooling. Such physical data, including the X-ray diffraction pattern, suggest that the molecule is a long fibrous double-stranded structure with the strands joined by hydrogen bonds between adenine and thymine bases.

A somewhat different polymer is formed when *E. coli* DNA polymerase I is incubated with high concentration of dGTP and dCTP in the presence of Mg^{2+}. Again, a lag period of several hours is found. This product contains guanine and cytosine, not necessarily in equal amounts, and nearest-neighbour sequence analysis shows that the frequencies of GpG and CpC are each 0.5. Mild acid hydrolysis releases all the dGMP but none of the dCMP. Sedimentation and viscosity measurements yield values similar to those found with the poly[d(A-T)] but the T_m value is much higher (83°C). These observations are consistent with the view that the molecule consists of two homopolymers, one containing only guanine and the other only cytosine, hydrogen-bonded throughout their lengths.

6.4.4 *E. coli* DNA polymerase II

The discovery of the polA⁻ mutant of *E. coli* which grows well and replicates its DNA in the usual manner in spite of the absence of the Kornberg enzyme gave an impetus to the search for another enzyme apparatus which can synthesize DNA. Two further polymerases, designated DNA polymerase II and III, were found in extracts of polA⁻ cells. These enzymes had not been detected previously because, in extracts of wild-type cells, they show little activity relative to DNA polymerase I when single-stranded or nicked DNA is used as template. DNA polymerase II and III show significant activity only with a 'gapped' DNA template (Table 6.2). The enzymes can be separated from one another by chromatography on phosphocellulose, DEAE-cellulose, or DNA–agarose [43].

Purified DNA polymerase II has a M_r of 120 000 and is homogeneous as judged by SDS/polyacrylamide gel electrophoresis. It synthesizes DNA in the $5' \rightarrow 3'$ direction

Table 6.2 DNA polymerases of *E. coli*

Polymerase (gene)	M_r	Molecules per cell	Nucleotides polymerized/s (a) per enzyme molecule (b) per bacterial cell	Direction of (a) polymerization (b) exonuclease action	Template (all require 3'-OH primer)
I (*polA*)	109 000	400	(a) 16–20 (b) 8000	(a) 5' → 3' (b) 3' → 5' and 5' → 3'	Denatured, nicked or gapped
II (*polB*)	120 000	17–100	(a) 2–5 (b) 500	(a) 5' → 3' (b) 3' → 5'	Gapped
III (*polC*)	180 000	10	(a) 250–1000 (b) 10 000	(a) 5' → 3' (b) 3' → 5' and 5' → 3'	Gapped

and for maximal activity it requires all four triphosphates, Mg^{2+}, NH_4^+, and a native DNA template containing single-stranded gaps 50–200 bases long. The rate of reaction falls off with longer gaps, but may be restored by addition of *E. coli* single-stranded DNA-binding protein (section 6.5.2). The enzyme also requires a 3'-OH primer. It is insensitive to sulphydryl reagents and is not affected by antiserum to DNA polymerase I. The purified enzyme, however, like polymerase I, only synthesizes DNA at rates a fraction of those found *in vivo*. The enzyme also possesses 3' → 5' exonuclease activity, but no 5' → 3' exonuclease activity [67].

The function of *E. coli* DNA polymerase II *in vivo* is unknown and mutants lacking the enzyme appear normal in all respects. However, double mutants lacking both polymerase I and polymerase II join Okazaki fragments even more slowly than do mutants lacking polymerase I alone [68].

6.4.5 *E. coli* DNA polymerase III

E. coli with a temperature-sensitive mutation in the gene for DNA polymerase III (*dnaE* or *polC*) are not viable at the restric-

tive temperature [69] and lysates prepared from them are defective in DNA synthesis [67] (section 6.7). Complementation of such lysates with DNA polymerase III purified from normal cells restores their DNA synthetic ability. This is strong evidence that, unlike DNA polymerase I and II, polymerase III is essential for DNA synthesis.

Although there are only about 10 molecules of DNA polymerase III per bacterial cell, its high rate of polymerization of nucleotides (Table 6.2) shows it to be capable of its proposed role in DNA synthesis. The best template for DNA polymerase III is double-stranded DNA with many small gaps containing 3'-OH priming ends.

DNA polymerase III exists in a complex holoenzyme form in the cell and it has proved difficult to determine the subunit composition of the enzyme. The α subunit (that coded for the *dnaE* or *polC* gene) of the holoenzyme has a M_r of 132 000 [43, 70–72] and in the DNA polymerase III core enzyme it is tightly associated with two small subunits: ε (M_r 27 000) and θ (M_r 10 000) [73] (Table 6.3). The core enzyme has both 3' → 5' exonuclease (which could be involved in proofreading) and 5' → 3' exonuclease activities, although the latter is only

Table 6.3 Subunit composition of *E. coli* DNA polymerase III holoenzyme [1–4, 54, 480]

Subunit	Gene	Size (kDa)	Name
α	*polC*	132	polymerase III
ε	*dnaQ*	27	$3' \rightarrow 5'$ exonuclease
θ		10	
τ	*dnaXZ*	71	
γ	–	52	EF II
δ	–	35	EF III?
δ'	–	33	
χ	–	15	
φ	–	12	
β	*dnaN*	37	EF I, copol III*

Bracket groupings: α, ε, θ = polymerase III core enzyme; core enzyme plus τ = pol III′; γ, δ, δ', χ, φ = γ complex; pol III′ plus γ complex = pol III*; pol III* plus β = DNA pol III holoenzyme.

manifest *in vitro* on duplex DNA with a single-stranded 5′-tail. For this reason the core enzyme cannot use a nicked primer/template but requires a gap. The $3' \rightarrow 5'$ exonuclease activity resides in the e subunit (the product of the *dnaQ* or *mutD* gene) and its role in proofreading will be considered in section 6.6.

DNA polymerase III core enzyme cannot use long single-stranded DNA molecules as template even when provided with a primer. It requires the addition of up to seven auxiliary subunits to increase its efficiency and to endow it with processivity, i.e. the ability to add nucleotides continuously without dissociation from the growing primer chain. In addition to a high degree of processivity required for rapid continuous DNA synthesis on the leading strand of the replication fork, the holenzyme must be able to recycle rapidly from the end of one Okazaki piece to the primer of the next. Only by such rapidly recycling can the 10 molecules of holoenzyme in a cell maintain an adequate rate of discontinuous synthesis on the lagging strand. To achieve these functions it has been proposed that the DNA polymerase III holoenzyme is an asymmetric dimer with twin active sites suitable for concurrent replica-

tion of both strands at the replication fork. Because of the extremely low cellular abundance of the subunits of DNA polymerase III holoenzyme it has only become possible to study their individual contributions as a result of their overproduction by genetic engineering techniques.

The *dnaZX* gene codes for the 71 000 dalton τ protein and the 52 000 dalton γ subunit comprising the *N*-terminal region of the τ protein [74, 75]. This is brought about by translational frameshifting generating a stop codon in a -1 frame which leads to formation of the γ protein [181]. The *N*-terminal region of τ and γ exhibits DNA-dependent ATPase activity whereas the *C*-terminal region of τ contains a potential DNA binding domain [178]. Complementation and reconstitution assays (section 6.7.5) have established that the holoenzyme requires both τ and γ for efficient activity, although τ can substitute for γ to some extent. However, a form of the enzyme known as polymerase III′ can be isolated having the constitution $(\alpha\varepsilon\theta\tau)_2$. The γ subunit can also be isolated as a subassembly containing two molecules of γ and one each of four other polypeptides (δ, δ', χ and φ) whose genetic loci and functional roles are as

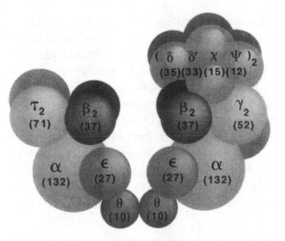

Fig. 6.15 Proposed V-shaped structure of *E. coli* DNA polymerase III holoenzyme. (Reproduced from [79] with permission of the authors and publishers.)

yet unknown [179]. This γ subassembly combines with polymerase III′ to form polymerase III*.

The interaction of polymerase with DNA can be broken down into several steps. The γ subassembly is crucial to the ATP-dependent transfer of the β subunit to the primed template which then interacts with the core enzyme to initiate processive DNA synthesis [180]. Addition of the τ subunit stabilizes the initiation complex and averts pausing in chain elongation at hairpin or duplex regions.

Formation of the initiation complex with a primed template results in a rearrangement of the holopolymerase such that the β subunit becomes inaccessible to antibody and the enzyme is unaffected by a concentration of KCl (150 mM) which inhibits the non-activated enzyme [76, 77]. The ATP-activated holoenzyme shows a much enhanced processivity, the one complex being capable of complete replication of a molecule of primed, circular, single-stranded phage DNA. On completing replication of such molecules the holoenzyme dissociates

and, in the presence of ATP, will reform on a new primed template [78]. This relocation takes a minute or more to complete and is too slow to manage discontinuous replication *in vivo*. The answer to this problem lies in the complex structure of the polymerase.

The holoenzyme can be viewed as a V-shaped complex, one arm of which contains τ and is responsible for continuous replication of the leading strand. The other arm contains the γ complex and is responsible for discontinuous synthesis on the lagging side of the fork (Fig. 6.15 [79]). For one enzyme complex to perform two polymerization reactions in different regions requires the DNA to loop back on itself. This has the big advantage that the site of termination of one Okazaki piece is adjacent to the site of initiation of the next which allows the rapid concerted transfer of the replicating complex from one primer to the next [80] (Fig. 6.15).

6.4.6 DNA polymerases of other prokaryotes

B. subtilis, a bacterium widely divergent from *E. coli*, also has three DNA polymerases which closely resemble those of *E. coli*. They differ in that DNA polymerase III is sensitive to 6-(p-hydroxyphenylazo)uracil (HPUra), an antibiotic active against Gram-positive bacteria. DNA synthesis in lysates of *B. subtilis* is also inhibited by HPUra [81].

Taq DNA polymerase is the enzyme present in the thermophylic bacterium *Thermus aquaticus*. It is of interest as the thermostable DNA polymerase is not denatured at the temperature at which DNA melts. It has therefore found a use in the *polymerase chain reaction* [82] (section A.7).

Many bacteriophages code for their own DNA polymerase. As already mentioned T4 DNA polymerase is encoded by the phage gene 43 and is a 104 kDa protein with both polymerase and $3' \rightarrow 5'$ exonuclease activity which plays a major role in phage DNA replication. As with *E. coli* DNA polymerase III it acts in a complex which ensures fidelity, speed and processivity [83].

Phage T7 gene 5 encodes a $3' \rightarrow 5'$ exonuclease/DNA polymerase of low processivity which combines with *E. coli* single-stranded DNA binding protein and with thioredoxin to stimulate both the rate of elongation and processivity [84, 85]. The gene 5 polymerase also associates strongly with the product of the phage gene 4 which has two functions – those of helicase (section 6.5.3) and primase. It is again quite possible that a dimeric complex can act at the replicating fork to promote both leading and lagging strand synthesis.

The smallest prokaryotic DNA polymerase so far analysed is the 63 kDa enzyme from phage PRD1. This enzyme shows significant sequence homologies with other polymerases, both prokaryotic and eukaryotic suggesting a common ancestral gene for all polymerases [86–88].

6.4.7 DNA polymerases in eukaryotes

Although the DNA polymerases of microorganisms have been studied more intensively, similar enzymes are present in eukaryotic cells. They resemble the bacterial enzymes in their requirement for a template, a $3'$-OH primer and four deoxyribonucleoside triphosphates [89, 90].

In higher eukaryotes there are at least four DNA polymerases known as α, β, γ, and δ and a fifth (ε) has recently been described [133, 811, 812]. In yeast DNA, polymerase I (POL1 or CDC17 gene) corresponds to DNA polymerase α, polymerase II to ε, polymerase III (CDC2 or POL3 gene) to δ and polymerase m (MIP1 gene) to γ [91–95], and they have recently been renamed accordingly [794].

On cell fractionation the bulk of DNA polymerase β is found in the nuclear fraction and there is considerable evidence that the location of DNA polymerase α *in vivo* is also intranuclear despite its recovery in high-speed supernatant fractions on cell disruption [96]. In those cells making DNA there is evidence for association of DNA polymerase α with the nuclear matrix [97, 98]. Indeed DNA polymerase α activity varies dramatically with the rate of cell division, being undetectable in non-growing cells and tissues [99].

(a) *DNA polymerase* α

This has a pH optimum in the range 6.5–8.0 and is highly susceptible to inhibition by thiol-active reagents (i.e. *N*-ethylmaleimide and *p*-chloromercuribenzoate). It is also inhibited strongly by araCTP and by aphidicolin (a tetracycline diterpenetetraol) [100]. In both instances inhibition is competitive with dCTP. Strangely, it is resistant to inhibition by $2',3'$-dideoxythymidine triphosphate (ddTTP) [101, 102] but is strongly inhibited by butylphenyl-dGTP and butylphenyl-dATP when present as a holoenzyme (see below) [103]. Use of these inhibitors together with consideration of the high enzymic activity found in growing cells has led to the conclusion that this enzyme plays a major role in DNA replication in eukaryotes. A temperature-sensitive mutant cell line which is defective in DNA synthesis at the non-permissive temperature has a heat-labile DNA polymerase α [104], and this enzyme has been shown to be involved in replication of SV40 DNA – further points illustrating the function of DNA polymerase α.

DNA polymerase α shows optimal activity with a gapped DNA template but shows a remarkable ability to use single-stranded DNA by forming transient hairpins [105]. It will not bind to duplex DNA. As normally isolated, polymerase α is only poorly processive, i.e. it adds about 11 nucleotides before it dissociates from the DNA [107]. The processivity can be dramatically increased in the presence of other proteins and ATP [108]. In contrast to the enzyme from *E. coli* no nuclease activity could initially be found associated with DNA polymerase α but this reflects the cryptic nature of the exonuclease (see below) [103, 111, 124].

Because of its extreme lability to proteolysis during purification and its apparent heterogeneity, DNA polymerase α has only recently been purified to the same extent as *E. coli* DNA polymerase I despite the fact that a 200-fold purification of the calf thymus enzyme was achieved in 1965 when it was separated from a terminal transferase (see below).

DNA polymerase α sediments in sucrose gradients as a broad peak in the 6–11S region and, on gel filtration, activity is found in a number of regions [105, 113]. It is probable, but by no means certain, that the native, undegraded enzyme consists of a 180 kDa polymerase together with three other subunits of about 70, 60 and 50 kDa. Association of the 180 kDa polymerase with the 70 kDa protein masks the $3' \rightarrow 5'$ exonuclease activity of the larger subunit [111, 114]. However, on isolation the bulk of the polymerase is recovered as an active 155 000-M_r protein together with several smaller active proteins which all share some primary sequence homology [111–115, 124]. The 60 and 50 kDa subunits comprise a *primase* activity which allows the enzyme to initiate replication on unprimed single-stranded cyclic DNAs [91, 108, 116–120]. The primase is a novel RNA polymerase (i.e. not RNA polymerase I, II or III – section 9.3) which synthesizes a short (8–12) oligoribonucleotide which, in the presence of deoxyribonucleotides, is extended by the DNA polymerase α activity. It can be isolated free of polymerase activity [121, 122]

but both the polymerase and primase activities are stimulated by accessory factors [106]. The association of primase with DNA polymerase α is restricted to the DNA-synthetic phase of the cell cycle and the complex forms part of the nuclear DNA replitase or replicase which also exhibits topoisomerase, RNase H and DNA-dependent ATPase activities [124].

An α-like polymerase/primase activity has also been isolated from an archaebacterium [125].

(b) *DNA polymerases δ and ε*

Polymerase δ was first described in 1976 [126] as a polymerase similar in properties to DNA polymerase α but having $3' \rightarrow 5'$ exonuclease activity. Thus both polymerase α and δ are present only in dividing cells and both are inhibited by aphidicolin (polymerase ε is similar). However, polymerase α is strongly inhibited by some dNTP analogues (e.g. butylphenyl dGTP) which scarcely inhibit polymerase δ or ε [127, 812]. Polymerase δ is dependent for activity on two auxiliary proteins: cyclin (or proliferating cell nuclear antigen, PCNA) and Activator 1, that dramatically increase processivity [109]. What was, initially, taken to be a PCNA-independent form of DNA polymerase δ is now known to be a separate enzyme known as polymerase ε (or two enzymes known as ε and ε^* with active polymerase subunits of 220 and 145 kDa, respectively; although the latter may be a proteolytic breakdown product of ε) [133, 812]. There is no antigenic cross reactivity between DNA polymerases α, δ (which consists of subunits of 125 and 48 kDa) and ε; nor between the equivalent polymerases I, III and II of yeast [94, 128–132, 811, 812].

Because of the presence of approximately equal activities of DNA polymerase α and δ it has been proposed that they act as a

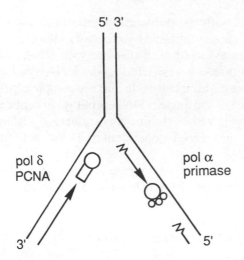

Fig. 6.16 A replication fork showing the combined action of DNA polymerase δ/PCNA catalysing leading strand synthesis and DNA polymerase α/primase catalysing the initiation and synthesis of Okazaki pieces.

dimer at the replication fork with the highly processive polymerase δ acting on the leading strand and the primase-associated polymerase α acting on the lagging strand (Fig. 6.16) [110, 134, 182]. DNA polymerase ε (and the yeast equivalent, DNA polymerase II) is also probably involved in replication [133] and it has been proposed that it may take over from DNA polymerase α in the synthesis of Okazaki pieces [812].

(c) *DNA polymerase* β

This is recovered in the nuclear fraction but shows little correlation between activity and the growth rate of the cells. This enzyme, which has been purified to homogeneity, has a pH optimum of about 8.6–9.0 and a M_r of 39 000 [135, 136]. DNA polymerase β shows optimal activity with native DNA activated by limited treatment with DNase I (to produce single-stranded nicks and short gaps bearing 3'-OH priming termini) and shows negligible activity with denatured DNA. It

is not, however, very efficient at gap filling [137]. It is relatively insensitive to thiol-active reagents, araCTP and aphidicolin but is strongly inhibited by ddTTP. The fact that DNA polymerase β never processes, i.e. it leaves the DNA primer after addition of a single nucleotide [53, 137], may account for its insensitivity to araCTP. It resembles DNA polymerase α in being devoid of overt nuclease activities but is found associated with a bidirectional, double-stranded exonuclease (DNase V) [138]. Thus neither enzyme has the overt, 'proofreading' capacity of the bacterial enzymes, and in fact both will add on to a mismatched primer terminus [139].

DNA polymerase β cannot use an oligoribonucleotide primer but, unlike polymerase α, given an oligodeoxyribonucleotide primer it is able to copy a ribonucleotide template (e.g. poly(A):oligo(dT)) [104]. DNA polymerase β is believed to play a role in repair of DNA and studies *in vitro* using nuclei from irradiated neuronal cells [140] or from non-dividing mouse L929 cells treated with nuclease show the repair synthesis to be sensitive to ddTTP. However, damage by ultraviolet light or alkylating agents, especially when it is severe, involves DNA polymerase α and DNA polymerases δ and ε in the repair reaction as well as DNA polymerase β [132, 141, 142, 812].

The use of an antibody to DNA polymerase β from calf thymus has demonstrated cross-reacting molecules of similar size in eukaryotes from protozoa to man indicating a lasting and essential role for this enzyme [143]. The gene for DNA polymerase β has been cloned and its regulation studied [136].

(d) *DNA polymerases of mitochondria, chloroplasts and viruses*

Animal cells also have a small amount of DNA polymerase γ, the mitochondrial DNA polymerase [104, 144]. It shows greatest activity using an oligdeoxyribonucleotide primer and a ribonucleotide template (e.g. poly(A).oligo(dT)). DNA polymerase γ from chick embryos is a tetramer having four identical subunits each of M_r 47 000 [145]. It is highly processive in contrast to DNA polymerases α and β [146] and this may be relevant to its function (section 6.9.5). It has a proofreading exonuclease [147].

Chloroplasts also possess the enzymic mechanisms for synthesizing their own DNA [148] and new enzymes are induced following viral infection [149–151].

The DNA polymerase coded for by herpes simplex virus (HSV) is very sensitive to inhibition by acyclovir and phosphonoacctic acid which makes these drugs suitable for use in the treatment of herpes virus infections [152] (section 3.7.7(c)). The HSV-coded enzyme is a 136 kDa protein with $3' \rightarrow 5'$ exonuclease activity which, in the presence of a DNA binding protein is highly processive [153]. The HSV-coded primase forms a complex with helicase activity [154] similar to that found with bacteriophage T7 (section 6.5.5).

(e) Terminal transferase

In addition to replicative DNA polymerase activities, which require the presence of all four deoxyribonucleoside triphosphates, animal cell extracts contain a separable enzyme responsible for the addition of nucleotidyl units to the ends of polynucleotide chains. This activity, which was originally described by Krakow in 1962 [155], does not require a template strand but catalyses the incorporation of nucleotide units from single triphosphates into terminal positions in the DNA primer molecule. It is not further stimulated by the addition of the other three triphosphates but it is stimulated by cysteine. It has been called 'terminal transferase enzyme' and may be used *in vitro*

in the biosynthesis of homopolymers of deoxyribonucleotides [156]. It is possible to distinguish between the terminal transferase and the replicative enzymes by the use of actinomycin D [48]. Terminal transferase shows strong amino acid homology to DNA polymerase β suggesting it may have arisen from a variant gene.

Unlike the replicative enzymes, which are stimulated by low concentrations of EDTA, the terminal transferase enzyme is completely inhibited by micromolar concentrations of EDTA which is believed to exert its effect by binding Zn^{2+}.

Terminal transferase has a M_r of 58 000– 62 000 [158] but the purified enzyme shows bands at 26 500 and 8000 M_r on electrophoresis in sodium dodecyl sulphate. This is a result of proteolysis during purification [159].

The activity of the terminal transferase is very low in all tissues except thymus and acute leukaemic lymphoblasts, i.e. lymphoid progenitor cells [160]. To what extent this reflects the peculiar physiological functions of such cells is unclear but the generation of short insertions at recombinational junctions formed during the rearrangements of immunoglobulin heavy chain genes is positively correlated with the presence of the enzyme [161] (section 7.5.7).

6.4.8 Reverse transcriptase or RNA-dependent DNA polymerase

The enzymes so far discussed copy DNA strands in the synthesis of DNA, but great excitement was created in 1970 by the announcement of the existence in certain RNA viruses of RNA-dependent DNA polymerases which use RNA as a template for the synthesis of DNA. These enzymes were discovered simultaneously by Temin [162] in the virus particles (virions) of Rous sarcoma virus (RSV) and by Baltimore [163] in Rauscher mouse leukaemia virus (R-MLV). The observation was confirmed for more than half a dozen RNA viruses by Spiegelman [164], and was seen as an important breakthrough in cancer research since the RNA viruses involved are oncogenic, i.e. capable of bringing about malignant change (chapter 3).

The virions of these, so-called, retroviruses contain a 70S RNA genome which, when treated with agents that disrupt hydrogen bonds, dissociates into two genetically identical RNA molecules and to several molecules of tRNA [165]. The unique RNA-directed DNA polymerase is also present in the virion. The enzyme uses one of the tRNA molecules (e.g. tRNA^{Trp}) [166] as a primer to synthesize DNA on the template 35S single-stranded RNA (section 6.14), and this distinguishes it from any cellular enzyme. Ribonucleoside triphosphates are without effect as substrates and the process is not susceptible to inhibition by actinomycin D.

The immediate product of the reaction is a double-stranded, RNA:DNA hybrid which is the result of the synthesis of a complementary strand of DNA using the single-stranded, viral RNA as template.

The *C*-terminal region of the enzyme shows ribonuclease H activity [167] which digests the RNA strand of the RNA:DNA hybrid molecule. A DNA-dependent DNA polymerase activity will then convert the hybrid into a duplex DNA. Cellular DNA polymerase can then replicate the duplex DNA so as to provide more copies which may be integrated into the genome of the host cell (section 6.14).

Reverse transcriptase is the product of the *pol* gene of retroviruses. It is synthesized as a gag-pol polyprotein, generally as the result of ribosomal frameshifting [314] (section 12.9.6(b)) during translation of the tran-

script from the integrated proviral DNA [168]. (Gag is a capsid protein.) This is necessitated where the gag and pol proteins are encoded in different frames in the message. (In other retroviruses these are in the same reading frame and the fusion protein is generated by reading through a termination codon – section 12.9.6(b).) The polyprotein is processed by a protease, present usually in the *N*-terminal region of the polymerase [169], to give the β subunit ($92\,000\,M_r$) and then the α subunit ($58\,000\,M_r$) which in the leukosis virus (ALV) combine to give the most active ($\alpha\beta$) form of the enzyme [170]. The enzyme from other retroviruses is a single polypeptide [171].

The enzyme shows only low fidelity and lacks associated exonuclease activity although it is capable of specifically hydrolysing the ribo strand of an RNA–DNA duplex [172, 173]. The 3′-end of the *pol* gene also encodes an endonuclease important for replication and integration (section 6.14) [170, 171] which altogether makes this a remarkable enzyme.

Reverse transcriptase may be used *in vitro* to make DNA copies of eukaryotic messenger RNA molecules. A thymidylate oligomer will anneal to the 3′-poly(A) tail of the mRNA and serve as a primer in a reaction which is of great importance in the production of cDNA (section A.8.2).

Reverse transcriptase is required not only for the replication of RNA retroviruses but also for the replication of DNA viruses such as hepatitis B virus and cauliflower mosaic virus [174, 175] indicating that these viruses replicate via an RNA intermediate. Again the mRNA contains information for the synthesis of both the polymerase and virion core polypeptide and the two are in different frames and the two genes overlap one another. In these cases, however, translation is initiated at an internal AUG and ribosomal frameshifting does not occur.

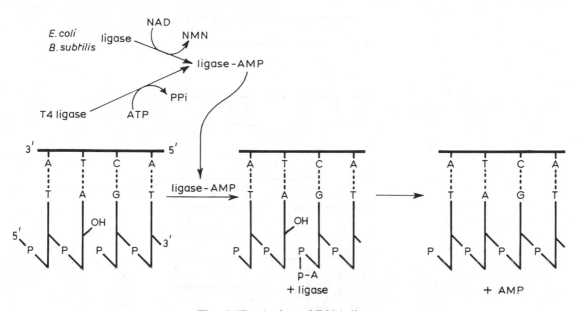

Fig. 6.17 Action of DNA ligases.

Reverse transcriptase is also involved in the mobility of some transposons (chapter 7) [176] and the telomerase (section 6.16 4) is, by definition, also a reverse transcriptase.

Reverse transcription is also found in bacteria where it functions in the production of multicopy, small, single-stranded DNA molecules [177].

6.5 OTHER PROTEINS INVOLVED IN REPLICATION

6.5.1 DNA ligases

Ligases catalyse the repair of single-stranded phosphodiester bond cleavages of the type introduced by endonuclease. They were first described in *E. coli* and have since been described in both animal and plant cells.

DNA ligases catalyse the formation of a phosphodiester bond between the free 5'-phosphate end of an oligo- or polynucleotide and the 3'-OH group of a second oligo- or polynucleotide next to it (Fig. 6.17). A ligase−AMP complex seems to be an obligatory intermediate and is formed by reaction with NAD in the case of *E. coli* and and *B. subtilis* [183] and with ATP in mammalian and phage-infected cells [184] (Fig. 6.17). The adenyl group is then transferred from the enzyme to the 5'-phosphoryl terminus of the DNA. The activated phosphoryl group is then attacked by the 3'-hydroxyl terminus of the DNA to form a phosphodiester bond. DNA ligase will close single-strand breaks in double-stranded DNA or in either strand of a polyribonucleotide:polydeoxyribonucleotide hybrid polymer; double-stranded ribopolymers are not substrates. Breakage of a single phosphodiester bond, without removal of a nucleotide, to give 5'-phosphoryl and 3'-OH termini is essential for repair by ligase activity (Fig. 6.18). Reaction is independent of the base composition around the cleavage point. The enzyme from *E. coli* or bacteriophage T4 will join short oligodeoxynucleotides in the presence of a long

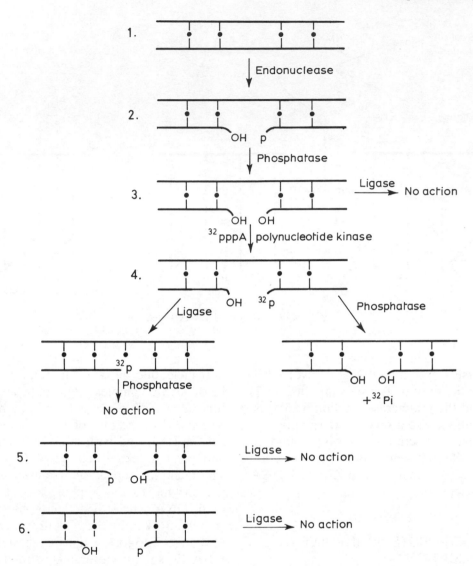

Fig. 6.18 An assay for DNA ligase. The substrate (4) is formed by treating double-stranded DNA (1) with DNase I to give single-stranded breaks (2), removing the 5′-phosphate residues with phosphatase (3) and replacing them with [^{32}P]phosphate groups with the aid of polynucleotide kinase and [γ-^{32}P]ATP. Forms (3), (5) and (6) are inactive as substrates for ligase.

complementary strand, e.g. d(T-G)$_3$, d(T-G)$_4$ or d(T-G)$_5$ can be joined in the presence of poly(dC-dA), and blunt end ligation (i.e. the joining of two duplex DNA molecules in the absence of any complementary overlapping sequence) can occur in the presence of excess ATP (section A.8.2).

(a) *Assay of DNA ligase*

DNA ligase has been assayed in a variety of systems, including the formation of covalently closed circles of double-stranded DNA, restoration of transforming activity of nicked DNA (e.g. [183]) and the formation of phosphatase-resistant radioactive phosphate [185] (Fig. 6.18).

(b) *Role of DNA ligase*

The nature of the reaction catalysed by DNA ligase makes it an important enzyme involved in DNA synthesis, DNA repair and genetic recombination. In all these cases, its postulated role is in re-establishing continuity by joining a stretch of newly synthesized DNA to pre-existing DNA by the formation of a phosphodiester bond.

It has, however, proved difficult to define a situation in which DNA ligase could be shown to be absolutely necessary for any one of these functions. Nevertheless, since *E. coli* ligase-deficient mutants selected in different ways exhibit abnormal uv sensitivity [186], it appears that ligase has some function in the repair process. Temperature-sensitive ligase mutants also accumulate Okazaki pieces at the restrictive temperature. That these pieces are eventually joined is probably a reflection of the presence of a few molecules of ligase still remaining in the mutant cells. There are normally about 300 molecules of ligase per cell, and it has been calculated that these are capable of sealing 7500 breaks per minute at 30°C [187]. Since only about 200 breaks per cell are formed on replication each minute there is a vast excess of enzyme.

Infection of *E. coli* by bacteriophage T4 leads to the synthesis of a T4-specified ligase which, in addition to joining two DNA chains, will also join RNA to DNA in an RNA–DNA hybrid [187]. Some ligase is essential for T4 development [186], and bacteriophage with a temperature-sensitive ligase show increased susceptibility to uv radiation (chapter 7).

There are two DNA ligases present in cells of higher eukaryotes. DNA ligase I (M_r 125 000) resembles DNA polymerase α in

being recovered partly in supernatant fractions on cell homogenization although it can be shown by immunocytochemical techniques to be a nuclear enzyme. It is present in much greater amounts in dividing than in resting tissues. The smaller, nuclear DNA ligase II does not change in activity with the growth rate of the cells [188–190].

6.5.2 Single-stranded DNA-binding proteins (SSB)

These are groups of proteins brought together under this heading because of a common property. They have been isolated from many sources, both prokaryotic and eukaryotic, but they do not necessarily all perform the same function *in vivo*.

The first of these to be characterized was that coded for by gene 32 of bacteriophage T4 [191]. This and similar proteins isolated from bacteriophage T7 and uninfected *E. coli* [192] have the ability to convert double-stranded DNA into single-stranded form at a temperature 40°C below the normal melting temperature (T_m) and for this reason they are also known as helix destabilizing (HD) proteins. The gene 32 protein has an M_r of 35 000 and binds co-operatively to single-stranded regions of DNA, each protein covering some 10 nucleotides [193]. Thus, if a limiting amount of gene 32 protein is mixed with excess single-stranded DNA some DNA molecules will become covered with protein and others remain free of protein. There are from 300 to 800 molecules of binding protein per uninfected *E. coli* cell, which is enough to cover about 1600 nucleotides at each replication fork [192]. A similar length of T4 DNA can be covered by the 10 000 molecules of gene 32 protein present in an infected cell. (There are about 60 phage T4 replication forks per infected cell.)

Purified polymerases are often stimulated

in a very specific manner by the presence of their complementary DNA-binding proteins. Thus the T4 DNA-binding protein will stimulate only the T4 polymerase and the T7 DNA-binding protein will stimulate the T7 polymerase [193], the *E. coli* DNA-binding protein stimulates *E. coli* polymerases II and holopolymerase III but does not stimulate polymerase I and III [194]. This probably arises as a result of formation of a specific complex between the polymerase and DNA-binding protein [129, 788]. The human SSB is a heterotrimer that stimulates unwinding of DNA and activity of DNA polymerase α and δ [198].

The SSBs from *E. coli* and T4-infected cells are believed to play a role in replication, recombination and repair [155, 156, 195] and an *E. coli* mutant (SSB-1) has been isolated showing a defective DNA-binding protein and extracts fail to convert phage G4 single strands into duplex form [196] (section 6.7.6). In contrast an SSB coded for by gene 5 of bacteriophage M13 plays an essential role in preventing further replication by stabilizing the single-stranded progeny viral DNA [197].

Thus SSBs serve a passive role. Although *in vitro* they can bring about denaturation of DNA, their role *in vivo* may simply be to stabilize single-stranded DNA.

6.5.3 DNA helicases (DNA-dependent ATPases) [204]

DNA helicases have the property of being able to unwind a DNA duplex, obtaining the necessary energy from the hydrolysis of ATP. They can be assayed by measuring the displacement of a radioactive oligonucleotide from a complementary single-stranded circle of M13 DNA [230].

In *E. coli* at least seven helicases have been partially characterized [199–201]; this multiplicity presumably implying multiple

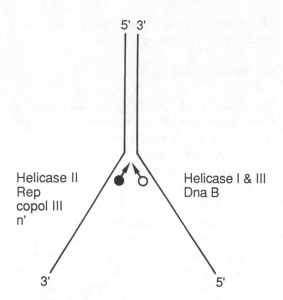

Fig. 6.19 DNA helicases acting at the replication fork. Some ($\bigcirc$) travel along the single strand in the $5' \rightarrow 3'$ direction (i.e. on the lagging side of the fork), whereas others ($\bullet$) travel in the $3' \rightarrow 5'$ direction (i.e. on the leading side).

functions for these proteins. All these proteins initially bind to single-stranded DNA and then translocate in either a $5' \rightarrow 3'$ ($\bigcirc$) or $3' \rightarrow 5'$ ($\bullet$) direction until they reach a duplex region when unwinding starts [204].

Helicase I ($\bigcirc$) is present only in F$^+$ strains of *E. coli* (i.e. strains carrying the F sex factor – section 3.6) and is the product of one of the transfer genes (*traI*) of the F factor [202]. Thus a role for helicase I in bacterial conjugation is implied and it plays no role in cellular DNA replication. Several molecules bind to a single-stranded region of DNA and travel down the single strand in a $5' \rightarrow 3'$ direction hydrolysing ATP. When a duplex region (e.g. a replication fork) is reached the DNA strands are unwound (Fig. 6.19) [203]. There are about 500 molecules per cell.

There are about 3000 molecules of helicase II ($\bullet$ M_r 82 000) per cell. It differs from helicase I in that it moves towards a replication fork in a $3' \rightarrow 5'$ direction (Fig. 6.19). However, it interacts specifically with holopolymerase III to promote processive unwinding of DNA in the presence of *E. coli* SSB [205, 206]. It is the product of the *E. coli uvrD* gene, mutants in which show increased rates of mutation and recombination [207, 208]. Immunological studies show that helicase II is involved in the replication of DNA of *E. coli*, phage lambda and plasmid Col E1 [209] although its main role may be in repair of DNA.

Rep (●) is so called because it is essential for the replication of small single-stranded phages (e.g. phage G4 and ϕX174 – section 6.9). Although not essential for *E. coli*, it speeds up replication and is essential in mutants defective in helicase II. Like helicase II it binds to a single-stranded region and translocates in a 3′ → 5′ direction until it reaches the replication fork (i.e. along the leading strand). For its unwinding action it requires a single-stranded binding protein to bind to the two arms of single-stranded DNA produced (Fig. 6.19) [192, 814]. Although Rep and helicase II share about 40% homology in amino acid sequence, they show functional differences. In particular, Rep acts processively whereas helicase II is required in stoichiometric amounts [210].

The β subunit of holopolymerase III (●, copolymerase III*) is also a helicase which travels down the leading strand and so is protein n′ (●, factor Y) [211, 212] (section 6.7.6).

Helicase III (○, $M_r = 2 \times 20\,000$) and the product of the *dnaB* gene (○) move along the bound strand of DNA in a 5′ → 3′ direction, i.e. along the lagging strand at the replication fork [213–215].

It is clear that helicases acting together on both sides of the fork would bring about efficient unwinding of the DNA duplex but it is not clear which pair of enzymes acts in any particular *in vivo* situation. However, both DnaB and protein n′ form part of the preprimosome (section 6.7.6) involved in lagging strand initiation; and, as they move in

opposite directions with respect to the bound strand, this would be expected to generate a loop of single-stranded DNA held together at its base by the preprimosome.

Two molecules of ATP are hydrolysed per nucleotide unwound [216]. This is a high price to pay for unwinding DNA as it involves the release of 60 kJ of energy to break hydrogen bonds, the energy of which varies from 5 kJ (A:T) to 21 kJ (G:C).

The gene 4 product of bacteriophage T7 is similar to Rep except that it forms a complex with the T7-coded DNA polymerase (the gene 5 product). Together they are able to extend a 3′-OH at a gap in the duplex, unwinding the DNA ahead of the complex and hydrolysing two ATP molecules for every dNMP incorporated [217, 218]. The gene 4 protein, in addition to acting as a helicase, is also able to synthesize a specific primer for initiation of Okazaki pieces (sections 6.3.4. and 6.11.1). A similar dual function is catalysed by the T4 gene 41/61 complex [219–221], and a second T4-coded helicase – the product of the non-essential *dda* gene – greatly speeds up the rate of replication fork movement *in vitro* [222, 223]. The Dda protein overcomes the inhibitory action exerted by RNA polymerase molecules which, when bound to the DNA template, can completely block fork movement [224]. Unwinding enzymes are also found in eukaryotes [225–227] and some DNA helicases are also RNA helicases. Indeed, some enzymes (e.g. the human p68 protein) unwind only RNA duplexes [228].

DNA helicases are also required to initiate transcription [204, 229].

6.5.4 Topoisomerases

As the replication fork moves forwards, the two DNA strands unwind. Either the whole DNA molecule has to rotate at 10 000 rev./min (an impossibility for closed cyclic DNA

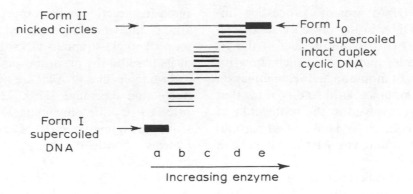

Fig. 6.20 Topoisomerase action. Diagram of agarose gel electrophoresis of SV40 DNA incubated with increasing amounts of topoisomerase I. Whereas an endonuclease would convert form I into form II directly, the topoisomerase produces DNA of intermediate linking numbers.

molecules) or some sort of swivel must be present. Cairns first proposed the idea of a swivel [3] which might act by alternating action of an endonuclease and a ligase. However, the two functions are present in a single enzyme, examples of which have been isolated from both prokaryotes and eukaryotes [231–236]. The activity was initially called a nicking closing enzyme but is now known as a topoisomerase as it catalyses the interconversion of different topological isomers of DNA (chapter 2). The difference between a topoisomerase and the earlier suggested mechanism is that with the topoisomerase the cut ends are not released from the enzyme although rotation of the two strands about one another is effectively achieved.

There are two types of topoisomerase. Topoisomerase I cuts only one strand of duplex DNA whereas both strands are cut by topoisomerase II. The eukaryotic enzymes produce a 5'-OH group (which can be phosphorylated by polynucleotide kinase) and a 3'-phosphate which is covalently linked to a tyrosine in the enzyme [233, 235, 237], although the *E. coli* topoisomerases bind to a 5'-phosphoryl group.

In vitro, topoisomerases are studied by following the result of their action on closed cyclic, supercoiled DNA molecules such as SV40 DNA or the plasmid Col E1 (Fig. 6.20). The topoisomerases will bring about a step by step relaxation of such supercoiled molecules by changing the linking number (section 2.5.1). Type I topoisomerases change the linking number in single steps indicating that they produce an enzyme-linked cut in one strand only of the DNA. The other strand is then passed through the cut which is resealed [238, 239]. This is equivalent to free rotation of the bond opposite to the nick.

Type II topoisomerases produce an enzyme-bridged break in both strands of DNA. In an ATP-requiring action, another region of duplex DNA passes through the gap thus changing the linking number two steps at a time (Fig. 6.21). In most instances both types of topoisomerase will act until all the supercoils are removed and the DNA is released as a fully relaxed, closed cyclic DNA molecule [240]. If the two regions of duplex DNA involved in topoisomerase II action are part of two different DNA molecules then these two molecules will become interlocked (Fig. 6.21). Because of this the enzyme is capable of producing and resolv-

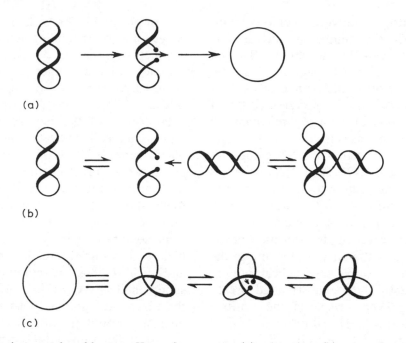

Fig. 6.21 Reactions catalysed by type II topoisomerases: (a) relaxation; (b) catenation and decatenation; (c) knotting and unknotting.

ing catenated and knotted molecules [235]. In a similar manner topoisomerase I can act opposite to a nick in a cyclic molecule and bring about the catenation and knotting of several supercoiled molecules to the nicked, cyclic molecule [238].

There are at least two types of topoisomerase II. The eukaryotic enzymes and the enzyme from bacteriophage T4 are only capable of relaxing supercoiled DNA. That is, they catalyse the conversion of supercoiled DNA into relaxed, cyclic duplex. They are homodimers having two subunits each of approximate Mr 170 000 [232, 235, 241–243]. Bacterial type II topoisomerases are known as DNA gyrases and they are additionally capable of introducing negative superhelical turns into covalently closed, relaxed, cyclic duplex DNA molecules. A third type of topoisomerase II has been reported in archaebacteria [244]. This reverse

gyrase is able to convert a relaxed cyclic DNA duplex into a positively supercoiled molecule.

E. coli DNA gyrase is a 400 kDa tetramer, the subunits of which are the products of two genes, *gyrA* and *gyrB* (also known as *nal* and *cou*). The *gyrB* product is a DNA-dependent ATPase (M_r 95 000) which is sensitive to the antibiotics coumermycin and novobiocin. The *gyrA* product is a 100 000–110 000 M_r, nicking-closing enzyme sensitive to nalidixic acid and oxolinic acid. In the presence of oxolinic acid or strong detergent these subunits remain bound to the 5′-ends of the cut [245] and hence the tetrameric enzyme holds the cut ends together. As we have said above, DNA gyrase, as well as relaxing supercoiled molecules can also reduce the linking number of cyclic duplex DNA molecules. Thus when incubated with ATP and DNA gyrase a relaxed duplex ring can be converted into a negatively supercoiled molecule with a linking number as much as 10% lower than that of the most stable B-DNA structure [246–248]. Although DNA gyrase is the only enzyme that can catalyse this reaction *in vitro* it has been suggested that it is the normal reaction catalysed *in vivo* by all type II topoisomerases [239].

A model for the action of DNA gyrase has been constructed [249]. The DNA first wraps itself around the protein to form a structure similar to a nucleosome (chapter 3). ATP then binds and causes a conformational change which drives the translocation of the DNA relative to the enzyme, leading to the formation of a positive superhelical loop on the DNA [250]. A double-stranded breaking/rejoining event, removes the positive superhelical turn [251] and on hydrolysis of ATP the enzyme returns to its original conformation leaving the DNA with a negative superhelical twist, i.e. partly unwound. The wrapping of the DNA around the gyrase produces overwound and underwound regions of duplex DNA such that when the DNA is left with an underwound region the overwound regions are selectively relaxed [739].

DNA gyrase may play a role either by removing positive superhelical turns induced into DNA during replication or, alternatively or additionally, by putting negative supertwists into the DNA ahead of the replication fork thus making unwinding easier. As treatment of *E. coli* with either coumermycin or nalidixic acid immediately stops DNA synthesis, the contribution of DNA gyrase is essential [252]. However, mutants in DNA gyrase are viable if compensatory mutations are present in the gene for topoisomerase I [253, 254]. Similar compensatory effects of topoisomerase I and II are seen on transcription [255], and DNA gyrase is also believed to be involved in recombination [256].

DNA gyrase is also required for replica-tion of phage T7 DNA [257], and phage T4 codes for a corresponding activity [242]. The products of T4 genes 39, 52 and 60 form a complex with topoisomerase activity which is essential for initiation of replication of T4 DNA.

Footprints of the binding of eukaryotic topoisomerase II indicate that binding takes place over only 20–30 bp, ruling out a mechanism which involves the DNA wrapping itself around the enzyme. This may explain the inability of the eukaryotic enzyme to introduce superhelical turns into DNA. Topoisomerase II interacts with two DNA helices at points where they cross (nodes), and such regions are commonly found in supercoiled DNA [813]. This explains how the enzyme can anchor loops of DNA to the nuclear scaffold (section 6.15.5). Topoisomerase II generates a double-stranded break in one duplex by sequentially cleaving the two strands, and this break may be religated with or without intervening strand passage. Strand passage is dependent on binding ATP and must involve massive conformational changes [813].

Decatenation (catalysed by topoisomerase II) is an essential step in the physical separation of cyclic DNA molecules after replication [234, 235, 258, 259] (section 6.16.2) whereas topoisomerase I acts near the replication fork [260].

In eukaryotic cells the activity of topoisomerase II fluctuates throughout the cell cycle. It is possible that activity is markedly affected by enzyme phosphorylation [261] or poly(ADP)-ribosylation [262]. The enzyme is associated with the axial domain at the base of chromatin loops in metaphase chromosomes [263] and is largely degraded as cells pass from mitosis into G1 [264]. From this it has been proposed that topoisomerase II has its major role in the regulation of chromatin decondensation and this is consistent with the finding that Top2 mutants in yeast die in anaphase [265] and accumulate catenated plasmids [266]. In contrast, topoisomerase I is associated with transcriptionally active chromatin [267] and inhibitors of human topoisomerase I rapidly block ribosomal RNA synthesis [268]. It appears not be be essential in the presence of topoisomerase II [266].

Treatment of topoisomerase/DNA mixtures with a denaturing agent such as SDS results in release of a cleaved DNA intermediate and has enabled the sequence specificity of the enzymes to be investigated. This has been helped by the finding that camptothecin – a cytotoxic alkaloid – stabilizes the cleavable complex of phosphorylated topoisomerase I and DNA causing the intermediates to accumulate [269, 270]. There is some evidence, however, that camptothecin stabilizes some complexes more than others and its use may give a false impression of the sequence specificity of the enzyme [271]. VP16 and Vm26 (teniposide) are two epipodophyllotoxins which specifically inhibit eukaryotic type II topoisomerases [272, 273] again by stabilizing the cleavable complex.

Although vertebrate topoisomerase I has a very degenerate consensus breakage site, the topoisomerase I from *Tetrahymena* binds with high affinity to a 20 bp region containing the cleavage site [274]. It cleaves preferentially in the following sequence

A(G/A)ACTT$\downarrow$ AGA(G/A)AAA(T/A)(T/A)(T/A)

This sequence occurs in two regions at both ends of the dimeric ribosomal DNA in *Tetrahymena*. In other situations, where the sequence is absent, topoisomerase I will cleave at random [275, 276]. DNA gyrase cleaves duplex DNA to give a four base pair stagger but there is controversy over whether or not it shows sequence specificity. Topoisomerase II from insects and from vertebrates cleaves DNA at different but imprecise consensus sequences including alternating purine/pyrimidine runs; and the yeast enzyme will act at oligo(A) tracts [277, 278].

6.5.5 Primase

A primase is an enzyme which makes the RNA primers required for initiation of Okazaki pieces (section 6.3). In *E. coli*, the *dnaG* gene encodes the primase which is a rifampicin-resistant RNA polymerase of M_r 64 000 [21, 279]. As well as polymerizing ribonucleotides it can also use deoxyribonucleotides [280, 281] and make a mixed primer although the first nucleotide is always rA.

This primase shows a very strong preference to initiate with adenosine followed by guanosine [280] and this suggests that initiation of Okazaki pieces may occur at particular sites on the lagging strand. However, the small phage P4 which requires only about 20 Okazaki pieces per round of replication shows no preferential initiation sites [282]. What appears to limit initiation of Okazaki pieces is the fact that the primase shows limited ability to bind to DNA. It will bind to rare sequences in single-stranded DNA when folded into a complex secondary structure [283, 694] but normally it can only bind to a complex of proteins already bound to DNA (sections 6.7.6 and 6.10.1) In contrast to this unresolved situation, the phage T7 primase always initiates at a very precise sequence (section 6.11.1).

The primase that is tightly associated with the eukaryotic DNA polymerase α is made up of two subunits (section 6.4.7(a)) and shows no stringent sequence requirements for the site of initiation. However, it does not act at random but primers are nearly always initiated with a GTP two to ten nucleotides 5′ to a hexanucleotide resembling the sequence $C_2A_2C_2$ [123].

6.6 FIDELITY OF REPLICATION

6.6.1 Introduction

To produce a sequence of bases in the DNA of a daughter cell which is *identical* to that in the parent, the process of DNA replication would need to be completely faithful. This is, obviously, not the case as mutants do arise spontaneously and this is not altogether a bad thing as it is on this varied counterpane that natural selection is able to act.

In chapter 7 we consider mechanisms which come into play to repair damage to DNA post-replicatively but even these repair mechanisms rely on the faithful production of DNA complementary to the template strand.

Fidelity may depend on a number of steps.

1. Even in the absence of enzyme, complementary base pairs are about 100-fold more stable than the non-complementary base pairs;
2. The polymerase must selectively bind the complementary deoxyribonucleoside triphosphate (K_m discrimination); thus a checking mechanism may exist which would allow rejection of incorrect triphosphates. Such enhanced discrimination may result from template-induced changes in enzyme conformation dictating the selection of the correct substrate, or increased binding of the enzyme to the template in the presence of the complementary nucleotide;
3. A kinetic proofreading model proposed by Hopfield [284] predicts that a high-energy [enzyme:template:dNMP] complex is produced with release of pyrophosphate. The polymerase can check the complex to see whether or not the correct base has been incorporated. The nucleotide (dNMP) may be released at this time but the chance of an incorrect nucleotide

being released is much higher than for the correct nucleotide. It is possible that incorrect dNMPs are released from the high-energy complex by a process that is the reversal of its formation (i.e. as dNTPs). Incorrect incorporation of nucleotides generally occurs 100 000-fold less frequently than does incorporation of the correct nucleotide [285].

4. Following incorporation, an exonucleolytic proofreading mechanism could remove incorrect 3′-terminal nucleotides which are mismatched with the template strand before elongation [286] (section 6.4.3(d)). These nucleotides are also released as monophosphates.

Both the kinetic and the exonucleolytic proofreading mechanisms rely on the reduced rate of reaction with a mismatched nucleotide. In the kinetic mechanism it is the joining of the bound nucleotide which is delayed; whereas with exonucleolytic proofreading it is the addition of the next nucleotide which is delayed allowing a greater opportunity for exonuclease action.

It is only if all these *error avoidance* mechanisms fail that post-replicative *error correction* mechanisms are required but, as explained above, these are also subject to the demands that a faithful complementary DNA strand is produced. Error correction is considered in section 7.3. Error avoidance mechanisms can increase the fidelity of replication to give an error rate of one mistake in 10 million base pairs replicated.

6.6.2 Fidelity assays and results

An assay that has proved particularly useful in assessing the fidelity of replication *in vitro* involves the copying of the single-stranded DNA of the phage ϕX174 [287, 288]. An amber mutant (ϕX174 am 16 – Table 6.4) is unable to grow on wild-type *E. coli* (but can

Table 6.4 Nucleotide substitutions possible at the am 16 codon of the ϕX174 genome

Gene A		Phe	Arg		
Mutant DNA sequence		5'-A G A T T T A G G C T G G-3'			
Gene B		Ile	am16	Ala	

Position at amber codon	Mispaired nucleotide		Gene A	Gene B
T	G		Phe-Arg	Glu
	T		Leu-Arg	Lys
	C (wt)		Leu-Arg	Gln
A	G		Phe-Arg	Ser
	A		Phe-Trp	Leu
	C		Phe-Gly	Trp
G	G		Phe-Thr	Tyr
	T		Phe-Lys	Ochre
	A		Phe-Met	Tyr

grow on the suppressor strain *E. coli* C62). Faithful copying by polymerase of the mutant ϕX174 DNA will reproduce mutant progeny; but a single base substitution at any one of the three bases of the amber codon will result in progeny viral DNA which can grow on wild-type *E. coli*. Using purified DNA polymerase α from calf thymus, Grosse *et al.* [288] showed a reversion frequency of about 1×10^{-4}. Of course all the base changes would not result in substitution with the original wild-type base and, as this region of the ϕX174 genome codes for two different proteins (section 3.7.7(a)), a variety of progeny phage can result. For instance, insertion of a guanine opposite the guanine of the amber codon (TAG) results in a change of an arginine into a threonine (AGG → ACG) in gene A and insertion of a tyrosine into gene B. (The genetic code is discussed in chapter 12.) This produces small phage plaques which show reduced growth at temperatures above 35°C.

From a study of the plaque morphology and related sequence data it can be estimated that the frequency of mispairing dATP with a guanine in the template (1 in 20 700) is an order of magnitude more common than the reciprocal mispairing of dGTP with an adenine in the template. Nonetheless, the major mutations are purine:purine mismatches (transversions) and G to T transitions [289].

An alternative assay system involves comparing the competition between the correct and incorrect nucleotide for incorporation in place of a dideoxynucleotide in the DNA sequencing reaction (section A.5.1). This system has the advantage that the effect of the environment of the substituted nucleotide can be studied [290].

Having established an assay system, the effect on mispairing of altering the relative concentrations of precursors can be studied. For instance, when the concentration of dATP is kept low and that of dCTP increased, the chances of mispairing with the first base of the amber triplet is increased.

However, other examples of an effect of pool bias were not found using DNA polymerase α.

The fidelity of the various eukaryotic DNA polymerases is very variable and it is only the high-M_r, 9S form of polymerase α which shows a fidelity comparable to that of the prokaryotic polymerases, both making less than 1 error in 10^5 nucleotides incorporated [103, 291, 292, 294]. This result suggests that a domain of polymerase α, which is lost on proteolysis, is vital for maintaining fidelity in a parallel way to that in which the proofreading exonuclease activity of the ε subunit of *E. coli* core polymerase III is used. This observation raised concern as it was thought that eukaryotic polymerases did not have an associated $3' \rightarrow 5'$ exonuclease. A human polymerase α–primase complex, consisting of four subunits of 195, 68, 55 and 48 kDa does, however, show $3' \rightarrow 5'$ exonuclease activity (section 6.4.7) and shows enhanced fidelity producing only one error in 10^5 bases incorporated [298]. However Roberts and Kunkel [299] only obtained enhanced fidelity using HeLa cell extracts and found that their polymerase α–primase complex showed no greater fidelity than polymerase α alone. These results indicate the involvement of some factor which readily dissociates from the complex and addition of the ε subunit of polymerase III to purified calf thymus polymerase α increases fidelity 7-fold even in the absence of physical interaction of the two proteins [295].

In the absence of exonucleolytic removal of a mismatched primer terminus, DNA polymerase α shows an elongation rate of only 5% of that with a fully matched terminus [296, 297]. It is not only the first base after the mismatch which is added more slowly, for the effect extends to additions five bases following the mismatch.

The prokaryotic DNA polymerases are as-sociated with more obvious proofreading exonucleases and in the ϕX174 am16 assay pool effects are more marked. Thus the frequency of a TAG $\rightarrow$ TGG transversion depends not only on the ratio of incorrect to correct triphosphate but also on the concentration of the next triphosphate, i.e. $[dGTP]^2/[dATP]$ [300]. By using deoxycytidine [1-thio]triphosphates which, once incorporated, cannot be removed by exonuclease action, Kunkel [301] calculated that proofreading can increase the fidelity of *E. coli* DNA polymerase I or phage T4 polymerase by up to 20-fold. The efficiency of proofreading depends on the adjacent nucleotide which is stacked together with the mismatched base [239] and the fidelity of these enzymes and *E. coli* DNA polymerase III holoenzyme is dependent on the divalent ions [302] and accessory proteins present [77, 303–306].

In an assay of fidelity which measures the production of dNMP by *E. coli* DNA polymerase I, Loeb's group have failed to find significant evidence for proofreading of either the kinetic or exonuclease variety [307] and claim that the rate of generation of dNMP by exonuclease action can account for only a minor increase in fidelity [308]. To what extent this relates to the role of DNA polymerase I in the repair, rather than replication of DNA is unclear and proofreading mechanisms are believed to be a major factor in limiting replication errors.

The use of the α subunit of *E. coli* DNA polymerase III (in the absence of the proofreading, ε subunit) indicates that there is a differential K_m of between 10^4 and 10^5 for incorporation of the correct, rather than an incorrect, nucleotide [309].

It is the post-replicative mismatch correction mechanisms which contribute to the increase in fidelity (up to one error per 10^9–10^{11} nucleotides incorporated) found *in vivo* (chapter 7).

The RNA-dependent DNA polymerases of retroviruses (reverse transcriptase) are very error-prone showing a fidelity similar to that of RNA polymerases, i.e. one error per 10^3–10^4 nucleotides incorporated [172, 173, 310, 311].

6.6.3 Frameshift errors

In addition to base mismatching, DNA polymerases can introduce an extra base, or omit a base, leading to frameshift errors. The frequency of such mutations is enhanced in the presence of intercalating agents (section 7.2.3) but they also occur spontaneously. Transient misalignment can occur along a repeated sequence leading to deletion or insertion of 5–8 bases, but the commonest type of deletion is of a single base which occurs at particular 'hot spots' dependent on the polymerase studied [312]. Complex deletions and insertions are believed to occur by the 3'-end of the primer becoming detached from the template strand and either folding back to use itself transiently as a template before reannealing in the correct position; or looping forwards to anneal with an upstream region of the template.

Some substitution errors actually occur by a misalignment mechanism. The polymerase is believed to skip a base but, before many nucleotides are added, realignment occurs to give a substitution mismatch [313]:

$$
\begin{array}{l}
5'\ \text{T T C G G} \\
3'\ \text{A A G C C T G G C C T}
\end{array}
$$

↓

$$
\begin{array}{l}
\text{T T C G G} \\
\text{A A G C C}\diagdown\diagup\text{G G C C T} \\
\hphantom{\text{A A G C C}}\text{T}
\end{array}
$$

↓

$$
\begin{array}{l}
\text{T T C G G C C C G} \\
\text{A A G C C T G G C C T}
\end{array}
$$

6.7 *IN VITRO* SYSTEMS FOR STUDYING DNA REPLICATION

6.7.1 *dna* mutants

Increased understanding of DNA replication has come about as a result of the intensive use of bacterial mutants, especially temperature-sensitive mutants, i.e. bacteria which grow normally at low temperatures but cease to grow at a higher, restrictive temperature.

The study of such conditional lethal mutants, defective in DNA synthesis, has demonstrated that a number of proteins are essential for DNA replication in *E. coli*. Some of these proteins together with the genes which code for them are listed in Table 6.5. Most of these proteins have now been purified and used with some success in attempts to reconstruct a system which will replicate DNA *in vitro*.

The mutant cells, when transferred to the restrictive temperature, may immediately cease to make DNA or they may continue to grow for some time. The former mutants are defective in the elongation of the replicating DNA molecule whereas the latter are defective in the process of initiation of new rounds of DNA replication.

The genetic approach thus indicates the involvement of a number of gene products in DNA replication but is able to say little about the detailed function of these proteins. The biochemist seeks to purify the proteins involved and to characterize their individual functions. A first step in this process involves the circumvention of the barrier of the cell membrane. This has been done in a number of ways [315, 316].

6.7.2 Permeable cells

Treatment of *E. coli* with lipid solvents (ether [317] or toluene [318]) although inhibiting cell division appears to leave the

Table 6.5 *E. coli* genes whose products play a role in DNA synthesis

Gene	Enzyme	Effect of mutation on DNA replication
dnaA	Initiation protein	No initiation
dnaB	Helicase	No synthesis
dnaC	Initiation protein	No initiation
dnaE (*polC*)	DNA polymerase III (α)	No synthesis
dnaF (*nrd*)	Ribonucleotide reductase	No synthesis
dnaG	Primase	No synthesis
dnaQ (*mutD*)	DNA polymerase III (ε)	Increased mutation
dnaS (*dut*)	dUTPase	Short Okazaki pieces
dnaT	i (primosome protein)	
dnaX/Z	DNA polymerase III (τ and γ)	Low processivity
polA	DNA polymerase I	Slow joining of Okazaki pieces, uv sensitive
polB	DNA polymerase II	Normal
lig	DNA ligase	Slow joining of Okazaki pieces
ssb	SSB	No synthesis
gyr	DNA gyrase	
top	DNA topoisomerase I	
uvrD	Helicase II	Increased mutation
priA	n′ (primosome protein)	
priB	n (primosome protein)	
priC	n″ (primosome protein)	

cells and constituent enzymes intact yet renders the membrane permeable to small molecules including nucleotide precursors. Such cells when incubated with the four deoxyribonucleoside triphosphates, ATP and Mg^{2+} continue semi-conservative replication. The products of such replication are the 10S Okazaki pieces which are only joined into high-molecular-weight DNA in the presence of NAD, the cofactor for *E. coli* DNA ligase. It was with toluenized cells and [α-^{32}P]deoxyribonucleotides that Okazaki was able to show the covalent linkage of RNA and DNA (section 6.3.4).

6.7.3 Cell lysates

The lysate formed on disintegration of the cell membrane is capable of maintaining DNA synthesis under certain conditions [319]. It is very important not to dilute the macromolecular constituents of the cell but to maintain them near their concentration *in vivo*. One way in which this has been achieved is by lysis of cells on a cellophane disc held on an agar surface. Small molecules are then allowed to diffuse through the disc into the concentrated lysate. Semiconservative DNA synthesis occurs in this and similar systems if the lysate is prepared from wild-type *E. coli*, but when it is prepared from *E. coli* cells defective in DNA synthesis then the lysate reflects the *in vivo* capabilities of the cells. Thus lysates prepared from mutants temperature-sensitive in the genes *dnaB*, *dnaD* or *dnaG* (which fail to extend growing DNA chains at the non-permissive temperature; Table 6.5) only synthesize DNA at low temperatures. In lysates prepared from mutants temperature-sensitive in the genes *dnaA* or *dnaC* (which fail to initiate DNA synthesis at non-permissive temperatures) DNA synthesis fails to occur only if the cells are maintained at the non-permissive temperature for one generation time before lysis [320].

This system has proved very useful in the purification of the products of these *dna* genes [315, 316]. Fractions containing the *dnaG* gene product from wild-type cells, when added to a lysate from a mutant temperature-sensitive in *dnaG*, are able to restore DNA synthesis at the non-permissive temperature. Although the products of the *dna* genes have all been at least partially purified by this *in vitro complementation assay*, functions for all these proteins are not readily established. However, they have been used with some success in reconstruction experiments using fully defined components (see below).

Permeablized cells [321] and lysates from eukaryotic cells [322–324] also show a limited ability to continue DNA synthesis when provided with the four deoxyribonucleoside triphosphates, ATP and Mg^{2+}. Eukaryotic cells can be rendered permeable to deoxyribonucleotides by treatment with hypo-osmotic buffers. DNA synthesis continues in these cells for about 30 minutes before slowing down. This pattern of synthesis is probably due to completion of pre-existing replicons coupled with an inability to initiate new replicons [325].

Cell lysates show similar characteristics [326] but isolated nuclei are much more limited in their capabilities unless supplemented with cytosol fractions [327]. In general, isolated nuclei appear capable of extending Okazaki pieces initiated *in vivo* but have only a limited capacity for ligating or initiating Okazaki pieces.

The disadvantage with lysates of eukaryotic cells is that very little is known about the nature of the DNA template or product. As with prokaryotes (see below), model systems have been used. Thus lysates of cells infected with SV40 or polyoma virus synthesize predominantly viral DNA *in vitro* (section 6.10.4) [322, 323], and soluble replication systems have also been prepared from adenovirus-infected cells and lysates made from cells infected with herpes simplex virus synthesize predominantly herpes simplex DNA *in vitro*.

6.7.4 Soluble extracts

Although useful in the identification and purification of some of the factors necessary for DNA synthesis, the cellophane-disc assay is of limited value in defining the function of the various gene products and, moreover, it relies almost entirely on possession of temperature-sensitive mutants for every protein involved.

A simplification involves the use of soluble extracts from *E. coli* free of the bacterial DNA. When cells are lysed with lysozyme the bacterial chromosome can be sedimented by centrifugation at $200\,000\,g$. Similar preparations can be made from animal cells, especially the eggs of *Drosophila* and *Xenopus*. The soluble preparation is capable of performing specific stages of DNA synthesis when supplemented with simple DNA substrates.

For instance, in order to study lagging strand synthesis, a soluble extract would be supplemented with a single-stranded DNA template with no ends to loop back, e.g. the DNA from the small, single-stranded, circular DNA phages ϕX174 or M13 (Fig. 6.22(a)).

In contrast, to study leading strand synthesis, a double-stranded DNA substrate with a nick in one strand is required. The 3'-OH at the nick can act as a primer to pro-

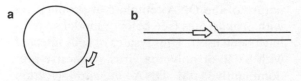

Fig. 6.22 Model substrates used to study DNA synthesis *in vitro*. A single-stranded cyclic DNA molecule (a) can be used to study initiation and elongation on the lagging strand. A nicked duplex (b) is used to study leading strand synthesis.

duce a fork (Fig. 6.22(b)). This situation is complicated because a single strand of DNA is displaced on fork movement and this may, in turn, be used as a template for lagging strand synthesis.

Once again, possession of mutants is an invaluable aid but the advantage of this system over the crude lysate is that specific steps in DNA replication may be studied and hence information on the function of particular proteins may be obtained.

6.7.5 Reconstruction experiments

The ultimate aim of these experiments is to achieve the ordered replication of a complex DNA molecule, such as the *E. coli* chromosome, using totally defined components. Initially simpler systems have been studied. The requirements for the conversion of single-stranded phage DNA into the double-stranded replicative form (RF) and the replication of RF have been delineated and the reactions carried out using purified components. Progress was not all plain sailing for, in addition to the components identified with the aid of mutants and purified by the *in vitro* complementation assay (see above), addition of soluble extract was required. From this extract, several further components have been purified for which, initially, no mutants were known.

The detailed results of these reconstruction experiments will now be considered.

The simplest situation to investigate is the elongation of a primer, hydrogen-bonded to a template DNA strand. When the components of such a system are known the synthesis of the primer can be investigated. Synthesis of a primer and its elongation as DNA is the reaction which occurs on the lagging strand of the replication fork, i.e. Okazaki piece synthesis (section 6.3). As indicated above, in order to ensure that a primer is being synthesized it is important to use a cyclic, single-stranded template molecule which has no free ends which can bend back and hybridize to the template strand (Fig. 6.11). Such templates are naturally provided by the DNA of the small bacteriophages fd (M13), G4 and ϕX174. The replication of these phage in *E. coli* takes place in three stages [42, 43].

1. The original infecting cyclic DNA molecule (known as the viral or plus strand) is used as a template on which a complementary minus strand is synthesized (SS → RF).
2. The resulting replicative form (RF) undergoes several cycles of replication by the so-called rolling circle mechanism (RF → RF) (section 6.9).
3. The RF molecules are used as templates from which progeny plus strands are synthesized. These are subsequently packaged into the viral coat as the mature virus leaves the cell (RF → SS).

Although of similar size the replication of the three small phage mentioned above differs in a number of significant steps which can be explained as a result of differences in DNA sequence.

6.7.6 Synthesis of Okazaki pieces (lagging strand synthesis)

At the replication fork, the lagging strand is present as single-stranded DNA upon which

Okazaki pieces have to be initiated and elongated.

(a) *Prokaryotes*

When single-stranded phage DNA is added to a cell-free extract it mimics the situation which occurs when these viruses infect a cell.

$$
\text{Phage DNA} \rightarrow \text{DS} \rightarrow
\begin{cases}
\text{DS} \rightarrow \text{SS} \\
\text{DS} \rightarrow \text{SS} \\
\text{DS} \rightarrow \text{SS} \\
\text{DS} \rightarrow \text{SS} \\
\text{DS} \rightarrow \text{SS}
\end{cases}
\rightarrow
\begin{array}{c}
\text{Many} \\
\text{new} \\
\text{phage}
\end{array}
$$

So a study of lagging strand synthesis is also a study of the first step in the replication of these phage, i.e. SS → DS or RF (replicative form) DNA.

In the cell, single-stranded DNA is always complexed with SSB and such a complex is formed as soon as the DNA from the small single-stranded phages (fd, M13, G4, ϕX174) is incubated with a host cell lysate or enters the host cell.

In fd DNA there remains exposed a hairpin loop which is able to bind *E. coli* RNA polymerase [328]. This enzyme synthesizes a short RNA which then acts as a primer for minus strand synthesis catalysed by DNA polymerase III holoenzyme [72, 329, 330]. This is able to process around the whole molecule until it comes back to the 5'-end of the primer. The primer can be removed by DNA polymerase I and the minus strand joined using DNA ligase (Fig. 6.23).

In G4 DNA a similar hairpin loop is exposed but this has a binding site for the DnaG protein – the primase [281, 331]. This enzyme is responsible for synthesizing the RNA primer in most prokaryotes including phage G4. The sequence of DNA around the origin of phage G4 and the sequence of the RNA primer synthesized by the primer is shown in Fig. 6.24 [21, 72, 331, 332]. Studies with phage ϕK have shown that the primase-binding site is more complex than shown in the figure and in fact two molecules of primase bind to a region of complex secondary structure having three hairpin loops and extending over 139 bases [333–335]. Although a primer of 28 nucleotides is shown in Fig. 6.24, in the presence of DNA polymerase as few as two ribonucleotides may be incorporated by primase before DNA polymerase takes over [336].

These two situations probably represent examples of viruses bypassing normal cellular mechanisms in order to enhance their own replication. These phage have introduced into their chromosome sequences that allow direct binding of RNA polymerase or DNA primase. These sequences occur only once per chromosome and so only one initiation event occurs every 5000 nucleotides or so. Such binding sequences do not occur regularly every 1500 nucleotides along the lagging strand in *E. coli* and are absent from the plus strand of phage ϕX174 DNA.

ϕX174 DNA may be more typical as no binding site for either RNA polymerase or primase is exposed. Comparison of the nucleotide sequences of the DNA of the closely related bacteriophages G4 and ϕX174 reveals a deletion in the latter covering the primase-binding site [331, 337]. In ϕX174 a complex is formed involving a series of proteins identified as DnaB, DnaC and proteins i, n, n' and n" (called X, Z, Y and W by Hurwitz) [43, 338, 339]. The name 'n' was given to a protein fraction (later separated into n, n' and n") which was sensitive to *N*-ethylmaleimide: 'i' is insensitive and is the product of the *dnaT* gene.

This complex, called the *preprimosome*, binds to a 55-nucleotide-long sequence which occurs once in ϕX174 DNA (around base

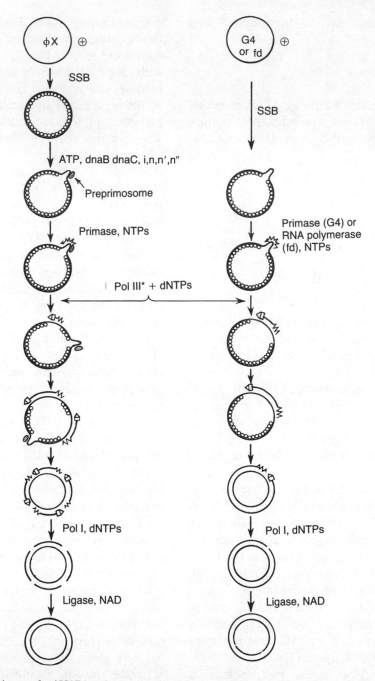

Fig. 6.23 Conversion of φX174, G4 and fd single-stranded DNA into duplex; ▷, DNA holopolymerase III; ⋀⋀, primer. (After Kornberg [42, 43, 344].)

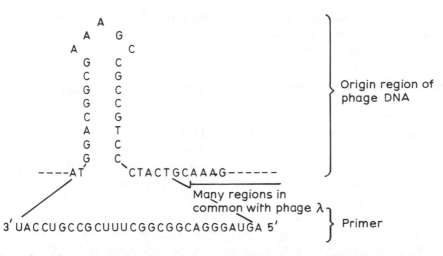

Fig. 6.24 Location of primer in phage G4 DNA (this untranslated region is absent from φX174 DNA [332, 336]).

2330) and forms a hairpin loop. It also occurs once on each strand of the duplex Col E1 plasmid as well as on *E. coli* DNA [340, 341] and this sequence, when inserted into fd DNA, renders this DNA capable of assembling the preprimosome. Protein n′ is a 76 000-M_r protein responsible for site selection [342, 343]. It recognizes a stem/loop structure in which the sequence AAGCGG is exposed in the loop and, once associated with the DNA, it binds proteins n, n″ and i and DnaC protein in an ATP requiring reaction. Six molecules of DnaB now bind and displace DnaC to form the preprimosome (Fig. 6.23) [344, 345].

The complex contains two helicases (n′ and DnaB) which migrate in opposite directions along the DNA causing it to form a loop around the preprimosome. This complex of several different proteins interacts with primase to form the primosome (Fig. 6.23) and an RNA primer is synthesized. As the primase moves off in the 3′ → 5′ direction along the template strand the DnaB helicase action causes the preprimosome to translocate in the opposite direction (5′ → 3′)

[338, 346, 347]. At various other sites the DnaB protein is able to affect the conformation of the DNA template thereby causing another molecule of primase to interact with the preprimosome and initiate synthesis of another primer [346, 348, 349].

In all, six to eight primers are made per round of replication on φX174 DNA [328]. They range in size from one to nine nucleotides long and, although they initiate with pppA-Pu, their internal sequence is random [347]. Their length is in part dependent on the availability of deoxyribonucleoside triphosphates and the DNA polymerizing enzymes.

In all cases the RNA primers formed are extended by DNA polymerase III holoenzyme which catalyses the addition of deoxyribonucleotides until the growing strand reaches the next RNA primer (Fig. 6.23). Because of the limited availability of DNA polymerase III holoenzyme several primers may be present awaiting extension, and rapid transfer of polymerase occurs from a completed Okazaki piece to the adjacent primer [350]. DNA polymerase I removes the RNA primer and completes the DNA chain in a nick-translation type of reaction (section 6.4.3(e)). Finally the chains are joined by DNA ligase to form the cyclic duplex molecule (RF). These reactions have been reviewed [351].

These are the reactions which occur on the lagging side of the replication fork (Fig. 6.10). It is important to establish a preprimosome on the lagging side of a fork but, once present, the preprimosome remains bound and can bind primase and initiate retrograde synthesis at multiple sites. These reactions are also those occurring during the initial (SS → RF) stages of replication of the small single-stranded phages; and following conjugal transfer of plasmids when a single strand of DNA is passed to the recipient bacterium (section 3.6). Even on completion of the synthesis of a double-stranded RF molecule, the preprimosome remains bound to the plus strand.

(b) *Eukaryotes*

In an attempt to characterize the enzymes involved in lagging-strand synthesis in eukaryotes similar closed cyclic single-stranded DNA molecules have been incubated with extracts of *Xenopus* eggs or *Drosophila* embryos [352–354, 364]. Such extracts were initially chosen as these tissues have the potential for very high rates of replication *in vivo*. These extracts, and extracts of cultured cells, catalyse the synthesis of an RNA primer which is extended to form a DNA chain of about 1000 nucleotides in length, complementary to the single-stranded template. The sequence of the RNA/DNA junction is random but the primers initiate preferentially with the sequence pppA. On fractionation of the extracts it appears that SSB and DNA polymerase α with its associated primase activity are the only proteins

required (section 6.4.7(a)) [91, 108, 116, 356, 357, 815].

6.7.7 Leading-strand synthesis

This reaction requires the extension of the growing daughter DNA strand and the concomitant unwinding of the duplex ahead of the point of polynucleotide synthesis. A suitable substrate to study this reaction is a nicked duplex molecule and this has been used in particular by Albert's group to investigate the proteins required for phage T4 DNA replication.

On a nicked DNA duplex a complex of six enzymes is able to displace the free 5'-end at the nick and extend the 3'-OH. The proteins required include a DNA polymerase (product of T4 gene 43); the archetypal SSB (gene 32 protein); polymerase accessory proteins (gene 44, 62 and 45 proteins) which increase the rate and processivity of polymerase action, and a helicase (gene 41 protein) which is required to unwind the displaced strand from the template [219, 303, 304, 358–361]. (Another helicase is required to allow the replicating complex to pass a bound RNA polymerase molecule [362].)

In conjunction with gene-61 protein, the gene-41 protein also has a primase function and the complex will synthesize and extend pentanucleotide primers on the displaced (lagging) strand [219, 362, 363]. This mixture of seven enzymes is thus able to catalyse the reaction going on at the replication fork. The Okazaki pieces synthesized on the lagging side are, however, some 10 000 nucleotides long as the complex lacks the capacity to remove RNA primers and join Okazaki pieces.

Further details as to how replication is thought to occur at the replication fork are given in section 6.4.5.

6.8 INITIATION OF REPLICATION – SIMPLE SYSTEMS

The results of the Meselson–Stahl experiment suggested that DNA replication would be found to be a sequential process and that successive rounds of replication would not begin at random positions on the chromosome. It is now clear that chromosomes from most viruses and prokaryotes, whether containing linear or cyclic DNA, initiate their replication at one specific site. For small DNA molecules, or for fragments of larger molecules, these sites can be visualized using the electron microscope when they appear as double-stranded 'bubbles' (Figs 6.5 and 6.7) [5, 365–368].

Cairns was able to visualize the whole of the replicating *E. coli* chromosome autoradiographically by growing the bacteria for several generations in the presence of [³H]-thymidine (Fig. 6.25), and he showed that the chromosome exists as a continuous piece of double-stranded cyclic DNA [3].

6.8.1 Methods of locating the origin and direction of replication

That the initiation of the replication bubble occurs at a unique site has been shown by a number of different techniques for viral, plasmid, bacterial and mitochondrial DNA. Electron microscopy studies have shown that initiation of replication of the linear T7 phage chromosome occurs 17% from one end [366] but cyclic chromosomes appear featureless in electron micrographs and so it is not immediately possible to define the position of replicating bubbles. However, Schnös and Inman [7] were able to show that initiation of replication occurred at a particular site with reference to the partial denaturation map of phage λ and in the plasmid Col E1 [368] and in the simian virus 40 [5]

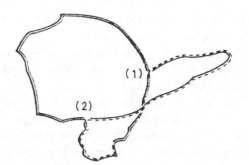

Fig. 6.25 Diagrammatic representation of an autoradiograph of the chromosome of *E. coli* labelled with [³H]thymidine for two generations [3]. The chromosome has been two thirds duplicated and the two replication forks are indicated as (1) and (2). (By courtesy of Dr John Cairns.)

chromosome replication is always initiated at a fixed distance from the site of action of the *Eco*R I restriction endonuclease (Figs 6.7 and 6.5).

In most cases (Col E1 is an exception) the midpoint of the replication bubble stays fixed with respect to the ends of the molecule. As the bubble gets larger it is thus clear that replication is occurring in both directions from the origin, i.e. replication is bidirectional.

As explained in chapter 3 those genes located near the origin of replication of the *E. coli* or *B. subtilis* chromosome are present in more than one copy per cell in growing cells (Fig. 3.15). It is possible to measure gene copy number in growing cells relative to non-growing cells (where all genes are present in one copy per cell) and hence, knowing the genetic map, locate the origin of replication [369, 370]. Early experiments to measure gene copy number relied on measuring transforming or transducing activity and required a large panel of mutant strains. More recent experiments measure the copy number or specific radioactivity of identifiable restriction fragments, an approach which has shown that replication of adenovirus DNA is initiated at both ends of the linear duplex molecule [371].

Initiation of the multiple, tandemly-arranged replicons of eukaryotes is considered in section 6.15.

The rate of replication of DNA is dependent on the temperature and the supply of nutrients, particularly deoxyribonucleotides. The minimum time required to replicate the *E. coli* chromosome is about 40 minutes, implying a rate of synthesis of about 1700 base pairs per second [372]. As replication is bidirectional the rate at each fork is about 850 base pairs per second or 14 μm per minute. (Compare this with the rate of transcription which proceeds at 35–40 nucleotides per second in *E. coli*.)

6.8.2 Initiation of replication of double-stranded DNA molecules

There is no single answer to the question 'what is the mechanism of initiation of replication, and how is it controlled?' On the face of it, each individual example appears different; yet there is a common problem; that is, how to produce a primer for DNA polymerase at a specific place and time? Primers may be of DNA (as in rolling circle replication) or of RNA, and the latter *may* be made by primase or RNA polymerase. Primers made by primase are usually short but those made by RNA polymerase are often several hundred nucleotides long.

We have seen how initiation of Okazaki pieces can occur on single-stranded molecules (lagging strand synthesis), and the problem which must be solved with a DNA duplex is how to generate a single-stranded region at the origin. One way in which this can be done is by transcription through the origin in which case the transcript *may* become the primer for DNA polymerase. Alternatively, specific proteins may bind to, and cause unwinding of, the origin region, thereby allowing primase to bind.

Duplex circles may replicate via a theta (θ) or Cairns structure in which the parental strands remain intact (Fig. 6.26(a)). Initiation involves production of a primer which is extended as DNA to produce the leading strand. Lagging-strand synthesis may start almost immediately or may be delayed for a short time to produce a D-loop (Fig. 6.26(c)) or until the replication fork has traversed the whole replicon, in which case synthesis leads to displacement of a single strand (displacement synthesis).

If lagging-strand synthesis proceeds through the origin region (Fig. 6.26(a)), or if separate initiation mechanisms exist for clockwise and anticlockwise replication (Figs 6.29 and 6.30), then replication will be bidirectional.

Linear duplex molecules may initiate internally by similar mechanisms (Fig. 6.26(b)) but some linear duplex molecules initiate replication at the ends leading to displacement synthesis (Fig. 6.26(e)). Initiation may occur at both ends or a linear single-stranded molecule may be produced alongside one daughter duplex.

Duplex circles often replicate by the rolling circle mechanism when initiation requires the action of a sequence-specific nuclease to generate a nick with a 3'-OH (Fig. 6.26(d)). (Excision repair (section 7.3.2) is initiated by a comparable nuclease.) DNA polymerase can use the 3'-OH as a primer in a strand displacement reaction.

A comprehensive review of prokaryotic DNA replication was published in 1985 [373].

6.9 ROLLING CIRCLE REPLICATION

In section 6.7.6 we saw the first stage in the replication of small, single-stranded DNA phages such as M13 and ϕX174, in which the

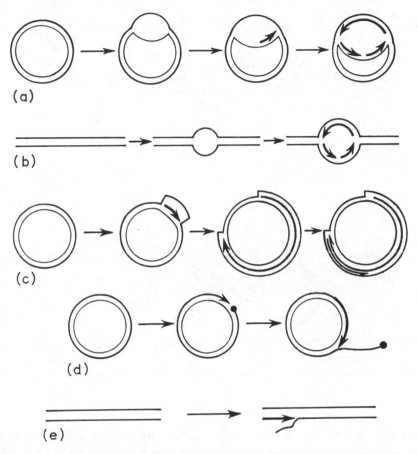

Fig. 6.26 Different ways in which initiation of replication can occur. (a) and (b) represent Cairns (θ) structures formed by cyclic and linear chromosomes, respectively; (c) is a D-loop and (d) a rolling circle; (e) represents an alternative method of initiation of replication of linear chromosomes.

single-stranded DNA was converted into a duplex circle: the replicative form (RF). These RFs now replicate via a rolling circle mechanism [351]. To initiate replication of the RF requires the intervention of a specific bacteriophage-coded protein (coded for by gene A of ϕX174). This protein makes an endonucleolytic cut in the plus (viral) strand of the RF DNA (Fig. 6.27). The site of the cut is specific and occurs between nucleotides 4305 and 4306 of ϕX174 RF DNA, within the region coding for the endonuclease. There is evidence that the endonuclease acts preferentially on the phage DNA mol-

ecule by which it was coded. The enzyme from phage ϕX174 is called the A (or cisA) protein and it recognizes a site of 2-fold symmetry which may be present as a hairpin structure [374].

The unique site of action of the gene II nuclease from bacteriophage fd (very closely related to M13) is between bases 5781 and 5782 which is adjacent to (24 bases 3', on the plus strand) the unique binding site of RNA polymerase (f in Fig. 6.27). In bacteriophage G4 the A protein acts on the opposite side of the DNA molecule (between nucleotides 506 and 507) to the primase-binding site (nucleo-

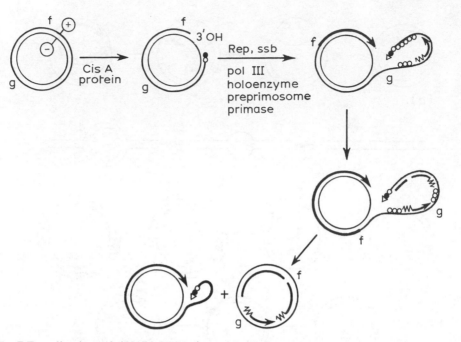

Fig. 6.27   RF replication of φX174 DNA (g and f correspond to the primase and RNA polymerase-binding sites on single-stranded DNA from phage G4 and fd, respectively, and, hence, indicate the position of initiation of minus strand synthesis with these phages) (based on the work of Kornberg and others [42, 43, 355, 376–379, 380].)

tides 3972–3992) (g in Fig. 6.27). The sequences recognized by the A protein of G4 and φX174 are identical and can be inserted into any plasmid. This makes plasmid replication dependent on the A protein and allows packaging of single-stranded plasmid DNA into preformed phage coats (see below) [341, 350].

Once the A protein has cut the plus strand, it remains attached to a dAMP residue at the 5′-terminus by means of a phosphotyrosine bond [375]. The 3′-end can act as a primer for plus strand synthesis by the DNA polymerase III holoenzyme which, together with the primosome, has remained attached to the DNA following SS → RF synthesis [350] (Fig. 6.27). Unwinding of the DNA is accomplished by combined action of Rep helicase, the A protein, and the SSB [293, 814]. DNA gyrase may also help to unwind the

DNA duplex ahead of the replicating fork but it is not required *in vitro* [293].

With phage fd the gene II protein does not remain bound to the 5′-end but rather there is an additional requirement for the host DnaA protein [376, 377].

As plus strand synthesis proceeds in the 5′ → 3′ direction, the 5′-end of the plus strand is peeled off the rolling, minus strand and it associates with SSB. The single plus strand, with its associated binding protein, thus resembles the initial bacteriophage DNA. In phage fd it remains unaltered until plus strand synthesis has gone almost a complete cycle when the RNA polymerase-binding site is exposed and minus strand synthesis is initiated [355]. In phage G4 the primase-binding site is exposed when plus strand synthesis has gone half way round the molecule [378]. With φX174 it is not certain whether or not an intact single-stranded plus strand is ever produced before the preprimosome reinitiates lagging-strand synthesis. However, as the preprimosome is retained on completion of SS → RF synthesis it appears likely that minus strand synthesis is initiated at several, non-random sites leading to the production of Okazaki pieces (Fig. 6.27) [379, 380].

When one round of plus strand synthesis is complete the A protein at the 5'-end of the plus strand is again brought adjacent to its site of action. It now recognizes the left half of the origin sequence, nicks it, and circularizes the displaced plus strand, i.e. the origin sequence is both an initiation and a termination sequence [380, 381]. A second round of plus strand synthesis is immediately reinitiated using the same complex of A protein and Rep [110].

As the ϕX174 infectious cycle proceeds, viral capsid proteins accumulate in the cell in the form of a prohead. This involves the products of viral genes F (major capsid protein), G (spike protein), H (spike-tip protein), B and D (non-structural proteins).

The prohead interacts with the replicating DNA–gene A protein complex in a reaction requiring phage gene C and J proteins, and this prevents initiation of further minus strand synthesis. Thus the single-stranded tail of the rolling circle is no longer available for conversion into duplex but rather is packaged into the phage prohead. Synthesis of mature phage particles by a reconstituted mixture of purified proteins has been achieved [42, 43, 382].

ϕX174 and G4 are lytic bacteriophage. However, the M13 family are non-lytic, filamentous bacteriophages and in these cases the single-stranded DNA tail produced late in RF replication becomes associated with the gene V protein. This binds DNA very tightly, displacing *E. coli* SSB, but is exchanged for coat proteins when the viral

DNA leaves the host cell [383]. It is the balance between gene V protein and gene II protein which determines the proportion of DNA molecules which undergo further rounds of DNA replication [384].

Rolling circle replication is not restricted to these single-stranded phages but often forms part of more complicated reactions. Thus it is involved in transfer of plasmid DNA from one bacterial cell to another [385] and in the transfer of DNA from the soil bacterium *Agrobacterium tumefaciens* to the plant genome [386, 387]. It is involved in the later stages of replication of phage λ [388] and viroids (section 3.7.10) are also believed to replicate by a rolling circle mechanism [389].

6.10 INITIATION OF REPLICATION USING PRIMASE

6.10.1 *E. coli* DNA replication

The origin of replication of the *E. coli* chromosome (*oriC*) has been located on a 245 bp sequence which can be cloned into small circular DNA molecules to give *oriC* plasmids. In such plasmids replication is bidirectional, dependent on DnaA protein and sensitive to rifampicin. The *oriC* region binds with high affinity to membrane proteins [390].

The *oriC* sequence is conserved in a number of Gram-negative species although it has been difficult to closely define the essential features [391–393]. In this region some base substitutions are deleterious; others not; but deletions are harmful suggesting some of the sequences may serve as spacers. The origin region probably provides multiple sites of interaction with replication initiation factors and must be present in a precise structure for activity.

In the *oriC* region there are four 9 bp

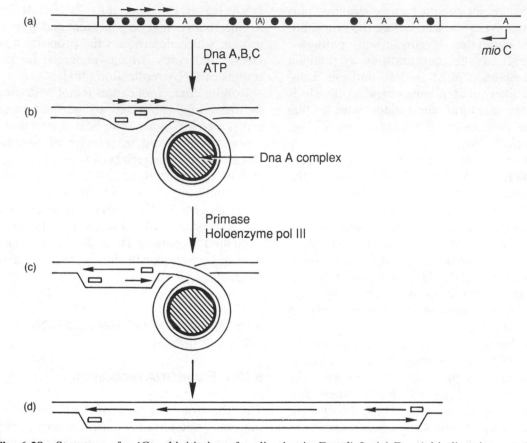

Fig. 6.28 Structure of *oriC* and initiation of replication in *E. coli*. In (a) DnaA binding sites are shown by an A and GATC sites by ●. The promoter of the *myoC* gene is indicated. Binding of a multimer of DnaA protein allows the melting of an AT-rich region to which DnaB can bind. (b) Primase and holopolymerase III initiate bidirectional replication to the left of the origin (c) and the rightward strand becomes the leading strand for clockwise replication (d).

binding sites for DnaA protein. Their sequence is TT(A/T)T(A/C)CA(A/C)A [393, 394]. Similar DnaA protein binding sites occur throughout the *E. coli* genome including within the *dnaA* gene. When they occur on the non-coding strand, the binding of DnaA protein terminates transcription [395, 698]. DnaA protein thus autoregulates its own synthesis and also links the expression of many genes to the replicative cycle. DnaA protein only binds to its recognition sequence when complexed with ATP and the ATP is

hydrolysed on release from DNA. There is then an exchange of ATP for bound ADP before rebinding can occur. This exchange is mediated by cAMP, the concentration of which, therefore, controls the binding of DnaA protein [396]. Although the concentration of DnaA protein can determine the time of initiation of DNA replication [397], there is little evidence that the concentration of either DnaA or cAMP normally play a major role in control of initiation.

Within *oriC* there are also eight or more

GATC sites which are methylated by the *dam* methylase (section 4.6). Methylation is important for initiation and indeed methylation of some sites may be essential [398, 399]. The promoter for the *dnaA* gene (which is adjacent to *oriC*) also contains GATC sites and methylation enhances expression. There is a lag of about 10 min between replication of origin DNA and its methylation and this transient hemimethylation following initiation is caused by the sequestration of the origin to membrane sites and is probably essential for chromosome segregation [400, 795, 796]. In addition, this sequestration blocks the initiation of new rounds of replication which will not start while the DNA remains hemimethylated [401, 789, 790]. The hemimethylated promoter of the *dnaA* gene is also inactivated by sequestration leading to inhibition of expression for a period of about 10 min following initiation of replication [790].

As well as depending on DnaA protein, initiation of replication of *E. coli* and the *oriC* plasmids depends on prior action of DNA gyrase and RNA polymerase and takes place preferentially in what might be a nucleosome-like complex produced by interaction of the DNA with HU protein (section 3.5.1).

The structure of part of *oriC* is shown in Fig. 6.28. Although there are two promoters for RNA polymerase within the origin region there is a strong promoter (*mioC* or p16 kDa) just outside the origin region and transcripts from this promoter enter *oriC* in a counterclockwise direction and some terminate at RNA/DNA transition sites associated with the GATC and DnaA protein binding sites (● and A in Fig. 6.28). Expression from the *mioC* promoter is under stringent control (i.e. it does not occur in the absence of protein synthesis – section 12.10.1), is inhibited when DnaA protein binds to a region near the promoter and is stimulated by *dam* methylation [402–404]. These RNA transcripts may form primers to initiate replication but this is not their only, nor their major, function. The principal function of transcription is to alter local DNA structure and allow an initiation complex to be formed at the origin (transcriptional activation) [405].

DnaA protein, in an ATP-activated form [406], interacts with origin DNA covered with HU protein and the DNA wraps itself around a core of 20–40 DnaA protein molecules. This leads to a stepwise opening up of the three 13 bp AT-rich regions which also bind DnaA protein [407] and this allows interaction of both strands with a DnaB/DnaC/ATP complex [408]. On binding to single-stranded DNA, DnaC is released and the helicase action of DnaB protein coupled with gyrase activity leads to opening up of the whole *oriC* region. This could allow the RNA transcripts to be used as primers for DNA synthesis by holopolymerase III [409, 410] in a reaction mediated by RNase H (section 6.11.3), but this is unlikely to occur *in vivo*.

Rather, DnaB is believed to act as a guide for primase to form primosomes on both strands of DNA. This allows initiation of replication in both the clockwise and counterclockwise direction [405, 411].

In this way the two leading arms of the fork are initiated for bidirectional replication (Fig. 6.28) [412, 413].

A system has been reconstructed from purified proteins which has enabled initiation to be separated into two stages [414–416]. In the absence of DNA polymerase III holoenzyme an RNA primer is made but not extended. If an inhibitor of *E. coli* RNA polymerase is now added along with DNA polymerase III holoenzyme, no further primers are made but those already present can now be extended. Topoisomerase I, RNase H and HU protein are required to prevent initiation at sites other than *oriC*. In addition to RNA polymerase, the production of a specific primer requires DNA gyrase and the products of the *dnaA*, *dnaB* and *dnaC* genes. Primase is not essential but when present it will interact with the preprimosome to synthesize a primer [417]. This shows that the *mioC* transcript can provide primers in this *in vitro* system. The second stage requires gyrase, single-stranded DNA-binding protein, and the products of the *dnaB* and *dnaC* genes in addition to the polymerase III holoenzyme. *In vivo* there is evidence that the products of genes *dnaI* and *dnaP* (and possibly also *dnaJ*, *dnaK* and *dnaL*) also play some part in initiation of replication.

6.10.2 Replication of phage lambda DNA

Initiation of replication at the λ origin is very similar to that at the bacterial origin but requires the products of phage genes *O* and *P* in addition to host functions [418].

The λ DNA molecule is linear but has sticky ends (section 3.7.6(a)) and cyclizes immediately on entry into the cell and this is essential for replication [419]. Replication is initiated at a unique origin and proceeds bidirectionally to give theta (θ) forms (cf. *E. coli* replication). However, at later stages of infection rolling circles are found.

Ori λ is a 65 bp region lying in the coding region for protein O [418]. It has four 18 bp repeats and a very AT-rich region (Fig. 6.29). Dimeric O protein molecules bind to the 18 bp repeats [420] and P protein then acts as a bridge to bind DnaB helicase, i.e. O and P proteins take the place of DnaA and DnaC proteins of the host cell.

The products of the *E. coli* genes *dnaJ* and *dnaK* (which are heat shock proteins) then bind to form an initiation complex [421, 422]. In the presence of ATP, SSB and

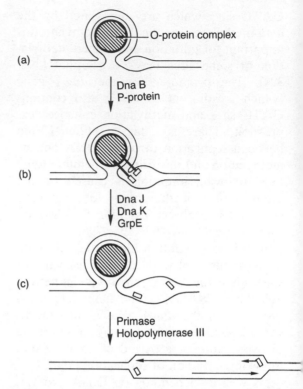

Fig. 6.29 Initiation of phage lambda DNA replication. The O protein forms a complex at the origin leading to melting of an AT-rich region (a) to which DnaB and P protein bind (b). The heat shock proteins DnaJ, DnaK and GrpE are required to dissociate this complex and allow DnaB to act as a helicase (c). Primase and holopolymerase III lead to initiation of bidirectional replication (d).

another heat shock protein, GrpE, there is a partial disassembly of the complex with release of the P and DnaJ proteins [422–424]. This activates the helicase activity of the DnaB protein in the same way as happens when the the DnaC protein is released from the *oriC* complex, i.e. proteins P and DnaC suppress the helicase activity of DnaB and the heat shock proteins are required as 'chaperonins' to catalyse protein disassembly reactions. The supercoiled λ DNA molecule becomes unwound at an AT-

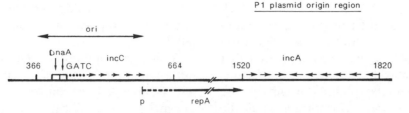

Fig. 6.30 The origin region of plasmid P1. DnaA binds to two regions in *ori* and there are five GATC sites (●). The *incC* and *incA* repeats are sites of binding RepA whose promoter lies within the origin region.

rich region to the right of the origin [425–427].

Primase (DnaG) can then bind to DnaB to form primosomes. If these form on both strands, bidirectional replication catalysed by DNA polymerase III holoenzyme is initiated. However, *in vitro* only rightward replication is initiated implying that leftward progression back through the origin is inhibited. Although not required *in vitro*, *E. coli* RNA polymerase is required *in vivo*, but whether it catalyses synthesis of a transcript which facilitates opening up the DNA duplex, or a transcript that can act as a primer for leftward replication is not clear [422, 428–431]. DNA gyrase is also required to release the tension arising from unwinding the DNA.

See sections 3.7.8 and 10.1.5 for control of lysogeny of phage λ [432, 433].

6.10.3 Low copy number plasmids

Replication of plasmids has been studied extensively as it was thought that they were a simple system which would reveal the secrets of bacterial chromosome replication. Simple they are not, and no two plasmids are alike. However, with the use of plasmids as DNA cloning vectors, control of copy number has assumed major importance [434].

A number of plasmids e.g. P1 (derived from a prophage), mini F and the R plasmids

are present in *E. coli* at one or two copies per cell. For replication they require DnaA protein but also a plasmid-coded initiator protein known as RepA. Binding of RepA to the origin region is required to enable DnaA to bind (i.e. binding of RepA takes the place of transcriptional activation seen in *E. coli*) but synthesis of RepA is autoregulated in that it also binds to its own promoter. RepA also binds to a separate region of the plasmid to regulate copy number.

Plasmid P1 will be used as an example of this sort of control [435, 436]. Miniplasmid P1 is a 1500 bp segment of prophage P1 that replicates as a plasmid. It has a 245 bp origin region and a 959 bp region coding for RepA. In addition there is a 285 bp region known as *incA* which regulates copy number (Fig. 6.30). The origin region consists of two DnaA binding sites followed by a region containing five GATCs which must be methylated for replication to be initiated, and five 19 bp repeats to which RepA binds (the *incC* region) – the last of these covers the promoter for *repA*. Although methylation of the GATCs may regulate segregation, other possible functions are more likely as plasmid P1 will not replicate in Dam⁻ cells. A composite of parts of plasmids P1 and F does replicate in Dam⁻ cells using the F plasmid origin but when this is inhibited the P1 plasmid origin still cannot function. Both origins function in Dam⁺ cells [437]. Plasmid P1 origin thus differs from *oriC* and this is confirmed by *in vitro* studies which show unmethylated P1 origins will not replicate. Phage P1 uses a different origin from the plasmid and it has been suggested that methylation may be the switch that determines whether the plasmid origin or the phage origin is used. The former would be switched off when undermethylated in the rapidly replicating, lytic cycle.

When RepA is synthesized it is complexed with DnaJ and is only activated by the action of the chaperonins DnaK and GrpE [801] (section 6.10.2). It then binds to the five sites in *incC* in a certain order. As it binds to its own promoter it switches off further synthesis. Binding of RepA to *incC* changes the DNA conformation and seeds binding of DnaA. This initiates replication in a manner similar to that described for the *E. coli* chromosome [699] (section 6.10.1). Any spare copies of RepA not bound to *incC* are mopped up by the nine RepA binding sites in *incA* thus regulating reinitiation. Deletion of *incA* leads to an increased copy number whereas extra copies of RepA binding sites introduced into *incA* inhibit replication by sequestering RepA. The reverse is true for *incC* which is essential for replication. Plasmids with interchangeable RepA are incompatible.

There is a paradox in this situation. If the RepA protein is sequestered by *incA* then the RepA gene will be activated as repression requires a finite concentration of RepA to bind to its promoter. Differential binding of RepA to its various binding sites could help to solve the dilemma. However, it has been shown that by DNA looping, the RepA bound to *incA* can still act as a repressor but not as an initiator protein, i.e. RepA can bind simultaneously to its promoter and to the *incA* control region thus sequestering the origin into a non-productive state[438]. It is also believed that RepA can bind to two different plasmid molecules and that either

intra- or interplasmid binding will reduce initiation of replication by steric hindrance [802].

A similar looping of DNA is involved in initiation of replication of plasmid R6K [439] and is involved in control of lysogeny of phage λ (section 3.7.8).

In plasmid R1, RepA only acts efficiently in *cis* (i.e. on the DNA molecule by which it was coded). In a *rho*-dependent termination (section 9.2.4) the 3′-end of the RepA mRNA remains associated with the DNA until translation is complete. RepA then associates with *ori* either by sliding along the DNA or by a looping mechanism [440].

In plasmids R1 and R6K, RepA synthesis can also be inhibited by another mechanism. A region (*copT*) in the leader of the RepA transcript hybridizes to a short complementary RNA (*copA*) which is made constitutively and this blocks translation of RepA mRNA. The formation of this hybrid is enhanced by RepA providing a further level of autoregulation [441–444]. This theme is extended in the case of plasmid Col E1 (section 6.11.3).

Some plasmids replicate via a rolling circle mechanism [445–447].

6.10.4 Papovavirus DNA replication

These similar, small, animal tumour viruses contain a cyclic DNA molecule which replicates as a θ-form [448], the parental strands remaining intact. They have been studied, not only for their own sake but as models for animal cell DNA replication. This is only poorly justified in the case of SV40 and polyoma for, although replication involves mainly host cell proteins and the DNA is present in the nucleus complexed with histones (i.e. as a minichromosome), replication is not restricted to one round per cell division. However, papilloma virus may be a

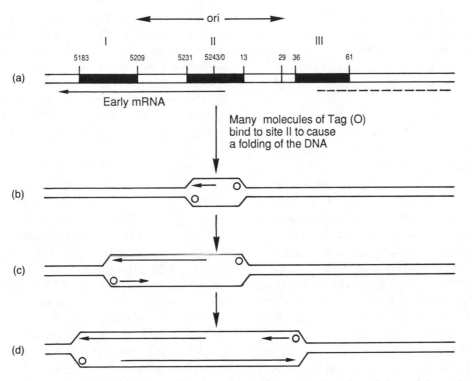

Fig. 6.31 Initiation of SV40 DNA replication. (a) indicates the location of the origin region and the binding sites for T antigen (I, II, and III). Many molecules of Tag bind to the origin to promote unwinding and Tag is then able to work as a helicase. DNA polymerase α/primase initiates leftward synthesis at one of several sites. Rightward synthesis is only initiated once the fork has left the essential origin region. DNA polymerase δ/PCNA soon takes over leading strand synthesis.

better example as it maintains a low copy number in mouse cells.

In SV40 and polyoma, initiation occurs at a unique site and proceeds bidirectionally. The origin region is also the site of initiation of transcription which also occurs in both directions (Fig. 3.19 and Fig. 6.31). One of the gene products (T antigen or Tag), synthesized early in infection, is essential for initiation of replication and its synthesis is autocontrolled [449]. It is an 82 kDa protein which must be phosphorylated at threonine 124 by the *cdc2* protein kinase for activity [450, 700]. There are three 26 bp palindromic sites in and around the origin region of SV40 to which Tag can bind and it binds most strongly to site I to block early mRNA synthesis. Only one of the Tag binding sites (site II) falls within the 65 bp essential origin region but elimination of both sites I and III reduces replication by 95% showing that two sites are required for efficient initiation [701].

Just outside the 65 bp origin region are two 72 bp enhancers and six Sp1 binding sites essential for early transcription and Tag production (Fig. 6.31 and section 10.4.4). Transcription is not essential for replication as it can be blocked with α amanitin if Tag is provided in *trans*, a situation that can be achieved by infecting *cos* cells (i.e. cells previously transformed with the *Tag* gene). Elimination of both enhancers and Sp1 binding sites, however, reduces replication dramatically showing them to play a role other than in the production of transcripts and the enhancer is absolutely required for polyoma virus DNA replication [451–453].

Tag, phosphorylated at threonine 124, binds to site II. Phosphorylation is not required for binding to site I and phosphorylation of serine interferes with binding to site II [450]. Hence, binding is stimulated by serine dephosphorylation of Tag which is dependent on a phosphoserine phosphatase [470]. Twelve molecules of Tag bind sequentially to four GAGGC sequences in site II (two on each strand) using the *N*-terminal domain of the protein to produce a change in DNA conformation and a stepwise untwisting of the DNA by 2–3 turns [454–459]. This opens up the SV40 duplex on both sides of the origin (including at the AT-rich region) in a reaction which involves ATP hydrolysis. In the presence of host-coded SSB (RP-A), the *C*-terminus of Tag acts as a helicase that moves bidirectionally from the origin ($3' \rightarrow 5'$ on the bound strand) [460–464, 662]. Thus Tag combines the function of *E. coli* proteins DnaA and DnaB.

The opened up origin is believed to be a suitable site at which the primase associated with DNA polymerase α can act. Isolation of nascent DNA from the origin region showed initiation to occur at many sites, each time with a short (7–8 base) ribonucleotide primer. Some 70% of primers start with adenosine, the remainder with guanosine [449, 465, 466]. On the early strand (i.e. the template for early mRNA synthesis) DNA primase will initiate at four major sites around nucleotide 5215 between T antigen-binding sites I and II, and the transition to continuous (leading strand) DNA synthesis occurs at about nucleotide 5210 (i.e. at the beginning of T antigen-binding site I). There are no initiations for late-strand synthesis in

the minimal origin region and it is proposed that only when leading-strand synthesis exposes primase-binding sites does initiation of late-strand synthesis occur [449]. *In vitro* results suggest that late-strand synthesis may also be initiated in any of the six repeated Sp1-binding sequences, removal of which reduces the efficiency of DNA replication *in vivo* [565, 467].

In vitro systems have shown that DNA polymerase α and its associated primase and SSB are involved in both initiation and continued SV40 DNA replication [102, 451, 465, 468, 470] and certainly host range specificity is partly a result of the ability of Tag to interact with primate polymerase α and not with murine polymerase α [469]. However, recent studies have also implicated DNA polymerases ε and δ (with its associated PCNA) in strand elongation (section 6.4.7(b)) [471–474, 812], and it is likely, once polymerase α/primase has initiated synthesis, that complexes containing both polymerase α and δ move bidirectionally away from the origin, catalysing lagging and leading strand synthesis, respectively [702, 807–810]. Later, one of these two polymerases may be replaced by polymerase ε [812]. *In vitro* systems have shown that elongation also requires topoisomerase, RNase H, DNA ligase, another DNA-dependent ATPase (RF-C), protein phosphatase PP2A and a double-stranded $5' \rightarrow 3'$ exonuclease and that it is inhibited by poly(ADP-ribose) polymerase [470, 475–477, 810]. RF-C (or Activator 1 – section 6.4.7(b)) is a complex of several proteins which function rather like the γ complex of *E. coli* DNA polymerase III to increase processivity (section 6.4.5) [807–809].

As well as binding to primate DNA polymerase α, SV40 Tag also binds to the cellular phosphoprotein p53 and this may explain the effect of SV40 on cell transformation and cell cycle control [478, 479, 797].

One characteristic of cellular DNA replication (section 6.15.1) is that it occurs once per cell cycle. Replication of SV40 and polyoma is different in that initiation of viral DNA synthesis occurs many times throughout the cell cycle (runaway replication). The closely related bovine papilloma virus (BPV), however, replicates only once per cell cycle in mouse cells and, like a plasmid, maintains a stable number (100–200) of molecules per cell [480]. The *C*-terminal region of BPV-E1 protein is essential for replication and binds to two regions at the origin of replication (cf. Tag, RepA and DnaA). However the *N* terminal region of E1 protein is required to maintain a stable copy number and mutants with deletions in the 5′-end of the E1 gene replicate extensively and kill the host cell. This is the runaway replication typical of SV40 growing in primate cells compared with the controlled replication typical of host cell replicons and wild type BPV.

Coinfection of cells with BPV and SV40 does not inhibit runaway SV40 replication but when cells are transfected with a composite SV40–4PV DNA molecule controlled replication is dominant, even in the presence of excess Tag bound to the SV40 origin. This shows that E1 protein does not exert its effect as an inhibitor of replication in *trans* and that control of copy number does not act at random on the pool of plasmid molecules. Rather, each individual plasmid replicon is allowed to replicate once per cell cycle by a *cis* acting regulator which distinguishes replicated plasmid from unreplicated plasmid. The relative position and orientation of the SV40 and BPV origins is of no consequence [481–483]. The fact that BPV replicates in synchrony with host replicons (section 6.15.1) implies it responds to factors which control host cell replication. These factors may interact with the *N*-terminus of the E1 protein DNA complex to delay chromosome segregation and prevent reinitiation of replication.

6.11 INITIATION USING RNA POLYMERASE

6.11.1 Replication of phage T7 DNA [484]

Initiation of bacteriophage T7 DNA synthesis occurs 17% from the so-called left end of the molecule. A replication bubble forms which expands bidirectionally. When the short side is completely synthesized a Y-shaped replicating chromosome is formed (Fig. 6.42). Concatamers, i.e. double length molecules, are found later in infection [485] (section 6.16.3).

T7 encodes most of its own replication protein in phage genes 1–6 [486, 487]. Gene 1 product is an RNA polymerase essential for the transcription of genes 2–6 and gene 2 protein inactivates the host RNA polymerase [488]. Gene 3 protein is an endonuclease and gene 6 protein an exonuclease involved among other things in the breakdown of host DNA.

There are two strong promoters for the T7 RNA polymerase in the origin region, closely followed by an AT-rich region (Fig. 6.32) [489, 490]. Transcription causes melting of the origin region which can be detected by treatment with a single-stranded endonuclease [491]. Transcripts of these promoters serve as primers for T7 DNA polymerase. At several sites within the AT-rich region there is a change over from RNA synthesis to DNA synthesis which proceeds unidirectionally towards the right [492].

In the presence of T7 gene 5 protein and the host protein thioredoxin (which together form the T7 DNA polymerase [493–495]) the T7 gene 4 protein catalyses an ATP-dependent unwinding of duplex DNA thereby promoting the DNA polymerase action

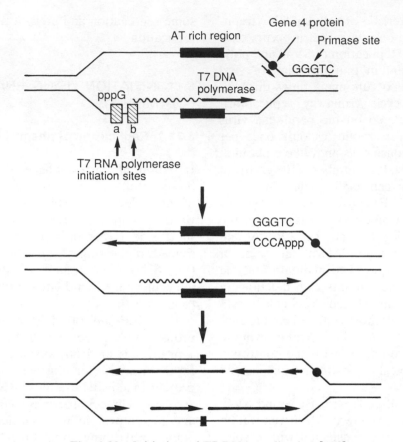

Fig. 6.32   Initiation of T7 DNA replication [490].

[218]. Gene 4 protein is thus a highly processive helicase and this complex is able to catalyse leading-strand synthesis; DNA-binding protein and gyrase further promote this action. Gyrase either relieves the positive supertwist generated by replication or it may actively help unwinding by imposing a negative supercoiling ahead of the fork.

A slightly different form of gene 4 protein has a second function in addition to its action as a DNA helicase [484]. At the replication fork, the two forms are present as a dimer, one subunit of which acts in a distributive manner as a primase for lagging strand synthesis [496]. This activity is initiated by the synthesis of a tetraribonucleotide of sequence pppACCC or pppACCA. Thus when a primase site is exposed by the rightward movement of the leading strand, synthesis of the lagging strand is initiated using the T7 gene 4 protein. The frequency of primase action is increased by a single-stranded DNA binding protein coded for by T7 gene 2.5 [497]. The tetraribonucleotide primer is extended by T7 DNA polymerase and the primase reassociates with the helicase at the fork. DNA synthesis now proceeds leftwards on the lagging strand through the origin region. With the help of the helicase action of gene 4 protein and single-stranded DNA-binding protein a fork is now established moving leftwards (Fig. 6.32) and replication is now bidirectional.

Thus as well as catalysing leading-strand

synthesis the gene 4-gene 5 protein complex can also catalyse initiation and elongation of Okazaki pieces on the lagging strand. It thus combines the functions of *E. coli* DnaB and DnaG proteins [487, 498–500].

Gene 6 exonuclease or host DNA polymerase I $5' \rightarrow 3'$ exonuclease can catalyse removal of RNA primers and the Okazaki pieces can be joined using either the host ligase or a ligase coded for by T7 gene 1.3.

6.11.2 Replication of phage T4 DNA

Replication of bacteriophage T4 DNA is far more complex and initiation of DNA synthesis appears to occur at several points along the long, circularly permuted, linear molecule [501, 502]. It has been proposed [503] that, as with T7, primary initiation occurs following synthesis of an RNA primer but in this case it is supposed that only the host RNA polymerase can act in this manner. Mutations in T4 genes 39, 52 and 60 lead to a delay in the onset of T4 replication and the DNA synthesis that does occur is dependent on the host cell gyrase [504]. These three genes code for an ATP-dependent topoisomerase [505] which may help to open up the duplex molecule. However, shortly after infection the host RNA polymerase is modified by interaction with the products of viral gene 33 and gene 55 and this affects promoter recognition rendering it unable to initiate DNA replication. Initiation now becomes dependent on genes 46 and 47 which code for enzymes involved in recombination (chapter 7). The model which has been proposed for these secondary initiations involves recombination between partially replicated molecules which contain single-stranded regions and the circularly permuted duplex molecules. This model explains the finding of complex branched molecules later in the T4 infectious cycle.

When cells of *E. coli* are infected with the T-even bacteriophages, the economy of the cells is completely altered so as to lead to the production of new phage DNA which differs from the host DNA in containing hydroxymethylcytosine in place of cytosine. These changes result in the production, in the infected cell, of a series of new and interesting enzymes [506–508].

1. Within a few minutes of infection a hydroxymethylase (the product of phage T4 gene 42) appears which brings about the conversion of dCMP into hydroxymethyldeoxycytidine monophosphate (dhmCMP).
2. At about the same time, a kinase is produced which phosphorylates dhmCMP to the corresponding triphosphate dhmCTP. Neither of these new enzymes is found in cells infected with phage T5 which does not contain hmC.

   The kinases for dTMP and dGMP are also greatly increased but not that for dAMP. This increase is due to the production of new enzymes which can be distinguished from the kinases present in the host cell before infection.
3. The formation of host DNA is prevented by the appearance of a pyrophosphatase (the product of phage T4 gene 56) which converts dCTP into pyrophosphate and dCMP which then acts as a substrate for the hydroxymethylase.
4. Five distinct glucosyltransferases are known to be induced after infection with T-even phages for the purpose of transferring glucose residues from uridine diphosphate glucose to the hmC of phage DNA in the proportions shown in Table 3.4. The glucosyltransferase found in T2 phage-infected cells transfers a glucose residue to hmC in the α-configuration. Two glucosylating enzymes are produced after T4 infection; one adds a glycosyl

group in α-linkage to hmC whereas the second also adds a glucose group but in the β-configuration. After T6 infection two glucosyltransferases are also produced. One adds a monoglucosyl residue to hmC in the α-linkage whereas the other reacts with the monoglucosylated groups on hmC to add a second glucose residue, the linkage between the residues being of the β-configuration.

5. A nuclease is produced which degrades DNA containing cytosine. T4 DNA which contains hydroxymethylcytosine is resistant.

In vivo the enzymes synthesizing the precursors for DNA synthesis are closely associated with the DNA polymerase [504, 509, 510] and *in vitro* the complex of enzymes preferentially uses deoxynucleoside monophosphates to make T4 DNA [511].

6.11.3 Plasmid Col E1 replication

This small double-stranded cyclic DNA molecule, from which the cloning vector pBR322 was derived, replicates from a unique origin but in one direction only [328].

Theta (θ) structures are formed and initiation does not involve permanent nicking of either strand [512, 513]. It is normally maintained with 10–20 copies per cell.

Up to three proteins bind near the origin and bring about supercoiling of the cyclic molecule [513] thereby exposing single-stranded regions to which *E. coli* RNA polymerase binds [514]. An RNA transcript (RNA II) is initiated 555 base pairs upstream from the origin (−555). This transcript, whose 5'-end adopts a complicated structure with at least 10 stems and loops, is not released from the origin region but a D-loop is formed with a DNA/RNA hybrid and single-stranded DNA [515–517] (Fig. 6.33). This

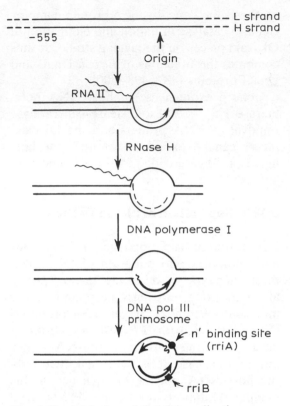

Fig. 6.33 Initiation of Col E1 replication.

hybrid is sensitive to RNase H (chapter 4) which fragments the RNA producing several primers which can be extended by *E. coli* DNA polymerase I with the help of DnaB helicase. This leads to the synthesis of a 6S piece of DNA (400 nucleotides) joined to a 20-long ribonucleotide primer, the whole remaining associated in the D-loop configuration [515, 516, 518, 519]. The DNA polymerase III replicating complex now takes over to extend the fork [325, 513, 515]. On the lagging side of the fork (on the L strand) there is an n' binding site (*rriA*) (section 6.7.6) at which lagging-strand synthesis can be initiated [325, 520].

However, lagging-strand synthesis cannot pass the origin region if a tail of unhybridized RNA II remains and so Col E1 replication remains unidirectional. In mutants lacking RNase H no leading strand synthesis is initiated at *ori* but the opening of the duplex allows initiation to occur on the lagging strand i.e. transcriptional activation [521, 522].

There are three GATC sites in the Col E1 primer promoter, methylation of which may affect membrane association (section 6.10.1) or which may play a role in the control of primer transcription [523].

There are two further complications to this story. As well as RNA II a second RNA (RNA I) is transcribed from the region −455 to −555 from the opposite strand of the DNA. This RNA, which can also assume a complex secondary structure, interacts with loops I–IV at the 5'-end of RNA II and prevents formation of loops VI–X and thereby interferes with primer formation. Once RNA II has been extended to longer than 360 nucleotides, loops VI–X can no longer be disrupted and so interaction of RNA I and RNA II must take place quickly for inhibition to occur. RNA I thus acts in *trans* to limit initiation and hence copy number of Col E1 and other incompatible plasmids (section 6.10.3).

A second *trans*-acting inhibitory element exists downstream from the origin (184 to 806 bp). The product of this region is the *rop* or *rom* protein (63 amino acids long) which enhances the binding of RNA I to RNA II [524–526]. It appears to bind to the 3'-end or mid region of RNA II thereby extending the period when RNA I and RNA II can interact. It also helps align RNAs I and II [520]. This protein appears to play no role in incompatibility but may regulate the balance between replication of duplex plasmid molecules (as described here) and the replication involved in the single-strand transfer reaction of conjugation (section 3.6).

In conjugal transfer a nick is introduced into one strand (the H-strand) and the endo-nuclease remains bound to the 5'-side of the nick. Rolling circle replication now leads to the synthesis of a new duplex, cyclic molecule and the parental H-strand can be passed to another bacterium through a conjugation tube in the form of a single strand (section 3.6). Cyclization follows transfer and conversion into a duplex molecule is initiated by the binding of n' protein to a recognition site (*rriB*) [527, 528]. Thus in conjugal transfer, replication of Col E1 closely resembles replication of ϕX174 except that the order of reactions is:

$$RF \rightarrow SS \rightarrow RF$$

A similar reaction is involved in the transfer of T-DNA from *Agrobacterium tumefaciens* to the plant cell [386].

6.11.4 Mitochondrial DNA replication

Animal cell mitochondria contain a closed circular duplex DNA molecule of about 16 kb that replicates unidirectionally from a unique origin. This origin region also contains the start site for transcription of both heavy and light strands of mitochondrial DNA [529, 530].

Transcription from both heavy and light strand promoters requires mitochondrial RNA polymerase and a mitochondrial transcription factor and the mouse mitochondrial heavy (H) strand transcript has a unique 5'-end at nucleotide 16 183 [530, 536, 537]. This transcript usually continues all the way around the mitochondrial genome but it can be processed by an RNase H-like activity. This enzyme (RNase MRP) is a ribonucleoprotein containing a nuclear-coded RNA of 265–275 nucleotides and it cleaves some copies of the transcript near a conserved sequence block (CSB) to provide a primer for DNA synthesis [529, 538]. Three such blocks (CSBI, II and III) are found in mouse mitochondrial DNA. Primers are from 80 to 155 nucleotides long, that is, similar in size to the primers found on phage T7 replication and the two RNA polymerases involved show some areas of homology [539].

Before initiation, the duplex undergoes partial unwinding in the presence of a topoisomerase to remove about 48 Watson–Crick turns [531]. Primers are extended by mitochondrial DNA polymerase γ [683] to produce a D-looped structure containing a 450-base length of newly synthesized DNA (the leading or heavy strand) (Fig. 6.26(c)) [529, 531, 532]. This small piece of newly synthesized 7S DNA is unstable and may turn over several times [533, 534] before it is extended. Extension occurs asymmetrically and unidirectionally. Thus although θ-forms are seen in the electron microscope one loop is a single-stranded parental H-strand displaced by synthesis of daughter H-strand.

D-loop DNA is extended and when H-strand synthesis has proceeded two-thirds of the way round the genome (99% in *Drosophila* mitochondria [535]) a sequence (O_L) is exposed which can form a stable hairpin structure [531, 532]. A 15–25 nucleotide RNA primer is synthesized here using mitochondrial DNA primase (another RNA containing enzyme [540]) which leads to initiation of L-strand synthesis [541–544]. Both strands are now extended and the two daughter molecules separate with one being still almost half single-stranded.

In sea-urchin oocytes mitochondrial DNA synthesis is similar to that found in mouse mitochondria but duplex synthesis occurs early with multiple initiations [545].

Yeast mitochondrial DNA contains several origins of replication. In petite strains in which a short stretch of DNA is excised and amplified, surrogate origins are used. In one petite strain the mitochondrial DNA contains only adenine and thymine nucleotides [546, 547].

Mitochondrial DNA in *Tetrahymena* and *Paramecium* is not cyclic but is a linear duplex [548–550]. In the former, replication is initiated near the centre of the molecule and 'eye-shaped' replication intermediates are seen in the electron microscope. Replication is bidirectional. In *Paramecium* replication proceeds unidirectionally from a cross-linked terminus [549, 550].

In kinetoplasts, the minicircle DNA may have more than one origin of replication of heavy strand synthesis and initiation of lagging strand synthesis appears to take place at multiple, random sites [551, 552].

The chloroplast DNA of higher plants and *Chlamydomonas* replicates via a D-loop structure [553].

6.12 TERMINAL INITIATION

Adenoviruses are animal viruses containing a linear duplex DNA molecule of about 36 000 bp. This replicates by a strand-displacement mechanism, initiation occurring at either end of the molecule (Fig. 6.34). The adenovirus DNA contains an inverted terminal repetition extending to the very ends of the genome. In different strains of adenoviruses the length of the terminal repeat varies from 103 to 162 bp [554, 555]. In adeno 2, three regions in this terminal repeat are required for initiation of replication. An AT-rich stretch from positions 10 to 18 is essential for binding of a 'terminal protein'–DNA polymerase complex (see below) and the regions 19–39 and 40–51 are involved in binding cellular transcription factors NFI and NFIII [555–563]. Covalently bound to both 5'-termini is a 55 000-M_r protein. Binding is by a phosphodiester link to the β-OH of a serine residue [554, 564].

Initiation requires host cell DNA topoisomerase I [565] and at least three virus-coded proteins: a 72 kDa SSB that may also bind to the ends of duplex molecules [554]; a

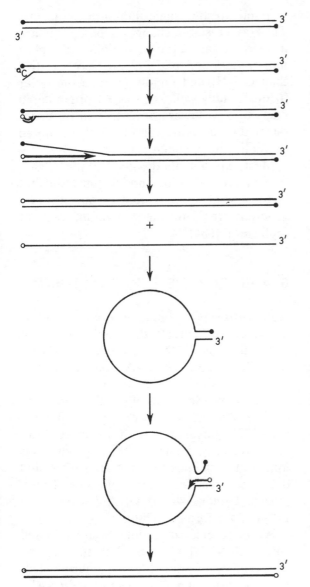

Fig. 6.34 Replication of adenovirus DNA. The terminal protein is represented by a circle (●, 55 kDa; ○, 80 kDa).

140 kDa DNA polymerase [149, 150, 558]; and an 80 kDa 'terminal protein' [554, 566–568]. The SSB associates with the ends of the DNA duplex to promote unwinding [569]. The terminal protein forms a complex with the DNA polymerase and in the presence of

NFI, it interacts with dCTP so that a dCMP residue becomes covalently linked to the terminal protein which now associates with the 55 kDa protein covalently bound to the 5′-ends of the adeno DNA. The dCMP then serves as a primer to initiate displacement synthesis (Fig. 6.34). Electron microscopy has been used to show that initiation usually occurs at only one end of the molecule so that one round of replication results in the formation of a duplex and a linear single-stranded molecule [554]. However, by pulse-labelling with [³H]thymidine and analysis of completed molecules it has been shown that termination must also occur at either end of adenovirus DNA. This shows that initiation can occur at either end and, occasionally, molecules are seen in the electron microscope which are replicating from both ends [554]. When a single strand of DNA is produced, the terminal repeat will allow it to form a panhandle structure to initiate conversion into a duplex DNA molecule [554]. Associated with separation of the new duplex from the displaced single strand, the $80\,000\,M_r$ terminal protein is reduced in size to $55\,000\,M_r$ but this processing does not appear to be essential [570].

Adenovirus is not unique in adopting this strategy for initiation of replication. Thus phage $\phi29$ initiates symmetrical DNA replication using a dAMP covalently linked to a $30\,000\,M_r$ terminal protein [571, 572] and Pearson *et al.* [573] have pointed out the similarity between this strategy and that adopted by ϕX174 which initiates replication of duplex circles by nicking one strand and covalently attaching a protein to the exposed end (section 6.9).

6.13 POSITIVE OR NEGATIVE CONTROL OF INITIATION

DNA replication and, it follows, cell division is initiated by a single event which occurs at the origin of replication of each replicon. This information led Jacob *et al.* [626] to propose a theory of positive control of replication similar to the induction/repression model for the control of protein synthesis in prokaryotes (chapter 10). This model proposes that a product of a replicon is a (protein) molecule which interacts with the origin to induce replication. The inducer molecule is trans-acting and its concentration dictates whether or not replication will be initiated.

An alternative model of Pritchard *et al.* [627] proposes negative control over initiation. These authors suggest that there is an inhibitor of replication present in cells. The inhibitor is synthesized by the replicon and only when its concentration is diluted (by growth of the cells) will initiation occur. The inhibitor may be RNA or protein or may involve iterated DNA sequences which compete for an activator protein [802] (section 6.10.3).

Autoregulation of an effector has elements of both positive and negative control and experiments involving fused plasmids are open to dual interpretation. Cabello *et al.* [629] fused Col E1 with the related plasmid pSC101 to form plasmid pSC134 which used only the Col E1 origin and maintained the copy number of Col E1. However, Col E1 cannot grow in *polA* mutants (section 6.11.3) and under these conditions pSC134 used the pSC101 origin and assumed the copy number of pSC101. These results are what would be expected of incompatible plasmids responding to identical repressors where the pSC101 origin is more sensitive to inhibition and hence is only induced in large cells in which the repressor has been diluted. Normally, the Col E1 origin is activated first and maintains the concentration of repressor. Only when the Col E1 origin is inactive does the concentration of repressor fall sufficiently for the pSC101 origin to be activated. Related plasmids are incompatible if they

respond to the same inhibitor molecule. Thus, as a result of changes in base sequence around the target for the inhibitor, the plasmids pMB9 and pBR322 are less sensitive than Col E1 and rapidly exclude the latter [630]. Equally well, the Col E1 origin could respond to a lower concentration of an inducer the concentration of which is limited by replication (section 6.11.3).

Most experiments designed to investigate the inhibitor dilution model conclude that factors other than inhibitor concentration generally regulate the maximum rate of replication [684].

6.14 RETROVIRUS REPLICATION [574]

Retroviruses pose a unique problem, for, as described in chapter 3, they replicate by first converting their RNA genome into DNA using RNA-directed DNA polymerase (reverse transcriptase – section 6.4.8). Models of reverse transcription of the RNA genome of retroviruses have been proposed [195, 575–577] to explain certain unexpected observations, i.e. the duplex DNA produced is longer at both ends than the viral RNA and during the reaction short molecules of both plus and minus strand DNA ('strong stop DNAs') are formed [578].

Avian reverse transcriptases are initiated with tRNA^{Trp} [166, 579] but the reverse transcription from Moloney murine leukaemia virus starts by using a tRNA^{Pro} which attaches to the viral RNA at about 150 nucleotides from the 5′-end. This tRNA^{Pro} serves as a primer for synthesis of minus strand DNA. Synthesis stops when the DNA strand reaches the 5′-end of the template, i.e. after 100–150 nucleotides have been added [580]. This is 'minus strand strong stop' DNA (Fig. 6.35).

The RNA genome is terminally redundant and digestion of the 5′-end by ribonuclease

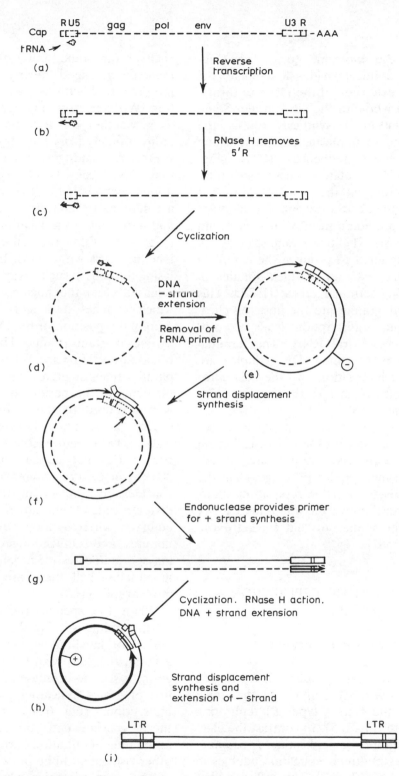

Fig. 6.35 Postulated mechanism of conversion of retroviral RNA into duplex DNA. See text for details.

H allows the molecule to cyclize (Fig. 6.35(d)) and this provides a template for the further extension of the DNA to form a copy of the whole of the viral strand. Some strand-displacement synthesis causes the minus strand to duplicate part of the long terminal repeat sequences (LTR, Fig. 6.35(f)). It is speculated that specific endonuclease action on the viral RNA by yet another domain of the reverse transcriptase could now provide a primer for plus strand DNA synthesis. The 'plus strand strong stop' DNA is extended to produce the complete duplex DNA each end of which contains an identical long terminal repeat (LTR). This duplex is integrated into the host cell chromosome and transcription leads to production of new viral RNA. The terminal nucleotides of the linear DNA duplex are required for integration into the host chromosome and are lost in the process [171, 577] (section 7.7.3).

Cauliflower mosaic and hepatitis B viruses are both cyclic, duplex DNA viruses but, in each case, both DNA strands are nicked. An RNA transcript slightly longer than the genomic DNA forms a replication intermediate, and is converted into duplex DNA in a manner similar to that for retroviral RNA [574, 581].

6.15 INITIATION OF REPLICATION – COMPLEX SYSTEMS

6.15.1 Replicons in eukaryotes

The electron micrograph of replicating polyoma virus DNA emphasizes the single origin and the Cairns type (θ) replicating intermediate (Fig. 6.5). In contrast the electron micrograph of replicating *Drosophila* DNA shows multiple replication bubbles in tandem array [367, 582]. This indicates that there is not a single origin of replication for each chromosome, but rather a series of *replicons* arranged linearly along the chromosome, each with its own origin of replication. What electron micrographs cannot tell us is whether the replicons are replicated bidirectionally from their origins.

Fibre radioautography studies also suggest that DNA consists of several, tandemly joined sections [583, 584] (replicons) which are separately replicated from their origins and show that this replication is bidirectional (Fig. 6.36). Fibre radioautography involves labelling cells with [^{3}H]thymidine and then lysing the cells in such a way that the DNA is released from the nucleus as a long fibre which attaches to a slide. On radioautography the position of the DNA is indicated by lines of grains produced by the β-particles released on ^{3}H decay. What is seen are two parallel lines of grains whose length is dependent on the period for which the cells were exposed to the [^{3}H]thymidine. The spacing between neighbouring lines gives the distance between replicons. The size of replicons varies between 15 and 60 μm (40–200 kb) in cultured mammalian cells but is considerably shorter in the very rapidly dividing cells of amphibian and insect early embryos, where S phase may last for only 20 minutes, and is much longer in slowly dividing spermatocytes [585, 586]. In a somatic, mammalian cell there are typically about 60 000 replicons.

When the specific radioactivity of the [^{3}H]thymidine is reduced after a short initial period of labelling at high specific radioactivity, the density of the radioautographic grains is seen to decrease at both ends of the line of grains indicating that replication is bidirectional (Fig. 6.36). It is important in these experiments to start labelling the cells at the time of initiation of replicon synthesis, otherwise gaps will be present in the middle of the lines of grains which may lead to misinterpretation of the results.

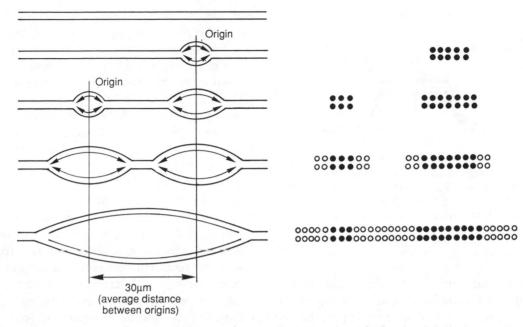

Fig. 6.36 Replication of mammalian DNA (two replicons are shown). On the right is the result of the fibre autoradiograph experiment described in the text (shown diagrammatically) and on the left is the interpretation originally given by Huberman and Riggs [583].

With an average size of 30 μm, the replicons of eukaryotes are very much smaller than the single replicon which is the *E. coli* chromosome (1300 μm). They are similar in size and may be physically identical with the loops of DNA seen in electron micrographs of dehistonized chromatin (chapter 3). The loops of DNA in dehistonized chromatin are essentially free of protein and are quickly digested by nucleases leaving behind the fragments of DNA which anchor the loops to the chromosome matrix (section 6.15.5).

By carrying out fibre radioautography following [³H]thymidine pulses of various durations, the rate of fork movement in mammalian cells is found to be only 0.5–1.2 μm per minute or about 60 base pairs per second [587]. However, in eukaryotes this slower rate is compensated for by the smaller size of the replicon which is completely replicated in about 15 minutes. It is clear that only about 5% of replicons in a eukaryotic cell are simultaneously active [588] and some control must be exercised over the order in which replicons initiate DNA synthesis (section 6.15.5).

The regulation of the overall initiation of replication (i.e. the passage of a cell from G1 into S-phase) is considered in sections 6.15.5 and 3.1.2.

6.15.2 Origins of replication in repeated genes

In the situation where genes occur in long tandem repeats it has been possible to identify an origin of replication in each repeat [589–591]. For example, electron microscopy has been used to localize the centre of replication bubbles of sea urchin and yeast ribosomal DNA to the non-transcribed

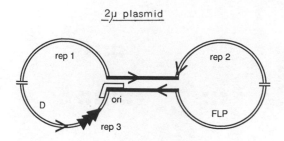

Fig. 6.37 The yeast 2 micron plasmid showing the origin region (open box) and the identical repeats across which recombination can occur (solid boxes). The lines (|) and arrows indicate the sites of initiation and termination of the genes indicated.

spacer and, in yeast, these replication bubbles are coincident with ARS sequences (section 6.15.3). Furthermore, at the beginning of S phase, two particular restriction fragments of the amplified dihydrofolate reductase (*dhfr*) gene in hamster cells are replicated first [592]. As a result of the use of different methods, there is at present some disagreement about the size and location of the origins in the *dhfr* gene, although the changeover from continuous (leading) strand synthesis to discontinuous (lagging) strand synthesis occurs within a region of 450 bp [791, 793].

6.15.3 ARS elements and yeast plasmids

Yeast cells contain a small 2 μm plasmid, packaged in nucleosomes, which replicates autonomously as a θ structure. For this to occur the plasmid requires a short origin-containing region at which bidirectional replication is initiated [593–596]. It also contains a region, *rep3*, of direct repeats forming a 62 bp centromere-like element adjacent to the origin, and genes for two proteins Rep1 and Rep2 (Fig. 6.37) (section 6.15.6). Other plasmids have been constructed which

lack the original plasmid origin but which can still replicate in yeast provided they contain one of up to 400, short regions from the yeast chromosomal DNA. Such regions are known as autonomous replicating sequences (ARS) as they confer this property on the plasmid [597–601]. ARS elements on yeast chromosomes are sites of matrix attachment (section 3.3.5) [602] and may bind initiator proteins [603].

Several ARS elements have been sequenced and they all contain a very AT-rich core region which has the consensus sequence; (A/T)TTTAT(A/G)TTT(T/A) [601, 604, 686]. Similar sequences obtained from non-yeast cells will also confer autonomous replication on yeast plasmids but we have no evidence yet that they act as origins of replication in their normal environment [605].

In addition to this 11 bp consensus sequence, flanking sequences of 24–109 bp are also required. These must be very AT-rich and for efficient replication the plasmid must be supercoiled under which conditions this region becomes single-stranded [606, 607]. It has, therefore, been proposed that ARS elements are analogous to the origins of *E. coli* or phage lambda [604]. Confirmation that ARS elements are, indeed, origins of replication has come from two-dimensional gel electrophoretic analysis which allows identification of non-linear (i.e. forked) DNA molecules. A surprising finding made when this technique was applied to the tandemly repeated yeast ribosomal RNA genes was that, although replication was initiated at (or very near to) an ARS element, the leftward fork stopped prematurely and the rightward fork (which moves in the direction of transcription) proceeded through several repeats (and ARS elements) before terminating at a stalled leftward fork; that is, not all ARS elements are used in each S-phase [604, 608] (section 6.16.1).

Using the technique found helpful in

yeast, autonomously replicating plasmids have been sought in mouse cells. One report [610] claiming success, indicated that many such plasmids contain the consensus sequence CTC(A/T)GAGA(C/G)(C/G)AA. However, it was later shown that the DNA was not present as ARS plasmids but as integrated DNA [609, 611–613]. In other cases [618–620], autonomously replicating plasmids have been obtained and, in one instance, initiation of replication was shown to occur within an *Alu* sequence (section 3.2.2(c)). In other cases, initiation of replication was found to occur at multiple sites and the chance of any of these sites being present increased with the size of the plasmid insert [805, 806].

Plasmids containing the origin of replication of the Epstein Barr Virus (EBV) together with certain other features are maintained as stable plasmids in cultured human cells, and experiments are in progress to search for human origins which can replace the EBV origin of replication [635]. Plasmids have been obtained containing a region upstream of the c-*myc* gene which confers the ability to replicate as an autosome which is stably partitioned in somatic mouse cells [697].

6.15.4 Other methods of identifying origins

In contrast to the results reported above, it now appears clear that any DNA sequence, even of prokaryotic origin, when injected into *Xenopus* eggs will replicate under the strict cell cycle control pertaining [614–617]. As described above (section 6.15.1), the *Xenopus* egg has very small replicons and replicates its DNA at an extremely rapid rate: S-phase lasts for only 20 min. For these reasons it may be an atypical system (section 6.15.5).

Handeli *et al.* [609] have identified origins in hamster cells by taking advantage of the finding that, in the absence of protein synthesis, parental nucleosomes segregate with the leading side of the replication fork. The lagging strand, being unprotected, is sensitive to nucleases (section 6.17). By mapping diverging leading strands, origins of replication can be located; and, by mapping converging leading strands, termini can be identified. They have shown that, when primary origins are deleted, secondary (previously inactive) origins are brought into play.

The replicating fork contains Y-shaped DNA molecules and restriction fragments containing a fork migrate anomalously on gel electophoresis. By running a second dimension under denaturing conditions and probing with radioactive sequences (section A 3 and A.4) the positions of particular forks and hence their point of origin can be determined [596, 604, 637].

Several other methods of locating origins have been devised [638–642, 662].

6.15.5 Matrix attachment and temporal replication

Labelling experiments show that DNA replication is initiated and continues in association with the nuclear matrix or scaffold; that residual nucleoprotein complex which remains when histones and other soluble proteins are removed from nuclei (often by treatment with high concentrations of NaCl) and the exposed DNA removed by deoxyribonuclease digestion [112, 621–625, 799] (section 3.3.5). That is, the replicating DNA is always at the bottom of the loops. It is proposed that, at the bottom of each loop, there is the complex of enzymes involved in DNA replication and the DNA threads through this complex as it is replicated. To

(a) Replicon cluster on linear DNA

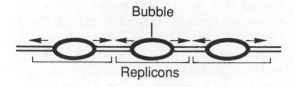

(b) Replicon cluster on DNA loops

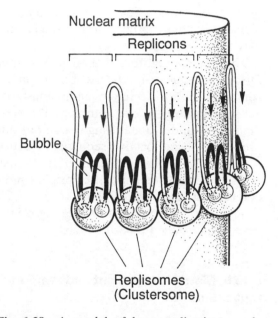

Fig. 6.38 A model of how replication can be initiated and continue on replisomes attached to the nuclear matrix or scaffold. As replication proceeds the fork is drawn through an immobilized complex of enzymes (the replisome). (Reproduced from [685] with permission of the authors and publishers.)

initiate replication the origin of a replicon must be associated with a 'replisome' on the nuclear matrix [685] (Fig. 6.38). Scaffold-attached regions of DNA isolated from yeast and from *Drosophila* cells can act as ARS elements in yeast [636, 675]. The regions of DNA associated with the scaffold or matrix

from a variety of species show certain similarities but are much longer than would be needed to bind one or two proteins [692]. They are enriched in binding sites for topo-isomerase II [689], an enzyme known to be involved in replicon fusion (section 6.16.2), and capable of holding together the two ends of a loop of DNA (section 6.5.4).

Most of the methods described in sections 6.15.2–6.15.4 provide results implying that replication is initiated at specific sequences along eukaryotic chromosomes, but that there is a hierarchy that controls their order of use or whether they are used at all [604]. Without this proviso, it is difficult to reconcile observations of specific origin sequences with the changes in size of replicons found in rapidly dividing embryos and very slowly growing cells [585, 586, 617], observations which imply that a particular origin may be used in some situations, but not in others. The dividing *Xenopus* egg may provide an environment of very low selectivity for origin sequences due to a superabundance of initiation factors. Thus initiation may require the binding of short AT-rich sequences to the nuclear matrix. Certain sequences may be more susceptible to binding than others and hence may initiate earlier in S phase. Other sequences may fail to initiate at all in slowly growing cells leading to an increased replicon size.

The working model assumes that origins of replication in eukaryotic chromosomes bind a specific initiation factor (or, perhaps, a series of factors), although none has yet been identified. This would make initiation formally similar to that occurring in the simple systems described in sections 6.8 to 6.12 and also in herpes viruses [631–633]. Initiation of replication in yeast and in *Xenopus* extracts depends on the activity of a protein kinase (cdc2 protein of *S. pombe* or the p34 of *Xenopus*) and this may act in an analogous manner to phosphorylation of DnaA or Tag [792] (sections 6.10.1 and 6.10.4).

Experiments of Rao and Johnson [628], where animal cells in the G1 phase of the cell cycle were fused with S-phase cells, strongly supported the idea of a *trans*-acting factor exerting control over initiation of DNA synthesis but did not distinguish between positive and negative control mechanisms.

Possible evidence for such a factor comes from work with extracts of *Xenopus* eggs which will replicate sperm chromatin [634]. The chromatin is packaged into nuclei and a second round of replication occurs only if the nuclear membrane is disrupted. It is proposed that, when this happens naturally at mitosis, a cytoplasmic factor gains access to the chromatin. Excess factor is removed in G1-phase thus limiting replication to one round per cell cycle.

One factor which regulates the timing of initiation of a replicon is its transcriptional activity. Transcriptionally inactive genes are present in condensed chromatin and replicate late in S-phase [640, 643–645]. For both transcription and replication, access of the DNA to initiation factors is required, and the availability of genes for transcription may determine their accessibility to replication factors. Indeed, in some cases identical factors are required for transcription and replication [662, 687, 688] (e.g. sections 6.10.4 and 6.12) and their function may be to help to unwind the DNA duplex at the site of initiation. When the transcriptional activity of a gene changes this is typically associated with a change in the time when it is replicated within S-phase [640, 644].

6.15.6 Amplification

The discussion so far has emphasized the point that replication must occur once only within each cell cycle in order to maintain the constancy of the amount of DNA per cell. Although normally the case [646], there are exceptions to this generality. The selective replication of specific sequences of DNA is known as amplification. It can occur as part of the developmental programme of an organism or in response to treatment with a drug or similar selection agent [800].

(a) *Developmental amplifications*

During oogenesis in *Xenopus* the ribosomal RNA genes are amplified many thousandfold such that the ribosomal DNA content of an oocyte may be as much as 2500 times the normal haploid value [647]. Amplification occurs in two phases. The first phase results in the 40-fold amplification of selected ribosomal repeat units and occurs in both sexes when only 9 to 16 primordial germ cells are present [648]. In the male these amplified sequences are lost at meiosis but in the female oocyte there is a further amplification; presumably to satisfy the special demand of the oocyte for large amounts of ribosomal RNA [647–649]. The amplified ribosomal DNA is produced in the form of extrachromosomal circles each containing several copies of the ribosomal DNA repeat and each capable of autonomous replication [650].

Ciliates contain two nuclei; a macronucleus which contains a small selection of genes which are transcriptionally active, and a micronucleus or germinal nucleus which gives rise to both nuclei in the next generation (section 7.5.6). In *Tetrahymena* the diploid micronucleus contains a single integrated rDNA gene per haploid genome. For the formation of the macronucleus, chromosome breakage occurs at specific sites on each side of the rDNA to produce a linear duplex containing a short inverted repeat at the 5′-end of the molecule. By intramolecular rearrangement this repeat acts as a primer to produce a double-length molecule to the ends of which are added from 20 to 70 copies of the telomeric sequence CCCCAA [651] (section 6.16.4). Similar dimers are found in *Physarum* and each copy of the rDNA in the dimer has an origin of replication but usually only one is active: the inactive origin carries a modified cytosine [652].

A similar selective amplification occurs to the chorion genes during *Drosophila* development. These genes which code for egg-shell proteins are amplified 16-fold in follicle cells. However, the amplified genes are not present as extrachromosomal circles but arise by multiple initiation of replication of specific origin regions within the gene [653–655]. The actual region amplified (80–100 kb) extends well beyond the centrally located chorion genes but it is the central region which shows the highest degree of amplification. A short segment of the amplified sequence acts in *cis* to control amplification [654]. This process of chorion gene amplification may be typical of a number of insect genes which can be seen to 'puff' in the polytene chromosomes of larval salivary glands under the influence of the steroid hormone ecdysone [655].

(b) Amplification by chemical selection

When undergoing selection for drug resistance, animal cells in culture may respond by overproducing a gene product. For example, cells treated with methotrexate, an inhibitor of dihydrofolate reductase (DHFR) (section 5.6), may respond by overproducing the inhibited enzyme. This is often a result of selective amplification of the gene coding for dihydrofolate reductase (*dhfr*) [653, 656]. Corresponding results are found with other drugs but it is not always the gene for the target enzyme which is amplified and chromosomal location markedly affects the frequency of amplification (657–659). Similar

amplification probably occurs in whole animals and has also been reported in *Leishmania* and bacteria [660, 661]. The mechanisms which give rise to the amplifications are believed to be active in the absence of the drug [646, 663] which acts largely as a selecting agent. Thus amplification of chromosomal segments is occurring spontaneously (see below) and does not simply involve one gene but rather includes a region up to 1000 kb in size. Under selective pressure, this whole region may be amplified as many as a thousand-fold when it can exist in one of two forms: (a) as a homogeneously staining region (*hsr*) integrated into a chromosome, when it remains stable for more than 12 months in the absence of the drug; or (b) as small acentric chromosome fragments (known as double minutes) which are rapidly lost (within two weeks) following removal of the drug. In the first case multiple rearrangements and translocations occur and the *hsr* is found at variable chromosomal locations.

(c) *Mechanism of amplifications*

The frequency of spontaneous twofold amplification of a gene has been estimated at one in a thousand per cell generation [646, 663] and this rate can be increased by a number of factors. For instance, the *dhfr* gene, along with several others whose products are required for DNA synthesis, is replicated and transcribed early in S phase but replication can be reinitiated if cells are treated with inhibitors of DNA synthesis early in S-phase [656, 664]. Excessive transcription under these conditions may provide an environment particularly suitable for re-initiation of replication of these genes. Further, if selection pressure is applied the amplification becomes apparent.

Although the actual mechanism of the initial amplification is not known in these cases, a model which is favoured involves so-called 'onion skin' replication and this is be-lieved to be the mechanism whereby copies of integrated viral DNA are produced linked to adjacent host sequences when trans-formed cells are induced [665–667] (Fig. 6.39). In some of the developmental ampli-fications *cis*-acting elements are known to be involved in controlling the amplification pro-cess [800].

Once produced, extra copies of DNA regions undergo rearrangement and recom-bination events via short repeated sequences in a manner similar to DNA transfected into cells [653, 656, 659] (section A.9.3). The consequences of over-replication depend on the type of rearrangements which occur sub-sequently. Reintegration into the chromo-some followed by further recombination events will lead to homogeneously staining regions containing multiple tandemly ar-ranged copies of the gene region. Similar duplicative recombination events are be-lieved to have been involved in the produc-tion of satellite DNA and in the production of the minisatellites which are a major source of the heterozygosity found in human DNA (section 3.2.2(b)).

The 2μ plasmid of yeast is maintained at 100 copies per cell by an intricate mechan-ism. The plasmid contains an inverted repeat where recombination can occur catalysed by the plasmid-coded recombinase, FLP (Fig. 6.37). Recombination results in inversion of the two halves of the plasmid and when this occurs in a replicating molecule it changes the θ structure into a double rolling circle. This leads to amplification of the copy num-ber which only stops when recombination occurs again. Rep 1 and Rep 2 (products of two plasmid genes) normally combine to repress the FLP recombinase gene (and *rep* 1). However, if the copy number falls the ratio of Rep 1 to Rep 2 falls and FLP recom-binase is induced leading to amplification. Rep 1 is also induced again leading to repres-

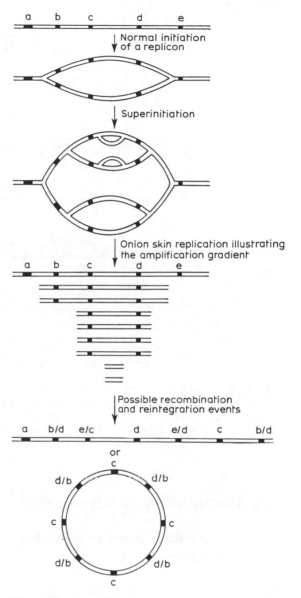

Fig. 6.39 Amplification by means of onion skin replication followed by recombination.

sion of FLP recombinase [668]. This ampli-fication mechanism may be of more general application and is considered, along with several others, by Stark *et al.* [800].

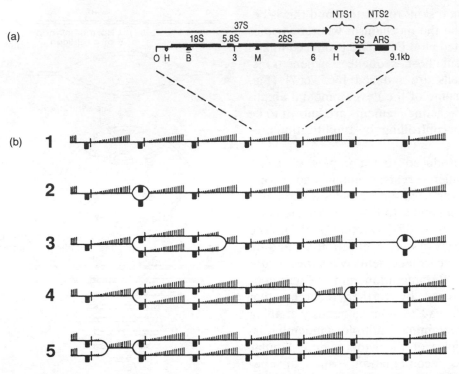

Fig. 6.40 Model for the replication of the rDNA of *S. cerevisiae*. (A) shows a diagram of a single 9.1 kb repeat unit (NTS is non-transcribed spacer and the arrows indicate the direction of transcription of the 37S and 5S RNAs). (B) is a diagram of the replication of several repeat units (the ARS elements are indicated by black boxes and the transcripts are indicated by short vertical lines increasing in length from left to right). (Reproduced from [669] with the permission of the authors and publisher.)

6.16 TERMINATION OF REPLICATION

6.16.1 Competition between replication and transcription

When a replication fork is travelling along the DNA in the same direction as an RNA polymerase molecule, the progress of one or the other may be slowed but there appears to be no reason that either replication or transcription be aborted. On the other hand, when a replicating complex encounters an RNA polymerase molecule travelling in the opposite direction the former is derailed.

This conclusion is reached from a consideration of the replication of the amplified rDNA genes of yeast (section 6.15.3). Replication, initiated within a non-transcribed spacer (NTS), proceeds bidirectionally but the leftward-moving fork soon encounters a transcription complex travelling towards it. This causes termination of replication [608, 669]. The rightward-moving fork will continue replication in the direction of transcription until it meets a stalled leftward-moving fork. This may involve replication through several NTSs in which no initiation has occurred (Fig. 6.40).

The derailment of replication complexes by RNA polymerase is obviously very important for the frequently transcribed genes such as that for ribosomal RNA, but it is probably one reason why the points of initiation of transcription and replication are close together in the genomes of SV40 and adenovirus and in mitochondrial DNA (sections 6.10.4, 6.12 and 6.11.4, respectively). Some genes (e.g. the avian histone H5 gene [641] and a c-*myc* gene [642]) use different origins of replication depending on whether or not they are being transcribed.

It is obvious that such problems could be even more significant in the larger replicons of bacteria. To minimize the difficulty, 85% of frequently transcribed genes in *E. coli* are transcribed in their direction of replication [690]. There are, however, some helicases that are able to displace a transcription complex to allow progression of the replication fork [670].

6.16.2 Replicon fusion

(a) *Termination signals*

The site of termination of replication, like the site of initiation, is characterized by abnormal DNA structures [676, 677] and specific termination signals (*ter*) have been characterized in bacteria and some plasmids [678–680]. *Ter* sequences are not palindromic and block replication forks approaching from their 5′-end and so a pair of sequences is required to block the two forks approaching from opposite directions. They are arranged in such a manner that the fork passes through the first *ter* sequence it encounters (from its 3′-end) and is arrested at the second. This means that the approaching forks become trapped between the pair of *ter* sequences, thus ensuring that termination occurs in this limited region of the chromosome [691].

The *ter* sequence binds a protein (Tus) which blocks helicase action and so prevents the unwinding of the parental duplex [681, 682, 693].

(b) *Unlinking the parental DNA*

A complication arises when the two replication forks moving in opposite directions around a cyclic chromosome approach one another. Unwinding takes place ahead of the fork for a distance of about 100 bp and when the unwound regions meet the result is two gapped duplex molecules, interlinked (i.e. catenated) 20–30 times (Fig. 6.41). If replication of the two daughter molecules continues to completion, then ligase will finally generate two interlinked, supercoiled molecules. This structure is the preferred substrate for the decatenase reaction catalysed by DNA gyrase [671]. In SV40 replication, the use of selective inhibitors has shown that topoisomerase I normally acts ahead of the fork to unwind the parental strands of DNA but, as replication nears completion, topoisomerase I is excluded and the rate of replication slows as it acts against the torsional strain. Replication of one of the daughter molecules appears to go to completion before the other, and DNA topoisomerase II is required to decatenate the dimer [469, 672–674].

6.16.3 Small linear viral chromosomes

These pose a different problem which has been solved in a variety of ways. The situation is best illustrated with T7 DNA replication (Fig. 6.42). The last Okazaki piece to be made on the lagging strand might start at the extreme 3′-end of the template but, when the primer is removed, this will leave a short single-stranded tail at each end of the linear molecule – a tail which cannot be replicated by any mechanism considered so far.

James Watson [722] postulated a mechanism whereby the problem might be solved. T7 DNA has an identical sequence of about 260 nucleotides at each end of the molecule,

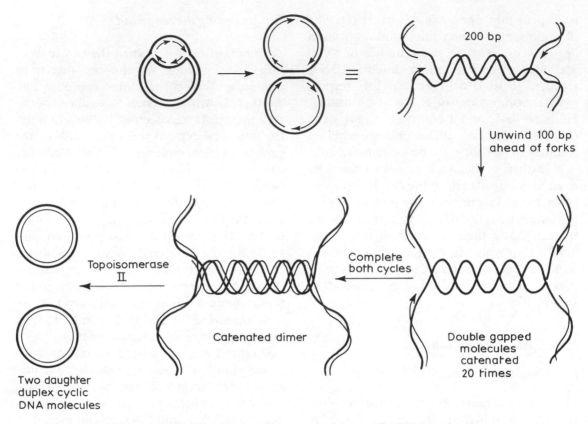

Fig. 6.41   Replicon fusion illustrated with a cyclic DNA molecule.

i.e. the molecule is terminally redundant. Thus the unreplicated single-stranded tail of one molecule could hydrogen-bond to the similar region of another molecule giving rise to a concatamer of almost two unit lengths (Fig. 6.42).

The two pieces can be joined together by ligase action and two new nicks introduced several nucleotides 5′ to the original nick. The gene 3 endonuclease is a candidate for the enzyme capable of making these specific cuts. Nick translation (section 6.4.3(e)) from the newly introduced 3′-OH groups will allow the concatamer to separate into two halves and the replication to be completed.

This model now appears to be an over-simplification as processing of concatamers is

closely linked to packaging and production of left and right hand ends is dependent on different factors. A model involving double-stranded breaks has been proposed [723].

Other small linear viral genomes cope differently with the problem of end replication. We have seen (section 6.12) that adenovirus DNA uses an unusual mechanism for initiation at the very ends of the DNA duplex which removes the extra stages required for termination by phage T7 DNA.

Phage lambda DNA has 'sticky' or cohesive ends of 12 or 19 nucleotides which enable it to form a cyclic molecule on entering its host cell. At the early stages of replication, phage lambda DNA replicates as a θ structure (section 6.10.2) but, at the later stages

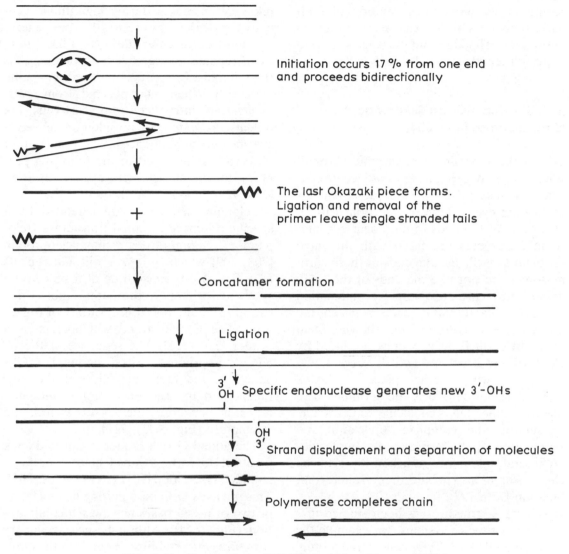

Initiation occurs 17 % from one end and proceeds bidirectionally

The last Okazaki piece forms. Ligation and removal of the primer leaves single stranded tails

+

Concatamer formation

Ligation

Specific endonuclease generates new 3'-OHs

Strand displacement and separation of molecules

Polymerase completes chain

Fig. 6.42 Replication of T7 DNA. (After Watson [722].)

of replication, long concatameric molecules are formed by a rolling circle mechanism (section 6.9). It is not clear how the switch in mechanism is effected. Unit length DNA molecules with sticky ends are regenerated from the concatamer during packaging. The lambda terminase cleaves the DNA at the *cos* site and remains bound to the left end so generated. In an ATP-dependent reaction, terminase-bound left ends are inserted into the capsid shell and the remainder of the DNA molecule follows until the next *cos* site is encountered when cleavage again occurs [724].

Different mechanisms are believed to be important in the replication of the small linear single-stranded DNA of the parvoviruses [721]. In these cases terminal redundancies allow the formation of fold-back sequences which can act as primers. Subsequent slippage and cleavage can generate full length molecules [695, 696].

6.16.4 Telomeres on eukaryotic chromosomes [803, 804]

Within the long linear eukaryotic chromosome, the converging forks of adjacent replicons can fuse in the manner described for replicating cyclic duplexes (section 6.16.2). However, at the ends of the linear chromosome eukaryotes are faced with the same problem as is T7 but the ends of these chromosomes are not like the ends of the small linear viral chromosomes. Recombination does not readily take place between the normal ends of chromosomes showing them to be different from new ends produced by chromosome breakage (which very rapidly recombine with another piece of duplex DNA). The telomere is the terminus of a linear, cellular DNA molecule which is able to support the complete replication and segregation of the two daughter molecules [725] and telomeres attached to artificial chromosomes also protect them from recombination [726, 727].

Telomere structure has been studied in a number of simpler systems, for example the ribosomal DNA of *Tetrahymena*, the termini of trypanosome chromosomes and yeast chromosomes [725, 728–733].

All telomeres so far studied have multiple (30–70) repeats of short, species-specific sequences such as TTGGGG (the major human telomere sequence is TTAGGG) [734, 735, 804] and have either the two strands of DNA covalently linked with a subterminal nick or have a single-stranded 3'-end, i.e. the G-rich strand extends by 12–16 nucleotides beyond the C-rich strand [731]. The actual number of repeats in a telomere increases during logarithmic vegetative growth by from 3 to 10 bp per generation but as cells enter stationary phase these extra repeats are removed. In human cells the number of repeats is reduced on ageing [736, 737]. When linear plasmids from *Tetrahymena* are introduced into yeast cells, the repeats are extended by addition of yeast-specific repeat sequences. These and other observations have led to the following proposals for telomere structure and replication [728, 729, 703–707, 738].

Telomere sequences are extended by a specific telomere terminal transferase (telomerase). Telomerase is a ribonucleoprotein [708, 709] which, *in vitro*, will add repeats (e.g. T_2G_4) one nucleotide at a time, onto synthetic primers representing the G-rich strand of telomeres from a variety of organisms. The C-rich strand will not act as a primer [710, 711]. The sequence at the 3'-end of the primer dictates whether T or G is added as the first nucleotide of the new repeat and the sequence of the telomeric repeat is dictated by the RNA present in the telomerase (Fig. 6.43) [712].

The added G-rich sequence can fold back on itself to form a novel hairpin or four-stranded DNA loop [707, 713–717, 731, 803]. This involves G:G base pairing in which one or two of the G bases has the *syn* configuration (Fig. 6.44). Although such secondary structures may stabilize the ends of the chromosomes, they may also interfere with telomerase action [157].

At this time it is not clear how the C-rich strand of telomeres is produced. One possibility is that, as the G-rich strand is extended, it may enable DNA polymerase and primase to initiate retrograde synthesis. Alternatively, although less likely, the folded back G-rich strand may act as a primer. Extending the 3'-end of the chromosome using telomerase means that initiation at the extreme 5'-end of the complementary strand is no

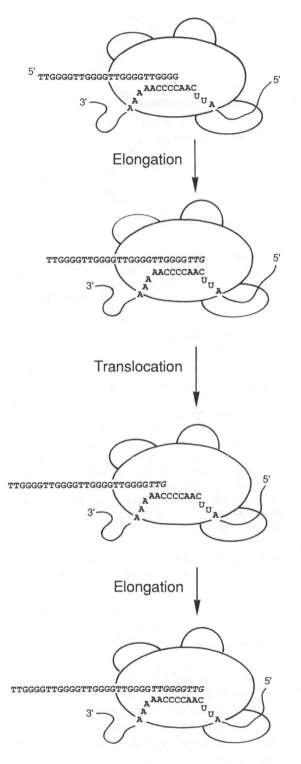

longer important. A process of continual telomere extension compensates for the loss of terminal sequences on DNA replication.

A somewhat different mechanism has been proposed for the replication of telomeres in *Oxytrichia* [718] where all the macronuclear telomeres are identical, with

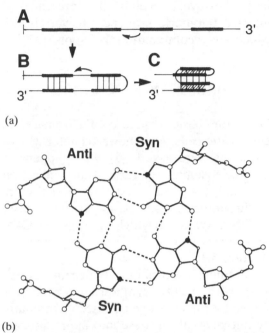

(a)

(b)

Fig. 6.44 (a) A model of how the G-rich strand could fold upon itself to form two- or four-stranded structures. (b) Shows a G tetrad involving Hoogsteen as well as Watson Crick hydrogen bonds; alternate nucleotides are in the *syn* and *anti* configuration. (Based on [705] with permission from the authors and publishers.)

Fig. 6.43 Model for telomerase action in *Tetrahymena*. The 3'-end of the telomere is shown binding to the RNA template within the enzyme. Three nucleotides (italics) are added and the enzyme translocates to the end of the telomere before another round of addition occurs. (Reproduced from [709] with permission of the authors and publisher and based on the model originally presented in [798].)

20 base pairs of G_4T_4/C_4A_4 and a 16-base 3′-single-stranded tail.

An alternative mechanism has been proposed for telomere extension in yeast [730]. This involves invasion of one telomere by the 5′-overhanging region of another (section 7.4.1), but evidence for such recombination is limited [719]. Such a mechanism may, however, explain the presence of internal telomeric sequences in vertebrates which are recombination 'hot spots' [720].

6.17 CHROMATIN REPLICATION

Once the basic structure of chromatin in eukaryotic cells was known (chapter 3) two questions were asked: (1) what happens to the nucleosomal structures at the replication fork and, (2) to what extent is the structure of chromatin responsible for the mechanism of DNA synthesis which involves (a) Okazaki pieces (section 6.3) and (b) replicons (section 6.15.1)?

That nucleosomes quickly become associated with newly synthesized DNA was shown by the cleavage of nascent chromatin by micrococcal nuclease into approximately 200 bp fragments [740, 741]. However, the rate and extent of cleavage of nascent chromatin is greater than that of mature chromatin indicating a somewhat different structure which requires 15 minutes to mature, i.e. a time sufficient to synthesize 22 000–54 000 bp of DNA or approximately one replicon [583, 742, 743].

As the replicating fork progresses along a region of nucleosome-associated DNA a number of different situations might result, some of which are illustrated in Fig. 6.45. Approaches used to distinguish between the possibilities include: (a) electron microscopic studies; (b) isolation and analysis of replicating chromatin; (c) cross-linking studies usually associated with attempts to separate

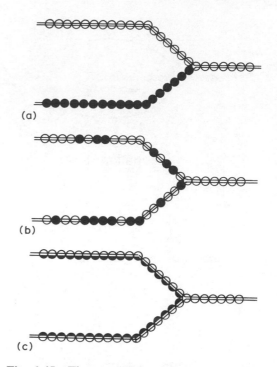

Fig. 6.45 Three possible mechanisms are shown for the distribution of nucleosomes at a replication fork (after Weintraub [757]). ○, parental nucleosome; ●, newly synthesized nucleosome. (a) All the old nucleosomes remain with one of the parental strands when they separate at the replication fork − cooperative segregation; (b) the old nucleosomes are dispersed at random between the two strands of DNA; (c) the old nucleosomes dissociate into half nucleosomes, one half going to each side of the fork.

the cross-linked species [751–756, 761]; (d) nuclease digestion experiments [757, 758]; and (e) the above approaches carried out with cells exposed to inhibitors of protein synthesis in an attempt to visualize the histone-free nascent DNA [741, 756, 757, 759–761].

The major controversy concerns the question whether old, parental nucleosomes are found adjacent to one another (as required for conservative segregation) or whether they are dispersed among newly synthesized nucleosomes.

Although some conservative segregation definitely appears to occur, the evidence is not yet strong enough to conclude that it represents the only mechanism [754, 759–761]. If segregation of nucleosomes is conservative the evidence favours the continued association of the old nucleosomes with the leading strand at the replication fork, with the newly synthesized nucleosomes being added to the lagging side [760].

Thus DNA fork movement may be considered to occur as a series of steps caused by relocation of intact, old nucleosomes on the leading side of the fork which is always made up of duplex DNA. Either the parental nucleosome remains associated with the leading strand [775] or is only transiently displaced and rapidly rebinds to duplex DNA on the leading strand but not to single-stranded DNA on the retrograde arm [761].

At each relocation a region of about 200 bases of DNA on the lagging side of the fork is rendered single-stranded providing a target for random initiation of RNA primers. For this reason the resulting Okazaki pieces have a mean size somewhat shorter than the 200 bases of nucleosomal DNA [327]. It is only when the DNA on the lagging side of the fork has been rendered double-stranded and been left behind by the replicating machinery that new nucleosomes form on the retrograde arm.

Newly synthesized nucleosomes consist predominantly of eight newly synthesized histones rather than a mixture of new and old histones. Histones, as well as DNA, are made predominantly in S-phase and inhibition of DNA synthesis inhibits synthesis of histones and causes pre-existing histone mRNA to break down with a much reduced half-life [744–746]. In yeast, co-ordinate control of histone gene transcription and replication appears to involve the location of the promoter for the H2A and H2B genes adjacent to an ARS element [747]. Some histone synthesis is not dependent on DNA synthesis, however, and, for instance, histone H3.3 is synthesized in quiescent cells and during G1 and G2 phase together with four H2A variants and H1a and $H1^0$ [748, 749]. Regulation of histone synthesis appears to take place at the level of transcription and translation [750, 762].

New nucleosomes form by initial addition of histones H3 and H4 to DNA (within 1000 bp of the replication fork) followed by histones H2A and H2B between 1000 and 10000 bp of the fork [786]. Histone H1 is added last and turns over quite readily [763]. DNA injected into *Xenopus* oocytes rapidly assumes a nucleosomal conformation, and similar, *in vitro*, studies using extracts of oocytes or SV40 infected cells [764, 765] have implicated at least two proteins in the assembly process. Nuclear factor N1 is involved in the binding of DNA to a tetramer of H3 and H4 and the heat-stable protein, nucleoplasmin, is involved in addition of H2A and H2B to the faces of the immature nucleosome [766–775].

As adjacent replicons tend to replicate at the same time in the cell cycle [583–585, 778–780], it is possible that a region of a chromosome becomes partially expanded and bidirectional replication is then initiated by a replisome [688]. Replicons have a length approximately equal to that of chromatin loops [776] or the equivalent of 400 nucleosomes or about 70 turns of a solenoid [777] and initiation of replication occurs at the point of attachment of a loop to the matrix (i.e. each loop is part of two adjacent replicons – section 6.15.5). As replicon size varies dramatically in developing amphibia and insects from $5\,\mu$m long in early cleavage to up to $350\,\mu$m long during spermatogenesis [585], chromatin structure would be expected to show parallel changes and this is indeed the case [743].

The altered structure associated with transcribing chromatin (section 10.6) requires between 1 and 3 minutes to re-form following replication [781, 782]. During this time, from 18 to 54 nucleosomes will have been added to the newly synthesized DNA and these nucleosomes consist of both newly synthesized and preexisting histones combined within a single octamer [783]. This implies that parental nucleosomes present on transcribing DNA dissociate to their individual monomers before reassembly. Nucleosomes present on transcriptionally active chromatin lack histone H1 which is replaced by several other proteins [784] (section 10.6.3). The promoter regions of active genes are often free of nucleosomes, and there is competition after replication between the binding of nucleosomes or transcription factors to this region [785, 787].

REFERENCES

1 Watson, J..D. and Crick, F. H. C. (1953) *Nature (London)*, **171**, 737.

2 Meselson, M. and Stahl, F. W. (1958) *Proc. Natl. Acad. Sci. USA*, **66**, 671.

3 Cairns, J. (1966) *Sci. Am.*, **214** (1), 36.

4 Taylor, J. H. (1963) in *Molecular Genetics* (ed. J. H. Taylor), Academic Press, New York, Part I, p. 65.

5 Fareed, G. C., Garon, C. F. and Salzman, N. P. (1972) *J. Virol.*, **10**, 484.

6 Okazaki, R., Okazaki, T., Sakabe, K., Sugimoto, K., Kainuma, R. *et al.* (1968) *Cold Spring Harbor Symp. Quant. Biol.*, **133**, 129.

7 Schnös, M. and Inman, R. B. (1970) *J. Mol. Biol.*, **51**, 61.

8 Magnussen, G., Pigiet, V., Winnacker, E. L., Abrarns, R. and Reichard, P. (1973) *Proc. Natl. Acad. Sci. USA*, **70**, 412.

9 Schlomai, J. and Kornberg, A. (1978) *J. Biol. Chem.*, **253**, 3305.

10 Tamanoi, F., Machida, Y. and Okazaki, T. (1978) *Cold Spring Harbor Symp. Quant. Biol.*, **43**, 239.

11 Tye, B-K., Nyman, P. O., Lehman, I. R., Hochhauser, S. and Weiss, B. (1977) *Proc. Natl. Acad. Sci. USA*, **74**, 154.

12 Tye, B-K. and Lehman, I. R. (1977) *J. Mol. Biol.*, **117**, 293.

13 Warner, H. R. and Duncan, B. K. (1978) *Nature (London)*, **272**, 32.

14 Olivera, B. M., Manlapaz-Ramos, P., Warner, H. R. and Duncan, B. K. (1979) *J. Mol. Biol.*, **128**, 265.

15 Okazaki, R., Sugimoto, K., Okazaki, T., Imac, Y. and Sugino, A. (1970) *Nature (London)*, **228**, 223.

16 Lehman, I. R., Tye, B-K. and Nyman, P. O. (1978) *Cold Spring Harbor Symp. Quant. Biol.*, **43**, 221.

17 Sugino, A. and Okazaki. R. (1972) *J. Mol. Biol.*, **164**, 61.

18 Brutlag, D., Schekman, R. and Kornberg, A. (1971) *Proc. Natl. Acad. Sci. USA*, **68**, 2826.

19 Wickner, W., Brutlag, D., Schekman. R. and Kornberg, A. (1972) *Proc. Natl. Acad. Sci. USA*, **69**, 965.

20 Bagdasarian, M. M., Izakawska, M. and Bagdasarian, M. (1977) *J. Bacteriol.*, **130**, 577.

21 Bouché, J. P., Zeckel, K. and Kornberg, A. (1975) *J. Biol. Chem.*, **250**, 5995.

22 Kurosawa, Y., Ogawa, T., Hirose, S., Okazaki, T. and Okazaki, R. (1975) *J. Mol. Biol.*, **96**, 653.

23 Okazaki, R., Okazaki, T., Hirose, S., Sugino, A., Ogawa, T. *et al.* (1975) *DNA Synthesis and its Regulation* (eds M. Goulian and P. Hanawalt), vol. III (Series ed. F. Fox) ICN-UCLA Symposium on Molecular and Cellular Biology, Benjamin, California.

24 Geider, K. (1976) *Curr. Top. Microbiol. Immunol.*, **74**, 58.

25 Seidel, H. (1967) *Biochim. Biophys. Acta*, **138**, 98.

26 Thomas, K. R., Manlapaz-Ramos, P., Lundquist, R. and Olivera, B. M. (1978) *Cold Spring Harbor Symp. Quant. Biol.*, **43**, 231.

27 Sugino, A. and Okazaki, R. (1973) *Proc. Natl. Acad. Sci. USA*, **70**, 88.

28 Huberman, J. A. and Horwitz, H. (1973) *Cold Spring Harbor Symp. Quant. Biol.*, **38**, 233.

29 Miyamoto, C. and Denhardt, D. T. (1977) *J. Mol. Biol.*, **116**, 681.

30 Anderson, M. L. M. (1978) *J. Mol. Biol.*, **118**, 277.

31 Kowalski, J. and Denhardt, D. T. (1979) *Nature (London)*, **281**, 704.

32 Tseng, B. Y., Erickson, J. M. and Goulian, M. (1979) *J. Mol. Biol.*, **129**, 531.

33 Blair, D. G., Sherratt, D. J., Clewel, D. B. and Helinski, D. R. (1972) *Proc. Natl. Acad. Sci. USA*, **69**, 2518.

34 Martens, P. A. and Clayton. D. A. (1979) *J. Mol. Biol.*, **135**, 327.

35 Ogawa, T., Hirose, S., Okazaki, T. and Okazaki, R. (1977) *J. Mol. Biol.*, **112**, 121.

36 Pigiet, V., Eliasson, R. and Reichard, P. (1974) *J. Mol. Biol.*, **84**, 197.

37 Olivera, B. M. and Bonhoeffer, F. (1972) *Nature (London) New Biol.*, **240**, 233.

38 Kurosawa, Y. and Okazaki, R. (1975) *J. Mol. Biol.*, **94**, 229.

39 Kuroda, R. K. and Okazaki, R. (1975) *J. Mol. Biol.*, **94**, 213.

40 Brynolf, K., Eliasson, R. and Reichard, P. (1974) *Cell*, **13**, 573.

41 Olivera, B. M. (1978) *Proc. Natl. Acad. Sci. USA*, **75**, 238.

42 Kornberg, A. (1974) *DNA Synthesis*, Freeman, San Francisco.

43 Kornberg, A. (1980) *DNA Replication*, Freeman, San Francisco; and supplement.

44 Ollis, D. L., Brick, P., Hamlin, R., Xuong, N. G. and Steitz, T. A. (1985) *Nature (London)*, **313**, 762.

45 Argos, P. (1988) *Nucleic Acids Res.*, **16**, 9909.

46 Kornberg, A. (1968) *Sci. Am.*, **21** (4) 64.

47 Josse, J., Kaiser, A. D. and Kornberg, A. (1961) *J. Biol. Chem.*, **236**, 864.

48 Becker, A. and Hurwitz, J. (1971) *Prog. Nucleic Acid Res. Mol. Biol.*, **11**, 423.

49 Jovin, T. M., Englund, P. T. and Bertsch, L. L. (1969) J. Biol. *Chem.*, **244**, 2996.

50 Que, B. G., Downey, K. M. and So, A. G. (1979) *Biochemistry*, **18**, 2064.

51 Joyce, C. M. and Steitz, T. A. (1987) *Trends Biochem Sci.*, **12**, 288.

52 Joyce, C. M. (1989), *J. Biol. Chem.*, **264**, 10 858.

53 Bambara, R. A., Vyemura, D. and Choi, T. (1978) *J. Biol. Chem.*, **253**, 413.

54 Matson, S. W., Capaldo-Kimball, F. N. and Bambara, R. A. (1978) *J. Biol. Chem.*, **253**, 7851.

55 Brutlag, D. and Kornberg, A. (1972) *J. Biol. Chem.*, **247**, 241.

56 Galas, D. J. and Branscomb, E. W. (1978) *J. Mol. Biol.*, **154**, 653.

57 Kunkel, T. A. (1988) *Cell*, **53**, 837.

58 Reha-Kranz, L. J. (1988) *J. Mol. Biol.*, **202**, 711.

59 Reha-Kranz, L. J. and Bessman, M. J. (1977) *J. Mol. Biol.*, **116**, 99.

60 Lo, K. Y. and Bessman, M. J. (1976) *J. Biol. Chem.*, **251**, 2480.

61 Cozzarelli, N. R., Kelly, R. B. and Kornberg, A. (1969) *J. Mol. Biol.*, **45**, 513.

62 Kelly, R. B., Cozzarelli, N. R., Deutscher, M. P., Lehman, I. R. and Kornberg, A. (1970) *J. Biol. Chem.*, **245**, 39.

63 de Lucia, P. and Cairns, J. (1969) *Nature (London)*, **224**, 1164.

64 Okazaki, R., Arisawa, M. and Sugino. A. (1971) *Proc. Natl. Acad. Sci. USA*, **68**, 2954.

65 Kato, T. and Konda, S. (1970) *J. Bacteriol.*, **104**, 871.

66 Emmerson, P. T. and Strikc, P. (1974) *Mechanism and Regulation of DNA Replication* (eds A. R. Kolber and M. Kohiyama), Plenum Press, New York, p. 47.

67 Gefter, M. L. (1974) *Prog. Nucleic Acid Res. Mol. Biol.*, **14**, 101.

68 Tait, R. C. and Smith, D. W. (1974) *Nature (London)*, **249**, 116.

69 Gefter, M. L., Hirota, Y., Kornberg, T., Wechsler, S. A. and Barnoux, C. (1971) *Proc. Natl. Acad. Sci. USA*, **68**, 3150.

70 Livingstone, D. M., Hinckle, D. C. and Richardson, C. C. (1975) *J. Biol. Chem.*, **250**, 461.

71 Lehman, I. R. (1974) in *The Enzymes* (ed. P. D. Boyer) Academic Press, New York, vol. 10, p. 237.

72 Kornberg, A. (1977) *Biochem. Soc. Trans.*, **5**, 359.

73 McHenry, C. S. and Crown, W. (1979) *J. Biol. Chem.*, **254**, 1748.

74 McHenry, C. S., Tomasiewicz, H., Griep, M. A., Fürste, J. P. and Flower, A. M. (1988) *DNA Replication and Mutagenesis* (eds R. E. Moses and W. C. Summers), American Society for Microbiology, Washington, p. 14.

75 Lee, S-H., Spector, M. P. and Walker, J. R. (1988) *DNA Replication and Mutagenesis* (eds R. E. Moses and W. C. Summers), American Society for Microbiology, Washington, p. 27.

76 Crute, J. J., La Duca, R. J., Johanson, K. O., McHenry, C. S. and Bambara, R. A. (1983) *J. Biol. Chem.*, **258**, 11 344.

77 Biswas, S. B. and Kornberg, A. (1984) *J. Biol. Chem.*, **259**, 7990.

78 Johanson, K. G. and McHenry, C. S. (1984) *J. Biol. Chem.*, **259**, 4589.

79 Maki, H., Maki, S. and Kornberg, A. (1988) *J. Biol. Chem.*, **263**, 6570.

80 Kornberg, A. (1988) *J. Biol. Chem.*, **263**, 1.

81 Gass, K. B. and Cozzarelli, N. R. (1974) *Methods Enzymol.*, **29**, 17.

82 White, T. J., Arnheim, N. and Erlich, H. A. (1989) *Trends Genet.*, **5**, 185.

83 Spicer, E. K., Rush, J., Fung, C., Reha-Kranz, L. J., Karam, J. D. *et al.* (1988) *J. Biol. Chem.*, **263**, 7478.

84 Tabor, S., Huber, H. E. and Richardson, C. C. (1988) *J. Biol. Chem.*, **263**, 16 212.

85 Huber, H. E., Tabor, S. and Richardson, C. C. (1988) *J. Biol. Chem.*, **263**, 16 274.

86 Jung, G., Leavitt, M. C., Hsieh, J-C. and Ito, J. (1987) *Proc. Natl. Acad. Sci. USA*, **84**, 8287.

87 Wang, S. W., Wahl, A. F., Yuan, P-M., Arai, N., Pearson, B. E., Arai, K., Korn, D., Hunkapiller, M. W. and Wang, T. S-F. (1988) *EMBO J.*, **7**, 37.

88 Pizzagalli, A., Valsasnini, P., Plevani, P. and Lucchini, G. (1988) *Proc. Natl. Acad. Sci. USA*, **85**, 3772.

89 Bollum, F. J. (1974) in *The Enzymes* (ed. P. D. Boyer) Academic Press, New York, vol. 10, p. 145.

90 Kaguni, L. S. and Lehman, I. R. (1988) *Biochim. Biophys. Acta*, **950**, 87.

91 Plevani, P., Badaracco, G., Augl, C. and Chang, L. M. S. (1984) *J. Biol. Chem.*, **259**, 7532.

92 Morrison, A., Araki, H., Clark, A. B., Hamatake, R. K. and Sugino, A. (1990) *Cell*, **62**, 1143.

93 Bauer, G. A., Heller, H. M. and Burgers, P. M. J. (1986) *J. Biol. Chem.*, **263**, 917.

94 Burgers, P. M. J. and Bauer, G. A. (1986) *J. Biol. Chem.*, **263**, 925.

95 Sitney, K. C., Budd, M. E. and Campbell, J. L. (1989) *Cell*, **56**, 559.

96 Nakamura, H., Monta, T., Masaki, S. and Yoshida, S. (1984) *Exp. Cell Res.*, **151**, 123.

97 Smith, H. C. and Berezney, R. (1983) *Biochemistry*, **22**, 3042.

98 Jones, C. and Su, R. T. (1982) *Nucleic Acids Res.*, **10**, 5517.

99 Wang, H. F. and Popenoe, E. A. (1977) *Biochim. Biophys. Acta*, **474**, 98.

100 Spadari, S., Sala, F. and Pedrali-Noy, G. (1982) *Trends Biochem. Sci.*, **7**, 29.

101 Waqar, M. A., Evans, M. J. and Huberman, J. A. (1978) *Nucleic Acids Res.*, **5**, 1933.

102 Edenberg, H. T., Anderson, S. and de Pamphilis, M. L. (1978) *J. Biol. Chem.*, **253**, 3273.

103 Reyland, M. E., Lehman, I. R. and Loeb, L. A. (1988) *J. Biol. Chem.*, **263**, 6518.

104 Murakami, Y., Yasuda, H., Miyazawa, H., Hanaoka, F. and Yamada, M. (1985) *Proc. Natl. Acad. Sci. USA*, **82**, 1761.

105 Mechali, M., Abadiedebat, J. and de Recondo, A. M. (1980) *J. Biol. Chem.*, **255**, 2114.

106 Goulian, M. and Heard, C. J. (1990) *J. Biol. Chem.*, **265**, 13 231.

107 Fisher, P. A., Wang, T. S-F. and Korn, D. (1979) *J. Biol. Chem.*, **254**, 6128.

108 Konig, H., Riedel, H. D. and Knippers, R. (1983) *Eur. J. Biochem.*, **135**, 435.

109 Lee, S-K. and Hurwitz, J. (1990) *Proc. Natl. Acad. Sci. USA*, **87**, 5672.

110 Brown, D. R., Roth, M. J., Reinberg, D. and Hurwitz, J. (1984) *J. Biol. Chem.*, **259**, 10 545.

111 Cotterill, S. M., Reyland, M. E., Loeb, L. A. and Lehman, I. R. (1987) *Proc. Natl. Acad. Sci. USA*, **84**, 5635.

112 Vaughn, J. P., Dijkwel, P. A., Mullenders, L. H. F. and Hamlin, J. L. (1990) *Nucleic Acids Res.*, **18**, 1965.

113 Pritchard, C. G. and De Pamphilis, M. L. (1983) *J. Biol. Chem.*, **258**, 9801.

114 Lehman, I. R. and Kaguni, L. S. (1989) *J. Biol. Chem.*, **264**, 4265.

115 Masaki, S., Tanabe, K. and Yoshida, S. (1984) *Nucleic Acids Res.*, **12**, 4455.

116 Gronostajski, R. M., Field, J. and Hurwitz, J. (1984) *J. Biol. Chem.*, **259**, 9479.

117 Grosse, F. and Krauss, G. (1985) *J. Biol. Chem.*, **260**, 1881.

118 Cotterill, S., Chui, G. and Lehman, I. R. (1987) *J. Biol. Chem.*, **262**, 16 100.

119 Cotterill, S., Chui, G. and Lehman, I. R. (1987) *J. Biol. Chem.*, **262**, 16 105.

120 Hirose, F., Yamamoto, S., Yamaguchi, M. and Matsukage, A. (1988) *J. Biol. Chem.*, **263**, 2925.

121 Suzuki, M., Enomoto, T., Masutani, C., Hanaoka, F., Yamada, M. and Ui, M. (1989) *J. Biol. Chem.*, **264**, 10 065.

122 Brooks, M. and and Dumas, L. B. (1989) *J. Biol. Chem.*, **264**, 3602.

123 Faust, E. A., Nagy, R. and Davey, S. K.

(1985) *Proc. Natl. Acad. Sci. USA*, **82**, 4023.

124 Ottiger, A. P. and Hubscher, U. (1984) *Proc. Natl. Acad. Sci. USA*, **81**, 3993.

125 Nakayami, M., Ben-Mahrez, K. and Kohikama, M., (1988) *Eur. J. Biochem.*, **175**, 265.

126 Byrnes, T. J., Downey, K. M., Black, V. L. and So, A. G. (1976) *Biochemistry*, **15**, 2817.

127 Wahl, A. F., Crute, J. J., Sabatino, R. D., Bodner, J. B., Marraccino, R. L. *et al.* (1986) *Biochemistry*, **25**, 7821.

128 Focher, F., Gassmann, M., Hafkemeyer, P., Ferrari, E., Spadari, S. *et al.* (1989) *Nucleic Acids Res.*, **17**, 1805.

129 Berg, J. M. (1990) *J. Biol. Chem.*, **265**, 6513.

130 Wong, S. W., Syvaoja, J., Tan, C-K., Downey, K. M., So, A. G. *et al.* (1989) *J. Biol. Chem.*, **264**, 5924.

131 Lee, M. Y. W. T. (1988) *DNA Replication and Mutagenesis* (eds R. E. Moses and W. C. Summers), American Society for Microbiology, Washington, p. 68.

132 Nishida, C., Reinhard, P. and Linn, S. (1988) *J. Biol. Chem.*, **263**, 501.

133 Syväoja, J., Suomensaari, S., Nishida, C., Goldsmith, J. S. Chui, G. S. J. *et al.* (1990) *Proc. Natl. Acad. Sci. USA*, **87**, 6664; and Syväoja, J. E. (1990) *BioEssays*, **12**, 533.

134 Focher, F., Ferrari, E., Spadari, S. and Hübscher, U. (1988) *FEBS Lett.*, **229**, 6.

135 Kunkel, T. A., Tcheng, J. E. and Meyer, R. R. (1978) *Biochim. Biophys. Acta*, **520**, 302.

136 Wilson, S., Abbotts, J. and Widen, S. (1988) *Biochim. Biophys. Acta*, **949**, 149.

137 Abbotts, J., SenGupta, D. N., Zmudzka, B. Z., Widen, S. G. and Wilson, S. H. (1988) *DNA Replication and Mutagenesis* (eds R. E. Moses and W. C. Summers), American Society for Microbiology, Washington, p. 55.

138 Randahl, H., Elliot, G. C. and Linn, S. (1988) *J. Biol. Chem.*, **263**, 12228.

139 Chang, L. M. S. (1973) *J. Biol. Chem.*, **248**, 6983.

140 Waser, J., Hubscher, V., Kuenzle, C. C. and Spadari, S. (1979) *Eur. J. Biochem.*, **97**, 361.

141 Dressler, S. L. and Lieberman, M. W. (1983) *J. Biol. Chem.*, **258**, 9990.

142 Mosbaugh, D. W. and Linn, S. (1984) *J. Biol. Chem.*, **259**, 10247.

143 Chang, L. M. S., Plevani, P. and Bollum,

F. J. (1982) *Proc. Natl. Acad. Sci. USA*, **79**, 758.

144 Spadari, S. and Weissbach, A. (1974) *J. Biol. Chem.*, **249**, 5809.

145 Yamaguchi, M., Matsukage, A. and Takahasi, T. (1980) *J. Biol. Chem.*, **255**, 7002.

146 Yamaguchi, M., Matsukage, A. and Takahashi, T. (1980) *Nature (London)*, **285**, 45.

147 Kunkel, T. A. and Soni, A. (1988) *J. Biol. Chem.*, **263**, 4450.

148 Tewari, K. K. and Wildman, S. G. (1967) *Proc. Natl. Acad. Sci. USA*, **58**, 689.

149 Stillman, B. W., Tamanoi, F. and Mathews, M. B. (1982) *Cell*, **31**, 613.

150 Field, J., Gronostajaski, R. M. and Hurwitz, J. (1984) *J. Biol. Chem.*, **259**, 9487.

151 Knopf, K. (1979) *Eur. J. Biochem.*, **98**, 231.

152 Furman, P. A., St. Clair, M. H., Fyfe, J. A., Rideout, J. L., Keller, P. M. *et al.* (1979) *J. Virol.*, **32**, 72.

153 O'Donnell, M. E., Elias, P. and Lehman, I. R. (1987) *J. Biol. Chem.*, **262** 4252.

154 Crute, J. J., Tsurumi, T., Zhu, L., Weller, S. K., Olivo, P. D. *et al.* (1989) *Proc. Natl. Acad. Sci. USA*, **86**, 2186.

155 Krakow, J. S., Coutsogeorgopoulos, C. and Canellakis, E. S. (1962) *Biochim. Biophys. Acta*, **55**, 639.

156 Bollum, F. (1965) *J. Biol. Chem.*, **240**, 2599.

157 Zahler, A. M., Williamson, J. R., Cech, T. R. and Prescott, D. M. (1991) *Nature (London)*, **350**, 718.

158 Chang, L. M. S., Plevani, P. and Bollum, F. J. (1982) *J. Biol. Chem.*, **257**, 5700.

159 Chang, L. M. S. and Bollum, F. J. (1971) *J. Biol. Chem.*, **246**, 909.

160 Srivastava, B. I. S. (1974) *Cancer Res.*, **34**, 1015.

161 Desiderio, S. V., Yancopoulos, G. D., Paskind, M., Thomas, E., Boss, M. A. *et al.* (1984) *Nature (London)*, **311**, 752.

162 Temin, H. M. and Mizutani, S. (1970) *Nature (London)*, **226**, 1211.

163 Baltimore, D. (1970) *Nature (London)*, **226**, 1209.

164 Spiegelman, S., Burny, A., Das, M., Keydar, J., Schlom, J. *et al.* (1970) *Nature (London)*, **228**, 430.

165 Wang, R. M. (1978) *Annu. Rev. Microbiol.*, **32**, 561.

166 Harada, F., Sawyer, R. C. and Dahlberg, J. E. (1975) *J. Biol. Chem.*, **250**, 3487.

167 Johnson, M. S., McClure, M. A., Feng,

D-F., Gray, J. and Doolittle, R. F. (1986) *Proc. Natl. Acad. Sci. USA*, **83**, 7648.

168 Jacks, T., Powere, M. D., Masiarz, F. R., Luciw, P. A., Barr, P. J. *et al.* (1988) *Nature (London)*, **231**, 280.

169 Pearl, L. H. and Taylor, W. R. (1987) *Nature (London)*, **328**, 482.

170 Varmus, H. E. (1985) *Nature (London)*, **314**, 583.

171 Panganiban, A. T. and Temin, H. M. (1984) *Proc. Natl. Acad. Sci. USA*, **81**, 7885.

172 Seal, G. and Loeb, L. A. (1976) *J. Biol. Chem.*, **251**, 975.

173 Battula, N., Dube, D. K. and Loeb, L. A. (1975) *J. Biol. Chem.*, **250**, 8404.

174 Chang, L-J., Pryciak, P., Ganem, D. and Varmus, H. E. (1989) *Nature (London)*, **337**, 364.

175 Penswick, J., Hubler, R. and Hohn, T. (1988) *J. Virol.*, **62**, 1460.

176 Baltimore, D. (1988) *Cell*, **40**, 481.

177 Varmus, H. E. (1989) *Cell*, **56**, 721.

178 Maki, S. and Kornberg, A. (1988) *J. Biol. Chem.*, **263**, 6547.

179 Maki, S. and Kornberg, A. (1988) *J. Biol. Chem.*, **263**, 6555.

180 Maki, S. and Kornberg, A. (1988) *J. Biol. Chem.*, **263**, 6561.

181 Tsuchihashi, Z. and Kornberg, A. (1990) *Proc. Natl. Acad. Sci. USA*, **87**, 2516.

182 Downey, K. H. Tan, C-K. and So, A. G. (1990) *BioEssays*, **12**, 231.

183 Laipis, P. J., Oliver, B. M. and Ganesan, A. T. (1969) *Proc. Natl. Acad. Sci. USA*, **62**, 289.

184 Tsukada, K. and Ichimura, M. (1971) *Biochem. Biophys. Res. Commun.*, **42**, 156.

185 Gefter, M. L., Becker, A. and Hurwitz, J. (1967) *Proc. Natl. Acad. Sci. USA*, **58**, 240.

186 Gellert, M. and Bullock, M. L. (1970) *Proc. Natl. Acad. Sci. USA*, **67**, 1580.

187 Lehman, I. R. (1974) in *The Enzymes* (ed. P. D. Boyer) Academic Press, New York, vol. 10, p. 237.

188 Soderhall, S. (1976) *Nature (London)*, **260**, 640.

189 Lasko, D. D., Tomkinson, A. E. and Lindahl, T. (1990) *J. Biol. Chem.*, **265**, 12 618.

190 Mezzina, M., Sarasin, A. Politi, N. and Bertazzoni, U. (1984) *Nucleic Acids Res.*, **12**, 5109.

191 Alberts, B. M. and Frey, L. (1970) *Nature (London)*, **227**, 1313.

192 Weiner, J. H., Bertsch, L. L. and Kornberg, A. (1975) *J. Biol. Chem.*, **250**, 1972.

193 Reuben, R. C. and Gefter, M. L. (1973) *Proc. Natl. Acad. Sci. USA*, **70**, 1846.

194 Geider, K. and Kornberg, A. (1974) *J. Biol. Chem.*, **249**, 3999.

195 Gilboa, E., Mitra, S. W., Goff, S. and Baltimore, D. (1979) *Cell*, **18**, 93.

196 Meyer, R. R., Glassberg, J. and Kornberg, A. (1979) *Proc. Natl. Acad. Sci. USA*, **76**, 1702.

197 Van Darp, B., Schneck, P. K. and Staudenbauer, W. L. (1979) *Eur. J. Biochem.*, **94**, 445.

198 Kenny, M. K., Schlegel, U., Furneaux, H. and Hurwitz, J. (1990) *J. Biol. Chem.*, **265**, 7693.

199 Kuhn, B., Abdel-Monem, M., Krell, H. and Hoffmann–Berling, H. (1978) *J. Biol. Chem.*, **254**, 11 343.

200 Lahue, E. E. and Matson, S. W. (1988) *J. Biol. Chem.*, **263**, 3208.

201 Abdel-Monem, M. and Hoffmann–Berling, H. (1980) *Trends Biochem. Sci.*, **5**, 128.

202 Abdel-Monem, M., Taucher-Scholz, G. and Klinkert, M-Q. (1983) *Proc. Natl. Acad. Sci. USA*, **80**, 4659.

203 Wessel, R., Müller, H. and Hoffmann-Berling, H. (1990) *Eur. J. Biochem.*, **189**, 277.

204 Matson, S. W. and Kaiser-Rogers, K. A., (1990) *Annu. Rev. Biochem.*, **59**, 289.

205 Georgi-Geisberger, P. and Hoffmann-Berling, H. (1990) *Eur. J. Biochem.*, **192**, 689.

206 Matson, S. W., (1986) *J. Biol. Chem.*, **261**, 10 169.

207 Oeda, K., Horiuchi, T. and Sakiguchi, M. (1982) *Nature (London)*, **298**, 98.

208 Kumura, K. and Sekiguchi, M. (1984) *J. Biol. Chem.*, **259**, 1560.

209 Klinkert, M-Q, Klein, A. and Abdel-Monem, M. (1980) *J. Biol. Chem.*, **255**, 9746.

210 Gilchrist, C. A. and Denhardt, D. T. (1987) *Nucleic Acids Res.*, **15**, 465.

211 Lee, M. S. and Marians, K. J. (1987) *Proc. Natl. Acad. Sci. USA*, **84**, 8345.

212 Lasken, R. S. and Kornberg, A. (1988) *J. Biol. Chem.*, **263**, 5512.

213 Yarranton, G. T., Das, R. H. and Gefter, M. L. (1979) *J. Biol. Chem.*, **254**, 12 002.

214 LeBowitz, J. H. and McMacken, R. (1986) *J. Biol. Chem.*, **261**, 4738.

215 Mok, M. and Marians, K. J. (1987) *J. Biol. Chem.*, **262**, 16 644.

216 Alberts, B. and Sternglanz, R. (1977) *Nature (London)*, **269**, 655.

217 Kolodner, R., Masamune, Y., Le Clere, J. E. and Richardson, C. C. (1978) *J. Biol. Chem.*, **253**, 566.

218 Matson, S. W., Tabor, S. and Richardson, C. C. (1983) *J. Biol. Chem.*, **258**, 14 017.

219 Alberts, B. M., Barry, J., Bedinger, P., Formosa, T., Jongeneel, C. V. *et al.* (1983) *Cold Spring Harbor Symp. Quant. Biol.*, **47**, 655.

220 Hinton, D. M. and Nossal, N. G. (1987) *J. Biol. Chem.*, **262**, 10 873.

221 Nossal, N. G. and Hinton, D. M. (1987) *J. Biol. Chem.*, **262**, 10 879.

222 Jongeneel, C. V., Formosa, T. and Alberts, B. M. (1984) *J. Biol. Chem.*, **259**, 12 925.

223 Jongeneel, C. V., Bedinger, P. and Alberts, B. M. (1984) *J. Biol. Chem.*, **259**, 12 933.

224 Bedinger, P., Hochstrasser, M., Jongeneel, C. V. and Alberts, B. M. (1983) *Cell*, **34**, 115.

225 Graw, J., Schlaeger, E-J. and Knippers, R. (1981) *J. Biol. Chem.*, **256**, 13 207.

226 Sung, P., Prakash, L., Matson, S. W. and Prakash, S. (1987) *Proc. Natl. Acad. Sci. USA*, **84**, 8951.

227 Arai, K., Yasuda, S. and Kornberg, A. (1981) *J. Biol. Chem.*, **256**, 5247.

228 Hirling, H., Scheffner, M., Restle, T. and Stahl, H. (1989) *Nature (London)*, **339**, 562.

229 Sapta, M., Burton, Z. F. and Greenblatt, J. (1989) *Nature (London)*, **341**, 410.

230 Smith, K. R., Yancey, J. E. and Matson, S. W. (1989) *J. Biol. Chem.*, **264**, 6119.

231 Wang, J. C. (1982) *Sci. Am.*, **247**, 84.

232 Sander, M. and Hsieh, T-S. (1983) *J. Biol. Chem.*, **258**, 8421.

233 Liu, L. F., Rowe, T. C., Yang, L., Tewey, K. M. and Chen, G. L. (1983) *J. Biol. Chem.*, **248**, 15 365.

234 Dugnet, M., Lavenot, C., Harper, F., Mirambeau, G. and de Recondo, A-M. (1983) *Nucleic Acids Res.*, **11**, 1059.

235 Liu, L. F. (1984) *CRC Crit. Rev. Biochem.*, **15**, 1.

236 Halligan, B. D., Edwards, K. A. and Liu, L. F. (1985) *J. Biol. Chem.*, **269**, 2475.

237 Champoux, J. J. (1981) *J. Biol. Chem.*, **256**, 4805.

238 Brown, P. O. and Cozzarelli, N. R. (1981) *Proc. Natl. Acad. Sci. USA*, **78**, 843.

239 Mhaskar, D. N. and Goodman, M. F. (1984) *J. Biol. Chem.*, **259**, 11 713.

240 Shure, M. and Vinograd, J. (1976) *Cell*, **8**, 215.

241 Shelton, E. R., Osheroff, N. and Brutlag, D. L. (1983) *J. Biol. Chem.*, **258**, 9530.

242 Liu, L. F., Liu, C-C. and Alberts, B. M. (1979) *Nature (London)*, **281**, 456.

243 Wang, J. C. (1984) *Nature (London)*, **309**, 669.

244 Kikuchi, A. and Asai, K. (1984) *Nature (London)*, **209**, 677.

245 Sugino, A., Higgins, N. P. and Cozzarelli, N. R. (1980) *Nucleic Acids Res.*, **8**, 3865.

246 Fisher L M (1981) *Nature (London)*, **294**, 607.

247 Sugino, A., Peebles, C. L., Kreuzer, K. N. and Cozzarelli, N. R. (1977) *Proc. Natl. Acad. Sci. USA*, **74**, 4767.

248 Mizuuchi, K., O'Dea, M. H. and Gellert, M. (1978) *Proc. Natl. Acad. Sci. USA*, **75**, 5960.

249 Denhardt, D. T. (1979) *Nature (London)*, **280**, 196.

250 Sugino, A., Higgins, N. P., Brown, P. O., Peebles, C. L. and Cozzarelli, N. R. (1978) *Proc. Natl. Acad. Sci. USA*, **75**, 4838.

251 Gellert, M., Mizuuchi, K., O'Dea, M. H., Ohmari, H. and Tomizawa, J. (1978) *Cold Spring Harbor Symp. Quant. Biol.*, **43**, 35.

252 Gellert, M., O'Dea, M. H., Itoh, T. and Tomizawa, J. (1976) *Proc. Natl. Acad. Sci. USA*, **73**, 4474.

253 Pruss, G. J., Manes, S. H. and Drlica, K. (1982) *Cell*, **31**, 35.

254 DiNardo, S., Voelkel, K. A., Sternglanz, R., Reynolds, A. E. and Wright, A. (1982) *Cell*, **31**, 43.

255 Pruss, G. J. and Drlica, K. (1989) *Cell*, **56**, 521.

256 Mizuuchi, K., Gellert, M. and Nash, H. A. (1978) *J. Mol. Biol.*, **121**, 375.

257 De Wyngaert, M. A. and Hinckle, D. C. (1979) *J. Virol.*, **29**, 529.

258 Holm, C., Goto, T., Wang, J. C. and Botstein, D. (1985) *Cell*, **41**, 553.

259 Marians, K. J. (1987) *J. Biol. Chem.*, **262**, 10 362.

260 Avemann, K., Knippers, R., Koller, T. and Sogo, J. M. (1988) *Mol. Cell. Biol.*, **8**, 3026.

261 Tse-Dinh, Y-C., Wong, T. W. and Goldberg, A. R. (1984) *Nature (London)*, **312**, 785.

262 Ferro, A. M., Higgins, N. P. and Olivera, B. M. (1983) *J. Biol. Chem.*, **258**, 6000.

263 Earnshaw, W. C. and Heck, M. M. S. (1985) *J. Cell Biol.*, **100**, 1716.

264 Heck, M. M. S., Hittelman, W. N. and Earnshaw, W. C. (1988) *Proc. Natl. Acad. Sci. USA*, **85**, 1086.

265 Uemura, T. and Yanagida, M. (1986) *EMBO J.*, **5**, 1003.

266 Yanagida, M. and Wang, J. C. (1987) *Nucleic Acids Research and Molecular Biology*, I (eds F. Eckstein and D. M. J. Lilley), Springer, p. 196.

267 Culotta, V. and Sollner-Webb, B. (1988) *Cell*, **52**, 585.

268 Zhang, H., Wang, J. C. and Liu, L. F. (1988) *Proc. Natl. Acad. Sci. USA*, **85**, 1060.

269 Kjeldsen, E., Bonven, B. J., Andoh, T., Ishii, K., Okada, K. *et al.* (1988) *J. Biol. Chem.*, **263**, 3912.

270 Pommier, Y., Kerrigan, D., Hartman, K. D. and Glazer, R. I. (1990) *J. Biol. Chem.*, **265**, 9418.

271 Kjeldsen, E., Mollerup, S., Thomsen, B., Bonven, B. J., Bolund, L. *et al.* (1988) *J. Mol. Biol.*, **202**, 333.

272 Chen, G. L., Yang, L., Rowe, T. C., Halligan, B. D., Tewey, K. M. *et al.* (1984) *J. Biol. Chem.*, **259**, 13 360.

273 Yang, L., Rowe, T. C. and Liu, L. F. (1985) *Cancer Res.*, **45**, 5872.

274 Stevnsner, T., Mortensen, U. H., Westergaard, O. and Bonven, B. J. (1989) *J. Biol. Chem.*, **264**, 10 110.

275 Been, M. D., Burgess, R. R. and Champoux, J. J. (1984) *Nucleic Acids Res.*, **12**, 3097.

276 Thomsen, B., Mollerup, S., Bonven, B. J., Frank, R., Blocker, H. *et al.* (1987) *EMBO J.*, **6**, 1817.

277 Spitzner, J. R. and Müller, M. T. (1988) *Nucleic Acids Res.*, **16**, 5533.

278 Spitzner, J. R., Chung, I. K. and Müller, M. T. (1990) *Nucleic Acids Res.*, **18**, 1.

279 Rowen, L. and Kornberg, A. (1978) *J. Biol. Chem.*, **253**, 758.

280 Rowen, L. and Kornberg, A. (1978) *J. Biol. Chem.*, **253**, 770.

281 Wickner, S. (1977). *Proc. Natl. Acad. Sci. USA*, **74**, 2815.

282 Kahn, M. and Hanawalt, P. (1979) *J. Mol. Biol.*, **128**, 501.

283 Sims, J. and Benz, E. W. (1980) *Proc. Natl. Acad. Sci. USA*, **77**, 900.

284 Hopfield, J. J. (1974) *Proc. Natl. Acad. Sci. USA*, **71**, 4135.

285 Radman, M. and Wagner, R. (1988) *Sci. Am.*, **259** (2), 24.

286 Kunkel, T. A. (1988) *Cell*, **53**, 837.

287 Kunkel, T A., Shaper, R. M., Beckman, R. A. and Loeb, L. A. (1981) *J. Biol. Chem.*, **256**, 9883.

288 Grosse, F., Krauss, G., Knill-Jones, J. W. and Fersht, A. R. (1983) *EMBO J.*, **2**, 1515.

289 Fersht, A. R. and Knill-Jones, J. W. (1983) *J. Mol. Biol.*, **165**, 633.

290 Lasken, R. S. and Goodman, M. F. (1985) *Proc. Natl. Acad. Sci. USA*, **82**, 1301.

291 Seal, G., Shearman, C. W. and Loeb, L. A. (1979) *J. Biol. Chem.*, **254**, 5229.

292 Brosius, S., Grosse, F. and Krauss, G. (1983) *Nucleic Acids Res.*, **11**, 193.

293 Shlomai, J., Polder, L., Arai, K. and Kornberg, A. (1981) *J. Biol. Chem.*, **256**, 5233.

294 Agarwal, S. S., Dube, D. K. and Loeb, L. A. (1979) *J. Biol. Chem.*, **254**, 101.

295 Perrino, F. W. and Loeb, L. A. (1989) *Proc. Natl. Acad. Sci. USA*, **86**, 3085.

296 Rechmann, B., Grosse, F. and Krauss, G. (1983) *Nucleic Acids Res.*, **11**, 7251.

297 Perrino, F. W. and Loeb, L. A. (1989) *J. Biol. Chem.*, **264**, 2898.

298 Bialek, G., Nasheuer, H-P., Goetz, H. and Grosse, F. (1989) *EMBO J.*, **8**, 1833.

299 Roberts, J. D. and Kunkel, T. A. (1988) *Proc. Natl. Acad. Sci. USA*, **85**, 7064.

300 Fersht, A. R. (1979) *Proc. Natl. Acad. Sci. USA*, **76**, 4946.

301 Kunkel, T. A. (1981) *Proc. Natl. Acad. Sci. USA*, **78**, 6734.

302 Hillebrand, G. G. and Beattie, K. L. (1984) *Nucleic Acids Res.*, **12**, 3173.

303 Bedinger, P. and Alberts, B. M. (1983) *J. Biol. Chem.*, **258**, 9649.

304 Topal, M. D. and Sinha, N. K. (1983) *J. Biol. Chem.*, **258**, 12 274.

305 Kunkel, T. A., Loeb, L. A. and Goodman, M. F. (1984) *J. Biol. Chem.*, **259**, 1539.

306 Schenermann, T., Tam, S., Burgers, P. M. J., Lu, C. and Echols, H. (1983) *Proc. Natl. Acad. Sci. USA*, **80**, 7085.

307 Loeb, L. A., Dube, D. K., Beckmann, R. A., Koplitz, M. and Gopinathan, K. P. (1981) *J. Biol. Chem.*, **256**, 3978.

308 Loeb, L. A., Weymouth, L. A., Kunkel, T. A., Gopinathan, K. P., Beckman, R. A. and Dube, D. K. (1978) *Cold Spring Harbor Symp. Quant. Biol.*, **43**, 921.

309 Sloane, D. L., Goodman, M. F. and Echols, H. (1988) *Nucleic Acids Res.*, **16**, 6465.

310 Reanney, D. (1984) *Nature (London)*, **307**, 318.

311 Roberts, J. D., Preston, B. D., Johnston, L. A., Soni, A., Loeb, L. A. and Kunkel, T. A. (1989) *Mol. Cell. Biol.*, **9**, 469.

312 Ripley, L. S. and Papanicolaou, C. (1988) in *DNA Replication and Mutagenesis* (eds R. E. Moses and W. C. Summers), American Society Microbiology, Washington, p. 227.

313 Boosalis, M. S., Mosbaugh, D. W., Hamatake, R., Sugino, A., Kunkel, T. A. and Goodman, M. F. (1989) *J. Biol. Chem.*, **264**, 11360.

314 Hatfield, D. and Oroszlan, S. (1990) *Trends Biochem. Sci.*, **15**, 186.

315 Wells, R. D. and Inman, R. B. (1973) *DNA Synthesis In Vitro*, University Park Press, Baltimore.

316 Wickner, R. B. (1974) in *Methods in Molecular Biology*, vol. 7, *DNA Replication* (ed. R. B. Wickner), Dekker, New York.

317 Vosberg, H. P. and Hoffmann–Berling, H. (1971) *J. Mol. Biol.*, **58**, 739.

318 Moses, R. E. and Richardson, C. C. (1970) *Proc. Natl. Acad. Sci. USA*, **67**, 674.

319 Schaller, H., Otto, B., Nusslein, V., Huf, J., Herrmann, R. and Bonhoeffer, F. (1972) *J. Mol. Biol.*, **63**, 183.

320 Nusslein, V. and Klein, A. (1974) in *Methods in Molecular Biology*, vol. 7 (ed. R. B. Wickner), Dekker, New York.

321 Burgoyne, L. A. (1972) *Biochem. J.*, **130**, 959.

322 Hunter, T. and Francke, B. (1974) *J. Virol.*, **13**, 125.

323 Winnacker, E. L., Magnussen, G. and Reichard, P. (1972) *J. Mol. Biol.*, **72**, 523.

324 De Pamphilis, M. L., Beard, P. and Berg, P. (1975) *J. Biol. Chem.*, **250**, 4340.

325 Berger, N. A., Petzold, S. J. and Johnson, E. S. (1977) *Biochim. Biophys. Acta*, **478**, 44.

326 Fraser, J. M. K. and Huberman, J. A. (1977) *J. Mol. Biol.*, **117**, 249.

327 De Pamphilis, M. L., Anderson, S., Bar-Shavit, R., Collins, E., Edenberg, H. *et al.* (1978) *Cold Spring Harbor Symp. Quant. Biol.*, **43**, 679.

328 Wickner, S. (1978) *Annu. Rev. Biochem.*, **47**, 1163.

329 Gray, C. P., Sommer, R., Polke, C., Beck, E. and Schaller, H. (1978) *Proc. Natl. Acad. Sci. USA*, **75**, 50.

330 Beck, E., Sommer, R., Averswald, E. A., Kurz, C., Zinc, B. *et al.* (1978) *Nucleic Acids Res.*, **5**, 4495.

331 Sims, J., Capon, D. and Dressler, D. (1979) *J. Biol. Chem.*, **254**, 12615.

332 Fiddes, J. C., Barrell, B. G. and Godson, G. N. (1978) *Proc. Natl. Acad. Sci. USA*, **75**, 1081.

333 Sims, J. and Benz, E. W. (1980) *Proc. Natl. Acad. Sci. USA*, **77**, 900.

334 Stayton, M. M. and Kornberg, A. (1983) *J. Biol. Chem.*, **258**, 13205.

335 Lambert, P. F., Waring, D. A., Wells, R. D. and Reznikoff, W. S. (1986) *J. Virol.*, **58**, 450.

336 Bouché, J. P., Rowen, L. and Kornberg, A. (1978) *J. Biol. Chem.*, **253**, 756.

337 Godson, G. N., Barrell, B. G., Staden, R. and Fiddes, J. C. (1978) *Nature (London)*, **276**, 236.

338 Sumida-Yasamoto, C., Ikeda, J-E., Benz, E., Marians, K. J., Vicuna, R. *et al.* (1978) *Cold Spring Harbor Symp. Quant. Biol.*, **43**, 311.

339 Wickner, S. and Hurwitz, J. (1976) *Proc. Natl. Acad. Sci. USA*, **73**, 1053.

340 Nomura, N., Low, R. L. and Ray, D. S. (1982) *Proc. Natl. Acad. Sci. USA*, **79**, 3153.

341 van der Ende, A., Teertstra, R., van der Avoort, H. G. A. M. and Weisbeek, P. J. (1983) *Nucleic Acids Res.*, **11**, 4957.

342 Shlomai, J. and Kornberg, A. (1981) *J. Biol. Chem.*, **256**, 6794.

343 Arai, K-I. and Kornberg, A. (1981) *Proc. Natl. Acad. Sci. USA*, **78**, 69.

344 Kobori, J. A. and Kornberg, A. (1982) *J. Biol. Chem.*, **257**, 13770.

345 Lee, M. S. and Marians, K. J. (1989) *J. Biol. Chem.*, **264**, 14531.

346 Arai, K-I., Low, R. L. and Kornberg, A. (1981) *Proc. Natl. Acad. Sci. USA*, **78**, 707.

347 Ogawa, T., Arai, K-I. and Okazaki, T. (1983) *J. Biol. Chem.*, **258**, 13353.

348 McMacken, R. and Kornberg, A. (1978) *J. Biol. Chem.*, **253**, 3313.

349 Arai, K-I. and Kornberg, A. (1979) *Proc. Natl. Acad. Sci. USA*, **76**, 4308.

350 Burgers, P. M. J. and Kornberg, A. (1983) *J. Biol. Chem.*, **258**, 7669.

351 Baas, P. D. (1985) *Biochim. Biophys. Acta*, **825**, 111.

352 Reidel, H-D, Konig, H., Stahl, G. and Knippers, R. (1982) *Nucleic Acids Res.*, **10**, 5621.

353 Mechali, M. and Harland, R. M. (1982) *Cell*, **30**, 93.

354 Yoda, K-Y. and Okazaki, T. (1983) *Nucleic Acids Res.*, **11**, 3433.

355 Meyer, T. F., Geider, K., Kurz, C. and Schaller, H. (1979) *Nature (London)*, **278**, 365.

356 Hu, S-Z., Wang, T. S-F. and Korn, D. (1984) *J. Biol. Chem.*, **259**, 2602.

357 Wang, T. S.-F., Hu, S-Z. and Korn, D. (1984) *J. Biol. Chem.*, **259**, 1854.

358 Piperno, J. R., Kallen, R. G. and Alberts, B. M. (1978) *J. Biol. Chem.*, **253**, 5180.

359 Nossal, N. G. and Peterlin, B. M. (1979) *J. Biol. Chem.*, **254**, 6032.

360 Sinha, N. K., Morris, C. F. and Alberts, B. M. (1980) *J. Biol. Chem.*, **255**, 4290.

361 Hibner, U. and Alberts, B. M. (1980) *Nature (London)*, **285**, 300.

362 Bedinger, P., Hochstrasser, M., Jongeneel, C. V. and Alberts, B. M. (1983) *Cell*, **34**, 115.

363 Muesing, M. A., Smith, D. H., Cabradilla, C. D., Benton, C. V., Lasky, L. A. and Capon, D. J. (1985) *Nature (London)*, **313**, 450.

364 Almouzni, G. and Mechali, M. (1988) *Biochim. Biophys. Acta*, **981**, 443.

365 Hirt, B. (1969) *J. Mol. Biol.*, **40**, 141.

366 Wolfson, J. and Dressler, D. (1972) *Proc. Natl. Acad. Sci. USA*, **69**, 2682.

367 Kriegstein, H. J. and Hogness, D. S. (1974) *Proc. Natl. Acad. Sci. USA*, **71**, 135.

368 Lovett, M. A., Katz, L. and Helinski, D. R. (1974) *Nature (London)*, **251**, 337.

369 Masters, M. and Broda, P. (1971) *Nature (London) New Biol.*, **232**, 137.

370 Hara, H. and Yoshikawa, H. (1973) *Nature (London) New Biol.*, **244**, 200.

371 Weingartner, B., Winnacker, E. L., Tolun, A. and Pettersson, U. (1976) *Cell*, **9**, 259.

372 Cairns, J. (1963) *J. Mol. Biol.*, **6**, 208.

373 Marians, K. J. (1985) *CRC Crit. Rev. Biochem.*, **17**, 153.

374 Sumida-Yasumoto, C., Yudelevich, A. and Hurwitz, J. (1976) *Proc. Natl. Acad. Sci. USA*, **73**, 1887.

375 Roth, M. J., Brown, D. R. and Hurwitz, J. (1984) *J. Biol. Chem.*, **259**, 10556.

376 Geider, K. and Meyer, T. F. (1978) *Cold Spring Harbor Symp. Quant. Biol.*, **43**, 59.

377 Mitra, S. and Stallions, D. R. (1976) *Eur. J. Biochem.*, **67**, 37.

378 Godson, G. N. (1977) *J. Mol. Biol.*, **117**, 353.

379 Eisenberg, S., Griffith, J. and Kornberg, A. (1977) *Proc Natl. Acad. Sci. USA*, **74**, 3198.

380 Denhardt, D. T. (1975) *J. Mol. Biol.*, **99**, 107.

381 Ende, A. van der, Langeveld, S. A., Teertstra, R., Arkel, G. A. van and Weisbeek, P. J. (1981) *Nucleic Acids Res.*, **9**, 2037.

382 Aoyama, A. and Hayashi, M. (1986) *Cell*, **47**, 99.

383 Fulford, W., Russell, M. and Model, P. (1987) *Annu. Rev. Cell Biol.*, **3**, 141.

384 Fulford, W. and Model, P. (1988) *J. Mol. Biol.*, **203**, 49.

385 Willets, M. and Wilkins, B. (1984) *Microbiol. Rev.*, **48**, 24.

386 Lichtenstein, C. (1986) *Nature (London)*, **322**, 682.

387 Buchanan-Wollaston, V., Passiatore, J. E. and Cannon, F. (1987) *Nature (London)*, **328**, 172.

388 Feiss, M. and Becker, A. (1983) in *Lambda II* (eds R. W. Hendrix, J. W. Roberts, F. W. Stahl and R. A. Weisberg), Cold Spring Harbor Press, p. 305.

389 Branch, A. D., Benenfeld, B. J. and Robertson, H. D. (1988) *Proc. Natl. Acad. Sci. USA*, **85**, 9128.

390 Hendrickson, W. G., Kusano, T., Yamaki, H., Balakrishnan, R., King, M. *et al.* (1982) *Cell*, **30**, 915.

391 Zyskind J. W., Harding, N., Clearly, M., Takeda, Y. and Smith D. W. (1982) *Proc. Natl. Acad. Sci. USA*, **80**, 1164.

392 Oka, A., Sasaki, H., Sugimoto, K. and Takanami, M. (1984) *J. Mol. Biol.*, **176**, 443.

393 Zyskind, J. W. and Smith, D. W. (1986) *Cell*, **46**, 489.

394 Fuller, R. S., Funnell, B. E. and Kornberg, A. (1984) *Cell*, **38**, 889.

395 Schaefer, C. and Messer, W. (1989) *EMBO J.*, **8**, 1609.

396 Hughes, P., Landoulsi, A. and Kohiyama, M. (1988) *Cell*, **55**, 343.

397 Løbner-Olesen, A., Skarsstad, K., Hansen, F. G., von Meyenburg, K. and Boye, E. (1989) *Cell*, **57**, 881.

398 Smith, D. W., Garland, A. M., Herman, G., Enns, R. E., Baker, T. A. *et al.* (1985) *EMBO J.*, **4**, 1319.

399 Yamaki, H., Ohtsubo, E., Nagai, K. and

Masda, Y. (1988) *Nucleic Acids Res.*, **16**, 5067.

400 Ogden, G. B., Pratt, M. J. and Schaechter, M. (1988) *Cell*, **54**, 127.

401 Russell, D. W. and Zinder, N. D. (1987) *Cell*, **50**, 1071.

402 Rokeach, L. A. and Zyskind, J. W. (1986) *Cell*, **46**, 763.

403 Schauza, M-A., Kücherer, C., Kölling, R., Messer, W. and Lotter, H. (1987) *Nucleic Acids Res.*, **15**, 2479.

404 Kohara, Y., Tohdoh, N., Jiang, X. and Okazaki, T. (1985) *Nucleic Acids Res.*, **13**, 6847.

405 Baker, T. A. and Kornberg, A. (1988) *Cell*, **55**, 113.

406 Sekimizu, K., Bramhill, D. and Kornberg, A. (1987) *Cell*, **50**, 259.

407 Yung, B. Y. and Kornberg, A. (1989) *J. Biol. Chem.*, **264**, 6146.

408 Wahle, E., Lasken, R. S. and Kornberg, A. (1989) *J. Biol. Chem.*, **264**, 2469.

409 Baker, T. A., Sekimuzi, K., Funnell, B. E. and Kornberg, A. (1986) *Cell*, **45**, 53.

410 Yamaki, H., Ohtsubo, E., Nagai, K. and Maeda, Y. (1988) *Nucleic Acids Res.*, **16**, 5067.

411 Bramhill, D. and Kornberg, A. (1988) *Cell*, **54**, 915.

412 Kohara, Y., Tohdoh, N., Jiang, X-W. and Okazaki, T. (1985) *Nucleic Acids Res.*, **13**, 6847.

413 Lasken, R. S. and Kornberg, A (1988) *J. Biol. Chem.*, **263**, 5372.

414 Kaguni, J. M. and Kornberg, A. (1984) *J. Biol. Chem.*, **259**, 8578.

415 Kaguni, J. M. and Kornberg, A. (1984) *Cell*, **38**, 183.

416 Funnell, B. E., Baker, T. A. and Kornberg, A. (1986) *J. Biol. Chem.*, **261**, 5616.

417 van der Ende, A., Baker, T. A., Ogawa, T. and Kornberg, A. (1985) *Proc. Natl. Acad. Sci. USA*, **82**, 3954.

418 Tsurimoto, T. and Matsubara, K. (1983) *Cold Spring Harbor Symp. Quant. Biol.*, **47**, 681.

419 Takahashi, S. (1975) *J. Mol. Biol.*, **94**, 385.

420 Tsuromoto, T. and Matsubara, K. (1981) *Nucleic Acids Res.*, **9**, 1789.

421 Liberek, K., Osipuik, J., Zylicz, M., Ang, D., Skorko, J. *et al.* (1990) *J. Biol. Chem.*, **265**, 3022.

422 Zylicz, M., Ang, D., Liberek, K. and

Georgopoulos, C. (1989) *EMBO J.*, **8**, 1601.

423 Alfano, C. and McMacken, R. (1989) *J. Biol. Chem.*, **264**, 10707.

424 Dodson, M., McMacken, R. and Echols, H. (1989) *J. Biol. Chem.*, **264**, 10719.

425 Dodson, M., Echols, H., Wickner, S., Alfano, C., Hensa-Wilmot, K. *et al.* (1986) *Proc. Natl. Acad. Sci. USA*, **83**, 7638.

426 Schnös, M., Zahn, K., Inman, R. B. and Blattner, F. R. (1988) *Cell*, **52**, 385.

427 Zahn, K and Blattner, F. R. (1987) *Science*, **236**, 416.

428 LeBowitz, J. H. and McMacken, R. (1984) *Nucleic Acids Res.*, **12**, 3069.

429 Tsurimoto, T. and Matsubara, K. (1984) *Proc. Natl. Acad. Sci. USA*, **81**, 7402.

430 Yodo, K., Yasuda, H., Jiang, X-W. and Okazaki, T. (1988) *Nucleic Acids Res.*, **16**, 6531.

431 Mensa-Wilmot, K., Seaby, R., Alfano, C., Wold, M. S., Gomes, B. and McMacken, R. (1989) *J. Biol. Chem.*, **264**, 2853.

432 Honigman, A., Hu. S-L., Chase, R. and Szybalski, W. (1976) *Nature (London)*, **262**, 112.

433 Walz, A., Pirrotta, V. and Ineichen, K. (1976) *Nature (London)*, **262**, 665.

434 Thomas, C. M. (1988) *Biochim. Biophys. Acta*, **949**, 253.

435 Pal, S. K., Mason, R. J. and Chattoraj, D. K. (1986) *J. Mol. Biol.*, **192**, 275.

436 Abeles, A. L. and Austin, S. J. (1987) *EMBO J.*, **6**, 3185.

437 Abeles, A. L. and Austin, S. J. (1988) in *DNA Replication and Mutagenesis* (eds R. E. Moses and W. C. Summers), American Society Microbiology, Washington, p. 103.

438 Chattoraj, D. K., Mason, R. J. and Wickner, S. H. (1988) *Cell*, **52**, 551.

439 Makherjee, S., Erickson, H. E. and Bastia, D. (1988) *Cell*, **52**, 375.

440 Masai, H. and Arai, K. (1988) *Nucleic Acids Res.*, **16**, 6493.

441 Masai, H. and Arai, K. (1988) in *DNA Replication and Mutagenesis* (eds R. E. Moses and W. C. Summers), American Society Microbiology, Washington, p. 113.

442 Persson, C., Wagner, E. G. H. and Nordström, K. (1990) *EMBO J.*, **9**, 3767 and 3777.

443 Patel, I and Bastia, D (1987) *Cell*, **51**, 455.

444 Öhman, M. and Wagner, E. G. H. (1989) *Nucleic Acids Res.*, **17**, 2857.

445 Puyet, A., del Solar, G. H. and Espinosa, M. (1988) *Nucleic Acids Res.*, **16**, 115.

446 Majunder, S. and Novick, R. P. (1988) *Nucleic Acids Res.*, **16**, 2897.

447 Gros, M. F., te Riele, H. and Ehrlich, S. D. (1987) *EMBO J.*, **6**, 3863.

448 Jaenisch, R., Mayer, A. and Levine, A. (1971) *Nature (London), New Biol.*, **233**, 72.

449 Hay, R. T. and De Pamphilis, M. L. (1982) *Cell*, **28**, 767.

450 McVey, D., Brizuela, L., Mohr, I., Marshak, D. R., Gluzman., Y. and Beach, D. (1989) *Nature (London)*, **341**, 503.

451 Ariga, H. and Sugano, S. (1983) *J. Virol.*, **48**, 481.

452 De Villiers, J., Schaffner, W., Tyndall, C., Lupton, S. and Kamen, R. (1984) *Nature (London)*, **312**, 242.

453 DePamphilis, M. L. (1988) *Cell*, **52**, 635.

454 Borowiec, J. A. and Hurwitz, J. (1988) *Proc. Natl. Acad. Sci. USA*, **85**, 64.

455 Stahl, H. and Knippers, R. (1987) *Biochim. Biophys. Acta*, **910**, 1.

456 Dean, F. B., Dodson, M., Echols, H. and Hurwitz, J. (1987) *Proc. Natl. Acad. Sci. USA*, **84**, 8981.

457 Mohr, I. J., Stillman, B. and Gluzman, Y. (1987) *EMBO J.*, **6**, 153.

458 Mastrangelo, I. A., Hough, P. V. C., Wall, J. S., Dodson, M., Dean, F. B. *et al.* (1989) *Nature (London)*, **338**, 658.

459 Roberts, J. M. (1989) *Proc. Natl. Acad. Sci. USA*, **86**, 3939.

460 Dean, F. B., Borowiec, J. A., Ishimi, Y., Deb, S., Tegtmeyer, P. *et al.* (1987) Proc. Natl. Acad. Sci. USA, **84**, 8267.

461 Wiekowsky, M., Schwarz, M. W. and Stahl, H. (1988) *J. Biol. Chem.*, **263**, 436.

462 Goetz, G. S., Dean, F. B., Hurwitz, J. and Matson, S. W. (1988) *J. Biol. Chem.*, **263**, 383.

463 Tsurimoto, T., Fairman, M. P. and Stillman, B. (1989) *Mol. Cell. Biol.*, **9**, 3839.

464 Bullock, P. A., Seo, Y. S. and Hurwitz, J. (1989) *Proc. Natl. Acad. Sci. USA*, **86**, 3944.

465 Tseng, B. Y. and Ahlem, C. N. (1984) *Proc. Natl. Acad. Sci. USA*, **81**, 2342.

466 Hay, R. T., Hendrickson, E. A. and De Pamphilis, M. L. (1984) *J. Mol. Biol.*, **175**, 131.

467 Bergsma, D. J., Olive, D. M., Hartzell, S. W. and Subramanian, K. N. (1982) *Proc. Natl. Acad. Sci. USA*, **79**, 381.

468 Li, J. J. and Kelly, T. J. (1984) *Proc. Natl. Acad. Sci. USA*, **81**, 6973.

469 Fairman, M. P. and Stillman, B. (1988) *EMBO J.*, **7**, 1211.

470 Weinberg, D. H., Collins, K. L., Simancek, P., Russo, A., Wold, M. S. *et al.* (1990) *Proc. Natl. Acad. Sci. USA*, **87**, 8692.

471 Prelich, G., Tan, C-K., Kostura, M., Mathews, M. B., So, A. G. *et al.* (1987) *Nature (London)*, **326**, 517.

472 Bravo, R., Frank, R., Blundell, P. A. and Macdonald-Bravo, H. (1987) *Nature (London)*, **326**, 515.

473 Blow, J. (1987) *Nature (London)*, **326**, 441.

474 Prelich, G. and Stillman, B. (1988) *Cell*, **53**, 117.

475 Ishimi, Y., Claude, A., Bullock, P. and Hurwitz, J. (1988) *J. Biol. Chem.*, **263**, 19 723.

476 Wold, M. S., Weinberg, D. H., Virshup, D. M., Li, J. J. and Kelly, T. J. (1989) *J. Biol. Chem.*, **264**, 2801.

477 Lee, S-H., Kwong, A. D., Ishimi, Y. and Hurwitz, J. (1989) *Proc. Natl. Acad. Sci. USA*, **86**, 4877.

478 Gannon, J. V. and Lane, D. P. (1987) *Nature (London)*, **329**, 456.

479 Braithwaite, A. W., Sturzbecher, H-W., Addison, C., Palmer, C., Rudge, K. *et al.* (1987) *Nature (London)*, **329**, 458.

480 Mecsas, J. and Sugden, B. (1987) *Annu. Rev. Cell Biol.*, **3**, 87.

481 Roberts, J. M. and Weintraub, H. (1986) *Cell*, **46**, 741.

482 Berg, L., Lusky, M., Stenlund, A. and Botchan, M. R. (1986) *Cell*, **46**, 753.

483 Roberts, J. M. and Weintraub, H. (1988) *Cell*, **52**, 397.

484 Nakai, H., Beauchamp, B. B., Bernstein, J., Huber, H. E., Tabor, S. *et al.* (1988) in *DNA Replication and Mutagenesis* (eds R. E. Moses and W. C. Summers), American Society Microbiology, Washington, p. 85.

485 Kolodner, R. and Richardson, C. C. (1978) *J. Biol. Chem.*, **253**, 574.

486 Masker, W. E. and Richardson, C. C. (1976) *J. Mol. Biol.*, **109**, 543.

487 Richardson, C. C., Romano, L. J., Kolodner, R., Le Clerc, J. E., Tamanoi, R. *et al.* (1978) *Cold Spring Harbor Symp. Quant. Biol.*, **43**, 427.

488 De Wyngaert, M. A. and Hinkle, D. C. (1979) *J. Biol. Chem.*, **254**, 11 247.

489 Fuller, C. W., Beauchamp, B. B., Engler, M. J., Lechner, R. L., Matson, S. W. *et al.* (1983) *Cold Spring Harbor Symp. Quant. Biol.*, **47**, 669.

490 Richardson, C. C. (1983) *Cell*, **33**, 315.

491 Strothkamp, R. E., Oakley, J. L. and Coleman, J. E. (1980) *Biochemistry*, **19**, 1074.

492 Sugimoto, K., Kohara, Y. and Okazaki, T. (1987) *Proc. Natl. Acad. Sci. USA*, **84**, 3977.

493 Nordstran, B., Randahl, H., Slaby, I. and Holmgren, A. (1981) *J. Biol. Chem.*, **256**, 3112.

494 Engler, M. J., Lechner, R. L. and Richardson, C. C. (1983) *J. Biol. Chem.*, **248**, 11 165.

495 Lechner, R. L., Engler, M. J. and Richardson, C. C. (1983) *J. Biol. Chem.*, **248**, 11 174.

496 Nakai, H and Richardson, C. C. (1988) *J. Biol. Chem.*, **263**, 9818.

497 Nakai, H and Richardson, C. C. (1988) *J. Biol. Chem.*, **263**, 9831.

498 Engler, M. J. and Richardson, C. C. (1983) *J. Biol. Chem.*, **248**, 11 197.

499 Wever, G. H., Fischer, H. and Hinkle, D. C. (1980) *J. Biol. Chem.*, **255**, 7965.

500 Fujiyama, A., Kohara, Y. and Okazaki, T. (1981) *Proc. Natl. Acad. Sci. USA*, **78**, 903.

501 Delius, H., House, C. and Lozinski, A. W. (1971) *Proc. Natl. Acad. Sci. USA*, **68**, 3049.

502 Alberts, B., Morris, C. F., Mace, D., Sinha, N., Bittner, M. *et al.* (1975) in *DNA Synthesis and its Regulation* (eds M. Goulian, P. Hanawalt and C. F. Fox), W. A. Benjamin, Menlo Park, p. 241.

503 Luder, A. and Mosig, G. (1982) *Proc. Natl. Acad. Sci. USA*, **79**, 1101.

504 Chiu, C-S., Tomack, P. K. and Greenberg, G. R. (1976) *Proc. Natl. Acad. Sci. USA*, **73**, 757.

505 Liu, L. F., Liu, C-C. and Alberts, B. M. (1980) *Cell*, **19**, 697.

506 Aposhian, H. V. and Kornberg, A. (1962) *J. Biol. Chem.*, **237**, 519.

507 Somerville, R., Ebisuyaki, K. and Greenberg, G. R. (1959) *Proc. Natl. Acad. Sci. USA*, **45**, 1240.

508 Prashad, N. and Hosoda, J. (1972) *J. Mol. Biol.*, **70**, 617.

509 Reddy, G. P. V., Singh, A., Stafford, M. E. and Mathews, C. K. (1977) *Proc. Natl. Acad. Sci. USA*, **74**, 3152.

510 Wirak, D. O. and Greenberg, G. R. (1980) *J. Biol. Chem.*, **255**, 1896.

511 Reddy, G. P. V. and Mathews, C. K. (1978) *J. Biol. Chem.*, **253**, 3461.

512 Reichard, P., Rowen, L., Eliasson, R., Hobbs, J. and Eckstein, F. (1978) *J. Biol. Chem.*, **253**, 7011.

513 Blair, D. G. and Helinsky, D. R. (1975) *J. Biol. Chem.*, **250**, 8785, and following papers.

514 Backman, K., Betlach, M., Boyer, H. W. and Yanofsky, S. (1978) *Cold Spring Harbor Symp. Quant. Biol.*, **43**, 69.

515 Itoh, T. and Tomizawa, J. (1978) *Cold Spring Harbor Symp. Quant. Biol.*, **43**, 409 and (1980) *Proc. Natl. Acad. Sci. USA*, **77**, 2450.

516 Bastia, D. (1977) *Nucleic Acids Res.*, **4**, 3123.

517 Masukata, H. and Tomizawa, J. (1984) *Cell*, **36**, 513.

518 Itoh, T. and Tomizawa, J. (1982) *Nucleic Acids Res.*, **10**, 5949.

519 Rosen, J., Ryder, T., Ohtsubo, H. and Ohtsubo, E. (1981) *Nature (London)*, **290**, 794.

520 Masai, H and Arai, K (1988) *J. Biol. Chem.*, **263**, 15 016.

521 Dasgupta, S., Masukata, H. and Yomizawa, J. (1987) *Cell*, **51**, 1113.

522 Masukata, H., Dasgupta, S. and Tomizawa, J. (1988) *Cell*, **51**, 1123.

523 Polisky, B. (1988) *Cell*, **55**, 929.

524 Tomizawa J. and Som, T. (1984) *Cell*, **38**, 871.

525 Lacatena R. M., Banner D. W., Castagnoli, L. and Cesareni, G. (1984) *Cell*, **37**, 1009.

526 Helmer-Citterich, M., Anceschi, M. M., Banner, D. W. and Cesareni, G. (1988) *EMBO J.*, **7**, 557.

527 Zipursky, S. L. and Marians, K. J. (1981) *Proc. Natl. Acad. Sci. USA*, **78**, 6111.

528 Nomura, N., Low, R. L. and Ray, D. S. (1982) *Proc. Natl. Acad. Sci. USA*, **79**, 3153.

529 Clayton, D. A. (1982) *Cell*, **28**, 693.

530 Chang, D. D., Fisher, R. P. and Clayton, D. A. (1987) *Biochim. Biophys. Acta*, **909**, 85.

531 Berk, A. J. and Clayton, D. A. (1976) *J. Mol. Biol.*, **100**, 85.

532 Robberson, D. L., Kasamasu, H. and Vinograd, J. (1972) *Proc. Natl. Acad. Sci. USA*, **69**, 736.

533 Bogenhagen, D. and Clayton, D. A. (1978) *J. Mol. Biol.*, **119**, 49.

534 Bogenhagen, D. and Clayton, D. A. (1978) *J. Mol. Biol.*, **130**, 69.

535 Goddard, J. M. and Wolstenholme, D. R. (1978) *Proc. Natl. Acad. Sci. USA*, **75**, 3886.

536 Yaginuma, K., Kobayashi, M., Taira, M. and Koike, K. (1982) *Nucleic Acids Res.*, **10**, 7531.

537 Chang, D. D., Hauswirth, W. W. and Clayton, D. A. (1985) *EMBO J.*, **4**, 1559.

538 Clayton, D. A. (1991) *Trends Biochem. Sci.*, **16**, 107.

539 Schinkel, A. H. and Tabak, H. F. (1981) *Trends Genet.*, **5**, 149.

540 Wong, T. W. and Clayton, D. A (1986) *Cell*, **45**, 817.

541 Wong, T. W. and Clayton, D. A. (1985) *Cell*, **42**, 951.

542 Tapper, D. P. and Clayton, D. A. (1982) *J. Mol. Biol.*, **162**, 1.

543 Wong, T. W. and Clayton, D. A. (1985) *Cell*, **42**, 951.

544 Hixson, J. E., Wong, T. W. and Clayton, D. A. (1986) *J. Biol. Chem.*, **261**, 2384.

545 Matsumoto L. Kasamatsu, H., Piko, L. and Vinograd, J. (1977) *J. Cell Biol.*, **63**, 146.

546 Bernardi, G. (1982) *Trends Biochem. Sci.*, **7**, 404.

547 Fangman, W. L. and Dujon, B. (1984) *Proc. Natl. Acad. Sci. USA*, **81**, 7156.

548 Arnberg, A. C., Van Bruggen, E. F. J., Clegg, R. A., Upholt, W. B. and Borst, P. (1974) *Biochim. Biophys. Acta*, **361**, 266.

549 Goddard, J. M. and Cummings, D. M. (1977) *J. Mol. Biol.*, **109**, 327.

550 Pritchard, A. E., Laping, J. L., Seilhamer, J. J. and Cumings, D. J. (1983) *J. Mol. Biol.*, **164**, 1.

551 Kitchin, P. A., Klein, V. A. and Englund, P. T. (1985) *J. Biol. Chem.*, **260**, 3844.

552 Birkenmeyer, L., Sugisaki, H. and Ray, D. S. (1987) *J. Biol. Chem.*, **262**, 2384.

553 Wu, M., Lou, J. K., Chang, D. Y., Chang, C. H. and Nie, Z. Q. (1986) *Proc. Natl. Acad. Sci. USA*, **83**, 6761.

554 Kelly, T. J. (1982) in *Organisation and Replication of Viral DNA* (ed. A. S. Kaplan) CRC press, Cleveland, p. 115.

555 Stillman, B. W. and Tamanoi, F. (1983) *Cold Spring Harbor Symp. Quant. Biol.*, **47**, 741.

556 Tamanoi, F. and Stillman, B. W. (1983) *Proc. Natl. Acad. Sci. USA*, **80**, 6446.

557 Challberg, M. D. and Rawlings, D. R. (1984) *Proc. Natl. Acad. Sci. USA*, **81**, 100.

558 Guggenheimer, R. A., Stillman, B. W.,

Nagata, K., Tamanoi, F. and Hurwitz, J. (1984) *Proc. Natl. Acad. Sci. USA*, **81**, 3069.

559 Hay, R. T. (1985) *EMBO J.*, **4**, 421.

560 O'Neill, E. A. and Kelly, T. J. (1988) *J. Biol. Chem.*, **263**, 931.

561 Pruij, G. J. M., van Driel, W., van Miltenburg, R. T. and van der Vliet, P. C. (1987) *EMBO J.*, **6**, 3771.

562 Pruij, G. J. M., van Driel, W., van Miltenburg, R. T. and van der Vliet, P. C. (1986) *Nature (London)*, **322**, 656.

563 Hay, R. T. (1985) *J. Biol. Chem.*, **186**, 129.

564 Nagata, K., Guggenheimer, R. A. and Hurwitz, J. (1983) *Proc. Natl. Acad. Sci. USA*, **80**, 6177.

565 Nagata, K., Guggenhelmer, R. A. and Hurwitz, J. (1983) *Proc. Natl. Acad. Sci. USA*, **80**, 4266.

566 Rijnders, A. W. M., van Bergen, B. G. M., van der Vliet, P. S. and Sussenbach, J. S. (1983) *Nucleic Acids Res.*, **11**, 8777.

567 Lichy, J. H., Field, J., Horwitz, M. S. and Hurwitz, J. (1982) *Proc. Natl. Acad. Sci. USA*, **79**, 5223.

568 Nagata, K., Guggenheimer, R. A., Enomoto, T., Lichy, J. H. and Hurwitz, J. (1982) *Proc. Natl. Acad. Sci. USA*, **79**, 6438.

569 Kenny, M. H. and Hurwitz, J. (1988) *J. Biol. Chem.*, **263**, 9809.

570 Challberg, M. D. and Kelly, T. J. (1981) *J. Virol.*, **38**, 272.

571 Watabe, K., Shih, M. F., Sugino, A. and Ito, J. (1982) *Proc. Natl. Acad. Sci. USA*, **79**, 5245.

572 Blanco, L., Bernard, A., Lazaro, J. M., Martin, G., Garmendia, C. *et al.* (1989) *J. Biol. Chem.*, **264**, 8935.

573 Pearson, G. D., Chow, K-C., Corden, J. L. and Harpst, J. A. (1981) in *The Initiation of DNA Replication* (ed. D. S. Ray), W. A. Benjamin, Menlo Park, p. 581.

574 Varmus, H. (1987) *Sci. Am.*, **257**, 48.

575 Swanstram, R., Varmus, H. E. and Bishop, J. M. (1981) *J. Biol. Chem.*, **256**, 1115.

576 Even, J., Anderson, S. J., Hampe, A., Galibert, F., Lowry, D., Khoury, G. and Sherr, C. J. (1983) *J. Virol.*, **65**, 1004.

577 Panganiban, A. T. and Temin, H. M. (1983) *Nature (London)*, **306**, 155.

578 Varmus, H. E., Heasley, S., Kung, H-J., Oppermann, H., Smith, V. C., Bishop, J. M. and Shank, P. R. (1978) *J. Mol. Biol.*, **130**, 55.

579 Flint, J. (1976) *Cell*, **8**, 151.
580 Bishop, J. M. (1978) *Annu. Rev. Biochem.*, **47**, 35.
581 Hohn, T., Hohn, B. and Pfeiffer, P. (1985) *Trends Biochem. Sci.*, **10**, 205.
582 Zakian, V. A. (1976) *J. Mol. Biol.*, **108**, 305.
583 Huberman, J. A. and Riggs, A. D. (1968) *J. Mol. Biol.*, **32**, 327.
584 Hand, R. (1978) *Cell*, **15**, 317.
585 Callan, H. G. (1972) *Proc. R. Soc. London Ser. B*, **181**, 19.
586 Woodland, H. R. and Pestell, R. Q. W. (1972) *Biochem. J.*, **127**, 597.
587 Housman, D. and Huberman, J. A. (1975) *J. Mol. Biol.*, **94**, 173.
588 Lewin, B. (1974) *Gene Expression*, vol. 2, Wiley, New York.
589 Bozzoni, I., Baldari, C. T., Armaldi, F. and Buongiorno-Nardelli, M. (1981) *Eur. J. Biochem.*, **118**, 585.
590 Botchan, P. M. and Dayton, A. I. (1982) *Nature (London)*, **299**, 453.
591 Heintz, N-H., Millbrandt, J. D., Greisen, K. S. and Hamlin, J. L. (1983) *Nature (London)*, **302**, 439.
592 Leu, T. H. and Hamlin, J. L. (1989) *Mol. Cell. Biol.*, **9**, 523.
593 Broach, J. R. (1982) *Cell*, **28**, 203.
594 Kikuchi, Y. (1983) *Cell*, **35**, 487.
595 Huberman, J. A., Spotila, L. D., Nawotka, K. A., El-Assouli, S. and Davis, L. R. (1987) *Cell*, **51**, 473.
596 Brewer, B. J. and Fangman, W. L. (1987) *Cell*, **51**, 463.
597 Tschumper, G. and Carbon, J. (1982) *J. Mol. Biol.*, **156**, 293.
598 Celniker, S. E. and Campbell, J. L. (1982) *Cell*, **31**, 201.
599 Fangman, W. L., Hice, R. H. and Chlebowicz-Sledziewska, E. (1983) *Cell*, **32**, 831.
600 Chan, C. S. M. and Tye, B-K. (1983) *Cell*, **33**, 563.
601 Kearsey, S. (1984) *Cell*, **37**, 299.
602 Amati, B. B. and Gasser, S. M. (1988) *Cell*, **54**, 967.
603 Campbell, J. (1988) *Trends Biochem. Sci.*, **13**, 212.
604 Umek, R. M., Linskens, M. H. K., Kowalski, D. and Huberman, J. A. (1989) *Biochim. Biophys. Acta*, **1007**, 1.
605 Gragerov, A. I., Danilevskaya, O. N., Didichenko, S. A. and Kaverina, E. N. (1988) *Nucleic Acids Res.*, **16**, 1169.
606 Palskill, T. G. and Newton, C. S. (1988) *Cell*, **53**, 441.
607 Umek, R. M. and Kowalski, D. (1988) *Cell*, **52**, 559.
608 Brewer, B. J. and Fangman, W. L. (1988) *Cell*, **55**, 637.
609 Handeli, S., Klar, A. S., Meuth, M. and Cedar, H. (1989) *Cell*, **57**, 909.
610 Holst, A., Müller, F., Zastrow, G., Zcntgraf, H., Schwendler, S. *et al.* (1988) *Cell*, **52**, 355.
611 Zastrow, G., Kochler, U., Müller, F., Klavinius, A., Wegner, M. *et al.* (1989) *Nucleic Acids Res.*, **17**, 1867.
612 Gilbert, D. and Cohen, S. N. (1989) *Cell*, **56**, 143.
613 Grummt, F. and Cohen, S. N. (1989) *Cell*, **56**, 143.
614 Harland, R. M. and Laskey, R. A. (1980) *Cell*, **21**, 761.
615 Harland, R. M. (1981) *Trends Biochem. Sci.*, **6**, 71.
616 Hines, P. J. and Benbow, R. M. (1982) *Cell*, **30**, 459.
617 Mechali, M. and Kearsey, S. (1984) *Cell*, **38**, 55.
618 Frappier, L. and Zannis-Hadjopoulos, M. (1987) *Proc. Natl. Acad. Sci. USA*, **84**, 6668.
619 Ariga, H., Itani, T. and Iguchi-Ariga, S. M. M. (1987) *Mol. Cell. Biol.*, **7**, 1.
620 Johnson, E. M. and Jelinek, W. R. (1986) *Proc. Natl. Acad. Sci. USA*, **83**, 4660.
621 Pardoll, D. M., Vogelstein, B. and Coffey, D. S. (1980) *Cell*, **19**, 527.
622 Aalen, J. M. A., Opstelten, R. J. G. and Waika, F. (1983) *Nucleic Acids Res.*, **11**, 1181.
623 Cook, P. R. and Lang, J. (1984) *Nucleic Acids Res.*, **12**, 1069.
624 Dijkwell, P. A., Wenink, P. W. and Poddighe, J. (1986) *Nucleic Acids Res.*, **14**, 3241.
625 Jackson, D. A. and Cook P. R. (1986) *EMBO J.*, **5**, 1403.
626 Jacob, F., Brenner, S. and Cuzin, F. (1963) *Cold Spring Harbor Symp. Quant. Biol.*, **28**, 329.
627 Pritchard, R. H., Barth, P. T. and Collins, J. (1969) *Symp. Soc. Gen. Microbiol.*, **19**, 293.
628 Rao, P. N. and Johnson, R. T. (1970) *Nature (London)*, **225**, 159.
629 Cabello, F., Timmis, K. and Cohen, S. N. (1976) *Nature (London)*, **259**, 285.
630 Lim, V. I. and Mazanov, A. L. (1978) *FEBS Lett.*, **88**, 118.

631 Elias, P. and Lehman, I. R. (1988) *Proc. Natl. Acad. Sci. USA*, **85**, 2959.

632 Olivo, P. D., Nelson, N. J. and Challberg, M. D. (1988) *Proc. Natl. Acad. Sci. USA*, **85**, 5414.

633 Gahn, T. A. and Schildkraut, C. L. (1989) *Cell*, **58**, 527.

634 Blow, J. J. and Laskey, R. A. (1988) *Nature (London)*, **332**, 546.

635 Krysan, P. J., Haase, S. B. and Calos, M. P. (1989) *Mol. Cell. Biol.*, **9**, 1026.

636 Amati, B. and Gasser, S. M. (1990) *Mol. Cell. Biol.*, **10**, 5442.

637 Huberman, J. A., Spotila, L. D., Nawotka, K. A., El-Assouli, S. M. and Davis, L. R. (1987) *Cell*, **51**, 473.

638 Vassilev, L. and Johnson, E. M. (1989) *Nucleic Acids Res.*, **17**, 7693.

639 Russev, G. and Vassilev, L. (1982) *J. Mol. Biol.*, **161**, 77.

640 Brown, E. H., Iqbal, M. A., Stuart, S., Hatton, K. S., Valinsky, J. *et al.* (1987) *Mol. Cell. Biol.*, **7**, 450.

641 Trempe, J. P., Lindstrom, Y. I. and Leffak, M. (1988) *Mol. Cell. Biol.*, **8**, 1657.

642 Leffak, M. and James, C. D. (1989) *Mol. Cell. Biol.*, **9**, 586.

643 Goldman, M. A., Holmquist, G. P., Gray, M. C., Caston, L. A. and Nag, A. (1984) *Science*, **224**, 686.

644 Dhar, V., Skoultchi, A. I. and Schildkraut, C. L. (1989) *Mol. Cell. Biol.*, **9**, 3524.

645 Enver, T. Brewer, A. C. and Patient, R. K. (1988) *Mol. Cell. Biol.*, **8**, 1301.

646 Weight, J. A., Smith, H. S., Watt, F. M., Hancock, M. C., Hudson, D. L *et al.* (1990) *Proc. Natl. Acad. Sci. USA*, **87**, 1791.

647 Gall, J. G. (1969) *Genetics*, **61**, Suppl., 121.

648 Bird, A. P. (1978) *Cold Spring Harbor Symp. Quant. Biol.*, **42**, 1179.

649 MacGreggor, H. C. (1972) *Biol. Rev.*, **47**, 173.

650 Rochair, J-D. and Bird, A. P. (1975) *Chromosoma*, **52**, 317.

651 Yao, M-C., Zhu, S-G. and Yao, C-H. (1985) *Mol. Cell. Biol.*, **5**, 1260.

652 Cooney, C. A., Eykholt, R. L. and Bradbury, E. M. (1988) *J. Mol. Biol.*, **204**, 889.

653 Stark, G-R. and Wahl, C. M. (1984) *Annu. Rev. Biochem.*, **53**, 447.

654 DeCicco, D. V. and Spradling, A. C. (1984) *Cell*, **38**, 45.

655 Ish-Horowitz, D. (1982) *Nature (London)*, **296**, 806.

656 Schimke, R. T. (1984) *Cell*, **37**, 705.

657 Wahl, G. M., de Saint Vincent, B. R. and DeRose, M. L. (1984) *Nature (London)*, **307**, 516.

658 Roninson, I. B., Abelson, H. T., Housman, D. E., Howell, N. and Varshavsky, A. (1984) *Nature (London)*, **309**, 626.

659 Borst, P. (1984) *Nature (London)*, **309**, 580.

660 Beverley, S. M., Coderre, J. A., Santi. D. V. and Schimke, R. T. (1984) *Cell*, **38**, 431.

661 Anderson, R. P. and Roth, J. R. (1977) *Annu. Rev. Microbiol.*, **31**, 473.

662 Virshup, D. M. (1990) *Current Opinion in Cell Biology*, **2**, 453.

663 Johnston, R. N., Beverley, S. M. and Schimke, R. T. (1983) *Proc. Natl. Acad. Sci. USA*, **88**, 3711.

664 Schimke, R. T., Sherwood, S. W., Hill, A. B. and Johnston, R. N. (1986) *Proc. Natl. Acad. Sci. USA*, **83**, 2157.

665 Botchan, M. R., Topp, W. and Sambrook, J. (1979) *Cold Spring Harbor Symp. Quant. Biol.*, **45**, 709.

666 Barau, N., Neer, A. and Manor, H. (1983) *Proc. Natl. Acad. Sci. USA*, **80**, 105.

667 Stark, G. R., Debatisse, M., Giulotto, E. and Wahl, G. M. (1989) *Cell*, **57**, 901.

668 Som, T., Armstrong, K. A., Volkert, F. C. and Broach, J. R. (1988) *Cell*, **52**, 27.

669 Linskens, M. H. K. and Huberman, J. A. (1988) *Mol. Cell. Biol.*, **8**, 4927.

670 Bedinger, P., Hochstrasser, M., Jongeneel, C. V. and Alberts, B. M. (1983) *Cell*, **34**, 115.

671 Marians, K. J. (1987) *J. Biol. Chem.*, **262**, 10 362.

672 Sundin, O. and Varshavsky, A. (1981) *Cell*, **25**, 659.

673 Snapke, R. M., Powelson, M. A. and Strayer, J. M. (1988) *Mol. Cell. Biol.*, **8**, 515.

674 Richter, A. and Stausfeld, U. (1988) *Nucleic Acids Res.*, **16**, 10 119.

675 Brun, C., Dang, Q. and Miassod, R. (1990) *Mol. Cell. Biol.*, **10**, 5455.

676 Corrigan, C. M., Haarsma, J. A., Smith, M. T. and Wake, R. G. (1987) *Nucleic Acids Res.*, **15**, 8501.

677 Hsieh, C-H. and Griffith, J. D. (1988) *Cell*, **52**, 535.

678 Horiuch, T. and Hidaka, M. (1988) *Cell*, **54**, 515.

679 Hill, T. M., Pelletier, A. J., Tecklenburg, M. L. and Kuempel, P. L. (1988) *Cell*, **55**, 459.

680 Hidaka, M., Akiyama, M. and Horiuchi, T. (1988) *Cell*, **55**, 467.

681 Khatri, G. S., MacAllister, T., Sista, P. R. and Bastia, D. (1989) *Cell*, **59**, 667.

682 Lee, E. H., Kornberg, A., Hidaka, M., Kobayashi, T. and Horiuchi, T. (1989) *Proc. Natl. Acad. Sci. USA*, **86**, 9104.

683 Zimmerman, W., Chen, S. M., Bolden, A. and Weissbach A. (1980) *J. Biol. Chem.*, **255**, 11 847.

684 Thomas, C. M. (1988) *Biochim. Biophys. Acta*, **949**, 253.

685 Tubo, R. A. and Berezney, R. (1987) *J. Biol. Chem.*, **262**, 5857.

686 Van Houten, J. V. and Newlon, C. S. (1990) *Mol. Cell. Biol.*, **10**, 3917.

687 Jones, K. A., Kadonaga, J. T., Rosenfeld, P. J., Kelly, T. J. and Tjian, R. (1987) *Cell*, **48**, 79.

688 Cheng, L. and Kelly, T. J. (1989) *Cell*, **59**, 541.

689 Adachi, Y., Käs, E. and Laemmli, U. K. (1990) *EMBO J.*, **8**, 3997.

690 Brewer, B. J. (1988) *Cell*, **53**, 679.

691 Kuempel, P. L., Pelletier, A. J. and Hill, T. M. (1989) *Cell*, **59**, 581.

692 Jackson, D. A., Dickinson, P. and Cook, P. R. (1990) *Nucleic Acids Res.*, **18**, 4385.

693 Hill, T. M. and Marians, K. M. (1990) *Proc. Natl. Acad. Sci. USA*, **87**, 2481.

694 Hiasa, H., Sakai, H., Komano, T. and Gadson, G. N. (1990) *Nucleic Acids Res.*, **18**, 4825.

695 Chen, C. K., Tyson, J. J., Lederman, M., Stout, E. R. and Bates, R. C. (1989) *J. Mol. Biol.*, **208**, 283.

696 Im, D-S. and Muzyczka, N. (1990) *Cell*, **61**, 447.

697 Sudo, K., Ogata, M., Sato, Y., Iguchi-Ariga, S. M. M. and Ariga, H. (1990) *Nucleic Acids Res.*, **18**, 5425.

698 Georgopoulos, C. (1989) *Trends Genet.*, **5**, 319.

699 Wickner, S. (1990) *Proc. Natl. Acad. Sci. USA*, **87**, 2690.

700 Prives, C. (1990) *Cell*, **61**, 735.

701 Gutierrez, C., Guo, Z-S., Roberts, J. and DePamphilis, M. L. (1990) *Mol. Cell. Biol.*, **10**, 1719.

702 Tsurimoto, T., Melendy, T. and Stillman, B. (1990) *Nature (London)*, **346**, 534.

703 Larsen, D. D., Spangler, E. A. and Blackburn, E. H. (1987) *Cell*, **56**, 477.

704 Allshire, R. C., Gosden, J. R., Cross, S. H., Cranston, G., Rout, D. *et al.* (1988) *Nucleic Acids Res.*, **332**, 656.

705 Panyutin, I. G., Kovalsky, O. I., Budowsky, E. I., Dickerson, R. E., Rikhirev, M. E. *et al.* (1990) *Proc. Natl. Acad. Sci. USA*, **87**, 867.

706 Murray, A. (1990) *Nature (London)*, **346**, 797.

707 Williamson, J. R., Raghuraman, M. K. and Cech, T. R. (1989) *Cell*, **59**, 871.

708 Morin, G. B. (1989) *Cell*, **59**, 521.

709 Zakian, V. A., Runge, K. and Wang, S-S. (1989) *Trends Genet.*, **6**, 12.

710 Greider, C. W. and Blackburn, E. H. (1985) *Cell*, **43**, 405.

711 Greider, C. W. and Blackburn, E. H. (1987) *Cell*, **51**, 887.

712 Lamond, A. I. (1989) *Trends Biochem. Sci.*, **14**, 202.

713 Hendersen, E., Hardin, C. C., Walk, S. K., Tinoco, I. and Blackburn, E. H. (1987) *Cell*, **51**, 899.

714 Weiner, A. M. (1988) *Cell*, **52**, 155.

715 Cech, T. R. (1988) *Nature (London)*, **332**, 777.

716 Sundquist, W. I. and Klug, A. (1989) *Nature (London)*, **342**, 825.

717 Sen, D. and Gilbert, W. (1990) *Nature (London)*, **344**, 410.

718 Zahler, A. M. and Prescott, D. M. (1988) *Nucleic Acids Res.*, **16**, 6953.

719 Szostak, J. W. (1989) *Nature (London)*, **337**, 303.

720 Hastie, N. D. and Allshire, R. C. (1989) *Trends Genet.*, **5**, 326.

721 Berns, K. I. and Hauswirth, W. W. (1982) in *Organisation and Replication of Viral DNA* (ed. A. S. Kaplan), CRC press, Cleveland, p. 3.

722 Watson, J. D. (1972), *Nature (London) New Biol.*, **239**, 197.

723 White, J. H. and Richardson, C. C. (1987) *J. Biol. Chem.*, **262**, 8851.

724 Feiss, M and Becker, A. (1983) in Lambda II (eds R. W. Hendrix, J. W. Roberts, F. W.

Stahl and R. A. Weisberg), Cold Spring Harbor Press, p. 305.

725 Johnson, E. M., Bergold, P. and Campbell, G. R. (1984) in *Recombinant DNA and Cell Proliferation* (eds G. Stein and J. Stein), Academic Press, New York, p. 303.

726 Price, C. M. (1990) *Mol. Cell. Biol.*, **10**, 3421.

727 Wilkie, A. O. M., Lamb, J., Harris, P. C., Finney, R. D. and Higgs, D. R. (1990) *Nature (London)*, **346**, 868.

728 Blackburn, E. H. and Szostak, J. W. (1984) *Annu. Rev. Biochem.*, **53**, 163.

729 Blackburn, E. H. (1984) *Cell*, **37**, 7.

730 Pluta, A. F. and Zakian, V. A. (1989) *Nature (London)*, **337**, 429.

731 Henderson, E. R. and Blackburn, E. H. (1989) *Mol. Cell. Biol.*, **9**, 345.

732 Wang, S-S. and Zakian, V. A. (1990) *Nature (London)*, **345**, 456.

733 Blackburn, E. H. (1990) *J. Biol. Chem.*, **265**, 5919.

734 Moyzis, R. K., Buckingham, J. M., Cram, L. S., Dani, M., Deaven, L. L. *et al.* (1988) *Proc. Natl. Acad. Sci. USA*, **85**, 6622.

735 Allshire, R. C., Dempster, M. and Hastie, N. D. (1989) *Nucleic Acids Res.*, **17**, 4611.

736 Harley, C. B., Futcher, A. B. and Greider, C. W. (1990) *Nature (London)*, **345**, 458.

737 Hastie, N. D., Dempster, M., Dunlop, M. G., Thompson, A. M., Green, D. K. *et al.* (1990) *Nature (London)*, **346**, 866.

738 Shampay, J., Szostak, J. W. and Blackburn, E. H. (1984) *Nature (London)*, **310**, 154.

739 Hsieh, T. (1990) *Curr. Opinion Cell Biol.*, **2**, 461.

740 Seale, R. L. (1976) *Cell*, **9**, 423.

741 Annunziato, A. T. and Seale, R. L. (1984) *Nucleic Acids Res.*, **12**, 6179.

742 Burgoyne, L. A., Mobbs, J. D. and Marshall, A. J. (1976) *Nucleic Acids Res.*, **3**, 3293.

743 Buongiorno-Nardelli, M., Michele, G., Carri, M. T. and Marilley, M. (1982) *Nature (London)*, **298**, 100.

744 Stahl, H. and Gallwitz, D. (1977) *Eur. J. Biochem.*, **72**, 385.

745 Huchhauser, S. J., Stein, J. L. and Stein, G. S. (1981) *Int. Rev. Cytol.*, **71**, 95.

746 Stein, G. S., Plumb, M. A., Stein, G. L., Marashi, F. F., Sierra, L. F. *et al.* (1984) in *Recombinant DNA and Cell Proliferation* (eds G. S. Stein and G. L. Stein), Academic Press, New York, p. 107.

747 Hereford, L., Bromley, S. and Osley, M. (1982) *Cell*, **30**, 305.

748 Plumb, M., Stein, J. and Stein, G. (1983) *Nucleic Acids Res.*, **11**, 7927.

749 Djondjurov, L. P., Yancheva, N. Y. and Ivanova, E. C. (1983) *Biochemistry*, **22**, 4095.

750 La Bella, F., Gallinari, P., McKinney, J. and Heintz, N. (1989) *Genes and Devel.*, **3**, 1982.

751 Cremisi, C., Chestier, A. and Yaniv, M. (1977) *Cell*, **12**, 947.

752 Worcel, A., Han, S. and Wong, M. L. (1978) *Cell*, **15**, 969.

753 Leffak, I. M., Grainger, K. R. and Weintraub, H. (1977) *Cell*, **12**, 837.

754 Russev, G. and Hancock, R. (1982) *Proc. Natl. Acad. Sci. USA*, **79**, 3143.

755 Jackson, V. and Chalkley, R. (1981) *Cell*, **23**, 121.

756 Russev, G. and Tsanev, R. (1979) *Eur. J. Biochem.*, **93**, 123.

757 Weintraub, H. (1976) *Cell*, **9**, 419.

758 Annunziato, A. T. and Seale, R. L. (1983) *J. Biol. Chem.*, **258**, 12675.

759 Riley, D. and Weintraub, H. (1979) *Proc. Natl. Acad Sci. USA*, **76**, 328.

760 Seidman, M. M., Levine, A. J. and Weintraub, H. (1979) *Cell*, **18**, 439.

761 Sogo, J. M., Stahl, H., Koller, Th. and Knippers, R. (1986) *J. Mol. Biol.*, **189**, 189.

762 Hoffmann, I. and Birnstiel, M. (1990) *Nature (London)*, **346**, 665.

763 Louters, L. and Chalkley, R. (1985) *Biochemistry*, **24**, 3080.

764 Corthésy, B., Léonnard, P. and Wahli, W. (1990) *Mol. Cell. Biol.*, **10**, 3926.

765 Banerjee, S. and Cantor, C. R. (1990) *Mol. Cell. Biol.*, **10**, 2863.

766 Worcel, A., Han, S. and Wong, H. L. (1978) *Cell*, **15**, 969.

767 Jackson, V. and Chalkley, R. (1981) *Cell*, **23**, 121.

768 Dilworth, S. M., Black, S. J. and Laskey, R. A. (1987) *Cell*, **51**, 1009.

769 Laskey, R. A. and Earnshaw, W. C. (1980) *Nature (London)*, **286**, 763.

770 Ryoji, M. and Worcel, A. (1984) *Cell*, **37**, 21.

771 Glikin, G. C., Ruberti, I. and Worcel, A. (1984) *Cell*, **37**, 33.

772 Stillman, B. (1988) *Cell*, **45**, 555.

773 Fotedar, R. and Roberts, J. M. (1989) *Proc. Natl. Acad Sci. USA*, **86**, 6459.

774 Almouzni, G., Clark, D. J., Méchali, M. and

Wolffe, A. P. (1990) *Nucleic Acids Res.*, **18**, 5767.

775 Bonne-Andrea, C., Wong, M. L. and Alberts, B. M. (1990) *Nature (London)*, **343**, 719.

776 Marsden, M. P. F. and Laemmli, U. K. (1979) *Cell*, **17**, 849.

777 Finch, J. T. and Klug, A. (1976) *Proc. Natl. Acad. Sci. USA*, **73**, 1897.

778 Kowalski, J. and Cheevers, W. P. (1976) *J. Mol. Biol.*, **104**, 603.

779 Edenberg, H. J. and Huberman, J. A. (1975) *Annu. Rev. Biochem.*, **44**, 245.

780 Sheinin, R. and Humbert, J. (1978) *Annu. Rev. Biochem.*, **47**, 277.

781 Weintraub, H. (1979) *Nucleic Acids Res.*, **7**, 781.

782 Solomon, M. J. and Varshavsky, A. (1987) *Mol. Cell. Biol.*, **7**, 3822.

783 Kumar, S. and Leffak, M. (1986) *Biochemistry*, **25**, 2055.

784 Rodriguez-Campos, A., Shimamura, A. and Worcel, A. (1989) *J. Mol. Biol.*, **209**, 135.

785 Workman, J. L., Abmayr, S. M., Cromlish, W. A. and Roeder, R. G. (1988) *Cell*, **55**, 211.

786 Ellison, M. J. and Pulleyblank, D. E. (1983) *J. Biol. Chem.*, **258**, 13 321.

787 Svaren, J. and Chalkley, R. (1990) *Trends Genet.*, **6**, 52.

788 Giedroc, D. P., Khan, R. and Barnhart, K. (1990) *J. Biol. Chem.*, **265**, 11444.

789 Boye, E. and Lbner-Olesen, A. (1990) *Cell*, **62**, 981.

790 Campbell, J. L. and Kleckner, N. (1990) *Cell*, **62**, 967.

791 Burhans, W. C., Vassilev, L. T., Caddle, M. S., Heintz, N. H. and DePamphilis, M. L. (1990) *Cell*, **62**, 955.

792 Blow, J. J. and Nurse, P. (1990) *Cell*, **62**, 855.

793 Vaughn, J. P., Dijkwel, P. A. and Hamlin, J. L. (1990) *Cell*, **61**, 1075.

794 Burgers, P. M. J., Bambara, R. A., Campbell, J. L., Chang, L. M. S., Downey, K. M. *et al.* (1990) *Eur. J. Biochem.*, **191**, 617.

795 Landoulsi, A., Malki, A., Kern, R., Kohiyama, M. and Hughes, P. (1990) *Cell*, **63**, 1053.

796 Gayama, S., Kataoka, T., Wachi, M., Tamura, G. and Nagai, K. (1990) *EMBO J.*, **9**, 3761.

797 Diller, S., Kassel, J., Nelson, C. E., Gryka, M. A., Litwak, G. *et al.* (1990) *Mol. Cell. Biol.*, **10**, 5772.

798 Greider, C. W. and Blackburn, E. H. (1989) *Nature (London)*, **337**, 331.

799 Nakayasu, H. and Berezney, R. (1989) *J. Cell Biol.*, **108**, 1.

800 Stark, G. R., Debatisse, M., Wahl, G. M. and Glover, D. M. (1990) *Gene Rearrangement* (eds B. D. Hames and D. M. Glover), Oxford, p. 99.

801 Wickner, S., Hoskins, J. and McKenney, K. (1991) *Nature (London)*, **350**, 165.

802 Nordström, K. (1990) *Cell*, **63**, 1121.

803 Blackburn, E. H. (1991) *Nature (London)*, **350**, 569.

804 Greider, C. W. (1990) *BioEssays*, **12**, 363.

805 Krysan, P. J. and Calos, M. P. (1991) *Mol. Cell. Biol.*, **11**, 1464.

806 Heinzel, S. S., Krysan, P. J., Tran, C. T. and Calos, M. P. (1991) *Mol. Cell. Biol.*, **11**, 2263.

807 Thömmes, P. and Hübscher, U. (1990) *Eur. J. Biochem.*, **194**, 699.

808 Hurwitz, J., Dean, F. B., Kwong, A. D. and Lee, S-H. (1990) *J. Biol. Chem.*, **265**, 18 043.

809 Tsurimoto, T. and Stillman, B. (1991) *J. Biol. Chem.*, **266**, 1950.

810 Tsurimoto, T. and Stillman, B. (1991) *J. Biol. Chem.*, **266**, 1961.

811 Weiser, T., Gassmann, M., Thömmes, P., Ferrari, E., Hafkemeyer, P. and Hübscher, U. (1991) *J. Biol. Chem.*, **266**, 10 420.

812 Bambara, R. A. and Jessee, C. B. (1991) *Biochim. Biophys. Acta*, **1088**, 11.

813 Osheroff, N., Zechiedrich, E. L. and Gale, K. C. (1991) *BioEssays*, **13**, 269.

814 Yarranton, G. T. and Gefter, M. L. (1979) *Proc. Natl. Acad. Sci. USA*, **76**, 1658.

815 Kagani, L. S., Rossingol, J-M., Conaway, R. C., Banks, G. R. and Lehman, I. R. (1983) *J. Biol. Chem.*, **258**, 9037.

7

Repair, recombination and rearrangement of DNA

7.1 INTRODUCTION

This chapter is concerned with alterations in the structure of DNA, and considers several different facets of this topic.

It is important to protect the genome from inadvertent damage, and cellular mechanisms exist to repair such damage. Section 6.6 considered error avoidance mechanisms which limit the number of errors which occur during DNA replication. Not all errors are avoided, however, and error correction mechanisms exist to increase the final fidelity of replication and to remove errors arising as a result of the action of mutagens.

On the other hand genetic variation is the *sine qua non* for evolution, and mutation and recombination together allow an extent of genetic flux. Then there are examples of precisely controlled changes in DNA, involving amplifications (considered in section 6.15.6) or rearrangement, that are part of the developmental programme of particular cells or organisms.

7.2 MUTATIONS AND MUTAGENS

Damage to DNA resulting in the insertion of incorrect bases or distortion of the normal double-helical structure must be corrected if the cell is to survive. This chapter starts with a consideration of some of the ways in which damage may arise and the various strategies the cell may adopt to rectify the damage. If the damage is quickly rectified little harm is done, but if the DNA is replicated before repair is complete, alternative repair mechanisms must be called into play and these usually require recombination events.

Alterations in the base pattern of DNA may arise in various ways. For example, existing bases may be replaced by others, or they may be deleted, or new bases may be inserted in the DNA chain. Occasional mistakes in the normal duplication of DNA give rise to spontaneous mutations, but such mistakes are surprisingly rare [1] (section 6.6). Misincorporation may occur as a result of spontaneous tautomerization leading to mispairing of iminocytosine with adenine, iminoadenine with cytosine, *enol*guanine with thymine or *enol*thymine with guanine. In addition, guanine can mispair with adenine in several different forms [2]. The rare tautomer will rapidly revert to the more stable form resulting in a mismatch which, unless corrected, will result in the perpetuation of the mutation at succeeding rounds of DNA replication. The frequency

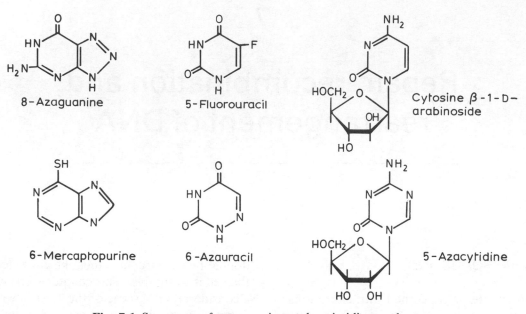

Fig. 7.1 Structures of some purine and pyrimidine analogues.

of mutations depends on conditions of temperature, pH, composition of growth medium, and the like, but it can be greatly increased by exposure of cells to ultraviolet and ionizing radiations (sections 7.2.4 and 7.2.5) or to certain types of chemical which are known collectively as mutagens. Such substances include base analogues, some dyes of the acridine series, alkylating agents, certain antibiotics, urethane, hydroxylamine and nitrous acid. This last substance may arise in the body from nitrates which are used in large amounts in fertilizers and which are reduced to nitrites by bacterial action in the mouth, stomach and bladder (section 2.7.2) [4]. Nitrous acid oxidizes the amino groups on nucleotide bases converting guanine into xanthine, adenine into hypoxanthine, cytosine into uracil and methylcytosine into thymine. The last three changes cause mispairing and, when not corrected, lead to transition mutations. Nitrous acid also causes the nitrosation of compounds such

as methylurea and methylnitroguanidine converting them into the potent alkylating agents MNU and MNNG (section 7.2.2). Hydroxylamine forms a derivative with the amino group of cytosine leading to transitions from G:C to A:T.

Mutagenic substances are the subject of an extensive literature [1, 3–9] which reflects the considerable effort now being devoted to attempts to inhibit cell division, especially in neoplastic tissues, by the use of compounds which might be expected to inhibit nucleic acid biosynthesis. Research in this field has been stimulated by the hope of finding a basis for an improved therapy for cancer. Reference has already been made (chapter 5) to the use of such compounds as azaserine and the folic acid antagonists in preventing the synthesis of the purine and pyrimidine nucleotides. Some of the other substances which have been used to prevent nucleic acid biosynthesis and to bring about mutations artificially are discussed below.

7.2.1 Base and nucleoside analogues

Some of the artificially produced base analogues are incorporated into RNA and DNA and may have powerful mutagenic effects [1, 8–10]. Among the most important analogues are the halogenated pyrimidines, and those bases where nitrogen has been substituted for a −CH= group (Fig. 7.1). The action of these unnatural bases seems, at least in some cases, to be twofold.

(1) They generally block some stage in the biosynthesis of the normal purine and pyrimidine nucleotides. Thus 8-azaguanine inhibits the biosynthesis of GMP and 6-mercaptopurine blocks the conversion of IMP into AMP [11]. In general, these inhibitions are brought about only after the inhibitor itself has been converted into its nucleotide. Thus 6-azauracil is converted first into its nucleoside (aza-U) then into its nucleotide (aza-UMP) which inhibits the action of orotidine 5′-phosphate decarboxylase (Fig. 5.6) and so prevents pyrimidine biosynthesis [5, 9]. 5-Fluorouracil, which has proved to be a potent inhibitor of the growth of certain tumours, is converted first into its ribonucleotide (F-UMP) and then into its deoxyribonucleotide (F-dUMP) which exerts its main effect by inhibiting conversion of dUMP into dTMP [8, 9] and hence inhibiting DNA synthesis (Fig. 5.8). 5F-dUMP is also formed when 5-fluorodeoxyuridine is added to cells in culture. As DNA synthesis is the only process affected, cells progress around the cycle and accumulate at the beginning of the S phase (section 3.2.1). The inhibition can be overcome by the addition of thymidine with the result that a population of cells in synchronized growth results [8, 12].

(2) They are themselves, after conversion into nucleotides, incorporated to varying degree into RNA and/or DNA although the incorporation may take an abnormal form.

Thus 8-azaguanine can be incorporated at the expense of guanine into the RNA of TMV [13] and, to a much larger extent, into the RNA of *B. cereus* [14]. Only very small amounts are incorporated into the DNA. 5-Azacytidine is incorporated largely into RNA and this rapidly interferes with protein synthesis [15]. Some is, however, incorporated into DNA and like 5-azadeoxycytidine it brings about changes in gene expression probably as a result of its inhibiting DNA methylation (chapter 4) [16, 17]. 2-Aminopurine is incorporated into DNA where it normally pairs with thymine, although it can also pair with cytosine. This capacity leads to frequent transitions both of A:T to G:C, and G:C to A:T. 5-Bromouracil can replace thymine in DNA where it normally base-pairs with adenine. However, in its rare *enol* state (which it assumes more readily than does thymine) it may pair with guanine instead of adenine so bringing about the base-pair transition of A:T into G:C [10]. DNA containing 5-bromouracil instead of thymine is very susceptible to breakage at light-induced bromouracil dimers [18] (see below). The main mutagenic effects of bromodeoxyuridine, however, appear not to arise from its incorporation into DNA in place of thymidine but rather from its effect on ribonucleotide reductase (section 5.5). This leads to an imbalance in the deoxyribonucleoside triphosphate pools [19] which may result in base misincorporation. In particular, the pool size of dCTP is reduced and some substitution of deoxycytidine with bromodeoxyuridine (in the *enol* form) occurs leading to transition from G:C to A:T as the misincorporated bromodeoxyuridine reverts to its more stable *keto* form. Addition of thymidine accentuates the mutagenic effect of bromodeoxyuridine by further decreasing the dCTP pool size whereas deoxycytidine antagonizes the mutagenic effects. 2-Aminopurine can also base-pair with either cytosine or thymine as described above, and, because it inhibits adenosine deaminase, it also increases the pool size of dATP which in turn increases the size of the dCTP pool relative to the dTTP pool leading to misincorporation of cytosine opposite adenine or 2-aminopurine [20]. The D-arabinosyl nucleosides are effectively analogues of deoxyribonucleosides (e.g. cytosine β-D-arabinoside is incorporated in place of deoxycytidine into DNA where it causes chain termination or a marked reduction in the rate of further chain extension [8, 9, 18, 21]).

7.2.2 Alkylating agents

The alkylating agents exert a variety of biological effects including mutagenesis, carcinogenesis and inhibition of tumour growth [1, 25–27]. They all carry one, two, or more alkyl groups in reactive form and include the well-known compounds sulphur mustard or di-(2-chloroethyl) sulphide and nitrogen mustard or methyl-di-(2-chloroethyl)amine. The action of the alkylating agents on DNA is complex. They are electrophilic reagents and react with nucleophilic centres, particularly with guanine at the N7 atom. The order of reactivity of different nucleophilic centres is N7–G $\gg$ N3–A > N1–A = N3–G = O6–G [3]. The bifunctional alkylating agents (i.e. those with two reactive alkyl groups) may bring about cross-linking between the opposing strands in the DNA molecule. Alkylation of purines in position 7 also gives rise to unstable quaternary nitrogens so that the alkylated purine may separate from the deoxyribose leaving a gap which might interfere with DNA replication or cause the incorporation of the wrong base [1, 22]. Alkylation of N7 of guanine with dimethylsulphate is the reaction used in Maxam–Gilbert sequencing to labilize the

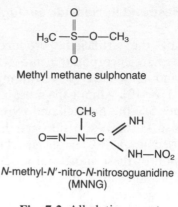

Fig. 7.2 Alkylating agents.

DNA chain in this position (section A.5). The phosphate groups may also be alkylated. The phosphate triester so formed is unstable and may hydrolyse between the sugar and the phosphate so that the DNA chain is broken.

The relationship between DNA alkylation and carcinogenesis (tumour induction) is tenuous. Although certain alkylations do cause miscoding (i.e. are mutagenic [6, 23]), this is not so for 7-methylguanine which is the major product of alkylation [24–26]. Good alkylating agents, e.g. dimethyl sulphate and methylmethanesulphonate, are poor carcinogens but the reverse is true for N-methyl-N'-nitro-N-nitrosoguanidine (MNNG) (Fig. 7.2). 2-O-Methylated bases may cause miscoding as may also 6-O-methylguanine and 3-N-methyladenine [24–26]. Dimethylnitrosamine is not itself an alkylating agent, but is converted into one in mammalian cells by cytochrome P450-dependent oxidation. Mitomycin C inhibits DNA synthesis by causing covalent cross-linking of the complementary DNA strands [9, 27]. The drug is reduced in the cell to produce an active bifunctional alkylating agent which cross-links guanine residues [28]. It is used to produce a feeder layer of dead cells on which sensitive mammalian

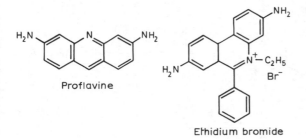

Proflavine

Ethidium bromide

Fig. 7.3 Structures of proflavine and ethidium bromide.

cells will grow. Phleomycin attaches co-valently to thymine residues in DNA to inhibit DNA polymerase action [9]. As with bleomycin, which binds in a similar way, this results in single-strand breaks in the DNA [9, 29].

7.2.3 Intercalating agents

The acridine dyes, e.g. proflavine, ethidium bromide and propidium di-iodide (Fig. 7.3),

are flat molecules which are able to insert themselves between the stacked bases of the DNA double helix. This brings about some unwinding of the DNA duplex which inter-feres with transcription [30]. The unwinding is resisted in closed circular DNA molecules (section 2.5) which therefore bind less dye and are therefore considerably denser than nicked cyclic molecules. This forms the basis of the separation of supercoiled molecules from non-supercoiled molecules on gradi-ents of caesium chloride containing the dye [31] (section A.2.1). The anthracenes and benzpyrenes (Fig. 7.4) present in soot, react with the 2-amino group of guanine before intercalating into the DNA duplex. Aflatoxins, which are potent liver carcino-gens produced by moulds (*Aspergillus* sp.), also preferentially intercalate next to purines while actinomycins react with deoxy-guanosine and intercalate between dG:dC nucleotide pairs. With actinomycin D the cyclic peptide (Fig. 7.4) lies in the groove in the double helix, hydrogen-bonded to the guanine [7]. Actinomycin D inhibits both

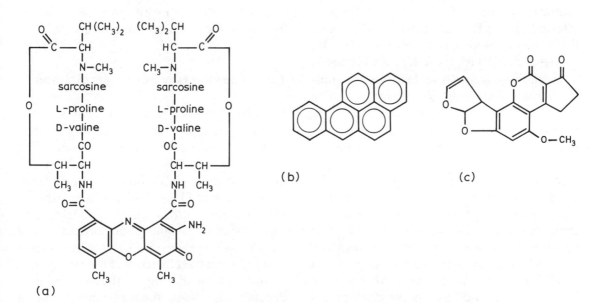

(a)

(b)

(c)

Fig. 7.4 Structure of (a) actinomycin D; (b) benzpyrene; (c) aflatoxin B1.

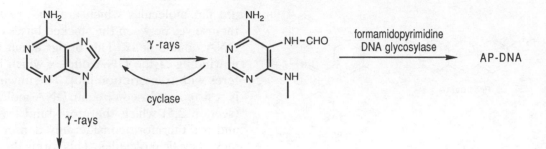

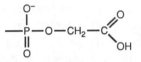

Fig. 7.5 Action of γ radiation on DNA.

DNA and RNA polymerases, although the former is much less sensitive [32]. At a concentration of actinomycin of $1\,\mu\text{M}$, for example, the DNA-dependent RNA polymerase is almost completely inhibited whereas the DNA polymerase is only slightly affected [33]. At lower concentrations the synthesis of the different sorts of RNA is affected to different extents, ribosomal RNA being particularly sensitive [34]. By partially unwinding the DNA double helix, intercalating agents can lead to the production of frameshift mutations in which at replication, extra bases are inserted or deleted, opposite to the intercalated molecule. Frameshift mutations can also arise in repeated sequences as a result of misalignment of one of the repeated units with the complement of another unit [36]. Acetylaminofluorene (AAF) is an intercalating agent which also becomes bound to the C8 of guanine. Particularly in runs of G this leads to a -1 frameshift; and in alternating CGs to a -2 frameshift. Where an AAF molecule is inserted into the DNA duplex it displaces the guanine to which it is bound. The guanine

is rotated out of the helix and so is not available at replication for base pairing with a dCTP and hence it is skipped by the polymerase [35]. Frameshift mutations can also arise in palindromic sequences where imperfect alignment occurs [37] (section 3.2.2(d)) and the association of potential hairpin structures with runs of guanine provides 'hotspots' for frameshift mutations [38].

7.2.4 The effects of ionizing radiations

In growing animal and plant cells there is an association between the onset of DNA synthesis and resistance to X-irradiation. Thereafter survival rates remain approximately constant throughout this period and then decline in G2 as damaged cells prepare for division with fragmented DNA and chromosomes. Single-strand breaks are a major lesion induced into DNA by ionizing radiation and the efficiency of repair depends in part on the nature of the termini produced, e.g. those breaks bounded by a 3'-OH and a 5'-phosphate can readily be

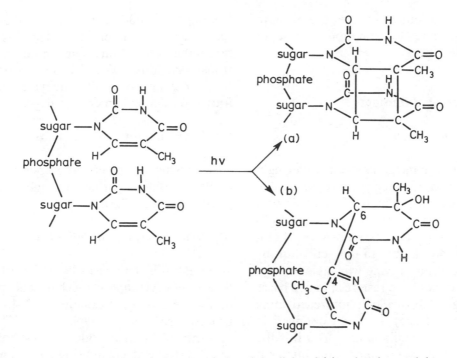

Fig. 7.6 The formation under the influence of ultraviolet light of (a) a thymine cyclobutane dimer and (b) a thymine:thymine (6:4) photoproduct.

rejoined by DNA ligase [39, 40]. The other major action of γ-rays is to break the imidazole ring of purines to produce a formamidopyrimidine ring (Fig. 7.5).

7.2.5 Ultraviolet radiation

Large doses of ultraviolet radiation can damage living cells by causing the formation of chemical bonds between adjacent pyrimidine nucleotides in the DNA. Two pyrimidine bases joined in this way in one strand form what is known as a dimer, and of the three possible types of pyrimidine dimer, the thymine dimer is formed most readily (Fig. 7.6). The presence of such dimers blocks the action of the DNA polymerase and so prevents replication [42]. The major product of ultraviolet irradiation is the cyclobutane:thymine dimer shown in Fig. 7.6 in which adjacent thymine bases are linked at the 5 and 6 atoms. However, it has been shown that the mutagenic effect of ultraviolet irradiation may result from a minor product in which the 5'-pyrimidine is linked between its 6 position and the 4 position of the 3'-pyrimidine (the 6:4 photoproduct – Fig. 7.6) [43].

7.3 REPAIR MECHANISMS

When DNA is damaged or when mistakes occur in its synthesis, it is important that it is repaired rapidly. Some damage can be directly reversed whereas other types of damage require that the faulty stretch of DNA be removed and replaced. If DNA replication intervenes between damage and repair the consequences are often more extreme, leading to mutations, most of which are lethal.

7.3.1 Reversal of damage

(a) *Photoreactivation*

When bacteria damaged by ultraviolet light are exposed to an intense source of visible light (wavelengths between 320 and 370 nm) a large proportion of the damaged cells recover. This process is known as photo-reactivation and is due to the activation by visible light of an enzyme which cleaves the pyrimidine dimers and restores the two bases to their original form. This photoreactivating enzyme (photolyase) has been obtained in pure form [3, 44, 45] and in *E. coli* it is the product of the genes *phrA* and *phrB*. It cleaves only the cyclobutane dimers and not the 6:4 adduct [45]. A similar enzyme occurs in eukaryotes [46, 47] and may be deficient in people who suffer from the disease xeroderma pigmentosum character-ized by extreme sensitivity to sunlight.

(b) *Removal of methyl groups*

An adaptive response to methylation of DNA with MNNG and other methylating agents is the dramatic increase in activity of a DNA methyltransferase which transfers methyl groups from O^6-methylguanine and methylphosphate triesters onto two cysteine residues in the protein [3, 25, 48, 49, 200]. The $18\,000$-M_r protein (the product of the *E. coli ada* gene) also removes other alkyl groups but shows greatest activity with methyl groups. When a methyl group is transferred from a methyltriester to cysteine 69, the Ada protein is converted into a positive transcriptional regulator which binds to the promoter of the *ada* operon

leading to a 100-fold increase in the trans-ferase activity per cell [201]. There is also a non-inducible O^6-methylguanine methyl-transferase in *E. coli* which is the product of the *ogt* gene. A similar enzyme as been found in mammalian cells [50].

(c) *Formamidopyrimidine cyclase*

This enzyme will reverse the opening of the imidazole ring of purines, brought about by γ-radiation (Fig. 7.5).

(d) *Joining of single-strand breaks*

Although this may occur by action of DNA ligase more complex mechanisms are re-quired following γ-irradiation as 3′-hydroxyl groups are not available [40, 41]. The majority of breaks produce a 3′-phosphate or 3′-phosphoglycolate (Fig. 7.5) and this must be removed (e.g. by exonuclease III action) to initiate repair [41].

(e) *Purine insertion*

An activity has been found which will insert purines into apurinic sites. The enzyme does not work at random but is reported to re-form the correct base-pairing [50–52].

7.3.2 Excision repair

This is the most common repair mechanism and is able to cope with most types of DNA damage. The first step in the reaction is unique to the *repair* of DNA and involves endonuclease action on one or both sides of the damage. In the classical example of phage T4 uv endonuclease V it is easy to consider excision repair as taking place in four stages. (i) A nick, with a 3′-OH terminus, is introduced on the 5′-side of a pyrimidine dimer (Fig. 7.7). (ii) A second

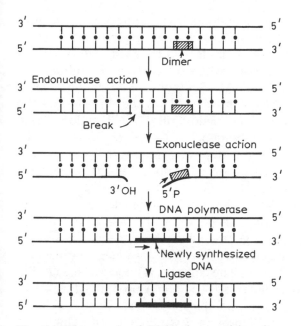

Fig. 7.7 The repair of DNA damaged by ultraviolet light. The action of uv has been to produce a dimer which is excised by sequential action of an endonuclease and exonuclease, leaving a short gap which is filled in by the action of DNA polymerase. The final join is effected by DNA ligase.

enzyme excises a short stretch of the DNA strand including the damaged region. (iii) DNA polymerase uses the intact complementary strand as template to synthesize a piece of DNA to fill the gap. (iv) The repair is completed by ligase action [3, 42, 53].

In *E. coli*, DNA polymerase I is the enzyme which first excises the portion of the affected strand (including, on average, 30 nucleotides [54]) by virtue of its $5' \to 3'$ nuclease action (section 6.4.3(e)) and then fills the gap [55, 56]. *polA$^-$* mutants show increased uv sensitivity [57, 58] and, although they do manage to repair lesions, the patch size is considerably bigger than in the wild-type [51, 52, 57, 59]. This may be because DNA polymerases II and III take over the role of polymerase I or that repair is occur-

ring by a recombination mechanism (see below). Depending on the nature of the damage, the endonucleolytic step is performed by one of a number of different enzymes.

Most commonly the damaged base is first recognized by a *glycosylase* which removes the base from the sugar phosphate backbone to produce apurinic or apyrimidinic DNA (known as AP DNA) [49, 50, 60–64]. Glycosylases are a family of small proteins (less than $30\,000\ M_r$) each one specific for a different unusual base. Glycosylases have been isolated which will remove uracil, hypoxanthine, 3-methyladenine, 7-methylguanine, 3-methylguanine, formamidopyrimidine, 5:6 hydrated thymine, urea or pyrimidine dimers. The role of *N*-glycosylases has been reviewed by Lindahl [65].

Some glycosylases appear to have intrinsic endonuclease activity. The *M. luteus* and the phage T4 uv endonucleases (considered above) first cleave the *N*-glycosyl bond between the 5'-pyrimidine of a pyrimidine dimer and its sugar and then cleave the phosphodiester bond 3' to the AP site leaving an aldehyde sugar [43] (Fig. 7.8). Endonuclease III of *E. coli* is a similar enzyme which recognizes reduced thymine rings in DNA and removes them at the same time as cleaving the phosphodiester backbone [66, 67]. N^7-methylguanine is spontaneously converted into 2:6 diamino-4-hydroxy–5-*N*-methylformamidopyrimidine (Fig. 7.9) by opening the imidazole ring [68]. This, together with formamidopyrimidines arising after γ-irradiation, is recognized by formamidopyrimidine glycosylase [69]. This glycosylase also cleaves the resulting AP DNA [70–72] by nicking to the side of the deoxyribose before the 3'-phosphate. The reaction, which is actually that of an AP lyase, occurs by β-elimination followed by δ-elimination resulting in a gap limited by 3'-phosphate and 5'-phosphate ends (Fig.

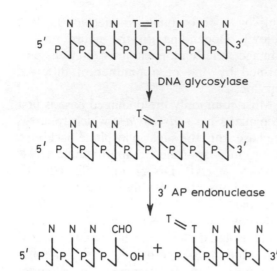

Fig. 7.8 The combined glycosylase and AP endonuclease action of the uv endonuclease of *M. luteus* and phage T4.

7.9). Subsequent gap filling requires the intervention of exonuclease III to remove the 3'-phosphate.

These enzymes that combine glycosylase

and AP endonuclease action are uncommon. More usually, following glycosylase action the DNA is cleaved by a separate AP endonuclease. Although cleavage usually occurs 5' to the AP site, there are AP endonucleases which can cleave 3' to the AP site and the combined action of the AP endonuclease and subsequent 3' → 5' or 5' → 3' exonuclease step determines the size of the patch produced – about 20–30 nucleotides long [3].

The major AP endonuclease of *E. coli* is that encoded in the *uvrA*, *uvrB*, and *uvrC* genes, which have been cloned and sequenced and their products purified [3, 73–75]. The complex is known as the (A)BC exinuclease. The UvrA protein (M_r 114 000) forms a complex {(UvrA)$_2$.UvrB} with the UvrB protein (M_r 84 000) that binds to damaged DNA. UvrA is released and the UvrC protein (M_r 70 000) binds, and nicks are made on *both* sides of the damaged region 12–13 nucleotides apart [76, 77] (Fig. 7.10). A 3'-hydroxyl is produced 5' to the damaged region and can serve as a primer

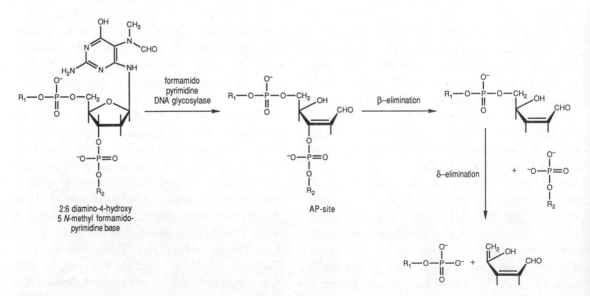

Fig. 7.9 Formamidopyrimidine DNA glycosylase acts to remove the damaged base and to cleave the phosphodiester backbone of DNA on both sides of the opened deoxyribose ring.

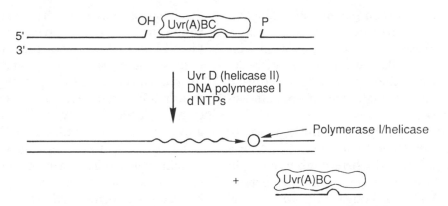

Fig. 7.10 The Uvr(A)BC exinuclease of *E. coli* recognizes AP sites and makes nicks on both sides of the damaged region. The 12–13 nucleotide-long fragment is released together with the exinuclease by helicase II action.

for DNA polymerase I which produces a patch only 12–13 nucleotides long. Helicase II (the product of the *uvrD* gene) is required to unwind the damaged region of DNA and release the UvrC protein [78, 79]. This mechanism may be particularly appropriate for bulky adducts as the cleavage sites are a short distance away from the distortion, however, the action of the (A)BC exinuclease is not dependent on gross distortion of the DNA duplex [80]. *E. coli* AP endonucleases are listed in Table 4.3.

A similar set of enzymes is present in eukaryotic cells [81], but in patients suffering from a number of diseases (e.g. xeroderma pigmentosum, ataxia telangiectasia, Fanconi's anaemia, Bloom's and Cockayne's syndrome) one of the repair enzymes is missing [82, 83]. Such people are therefore abnormally sensitive to exposure to sunlight and tend readily to develop skin cancer [3, 46, 50, 84–86].

7.3.3 Mismatch repair

When a normal but incorrect base is introduced into DNA at replication, it is usually

removed by a proofreading mechanism before the next nucleotide is added (section 6.4.3(d)). When this mechanism fails, the product is a duplex DNA molecule with a mismatch. A repair system which can correct the mismatch must discriminate between two normal bases, one of which is incorrect. A random process would accentuate the error in 50% of cases, but mismatch correction systems use the lack of methylation on the newly synthesized (incorrect) strand to determine which strand requires repair.

Such systems are present in both prokaryotes and eukaryotes, and although they are better understood at present in the former, significant advances are being made in the field of eukaryotic mismatch repair [336]. Methylation (section 4.6.1) occurs shortly after replication of DNA but, for a finite time, the daughter strand is unmethylated. It is during this time window, when the DNA is hemimethylated, that mismatch correction is activated. *Dam* mutants which have a much decreased level of DNA adenine methylation show increased rates of mutation [16, 87–89]. This is because in *Dam⁻* cells neither strand of DNA is methylated and the repair system cannot discriminate between parental and daughter strand and so repair occurs at random. Mismatch correction involves the Mut proteins and in *Mut*⁻ cells the repair system is defective, mismatches are not removed, and there is again a high rate of spontaneous mutagenesis.

Mismatch repair involves the products of the *mutH*, *mutL* and *mutS* genes. MutS is a 97 kDa protein which recognizes all eight possible mismatches (i.e. it binds to heteroduplex DNA) although not with equal efficiency [90, 91]. MutL is a 70 kDa dimer which, in the presence of ATP, binds to the MutS–heteroduplex complex to activate MutH. This is a 25 kDa endonuclease which cleaves DNA 5′ to the G in an unmethylated GATC sequence to produce a nick bounded by a 3′-OH and a 5′-phosphate. Fully methylated DNA is not cleaved and unmethylated DNA is cleaved on either one strand or the other strand at random [92].

How the signal passes from the mismatch (where MutS·and MutH are bound) to the GATC (which can be several kilobase pairs away) is not known. Neither is it known whether nicks are made at GATCs on both sides of the mismatch or whether a second nick is generated at the mismatch itself. One possibility is that helicase II (the product of the *mutU* or *uvrD* gene) binds at the mismatch and travels bidirectionally, opening up the duplex, until hemimethylated GATCs are encountered. This would lead to displacement of the incorrect strand. The single-stranded DNA is covered in SSB and the 3′-OH serves as a primer for DNA holopolymerase III. This mechanism leads to patches several kilobases long.

There is another mismatch repair system which neither depends on DNA methylation nor the *mutHLS* system and which produces shorter repair patches (up to 300 bases). It is specifically involved in converting A:G mismatches into C:G base pairs [93, 94].

Such mismatches occur frequently at replication and involve the *syn* form of dGMP (see [2] and [89] for the structures of some unusual base pairs). They appear not to be recognized by DNA polymerase III at proofreading but there is a supplementary enzyme (coded for by *mutT*) which probably degrades the *syn* form of dGTP [95, 96].

There is a third mismatch correction system which relies on nicks arising in the daughter strand at replication to discriminate between the correct and incorrect base [97] and this appears of more significance in eukaryotes [98, 334]. This mismatch correction involves DNA polymerase α or δ.

Not all mismatches arise at replication. Deamination of 5-methylcytosine to thymine can lead to certain cytosines (e.g. in *E. coli*, those methylated in the sequence CC(A/T)GG by the *dcm* methylase) being 'hot spots' for mutation if repair is not efficient [99]. The resulting T:G mismatches are specifically corrected to C:G pairs by a mechanism which produces very short patches (VSP repair) and which, in *E. coli*, requires MutL and MutS but not MutH or helicase II [100].

A similar, very short patch repair system is also present in eukaryotes [101]. Here the initial step appears to be the removal, by a glycosylase, of the thymine paired with guanine. The single nucleotide gap produced by subsequent AP endonuclease action, is filled using DNA polymerase β [335]. If not repaired before replication then the daughter molecules will have a

$$\begin{matrix} -\text{TG}- \\ -\text{AC}- \end{matrix} \text{ in place of the original } \begin{matrix} -\text{mCG}- \\ -\text{GCm}- \end{matrix}$$

Such mutations are frequent in eukaryotes [102, 223] and are believed to be one of the reasons why the vertebrate genome is deficient in the CG dinucleotide [103–105].

Mismatches can also arise by incorrect alignment of the DNA strands where several

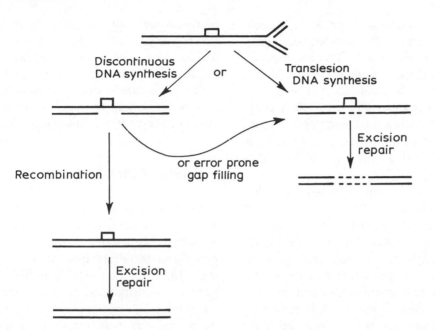

Fig. 7.11 Different repair processes occurring following replication of a region of damaged DNA.

repeating units are arranged in tandem or when imperfect palindromic sequences form cruciform structures. These often result in repair reactions leading to deletion or insertion of one or two nucleotides [37].

Mismatches also arise following recombination and in *E. coli* these are also subject to a correction mechanism involving the MutHLS system. This can effectively reverse the effects of recombination, but also acts as a barrier to recombination between the DNA of different species [117].

7.3.4 Bypass synthesis, translesion replication or error-prone repair

If repair is quickly and accurately effected before the damaged DNA is replicated the integrity of the genetic message is maintained. Such excision repair is very effective, of high fidelity and leads to the insertion of only small repair patches. Nonetheless,

occasionally the cell tries to replicate the damaged region of DNA, and this occurs more often when excision repair is slow or defective, as in *uvr⁻* cells or in patients with xeroderma pigmentosum. When the advancing polymerase reaches a lesion such as a thymine dimer two possibilities arise. Either the polymerase stops and a gap is left in the daughter strand opposite the lesion; or the polymerase replicates past the lesion, inserting a series of random nucleotides. Even if the gap is later filled by insertion of random nucleotides, then the bypass synthesis is error-prone. An alternative to this error-prone synthesis involves the insertion of a patch from an intact duplex (Fig. 7.11) but this involves recombination mechanisms which are discussed in section 7.4.

Normally DNA polymerase III is very processive (section 6.4.5) but it dissociates from the DNA when a nicked duplex is formed or when it encounters a lesion. Termination occurs because a mismatched terminus is produced and the $3' \rightarrow 5'$ exonuclease will remove the nucleotides, newly added opposite the lesion. The enzyme can rebind and perform a futile cycle of addition and removal of nucleotides. Occasionally, the highly processive nature of DNA polymerase III may allow it to completely bypass the lesion and this bypass synthesis is accentuated during the, so-called, SOS response [106].

In *E. coli* the presence of single-stranded regions produced in DNA as a result of damage induces the SOS functions [47, 107, 108]. The RecA protein binds to the single-stranded regions and is activated to form a proteolytic enzyme which causes the cleavage of LexA protein, the common repressor of at least 17 genes. Only following repair of the damage is the RecA protease inactivated and the level of LexA protein allowed to build up to again repress this pleiotropic effect. Among the proteins under RecA/LexA control is the phage lambda repressor (explaining the induction of prophage by agents which damage DNA – section 3.7.8) and the *uvrA*, *uvrB* and *uvrC* genes.

In addition, the *umuC* and *umuD* genes are activated, and various suggestions have been made as to how the induced proteins may act to promote translesion synthesis. (UmuD itself is proteolytically activated by RecA [109, 110].) They may act to increase the processivity of DNA polymerase III [111]. This may occur via modification of the β subunit of DNA polymerase III [112] or by inhibition of the $3' \rightarrow 5'$ proofreading exonuclease activity of the ε subunit. An alternative form of DNA polymerase III is found in induced cells [113] and this may involve the substitution of DNA polymerase I (or a variant form of it or of DNA polymerase II) for the α subunit in the DNA polymerase holoenzyme [107, 113, 114]. A decreased fidelity of the replicative poly-

merase would not, however, be restricted to the regions of DNA actually damaged but would occur generally in replicated DNA [3, 46, 47, 115, 116]. To what extent this is actually the case is still open to debate but it may be the origin of many of the mutations arising as a result of damage to DNA.

7.4 RECOMBINATION

Recombination is the production of new DNA molecule(s) from two parental DNA molecules. It usually occurs by exchange of genetic material (reciprocal recombination) such that the new DNA molecules carry genetic information derived from both parental DNA molecules, but it can occur by a one-way transfer (gene conversion). Recombination involves a physical rearrangement of the parental DNA and can be observed during (1) a mixed infection of two viruses or plasmids carrying different genetic markers, (2) crossing over at meiosis in eukaryotic cells, (3) integration of bacteriophage, viral or plasmid DNA into a host cell chromosome, (4) transformation of cells with foreign DNA, (5) gene conversion, (6) post-replication repair or (7) a variety of other processes considered later in this chapter. If it were not for recombination, genes would forever remain linked on the same chromosome.

In higher eukaryotes recombination between chromosomes leads to chiasma formation at meiosis in which two non-sister chromatids have become linked together at corresponding sequences. Each chromatid is a duplex DNA molecule (with associated protein) and, following replication, a chromosome consists of two sister chromatids. At chromosome pairing in meiosis the homologous chromosomes become associated to form a synaptonemal complex and it is within this complex that crossing over

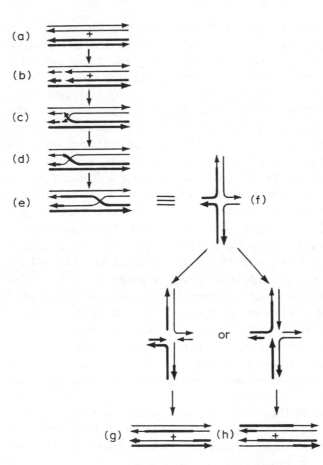

Fig. 7.12 The Holliday model of recombination. See text for details. Structures (e) and (f) are equivalent.

occurs. However, there is a conceptual gap between what is observed to occur between these huge nucleoprotein complexes and what is presumed to occur at the DNA level. Nucleoprotein complexes are also formed during recombination in *E. coli* and these are more amenable to study (section 7.4.2). General recombination occurs between any regions of DNA which are wholly or largely complementary in sequence. In site-specific recombination (section 7.4.4) complementarity exists over only a specific limited region but the reaction is catalysed by enzymes which recognize this complementary site.

7.4.1 Models of general recombination

The original Holliday model [118] was proposed to explain genetic results obtained in fungi but applies equally well to Rec-mediated recombination in *E. coli* (section 7.4.2). Indeed it is this latter system which has provided the experimental evidence to support the model. Holliday proposed that recombination between homologous duplex molecules is initiated by an endonuclease nicking corresponding regions of homologous strands of the paired duplexes (Fig. 7.12(b)). The nicked ends dissociate from their complementary strands and reciprocal strand invasion (Fig. 7.12(c)) produces a joint molecule in which each duplex has a heteroduplex region derived from one strand from each parental duplex. Ligase action generates stable joint molecules (Fig. 7.12(d)). Branch migration leads to the formation of extensive heteroduplex regions (Fig. 7. 12(e)). The joint molecule is known as a Holliday or Chi (χ) structure and can be visualized in electron micrographs, so providing proof of the existence of joint molecules [119].

The resolution of a Holliday structure requires two nicks which may occur in one of two ways leading to either a short length of heteroduplex DNA in the otherwise undisturbed parental molecules (patching) (Fig. 7 12(g)); or the production of duplex molecules derived from different halves of the two parental molecules (Fig. 7.12(h)). The Holliday structure can be drawn more simply as a linear alignment of four DNA strands (Fig. 7.13(a)). Resolution then requires the cleavage of the two 'Watson' strands or the two 'Crick' strands at the boundary of the heteroduplex region [138]. These diagrams hide the information that, in the 4-way junction, the two duplexes adopt a planar, stacked X-shaped structure in which the most energetically favourable conformation appears to be that where the exchanging strands do not cross but enter and leave the junction on the same side (Fig. 7.13(b)) [41, 333].

When the two parental molecules are cyclic duplexes then the joint molecule is a figure 8 which, on resolution, will give either two cyclic molecules each with a heteroduplex region or alternatively a double length cyclic molecule (the cointegrate) made up from both parental molecules. Both the figure 8 and the double-length cyclic molecules can be seen in electron micrographs [120, 121].

A variation on the Holliday model was proposed by Meselson and Radding [122] to explain formation of heterozygous regions which are not perfectly reciprocal. They suggested that recombination is initiated by a nick in only one of the parental duplexes. In a strand displacement reaction, DNA polymerase extends the 3'-OH and the 5'-end invades the homologous duplex (the recipient) to form a D-loop. A second nick leads to digestion of the displaced D-loop strand and ligase action generates Chi structures. Such a gene conversion mechanism can account for the concerted evolution of gene families [249]. Potter and Dressler [119, 123] have suggested that no nicking is required if strand invasions can occur at transiently single-stranded regions. More recently, as a result of studies with the RAD 52 mutant in yeast, it has been proposed that recombination is initiated with a double-stranded break in one of the aligned duplexes [124, 125]. Nucleases extend the break to form a gap with 3'-extensions which can invade the homologous region on the other parental duplex. Repair synthesis leads to production of two Holliday junctions which can be resolved in two different ways.

It is probable that different models reflect different pathways of recombination, three of which are known in *E. coli* [126]. In all cases it is likely that the site of the initiating nick or cut is at a preferred sequence [127].

Where the heteroduplex contains mismatched base pairs the mismatch correction system (section 7.3.3) will come into play in a random manner. In the Meselson and Radding model this may restore the recipient to its original sequence or convert it into the sequence of the donor, and a similar outcome has been shown to occur when recombination is initiated at a double-stranded break [127–129]. Where one of the strands of the heteroduplex contains an abnormal or damaged base, excision repair will use the complementary strand as a template to

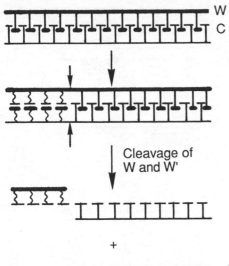
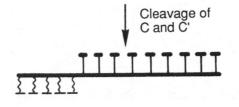

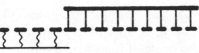
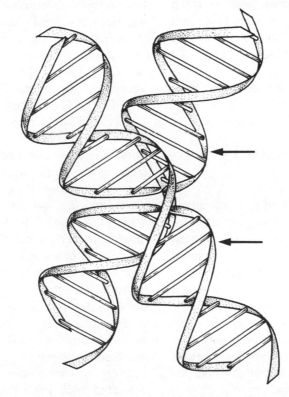

Fig. 7.13 The Holliday junction can be represented as a linear 4-stranded structure (a) which is resolved by rearrangement of hydrogen bonds and cleavage of the two 'Watson' strands or the two 'Crick' strands [138]. The junction has been shown to resemble an X where the two cleaved strands are not crossed (b). The arrows indicate the position of cleavage by T4 endonuclease VII. From [333] with permission of the authors and publishers.

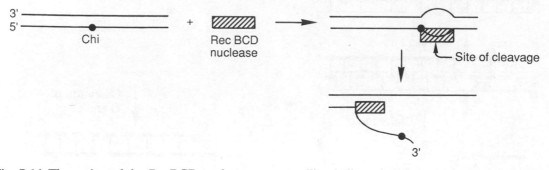

Fig. 7.14 The action of the RecBCD nuclease as a travelling helicase is changed to a nuclease when a Chi site is encountered.

replace the damage with the order of the new bases being dictated by the paired duplex molecule (Fig. 7.11).

Phage-encoded endonucleases (e.g. T4 endonuclease VII) have been isolated which will resolve Chi structures *in vitro* [140] and they will also cleave cruciform structures which are formally analogous [141]. T4 endonuclease VII acts exclusively at the junction point . and similar enzymes have been isolated from yeast [139].

7.4.2 The *E. coli* rec system and single-strand invasion

General recombination [130] can take place anywhere along the length of two complementary DNA molecules but may be initiated at preferred sequences. It has been studied in *E. coli* infected with two bacteriophages carrying different genetic markers when multiple recombination events may occur; or following conjugation of *hfr* strains (section 3.6). In *E. coli*, general recombination is dependent on one of three host *rec* systems as well as other enzymes and proteins involved in DNA replication.

In wild-type cells, recombination uses the RecBCD pathway. RecBCD nuclease (exonuclease V, the product of genes *recB*, *recC*

and *recD*) has multiple enzymic activities. Commencing at an end of a linear, duplex DNA molecule, it can travel along the DNA at about 300 bp per second, acting as a helicase and unwinding the DNA. Rewinding occurs behind the enzyme but single-stranded loops form transiently and these are sensitive to exo- and endonuclease cleavage [131–134]. Endonucleolytic cleavage is greatly enhanced when the travelling RecBCD nuclease approaches a Chi site [135]. The Chi site has the sequence 5′-GCTGGTGG–3′ and the RecBCD nuclease, as it approaches from the right, cleaves the single-stranded loop of DNA (produced by its unwinding action) four to six nucleotides on the 3′-side of Chi (Fig. 7.14) [132–134]. The enzyme continues to unwind the DNA but this now leads to the production of a single-stranded tail of DNA. The Chi sites are therefore recombination hotspots and they occur, on average, once per 4 kb in the *E. coli* genome. The presence of enzymes such as the RecBCD nuclease also explains why ends of DNA molecules are recombinogenic.

In *recBC* mutants which also have an activated *recE* gene (on a cryptic plasmid) the encoded exonuclease VIII digests double-stranded, linear DNA in the 5′ → 3′ direction, producing 3′-tails. The RecE exo-

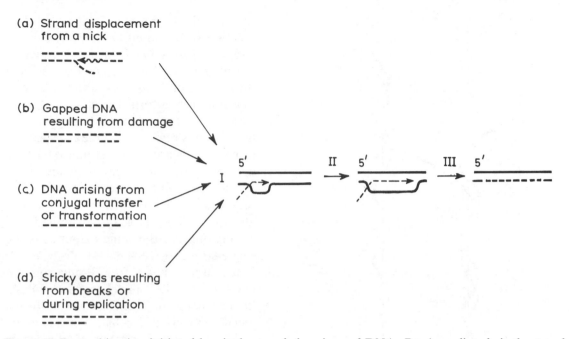

(a) Strand displacement from a nick

(b) Gapped DNA resulting from damage

(c) DNA arising from conjugal transfer or transformation

(d) Sticky ends resulting from breaks or during replication

Fig. 7.15 Recombination initiated by single-stranded regions of DNA. RecA-mediated single-strand invasion leads to formation of D-loops (I); branch migration occurs over the region of homology (II); nuclease and ligase action generate the recombinant molecule (III). This series of reactions takes place within a sheath of RecA protein filaments.

nuclease is similar to the Red exonuclease of phage lambda which acts preferentially on 5′-phosphate ends of duplex DNA to produce molecules with single-stranded 3′-tails. Annealing of complementary single-stranded regions from two different molecules, followed by exonuclease action leads to formation of a hybrid molecule by a non-reciprocal method, i.e. only one complete DNA molecule results. If the tails are not removed by exonuclease [136] but endo-nuclease and polymerase intervene (as in bacteriophage T7 replication, section 6.11.1) a reciprocal recombination can occur [137].

In *recBC* mutants which have an active *recF* gene and an inactive exonuclease I, intermediates with 3′-tails are again formed. In all three pathways the single-stranded tail is thought to be involved in the RecA-mediated strand invasion [126].

The 38 kDa RecA protein, among other functions, is able to catalyse the annealing of single-stranded DNA to a homologous duplex at the expense of ATP hydrolysis. The single strands may arise as described above or by a variety of other mechanisms. Thus RecA-mediated recombination is in-volved in the integration of the single-stranded DNA arising from conjugal transfer (section 3.6) or bacterial transformation (section 3.5.3). Alternatively single strands may be produced during replication of damaged DNA or by a strand-displacement mechanism acting from a nick in duplex DNA (Fig. 7.15). Double-stranded breaks may produce DNA molecules with over-lapping (sticky) ends [124, 125] and sticky ends may be important as sites of secondary initiation of replication of phage T4 DNA (section 6.11.2). High rates of recombin-

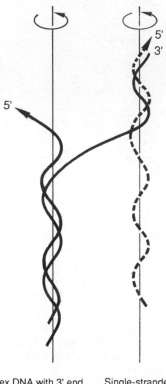

Duplex DNA with 3' end
complementary to a
region of SS-DNA

Single-stranded DNA
associated with RecA
protein

Fig. 7.16 Single-strand invasion takes place by rotation of both the duplex DNA and the RecA-coated single strand. From [143] with permission of the author and publishers.

ation are found in mutants defective in ligase, DNA polymerase I or dUTPase, situations which lead to an increased frequency of nicks or gaps in DNA.

RecA promotes the invasion, by a single-stranded DNA molecule, of a homologous DNA duplex, resulting in the displacement of one of the duplex strands and its replacement with the invading single strand. The reaction takes place in three stages [142–145].

1. RecA protein polymerizes onto single-stranded DNA in the presence of ATP, starting at or towards the 5'-end, to form

the, so-called, presynaptic filament. This reaction is promoted by SSB. The single strand of DNA lies in a deep groove in the filament and is stretched out so that the spacing of the bases is about 1.5 times that found in B-form DNA.

2. The presynaptic filament aligns with homologous regions in duplex DNA. The three strands are held together over long distances and alignment may involve a paranemic joint. Free ends are not required in the duplex DNA for this alignment. The duplex is extended and underwound to the same extent as is the single-stranded DNA.

3. If the duplex has a free 3'-end complementary to the single-stranded DNA, then exchange of strands occurs to form a plectonemic joint. Exchange occurs in the $5' \rightarrow 3'$ direction relative to the single strand and requires ATP hydrolysis. It is probable that both the RecA nucleo-protein filament and the duplex DNA rotate about their longitudinal axes to affect transfer (Fig. 7.16) [146]. The product is a heteroduplex in which one strand of the duplex was the original single-stranded DNA molecule. A similar activity is found in eukaryotic cells [247].

One of the ways in which these reactions have been studied *in vitro* is using a cyclic, single-stranded molecule and linear duplex DNA and the reaction occurs more readily if one end of the duplex is complementary to the single strand as topological problems are thereby avoided. Annealing is initiated with the 3'-end of the complementary strand, i.e. the single strand invades with $5' \rightarrow 3'$ polarity [147–149]. Strand transfer is blocked when a region of non-homology arises and three mismatches in a row are sufficient [150]. It can, however, restart even after an insertion of 1000 bases in the single-stranded DNA.

Another model system is the interaction of linear, single-stranded DNA with homologous, supercoiled, cyclic duplex DNA molecules. In this reaction a D-loop forms with relaxation of the supercoiled molecule, but the lack of a free end in the duplex blocks any net transfer. The D-loop travels around the duplex from the 5'-end of the single-stranded DNA until the 3'-end is reached, when the original single strand is again released. RecA protein remains associated with the duplex which is underwound and this prevents the reinitiation of a second cycle [145].

RecA protein is also able to catalyse reciprocal exchange of DNA strands between two duplex molecules [119, 151] but only if one of them contains a single-stranded gap to allow nucleation [152, 153].

7.4.3 Illegitimate recombination

Certain recombination events appeared to occur in regions where no homology was apparent. However, on further analysis, it became clear that, in many instances, short homologous regions were present and it was between these regions that the recombination events were taking place. In other cases, recombination occurs as a consequence of breakage of DNA molecules (e.g. by restriction enzyme or topoisomerase action) or as a result of replication errors and various examples are considered by Berg [250]. When a double-stranded break occurs somewhere between repeats, resection of the intervening DNA can allow annealing of the complementary strands of the adjacent repeats. Ligation then results in the joining of the two fragments at a common sequence [248]. Alternatively, a copy-choice mechanism involves errors in replication in which the polymerase jumps from one repeat to the other. The repeats may even be present on

different chromosomes, and the resulting errors are similar to those involved in the generation of frameshift mutations [250].

7.4.4 Site-specific recombination

Whereas general recombination can occur between any homologous sequences, the enzymes which catalyse site-specific recombination act only at specific, short sequences which must be present in both parental duplex DNA molecules. The mechanism of site-specific recombination has been studied primarily in the lambdoid phages but it also occurs in other systems, both prokaryotic and eukaryotic. It is required for many of the gene rearrangements considered in section 7.5.

7.5 GENOME REARRANGEMENTS

This section considers a number of selected examples where recombination mechanisms are involved in regulated changes in the genome. In few cases are the molecular mechanisms clear, although most involve site-specific recombination.

7.5.1 Integration of phage lambda DNA

In the lysogenic state phage lambda (λ) is integrated at a specific site in the bacterial chromosome (section 3.7.8). Integration does not require the *rec* or *red* systems of general recombination but does require specific sequences on both the phage and bacterial DNA (the *att* sites) as well as phage-coded proteins. Phage ϕ80 integrates into the *E. coli* chromosome at the site *att80* (near *lac*) and phage λ at the site *attλ* (between *gal* and *bio*). The site on the phage DNA and the site on the bacterial DNA are

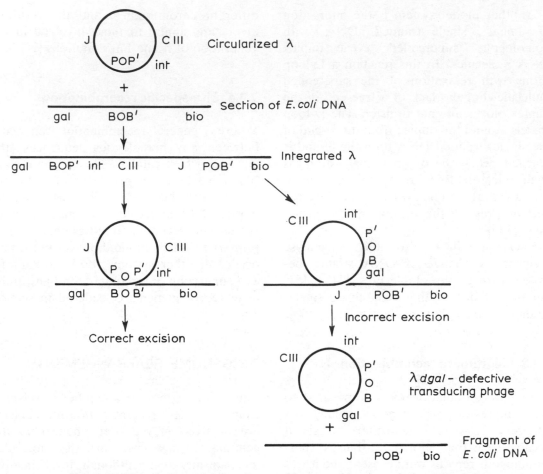

Fig. 7.17 Campbell model for site-specific recombination of phage lambda.

known as POP' (*attP*) and BOB' (*attB*), respectively. Both are made of three parts. The O (or core) region is a sequence of 15 bases which is identical in the two regions. During recombination staggered cuts are made in the two core regions to give a seven-base 3'-extension. These then cross-hybridize so that recombination results in formation of BOP' (*attL*) and POB' (*attR*) (140, 154–156, 250) (Figs 7.17 and 7.18).

Formation of this structure requires the phage-coded Int protein and the host-coded IHF (integration host factor). The Int protein is a 42 000-M_r topoisomerase I which

binds to seven sites in POP'; two in the 152 bp P region; a pair in O; and three adjacent sites in the P' region [157–160]. It also binds to a pair of sites in the core region of *attB*. It is the *N*-terminal domain of Int which binds sites in P and P' but the *C*-terminal domain binds the inverted repeats in the core. Int thus has the ability to bridge two regions in *attP*. *In vitro* Int cleaves the DNA at the positions indicated in Fig. 7.18 and becomes covalently bound to the 3'-phosphate of the DNA at the site of breakage.

IHF is a DNA-binding protein which is

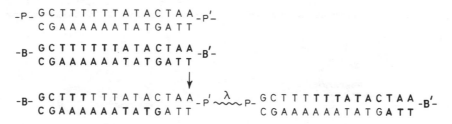

Fig. 7.18 Details of the cleavage and rejoining of the core regions which occur on the site-specific recombination of phage lambda.

required to form the Holliday-type intermediates but not for their resolution. IHF is a 20-kDa $\alpha\beta$ dimer that binds to three regions in *attP* and causes the bending of the DNA in the complex nucleoprotein array. This bending is essential for the Int protein to act at two sites simultaneously and requires that the phage DNA be supercoiled for integration to occur [161–163]. This complex structure (known as an intasome) is presumably required to ensure unidirectional recombination (i.e. integration). Most other examples of site-specific recombination involve multiple rounds and are simpler in their protein requirements [160].

Excision of phage DNA requires a second phage-coded protein (Xis) which serves as a directional switch by inhibiting the integration and allowing only the excisional recombination event [164]. Incorrect excision leads to defective bacteriophage carrying genetic markers adjacent to *att*, i.e. for bacteriophage λ either *gal* or *bio*. Such defective transducing bacteriophages can integrate into rec^+ cells at the site corresponding to their markers (e.g. λd*gal* can integrate at a *gal* site) and have proved very useful to bacterial geneticists.

7.5.2 Nitrogen-fixing genes

Excision of a lysogenic phage, where a site-specific recombination joins together two previously distant regions of the genome, leads to the loss of the intervening (phage) DNA. A similar loss of DNA occurs during the development of the cyanobacterium *Anabaena* from a vegetative cell to a heterocyst when there is a rearrangement of the genes required for nitrogen fixation (*nif*). There are two nitrogen fixation operons, known as *nifH* (containing the genes *nifH*, *nifD* and *nifK*) and *nifB* (containing the genes *nifB*, *fdxN*, *nifS* and *nifU*) together with the genes *nifQ* and *nifV*. In the vegetative cell the *nifK* gene is separated by 11 kb from the *nifD* and *nifH* genes, but during heterocyst formation the intervening DNA is excised to form a circular molecule which remains in the heterocyst. Excision occurs as a result of site-specific recombination between two 11 bp directly repeated sequences and requires the products of the *xisA* gene [171]. The rearrangement allows the three genes to function as a single transcriptional unit [172].

A second excision event, occurring independently as a result of recombination between two 5 bp directly repeated sequences, deletes a 55 kb element from the *fdxN* gene, thereby juxtaposing the *nifB*, *fdxN*, *nifS* and *nifU* genes. These two excisions lead to the two nitrogen fixation operons being adjacent to each other in the heterocyst [173].

Heterocysts develop at regular intervals along a filament of vegetative cells and, as with the antibody-producing cells (section

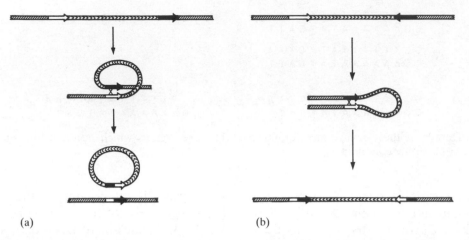

(a)   (b)

Fig. 7.19 Recombination between signal sequences arranged in tandem on the chromosome can lead to deletion (a) or inversion (b) depending on the relative orientation of the signal sequences.

7.5.7), such differentiation is irreversible. Here we have an example of a colony of bacteria in which some cells sacrifice their potential for multiplication in order to supply fixed nitrogen compounds to the remaining cells.

Because recombination results in loss of the intervening DNA, such rearrangements differ from the situation where the *inversion* of an intervening stretch of DNA can bring together two previously separated regions of DNA (e.g. flagellar phase variation in *Salmonella* [174] and *E. coli* [175], and genome isomerization in herpes simplex virus [176]) which is a potentially reversible reaction in which no DNA is lost. The molecular mechanisms of segment inversion and elimination events are not, however, fundamentally different. Rather they depend only on the relative orientation of the two sequences undergoing recombination [177, 178] (Fig. 7.19).

7.5.3 Pilus phase variation

This, and the next two examples from eukaryotes, involve the transfer of a gene from a site in which it is inactive into an expression-linked site [250].

In *Neisseria gonorrhoeae* (causes gonorrhoea) and *N. meningitidis* (causes menigococcal meningitis) the presence and nature of the pilus (required in bacterial conjugation – section 3.6) is variable. Thus, some cells lack a pilus (i.e. fail to produce an effective monomer – pilin – molecule) and others have pili containing one of a series of different pilin molecules. The active pilin gene is expressed from the expression-linked site *pilE*, whereas other genes are available at *pilS* (silent) or on the genomes of other *Neisseria* present in the population and available as a result of efficient transformation (section 3.5.3). Exchange of pilin genes at the *pilE* site is *recA* dependent and involves recombination between the incumbent *pilE* gene and one of the other available *pil* genes. The loss of an effective pilin (P^+ to P^-) occurs at a frequency of about 10^{-3}, whereas reversion to P^+ which requires production of effective pilin (i.e. one that forms an effective pilus rather than a soluble protein) is less than one tenth as frequent [165–168].

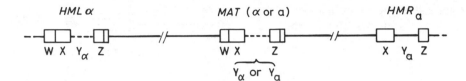

Fig. 7.20 Yeast mating type genes.

It should be noted that this is not the only mechanism in which changes in DNA can lead to the loss of gene expression in these and related bacteria. A number of situations exist in bacteria where insertions or deletions of one or more nucleotides can occur in the DNA by 'slipped-strand' mispairing in a region containing multiple, short tandem repeats (e.g. $(CTCTT)_n$) or runs of a single base (e.g. C_n) [165–170].

7.5.4 Yeast mating-type locus

In *Saccharomyces* there are two mating types: a and α. Fusion occurs only between an a and an α haploid cell and leads to formation of a diploid cell (a/α) which can sporulate to form haploid spores. However, in a clone derived from a single haploid cell (a or α) some cells will switch their mating type [179]. (Fission yeast undergo similar switches in mating type [332].)

Mating type is conferred by a specific region (Ya or Yα) in the mating-type locus (MAT). Inactive copies of Ya and Yα also occur at other loci to the right (*HMRa*) and left (*HMLα*), respectively of *MAT* (Fig. 7.20). The inactive copies are repressed by a *cis*-acting silencer whose action is mediated by four *SIR* genes. However, one of the inactive genes may duplicate itself and the copy then moves from the influence of the silencer to replace the gene in the mating-type locus and this may bring about a change from an a-type to an α-type cell, or vice versa [180]. This is a gene conversion

event involving what is known as the cassette mechanism as different loci can be inserted into the 'play position'.

Ya (642 bp) is not homologous to Yα (747 bp) but in all three sites (*MAT*, *HMRa* and *HMLα*) the Y sequence is flanked by homologous regions within which recombination is thought to occur, e.g. XYaZ or XYαZ (Fig. 7.20). The mating-type switch is initiated when a double-stranded cleavage occurs in a loop region to the 3'-side of the gene in the Z region at the *MAT* locus. Although similar sequences are also present at the YZ junction in the *HMR* and *HML* loci, they are not cleaved, perhaps as a result of their being in a different chromatin conformation. This cleavage is brought about by an endonuclease (the product of the *HO* gene) that generates the required site-specific, double-stranded break with four-base 3'-extensions [182]. Exonuclease action leads to degradation of the Y-region at *MAT*, and the DNA is repaired by taking information from one of the inactive copies. This probably involves an invasion of the inactive copy by the free DNA ends followed by extension of the 3'-ends using the complementary strands of the inactive gene as template. In 90% of cases this leads to a switch in mating type.

The frequency of mating-type switching is, therefore, dependent on the level of the HO endonuclease. Yeast strains lacking an active *HO* gene, switch mating types at a frequency of only 10^{-6} whereas HO^+ cells may switch every cell division. Control of transcription of the *HO* gene is complex and differs with cell type, cell age and position in the cell cycle [181].

7.5.5 Variant surface glycoprotein (VSG) genes in trypanosomes [251]

Trypanosomes are protozoan parasites responsible for sleeping-sickness and nagana [183–186]. The immune response engendered by infection is directed against the surface glycoprotein of the protozoan. However, the parasite is able to change the surface glycoprotein and so avoid the host's defence mechanisms. There are about a thousand VSG genes in each trypanosome making up about 10% of the genome, but only one – the expression-linked copy (ELC) – is active at any one time. Most of the genes (the basic copy or BC genes) are arranged in tandem, separated by several copies of incompletely homologous 70 bp repeats and all the genes have a common sequence of 6–8 bp at their 3'-ends. In addition to the normal chromosomes, trypanosomes have about 100 minichromosomes each one of which has a VSG gene towards each end. These subtelomeric VSG genes have copies of the 70 bp repeats in their upstream region and towards the end of the chromosomes they have tandem repeats of the telomeric sequence 5'-CCCTAA–3' (section 6.16.4). Up to 20 of these subtelomeric copies have the potential to be active (i.e. have a functional promoter) but only one is active at a time. Which of the expression-linked copies is transcribed appears to be determined at random but the outcome is that parasites with a series of different surface glycoproteins are present in the bloodstream early in infection.

Expression-linked copies are always subtelomeric and differ from the other genes by the presence of a 5'-exon of 35 bp. This

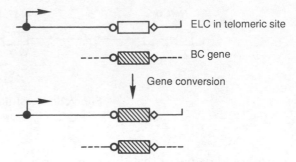

Fig. 7.21 Gene conversion is involved in the replacement of a subtelomeric ELC (expression-linked copy) of the trypanosome variant surface glycoprotein gene by a copy of a basic copy (BC) gene. Transcription starts from the miniexon (●).

miniexon, which is absent from all non-functional copies of the gene, is present in 200 or more copies as part of a tandem repeat which may form part of a complex promoter (section 10.4) [186]. As most VSG genes are not present in expression-linked sites, their expression requires the transposition of an inactive VSG gene to the expression-linked site to replace the resident gene. This takes place most readily between subtelomeric genes and, although reciprocal recombination events are known, transfer more commonly requires the basic copy gene to be duplicated and for one copy to replace the ELC (i.e. gene conversion, Fig. 7.21).

Recombination involves the 70 bp repeats on the 5'-side of the gene and the common 6–8 bp homologous region at the 3'-ends of the genes [188, 189]. The 5'-miniexons are not lost from the ELC. Transposition of a subtelomeric gene into the EL site may involve recombination only at the 5'-end of the gene so that the whole of the end of the chromosome is replaced. Alternatively, activation of a subtelomeric VSG may occur by a non-duplicative event involving reciprocal recombination between the ends of two chromosomes – chromosome end exchange [186, 190]. Occasionally recombination

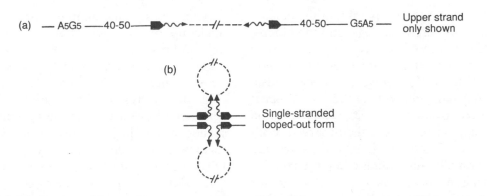

Fig. 7.22 Elimination of internal sequences on formation of the macronucleus in ciliates may involve the formation of single-stranded loops of AT-rich DNA.

occurs at sites within the gene to generate chimeric VSGs.

Recombination involving the ends of chromosomes is also common on the malarial parasite where, again, genes for common antigens are found in subtelomeric regions [191]. Antigenic variation in the vector-borne relapsing fever (caused by *Borrelia hermsii*) may involve similar DNA rearrangements [192].

7.5.6 Macronucleus formation

Ciliated protozoa have two sorts of nuclei. The micronucleus contains a full complement of DNA and is retained in the germ line. The macronucleus contains multiple (about 1000) copies of a small fraction (5–70%) of the genes present in the micronucleus. One macronucleus may contain 20 000 distinct DNA molecules each of average size, 2 kb. Both nuclei replicate in vegetative cells but only the macronucleus is transcriptionally active.

Following conjugation the macronucleus is destroyed and a new one generated from a mitotic copy of the micronucleus. This generation phase starts with a number of rounds of replication generating polytene chromosomes which are then broken down to gene-size pieces (0.4–15 kb long) to the ends of which telomeres are attached [193] (section 6.16.4). This processing involves many, site-specific, deletion–ligation events throughout the genome, some of which are essential to remove sequences from within gene-coding regions. The internally eliminated sequences (IESs) which are released as free circular DNA molecules, can be unique but also include a number of highly repetitive elements [196–199].

The size of IESs range from one of 14 bp in *Oxytricha*, to some of several thousand bp in *Tetrahymena*. A common feature of IESs is a pair of direct repeats of 2–6 bp at the end of the deleted segment, The sequence of the repeats differs between different IESs and following elimination of the IES, one copy of the repeat is retained. In addition, just inside the direct repeats of some, but not all, IESs is a pair of inverted repeats (Fig. 7.22). A polypurine tract (A_5G_5) situated 40–50 bp distal to each terminal repeat may also be essential for elimination.

Several features of the *in vitro* reaction appear inconsistent with a mechanism involving general recombination, not least being the extremely small size of some of the deleted regions. The AT-rich nature of

IESs has led to the proposal that the deleted regions are present as single-stranded loops, held together by the inverted repeat, thereby bringing together the direct repeats (section 7.5.7) (Fig. 7.22). However, a mechanism involving staggered cuts in the direct repeat region, followed by intramolecular ligation would appear to achieve the observed result (section 7.5.2 and Fig. 7.19) and a staggered cut in homologous strands of a paired duplex has been proposed as the mechanism for generating a dimer of the ribosomal DNA in the macronucleus of *Tetrahymena* [195] (section 8.4.3).

7.5.7 Immunoglobulin and T-cell receptor genes

B lymphocytes have the ability to recognize a wide range of foreign antigens in solution and bind them by means of a cell surface antibody (immunoglobulin) molecule. Later, these lymphocytes secrete soluble immunoglobulins. In contrast T lymphocytes interact with cellular antigens and this is achieved by the T-cell receptor only recognizing the antigen when it is presented along with a major histocompatibility (MHC) protein [202, 203]. For the interaction of the antigen with the antibody or T-cell receptor to be specific requires millions of different antibodies and T-cell receptors to react with the millions of different antigens which may possibly be encountered in the life of an animal. The ability of animals to generate a multiplicity of antibodies to novel antigenic stimuli posed a problem to scientists for many years. Dreyer and Bennett [204] proposed in 1965 that a single polypeptide might be synthesized using combined information present in two genes and since then numerous experiments have elucidated the details showing that this theory for the generation of antibody diversity is broadly correct [205–207].

Antibodies (or immunoglobulins) are made up of four polypeptide chains: two light and two heavy chains. Each chain has a region (the constant region) which is of similar sequence for all antibodies of a given class and a region (the variable region) which can have a wide range of amino acid sequences [202]. The constant and variable regions are encoded separately in germ-line DNA but the gene segments are brought together in antibody-producing cells by a somatic recombination event [203, 205–207].

(a) *Light chains*

In the mouse and man (two species extensively investigated) there are two classes of light chains (κ – the more commonly expressed – and λ) on different chromosomes, and each with a characteristic constant region gene (Cκ and Cλ). A short distance upstream from the Cκ region is a series of joining (Jκ) sequences each of about 30 nucleotides in length. Much further upstream are a large number of Vκ sequences (Fig. 7.23(a)). The Cλ complex consists of several, closely related genes (Cλ1, Cλ2 etc.) each associated with a single Jλ sequence and a small number of Vλ genes [207]. The somatic cell gene rearrangement involves the joining of a particular V region to one of the J regions to form a VJ region with elimination of the intervening DNA. In different cells the choice of V and J regions probably occurs at random to produce a battery of lymphocytes each with a different VJ combination. Transcription leads to the synthesis of a premessenger RNA molecule which is processed to eliminate the intervening sequences within the V region and between the selected J region and the C region and thus produce the mRNA for the immunoglobulin light chain (section 9.3 and Fig. 7.23(a)).

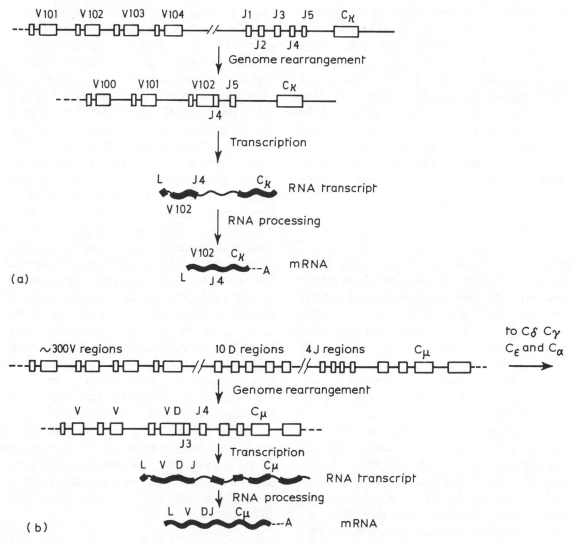

Fig. 7.23 Structure, rearrangement and expression of immunoglobulin genes; for details see text. The letters refer to the leader (L), variable (V), joining (J), diversity (D) and constant (C) regions of the kappa light chain genes (a) or the heavy chain genes (b).

(b) *Heavy chains*

The heavy-chain genes are also made up of variable (V), joining (J) and constant (C) regions but the diversity of heavy chain sequences is further increased by the presence of another group of gene segments (the diversity or D regions) which lie be-

tween the V and J regions (Fig. 7.23(b)). The first step in the assembly of an active heavy chain gene is the joining of a D and J_H region and, early in development, the 3'-D segment is used preferentially. (Later, secondary rearrangements involve more 5'-D segments.) The second step involves the joining of a V_H region to the rearranged

DJ$_H$ segment. There are several different C regions (e.g. Cμ, Cγ, Cα) expressed in turn during the life of a single lymphocyte. Additionally each C region is made up of several exons representing the different domains of the resulting heavy chains [208, 209]. Initially the Cμ gene is active to produce IgM antibodies in immature lymphocytes. Later IgG (from Cγ) and IgA (from Cα) are produced by subsequent DNA rearrangements. This is known as class switching (section 7.5.7(f)).

(c) *Allelic exclusion*

There are two alleles of the heavy chain genes (and of the various light chain genes) but only one functional rearranged gene is produced [202, 206, 207]. The other genes may not have rearranged or may have undergone a non-productive rearrangement to produce an aberrant gene. This process is called allelic exclusion and implies a feedback regulation such that only when a functional rearrangement has occurred are further rearrangements inhibited [206, 207, 210]. It is the presence of a membrane bound IgM protein which appears to be instrumental in inhibiting a second joining of V$_H$ to DJ$_H$ and of V$_\kappa$ to J$_\kappa$. As the κ genes rearrange before the λ genes, these latter are only expressed if the initial rearrangements fail to lead to the production of a functional IgM. The ordered rearrangement of the various genes during development, which forms the basis for allelic exclusion, is believed to depend on their accessibility to the recombinase (see below) [206].

(d) *T-cell receptor genes*

T-cell receptors are dimeric molecules containing an α and a β chain both of which are divided into a variable (V) and a constant (C) region. The gene for the β chain resembles the gene for the immunoglobulin H chain in that it is made up of Vβ, Dβ (diversity), Jβ (joining) and Cβ segments that rearrange early during T-cell differentiation. There are two adjacent Cβ genes, each associated with seven Jβ segments and one Dβ segment. In the mouse there are eleven Vβ genes but some of these are closely related to each other (subfamilies) [211, 212]. The α chain genes, which rearrange after the β chain genes, consist of V, J and C regions only. There are at least ten subfamilies of Vα genes each containing from one to ten members. There is only one Cα gene associated with 19 Jα segments [213–216]. Rearrangement of the T-cell receptor genes is required for expression and takes place in a manner strictly analogous to the rearrangement of the immunoglobulin genes and there is the same scope for generating diversity [209]. Some (possibly cytotoxic) T cells contain, in place of the $\alpha\beta$ receptor, a $\gamma\delta$ receptor. The human γ gene contains two C regions, five J regions and 14 V regions, although six of these are pseudogenes. In this case diversity seems to be restricted to the V–J join. The δ genes are found upstream of the J segments in the α locus. There is only limited expression of γ chain genes in the early stages of development of helper T cells [209, 217, 237].

(e) *The joining reactions*

The details of the joining reactions are not yet known but each of the rearrangements leading to a functional immunoglobulin gene probably occurs by the same mechanism of site-specific recombination [205, 218]. The enzyme activity involved (the V(D)J recombinase) is present only in pre–B and T cells and acts on the appropriate sequences, when they are accessible, independently of their position in the chromosome. The reactions have been studied by use of recombinant

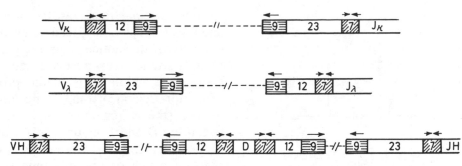

Fig. 7.24 Details of the arrangement of the heptamer and nanomer sequences flanking the V, D and J regions of immunoglobulin genes.

DNA molecules containing, for instance, V and J regions flanking a reporter gene introduced into transformed pre–B or T cells [206, 207, 219, 246]. It has been possible to express recombinase activity in fibroblasts and the putative gene (RAG–1) encoding the enzyme has recently been isolated [220].

In the germ line, each segment is bounded by a palindromic heptameric oligonucleotide of consensus sequence 5'-CACAGTG–3' followed by a spacer of 12 bp or 23 bp and a nanomeric oligonucleotide of consensus sequence ACAAAAACC. The nonamer always has the orientation in which the terminal CC is directed away from the gene segment (Fig. 7.24). There are several possible ways to bring together the two regions which are to undergo recombination. Intrastrand recombination could occur by looping out the intervening sequences and aligning the short consensus sequences (Fig. 7.25). This could involve deletion of the intervening sequences or the inversion of a region of DNA depending on the relative orientation of the genes, and both have been observed [178, 221, 252]. The two regions which undergo the recombination always consist of one with a 12 bp spacer and the other with a 23 bp spacer (one and two turns of the double helix, respectively). This is known as the 12/23 rule. Vκ regions all end with a 12 bp spacer and Vλ regions with a 23 bp spacer whereas the corresponding J regions begin with a 23 bp spacer (κ) or a 12 bp spacer (λ). This thus ensures that a V region always joins to a J region and not, for instance, to another V region. Similar rules apply to the rearrangements of the heavy-chain genes and the T-cell receptor genes [205].

In order to bring about the observed recombination, Alt and Baltimore [221] point out that the reaction must be initiated by breaks involving all four strands when they are aligned at the signal sequences (Fig. 7.25(a)). The breaks occur at the boundaries of the heptamers and then the two heptamers are precisely ligated, back to back (Fig. 7.25(b)). This leads to deletion or inversion of the intervening DNA. In the second step the coding elements are joined together; but this is an imprecise event resulting from the loss of nucleotides from the two ends by exonuclease action. This leads to a series of different junctions for every combination of V and J segments and hence a further increase in sequence diversity. If the recombination event leads to a change in reading frame (section 12.2) then J regions with entirely new amino sequences could be generated but when this occurs the joining is non-productive. In the D/J joining reaction of

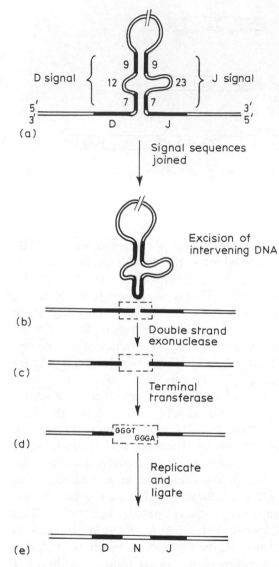

(a)

Signal sequences
joined

(b)

Excision of
intervening DNA

Double strand
exonuclease

(c)

Terminal
transferase

(d)

Replicate
and
ligate

(e)   D   N   J

Fig. 7.25 A model for the D–J_H recombination involved in formation of heavy-chain genes, illustrating the involvement of terminal transferase in the generation of the hypervariable N (for nucleotides) region (based on [221] and reproduced with permission of the authors and publishers).

heavy chain genes, in addition to the deletion of nucleotides at the joining site, nucleotides are inserted in an apparently random fashion. This insertion which may be cata-

lysed by terminal transferase (section 6.4.7(v) and Fig. 7.25) obviously leads to even greater diversity of nucleotide sequence [221, 246]. Double-stranded exonuclease and terminal transferase action leads to the production of a short region of random nucleotide sequence (the N region in Fig. 7.25(e)) between the D and J regions [221]. Formation of the N region is restricted to D/J joining, a reaction which occurs at a time when terminal transferase is active in immature lymphocytes, before joining of the V regions or the rearrangement of the light chain genes (section 7.5.7(c)).

(f) *The mechanism of class switching*

Because the $C\mu$ gene is closest to the J_H segments it is this constant region which is initially expressed following the rearrangements of the variable region discussed above (Fig. 7.23(b)). When class switching occurs the $C\mu$ gene region is replaced by another C_H region but no change occurs in the variable region at this time, i.e. a lymphocyte only expresses one variable region but during its lifetime the constant region changes. Switching is brought about by recombination reactions which lead to the deletion of stretches of DNA. Recombination occurs at or near S (switch) sequences which comprise multiple copies of partly conserved pentameric sequences (e.g. TGAGC, GAGCT and TGGGG) which occur in the 5'-flanking region of each C_H gene except the $C\delta$ gene [224–226]. These sequences resemble the Chi sequences which are recombination hotspots in *E. coli* (section 7.4.2).

Recombination between S sequences occurs on the same chromosome by a looping out mechanism which normally leads to the deletion of the intervening DNA [227–229]. (Some abortive switches have been shown to result from inversion [228].) The rearranged switch regions contain multiple point mutations, small deletions and duplications of the switch signals. To account for these it has been proposed that switch recombination involves a copy-choice mechanism in which DNA synthesis initiated in Sμ transfers to the aligned Sα region [233, 234] (but see also section 7.5.7(e)).

The change to expression of the Cδ gene occurs not by DNA rearrangement (there is no S sequence) but by readthrough transcription and differential processing of the transcript (section 11.3) at a stage in development before class switching occurs [230, 231]. Such differential splicing may also lead to the transient coexpression of Cμ and Cγ genes (and even Cϵ or Cα genes) at a stage before DNA rearrangement [226, 232].

(g) *Somatic mutation*

A comparison with germ-line sequences shows the presence of multiple single-base mutations in the V, J and D segments of both light- and heavy-chain genes [205, 235]. It has been proposed that these mutations arise in combination with the joining and switching rearrangements discussed above. The mechanism envisaged involves error-prone repair of gaps generated by nicking of the DNA followed by exonuclease action (section 7.3.4). When such a mutation generates an antibody with increased affinity for the antigen it will be selected during the clonal expansion phase (see below).

(h) *Clonal selection*

It is the variable region of the antibody molecule which represents the antigen-combining site. A particular antigen will interact with only a small number of lymphocytes each carrying on their surface an IgM with affinity for the antigen. When an antigen interacts with a particular lymphocyte that lymphocyte is induced to divide. This leads to the production of a clone of lymphocytes each having the particular DNA rearrangement coding for the active immunoglobulin (which may now be secreted as an IgG). This clonal selection was initially postulated to occur by MacFarlane Burnet [236] to explain how the immune system works, not by inducing change, but by selecting for and amplifying those lymphocytes carrying the most appropriate antibody. It has taken 30 years to elucidate the mechanisms used to generate the diversity of antibodies and B-lymphocytes on which clonal selection may act and in so doing a completely new insight has been obtained into the plasticity of DNA. The situation with T-lymphocytes is more complex and involves both positive and negative selection mechanisms, operating in the thymus, to generate T cells with useful receptors and to eliminate those which might damage the host [238].

7.5.8 Chromosomal translocations

A chromosomal translocation is the reciprocal exchange of material between two chromosomes and, although translocations have been observed for many years, they have recently attracted greater interest as a result of their prevalence in certain cancers [337]. Thus in chronic myeloid leukaemia (CML) there has been (in at least 90% of cases) an exchange between the end of the long arms of human chromosomes 9 and 22 such that chromosome 22 is shortened to form the characteristic Philadelphia chromosome. Other translocations associated with cancers are between chromosomes 8 and 21 (acute myeloid leukaemia: AML); 8 and 2, 8 and 14 or 8 and 22 (Burkitt's lymphoma); 15 and 17 (acute non-lymphocytic leukaemia); 11 and 14 (acute lymphoblastic and chronic lymphocytic leukaemia); and 18 and 14 (follicular lymphoma) (Fig. 7.26). These dis-

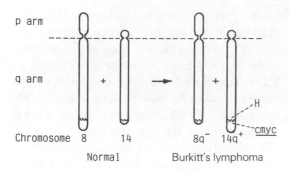

Fig. 7.26 A chromosomal translocation involved in the initiation of Burkitt's lymphoma brings the c-*myc* oncogene from chromosome 8 under the control of the immunoglobulin heavy chain promoter on chromosome 14.

eases are all leukaemias, i.e. occur in the B or T lymphocytes, and many of the translocations occur within or close to the immunoglobulin genes on chromosomes 14 (heavy-chain genes), 22 (lambda light-chain genes) and 2 (kappa light-chain genes). Others occur near to the T-cell receptor genes (e.g. the α chain gene on chromosome 14) [239–241]. It would therefore appear that these translocations occur as aberrations of events normally involved in the rearrangement of the immunoglobulin genes. The aberrant rearrangements bring together an oncogene, e.g. c-*myc* on chromosome 8 (section 3.7.9), and part of the antibody gene leading to the production of an altered oncogene product or enhanced oncogene transcription [242, 243]. In patients with acute promyelocytic leukaemia a rearrangement has disrupted the retinoic acid receptor gene (an antioncogene) [253].

Not all translocations involve aberrant immunoglobulin gene rearrangements. For instance, the small number of females who suffer from Duchenne muscular dystrophy have an X:autosome translocation [244], and large deletions covering the human steroid sulphatase locus on the X chromosome appear to arise by recombination between

directly repeated sequence elements [245]. The site of the initial pairing at meiosis between the X and Y chromosomes is a common sequence near the telomeres. This is also the site at which obligatory recombination occurs between one chromatid of the X and one chromatid of the Y chromosomes. The effect of this recombination is to make it appear that genes distal to the site of recombination are not sex-linked, i.e. they are pseudoautosomal [246–248].

7.6 GENE DUPLICATION AND PSEUDOGENES

7.6.1 Multiple related copies of eukaryotic genes

Hybridization experiments with eukaryotic genomic DNA have shown that for many of the genes encoding proteins there exist multiple copies of related DNA. These differ from multiple copies produced by gene amplification (section 6.15.6) in the fact that they are not transient and generally not present in such large numbers, although a notable exception to the latter characteristic is exhibited by the genes encoding the stable RNAs (section 8.4). The mechanism by which some of these related sequences have arisen involves a retroviral-like transposition, which results in the non-functional 'processed pseudogenes' described in section 7.7.4 below. In this section, however, we are concerned with related sequences that are thought to have arisen by the tandem duplication of an original sequence. Such gene duplication is important in evolutionary terms, as it provides the potential for one copy to diverge while the other still retains its original function. Nevertheless, in some cases the products of such duplication have been maintained as identical or near-identical copies (e.g. members of the histone gene

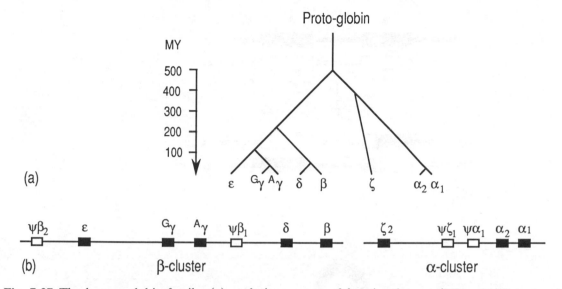

Fig. 7.27 The human globin family: (a) evolutionary tree of functional genes (MY = million years), after [258], with permission; (b) organization of functional genes (filled blocks) and pseudogenes (open blocks) in the human α- and β-globin clusters.

cluster – section 8.4.1), whereas in others they have diverged to produce different isoforms of the same protein (e.g. the tissue-specific actin and myosin isoforms [254]). In some instances much greater divergence occurs, and this may result in the relationship between the genes being undetectable by DNA hybridization. Such distant relationships have been revealed by computer-assisted comparison of nucleotide or protein sequences, an extreme example being the genes coding for epidermal growth factor and its receptor [255]. Divergence does not, however, always result in new functional genes, but can instead give rise to pseudogenes (section 7.6.3).

7.6.2 Mechanism of tandem gene duplication

A gene family that provides clear evidence of repeated duplication events is the human globin gene family, which is described in greater detail in section 8.4.2. Extensive nucleotide sequence analysis, together with a knowledge of the number and organization of globin genes in other species, indicates the evolutionary relationships shown in Fig. 7.27 [256, 257]. It is thought that an initial duplication gave rise to the prototypes of two families, the alpha and beta families. Although in many species these are now on separate chromosomes, the result, it is presumed, of a translocation event, their common origin is indicated by the fact that in amphibians they are still linked on the same chromosome [259]. The human α-globin family (on chromosome 16) contains α and ζ genes, and the β-globin family (on chromosome 11) contains β, γ, δ and ε genes (Fig. 7.27). There are two similar functional forms of the α- and γ-genes and, in addition, certain non-functional pseudogenes. These latter will be discussed in section 7.6.3, below.

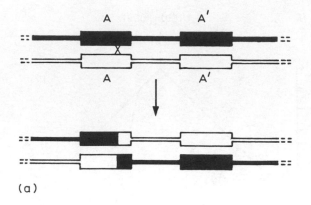

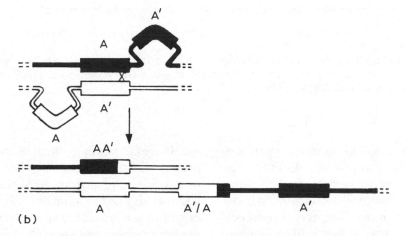

Fig. 7.28 Increase in gene number by homologous recombination involving unequal crossing over. The diagram shows a representation of two tandem copies, A and A′, of a gene (or region of DNA) on two recombining non-sister chromatids (distinguished by light and dark shading) at meiosis. Reciprocal crossing-over on copy A of properly aligned chromatids (a) does not affect gene number, whereas unequal crossing over between A and A′ on misaligned chromatids (b) generates different products containing one and three copies of the gene, respectively.

The best understood mechanism in which tandem copies of a gene can be generated is homologous reciprocal recombination at meiosis (section 7.4), but involving unequal crossing-over. This is most easily envisaged in a situation where two similar tandem copies of the gene already exist, in which case one of the recombinant chromosomes produced will have three copies of the gene (Fig. 7.28). Evidence that such unequal crossing-over

can occur among globin genes is provided by the analysis of certain haemoglobinopathies [260]. In haemoglobin anti-Lepore an extra gene has arisen between the δ and β genes and this has a β/δ fusion structure, as would be predicted for such a crossing-over mechanism (cf. Fig. 7.28 where the genes A and A′ would be formally equivalent to δ and β, respectively). The other predicted possible result of unequal crossing-over, contraction of gene number, is illustrated in haemoglobin Lepore, where the β and δ genes have been replaced by a δ/β fusion gene. Another, more ancient, example of such contraction is the fusion of the $\psi\beta$ and δ genes (Fig. 7.27) in lemur [261].

The model for gene duplication by unequal crossing-over presented in Fig. 7.28 cannot, as such, be applied to the duplication of a single gene (e.g. the original ancestral globin gene in Fig. 7.27). One possibility is that a single gene might undergo 'incidental' duplication when it becomes flanked by similar copies of repeated DNA, which then undergo recombination with unequal crossing-over. Among the candidates for repeated DNA that might arise by chance in such a flanking position are the highly repetitive eukaryotic LINEs and SINEs (sections 3.2.2(c) and 7.7.4). This is because their mode of duplication and dispersal is independent of recombination, being thought to involve retroviral-like reintegration of DNA copies of RNA transcripts (section 7.7.4). (Such a mechanism is excluded for the duplication of the protein-coding genes themselves as it leads to the formation of the processed pseudogenes described in section 7.7.4.)

The arrangement and copy number of the predicted products of homologous recombination — the three copies of the target DNA and the two copies of the unique DNA between the original double-copy target — should allow one to determine whether such events were responsible for a given gene duplication. Except in the case of relatively recent events, however, the accumulation of mutations in such non-functional spacer DNA, and the occurrence of gene correction or conversion events (section 7.6.4), generally obscure the picture. Nevertheless it is still of interest to examine the structures of the more recently duplicated globin genes. In the case of the human $^{G}\gamma/^{A}\gamma$, δ/β, $\psi\zeta_1/\zeta_2$

and α_1/α_2 pairs (Fig. 7.27), duplicated regions of c. 5, 7, 12 and 4 kb DNA are still evident [257, 262–264]. These compare with c. 2 kb for the length of the globin gene, including introns and the 5'- and 3'-untranslated regions. In the case of the $^{G}\gamma/^{A}\gamma$ pair, Smithies and coworkers found a triply repeated sequence of 110–170 bp that could theoretically have flanked a single proto-γ gene to produce the duplication [263]. Nevertheless these workers were not prepared to conclude that this repeated sequence was the cause of the γ-globin duplication, as they were able to construct alternative models of gene duplication via non-homologous recombination that could generate such triply repeated DNA sequences from an original single sequence [263]. In the case of other duplicated genes, such as the human α_1/α_2 globin pair [265] or the human haptoglobin pair [266], no such triply repeated sequences are apparent.

Although this section is concerned with duplication events involving several kilobases of DNA, it should be mentioned that the duplication of much smaller stretches of DNA can occur giving rise to internal repeated units within genes, and to simple tandemly repeated DNA such as satellite DNA [267] (Fig. 3.4).

7.6.3 Pseudogenes

Whereas certain mutations might lead to new and useful functions in a duplicated gene, it is clear that others could be so deleterious as to render the gene functionally useless. Such genes in the latter category are termed *pseudogenes* [268], and there are several examples of them in the globin gene family [269] where they are designated by the greek letter ψ (psi) preceding the symbol for their active counterpart (Fig. 7.27). For example, the human α-globin pseudogene, $\psi\alpha_1$, contains both a mutation in the initiation codon and frameshift deletions that render it incapable of producing a globin product [265].

The evolutionary history of globin genes can be deduced in two different ways, both of which provide evidence that contemporary pseudogenes have been active for part of their lives. One method involves comparison of mutations in the coding region of the gene at silent sites (i.e. those that would not produce a change in the encoded amino acid) and at replacement sites (those that would produce such a change). For an inactive gene the frequency of mutations at these sites should be equal, whereas in an active gene mutations at replacement sites are selected against. By these criteria $\psi\alpha_1$ is judged to have been active for part of its history [265]. The other method involves phylogenetic comparisons. For example, the goat equivalent of the human pseudogene $\psi\beta_1$ (now more generally referred to as $\psi\eta$) is still functional [261], implying that the human $\psi\beta_1$ is the descendant of a relatively ancient active gene. Conversely, although the δ globin gene is still active in man, it has become a pseudogene in the Old World monkeys [264]. It is interesting that the relatively recent silencing of this gene has allowed the mutation responsible to be identified as involving the region of DNA 5′ to the site for initiation of transcription (chapter 10).

Finally, it should be pointed out that there are also pseudogenes corresponding to genes that do not encode proteins: indeed those related to the 5S rRNA genes of X. *laevis* (Fig. 8.9) were the first pseudogenes to be discovered [270].

7.6.4 Concerted evolution of duplicated genes

After gene duplication has occurred it might be expected that the two copies would accumulate mutations independently and thus diverge. Although this has, of course, happened in many instances, in others (e.g. the repeated histone gene cluster and the repeated 5S rRNA genes of X. *laevis*) the different copies have been maintained in a remarkably similar state. In order to explain such concerted evolution it is necessary to postulate that a mechanism or mechanisms exist to correct divergences as they occur. Once again the intensely studied globin gene family provides an opportunity to explore this problem. Thus, there is clear evidence for the operation of such a correction mechanism in the case of the human α_1/α_2 [271] and $^G\gamma/^A\gamma$ [263] globin pairs. This evidence is the discrepancy between the time of the gene duplication based on phylogenetic comparison and that deduced from the apparent rate of accumulation of mutations in amino acid or nucleotide.

One proposal is that a series of unequal homologous crossing-over events of the type depicted in Fig. 7.28 can lead to homogenization of the structure of the members of a pair [262, 271]. What is envisaged is a series of successive expansions and contractions involving the interchange of genetic material. That such expansions and contractions are continually occurring is indicated by polymorphisms in α-globin number in human populations [271]. Furthermore, the human δ-globin gene has a structure that indicates it may be derived from the rescue of a silenced member of the β-globin pair by its active counterpart, and that this could have occurred through unequal crossing-over [264].

Although there is good evidence that gene correction by unequal crossing-over operates in the case of rRNA genes in *Drosophila* [272] and yeast [273], there is another mechanism that can operate on multigene families. This mechanism, gene conversion, involves the replacement of one of the genes by a copy of the other in a non-reciprocal process [274]. Gene conversion has already been considered in section 7.5.5 in relation to situations where it is employed to control specific gene expression. It has been suggested that this mechanism has been operative in the case of the human $^G\gamma/^A\gamma$ pair because of a greater similarity between the $^A\gamma$ and $^G\gamma$ genes on one chromosome than the $^A\gamma$ alleles on different chromosomes [275]. Although the mechanism of such conversion in this case is purely a matter of speculation, it was suggested that certain simple sequences might act as recognition signals [263]. Such sequences do not, however, appear to be a universal feature of genes undergoing correction.

Any mechanism for gene correction must also explain how genes can escape from such mechanisms to undergo divergence. In the case of a mechanism involving recombination, it has been proposed that the more rapid accumulation of extensive mutations (e.g. involving deletions and duplications) in the introns (section 8.2) might allow escape. The larger size of the introns in the β-globin family compared with those in the α-globin family has been suggested as the reason why members of the β-globin family appear to have become more diverged [271]. Another way in which the duplicated regions could diverge rapidly is if one member acquired an insertion of transposed mobile DNA. An example of this might be the *Alu* sequences (section 7.7.4) in the flanking regions of the human α-globin genes [276]. Finally, an extreme mechanism of severely decreasing (although not entirely eliminating) the probability of gene correction through homologous crossing-over is separation of the pair through chromosome translocation.

7.7 TRANSPOSITION OF DNA

7.7.1 Transposable elements

The dogma that the genome is stable arose from the observation of stable phenotypes in individual cells and organisms, and the success of genetic analysis based on the concept of fixed positions for genetic loci within the genome. It was the observation by Barbara McClintock of a non-stable phenotype in maize (reviewed in [277]) that led to the first proposal of the existence of the mobile genetic elements. She named the elements in maize 'controlling elements', but such mobile elements are now more generally termed 'transposable elements', or transposons. Transposable elements were later detected in bacteria by the mutations they produced preventing expression of certain genes [278, 279]. Because of the wealth of knowledge regarding the genome of *E. coli*, and the power of classical bacterial genetics, it was natural that the bacterial transposable elements should initially have come under most intense study. However, much is now known of eukaryotic transposable elements, especially those of *Drosophila* and yeast, where classical genetic analysis has complemented recombinant DNA techniques.

Transposable elements may be defined as segments of DNA that are bounded by specific terminal DNA sequences and that have the ability to move to new DNA sites with little or no specificity for the latter. This limited target-site specificity distinguishes transposition from site-specific recombination (e.g. of bacteriophage lambda − section 7.5.1). A diagnostic feature of transposition is the presence of short direct repeats flanking the element at its site of integration (target site). These direct repeats are not part of the element, but are generated from the original target site by the formation of a staggered endonucleolytic break in the DNA, to the ends of which the transposed DNA is attached. The filling of the remaining single-stranded gaps on either side generates the direct repeat (Fig. 7.29), the size of which (generally in the range 3–13 bp) is characteristic of a given transposable element.

Transposons of prokaryotes and eukaryotes exhibit differences as well as similarities, and will therefore be considered separately. However, one common feature of more general significance should be emphasized at this juncture. That is their ability to cause restructuring of the genome. On the one hand they can cause deletion or inversion of regions of DNA lying between them (Fig. 7.19); and on the other hand, at least in some cases, they have the ability to acquire and transpose other genomic DNA. In the latter case, recent evolutionary consequences can be seen in the spread of genes specifying antibiotic resistance to pathogenic organisms. It is possible that analogous transposition might have had a major role during evolution in the more distant past [280, 281].

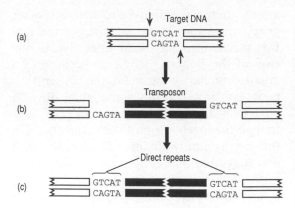

Fig. 7.29 Model to explain the generation of direct repeats of the target of transposon integration: (a) endonucleolytic cleavage of the target DNA generates a staggered break; (b) the transposable element is ligated to the protruding ends of the cleaved target site; (c) cellular enzymes repair the single-stranded gaps, generating flanking direct repeats.

7.7.2 Prokaryotic transposons [282–284]

(a) *Structural organization of bacterial transposons*

Historically, bacterial transposable elements have been assigned to one of three classes on the basis of structure and coding potential (Fig. 7.30). Although the two different mechanisms of transposition (see below) cut across this classification, it is still useful as an anatomical introduction. Class I contains elements that encode an enzyme that mediates the transposition process, the *transposase*, together with sequence information that governs the regulation of transposition. The simplest members of Class I are the insertion or IS elements (termed IS1, IS5 etc.) [285], the termini of which consist of *inverted repeat* sequences, the size and structure of which characterize a particular IS element. The lengths of the terminal inverted repeats of different IS elements usually lie in the range 15–25 bp. In fact, IS elements are generally thought to use the same reading frame to encode two proteins, the second being a *repressor* of transposition. In the case of the element IS1 of Tn9 they are produced from the same transcript by translational frameshifting (section 12.9.6); in the case of IS50 of Tn5, overlapping mRNAs are produced from different promoters.

Class I also contains what are termed *composite transposons* (e.g. Tn5, Tn9, Tn10) [286, 287] which comprise two IS or IS-like elements (in inverted or direct relative orientation) flanking a piece of 'passenger' DNA that is generally a gene specifying antibiotic resistance. In composite transposons both IS or IS-like elements need not be functional, and the extreme terminal repeat contributed by each member of the pair may be regarded as defining the limits of the transposon. Nevertheless, individual functional modules can transpose independently.

Class II elements (e.g. Tn3, $\gamma\delta$) [288] also carry various 'passenger' genes (e.g. the *bla* gene, encoding β-lactamase), but have only single terminal repeats (of *c.* 38 bp) at their

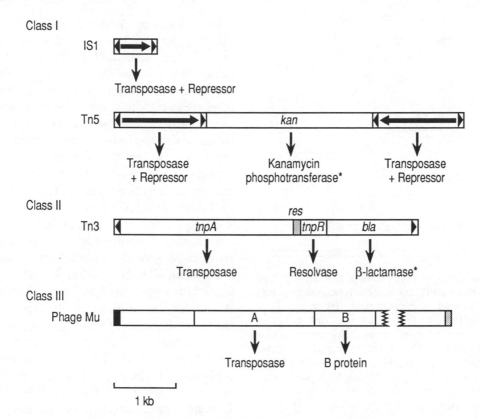

Fig. 7.30 The structures of some bacterial transposable elements. The terminal inverted repeats of class I and II elements (represented by solid arrowheads) and the termini of the 40-kb bacteriophage Mu (filled areas) are not to scale. The horizontal arrows indicate the directionality of the long terminal repeats of Tn5. *Products of 'passenger genes', not necessary for transposition.

extremities. In addition to a gene coding for a transposase (*tnpA*), they are typified by the fact that they contain a gene (*tnpR*) coding for a site-specific recombinase, the resolvase, and an internal site (*res*) which the latter protein recognizes. The resolvase can also act as a repressor of the transcription of the *tnpA* and *tnpR* genes.

A separate class, Class III, has been created to accommodate the bacteriophages such as Mu [289, 290], which replicate by a transpositional mechanism. In addition to encoding a considerable number of bacteriophage functions (coat proteins etc.), Mu possesses two genes, A and B, involved in transposition. Gene A is the transposase and

gene B greatly enhances the efficiency of transposition. It is, in fact, the high efficiency of transposition of Mu ($c.$ 100 transpositions per lytic infection) compared with that of other transposons (10^{-7} to 10^{-5} per generation) that has made it particularly suitable for studying the mechanism of transposition. Mu transposes by the same replicative mechanism employed by Class II elements (see below), and has served as a model for the transposition of the latter.

(b) *Mechanisms of transposition*

There are two types of mechanism for transposition: conservative and replicative. In

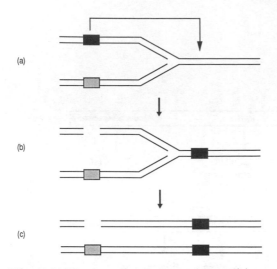

Fig. 7.31 Conservative transposition: (a) one copy of the newly replicated transposons (rectangular boxes) is excised and (b) moves to a position in front of the replication fork; (c) after the replication fork has passed through this one of the daughter DNAs has acquired a second copy of the transposon. The other daughter DNA has a gap in the position of the transposon that has migrated, and is destroyed in bacteria (although in eukaryotes this gap is repaired).

both cases the initial integrative phase of transposition has an absolute requirement for the expression of the transposase enzyme. Each transposase specifically recognizes the termini of the transposable element which encodes it, and is presumed to bring the element to its site of integration. The transposase then catalyses the staggered endonucleolytic cleavage at the target site, and the precise nicking at the ends of the element. The size of the target direct repeat, reflecting the length of the overhang in the staggered break, is characteristic of individual transposons.

Conservative transposition (also known as 'cut-and-paste') is exhibited by Class I transposons (e.g. Tn5 and Tn10), and is also an option available to certain Class II and III transposons. The transposon is lost from its

original location in the donor DNA during the transfer, and, as the double-stranded break in the donor DNA is not repaired, the donor DNA is, in fact, destroyed. Despite the loss of the donor, it is possible for a net increase in transposon number to result from conservative transposition. This can happen intramolecularly when transposition occurs from one of the two copies of the newly replicated transposon to a position ahead of the replication fork (Fig. 7.31). In the case of intermolecular transposition, e.g. from a transposon on a plasmid to a bacterial chromosome, compensatory replication of another copy of the plasmid can occur.

Replicative transposition is exhibited by Class II transposons and bacteriophage Mu. A model of replicative transposition is shown in Fig. 7.32 [283]. This derives from that of Shapiro [291], modified as proposed by Ohtsubo *et al*. [292] to allow for non-replicative formation of simple insertions, and owes much to the original ideas of Grindley and Sherratt [293]. Its key feature is the formation of an intermediate, evidence for the existence of which has been obtained by Mizuuchi and coworkers [294, 295] using bacteriophage Mu. The initial step (Fig. 7.32(a)) involves the staggered endonucleolytic break in the target DNA that produces 5'-protruding ends, and a single nick at each end of the transposable element leaving a free 3'-OH on the sides adjacent to the element. The 5'-ends of the target are then ligated to the 3'-OH groups on the element, resulting in a structure which resembles two replication forks (Fig. 7.32(b)). It is assumed that host enzymes attach to the forks of this intermediate and replicate through the transposable element, without necessarily replicating any other DNA. The product of this (Fig. 7.32(c)), containing two copies of the transposon and both donor and acceptor chromosomes, is termed a 'cointegrate'. In Class II transposons the resolvase encoded

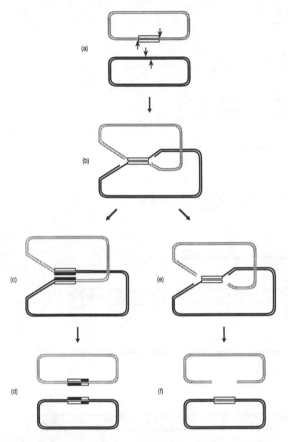

Fig. 7.32 Model for replicative transposition. The sites of nucleolytic cleavage are indicated by arrows, and the original and newly synthesized copies of the transposable element by open and solid areas, respectively. The donor replicon is shown in grey and the target replicon in black. For other details see text.

by the *tnpR* gene (Fig. 7.30) binds to the *res* site to promote homologous recombination regenerating host and target replicons (Fig. 7.32(d)). An enzyme to resolve cointegrates seems to have emerged in Class II transposons because of the intrinsic instability of two different fused genomes. In bacteriophage Mu there is no need for such resolution of the cointegrate, which is merely the bacterial genome containing Mu as prophage.

The mechanism by which the resolvase promotes site-specific recombination is thought to be fundamentally different from that of the integrase of bacteriophage lambda, involving Holliday junctions (section 7.4.1), and a detailed model for the formation and resolution of recombination synapses has been presented [330]. The active form of the resolvase is a dimer (allowing it to bridge the two recombining regions of DNA) and the three-dimensional structure of the catalytic portion of this has recently been determined [331].

The structure in Fig. 7.32(b) is also thought to be an intermediate in the alternative formation of non-replicative simple insertions by Mu and some Class II elements. It is assumed that some feature of the forks in this case prevents replication occurring from them, and allows nucleolytic attack on the donor replicon (Fig. 7.32(e)). Gap repair then yields the simple insertion of Fig. 7.32(f). It has been suggested that the 5'-exonuclease activity of DNA polymerase I might act on the intermediate in the absence of replication to effect such destruction of the donor DNA [283].

(c) *Regulation of transposition [338]*

There are several different ways in which transposition is subject to regulation, and these will be mentioned briefly. Both class II transposons, such as Tn3 [288], and bacteriophage Mu [289] are under transcriptional control by protein repressors (a second function of the resolvase, in the case of Tn3) which bind to the transposon DNA. The repressor proteins of class I transposons, such as those for IS1 and Tn5, mentioned above (section 7.7.2(a)), also bind DNA, presumably by virtue of their partial identity to the DNA-binding transposases [286]. DNA–protein interaction may also play a role in the phenomenon of *immunity*, shown by Tn3 [288]. This is the reduced tendency

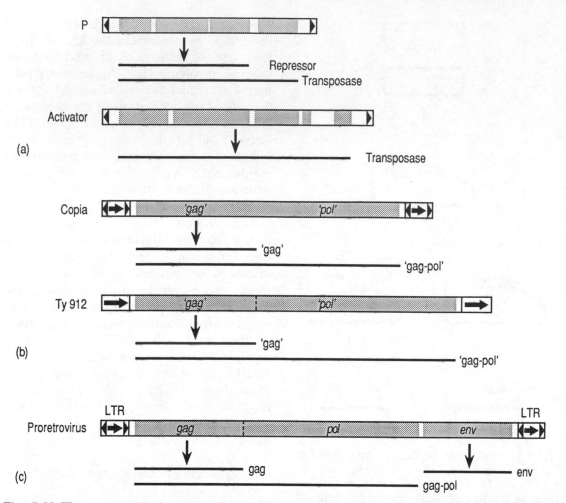

Fig. 7.33 The structures of some eukaryotic transposable elements: (a) elements that transpose conservatively; (b) integrating retroelements; (c) proviral DNA of a retrovirus. The shaded areas represent exons, the vertical broken lines represent termination codons, and the bold horizontal lines the primary translation products. Repeat sequences are represented as in Fig. 7.30. (The alternative splicing of *Copia* described in the text is not illustrated.)

for transposition to occur into DNA already harbouring the same transposon, and it has been proposed that specific binding to the ends of the transposon in the target DNA may cause sequestration of limiting amounts of transposase, preventing transposition from occurring.

A different mechanism of regulation is shown by Tn10, involving translational

inhibition by an antisense RNA [287]. This is described in more detail in section 10.7. Finally, the methylation state of the DNA may influence transposition: demethylation of transposons IS10 and IS50 increases their tendency to transposition, and hemimethylated transposons are also more mobile [287]. This may help to co-ordinate transposition to replication (Fig. 7.31), as, immediately after replication, the transposon DNA will be hemimethylated.

7.7.3 Eukaryotic transposons [284, 296]

The structures of some selected eukaryotic transposons are illustrated in Fig. 7.33. These can be divided into two classes, both of which differ in structure from prokaryotic transposons. Nevertheless there is a fundamental similarity in the mechanisms of transposition of both eukaryotic and prokaryotic transposons. This is most immediately apparent for those eukaryotic elements that transpose in a conservative manner, and these will be considered first. In accessing the protein-coding potential of eukaryotic transposons it should be borne in mind that eukaryotic mRNAs are functionally monocistronic (section 12.5.1), so that it is the nature of the spliced transcript(s) and their reading frames, rather than open reading frames in genes, which are of relevance.

(a) *Elements that transpose conservatively*

The two best characterized eukaryotic transposons in this category are the P factors of *Drosophila melanogaster* and the 'controlling element', *Activator*, of maize. These are bounded by short (31 bp and 11 bp, respectively) inverted repeat sequences, and in their central portion contain genetic information for a transposase (Fig. 7.33(a)). Thus, structurally they bear some resemblance to the Tn3 family of bacterial transposons, although mechanistically they are more like the non-replicative class I bacterial transposons.

P factors (and defective derivatives called P elements) [297, 298] are one of the elements involved in a complex genetic phenomenon called hybrid dysgenesis (the other type of element that can cause this is the

I element, section 7.7.4(a)) in which they have been shown to be responsible for mutagenesis of the progeny of certain interstrain crosses by transpositional insertion in which 8-bp target-site direct repeats are generated. A peculiarity of the transposition is that it only occurs in the germ line, and thus there must be a mechanism for the repression of transposition in somatic cells. The complete sequence of a 2.9-kb P factor shows the presence of four potential open reading frames (which are, in fact, exons) all of which are required for transposition. This is because in the germ line the portions of the transcript corresponding to these four reading frames are spliced to produce a single open reading frame in the mRNA, encoding the 87 kDa transposase. In somatic cells, however, only the first two introns are removed, and the open reading frame in the mRNA terminates before the RNA corresponding to the final open reading frame of the element. The smaller protein (66 kDa) encoded by this mRNA lacks transposase activity but is a repressor of transposition, acting by preventing the removal of the third intron. Although there is clear evidence that P factors transpose by a conservative cut-and-paste mechanism, this differs from that of the class I transposons of bacteria in that the gap left by the excised element is filled by double-stranded repair. This involves a copying mechanism, generally from a sister chromatid, and thus in flies homozygous for the P factor the original element can be regenerated, even though the transposition is not replicative.

The *Activator* (*Ac*) element of maize [299] and the defective element *Dissociation* (now known to be derived from *Ac* by deletion) were the first transposable elements studied by Barbara McClintock. *Ac* is 4.6 kb long, generates 8-bp target-site direct repeats, and has two potential open reading frames and five exons. Although the evidence is equivocal, it seems that *Ac* has a single transcript that encodes a transposase (a deletion in the transcript inactivates transposition in *Dissociation*). However, there is also evidence for a repressor function of *Ac*, and it has been suggested that this may be contained in the same polypeptide as the transposase. Like P factors, *Ac* transposes by a cut-and-paste mechanism with repair of the gap left behind. However this repair must be by a different mechanism from that in prokaryotic transposons, as it generally leaves both copies of the original flanking duplicated sequence at the empty donor site.

(b) *Integrating retro-elements*

The second category of eukaryotic transposons, exemplified by *copia* and Ty (Fig. 7.33(b)), have a replicative mode of transposition. This differs from that in prokaryotes in that it involves a reverse transcriptase similar to that found in retroviruses. In order to appreciate this one may consider how the double-stranded proviral DNAs of retroviruses (section 6.14) might be considered as transposable elements. From this point of view the single-stranded viral structural RNA is regarded as an intermediate in the genomic transposition of the proviral DNA to other sites within the same or a different genome.

The structure of the proviral DNA is presented in Fig. 7.33(c) in a manner which allows comparison with eukaryotic transposable elements (Fig. 7.33(b)). The proviral DNA is flanked by direct repeats (LTRs) of 250–1400 bp, each of which includes a short inverted repeat sequence (5–13 bp long) at its end. There are three immediately apparent viral genes, *gag*, *pol* and *env* (cf. Fig. 6.35), between these repeats, although these disguise a greater coding capacity. The *env* gene product is translated from a subgenomic mRNA, and the main product of translation of the complete mRNA is the *gag* gene product which is subsequently subject to proteolytic cleavage to four smaller species. The reverse transcriptase encoded by the *pol* gene is generated from a 180 kDa polyprotein translated from the *gag* initiation codon, involving either readthrough suppression or frameshifting (depending on the virus) to bypass the *gag* termination codon (section 12.9.6). Other proteins produced from the 180 kDa polyprotein are the 13 kDa viral protease, which is located at the 5′-end of the *pol* gene, and an endonuclease, IN, which is located at the 3′-end of the *pol* gene.

The mechanism of reintegration of the proviral DNA into the host chromosome [300, 301] will be considered briefly, as it is regarded as the paradigm for the transposition of *copia* and Ty. This involves reverse transcription of the 'intermediate' viral RNA to generate the linear blunt-ended double-stranded DNA molecule (i) of Fig. 6.35. The LTRs are recognized by the integration protein, IN, the endonuclease activity of which removes two base pairs from their 3′-OH ends. The DNA is then ligated to a staggered break produced by the IN protein at a random site in the target DNA. Gap-filling produces direct repeats (usually 5 bp in length) of the target DNA flanking the insertion. The enzymes responsible for these latter activities have not yet been identified. Although the involvement of reverse transcription is a point of striking difference between this proposed mode of eukaryotic transposition and replicative transposition in prokaryotes, the integration of the double-stranded DNA precursor into the target DNA can be regarded as being formally similar to that shown in Fig. 7.32 [302].

We now turn to those eukaryotic transposons for which the retrovirus model is thought to apply. *Copia* and the *copia*-like elements of Drosophila [303, 304] have more obvious structural similarity to retroviruses in the relatively large (*c.* 300 bp) terminal direct repeats, each of which is bounded by a short (17 bp) inverted repeat (Fig. 7.33(b)). The size of the target direct repeat is 5 bp. The *copia*-like element 17.6 [305] has open reading frames with similarities to the *gag*, *pol* and *env* genes, although in the shorter (5.1 kb) *copia* itself [306], in which there is a single extensive open reading frame, the homologue of the env gene is absent. There are two RNAs transcribed from *copia*: one a full-length transcript for a 'gag-pol' fusion protein, and a smaller spliced transcript encoding 'gag'. RNA sequences related to *copia* have been found associated with reverse transcriptase in virus-like particles, and extrachromosomal copies of *copia* have also been found, although there is only indirect evidence that they may be intermediates in transposition.

The Ty element of yeast [307, 308] is 5.9 kb in length and generates 5 bp long target-site direct repeats. It is bounded by long (*c.* 300 bp) terminal direct repeats, but unlike the retroviruses and *copia*, these do not contain terminal inverted repeats. Nevertheless there are open reading frames with clear similarities to *gag* and *pol*, and a much greater body of evidence for a retroviral-like mode of transposition than has been obtained for *copia*. This includes evidence for the synthesis of a *gag-pol* fusion protein which, as in the case of Rous sarcoma virus, involves translational frame-shifting and undergoes post-translational proteolysis. Furthermore, virus-like particles containing Ty RNA and several protein products, including reverse transcriptase, have been described. Perhaps the most convincing experiment to show that RNA is an intermediate in the transposition of Ty was one in which an intron was inserted into a Ty element under the artificial control of a yeast GAL 1 promoter. When transposition was

induced by galactose (i.e. inducing transcription), it was found that the intron had been correctly spliced out of the transposed copy [309].

The virus-like particles associated with *copia* and Ty differ from those of retroviruses in not being infectious as they lack the possibility of an extracellular phase. Although in most cases the provirus stage of a retrovirus is an intermediate in the horizontal transmission of the virus via somatic cells, in certain cases there is evidence for the vertical transmission of *endogenous retroviruses* via non-somatic cells. In these cases they are, in fact, acting as transposable elements analogous to *copia* and Ty [310]. One such endogenous retrovirus is responsible for the dilute coat colour mutation in mouse [311]. In addition there are defective elements related to retroviral genes but lacking the possibility of an extracellular phase. These are the genes that encode the *intracisternal A-particles* [312].

7.7.4 Other transposing retro-elements

The clearly defined eukaryotic transposons, *copia* and Ty, discussed above, are not the only families of DNA, the mobility of which depends on retroposition. Some of these other families (e.g. the SINEs and the processed pseudogenes) do not encode their own reverse transcriptase, and this was the justification for assigning them to a different class, *retroposons*, rather than transposons. It has subsequently emerged that members of one of the families designated as retroposons, the LINEs, do encode their own reverse transcriptase, so that they could equally well be considered as transposons. A rather arbitrary justification for classifying them differently is that they do not have long terminal repeats and lack regions predicted to encode proteins with homology to *gag* and

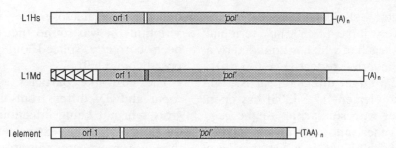

Fig. 7.34 The structures of some members of the LINE family. The lighter shaded areas represent open reading frames, and the darker area in L1Md represents a region where the two reading frames overlap. The triangles in L1Md represent direct repeat sequences.

the integrase, unlike *copia* and Ty. However, as their transposition is much less well characterized it is convenient to consider them here among a variety of what are thought to be transposing retro-elements.

(a) *LINEs*

LINEs are the long interspersed repetitive elements, previously mentioned in section 3.2.2(c), and initially thought to be peculiar to vertebrates, where they were first detected, especially in rodents and in man [313, 314]. The main difficulty in studying the vertebrate LINEs (e.g. the L1Md family of mouse and the L1Hs family of man) is identifying a functional member in the large number of copies ($c.$ 10^5) in the genome, most of which are defective. For example, many copies lack varying portions of the 5'-end present in the longest versions, consistent with their generation by different extents of reverse transcription. (The target-site direct repeats of 5–15 nucleotides imply that this truncation did not occur after transposition.) Indeed, initially the misimpression arose that the shorter members represented several different families (e.g. the *Bam*HI, MIF–1, *Bam*–5 and R families, which are all, in fact, derived from L1Md).

The full-length L1Md and L1Hs elements are $c.$ 6–7 kb long, and contain two open

reading frames, the second of which is homologous to and thought to encode reverse transcriptase (Fig. 7.34). They both terminate in a polyadenylation/processing signal followed by a poly(dA) tail at their 3'-end, but differ in their 5'-untranslated regions, which are nevertheless thought to contain the promoters for transcription by cellular RNA polymerase II. Because it has only recently been possible to isolate an active L1 element [315] there is limited information available on the patterns of transcription and the expression of the genetic potential of LINEs. It appears that transposition is restricted to the germ line, as transcripts have only been detected in cell lines such as those from teratocarcinomas, and in these cells particle-like intermediates have been detected. Evidence for the expression of the protein product of orf 1 (Fig. 7.34) has also been presented [339].

More recently it has emerged that LINEs are also present in *Drosophila* [316, 317], the first *Drosophila* LINE member identified being the *I element*, which had long been studied genetically as the cause of I–R hybrid dysgenesis. This enabled functional I elements to be identified genetically. I elements lacking 5'-portions also occur, although defective elements are not as numerous as in vertebrates. The structure of the 5.4 kb I element is compared in Fig. 7.34

with the vertebrate L1Hs and L1Md. It can be seen to share the two open reading frames with the latter, as well as an A-rich 3′-terminal sequence, although this is a repeating TAA sequence rather than poly(dA). It is uncertain which RNA polymerase transcribes the I elements, although the promoter appears to be downstream of the transcription start site, which may account for the greater number of functional elements.

Other mobile elements that have turned out to be LINEs are the Cin4 element of *Z. mays*, and the ingi3 element of the protozoan *Trypanosoma brucei*. The type I and II ribosomal insertion elements of insects are also related to LINEs, although these are clearly more specific in the sites into which they integrate. At least in the case of the type II elements, this can be ascribed to an additional function encoded by the element.

(b) *msDNA elements* [318]

Retro-elements are not confined to eukaryotes, but have also been detected in certain strains of bacteria. Although these are now known to originate from a bacterial retrovirus, it is most convenient to describe them in this section.

The product of these elements preceded their characterization. Initially discovered in myxobacteria, but also present in *E. coli*, these were first designated as multicopy single-stranded DNA (msDNA), but subsequently the DNA was found to be covalently joined to an RNA molecule by a 2′–5′ linkage to a guanosine residue (Fig. 7.35(b)). The DNA and RNA portions of the msDNA are encoded by overlapping DNA (Fig. 7.35(a)), adjacent to which is a gene for a reverse transcriptase thought to be responsible for synthesis of the DNA component of msDNA. Those isolates of *E. coli* that contain the element have only one copy, but comparison with isolates that lack

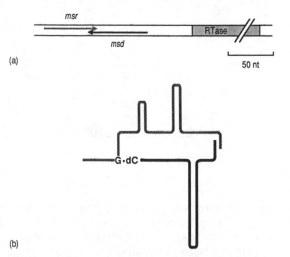

Fig. 7.35 Bacterial msDNA: (a) organization of the genes *msr* and *msd* encoding the RNA and DNA components of the msDNA, and the adjacent gene encoding a reverse transcriptase (RTase); (b) diagrammatic representation of the msDNA molecule, with the RNA component in grey and the DNA component in black. The overlap between these and the 'hairpin' loops represent hydrogen bonded regions.

the element shows that it is bounded by a 26-bp direct repeat of DNA present in the latter. The region flanked by these direct repeats is 34 kb in length and contains a copy of the *dam* methylase gene in addition to the region encoding the reverse transcriptase and the components of msDNA [329]. The bacteriophage responsible for producing these elements was recently described, and has been designated retronphage φR73 [222]. Although this retronphage is structurally quite different from eukaryotic retroviruses, it will be interesting to compare the modes of integration of the two.

(c) *SINEs*

SINEs are the short interspersed repetitive elements previously mentioned in section 3.2.2(c), and are typified by the *Alu* family in

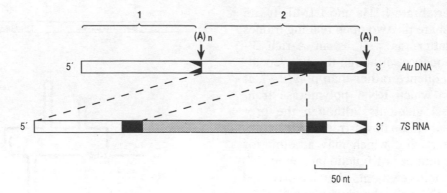

Fig. 7.36 Structural features of *Alu* DNA and relationship to 7S RNA. The two imperfectly repeated copies of the *Alu* dimer are numbered 1 and 2, the extra DNA in the latter being indicated by solid shading. The grey area represents the central sequences of 7S RNA absent from either of the *Alu* repeats and from other SINEs.

Man and the B1 family in rodents [319]. These SINEs have structures that are related to the 7S RNA component of the signal recognition particle (section 12.8.1), but are flanked by (target-site) direct repeats, lack the central section of the latter and have acquired a 3′-oligo(dA) tail (Fig. 7.36). In the case of *Alu*, the structure is an imperfect head-to-tail dimer. Other SINEs, such as the prosimian *galago*, appear to be derived from tRNA genes. The SINEs (like the 7S RNA and tRNA genes) are transcribed by RNA polymerase III, and are thus likely to terminate at oligo(dT) stretches in flanking 3′-regions of the DNA (section 9.5.2). Transcripts would thus have the potential for secondary structure interactions between the corresponding 3′-oligo(T) sequence and the intrinsic oligo(A) sequence which could act as a priming site for reverse transcription, leading to a cDNA copy that could be reintegrated into the genome. The source of the reverse transcriptase involved in this process is unknown, but, in contrast to the situation for the retro-elements already discussed, it is certainly not encoded by the SINE itself. Other details of the transposition process are also obscure. Transposed *Alu* sequences still

contain their (RNA polymerase III) internal promoters, and thus have the potential for further transposition. This may explain the extreme abundance of *Alu* sequences (*c.* 10^5 copies in the human genome). The accumulation of mutations may, of course, inactivate the promoters of some *Alu* sequences.

(d) *Processed pseudogenes*

The last type of DNA to be considered that arises from retroposition is the processed pseudogene [320, 321], which, in contrast to the examples above, lacks the capacity for further transmission. Processed pseudogenes generally differ in the following respects from the type of tandem duplicate pseudogene discussed in section 7.6.3: (i) they extend only from the cap site to the site of polyadenylation; (ii) they possess a poly(dA) tail (which is absent from the functional gene) in the expected position following a polyadenylation/processing signal (section 11.5.1); (iii) they clearly lack all introns present in the functional gene; (iv) they are flanked by direct-repeat sequences (typically 11–15 bp) immediately preceding the transcriptional start site and immediately following the poly(dA) tail; and (v) they reside at chromosomal locations distant from the functional gene. These characteristics clearly indicate that these pseudogenes originated from spliced polyadenylated mRNAs, DNA copies of which were inserted at sites in the chromosome created by a staggered endonucleolytic break. Most of such processed pseudogenes have been inserted at sites at which the lack of a promoter renders transcription impossible, and most have accumulated mutations that would render any transcript functionless. However, in the case of a chicken calmodulin processed pseudogene lacking frameshift or nonsense mutations there is evidence for expression [322]. A similar situation obtains for the functional rat insulin I gene, which, because of its location on the same chromosome as the rat insulin II gene and the fact that it contains one of the three introns of the latter, had been thought to have been derived from gene II by a tandem duplication. Subsequent analysis [323] demonstrated that the gene I has a poly(dA) tail, is flanked by direct repeats and is c. 100 000 kb from gene II. It appears that the insulin I gene arose from the reverse transcription of a partially processed transcript of gene II that had initiated upstream from the cap site and included the promoter.

Vertical transmission of processed pseudogenes can only occur if they arise in germ cells, and the most numerous processed pseudogenes correspond to mRNAs that one would expect to be abundant in undifferentiated egg or sperm cells. These include multiple processed pseudogenes corresponding to glycolytic enzymes and to cytoskeletal proteins such as actin. In actin, for example, it is mRNAs corresponding to the cytoplasmic (β and γ) isotypes, rather than the muscle-specific isotypes, that have given rise to processed pseudogenes. Nevertheless, processed pseudogenes corresponding to

α-globin and immunoglobulin λ, normally the products of differentiated cells, have been described. These, like the insulin I gene, mentioned above, are atypical in corresponding to transcripts extending upstream beyond the normal 5′-cap site. The possibility that such mRNAs are expressed (albeit transiently and at a low level) in the germ line (rather than being somehow transmitted horizontally from differentiated cells) is interesting in relation to the gene conversion thought to operate on globin (section 7.6.4) and immunoglobulin genes. It has been suggested [324] that this conversion, which would, of course, have to occur in the germ line, might possibly involve reverse transcripts of such transiently expressed mRNAs.

Two fundamental questions arise regarding the reverse transcription of the mRNAs that give rise to processed pseudogenes. The first is the source of the reverse transcriptase (and any other components of the transpositional machinery that may be required); the second is the nature of the primer for reverse transcription. One possibility is that the reverse transcriptase derives from retroviral infection, another is that it might be the reverse transcriptase encoded by endogenous transposable elements (e.g. endogenous retroviruses, intracisternal A-particles or LINEs), and yet another that it is a higher eukaryotic homologue of the reverse transcriptase involved in telomere synthesis in ciliates [325]. The mechanism of priming is also uncertain because, in the absence of specific binding sites, it is unlikely to involve a tRNA species as in the case of retroviruses (section 6.14), and possibly *copia* and Ty. One model [321] rests on the observation that the target sites of transposition frequently have a dA-rich segment at their 5′-ends. A staggered break at such a point would provide an oligo(dT) attachment site for the poly(A) tail of the mRNA, and this oligo (dT) segment would act as the primer for reverse-transcribing the mRNA. However, this mechanism, despite solving the problem of priming, is so different from that for retroviral integration that it is difficult to conceive that it could employ the same enzymes.

Types of RNA other than mRNA can generate pseudogenes by mechanisms which must involve reverse transcription. The small nuclear RNAs (U1 to U6) give rise to two structural classes of pseudogenes flanked by target-site direct repeats of similar size range to those flanking the processed pseudogenes described above [326]. One class of these are full length (or almost so) but terminate with poly(A) tails, absent from the parent RNA. This suggests a requirement for aberrant polyadenylation for retroposition, and would be consistent with priming models involving such poly(A) sequences. The other class are truncated to various extents at their 3'-ends. In the case of the severely truncated, non-polyadenylated U3 pseudogenes this clearly relates to the potential for self-hybridization at the 3'-end that could allow the self-priming of reverse transcripts of the size observed [327]. It also seems likely that self-priming is involved in the generation of less severely 3'-truncated non-polyadenylated pseudogenes of these small nuclear RNAs. Processed pseudogenes of 7S RNA have also been described, and it is important to emphasize that these are distinct from the SINEs such as *Alu* related to 7S RNA (section 7.7.4(c)). The processed pseudogenes of 7S RNA exhibit 3'-truncations, and are assumed to have been generated by self-priming facilitated by a suitable 3'-secondary structure [328].

REFERENCES

1 Singer, B. and Kusmievek, J. T. (1982) *Annu. Rev. Biochem.*, **51**, 655.

2 Kennard, O. (1987) in *Nucleic Acids and Molecular Biology*, **Vol 1** (eds F. Eckstein and D. J. M. Lilley), Springer-Verlag, p. 25.

3 Friedberg, E. C. (1985) *DNA Repair*, W. H. Freeman and Co., San Francisco.

4 Tannenbaum, S. R., Weisman, M. and Fett, D. (1976) *Food Cosmet. Toxicol.*, **114**, 549.

5 Stock, J. A. (1975) *Biology of Cancer* (2nd edn) (eds E. J. Ambrose and F. S. C. Roe), Ellis Horwood, Chichester, p. 279.

6 Zarbl, H., Sukumar, S., Arthur, A. V., Martin-Zanka, D. and Barbacid, M. (1985) *Nature (London)*, **315**, 382.

7 Sobell, H. M. (1973) *Prog. Nuceic Acid Res. Mol. Biol.*, **13**, 153.

8 Walker, R. T., De Clercq, E. and Eckstein, F. (1979) *Nucleoside Analogues*, Plenum Press, New York.

9 Kersten, H. and Kersten, W. (1974) *Inhibitors of Nucleic Acid Synthesis*, Chapman and Hall, London.

10 Heinrich, M. and Krauss, G. (1989) *J. Biol. Chem.*, **264**, 119.

11 Lasnitski, J., Matthews, R. E. F. and Smith, J. D. (1954) *Nature (London)*, **173**, 346.

12 Adams, R. L. P. (1990) *Cell Culture for Biochemists*, 2nd edn (eds R. H. Burdon and P. H. van Knippenberg), Elsevier, Amsterdam.

13 Matthews, R. E. F. (1953) *Nature (London)*, **171**, 1065.

14 Hilmoe, R. J. and Heppell, L. A. (1957) *J. Am. Chem. Soc.*, **79**, 4810.

15 Zain, B. S., Adams, R. L. P. and Imrie, R. C. (1973) *Cancer Res.*, **33**, 40.

16 Adams, R. L. P. and Burdon, R. H. (1985) *The Molecular Biology of DNA Methylation*, Springer-Verlag, New York.

17 Taylor, S. M. and Jones, P. A. (1979) *Cell*, **17**, 771.

18 Roy-Bowman, P. (1970) *Analogues of Nucleic Acid Components*, Springer-Verlag, New York.

19 Kaufman, E. R. and Davidson, R. L. (1978) *Proc. Natl. Acad. Sci. USA*, **75**, 4982.

20 Hopkins, R. L. and Goodman, M. F. (1980) *Proc. Natl. Acad. Sci. USA*, **77**, 1801.

21 Hunter, T. and Francke, B. (1975) *J. Virol.*, **15**, 759.

22 Lawley, P. D. and Brookes, P. (1961) *Nature (London)*, **92**, 1081.

23 Roberts, J. J. (1975) *Biology of Cancer* (eds E. J. Ambrose and F. J. C. Roe) Ellis Horwood, Chichester.

24 Lijinsky, W. (1976) *Prog. Nucleic Acid Res. Mol. Biol.*, **17**, 247.

25 Lindahl, T. (1981) *Chromosome Damage and Repair* (eds E. Seeberg and K. Kleppe), Plenum Press, New York, p. 207.

26 Eadie, J. S., Conrad, M., Toorchen, D. and Topal, M. D. (1984) *Nature (London)*, **308**, 201.

27 Iyer, V. N. and Szybalski, W. (1963) *Proc. Natl. Acad. Sci. USA*, **50**, 355.

28 Goldberg, I. H. and Friedman, P. A. (1971) *Annu. Rev. Biochem.*, **40**, 775.

29 Stern, R., Rose, J. A. and Friedman, R. M. (1974) *Biochemistry*, **13**, 307.

30 Neidle, S. and Abraham, Z. (1985) *CRC Crit. Rev. Biochem.*, **17**, 73.

31 Radloff, R., Bawer, W. and Vinograd, J. (1967) *Proc. Natl. Acad. Sci. USA*, **57**, 1514.

32 Reich, E. and Goldberg, I. H. (1964) *Prog. Nucleic Acid Res.*, **3**, 184.

33 Hurwitz, J. and August, J. T. (1963) *Prog. Nucleic Acid Res.*, **1**, 59.

34 Penrnan, S., Vesco, C. and Penman, M. (1968) *J. Mol. Biol.*, **34**, 49.

35 Fuchs, R. P. P., Freund, A-M., Bichara, M. and Koffel-Schwartz, N. (1988) in *DNA Replication and Mutagenesis* (eds R. E. Moses and W. C. Summers), American Society for Microbiology, Washington, p. 263.

36 Okada, Y., Shreisinger, G., Owen, J. E., Ncwton, J., Tsugita, A. and Inouye, M. (1972) *Nature (London)*, **236**, 338.

37 deBoer, J. G. and Ripley, L. S. (1984) *Proc. Natl. Acad. Sci. USA*, **81**, 5528.

38 Hampsey, D. M., Koski, R. A. and Sherman, F. (1986) *Mol. Cell. Biol.*, **6**, 4425.

39 Hutchinson, F. (1985) *Prog. Nucleic Acid Res. Mol. Biol.*, **32**, 115.

40 Henner, W. D., Grunberg, S. M. and Haseltine, W. A. (1983) *J. Biol. Chem.*, **258**, 15 198.

41 Lilley, D. J. M. (1990) *Curr. Opinion Cell Biol.*, **2**, 464.

42 Hanawalt, P. C. (1972) *Endeavour*, **31**, 83.

43 Haseltine, W. A. (1983) *Cell*, **33**, 13.

44 Sutherland, B. M., Chamberlin, M. J. and Sutherland, J. C. (1973) *J. Biol. Chem.*, **248**, 4200.

45 Sancar, A., Franklin, K. A. and Sancar, G. B. (1984) *Proc. Natl. Acad. Sci. USA*, **81**, 7397.

46 Lehman, A. R. and Bridges, B. A. (1977) *Essays Biochem.*, **13**, 71.

47 Hanawalt, P. C., Cooper, P. K., Ganesan, A. K. and Smith, C. A. (1979) *Annu. Rev. Biochem.*, **48**, 783.

48 Olsson, M. and Lindahl, T. (1980) *J. Biol. Chem.*, **255**, 10 569.

49 Warner, H. (1983) in *Enzymes of Nucleic Acid Synthesis and Modification* (ed. S. T. Jacob), CRC Press, Cleveland, vol. 1, p. 145.

50 Pegg, A. E. and Bennett, R. A. (1983) in *Enzymes of Nucleic Acid Synthesis and Modifcation* (ed. S. T. Jacob), CRC Press, Cleveland, vol. 1, p. 79.

51 Livneh, Z., Elad, D. and Sperling, J. (1979) *Proc. Natl. Acad. Sci. USA*, **76**, 1089.

52 Deutsch, W. A. and Linn, S. (1979) *J. Biol. Chem.*, **254**, 12 099.

53 Kushner, S. R., Kaplan, J. C., Ono, H. and Grossman, L. (1971) *Biochemistry*, **10**, 3325.

54 Setlow, R. B. and Canier, W. L. (1964) *Proc. Natl. Acad. Sci. USA*, **51**, 226.

55 Monk, M., Peacey, M. and Gross, J. D. (1971) *J. Mol. Biol.*, **58**, 623.

56 Kato, T. and Kondo, S. (1970) *J. Bacteriol.*, **104**, 871.

57 Cooper, P. K. and Hanawalt, P. C. (1972) *J. Mol. Biol.*, **67**, 1.

58 Cooper, P. (1977) *Mol. Gen. Genet.*, **150**, 1.

59 Dorson, J. W., Deutsch, W. A. and Moses, R. E. (1978) *J. Biol. Chem.*, **253**, 660.

60 Talpaert-parle, M., Clerici, L. and Campagnari, F. (1979) *J. Biol. Chem.*, **254**, 6387.

61 Wist, E., Uhjem, O. and Krokan, H. (1978) *Biochim. Biophys. Acta*, **520**, 253.

62 Tamanoi, F. and Okazaki, T. (1978) *Proc. Natl. Acad. Sci. USA*, **75**, 2195.

63 Karran, P. and Lindahl, T. (1978) *J. Biol. Chem.*, **253**, 5877.

64 Kirtikar, D. M., Dipple, A. and Goldthwait, D. A. (1975) *Biochemistry*, **14**, 5548.

65 Lindahl, T. (1979) *Prog. Nucleic Acid Res. Mol. Biol.*, **22**, 135.

66 Demple, B. and Linn, S. (1980) *Nature (London)*, **287**, 203.

67 Breimer, L. H. and Lindahl, T. (1984) *J. Biol. Chem.*, **259**, 5543.

68 Haines, J. A., Reese, C. B. and Lord Todd (1962) *J. Chem. Soc.* 5281.

69 Bailly, V., Verly, W. G., O'Connor, T. and Laval, J. (1989) *Biochem. J.* **262**, 581.

70 O'Connor, T. R. and Laval, J. (1989) *Proc. Natl. Acad. Sci. USA*, **86**, 5222.

71 Bailly, V. and Verly, W. G. (1987) *Bichem.*

J., **242**, 565.

72 Boiteux, S., O'Connor, T. R., Lederer, F., Gouyeffe, A. and Laval, J. (1990) *J. Biol. Chem.*, **265**, 3916.

73 Seeberg, E. (1981) *Prog. Nucleic Acid Res. Mol. Biol.*, **26**, 217.

74 Sancar, A. and Rupp, W. D. (1983) *Cell*, **33**, 249.

75 Seeberg, E. and Steinum, A-L. (1983) in *Cellular Responses to DNA Damage* (eds E. D. Friedberg and B. A. Bridges), Alan R. Liss, New York, p. 39.

76 Yeung, A. T., Mattes, W. B., Oh, E. Y. and Grossman, L. (1983) *Proc. Natl. Acad. Sci. USA*, **80**, 6157.

77 Orren, D. K. and Sancar, A. (1989) *Proc. Natl. Acad. Sci. USA*, **86**, 5237.

78 Caron, P. R., Kushner, S. R. and Grossman, L. (1985) *Proc. Natl. Acad. Sci. USA*, **82**, 4925.

79 Husain, I., van Houten, B., Thomas, D. C., Abdel-Monem, M. and Sancar, A. (1985) *Proc. Natl. Acad. Sci. USA*, **82**, 6774.

80 Pu, W. T., Kahn, R., Munn, M. M. and Rupp, W. D. (1989) *J. Biol. Chem.*, **264**, 20 697.

81 Matsumoto, Y. and Bogenhagen, D. F. (1989) *Mol. Cell. Biol.*, **9**, 3750.

82 Hanawalt, P. C. and Sarasix, A. (1986) *Trends Genet.*, **2**, 124.

83 Tanaka, K., Miura, N., Satokata, I., Miyamoto, I., Yoshida, M. C. *et al.* (1990) *Nature (London)*, **348**, 73.

84 Smith, C. A. and Hanawalt, P. C. (1978) *Proc. Natl. Acad. Sci. USA*, **75**, 2598.

85 Marx, J. L. (1978) *Science*, **200**, 518.

86 Park, S. D. and Cleaver, J. E. (1979) *Nucleic Acids Res.*, **7**, 1151.

87 Glickman, B. W. and Radman, M. (1980) *Proc. Natl. Acad. Sci. USA*, **77**, 1063.

88 Lu, A-L., Clark, S. and Modrich, P. (1983) *Proc. Natl. Acad. Sci. USA*, **80**, 4639.

89 Modrich, P. (1987) *Annu. Rev. Biochem.*, **56**, 435.

90 Modrich, P. (1989) *J. Biol. Chem.*, **264**, 6597.

91 Radman, M. and Wagner, R. (1988) *Sci. Am.*, **259** (2), 24.

92 Welsh, K. M., Lu, A-L., Clark, S. and Modrich, P. (1987) *J. Biol. Chem.*, **262**, 15 624.

93 Lu, A-L. and Chang, D-Y. (1988) *Cell*, **54**, 805.

94 Su, S-S., Lahue, R. S., Au, K. G. and Modrich, P. (1988) *J. Biol. Chem.*, **263**, 6829.

95 Schaaper, R. M. and Dunn, R. L. (1987) *J. Biol. Chem.*, **262**, 16 267.

96 Akiyama, H., Maki, H., Sekiguchi, M. and Horiuchi, T. (1989) *Proc. Natl. Acad. Sci. USA*, **86**, 3949.

97 Lahue, R. S., Su, S-S and Modrich, P. (1987) *Proc. Natl. Acad. Sci. USA*, **84**, 1482.

98 Hare, J. T. and Taylor, J. H. (1985) *Proc. Natl. Acad. Sci. USA*, **82**, 7350.

99 Coulondre, C., Miller, J. H., Farabaugh, P. J. and Gilbert, W. (1978) *Nature (London)*, **274**, 775.

100 Zell, R. and Fritz, H-J. (1987) *EMBO J.*, **6**, 1809.

101 Wiebauer, K. and Jiricny, J. (1989) *Nature (London)*, 339, 234.

102 Green, P. M., Montandon, A. J., Bentley, D. R., Ljung, R., Nilsson, I. M. and Giannelli, F. (1990) *Nucleic Acids Res.*, **18**, 3227.

103 Salser, W. (1977) *Cold Spring Harbor Symp. Quant. Biol.*, **42**, 985.

104 Bird, A. P., Taggart, M. H. and Smith, B. A. (1979) *Cell*, **17**, 889.

105 Adams, R. L. P. and Eason, R. (1984) *Nucleic Acids Res.*, **12**, 5869.

106 Livneh, Z., Schwartz, H., Hevroni, D., Shavitt, O., Tadmor, Y. and Cohen, O. (1988) in *DNA Replication and Mutagenesis* (eds R. E. Moses and W. C. Summers), American Society for Microbiology, Washington, p. 296.

107 Lackey, D., Krauss, S. W. and Linn, S. (1985) *J. Biol. Chem.*, **260**, 3178.

108 Witkin, E. M. and Kogoma, T. (1984) *Proc. Natl. Acad. Sci. USA*, **81**, 7539.

109 Shinagawa, H., Iwasaki, H., Kato, T. and Nakata, A. (1988) *Proc. Natl. Acad. Sci. USA*, **85**, 1806.

110 Burckhart, S. E., Woodgate, R., Scheuermann, R. H. and Echols, H. (1988) *Proc. Natl. Acad. Sci. USA*, **85**, 1811.

111 Shwartz, H., Shavitt, O. and Livneh, Z. (1988) *J. Biol. Chem.*, **263**, 18 277.

112 Shavitt, O. and Livneh, Z. (1989) *J. Biol. Chem.*, **264**, 11 275.

113 Goodman, M. F., Petruska, J., Boosalis, M. S., Randall, S. K., Sowers, L. C. and Mendelman, L. (1988) in *DNA Replication and Mutagenesis* (eds R. E. Moses and W. C.

Summers), American Society for Microbiology, Washington, p. 284.

114 Bryan, S., Hagensee, M. E. and Moses, R. E. (1988) in *DNA Replication and Mutagenesis* (eds R. E. Moses and W. C. Summers), American Society for Microbiology, Washington, p. 305.

115 Maenhaut-Michel, G. (1988) in *DNA Replication and Mutagenesis* (eds R. E. Moses and W. C. Summers), American Society for Microbiology, Washington, p. 332.

116 Eisenstadt, E. (1988) in *DNA Replication and Mutagenesis* (eds R. E. Moses and W. C. Summers), American Society for Microbiology, Washington, p. 403.

117 Rayssiguier, C., Thaler, D. S. and Radman, M. (1989) *Nature (London)*, **342**, 396.

118 Holliday, R. (1964) *Genet. Res.*, **5**, 282.

119 Dressler, D. and Potter, H. (1982) *Annu. Rev. Biochem.*, **51**, 727.

120 Potter, H. and Dressler, D. (1976) *Proc. Natl. Acad. Sci. USA*, **73**, 3000.

121 Thompson, B. J., Camien, M. N. and Warner, R. C. (1976) *Proc. Natl. Acad. Sci. USA*, **73**, 2299.

122 Meselson, M. S. and Radding, C. M. (1975) *Proc. Natl. Acad. Sci. USA*, **72**, 358.

123 Potter, H. and Dressler, D. (1979) *Cold Spring Harbor Symp. Quant. Biol.*, **43**, 969.

124 Szostak, J. W., Orr-Weaver, T. L. and Rothstein, R. J. (1983) *Cell*, **33**, 25.

125 Whitehouse, H. L. K. (1983) *Nature (London)*, **306**, 645.

126 Smith, G. R. (1989) *Cell*, **58**, 807.

127 Nicholas, A., Treco, D., Schultes, N. P. and Szostak, J. W. (1989) *Nature (London)*, **338**, 35.

128 Sun, H., Treco, D., Schultes, N. P. and Szostak, J. W. (1989) *Nature (London)*, **338**, 87.

129 Fincham, J. R. S. and Oliver, P. (1989) *Nature (London)*, **338**, 14.

130 Radding, C. M. (1978) *Annu. Rev. Biochem.*, **47**, 847.

131 Telander-Muskavitch and Linn, S. (1981) in *The Enzymes* (ed. P. D. Boyer) Academic Press, New York, vol. 14, p. 233.

132 Chaudhury, A. M. and Smith, G. R. (1984) *Proc. Natl. Acad. Sci. USA*, **81**, 7850.

133 Ponticelli, A. S., Schultz, D. W., Taylor, A. F. and Smith, G. R. (1985) *Cell*, **41**, 145.

134 Taylor, A. F., Schultz, D. W., Ponticelli, A. S. and Smith G. R. (1985) *Cell*, **41**, 153.

135 Smith, G. R. (1983) *Cell*, **34**, 709.

136 Cassuto, E. and Radding, C. (1971) *Nature (London) New Biol.*, **229**, 13.

137 Boon, T. and Zinder, N. D. (1971) *J. Mol. Biol.*, **58**, 133.

138 Müller, B., Jones, C., Kemper, B. and West, S. C. (1990) *Cell*, **60**, 329.

139 West, S. C. (1989) in *Nucleic Acids and Molecular Biology*, vol 3 (eds F. Eckstein and D. M. J. Lilley), Springer-Verlag, Berlin, p. 44.

140 Craig, N. L. (1984) *Nature (London)*, **311**, 706.

141 Lilley, D. M. J. and Kemper, B. (1984) *Cell*, **36**, 413.

142 Griffith, J. D. and Harris, L. D. (1988) *CRC Crit. Rev. Biochem.*, **23**, Suppl. **1**, 543.

143 Radding, C. M. (1989) *Biochim. Biophys, Acta*, **1008**, 131.

144 Dicapua, E. and Koller, Th. (1987) In *Nucleic Acids and Molecular Biology*, vol 1 (eds F. Eckstein and D. M. J. Lilley) Springer-Verlag, Berlin, p. 174.

145 Register, J. C. and Griffith, J. (1988) *J. Biol. Chem.*, **263**, 11 029.

146 Honigberg, S. M. and Radding, C. M. (1988) *Cell*, **54**, 525.

147 West, S. C., Cassuto, E. and Howard Flanders, P. (1981) *Proc. Natl. Acad. Sci. USA*, **78**, 6149.

148 Kahn, R., Cunningham, R. P., Das-Gupta, C. and Radding, C. M. (1981) *Proc. Natl. Acad. Sci. USA*, **78**, 4786.

149 Cox, M. M. and Lehman, I. R. (1981) *Proc. Natl. Acad. Sci. USA*, **78**, 6018.

150 Cox, M. M. and Lehman, I. R. (1981) *Proc. Natl. Acad. Sci. USA*, **78**, 3433.

151 West, S. C., Cassuto, E. and Howard Flanders, P. (1981) *Proc. Natl. Acad. Sci. USA*, **78**, 2100.

152 Conley, E. C. and West, S. C. (1989) *Cell*, **56**, 987.

153 Linsley, J. E. and Cox, M. M. (1990) *J. Biol. Chem.*, **265**, 10 164.

154 Campbell, A. (1969) *Episomes*, Harper and Row, New York.

155 Hsu, P-L, Ross, W. and Landy, A. (1980) *Nature (London)*, **285**, 85.

156 Mizuuchi, K. (1981) *Cold Spring Harbor Symp. Quant. Biol.*, **45**, 429.

157 Ross, W. and Landy, A. (1983) *Cell*, **33**, 261.

158 Hsu, P. L. and Landy, A. (1984) *Nature*

(London), **311**, 721.

159 Craig, N. L. and Nash, H. A. (1983) *Cell*, **35**, 795.

160 Nash, H. A. (1990) *Trends Bichem. Sci.*, **15**, 222.

161 Goodman, S. D. and Nash, H. A. (1989) *Nature (London)*, **341**, 251.

162 Snyder, U. K., Thompson, J. F. and Landy, A. (1989) *Nature (London)*, **341**, 255.

163 Travers, A. (1989) *Nature (London)*, **341**, 184.

164 Yin, S., Bushman, W. and Landy, A. (1985) *Proc. Natl. Acad. Sci. USA*, **82**, 1040.

165 Segal, E., Billyard, E., So, M., Starzbach, S. and Meyer, T. F. (1985) *Cell*, **40**, 293.

166 Seifert, H. S., Ajioka, R. S., Marchal, C., Sparling, R. F. and So, M. (1989) *Nature (London)*, **336**, 392.

167 Gibbs, C. P., Reimann, B-Y, Schultz, E, Kaufmann, A., Haas, R. and Meyer, T. F. (1989) *Nature (London)*, **338**, 655.

168 Saunders, J. R. (1989) *Nature (London)*, **338**, 622.

169 Murphy, G. L., Connal, T. D., Barritt, D. S., Koomey, M. and Cannon, J. G. (1989) *Cell*, **56**, 539.

170 Stibitz, S., Aaronson, W., Monack, D. and Falkow, S. (1989) *Nature (London)*, **338**, 266.

171 Golden, J. W. and Wiest, D. R. (1988) *Science*, **242**, 1421.

172 Golden, J. W., Robinson, S. J. and Haselkorn, R. (1985) *Nature (London)*, **314**, 419.

173 Mulligan, M. E. and Haselkorn, R. (1989) *J. Biol. Chem.*, **264**, 19 200.

174 Silverman, M. and Simon, M. (1983) in *Mobile Genetic Elements* (ed. J. A. Shapiro), Academic Press, New York, p. 622.

175 Abraham, J. M., Freitag, C. S., Clements, J. R. and Eisenstein, B. I. (1985) *Proc. Natl. Acad. Sci. USA*, **82**, 5724.

176 Chou, J. and Roizman, B. (1985) *Cell*, **41**, 803.

177 Craig, N. L. (1985) *Cell*, **41**, 649.

178 Baltimore, D. (1986) *Nature (London)*, **319**, 12.

179 Herskovitz, I. (1988) *Microbiol. Rev.*, **52**, 536.

180 Nasmyth, K. and Shore, D. (1987) *Science*, **237**, 1162.

181 Nasmyth, K. (1985) *Cell*, **42**, 213.

182 Kostriken, R., Strathern, J. N., Klar, A. J.

S., Hicks, J. B. and Heffron, F. (1983) *Cell*, **35**, 167.

183 Donelson, J. E. and Rice-Ficht, A. C. (1985) *Microbiol. Rev.*, **49**, 107.

184 Bernards, A. (1984) *Biochim. Biophys. Acta*, **824**, 1.

185 Donelson, J. E. and Turner, M. J. (1985) *Sci. Am.*, **252** (2), 32.

186 Borst, P., Bernards, A., Van der Ploeg, L. H. T., Michels, P. A. M., Lin, A. Y. C. *et al.* (1983) *Eur. J. Biochem.*, **137**, 383.

187 Pays, E. (1989) *Trends Genet.*, **5**, 389.

188 Borst, P. and Cross, G. A. M. (1982) *Cell*, **29**, 291.

189 Pays, E. (1985) *Prog. Nucleic Acid Res. Mol. Biol.*, **32**, 1.

190 Pays. E., Guyaux, M., Aerts, D., Van Meirvenne, N. and Steinert, M. (1985) *Nature (London)*, **316**, 562.

191 Foote, S. J. and Kemp, D. J. (1989) *Trends Genet.*, **5**, 337.

192 Meier, J. T., Simon, M. I. and Barbour, A. G. (1985) *Cell*, **41**, 403.

193 Yao, M-C. (1989) in *Mobile DNA* (eds D. E. Berg and M. M. Howe), American Society for Microbiology, Washington, p. 715.

194 Roth, M. and Prescott, D. M. (1985) *Cell*, **41**, 411.

195 Yao, M-C., Zhu, S-G. and Yao, C-H. (1985) *Mol. Cell. Biol.*, **5**, 1260.

196 Robinson, E. K., Cohen, P. D. and Blackburn, E. H. (1989) *Cell*, **58**, 887.

197 Tausta, S. L. and Klobutcher, L. A. (1989) *Cell*, **59**, 1019.

198 Jahn, C. L., Krikau, M. F. and Shyman, S. (1989) *Cell*, **59**, 1009.

199 Godiska, R. and Yao, M-C. (1990) *Cell*, **61**, 1237.

200 Lindahl, T., Sedwick, B., Sekiguchi, M. and Nakabeppu, Y. (1988) *Annu. Rev. Biochem.*, **57**, 133.

201 Teo, I., Sedwick, B., Kilpatrick, M. W., McCarthy, T. V. and Lindahl, T. (1986) *Cell*, **45**, 315.

202 Roit, I. (1988) *Essential Immunology* (6th edn), Blackwell, Oxford.

203 Tonegawa, S. (1985) *Sci. Am.*, **253** (4), 104.

204 Dreyer, W. J. and Bennett, J. C. (1965) *Proc. Natl. Acad. Sci. USA*, **54**, 864.

205 Tonegawa, S. (1983) *Nature (London)*, **302**, 575.

206 Blackwell, T. K. and Alt, F. W. (1988) in *Molecular Immunology* (eds B. D. Hames

and D. M. Glover), IRL Press, Oxford.

207 Yancopoulos, G. D. and Alt, F. W. (1986) *Annu. Rev. Immunol.*, **4**, 339.

208 Sakano, H., Rogers, J. H., Huppi, K., Brack, C., Trannecker, A. *et al.* (1979) *Nature (London)*, **277**, 627.

209 Cushley, W. (1986) in *Multidomain Protein Structure and Function* (eds D. G. Hardie and J. R. Coggins), Elsevier, Amsterdam.

210 Early, P. and Hood, L. (1981) *Cell*, **24**, 1.

211 Gascoigne, N. R. J., Chien, Y.-H., Becker, D. M., Kavaler, J. and Davis, M. M. (1984) *Nature (London)*, **310**, 387.

212 Sims, J. C., Tunnacliffe, A., Smith, W. J. and Rabbitts, T. H. (1984) *Nature (London)*, 312, 541.

213 Arden, N., Klotz, K. L., Siu, G. and Hood, L. E. (1985) *Nature (London)*, **316**, 783.

214 Yoshikai, Y., Clark, S. P., Taylor, S., Sohn, U., Wilson, B. I. *et al.* (1985) *Nature (London)*, **316**, 837.

215 Winoto, A., Mjolsness, S. and Hood, L. (1985) *Nature (London)*, **316**, 832.

216 Hayday, A. C., Diamond, D. J., Tanigawa, G., Heilig, J. S., Folsan, V. *et al.* (1985) *Nature (London)*, **316**, 828.

217 Heilig, J. S., Glimcher, L. H., Kranz, D. M., Clayton, L. K., Greenstein, J. L. *et al.* (1985) *Nature (London)*, **317**, 68.

218 Seidman, J. G., Max, E. E. and Leder, P. (1979) *Nature (London)*, **280**, 370.

219 Hesse, J. E., Lieber, M. R., Gellert, M. and Mizuuchi, K. (1987) *Cell*, **49**, 775.

220 Schatz, D. G., Oettiger, M. A. and Baltimore, D. (1989) *Cell*, **59**, 1035.

221 Alt, F. W. and Baltimore, D. (1982) *Proc. Natl. Acad. Sci. USA*, **79**, 4118.

222 Inouye, S., Sunshine, M. G., Six, E. W. and Inouye, M. (1991) *Science*, **252**, 969.

223 Rideout, W. M., Coetzee, G. A., Olumi, A. F. and Jones, P. A. (1990) *Science*, **249**, 1288.

224 Sakano, H., Maki, R., Kurosawa, Y., Roeder, W. and Tonegawa, S. (1980) *Nature (London)*, **282**, 676.

225 Kataoka, T., Miyata, T. and Honjo, T. (1981) *Cell*, **23**, 357.

226 Shimizu, A. and Honjo, T. (1984) *Cell*, **36**, 801.

227 Obata, M., Kataoka, T., Nakai, S., Yamagishi, H., Takahashi, N. *et al.* (1981) *Proc. Natl. Acad. Sci. USA*, **78**, 2437.

228 Jäck, H-M., McDowell, M., Steinberg, C. M. and Wabl, M. (1988) *Proc. Natl. Acad. Sci. USA*, **85**, 1581.

229 Schwedler, U. von, Jäck, H-M. and Wabl, M. (1990) *Nature (London)*, **345**, 452.

230 Liu, C-P., Tucker, P. W., Mushinski, J. F. and Blattner, F. R. (1980) *Science*, **209**, 1348.

231 Knapp, M. R., Liu, C-P, Newell, N., Ward, R. B., Tucker, P. W. *et al.* (1982) *Proc. Natl. Acad. Sci. USA*, **79**, 2996.

232 Yaoita, Y., Kumagai, Y., Okumura, K. and Honjo, T. (1982) *Nature (London)*, **297**, 697.

233 Dunnick, W., Wilson, M. and Stavnezer, J. (1989) *Mol. Cell. Biol.*, **9**, 1850.

234 Dunnick, W. and Stavnezer, J. (1990) *Mol. Cell. Biol.*, **10**, 397.

235 Baltimore, D. (1981) *Cell*, **26**, 295.

236 Burnet, F. M. (1959) *The Clonal Selection Theory of Acquired Immunity*, Vanderbilt University Press, Nashville.

237 LeFrank, M. P. and Rabbitts, T. H. (1989) *Trends Biochem. Sci.*, **14**, 214.

238 Blackman, M., Kappler, J. and Marrack, P. (1990) *Science*, **248**, 1335.

239 Adams, J. M. (1985) *Nature (London)*, **315**, 542.

240 von Lindern, M., Poustka, A., Lehrach, H. and Grosveld, G. (1990) *Mol. Cell Biol.*, **10**, 4016.

241 Lewis, W. H., Michalopoulos, E. E., Williams, D. L., Minden, M. D. and Mak, T. W. (1985) *Nature (London)*, **317**, 544.

242 Rabbitts, T. H. (1984) in *Molecular Biology and Human Disease* (eds A. Macleod and K. Sikora), Blackwell, Oxford, p. 164.

243 Croce, C. M. and Klein, G. (1985) *Sci. Am.*, **252** (3), 44.

244 Ray. P. N., Belfall, B., Duff, C., Logan, C., Kean, V. *et al.* (1985) *Nature (London)*, **318**, 672.

245 Yeu, P. H., Li, X-M., Tsai, S-P., Johnson, C., Mohandas, T. and Shapiro, L. J. (1990) *Cell*, **61**, 603.

246 Blackwell, T. K. and Alt, F. W. (1989) *J. Biol. Chem.*, **264**, 10 327.

247 Hsieh, P. and Camerini-Otero, R. W. (1989) *J. Biol. Chem.*, **264**, 5089.

248 Brunier, D., Michel, B. and Ehrlich, S. D. (1988) *Cell*, **52**, 883.

249 Amstutz, H., Munz, P., Heyer, W-D., Leupold, U. and Kohli, J. (1985) *Cell*, **40**, 879.

250 Berg, D. E. (1990) in *Gene Rearrangement* (eds B. D. Hames and D. M. Glover), Oxford University Press, Oxford, p. 1.

251 Van der Ploeg, L. H. T. (1990) in *Gene Rearrangement* (eds B. D. Hames and D. M. Glover) Oxford University Press, Oxford, p. 51.

252 Weichhold, G. M., Klobeck, H-G., Ohnheiser, R., Combriato, G. and Zachau, H. G. (1990) *Nature (London)*, **347**, 90.

253 de Thé, H., Chomienne, C., Lanotte, M., Degos, L. and Dejean, A. (1990) *Nature (London)*, **347**, 558.

254 Buckingham, M. E. (1985) *Essays Biochem.*, **20**, 77.

255 Pfeffer, S. and Ullrich, A. (1985) *Nature (London)*, **313**, 184.

256 Efstratiadis, A., Posakony, J. W., Maniatis, T., Lawn, R. M., O'Connell, C. *et al.* (1981) *Cell*, **26**, 653.

257 Proudfoot, N. J., Gil, A. and Maniatis, T. (1982) *Cell*, **31**, 553.

258 Jeffreys, A. J., Harris, S., Barrie, P. A., Wood, D., Blanchetot, A. *et al.* (1983) in *Evolution from Molecules to Man* (ed. D. S. Bendall), Cambridge University Press, Cambridge, p. 175.

259 Jeffreys, A. J., Wilson, V., Wood, D., Simons, J. P., Kay, R. M. *et al.* (1980) *Cell*, **21**, 555.

260 Lang, A. and Lorkin, P. A. (1976) *Br. Med. Bull.*, **32**, 239.

261 Goodman, M., Koop, B. F., Czelusniak, J., Weiss, M. L. and Slightom, J. L. (1984) *J. Mol. Biol.*, **180**, 803.

262 Lauer, J., Shen, C-K. S. and Maniatis, T. (1980) *Cell*, **20**, 119.

263 Shen, S., Slightom, J. L. and Smithies, O. (1981) *Cell*, **26**, 191.

264 Martin, S. L., Vincent, K. A. and Wilson, A. C. (1983) *J. Mol. Biol.*, **164**, 513.

265 Proudfoot, N. J. and Maniatis, T. (1980) *Cell*, **21**, 537.

266 Maeda, N. (1985) *J. Biol. Chem.*, **260**, 6698.

267 Smith, G. P. (1976) *Science*, **191**, 528.

268 Jeffreys, A. J. and Harris, S. (1984) *Bio-Essays*, **1**, 253.

269 Little, P. F. R. (1982) *Cell*, **28**, 683.

270 Ford, P. (1978) *Nature (London)*, **271**, 205.

271 Zimmer, E. A., Martin, S. L., Beverley, S. M., Kan, Y. and Wilson, A. C. (1980) *Proc. Natl. Acad. Sci. USA*, **77**, 2158.

272 Tartof, K. D. (1974) *Proc. Natl. Acad. Sci. USA*, **71**, 1272.

273 Petes, T. D. (1980) *Cell*, **19**, 765.

274 Baltimore, D. (1981) *Cell*, **24**, 592.

275 Slightom, J. L., Blechl, A. E. and Smithies, O. (1980) *Cell*, **21**, 627.

276 Hess, J. F., Fox, M., Schmid, C. and Shen, C-K. J. (1983) *Proc. Natl. Acad. Sci. USA*, **80**, 5970.

277 McClintock, B. (1984) *Science*, **226**, 792.

278 Jordan, E., Saedler, H. and Starlinger, P. (1968) *Mol. Gen. Genet.*, **102**, 353.

279 Shapiro, J. A. (1969) *J. Mol. Biol.*, **40**, 93.

280 Reznikoff, W. S. (1983) in *Gene Function in Prokaryotes* (eds J. Beckwith, J. Davies and J. A. Gallant), Cold Spring Harbor Monograph Series, p. 229.

281 Syvanen, M. (1984) *Annu. Rev. Genet.*, **18**, 271.

282 Shapiro, J. A. (1983) *Mobile Genetic Elements*, Academic Press, New York.

283 Grindley, N. D. F. and Reed, R. R. (1985) *Annu. Rev. Biochem.*, **54**, 863.

284 Berg, D. E. and Howe, M. (1989) *Mobile DNA*, American Society for Microbiology, Washington.

285 Galas, D. J. and Chandler, M. (1989) in *Mobile DNA* (eds D. E. Berg and M. Howe), American Society for Microbiology, Washington, p. 109.

286 Berg, D. E. (1989) in *Mobile DNA* (eds D. E. Berg and M. Howe), American Society for Microbiology, Washington, p. 185.

287 Kleckner, N. (1989) in *Mobile DNA* (eds D. E. Berg and M. Howe), American Society for Microbiology, Washington, p. 227.

288 Sherratt, D. (1989) in *Mobile DNA* (eds D. E. Berg and M. Howe), American Society for Microbiology, Washington, p. 163.

289 Pato, M. L. (1989) in *Mobile DNA* (eds D. E. Berg and M. Howe), American Society for Microbiology, Washington, p. 23.

290 Mizuuchi, K. and Craigie, R. (1986) *Annu. Rev. Genet.*, **20**, 385.

291 Shapiro, J. A. (1979) *Proc. Natl. Acad. Sci. USA*, **76**, 1933.

292 Ohtsubo, E., Zenilman, M., Ohtsubo, H., McCormick, M., Machida, C. and Machida, Y. (1981) *Cold Spring Harbor Symp. Quant. Biol.*, **45**, 283.

293 Grindley, N. D. F. and Sherratt, D. (1979) *Cold Spring Harbor Symp. Quant. Biol.*, **43**, 1257.

294 Mizuuchi, K. (1984) *Cell*, **39**, 395.

295 Craigie, R. and Mizuuchi, K. (1985) *Cell*,

41, 867.

296 Finegan, D. J. (1989) *Trends Genet.*, **5**, 103.

297 Craig, N. L. (1990) *Cell*, **62**, 399.

298 Engels, W. R. (1989) in *Mobile DNA* (eds D. E. Berg and M. Howe), American Society for Microbiology, Washington, p. 437.

299 Fedoroff, N. (1989) in *Mobile DNA* (eds D. E. Berg and M. Howe), American Society for Microbiology, Washington, p. 375.

300 Varmus, H. and Brown, P. (1989) in *Mobile DNA* (eds D. E. Berg and M. Howe), American Society for Microbiology, Washington, p. 53.

301 Grandgenett, D. P. and Mumm, S. (1990) *Cell*, **60**, 3.

302 Plasterk, R. H. A. (1990) *New Biologist*, **2**, 787.

303 Rubin, G. M. (1983) in *Mobile Genetic Elements* (ed. J. A. Shapiro), Academic Press, New York, p. 329.

304 Bingham, P. M. and Zachar, Z. (1989) in *Mobile DNA* (eds D. E. Berg and M. Howe), American Society for Microbiology, Washington, p. 485.

305 Saigo, K., Kugimiya, W., Matsuo, Y., Inouye, S., Yoshioka, K. and Yuki, S. (1984) *Nature (London)*, **312**, 659.

306 Mount, S. M. and Rubin, G. M. (1985) *Mol. Cell. Biol.*, **5**, 1630.

307 Adams, S. E., Kingsman, S. M. and Kingsman, A. J. (1987) *BioEssays*, **7**, 3.

308 Boeke, J. D. (1989) in *Mobile DNA* (eds D. E. Berg and M. Howe), American Society for Microbiology, Washington, p. 335.

309 Boeke, J. D., Garfinkel, D. J., Styles, C. A. and Fink, G. R. (1985) *Cell*, **40**, 491.

310 Jaenisch, R. (1983) *Cell*, **32**, 5.

311 Jenkins, N. A., Copeland, N. G., Taylor, B. A. and Lee, B. K. (1981) *Nature (London)*, **293**, 370.

312 Ono, M., Toh, H., Miyata, T. and Awaya, T. (1985) *J. Virol.*, **55**, 387.

313 Fanning, T. G. and Singer, M. F. (1987) *Biochim. Biophys. Acta*, **910**, 203.

314 Hutchison, C. A., Hardies, S. C., Loeb, D. D., Shehee, W. R. and Edgell, M. H. (1989) in *Mobile DNA* (eds D. E. Berg and M. Howe), American Society for Microbiology, Washington, p. 593.

315 Di Nocera, P. P. and Sakaki, Y. (1990) *Trends Genet.*, **6**, 29.

316 Finnegan, D. J. (1989) in *Mobile DNA* (eds D. E. Berg and M. Howe), American Society for Microbiology, Washington, p. 503.

317 Bucheton, A. (1990) *Trends Genet.*, **6**, 16.

318 Varmus, H. E. (1989) *Cell*, **56**, 721.

319 Deininger, P. (1989) in *Mobile DNA* (eds D. E. Berg and M. Howe), American Society for Microbiology, Washington, p. 619.

320 Vanin, E. F. (1985) *Annu. Rev. Genet.*, **19**, 253.

321 Wagner, M. (1986) *Trends Genet.*, **2**, 134.

322 Stein, J. P., Munjaal, R. P., Lagace, L., Lai, E. C., O'Malley, B. W. *et al.* (1983) *Proc. Natl. Acad. Sci. USA*, **80**, 6485.

323 Soares, M. B., Schon, E., Henderson, A., Karathanasis, S. K., Cate, R. *et al.* (1985) *Mol. Cell. Biol.*, **5**, 2090.

324 Baltimore, D. (1985) *Cell*, **40**, 481.

325 Boeke, J. D. (1990) *Cell*, **61**, 193.

326 Van Ardsall, S. W., Denison, R. A., Bernstein, L. B., Weiner, A. M., Manser, T. *et al.* (1981) *Cell*, **26**, 11.

327 Bernstein, L. B., Mount, S. M. and Weiner, A. M. (1983) *Cell*, **32**, 461.

328 Ullu, E. and Weiner, A. M. (1984) *EMBO J.*, **3**, 3303.

329 Hsu, M.-Y., Inouye, M. and Inouye, S. (1991) *Proc. Natl. Acad. Sci. USA*, **87**, 9454.

330 Stark, W. M., Sherratt, D. J. and Boocock, M. R. (1989) *Cell*, **58**, 779.

331 Sanderson, M. R., Freemont, P. S., Rice, P. A., Goldman, A., Hatfull, G. F. *et al.* (1990) *Cell*, **63**, 1323.

332 Egel, R., Nielsen, O. and Weilguny, D. (1990) *Trends Genet.*, **6**, 369.

333 Murchie, A. I. H., Clegg, R. M., von Kitzing, E., Duckett, D. R., Diekmann, S. *et al.* (1989) *Nature (London)*, **341**, 763.

334 Holmes, J., Clarke, S. and Modrich, P. (1990) *Proc. Natl. Acad. Sci. USA*, **87**, 5837.

335 Wiebauer, K. and Jiricny, J. (1990) *Proc. Natl. Acad. Sci. USA*, **87**, 5842.

336 Heywood, L. A. and Burke, J. F. (1990) *BioEssays*, **12**, 473.

337 Croce, C. M. (1987) *Cell*, **49**, 155.

338 Kleckner, N. (1990) *Annu. Rev. Cell Biol.*, **6**, 297.

339 Leibold, D. M., Swergold, G. D., Singer, M. F., Thayer, R.E., Dombroski, B. A. *et al.* (1990) *Proc. Natl. Acad. Sci. USA*, **87**, 6990.

8

The arrangement of genes

8.1 GENE NUMBERS AND SPACING

The most recently published linkage map for the *E. coli* chromosome contains the positions of over 1400 genes [1]. Work is now well advanced on the total sequencing of the genome and it seems likely that the estimated 4750 kb of *E. coli* DNA will contain 2500–4000 genes. Most of these genes are present as a single copy although there are seven sets of the genes that encode the ribosomal RNAs (section 8.4.3). Many of the genes occur in *operons*; groups of genes of related function which may, for instance, encode all the enzymes in a metabolic pathway. The genes of an operon are united by common controls of their expression (sections 10.1 and 10.2) and tend to be transcribed from common promoters into polycistronic mRNAs (section 11.1.1).

The 680 mapped genes of *Salmonella typhimurium* show a striking overall similarity in map order with those of *E. coli* [2, 3]. But those of three partially mapped *Psuedomonas* species have a substantially different arrangement and are not clustered but dispersed and non-contiguous [4].

Prokaryotic DNA is used efficiently with only short regions of non-coding DNA between genes. Thus, of the five polypeptide-encoding genes of the *E. coli*

tryptophan operon (section 10.1.2), two genes have initiation codons that overlap the termination codon of the previous gene, and the largest distance between one gene and the next is 14 nucleotides [5].

Viruses may have evolved from plasmids and probably arose when a plasmid acquired a protein coat. However, the capsid coats of viruses impose severe limitations on the amount of nucleic acid that can be enclosed so both RNA and DNA viruses also exhibit efficient use of the available coding potential. Some small viruses, such as ϕX174 take this economy to extremes with overlapping genes (section 3.7.7).

The mammalian genome with approximately 3×10^9 base pairs has over 600 times the coding potential of *E. coli* but it seems unlikely that there is a need for more than 50 times the prokaryotic complexity. Various features of the eukaryotic genome that account for the excess coding potential are considered in the relevant sections of this book. Thus, the common finding that eukaryotic genes often occur in multiple functional and non-functional copies is considered in section 8.4. The fact that many, and in some species most, genes of eukaryotic cells are discontinuous with the coding sequence interrupted by non-coding elements that may be much longer than the gene itself is discussed in sections 8.2 and 11.3. Much of the DNA in a cell is repetitive and of no known function. This includes both the highly repetitive, satellite DNA and the more dispersed repetitive elements (section 3.2.2).

8.2 GENES ARE OFTEN DISCONTINUOUS

The availability of purified mRNA species (section 11.1.2), the use of reverse transcriptase to make a radioactive probe by copying the messenger into a complementary DNA (cDNA) and the availability of restriction enzymes that cleave DNA at specific sites, led to the discovery that the genes encoding eukaryotic proteins are often discontinuous. Leder and colleagues [6, 7] digested mouse DNA with restriction enzymes, separated the fragments on agarose gels and transferred them to cellulose nitrate by Southern blotting (section A.3). They then searched the blotted DNA for fragments that contained globin-encoding sequences by hybridization with a ^{32}P-labelled β-globin cDNA. Their results showed that the β-globin gene did not exist as a contiguous stretch of DNA but as three blocks of coding sequence separated by sections of non-coding DNA which became known as *introns*, inserts or intervening sequences. The sections of coding sequence became known as *exons*. Leder and colleagues [7] further showed that the largest of the two globin introns could be visualized in the electron microscope. Hybrids between the mRNA and a cloned globin gene, formed under conditions in which RNA: DNA hybrids were more stable than the DNA:DNA double helix, had an unusual appearance. Instead of the expected single continuous loops of displaced DNA, two smaller loops of displaced single-stranded

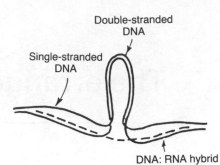

Fig. 8.1 Diagrammatic representation of an R loop. The pattern of hybridization that would be expected between a mRNA and a gene containing a single large intron.

DNA were separated by a loop of double-stranded DNA (Fig. 8.1). This so-called R loop was caused by the larger intron, having no complement in the cDNA and thus continuing to exist as a DNA duplex. Small introns cannot be detected by this method. An electron micrograph of a hybrid molecule formed between a cloned ovomucoid DNA and the ovomucoid mRNA is illustrated in Fig. 11.4. It shows the R loops formed by the seven introns of the gene [17].

It is now known that many, and in higher eukaryotes most, protein-encoding genes of eukaryotes and their viruses are discontinuous. Exceptions include most histone genes, the interferon gene, and adenovirus polypeptide IX and most herpes virus and orthopoxvirus genes [8–11, 30]. It is also known that the initial transcript of genes includes introns and exons and that the former are spliced out of the pre-mRNA. The mechanisms of this processing are discussed in section 11.3.1. The arrangement of the coding and non-coding blocks of some of the first genes to be characterized is illustrated in Fig. 8.2. In many genes, the combined length of the introns exceeds that of the exons. Thus the chicken ovalbumin gene contains 7564 base pairs whereas the

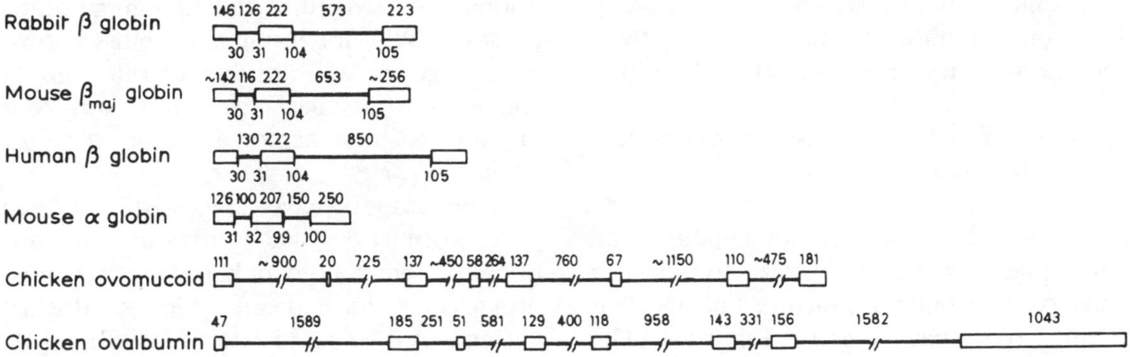

Rabbit β globin

146 126 222   573   223
       30 31 104   105

Mouse β_maj globin

~142 116 222   653   ~256
        30 31 104   105

Human β globin

130 222   850
     30 31 104   105

Mouse α globin

126 100 207 150 250
     31 32 99 100

Chicken ovomucoid

111 ~900 20 725 137 ~450 58 264 137 760 67 ~1150 110 ~475 181

Chicken ovalbumin

47   1589   185 251 51 581 129 400 118   958   143 331 156   1582   1043

Fig. 8.2 The organization of introns and exons in some of the first genes characterized. The boxes represent exons and the intervening lines, introns. The numbers above the diagrams indicate the numbers of nucleotides in the segments. The numbers below the globin diagrams represent the amino acid codons that occur on either side of the introns.

mRNA is only 1872 nucleotides long. The difference of 5692 nucleotides represents non-coding sequences arranged in seven introns [12]. The total length of the avian vitellogenin gene is approximately 23 000 base pairs; 25 large and 6–10 small introns are interspersed among the 6700 base pairs that encode the mRNA [13]. The chicken pro-α-2(1) collagen gene contains about 50 introns [14]. The most extreme examples known include the dystrophin gene that contains more than 65 exons spread over more than two million base pairs of DNA ([15] and references therein) and the *c-abl* gene, the first intron of which is more than 200 kb long [16].

8.2.1 Intron types

(a) *Classical introns*

The introns so far discussed, that occur in the nuclear genes of eukaryotic cells, have become known as classical introns. They are characterized as beginning with the dinucleotide GT and ending with the dinucleotide AG. Their transcripts in pre-mRNA are excised by a multi-ribonucleoprotein particle known as a spliceosome (section 11.3.1). Considerable controversy surrounds their origins, function and whether they are mobile (section 8.2.2).

(b) *Group I introns (also known as type I or class I introns)*

Group I introns share a highly conserved secondary structure and occur in mito-chondrial genes, chloroplast genes and in the nuclear genes of lower eukaryotes [18, 19]. They are also found in the genes of prokaryotic bacteriophage [20] and in cyanobacteria [21]. They are apparently not essential; often being present in one strain or species but not in another close relative. In some cases at least they are mobile. Thus, the omega intron of *Saccharomyces* contains an open reading frame that encodes a endonuclease. This site-specific enzyme is responsible for the mobility of the intron by introducing a double-stranded cut in the yeast DNA into which the intron is inserted [22, 23]. There is also evidence that group I introns can transpose to new sites by a reversal of their splicing mechanism [24].

The splicing of transcripts of some group I introns is autocatalytic whereas others require ancillary proteins (section 11.3.2).

(c) *Group II introns (also known as type II or class II introns)*

Found in the genomes of mitochondria and chloroplasts, group II introns also have a conserved secondary structure but one that differs from that of group I introns. The splicing mechanism of the intron transcripts most closely resembles classical introns but, like group I introns, the process can be autocatalytic (section 11.3.3). They have not been demonstrated to transpose but may encode a reverse transcriptase-like protein that could be responsible for mobility [25].

(d) *Group III introns*

The chloroplasts of *Euglena* contain short introns that are very AT-rich [26]. Nothing is known of their splicing or whether they exhibit mobility within the genome.

(e) *tRNA introns*

Some of the tRNA genes of eukaryotes and archaebacteria may contain short introns that in all known cases are inserted to the 3′-side of the anticodon. They are particularly common in yeast nuclear tRNA genes (40 of the 400 tRNA genes) and their transcripts are edited out of pre-tRNA by a two-stage process. The boundaries of the intron are first cut by an endonuclease and then the two tRNA half molecules are ligated together (section 11.3.5).

8.2.2 The origins and role of introns

The existence of introns provokes an obvious question: 'What possible evol-utionary or structural function might they possess?' Why for instance do most known histone genes lack introns yet other genes contain introns that have been conserved in their position across a billion years of evolution [27].

Some authors have suggested that they have no function; that introns are an evolutionary remnant in eukaryotes that most prokaryotes have been able to discard because of their rapid rate of evolution [28, 29]. That this might be occurring, though at a slow rate in eukaryotes is supported by the finding that simple, rapidly dividing eukaryotes, such as yeast, tend to have smaller introns than those of higher cells. Furthermore, many of their genes have no introns.

An extension of these ideas, that introns represent an early stage of gene evolution, is that split genes allow the rapid evolution of new genes [28]. An exon might simply represent a function-encoding protein domain able, for instance, to bind a particular coenzyme, span a membrane or locate in the main groove of DNA. Genetic recombinations might then bring together new combinations of exons and generate a protein with a new combination of functional domains. By this theory introns are seen as useless sequences trapped between reorganized exons. Without them, however, and the ability of the cells to remove their transcripts (section 11.3), recombination would have to join functional domains exactly so as to preserve both their identity and their reading frames. Such precise reorganization would be very rare and gene evolution would be much slower.

Support for this exon shuffling theory first came from the studies of Tonegawa and coworkers [31] who demonstrated a functional relationship between the domains of the heavy chain of immunoglobulin and the exons of its gene. Each of the six functional polypeptide units, including the 14-amino acid hinge region, is encoded in a separate exon. Similarly, Stein *et al.* [32] have shown that the intervening sequences of chicken ovomucoid gene separate coding portions that correspond to the functional domains of the protein. Other examples include pyruvate kinase [33] and the low density lipoprotein (LDL) receptor [34]. The latter example contains 18 exons most of which correlate with functional domains in the LDL receptor protein. Thirteen of the exons are homologous with functionally similar domains in other proteins. The central exon of globin genes encodes the haem-binding domain of the protein [35] and although the other exons do not encode such clear cut domains as in the examples above, they do divide the protein into separate compact structural regions [36].

Further support for the role of exons in the evolution of proteins comes from the study of gene families and structurally related genes. Thus, all known vertebrate globin and myoglobin genes have two introns at the same position within the coding sequence (Fig. 8.2). The relationship even extends to the plant leg haemoglobins [37] which have three introns, two of which correspond to those of vertebrate globins. However, not all the evidence supports the concept of exon shuffling. For instance, the nucleotide coenzyme binding domain of dehydrogenases is flanked by introns in several genes but they are not in conserved positions or even in the same reading frame [38]. Such introns cannot easily be accommodated in an exon shuffling theory in which an ancestral intron is involved in the evolution of the gene family. In fact, it is clear that in this case the introns must have been inserted into the genes after the divergence of multigene families. The same is true for the introns in the genes of the serine protease family [39], collagens, actins,

tubulins, calcium-binding proteins, zinc-finger proteins and a host of other examples (reviewed in [20]). These introns are known as *discordant introns* and it appears that most classical introns are of this type. Possible mechanisms by which discordant introns might be inserted into genes have been reviewed [20, 40] but they are largely hypothetical. Indeed the origins of introns are still shrouded in uncertainty.

All of the above ideas imply that introns are functionless sequences and multiple examples testify that genes with deleted introns can be equally well expressed. Furthermore, with the exception of the splicing regions, the nucleotide sequences of introns exhibit a much greater sequence divergence than those of exons, again indicating a low information content. Nevertheless, it is clear that introns can carry at least some information. Many examples are known of alternative splicing pathways by which more than one mRNA species can be created from a single pre-mRNA. This implies differential recognition by the splicing machinery of sequence elements within an intron. This is considered in detail in section 11.9.1. Many group I and group two introns also include open reading frames that encode proteins involved in splicing (section 11.3.2) or intron mobility [25].

8.3 MULTIGENE FAMILIES

As outlined in the preceding section, one potential way in which new genes may evolve is by the grouping of previously unrelated exons into a new transcriptional unit. Another mechanism, which appears to be common, is gene duplication. This may give rise to multicopy genes and/or allow the evolution of new genes. Detection of gene families usually follows C_0t analysis (section

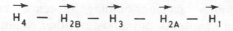

Fig. 8.3 The histone gene cluster of *Psammechinus*.

3.2.2) or the use of a cloned cDNA to isolate genomic clones. A family is indicated when the cDNA hybridizes to more than one area of the genome. Subsequent analysis may reveal that some members of a family are not expressed (pseudogenes) or are only expressed in specific tissues.

Multicopy genes are commonly found for RNA or protein species that require to be produced in large quantities, particularly when synthesis is concentrated into a short period of time. Thus, the multicopy histone genes (section 8.4.1) can be explained by the need, during the S-phase of the cell cycle, to synthesize sufficient histone to complex with a further copy of the complete genome. Such multicopy genes include those encoding rRNA (section 8.4.3), tRNA (section 8.4.5), 5S RNA (section 8.4.4), ribosomal proteins [41], histones (section 8.4.1) and abundant structural proteins such as collagen [42], actin [43], tubulin [43, 44], keratin [45], lens crystallins [46] and insect eggshell chorion [47].

If a gene duplicates and no evolutionary advantage results from the duplication, then the extra copy can mutate without adverse effects on the organism. In many cases, this will simply result in a pseudogene (section 7.6.3) but eventually advantageous new genes may result. Thus, the mouse α-foeto-protein and serum albumin genes exhibit considerable sequence homology and are assumed to have arisen by duplication of a common ancestral gene [48]. Many further examples could be listed among which the families of receptor proteins [49, 50], membrane transporter proteins [51], G-proteins [52] and many isoenzymes are

well-known examples. The globin genes are an example considered in section 8.4.2.

It should be noted that multiple related protein species are not always the products of a gene family; they can also be generated by differential processing of one gene transcript (section 11.9).

8.4 GENE CLUSTERING

8.4.1 Histone genes

The members of a gene family may occur closely coupled (arranged in tandem along the DNA), or they may be dispersed as individual genes or small clusters. Histone genes, which exhibit all three arrangements, will serve as an example (reviewed in [53, 54]).

Histones are small, highly conserved proteins that organize DNA into nucleosomes (section 3.3.2). Their genes were first analysed in detail in the sea urchin *Psammechinus miliaris* where they were found to be so highly repeated and clumped that, in caesium chloride–actinomycin density gradients, the genes migrated as a satellite, separable from the main band of DNA [55]. The histone genes so obtained, when digested with the restriction enzyme *Eco*R I or *Hin*d III gave rise to 6.3 kb fragments of DNA which contained the genes for all five histones [55]. Within the fragments, the genes were arranged in the order illustrated in Fig. 8.3. The arrows indicate the polarity of the genes and show that they are all transcribed in the same direction and therefore occur in sequence on the same strand of the DNA [56]. There is no evidence, however, that they are ever transcribed as a polycistronic mRNA. Each gene has its own TATA box promoter in its 5'-flanking sequences (section 9.4.1) and is separately transcribed by RNA polymerase

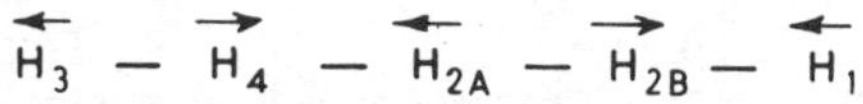

$$\overset{\leftarrow}{H_3} - \overset{\rightarrow}{H_4} - \overset{\leftarrow}{H_{2A}} - \overset{\rightarrow}{H_{2B}} - \overset{\leftarrow}{H_1}$$

Fig. 8.4 The histone gene cluster of *Drosophila*.

II. Each of the genes is separated from the next by non-transcribed AT-rich spacer and the quintets are arranged in tandem with 300–600 copies per haploid genome. Each cluster has one *Eco*R I and one *Hin*d III cleavage site; a fact that permitted their original isolation and cloning [55].

The availability of the sea urchin histone genes allowed their use as probes for the isolation and cloning of histone genes from other species. cDNA copies of purified histone mRNA were also employed as a probes. It was soon found that the histone genes of two other distantly related sea urchin species occur in very similar quintets [57] in which the main difference from *P. miliaris* was the size and sequence of the non-transcribed spacers. The spacers even show some heterogeneity (microheterogeneity) within a species.

The genes of the fruit fly, *Drosophila melanogaster*, are also in quintets [58] though here the order is different (Fig. 8.4). Furthermore, the *Drosophila* genes have a variable polarity within the cluster, some running from left to right and the others in the reverse direction. Since transcription is always in a $5' \rightarrow 3'$ direction (section 9.2), this means that the coding sequence of some genes is on one strand of the DNA whereas the rest are on the other. There are about 100 copies of the *Drosophila* histone genes.

At this point, it appeared that histone genes and their transcripts were characterized by a number of features that set them apart from other known protein-encoding genes. They were highly reiterated, clustered into tandem repeated quintets, contained no introns and their transcripts were not polyadenylated (section

11.5.2). It is now known that none of these generalizations are universally true. Particularly in vertebrates, some histone genes do contain introns [28], some of their transcripts are polyadenylated (section 11.5.1) and many do not occur as highly reiterated, tandemly repeated quintets.

Schümperli [59] has grouped the histones of higher eukaryotes into three classes:

1. *Replication-dependent variants.* A family of 5–20 genes per histone protein that are clustered but not organized into conserved repeats. Thus, within the amphibian, *Xenopus borealis*, at least 70% of histone genes occur in a major cluster that has the same gene order as *Drosophila* but is in a large, 16 kb, segment of DNA for which tandem linkage has not been demonstrated. The closely related *Xenopus laevis*, on the other hand, has at least two major gene cluster types each of which shows considerable length and sequence heterogeneity [60, 61]. Mammals and birds show even less regularity and isolated clones indicate that histone genes are clustered in a largely random fashion except that H2A and H2B tend to be paired [28, 54]. The genes are moderately conserved and contain no introns. Their expression is tightly linked to the S phase of the cell cycle [59] and the mRNAs are not polyadenylated. Instead they contain a palindromic 3'-sequence that is involved in the termination and processing of transcripts (sections 9.5(c) and 11.5.2).

2. *Replication-independent or replacement variants.* These histones accumulate in non-dividing, differentiated cells, are encoded in single copy genes and are transcribed into polyadenylated mRNA [62, 63].

3. *Tissue-specific variants.* The best known of these is histone H5, a histone found

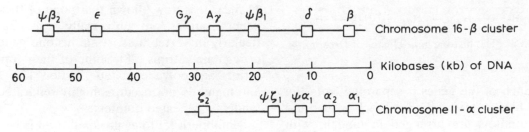

Fig. 8.5   The human globin gene clusters.

only in the nucleated erythrocytes of birds and amphibia.

8.4.2 Globin genes

Other much-studied gene clusters are those for the α and β globins (reviewed in [64]). In the adult mammal, haemoglobin consists of two alpha (α) and two beta (β) polypeptides. In the foetus, the β chain is replaced by gamma (γ) chains which, since they give haemoglobin a higher affinity for oxygen, facilitate the passage of the gas across the placenta from maternal to foetal blood. In human embryos of less than eight weeks two further polypeptides occur in haemoglobin. The β chain is replaced by epsilon (ε) polypeptides and the α chains are replaced by zeta (ζ) polypeptides. From eight weeks the embryonic polypeptides are gradually replaced by foetal chains and hybrid molecules occur.

There is then a family of α-like globins and a family of β-like globins and, in the genome, these are represented as clustered genes. The arrangement of the human globin genes is illustrated in Fig. 8.5. It can be seen that separate genes encode each of the embryonic, foetal and adult polypeptides and that some (α, ζ and γ) are encoded in duplicated genes. All members of each family have two introns inserted at the same position in the coding sequence as α and β

respectively. In both families, the genes are arranged in the order in which they are expressed and the control of β-globin gene expression is considered in detail in section 10.4.5. Both families also include pseudogenes (identified with the symbol ψ) which have arisen by duplication but are defective and do not produce a functional polypeptide. (Pseudogenes are discussed in sections 7.6.3 and 7.6.4.)

The globin genes of other primates are very similar to those of man and, although those of other vertebrates have their own embryonic variants, in mammals at least, it appears that the active genes are always arranged in a $5' \to 3'$ direction in the order in which they are expressed.

8.4.3 rRNA genes

(a) *The rRNA genes of prokaryotes*

In *E. coli* there are seven rRNA transcription units dispersed through the genome and arranged in operons that are termed *rrnA* to *E*, *G* and *H*. Each operon contains the genes for the three ribosomal RNAs in the order 16S, 23S, 5S together with one or more genes for tRNA. Each gene is separated from the next by a transcribed spacer. The number and location of the tRNA genes varies. In four of the seven operons the spacer between the 16S and 23S RNA

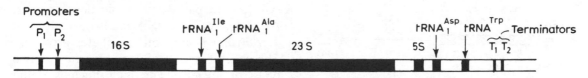

Fig. 8.6 A schematic representation of the *E. coli* ribosomal transcription unit, *rrnC*.

contains a tRNA^{Glu} gene whereas in the remaining three operons the spacer contains genes for tRNA^{Ile} and tRNA^{Ala} [65–68]. There may also be tRNA genes following the 5S RNA (trailer genes) and in the *E. coli* operon, *rrnC*, this region contains the only tRNA^{Trp} gene of the genome together with a gene for tRNA^{Asp} (Fig. 8.6). The operon *rrnD* contains two 5S RNA genes with a tRNA^{Thr} gene sandwiched between them [69]. All the operons are flanked by typical prokaryotic promoters (section 9.2.2) and rho-independent termination signals (section 9.2.4). The regulation of rRNA gene expression by the so-called stringent control in which limited availability of charged tRNA inhibits *rrn* operon transcription in relA^{+} (stringent strains) but not in relaxed strains is considered in section 12.10.1.

(b) *The rRNA genes of eukaryotes*

Ribosomal RNA represents more than 80% of the cellular RNA and, in order that the cell can produce the quantities required, the rRNA species are encoded in multiple genes (rDNA). Furthermore the genes are transcribed at a rapid rate with one RNA polymerase I molecule per 100 bp of rDNA each transcribing at 30 nucleotides per second (an electron micrograph of rRNA genes being transcribed is illustrated in Fig. 11.29).

The clusters of ribosomal genes occur at a few locations in the genome which are known as nucleolar organizers because of their ability to associate into nucleoli. It was

this clustering of ribosomal genes, and the high GC-content of the rDNA of some species, that permitted their early isolation as satellites on CsCl density gradients [70]. Clustering also explains why a single mutation in an anucleolate *Xenopus laevis* [70] or the so-called bobbed mutants in *Drosophila* [71, 72] can lead to the deletion of all or most rRNA gene copies. The nucleolus also contains the specific transcribing enzyme, RNA polymerase I (section 9.3) and other proteins that are involved in the transcription and processing of rRNA [73]. For reviews of the structure and, where relevant, the amplification of rDNA from a range of species see many of the chapters in [74] and [75].

Each nucleolar rRNA transcription unit contains the genes for three of the four ribosomal RNA species. These are known from their sedimentation constants in higher animals as 18S, 28S and 5.8S RNA but in fact their size is variable and in lower eukaryotes the major species may have sedimentation constants as low as 17S and 25S. Within the transcription unit, the genes for 18S, 28S and 5.8S RNA are separated from each other by internal transcribed spacers (ITS) and they are flanked on their 5'-side by an external transcribed spacer (ETS). The transcription units themselves are clustered in tandem repeats; each unit being separated from the next by the so-called non-transcribed spacer (NTS). Increasingly, however, it is being realized that, in some species at least, this spacer can be partially or almost wholly transcribed

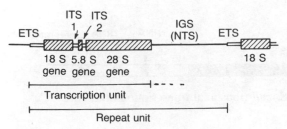

Fig. 8.7 The organization of eukaryotic rDNA (*Xenopus laevis*).

(sections 9.5.3 and 11.6.2) and it is more accurately named the intergenic spacer (IGS) (Fig. 8.7).

Considerable variation occurs in the number, arrangement, size and chromosomal location of the rDNA clusters (reviewed in [76–78]). Man has five clusters located on chromosomes 13, 14, 15, 21 and 22 [79]. Maize contains all of its several thousand genes in one organizer on chromosome 6 and *Drosophila* has organizers on the X and Y chromosomes [72].

Most lower eukaryotes have several hundred tandem rDNA repeated genes; mammals have 100–300; plants and amphibia have several thousand or even tens of thousands. There is, however, considerable species variation and Long and Dawid [76] have catalogued numerous instances in which closely related species or even strains exhibit large variations in the repetitiveness of their rRNA genes. It would appear that there are often more genes than are required for viability. In addition, some organisms are able to amplify their ribosomal rRNA genes. Amplification is a common feature of oogenesis and, in *Xenopus*, occurs by the production of large circles of extrachromosomal DNA each containing hundreds or even thousands of rDNA repeating units [80]. The amplification allows the accumulation of ribosomes and prepares the oocyte for the massive increase in protein

synthesis in early embryonic development. Amplification of rDNA is also a feature of *Tetrahymena* and other ciliate protozoa. The vegetative cell of *Tetrahymena* has two types of nucleus. The micronucleus is the germinal nucleus that undergoes meiosis in sexual conjugation. However, it is only a repository of genetic information and is not expressed. It contains a single ribosomal transcription unit [81]. The macronucleus is derived from the micronucleus, is polyploid and controls normal cell metabolism. Within it, the rDNA is amplified to 400 copies per haploid genome; equivalent to approximately 20 000 copies per polyploid nucleus [82, 83]. The mechanism of amplification is discussed in section 7.5.6.

A further source of variability in rDNA is in the length and arrangement of both the coding and the non-coding regions of the transcription units (Fig. 8.8). The genes for the major rRNA species are always arranged on the bacterial pattern, i.e. ETS-18S-ITS-28S and the 5.8S gene is inserted into the ITS. However, there is considerable variation on this theme. Largely due to differences in the transcribed spacers, the transcription unit can vary in length from 6–8 kb in lower eukaryotes [84] to 13 kb in mammals [85]. Furthermore, the intergenic spacers show even greater length diversity so that the length of the total repeating unit of rDNA varies from 9 kb in yeast [86] to 34–44 kb in mammals [87–89].

In the mature ribosome, 5.8S RNA is base paired to the 5′-end of 28S RNA and, in sequence and secondary structure the two are the evolutionary and structural equivalents of the 5′-end of prokaryotic 23S RNA. One can therefore think of the ITS between the 5.8S and 28S genes (ITS-2) as a piece of DNA inserted 150 nucleotides from the end of large ribosomal subunit RNA (5.8S RNA is reviewed in [90]). Similar arguments can be made for other variations

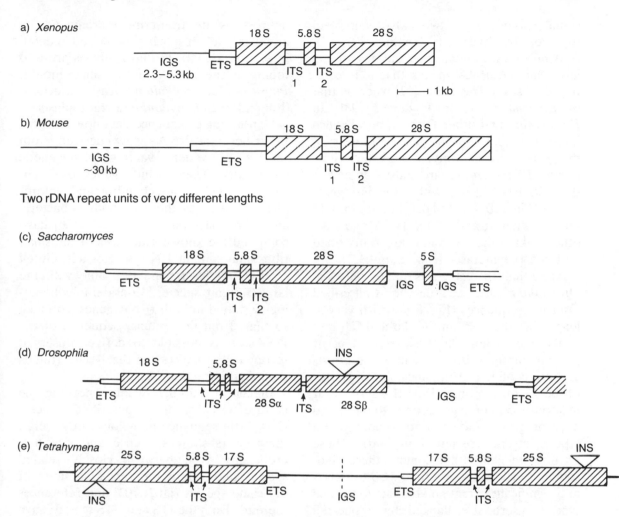

a) *Xenopus*

b) *Mouse*

Two rDNA repeat units of very different lengths

(c) *Saccharomyces*

(d) *Drosophila*

(e) *Tetrahymena*

Three unusual rDNA arrangements

Fig. 8.8 Some eukaryotic rRNA gene arrangements. (a) and (b) compare the very different repeat unit lengths of two vertebrates, (a) an amphibian (*Xenopus*) and (b) a mammal (mouse). (c), (d) and (e) illustrate three unusual gene arrangements. The *Saccharomyces* rDNA unit (c) includes the 5S genes that are transcribed from the other strand of the DNA by RNA polymerase III. In *Drosophila* (d), both the 5.8S and the 28S genes are fragmented by extra transcribed spacers and some 28S genes contain insertion sequences (INS) that are not transcribed. The extrachromosomal rRNA genes of *Tetrahymena* (e) occur as palindromes and the dotted line indicates the axis of symmetry. They may also contain insertion sequences (INS) that are group I introns.

that are found in the arrangement of rRNA genes. Thus the 28S rRNA gene of some eukaryotes is divided into two fragments, α and β, by another internal transcribed spacer (Fig. 8.8). As with the transcripts of other transcribed spacers the extra sequences are removed during processing (section 11.6) but as a result the RNA of the large ribosomal subunit is in two halves and held together by hydrogen bonding [91]. In trypanosomes, similar interruptions to the large subunit RNA means that it is composed of seven fragments [92] whereas that of *Euglena* is in 16 fragments [93]. In *Drosophila*, and other Diptera, an ITS also interrupts the 5.8S RNA gene [94]. Yeast of the genus *Saccharomyces* and the slime mould, *Dictyostelium*, are unusual in having their 5S RNA genes within the intergenic spacer (Fig. 8.8) [86, 95] but they are located on the other strand of the DNA and, like other eukaryotic 5S genes, are transcribed by RNA polymerase III (section 9.4.2).

About half the 250 genes for 26S rRNA in *Drosophila* also contain non-transcribed insertion sequences (Fig. 8.8) which vary in length between 0.5 and 6.0 kb and fall into two classes. Those known as type I occur predominantly as 0.5, 1.0 and 5.0 kb inserts and are confined to the nucleolar organizer of the X chromosome. Type II inserts occur in about 15% of the genes on both the X and Y chromosome and show no homology with type I inserts (reviewed in [96]). Those *Drosophila* genes that contain these ribosomal insertions are not processed to rRNA to a significant extent and the fact that some type I inserts are flanked by duplicated sequences suggests that they arose as transposable elements [96]. Similar non-transcribed inserts, which should perhaps be called interruptions have been found in a small percentage of the rDNA genes of a number of other insects (e.g. [97]).

Some unicellular protozoa have extra-chromosomal copies of rRNA genes. Those of *Tetrahymena* occur as palindromes each containing two transcriptional units usually in a linear molecule (Fig. 8.8) but in some cases as circles [98, 99]. In some strains and species of *Tetrahymena* these extrachromosomal genes are interrupted with a group I intron. This intron is of considerable current interest as its transcript is removed by autocatalytic self splicing (described in detail in section 11.3.2). Group I introns are also found in the rDNA of the slime mould, *Physarum polycephalum*, yeast mitochondria and in *Chlamydomonas* chloroplasts.

Considerable sequence data now exist for the ribosomal RNA genes derived from species of widely varying phylogenetic complexity. They include genes from eukaryotic nuclei, mitochondria and chloroplasts as well as those of both eubacteria and archaebacteria. Neefs *et al*. [100] have compiled the known sequences of the small ribosomal subunit RNA species and Gutell *et al*. [101] have done the same with the large subunit species. Considerable homology is found in both rRNA genes; so much so that, from the primary sequence of the DNAs, it is possible to derive conserved secondary structures for the RNAs (this is described in section 12.6.1).

In contrast to the regions encoding the mature rRNAs, the transcribed spacers show little sequence homology. In general, the spacers show sequence and size heterogeneity even between closely related species and within the repeating units of the same species ([102, 103] and references therein). Both the ITS and ETS tend to have regions rich in one or more nucleotides, lengths of homopolymers, dinucleotide polymers and repeated elements but again these are characteristic of the species and are not conserved between species. The intergenic spacers show even greater heterogeneity [103] and commonly vary in length in a single gene cluster. Since it is within this spacer that many of the sequence elements associated with the control of transcription occur, their variability explains in part the species specificity of rRNA transcription that is described in section 9.4.3.

8.4.4 5S rRNA genes

The sequences of over 600 prokaryotic and eukaryotic 5S rRNA genes have been compiled by Specht *et al.* [104]. The multiple genes that encode 5S RNA are not normally linked to those of other ribosomal RNA species although they are in the yeast, *Saccharomyces* (section 8.4.3 and Fig. 8.8) and in the slime mould, *Dictyostelium discoidium* [86, 95]. In some lower eukaryotes, including *Neurospora* [105] and the fission yeast, *Schizosaccharomyces pombe* [106] they are dispersed throughout the genome but in all higher eukaryotes so far investigated, including plants [107, 108], insects [109–111] amphibia (reviewed in [72]) and mammals [112], they are clustered through the genome in tandem arrays. Chromosome location is quite variable. For instance, the approximately 160 repeats of the *Drosophila* 5S gene appear to be arranged in a single cluster on chromosome 2 [109–111] whereas those of *Xenopus* occur at the telomeres of most if not all chromosomes [113].

The two genera, *Drosophila and Xenopus* have historically provided the favoured system for the study of 5S RNA genes. Those of *Drosophila* are simpler than *Xenopus* and the repeating unit consists of the 135 bp gene separated from its neighbour by an AT-rich spacer. This spacer varies in length depending on the number of copies of a hepatomeric base sequence [114] but in most units there are five copies, giving a 238 bp spacer and a repeat length of 373 bp. The genes of *Xenopus* are more complex and occur as two major types encoding oocyte and somatic 5S RNAs. The oocyte genes are only expressed during oogenesis whereas the somatic genes are expressed in oocytes and somatic cells (reviewed in [115]). Control of the process is discussed in section 9.4.2(c). The two RNA variants

differ from each other in six nucleotides [116]. In *X. laevis* there are 20 000 copies of the oocyte gene [117] compared with 400 copies of the somatic gene. It is assumed that, during oogenesis, the large number of the oocyte 5S genes allows their transcription to keep pace with that of the amplified genes for 18S and 28S rRNA. The repeat units of *X. laevis* oocyte 5S DNA are larger than those of *Drosophila* and each contains a pseudogene [118] as well as the 5S gene (pseudogenes are discussed in section 7.6.3). They are also heterogeneous in length, again largely because of variation in the length of an AT-rich spacer [119–121]. Figure 8.9 illustrates the organization of the repeating unit.

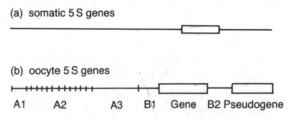

(a) somatic 5 S genes

(b) oocyte 5 S genes

A1   A2   A3   B1   Gene   B2 Pseudogene

Fig. 8.9 (a) The structure of the 880 bp repeat unit of the tandemly repeated *Xenopus* somatic 5S RNA genes. (b) The structure of the 650–850 bp repeat unit of the tandemly repeated *Xenopus* oocyte 5S RNA genes. A1, first 40 bp of the AT-rich portion of the spacer, constant in length and highly conserved. A2, a 15 bp repeating sequence; variation in the number of repeats is the main reason for heterogeneity in the length of the repeat unit. A3, a 24/25 bp unit repeated eight times with some variation at nucleotides 4, 13, 14, 18 and 20 and in the presence or absence of base pair 25. B1, 49 bp, GC-rich and much less repetitive than the remainder of the spacer but containing palindromes. Gene, 121 bp containing an internal promoter for RNA polymerase III (section 9.4.2). B2, 73 bp showing considerable homology with B1 plus the 3′ 24 bp of A3. Pseudogene, 101 bp, identical to the first 101 bp of the gene in all but 10 nucleotides.

The 5S DNA of mammals is also heterogeneous in length, sequence and possibly function and has very long repeat lengths which range from 6 kb in mouse, 2.2 kb in hamster and 2.8 kb in man [112, 119, 120]. Again the variation is due to sequences other than the gene which, like the RNA that it encodes, is highly conserved across the phylogenetic spectrum [122].

The 5S genes are transcribed by RNA polymerase III. The enzyme, its promoters, need for *trans*-acting transcription factors and termination sites are considered in chapter 9.

8.4.5 tRNA genes

(a) *Prokaryotic tRNA genes*

The structure and organization of the tRNA genes of *E. coli* and *B. subtilis* have been the subject of reviews [123, 124].

At least 54 tRNA genes, encoding 35 subspecies and representing an estimated two thirds of the total tRNA genes, have been mapped in *E. coli*. Many were located as a result of the mapping of suppressor mutations; most of which result from changes in the anticodon-encoding region of tRNA genes. Others have been mapped by demonstrating the enhanced synthesis of specific tRNAs after the insertion of a prophage or an F′ factor caused the amplification of expression. Still others were characterized by the analysis of tRNA gene clusters and of rRNA operons that contained tRNA genes (section 8.4.3). Fournier and Ozeki [123] have produced a chromosome map of *E. coli* showing that the 54 genes are dispersed throughout the genome with no correlation between their position and that of the genes for aminoacyl tRNA synthetases. Of the 54, however, 37 occur in polycistronic operons of which 14 are in the

seven rRNA operons (section 8.4.3 and Fig. 8.6). The remaining clustered genes occur in six operons that encode from 3 to 7 tRNAs per cluster. Fournier and Ozeki [123] have catalogued the sequences of 15 *E. coli* tRNA genes.

The analysed genes of the coliform bacteria *Salmonella typhimurium* reveal its close relationship with *E. coli*. Thus, one four-tRNA gene operon is almost identical in sequence and chromosomal position to its counterpart in *E. coli* [124, 125]. The more distantly related *B. subtilis*, however, differs considerably [126]. Its tRNA genes are heavily clustered into 10 known operons that are concentrated in two regions of the chromosome. One operon contains 21 genes and another 16 genes. All *E. coli* tRNA genes encode the CCA 3′-terminus, 12 of the 51 analysed *B. subtilis* genes do not. Repeated genes of a specific tRNA species occur in several *E. coli* operons but they have not been found in the operons of *B. subtilis*. Other differences between these examples of Gram-negative and Gram-positive eubacteria are listed by Vold [126].

(b) *Eukaryotic tRNA genes*

The structure and organization of eukaryotic tRNA genes has been reviewed by Clarkson [127]. Whereas prokaryotes contain one or a few copies of the gene encoding any given tRNA species, eukaryotic tRNAs are encoded in multiple copy genes. Presumably this is necessary in order that the cell can produce the required quantities of tRNA (20% of total cell RNA). Nevertheless, a tabulation of tRNA gene numbers by Long and Dawid [76] reveals that abundance is poorly correlated with genetic complexity. For instance, among the fungi, *Saccharomyces* has 360 genes for total tRNA whereas *Neurospora* has 2640.

The most striking feature of the arrangement of tRNA genes in eukaryotic cells is the species to species variation. Some, like many yeast genes, are dispersed throughout the genome [128, 129]. Others occur in clusters. The 600–800 *Drosophila* genes are arranged in approximately 60 such clusters spread over most of the chromosomes. The arrangement within a cluster is totally irregular, however, and any given cluster may contain several types of tRNA gene, some in multiple copies. Furthermore, they are irregularly spaced and transcribed from both strands of the DNA. One cluster that has been fully analysed contained eight tRNA^{Asn} genes, four tRNA^{Arg} genes, five tRNA^{Lys} genes and one tRNA^{Ile} gene [130]. The tRNA genes of the amphibian, *Xenopus*, occur in tandemly repeating clusters. One such cluster consisted of approximately 150 copies of a 3.18 kb repeating fragment that contained eight tRNA genes including two copies of tRNA^{Met}, and six single copy genes [131, 132]. In mammals, tRNA genes appear to be solitary or arranged in small clusters (1–2.5 kb DNA) with each cluster separated by a larger region (8–20 kb) of DNA that does not encode tRNA. A 13.4 kb rat DNA fragment containing two such clusters has been analysed by Rosen *et al.* [132]. One cluster contained tRNA^{Leu}, tRNA^{Asp}, tRNA^{Gly} and tRNA^{Glu} and the second cluster contained the same genes except that tRNA^{Leu} was a pseudogene. Blot analysis revealed that the clusters were part of a unit that was repeated about ten times.

The enzyme that transcribes eukaryotic tRNA genes, RNA polymerase III, together with the promoters and termination signals that it recognizes are described in chapter 9. With one known exception, each tRNA gene, whether solitary or part of a cluster is organized as a single transcription unit. The exception is a pair of yeast genes, tRNA^{Arg} and tRNA^{Asp} that are separated by only 10 bp and are transcribed as a 170-nucleotide dimeric precursor that is processed to the mature tRNAs [133]. Transcription depends on a promoter to the 5′-side of the gene pair [134].

8.5 THE GENES OF MITOCHONDRIAL AND CHLOROPLAST DNA

The genes of subcellular particles are organized and expressed in ways which differ both from those of prokaryotes and those of the eukaryotic nucleus (reviewed in [135–140]). Even within themselves they are very variable. In general, animals have small circular mitochondrial genomes of 14–18 kb, fungi have larger circles of 18–78 kb and plant mitochondria and chloroplasts have much larger circles. Higher plant mitochondrial DNAs vary from 200 to 2500 kb and even in a single family of flowering plants, the *Cucurbitaceae*, can vary over a sevenfold size range (330 kb to 2500 kb). The reason for this variation is the presence of repeated sequences that recombine, even within a single species to give circular molecules of various sizes. Thus, in *Brassica campestris*, the 218 kb, so called master chromosome, contains two copies of a 2 kb repeat that separate 83 kb and 135 kb segments of DNA. However, direct recombination between the repeats gives rise to lower numbers of smaller circles; one containing the 83 kb segment and the other the 135 kb segment [141]. The 570 kb master chromosome of maize mitochondria contains six repeats that recombine to give multiple-sized genomes [142]. In addition to this source of variability, many plant mitochondria contain DNA sequences derived from chloroplast DNA. Table 8.1 compares the structure and expression of the mitochondrial genomes of yeast and man.

Table 8.1 A comparison of the structure and expression of the yeast (*Saccharomyces*) and human mitochondrial genomes.

Genome structure and expression	Yeast	Human
Size	73–78 kb	16–57 kb
Shape	Circular	Circular
Encoded rRNA	15S and 21S	12S and 16S
Transcription	Asymmetric	Symmetric
Promoters	Five	One
Introns	In some genes	None
Processing	No polyadenylation	mRNAs polyadenylated

8.5.1 Protein-encoding genes of mitochondria and chloroplasts

(a) *Mitochondria*

Most mitochondrial proteins are synthesized from the transcripts of nuclear genes on cytosolic ribosomes and are then imported post-translationally into the mitochondria (section 12.8.3). However, some proteins, largely constituents of the inner mitochondrial membrane, are encoded in the mitochondrial genome. For instance, the 60 kb mitochondrial DNA of *Neurospora* [143] encodes three subunits of cytochrome oxidase, apocytochrome *b*, two or three subunits of the mitochondrial ATPase and seven subunits of NADH dehydrogenase. There appears, however, to be wide variation in the degree to which these electron transport proteins are mitochondrial or nuclear genes. Thus, in yeast, none of the NADH dehydrogenase subunits are mitochondrial genes. The α-subunit of mitochondrial ATPase is a nuclear gene product in mammals and fungi but is encoded in the mitochondrial genome of higher plants. The mitochondrial DNA of at least some plants encode genes for some of the proteins of the subcellular ribosomes [140] and many mitochondria have open reading frames that have not been assigned to known proteins.

Group I introns are found in some mitochondrial genes of fungi and higher plants and the processing of their transcripts is considered in section 11.3.2. Within the introns of fungal mitochondria there may be genes for endonucleases or maturases that are involved in the splicing (section 11.3.2) or mobility of the introns (section 8.2.1). The variability within mitochondria from various sources extends to the presence or absence of introns. Thus, the so-called *cob* gene, encoding apocytochrome *b*, exists in 2 and 5 intron versions in yeast but in mammals no mitochondrial genes contain introns and the *cob* gene is flanked by tRNA genes.

The mRNA transcripts of mammalian mitochondria are polyadenylated and the economy with which the genomic sequence is utilized is such that many of the gene transcripts require the poly(A) tail to complete the translation stop codon [135]. No obvious polyadenylation signals have been identified in mitochondrial transcripts and mitochondrial mRNAs are not capped [144].

(b) *Chloroplasts*

There are many more genes encoded in chloroplast genomes than in those of mitochondria. The complete nucleotide sequence has been determined for the chloroplast

DNA of the liverwort, *Marchantia polymorpha* [145] as has that of tobacco, *Nicotiana tabacum* [146]. The gene composition of the two species has been compared [138]. The liverwort chloroplast genome is 121 024 bp whereas that of tobacco is 155 844 bp. The size difference is largely accounted for by the length of the inverted repeat elements (IR_A and IR_B) that separate the large and small single copy regions (LSC and SSC) of all known chloroplast genomes except those of some Leguminosae. The gene composition of the two genomes is very similar. The liverwort has 128 genes, 9 of which are duplicated in the repeat elements. Tobacco has 146 genes, 24 of which are duplicates. There are thus 119 gene types in *Marchantia* and 122 in tobacco and they can be divided as follows. Approximately 60 genes are involved in transcription and, besides those for rRNA and tRNA, they include genes encoding some ribosomal proteins, subunits for chloroplast RNA polymerase and an initiation factor. A further 20 genes encode components of the photosynthetic electron transport chain; including some of those of photosystems I and II, cytochrome *b/f*, ATPase subunits and the large subunit of ribulose 1,5-bisphosphate carboxylase. Ohyama *et al.* [138] have reviewed the possible nature of the proteins encoded in some of the other open reading frames. They include probable components of the chloroplast nucleotide-driven transport system, genes with homology to 4 Fe−4 S-type ferredoxin and additional putative components of the electron transport system.

8.5.2 Mitochondrial and chloroplast rRNA genes

Mitochondria and chloroplasts have the capacity to make some of their own proteins and to do so produce their own ribosomes, the RNAs for which are encoded in their genomes. The sequences of the rRNA genes of both mitochondria and chloroplasts are known for a wide variety of species (listed in [100, 101]). In general they support the concept that these organelles may have arisen by endosymbiotic capture of bacteria. The argument is complicated however by the very variable gene arrangement, particularly in mitochondria. The small $5 \mu m$ circular mitochondrial DNA of mammals have their genetic information highly compressed. Their rRNAs are very small (12S and 16S) and the rRNA genes are interspersed with tRNA genes rather than spacers [147]. The large $25 \mu m$ circular DNA of yeast mitochondria, on the other hand, has the genes for the 15S and 21S rRNAs separated by a segment of DNA which is at least 25 kb long and contains two cytochrome oxidase genes and most of the mitochondrial tRNA genes [148]. Furthermore, the genes for the 21S rRNA contain an insert [148] and the two ribosomal genes are separately transcribed from different promoters [149]. The mitochondrial genes of *Aspergillus* [150] and the linear mitochondrial genome of ciliate protozoans [151] appear to present an intermediate degree of complexity. The mitochondrial rDNA of the algae, *Chlamydomonas reinhardtii* on the other hand has a much more scrambled arrangement. The 16S rRNA gene is split into four segments and the 23S gene is split into at least eight. The segments or modules are interspersed with protein-encoding and tRNA genes through a 6 kb stretch of the genome [152]. Higher plant mitochondrial rRNAs and their genes are larger than those of fungi and animals (26S equivalent to 3.5 kb and 18S equivalent to 2.0 kb) and, unlike the mitochondrial RNA of other organisms, they possess 5S genes. They exhibit much

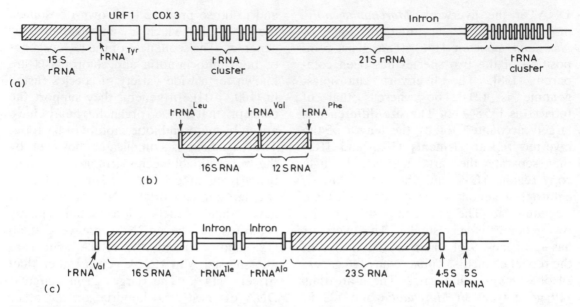

Fig. 8.10 Ribosomal DNA of subcellular particles. (a) mitochondrial rDNA of *Aspergillus*; (b) mitochondrial rDNA of mouse; (c) chloroplast rDNA of maize.

closer sequence homology to chloroplast and eubacterial rRNA genes than to other mitochondrial rRNAs and it has been suggested that they evolved separately [153]. Their 18S and 5S genes are tightly linked whereas those of 26S rRNA are located elsewhere in the repeat element.

Figure 8.10 illustrates the gene arrangement of the mitochondrial rDNA of *Aspergillus* and mouse (rat, human and bovine genes share the same arrangement as mouse).

The genes for the four rRNAs of higher plant chloroplasts are clustered in the order 16S, 23S, 4.5S and 5S [154, 155] and have intron-containing tRNA genes in their intervening sequences (Fig. 8.10c). In all known chloroplasts except those of a few legumes, the rDNA is part of the inverted repeat and there are therefore two sets of genes [145, 146, 156].

8.5.3 Mitochondrial and chloroplast tRNA genes

The human, mouse and bovine mitochondrial genome each encode 22 tRNA genes. They are located as singles, doubles, triplets and one five-gene sequence such that they separate from each other nearly all of the other mitochondrial genes (Fig. 8.10) (reviewed in [135, 136]). They occur on both strands of the DNA, the encoded tRNAs are smaller than their cytoplasmic counterparts (59−75 nucleotides) and they are relatively undermethylated. The 59 nucleotide $tRNA^{Ser}$ has lost the D stem and loop portion of the clover leaf secondary structure that is characteristic of all tRNAs (section 12.3.1).

As with rRNA genes, the tRNA genes of plant mitochondria show greater homology with chloroplast and bacterial tRNA genes than with those of other mitochondria (reviewed in [140]) and it thus appears that the chloroplast-derived sequences found in plant mitochondria can include active genes. Plant mitochondrial tRNA genes are not clustered and like yeast but unlike mammals they include a separate initiator tRNA^{Met}. It appears that the mitochondrial genome of many organisms includes genes for all of the tRNAs required for mitochondrial protein synthesis. However, some mitochondrial tRNAs of *Chlamydomonas*, *Tetrahymena* and higher plants are encoded in the nucleus and imported into the mitochondria [157–159].

The completely sequenced chloroplast genomes of the liverwort, *Marchantia* and of tobacco [145, 146], together with the substantially characterized genomes of *Euglena*, pea and geranium encode a complete set of tRNAs. Some of the genes are clumped whereas others are isolated. Some are duplicated in the inverted repeats and a few have been shown to contain introns.

REFERENCES

1 Bachmann, B. J. (1990) *Microbiol. Rev.*, **54**, 130.
2 Sanderson, K. E. and Roth, J. R. (1988) *Microbiol. Rev.*, **52**, 485.
3 Riley, M. and Krawiec, S. (1987) in *E. coli and Salmonella typhimurium: Cellular and Molecular Biology* (eds M. Schaechter and H. E. Umbarger), American Society Microbiology, Washington.
4 Holloway, B. W. and Morgan, A. F. (1986) *Annu. Rev. Microbiol.*, **46**, 79.
5 Platt, T. (1981) *Nucleic Acids Res.*, **9**, 6659.
6 Tilghman, S. M., Tiemeier, D. C., Polsky, F., Edgell, M. H., Seidman, J. G. *et al.* (1977) *Proc. Natl. Acad. Sci. USA*, **74**, 4406.
7 Tilghman, S. M., Tiemeier, D. C., Seidman, J. G., Peterlin, B. H., Sullivan, M. *et al.* (1978) *Proc. Natl. Acad. Sci. USA*, **75**, 725.
8 Schaffner, W., Gross, K., Telford, J. and

Birnstiel, M. (1976) *Cell*, **8**, 471.
9 Nagata, S., Mantei, N. and Weissmann, C. (1980) *Nature (London)*, **287**, 401.
10 McGeoch, D. J., Dalrymple, M. E., Davison, A. J., Dolan, A., Frame, M. C. *et al.* (1988) *J. Gen. Virol.*, **69**, 1531.
11 Aleström, P., Akasjärski, G., Perricandet, M., Mathews, M. B., Klessing, D. F. and Petterson, U. (1980) *Cell*, **19**, 671.
12 Woo, S. L., Beattie, W. G., Catterall, J. F., Dugaiczyk, A., Staden, R. *et al.* (1981) *Biochemistry*, **20**, 6437.
13 Wilks, A., Cato, A. C., Cozens, P. J., Mattaj, I. W. and Jost, J-P. (1981) *Gene*, **16**, 249.
14 Wozney, J., Hanahan, D., Tate, V., Boedtker, H. and Doty, P. (1981) *Nature* **294**, 129.
15 Feener, C. A., Koening, M. and Kunkel, L. M. (1989) *Nature (London)*, **338**, 509.
16 Bernards, A., Rubin, C. M., Westbrook, C. A., Paskind, M. and Baltimore, D. (1987) *Mol. Cell. Biol.*, **7**, 3231.
17 Lai, E. C. *et al.* (1979) *Cell*, **18**, 829.
18 Lambowitz, A. (1989) *Cell*, **56**, 323.
19 Scazzocchio, C. (1989) *Trends Genet.*, **5**, 168.
20 Rogers, J. H. (1989) *Trends Genet.*, **5**, 213.
21 Xu, M-Q., Kathe, S. D., Goodrich-Blair, H., Nierzwicki, S. A. and Shub, D. A. (1990) *Science*, **250**, 1566.
22 Jacquier, A. and Dujon, B. (1985) *Cell*, **41**, 383.
23 Macreadie, I. G., Scott, R. M., Zinn, A. R. and Rutow, R. A. (1985) *Cell*, **41**, 395.
24 Woodson, S. A. and Chech, T. R. (1989) *Cell*, **57**, 335.
25 Michel, F. and Jacquier, A. (1987) *Cold Spring Harbor Symp. Quant. Biol.*, **52**, 201.
26 Christopher, D. A. and Hallick, R. B. (1989) *Nucleic Acids Res.*, **17**, 7591.
27 Kuhsel, M. G., Strickland, R. and Palmer, J. D. (1990) *Science*, **250**, 1570.
28 Gilbert, W. (1978) *Nature (London)*, **271**, 501.
29 Darnell, J. E. (1978) *Science*, **202**, 1257.
30 Goebel, S. J., Johnson, G. P., Perkus, M. E., Davis, S. W., Winslow, J. P. *et al.* (1990) *Virology*, **179**, 247.
31 Sakano, H., Rogers, J. M., Hüppi, K., Brack, C., Traunecker, A. *et al.* (1979) *Nature (London)*, **277**, 627.
32 Stein, J. P., Catterall, J. F., Kristo, P.,

Means, A. R. and O'Malley, B. W. (1980) *Cell*, **21**, 681.

33 Lonberg, N. and Gilbert, W. (1985) *Cell*, **40**, 81.

34 Südhof, T. C., Goldstein, J. D., Brown, M. S. and Russell, D. W. (1985) *Science*, **228**, 8l5.

35 Craik, C. S., Bachman, S. R. and Beychok, S. (1980) *Proc. Natl. Acad. Sci. USA*, **77**, 1384.

36 Go, M. (1981) *Nature (London)*, **291**, 90.

37 Jensen, E. O., Paludan, K., Hyldig-Nielsen, J. J., Jorgensen, P. and Marcker, K. A. (1981) *Nature (London)*, **291**, 677.

38 Traut, T. W. (1988) *Proc. Natl. Acad. Sci. USA*, **85**, 2944.

39 Rogers, J. (1985) *Nature (London)*, **315**, 458.

40 Rogers, J. H. (1990) *FEBS Lett.*, **268**, 339.

41 Mager, W. H. (1988) *Biochim. Biophys. Acta*, **949**, 1.

42 Vuorio, E. and deCrombrugghe, B. (1990) *Annu. Rev. Biochem.*, **59**, 837.

43 Firtel, R. A. (1981) *Cell*, **24**, 6.

44 Gwo-Shu Lee, M., Lewis, S. A., Wilde, C. D. and Cowan, N. J. (1983) *Cell*, **33**, 477.

45 Fuchs, E. V., Coppock, S. M., Green, H. and Cleaveland, D. W. (1981) *Cell*, **27**, 75.

46 Piatigorsky, J. (1984) *Cell*, **38**, 620.

47 Orr-Weaver, T. C. (1991) *BioEssays*, **13**, 97.

48 Ingram, R. S., Scott, R. W. and Tilghman, S. M. (1981) *Proc. Natl. Acad. Sci. USA*, **78**, 4694.

49 Allan, G. F., Tsai, S. Y., O'Malley, B. W. (1991) *BioEssays*, **13**, 73.

50 Bonner, T. I. (1987) *Trends Neurosci.*, **12**, 148.

51 Trimmer, J. S. and Agnew, W. S. (1989) *Annu. Rev. Physiol.*, **51**, 401.

52 Libert, F., Parmentier, M., Lefort, A., Dinsart, C., Van Sande, J. *et al.* (1989) *Science*, **244**, 569.

53 Hentschel, C. C. and Birnstiel, M. L. (1981) *Cell*, **25**, 301.

54 Maxson, R., Cohn, R. and Kedes, L. H. (1983) *Annu. Rev. Genet.*, **17**, 239.

55 Schaffner, W., Gross, K., Telford, J. and Birnstiel, M. (1976) *Cell*, **8**, 471.

56 Gross, K., Schaffner, W., Telford, J. and Birnstiel, M. (1976) *Cell*, **8**, 479.

57 Wu, M., Holmes, D. S., Davidson, N., Cohn, R. H. and Kedes, L. H. (1976) *Cell*, **9**, 163.

58 Lifton, R. P., Goldberg, M. L., Karp, R. and Hogness, D. S. (1977) *Cold Spring Harbor Symp. Quant. Biol.*, **42**, 1047.

59 Schümperli, D. (1986) *Cell*, **45**, 471.

60 Old, R. W., Woodland, H. R., Ballentine, J. E. M., Aldridge, T. C., Newton, C. A. *et al.* (1982) *Nucleic Acids Res.*, **10**, 7561.

61 Turner, P. C. and Woodland, H. R. (1983) *Nucleic Acids Res.*, **11**, 971.

62 Kreig, P. A., Robins, A. J., D'Andrea, R. and Wells, J. (1983) *Nucleic Acids Res.*, **11**, 619.

63 Brush, D., Dodgson, J. B., Choi, O-R., Stevens, P. W. and Engel, J. D. (1985) *Mol. Cell. Biol.*, **5**, 1307.

64 Maniatis, T., Fritsch, E. F., Lauer, J. and Lawn, R. M. (1980) *Annu. Rev. Genet.*, **14**, 145.

65 Lund, E., Dahlberg, J. E., Lindahl, L., Jaskunas, S. R., Dennis, P. P. and Nomura, M. (1976) *Cell*, **7**, 165.

66 Lund, E. and Dahlberg, J. E. (1977) *Cell*, **11**, 247.

67 Young, R. A., Macklis, R. and Steitz, J. A. (1979) *J. Biol. Chem.*, **254**, 3264.

68 Brosius, J., Dull, T. J., Sleeter, D. D. and Moller, H. F. (1981) *J. Mol. Biol.*, **148**, 107.

69 Duester, G. L. and Holmes, W. M. (1980) *Nucleic Acids Res.*, **8**, 3793.

70 Wallace, H. and Birnstiel, M. L. (1966) *Biochim. Biophys. Acta*, **114**, 296.

71 Marrakechi, M. (1974) *Mol. Gen. Genet.*, **135**, 213.

72 Ritossa, F. (1976) in *The Genetics and Biology of Drosophila* (eds M. Ashburner and F. Novilski), Academic Press, New York, vol. 1B, p. 801.

73 Reeder, R. H. (1990) *Trends Genet.*, **6**, 390.

74 Busch, H. and Rothblum, L. (1982) *The Cell Nucleus* vols 10, 11 and 12, Academic Press, New York and London.

75 Sollner-Webb, B. and Tower, J. (1986) *Annu. Rev. Biochem.*, **55**, 801.

76 Long, E. D. and Dawid, T. B. (1980) *Annu. Rev. Biochem.*, **49**, 727.

77 Hadjiolov, A. A. (1980) in *Subcellular Biochemistry* (ed. D. B. Roodyn), Plenum Press, New York, vol. 7, p. 1.

78 Mandal, R. K. (1984) *Prog. Nucleic Acid Res. Mol. Biol.*, **31**, 115.

79 Henderson, A. S., Warburton, D. and Atwood, K. C. (1972) *Proc. Natl. Acad. Sci. USA*, **69**, 3394.

80 Miller, O. L. and Beatty, B. R. (1969) *Genetics*, **61**, (Suppl.), 133.

81 Yao, M-C., and Gall, J. G. (1977) *Cell*, **12**, 121.

82 Yao, M-C., Blackburn, E. and Gall, J. G. (1978) *Cold Spring Harbor Symp. Quant. Biol.*, **43**, 1293.

83 Yao, M-C., Blackburn, E. and Gall, J. (1981) *J. Cell Biol.*, **90**, 515.

84 Klootwijk, I., de Jonge, P. and Planta, R. J. (1979) *Nucleic Acids Res.*, **6**, 27.

85 Wellauer, P. K. and Dawid, I. B. (1973) *Proc. Natl. Acad. Sci. USA*, **70**, 2827.

86 Philippsen, P., Thomas, M., Kramer, R. A. and Davis, R. W. (1978) *J. Mol. Biol.*, **123**, 387.

87 Meunier-Rotival, M., Cortadas, I., Macaya, G. and Bernardi, G. (1979) *Nucleic Acids Res.*, **6**, 2109.

88 Kominami, R., Urano, Y., Mishima, Y. and Maramatsu, M. (1981) *Nucleic Acids Res.*, **9**, 3219.

89 Wellauer, P. K. and Dawid, I. B. (1979) *J. Mol. Biol.*, **128**, 289.

90 Walker, T. H. and Pace, N. R. (1983) *Cell*, **33**, 320.

91 Wellauer, P. K. and Dawid, I. B. (1977) *Cell*, **10**, 193.

92 Spencer, D. F., Collings, J. C., Schnare, M. N. and Gray, M. W. (1987) *EMBO J.*, **6**, 1063.

93 Schnare, M. N. and Gray, M. W. (1990) *J. Mol. Biol.*, **215**, 73.

94 Jordan, B. R., Latil-Damotte, M. and Jourdan, R. (1980) *Nucleic Acids Res.*, **8**, 3565.

95 Maizels, N. (1976) *Cell*, **9**, 431.

96 Glover, D. M. (1981) *Cell*, **26**, 297.

97 Lecanidou, R., Eickbush, T. H. and Kafatos, F. C. (1984) *Nucleic Acids Res.*, **12**, 4703.

98 Karrer, K. M. and Gall, J. G. (1976) *J. Mol. Biol.*, **104**, 421.

99 Engberg, J., Andersson, P. and Leick, V. (1976) *J. Mol. Biol.*, **104**, 455.

100 Neefs, J-M., Van de Peer, Y., Hendriks, L. and De Wachter, R. (1990) *Nucleic Acids Res.*, **18** (Suppl), 2237.

101 Gutell, R. R., Schnare, M. N. and Gray, M. W. (1990) *Nucleic Acids Res.*, **18** (Suppl), 2319.

102 Bourbon, H., Michot, B., Hassouna, N., Feliu, J. and Bachellerie, J-P. (1988) *DNA*, **7**, 181.

103 Gonzalez, I. L., Chambers, C., Gorski, J. L., Stambolian, D., Schmickel, R. D.

et al. (1990) *J. Mol. Biol.*, **212**, 27.

104 Specht, T., Wolters, J. and Erdmann, V. A. (1990) *Nucleic Acids Res.*, **18** (suppl), 2215.

105 Selker, E. U., Yanofsky, C., Driftmeier, K., Metzenberg, R. L., Alzner-DeWeerd, B. et al. (1981) *Cell*, **24**, 819.

106 Mao, J., Appel, B., Schaack, J., Sharp, S., Yamada, H. and Söll, D. (1982) *Nucleic Acids Res.*, **10**, 487.

107 Rafalski, J. A., Wiewiorowski, M. and Söll, D. (1982) *Nucleic Acids Res.*, **10**, 7635.

108 Goldbrough, P. B., Ellis, T. and Lomonossoff, G. P. (1982) *Nucleic Acids Res.*, **10**, 4501.

109 Hershey, N. D., Conrad, S. E., Sodja, A., Yen, P. H., Cohen, M. *et al.* (1977) *Cell*, **11**, 585.

110 Procunier, J. D. and Dunn, R. J. (1978) *Cell*, **15**, 1087.

111 Artavanis-Tsakonas, S., Schedl, P., Tschudi, C., Pirotta, V., Stewart, R. *et al.* (1977) *Cell*, **12**, 1057.

112 Emerson, B. M. and Roeder, R. G. (1984) *J. Biol. Chem.*, **259**, 7916.

113 Pardue, M. L., Brown, D. D. and Birnstiel, M. L. (1973) *Chromosoma*, **42**, 191.

114 Tschudi, C. and Pirrotta, V. (1980) *Nucleic Acids Res.*, **8**, 441.

115 Wolffe, A. P. and Brown, D. D. (1988) *Science*, **241**, 1626.

116 Ford, P. J. and Brown, R. D. (1976) *Cell*, **8**, 485.

117 Brown, D. D., Wensink, P. C. and Jordan, E. (1971) *Proc. Natl. Acad. Sci. USA*, **68**, 3175.

118 Jacq, C., Miller, J. R. and Brownlee, G. G. (1977) *Cell*, **12**, 109.

119 Hart, R. P. and Folk, W. R. (1982) *J. Biol. Chem.*, **257**, 11 706.

120 Krol, A., Gallinaro, H., Lazar, E., Jacob, M. and Branlant, C. (1981) *Nucleic Acids Res.*, **9**, 769.

121 Birkenmeier, E. H., Brown, D. D. and Jordon, E. (1978) *Cell*, **15**, 1077.

122 Erdmann, V. A., Wolters, J., Huysmans, E. and Dewachter, R. (1985) *Nucleic Acids Res.*, **13**, r105.

123 Fournier, M. J. and Ozeki, H. (1985) *Microbiol. Rev.*, **49**, 379.

124 Hsu, L. M., Klee, H. J., Zagorski, J. and Fournier, M. J. (1984) *J. Bacteriol.*, **148**, 934.

125 Bossi, L. (1983) *Mol. Gen. Genet.*, **192**, 163.

126 Vold, B. S. (1985) *Microbiol. Rev.*, **49**, 71.

127 Clarkson, S. G. (1983) in *Eukaryotic Genes: Their Structure, Activity and Regulation* (eds N. McLean, S. P. Gregory and R. A. Flavell), Butterworth Press, London, p. 239.

128 Olson, M. V., Montgomery, D. L., Hooper, A. K., Page, G. S., Horodski, F. *et al.* (1977) *Nature (London)*, **267**, 639.

129 Valenzuela, P., Venegas, A., Weinberg, F., Bishop, R. and Rutter, R. J. (1978) *Proc. Natl. Acad. Sci. USA*, **75**, 190.

130 Yen, P. H. and Davidson, N. (1980) *Cell*, **22**, 137.

131 Müller, F. and Clarkson, S. G. (1980) *Cell*, **19**, 345.

132 Rosen, A., Sarid, S. and Daniel, V. (1984) *Nucleic Acids Res.*, **12**, 4893.

133 Schmidt, O., Ma, O. J., Ogden, R., Beckman, J, Sakano, H. *et al.* (1980) *Nature (London)*, **287**, 750.

134 Kjellin-Straby, K., Engelke, D. R. and Abelson, J. (1983) International tRNA Workshop, Hakone, Japan.

135 Clayton, D. A. (1984) *Annu. Rev. Biochem.*, **53**, 573.

136 Tzagoloff, A. and Myers, A. M. (1986) *Annu. Rev. Biochem.*, **55**, 249.

137 Whitfield, P. R. and Bottomley, W. (1983) *Annu. Rev. Plant Physiol.*, **34**, 279.

138 Ohyama, K., Kohchi, T., Sano, T and Yamada, Y. (1988) *Trends Biochem. Sci.*, **13**, 19.

139 Schinkel, A. H. and Tabak, H. F. (1989) *Trends Genet.*, **5**, 149.

140 Levings, C. S. and Brown, G. G. (1989) *Cell*, **56**, 171.

141 Palmer, J. D. and Shields, C. R. (1984) *Nature (London)*, **307**, 437.

142 Lonsdale, D. M., Hodge, T. P. and Fauron, C. (1984) *Nucleic Acids Res.*, **12**, 9249.

143 Collins, R. A. (1987) in *Genetic Maps* (ed.

S. J. O'Brien.) vol. 4, Cold Spring Harbor Laboratory, Cold Spring Harbor, New York, p. 322.

144 Grohmann, K., Amalric, F., Crews, S. and Attardi, G. (1978) *Nucleic Acids Res.*, **5**, 637.

145 Ohyama, K., Fukuzawa, H., Kohchi, T., Shirai, H., Sano, T. *et al.* (1986) *Nature (London)*, **322**, 572.

146 Shinozaki, K., Ohme, M., Tanaka, M., Wakasugi, T., Hayashida, N. *et al.* (1986) *EMBO J.*, **5**, 2043.

147 Van Etten, R. A., Walberg, M. W. and Clayton, D. A. (1980) *Cell*, **22**, 157.

148 Borst, P. and Grivell, L. A. (1978) *Cell*, **15**, 705.

149 Levens, D., Ticho, B., Ackerman, E and Rabinowitz, M. (1981) *J. Biol. Chem.*, **256**, 5226.

150 Kochel, H. and Kuntzel, H. (1981) *Nucleic Acids Res.*, **9**, 5689.

151 Seithamer, J. J. and Cummings, D. J. (1981) *Nucleic Acids Res.*, **9**, 6391.

152 Boer, P. H. and Gray, M. W. (1988) *Cell*, **55**, 399.

153 Gray, M. W., Sankoff, D. and Cedergren, R. J. (1984) *Nucleic Acids Res.*, **12**, 5837.

154 Bedbrook, J. R. and Kolodner, R. (1979) *Annu. Rev. Plant Physiol.*, **30**, 593.

155 Rochaix, J. D. and Darlix, J. L. (1982) *J. Mol. Biol.*, **159**, 383.

156 Palmer, J. D. and Thompson, W. F. (1982) *Cell*, **29**, 537.

157 Suyama, Y, (1986) *Curr. Genet.*, **10**, 411.

158 Gray, M. W. and Boer, P. H. (1988) *Phil. Trans. Roy. Soc. (London)*, **B319**, 135.

159 Green, G. A., Marechal, L., Weil, J. H. and Guillemaut, P. A. (1987) *Plant Mol. Biol.*, **10**, 13.

9

RNA biosynthesis

Gene expression involves a sequential flow of information. In the first stage of this process, known as transcription, one strand of double-stranded DNA is copied into an RNA. In contrast to DNA replication, the genome is not copied in its entirety. Rather, defined units of genetic information are copied into RNA molecules which may function as messenger (mRNA), form a component of ribosomes (rRNA) or have an adapter function (tRNA). All three of these RNA products take part in the subsequent stage of gene expression, known as translation, in which mRNAs are used as coded messages for the synthesis of protein. Translation is the subject of chapter 12.

9.1 CONVENTIONS AND TERMS ASSOCIATED WITH RNA BIOSYNTHESIS

RNA biosynthesis is called *transcription*. A *transcription unit* is a segment of DNA (gene or genes) that is *transcribed* (copied) into a single RNA molecule by the enzyme RNA polymerase. Associated with this segment of the genome are various non-transcribed DNA elements such as *promoters*, *enhancers* and *operators* that in-fluence the frequency with which any transcription unit is transcribed (they are to be fully described in forthcoming sections). These elements may include nucleotide sequences that are recognized by RNA polymerase or some other DNA-binding protein that modulates RNA synthesis. A given recognition sequence, or one like it, may thus be found associated with all the genes influenced by a particular protein. Such sequences are known as *consensus sequences*. Typically, they are not totally conserved but are a consensus from which any similar sequence deviates only marginally.

Associated with the controlled transcription of a gene, there may also be elements, such as *regulator genes*, that are independently transcribed and cause the formation of proteins that influence the frequency of transcription. Segments of DNA that influence the activity of genes with which they are contiguous are described as *cis-acting*. This differentiates them from *trans-acting elements* that give rise to diffusible products that can act at many sites whether on the same or different chromosomes. *Trans*-acting elements are transcribed and translated into proteins called *trans-acting factors* or *transcription factors*. These influence transcription by interacting with promoters, enhancers and other *cis*-acting elements.

RNA is synthesized in a $5' \rightarrow 3'$ direction from the complementary strand of the DNA. However, when describing the nucleotide sequence of genes it is conventional to describe the *coding strand* as that having the same sequence as the RNA transcript (except that in the RNA, uridines replace thymidines). The point at which transcription starts is called the *initiation site* or *start site*. Sequences conventionally written to the left of the transcription unit are described as being 5' to the start site. Such elements are also described as being *upstream* whereas those on the 3'-side of the start site are described as *downstream*. Individual nucleotides are numbered away from the start site and are given positive values when downstream and negative values when upstream. Thus, a nucleotide 20 base pairs before the start site is −20 and one 10 base pairs after the start site is at position +10.

9.2 TRANSCRIPTION IS BEST UNDERSTOOD IN PROKARYOTES

9.2.1 *E. coli* RNA polymerase

The enzymes that catalyse the synthesis of RNA from a DNA template are known as DNA-dependent RNA polymerases. These enzymes require ribonucleoside triphosphates as substrates and transfer nucleoside monophosphates onto the 3'-OH terminus of the growing RNA chain. Polymerization is thus in a $5' \rightarrow 3'$ direction and produces an RNA chain the sequence of which is determined by Watson and Crick base-pairing with the DNA template. RNA polymerase activity was first detected in rat liver nuclei by Weiss and Gladstone [1] but it

is the bacterial enzymes, particularly that of *E. coli*, that are most fully characterized (reviewed in [2–4]). The complete *E. coli* enzyme or holoenzyme, of M_r 449068, contains five polypeptide chains: two α-chains of M_r 36512; one β-chain of M_r 150619; one β'-chain of M_r 155162 and one σ-chain of M_r 70236.

The β and β'-subunits each contain a Zn^{2+} ion [5] which is essential for catalytic activity. The enzyme requires all four ribonucleoside triphosphates and a divalent metal ion which *in vivo* is Mg^{2+} but *in vitro* can be Mn^{2+}. When isolated, the enzyme normally consists of a mixture of the holoenzyme and the core enzyme from which the σ-subunit, commonly known as the sigma factor, is missing. The core enzyme is the catalytic component of the enzyme. It has a general ability to bind to DNA and, if the DNA is nicked, can catalyse the synthesis of RNA from the nicks. It is not, however, able to catalyse specific initiation. The addition of a sigma factor to the core enzyme reconstitutes a fully active holoenzyme [6] that now has reduced affinity for non-specific DNA sequences but a considerably increased affinity for specific recognition sites near the beginnings of the sequences to be transcribed. These sequences are known as *promoters*. Their nature and role in the initiation of transcription by RNA polymerase is further discussed in the following section. The transcription of most genes requires the predominant or primary sigma factor. In many species, however, there are alternative sigma factors that recognize the promoters of different sets of genes. Alternative RNA polymerase sigma factors are thus a means for the selective expression of specific genes and this is considered in section 10.3.1.

The function and mode of action of the subunits within the core enzyme are incompletely understood. Numerous studies employing inhibitors, affinity labelling, reconstitution and genetic methods [7–9] confirm that α, β and β' are all necessary components of the core enzyme. They also show that the β'-subunit appears to be largely responsible for DNA binding. It is basic as would be expected of a protein with such a function and it binds the polyanion heparin that inhibits transcription *in vitro*. The β subunit has been implicated in binding both the substrates and the products of catalysis and it contains at least a portion of the catalytic site for RNA synthesis. It is also the site of action of antibiotics such as the streptolydigins which inhibit RNA chain elongation and rifamycins which inhibit initiation by preventing the formation of the first phosphodiester bond.

There are approximately 7000 RNA polymerase molecules in an *E. coli* cell and, depending on the rate of growth, about 2000–5000 of these will ·be actively transcribing RNA at any given time. The RNA polymerases of other bacteria, as well as those of blue–green algae, are similar to those of *E. coli* and contain subunits that are homologous to the α, β, β' and σ-chains. Indeed, in many though not all cases, the homology is such that reconstituted enzymes containing heterologous mixtures of subunits derived from different species are still catalytically active [3].

9.2.2 Initiation of prokaryotic transcription

The mechanism by which RNA polymerase initiates transcription has been most extensively studied with the enzyme from *E. coli* and has been shown to occur in three stages. First, the holoenzyme recognizes and binds to specific promoters. Second, this binding leads to the formation of the so-called 'open complex' in which a portion of the double helix is unwound. Finally, RNA synthesis is initiated (reviewed in [10]).

(a) *Binding of RNA polymerase to prokaryotic promoters*

The binding of RNA polymerase to transcriptional promoters appears to require an initial non-specific association with DNA. There are calculated to be 4×10^6 of these non-specific binding sites in the *E. coli* genome and the fact that this corresponds closely to the total number of base pairs (4.5×10^6) emphasizes the non-specific nature of the interaction. The non-specific binding constant of approximately $2 \times 10^{11} \, \text{M}^{-1}$ is strong when set against the binding of many proteins to their ligands. Nevertheless, it is relatively weak compared with the binding to specific promoters and the DNA remains in the double-stranded (closed) form.

The precise role of non-specific binding and the mechanism by which the enzyme moves from them to specific promoters is not fully understood. Von Hippel and coworkers [11, 12] have shown that three-dimensional diffusion processes are not fast enough to account for the transfer. Furthermore, they have reviewed evidence that suggests that non-specific binding is electrostatic and that, in this mode, facilitated transfer may occur by sliding along the DNA [13, 14] or by hopping [15] or a combination of both processes [16]. Heumann and coworkers [17, 18] have presented data favouring facilitated diffusion and have described the non-specific DNA as an 'antenna' along which RNA polymerase moves to the promoter. They reasoned that if such an effect occurred, occupancy of the promoter should be positively influenced by an increasing length of flanking sequence. Such an effect was demonstrated with downstream but not upstream sequences.

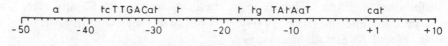

Fig. 9.1 Consensus sequence for the *E. coli* promoter.

Transfer from non-specific binding to a tight binding complex with a specific promoter depends on the presence of sigma factor. The binding constant for the association varies from 10^{12} to $10^{14}\,M^{-1}$ and the strength of the complex, referred to as *promoter strength*, is strongly related to the efficiency with which the associated gene is transcribed.

E. coli RNA polymerase holoenzyme will protect promoter regions from nuclease attack. The length of the protected segment varies with the nuclease and digestion conditions used but it is clear that the enzyme binds asymmetrically to the start site of transcription such that it covers up to 50 nucleotides on the 5′-side of the gene and up to 20 nucleotides into the gene [13]. Within the 5′-side of this protected area, there are two segments, centred around nucleotides −10 and −35, the sequences of which are strongly conserved. It is these segments that are the promoters to which the holoenzyme binds. The −10 sequence, often known as the Pribnow Box, has the consensus sequence TATAAT and the −35 region has a consensus sequence TTGACA [19, 20]. It should be emphasized that not all promoters have these sequences. However, any given promoter will vary from the consensus at no more than a few positions and this variation is more likely to occur in some nucleotides than in others. Thus, in a compilation of 112 well-defined *E. coli* promoter regions [21], a consensus sequence for the region −50 to +10 was derived. This is illustrated in Fig. 9.1, where upper case letters indicate strongly conserved nucleotides, lower case letters

represent weakly conserved nucleotides and no letters are inserted where there is no evidence for conservation. The importance of the consensus sequences has also been investigated in promoter mutants [21, 22] which are described as *down* or *up mutants* depending on whether they decrease or increase transcription from the promoter. In an analysis of 98 *E. coli* mutants, nearly all followed a general rule that down mutants showed decreased homology with the consensus sequence whereas up mutants had increased homology [22]. The length of DNA between the −10 and −35 regions is also fairly stable; 100 of the 112 promoter sequences compiled by Hawley and McClure [21] had 17±1 intervening nucleotides, however sequences ranging from 15 to 20 nucleotides are known. Similarly the distance between the −10 region and the transcription start site can vary from 5 to 9 nucleotides.

An overall picture of *E. coli* promoters is thus of fairly tightly constrained sequence and spatially defined elements but nevertheless with a substantial degree of variability. Many promoters are in fact weak; their function relies on and is controlled by other proteins that recognize regulatory elements and assist initiation (These controlling factors are considered in chapter 10.) Notwithstanding this, there are still problems. First, it is not clear that there is sufficient information in what is understood of promoter elements to define their unambiguous recognition by RNA polymerase. Second, there is little correlation between initial promoter recognition (rate of complex formation) and promoter strength defined

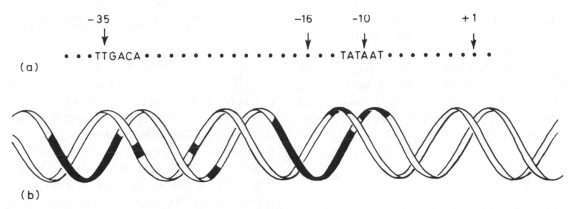

Fig. 9.2 Contact points for *E. coli* RNA polymerase on the promoter of the *β*-galactosidase gene. The sequence of the promoter coding strand (a) is aligned with the distribution of contact points (darkened areas) on one face of the DNA double helix (b).

as rate of RNA synthesis [23]. O'Neil [24], in an analysis of 52 well-characterized promoters, found that on average they conserved only 3.9 bases in the −35 consensus sequence and 4.2 bases in the −10 consensus sequence. This level of degeneracy could lead on average to promoter recognition every 200 base pairs of DNA; obviously with many false positives. Clearly, additional information must specify an RNA polymerase recognition site.

Some further information comes from the analysis of contact points between RNA polymerase and its promoter. Such experiments are performed by examining the extent to which the polymerase can protect specific nucleotides from base-modifying reagents. Alternatively, mutations and base-specific reagents can be tested for their effect on enzyme binding. Contact points are heavily concentrated in the −35 and −10 consensus regions of the DNA and are also clustered in the −16 area [20]. In all three areas, the contact nucleotides are not always those that are most strongly conserved (Fig. 9.2) so it would appear that features additional to base sequence, such as three-dimensional structure and

hydrogen-bond pattern [12], may be important in enzyme recognition. That three-dimensional structure must be important is indicated by the finding that the length of the DNA segment between the −10 and −35 regions has considerable influence on promoter strength. In the B form of DNA, the −35, −10 and −16 regions all lie on the face to which RNA polymerase binds; a change in the length of the intervening segment would skew this arrangement [25] which might also be influenced by supercoiling [26]. One recent suggestion is that there are three classes of promoter depending on whether there are 16, 17 or 18 nucleotides between the −10 and −35 regions [24].

Experiments with mutant sigma factors and with mutant promoters demonstrate that the factor recognizes both the −10 and the −35 regions on the promoter ([27] and references therein). Two out of four highly conserved regions of the polypeptide are involved in the recognition.

The promoters of bacteriophage that rely on the RNA polymerase of their host compete very effectively for the enzyme. It is perhaps surprising, therefore, that the consensus sequence of 29 bacteriophage T4 promoters differs significantly from that of *E. coli* [28]. The difference raises the possibility that additional proteins are involved in their recognition. Variant prokaryotic RNA polymerase promoters associated with the expression of complex gene sets are discussed in section 10.3.1.

(b) *Formation of an open promoter complex and the initiation of RNA synthesis*

With the tight binding of RNA polymerase to a promoter site, there is a rapid transition to the so-called *open-complex* in which the DNA is partially unwound (Fig. 9.3). Unwound regions of DNA are accessible to base-modifying reagents. Siebenlist *et al.* [20] exploited this and used dimethyl sulphate to methylate the N1 and N3 positions of adenines and cytosines, respectively, in the unwound segments of the open complex. They showed that a 12 bp region of the promoter is unwound and stretches from the middle of the −10 region to just beyond the RNA start site (specifically from −9 to +3). The fact that this region overlaps with a contact point between one strand of the DNA and sigma factor [27] indicates that the essential role of this subunit in correct initiation may be associated with the formation of open complexes. Buc and colleagues [29−32] have investigated both the opening and the associated unwinding of strong and weak promoters. They find that although the kinetics of opening are different, both open to the same extent, i.e. about 12 bp. They have further investigated the role of supercoiling in the process and have suggested that initially the DNA is wrapped around the polymerase in a nucleosome-like form [32]. In at least some promoters, the open complex is then believed to form via an intermediate step:

$$R + P \xrightarrow{\textit{fast}} RP_c \xrightarrow{\textit{rate-limiting step}} RP_i$$

$$\xrightarrow{\textit{major induced-fit}} RP_0$$

R is RNA polymerase and P is the promoter. RP_c is the closed enzyme promoter complex in which the DNA is still double-stranded and the −10 and −35 regions are not fully registered with the enzyme. RP_i is envisaged as an intermediate stage in which the DNA is still double-stranded but the −10 and −35 regions are in register and the enzyme is strictly positioned with respect to the DNA backbone. During the transition between RP_c and RP_i, the DNA is topologically unwound and the unwinding (untwisting or negative writhing) is converted into strand separation in the final step; the formation of the open complex, RP_0.

Transcription in prokaryotes is most commonly initiated with a purine nucleotide at a site 6 or 7 bp downstream from the −10 region. Of 88 promoter start sites examined by Hawley and McClure [21], A and G occurred at position +1 of the coding strand 45 and 37 times, respectively, and pyrimidines occurred in only six sequences. They also demonstrated that pyrimidines were preferred at positions −1 and +2 with CAT as the overall consensus sequence for the triplet −1, +1, +2 (Fig. 9.1). Thus, a consensus transcript would begin pppApU. The binding of the initiation nucleotide to RNA polymerase is an order of magnitude stronger than that of succeeding nucleotides and it has been suggested that it occupies a separate initiating site on the enzyme [33].

At an early point in the incorporation of nucleoside phosphates into new transcripts, two further changes occur in the initiation complex. In the first of these, the open complex is transformed into a ternary complex which is much more resistant to dissociation by high salt concentrations, is resistant to inhibition by rifampicin and in which RNA

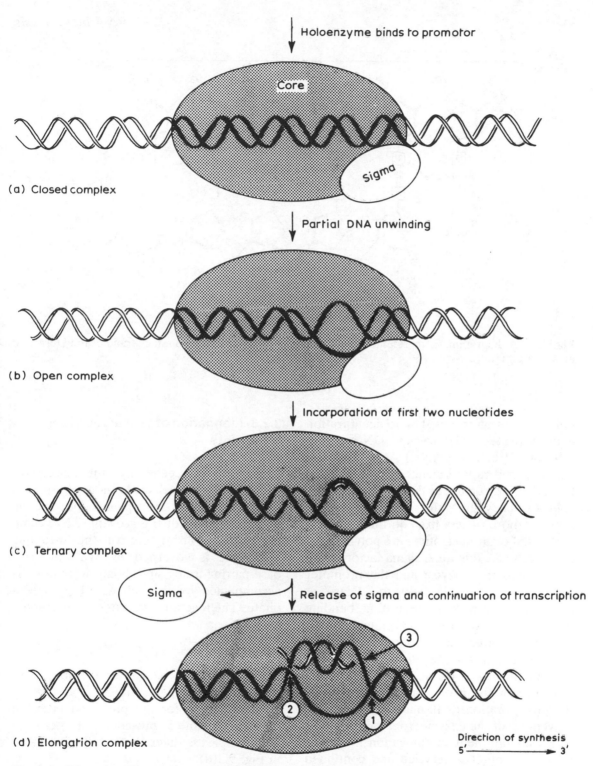

Fig. 9.3 Diagrammatic representation of the stages in the initiation of transcription. On the elongation complex (d) 1, 2 and 3 represent the active centres for the removal of positive supercoiling, the removal of negative supercoiling and elongation, respectively.

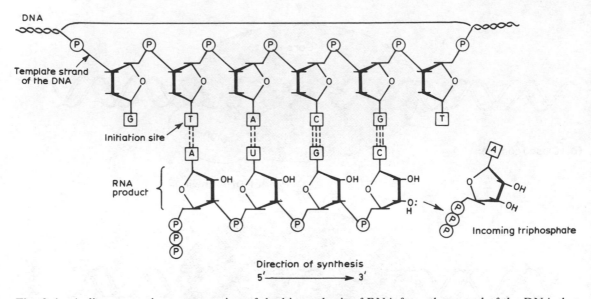

Fig. 9.4 A diagrammatic representation of the biosynthesis of RNA from the strand of the DNA that acts as a template.

polymerase is more resistant to denaturation and to proteolytic enzymes. DNase footprinting (section A.9.2), photoaffinity labelling and rapid kinetics have been used to study the structure of this *ternary complex* [34–36]. They show that RNA polymerase remains more or less fixed on the promoter as the first eight nucleotides are polymerized into RNA. At this time, sigma factor is still attached to the enzyme and the promoter shows increased sensitivity to DNase at position −25, perhaps indicating bending of the DNA in this region [34]. Sigma factor is released after the polymerization of 10 nucleotides [34] and the release is apparently dependent on and driven by the nucleotide sequence of the promoter [35]. Only after the incorporation of 10 nucleotides is there movement of the enzyme from the promoter and a commitment to elongation [34]. The released subunit is recycled and continued elongation of the transcript is catalysed by the core enzyme (Fig. 9.3).

9.2.3 Elongation of prokaryotic transcripts

Elongation proceeds by the successive addition of ribonucleoside monophosphates from substrate triphosphates on to the 3′-OH terminus of the growing RNA chain (Fig. 9.4). The nature of the incoming nucleotide is governed by Watson–Crick base-pairing rules and bond formation is accompanied by the release of pyrophosphate. The reaction can thus be represented as:

$$\text{pppXpY} \xrightarrow{n\text{ZTP}} \text{pppXpY}(\text{pZ})_n + n\text{PP}_i$$

X, Y and Z can be any of the four ribonucleotides although evidence has already been presented for the preferred insertion of purines as the 5′-nucleotide. The elongation complex is diagrammatically illustrated in Fig. 9.3(d).

There is a mechanical problem posed by the progress of RNA polymerase that is best conceptualized by a simple model. Take a double stand of fixed cord, such as the double string employed to draw a blind. If this is rotated repeatedly clockwise, a double helix is produced that can serve as our model of DNA. Now insert a pencil through the helix and move it away from you between the strings. You have, in effect, a mechanistic model of RNA polymerase moving along the DNA helix and, as you move the pencil, the string will overwind in front of the pencil and underwind behind. This is exactly what is considered to happen in transcription. The DNA will become positively supercoiled in front of RNA polymerase and negatively supercoiled behind it [37]. If we now continue our experiment, we rapidly reach a point at which we cannot push the pencil any further; there is too much resistance from the overwound string and it cannot be overwound anymore. In a cell, such overwinding must be released. One way to do this would be to allow the DNA and transcribing machinery to rotate around each other. This immediately appears unlikely. The enormously long filament of DNA could hardly rotate and the concept of the growing nascent RNA chain rotating around the DNA is almost as bizarre. The solution proposed by Liu and Wang [37], in their *twin-supercoiled-domain model*, was that waves of supercoiling generated by the transcribing polymerase would be released enzymically. DNA gyrase (topoisomerase II) would remove the positive supercoiling ahead of the RNA polymerase and DNA topoisomerase I would remove negative supercoiling behind it. Wang and coworkers [38, 39] have since used inhibitors of the different topoisomerases to confirm the existence of waves of supercoiling *in vivo*. Studies *in vitro* [40] have also provided supporting data.

RNA polymerases from a wide variety of bacteria catalyse the synthesis of RNA from a natural template at 12–19 nucleotides per second [41]. This is slower than the theoretical rate of up to 60 nucleotides per second because transcription from natural templates is not uniform. Many groups have shown that elongation is a discontinuous process in that there are pause sites along the DNA at which RNA polymerase slows or stops. Elongation is rapid between these sites (e.g. [42]). Pausing occurs at GC-rich regions and at points 16–20 nucleotides downstream of regions of dyad symmetry. At these latter regions the delay is associated with the formation, by the nascent RNA, of stem–loop structures either because it base pairs with itself or with the coding strand of the DNA [43, 44]. Pausing is a necessary preliminary to termination (section 9.2.4) and premature termination can occur at internal pausing sites (section 10.2). Viral RNA polymerases are apparently not delayed by the DNA of their hosts and have been reported to elongate at up to 200 nucleotides per second [45].

A number of inhibitors have proved useful for the study of elongation [41–45]. The antibiotics rifampicin and streptovaricin bind to the β subunit and block initiation whereas streptolydigin also binds to the β subunit but interferes with the elongation steps. Another agent which prevents elongation is actinomycin D. However, it does so, not by binding to the enzyme, but by complexing with deoxyguanosine residues in the DNA template and thus preventing movement of the core enzyme along the template (section 7.2.3).

9.2.4 Termination of prokaryotic transcripts

Three events are required when the transcription of a gene is terminated; elongation ceases, the transcript is released and RNA polymerase is released. Studies with pro-karyotes have demonstrated that these events occur by at least two mechanisms that are known as factor-independent termination and factor-dependent termination. Both of these have been the subject of reviews [29, 46–48]. Before discussing each in turn it should be stressed that the precise identification of termination sites is not easy. The 5′-end of a prokaryotic transcript can be unambiguously identified by its triphosphate group. No such marker defines the terminus of transcription and an observed 3′-end may be the result of post-transcriptional cleavage.

Termination has largely been studied *in vitro* using methods that involve the synthesis, by purified RNA polymerase, of specific transcripts from bacteriophage or from defined bacterial genes. Commonly used systems include the bacterial *trp* operon (genes encoding the enzymes that synthesize tryptophan), the DNA coliphages T7, T3, ϕX174 and bacteriophage lambda. Termination occurs at defined sites in these systems and, in a number of them, concern that the sites might be artifacts of *in vitro* incubation conditions has been dispelled by confirmation of their use *in vivo*.

(a) *Factor-independent termination*

Sites on the DNA at which independent termination occurs have a characteristic structure that comprises a GC-rich inverted repeat followed, on the template strand, by a run of adenylate residues. The former of these regions results in the formation of GC-rich regions in the transcript, which are able to base-pair into a 'hairpin' or stem–loop structure (Fig. 9.5). Such loops, which typically contain seven to ten G:C base pairs, have been shown to cause RNA polymerase to pause [49]. Mutations that lengthen the region of dyad symmetry, over

a minimum of 6 bp, strengthen the efficiency of termination as does the incorporation into the transcript of nucleotide analogues, such as bromo- or iodo-CMP, that stabilize G:C pairs. Conversely, weakening of G:C pairs, such as occurs when inosine is substituted for guanosine, decreases termination efficiency [46].

Shortly after the GC-rich region, a run of four to eight adenylates in the template will dictate that a string of uridylate residues are transcribed into the nascent RNA. Decreasing the number of these residues has been shown to reduce the efficiency of termination [50] and it appears that the importance of the sequence resides in the very unstable nature of the rU:dA hybrid [51]. Thus three features, the disruption of the RNA:DNA hybrid caused by G:C base-pairing, the pausing of RNA polymerase and instability of the rU:dA region, all combine to facilitate the release of the transcript from the template. O'Hare and Hayward [52], working with a termination site, T1, of coliphage T7, found that *in vitro* recognition of the stop signal and release of RNA occurred with a $t_{1/2}$ of 3 min whereas release of RNA polymerase took 12 min. However, more recent analyses indicate that the enzyme is released within 13 s of the cessation of elongation [53]. Platt [271] has suggested that RNA polymerase undergoes a conformational change to a termination complex at some time in this process and the concept is supported by the isolation of polymerase mutants that are modified in termination efficiency. Arndt and Chamberlin [54] have suggested that the RNA hairpins bring about this conformational change by converting the ternary elongation complex into what they call a 'release mode'.

Some recent data [55, 56] suggest that the above picture of factor-independent termination may be somewhat simplistic. It

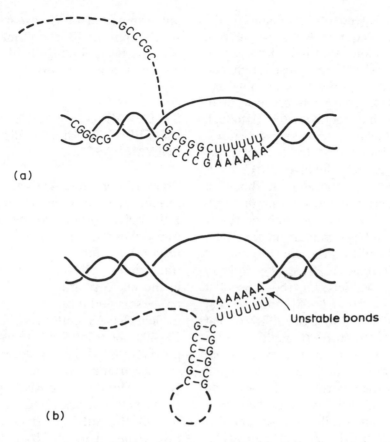

(a)

(b)

Fig. 9.5 Independent termination. RNA polymerase is omitted from the diagram for reasons of clarity (adapted from Platt [271]). (a) point at which elongation stops. (b) formation of hairpin followed by termination.

was demonstrated that sequences associated with the promoter can have a profound influence on termination but the way in which this occurs is totally unknown.

(b) *Factor-dependent termination*

Roberts [57] discovered a protein, which he called rho (ρ), that when added to an *E. coli* transcription system *in vitro* causes the generation of transcripts with discrete 3'-ends. The protein is basic in character and in solution consists of six identical, 419 amino acid subunits arranged in a ring

structure. It recognizes and binds to the nascent RNA transcript and the RNA-binding domain is in the first 151 amino acids of the subunit [58]. It binds to 13 ± 1 nucleotides per monomer and 78 ± 6 nucleotides per hexamer [59] such that the polynucleotide appears to be wrapped around or condensed within the protein oligomer [60]. The protein also exhibits RNA-dependent ATPase activity that is located in the carboxyl two thirds of the polypeptide [61] and is required for the helicase activity that brings about the unwinding of RNA:DNA hybrids [62–64].

As with independent termination, the activity of rho requires RNA polymerase to pause in its elongation [65]. Rho-dependent sites (*rut sites*) are cytosine-rich and guanosine-poor [260] but they do not have the degree of sequence conservation associated with rho-independent sites. Although many can be drawn as stem–loop structures, the base-pairing tends to be relatively unstable and largely involves A:U base-pairing. Indeed, a considerable body of evidence has accumulated showing that rho-dependent termination is enhanced by decreased secondary structure in the RNA transcript [65] and that rho binds to nascent mRNA that is relatively free of secondary structure [66]. Models for rho-dependent termination [46, 50, 67, 68] propose that the factor first binds to nascent RNA that lacks secondary structure. It may then move along the transcript until it finds a paused RNA polymerase and this movement could require the hydrolysis of nucleoside triphosphate [46]. Alternatively, the hydrolysis of NTP may result from a conformational change in rho brought about by the interaction of the nucleotide with the ternary complex. This in turn could cause the release of the nascent RNA chain ([62] and references therein).

It seems likely that there are a number of factors that bring about termination in prokaryotes. Nus A protein, which is considered in chapter 10 under the control of termination (section 10.2.2), is a termination factor that, in bacteriophage lambda, interacts with an anti-termination factor N. Antitermination is a process whereby RNA polymerase is rendered insensitive to some termination mechanisms. The best-known systems in which it has been studied are; the regulation of gene expression in bacteriophage lambda (section 10.2.2); the attenuation of amino acid biosynthetic operons (section 10.2.1) and

antitermination of *E. coli* rRNA synthesis which appears to share many similarities with the bacteriophage lambda system ([69] and references therein).

9.3 EUKARYOTES HAVE THREE DIFFERENT NUCLEAR RNA POLYMERASES

Transcription in eukaryotic nuclei is performed by three separate enzymes [70–74]. RNA polymerase I is located in the nucleoli, transcribes the genes for rRNA and is responsible for 50–70% of total RNA transcription. RNA polymerase II occurs in the nucleoplasm and synthesizes mRNA precursors and most of the U-series of small nuclear RNAs (snRNAs). It accounts for 20–40% of total RNA transcription. RNA polymerase III is also located in the nucleoplasm and transcribes a series of small RNAs. These include RNA species required for protein synthesis (tRNA and 5S RNA), one of the snRNAs required for mRNA processing (U6 snRNA), a small RNA required for protein transport (7SL RNA), RNAs required for the regulation of viral gene expression (adenovirus VA RNA and Epstein Barr Virus EBER RNA) and other small RNA species of unknown function, such as 7SK RNA. RNA polymerase III synthesizes approximately 10% of total transcribed RNA.

All three enzymes catalyse RNA synthesis on a DNA template as previously described for the *E. coli* enzyme and they require a divalent metal ion which can be Mg^{2+} or Mn^{2+}. When assayed *in vitro*, polymerase II is much more active with Mn^{2+} and polymerase III is slightly so. However, concentrations of Mn^{2+} *in vivo* are low and there is evidence that the use of Mn^{2+} *in vitro* may alter the binding properties of the enzyme [75]. In the analysis of enzyme activity, use is also made of their differing sensitivity to the fungal amatoxins (α-, β- and γ-amanitin, amanin and amanullin) of which the most commonly used is α-amanitin. RNA polymerase II is inhibited by α-amanitin at 50 ng/ml. RNA polymerase III is much less sensitive but is inhibited at concentrations of 5 μg/ml whereas RNA polymerase II is insensitive. De Mercoyrol [76] has demonstrated that α-amanitin does not block the formation of the first phosphodiester bond but inhibits the catalytic accumulation of trinucleotide, perhaps because it blocks an isomerization step required for translocation after phosphodiester bond formation. As with prokaryotic RNA polymerase, all three eukaryotic enzymes contain bound zinc, probably associated with a conserved amino acid sequence in the largest subunits [77].

Sedimentation analyses show that all three enzymes are very large with M_r values of 500 000–600 000 and they contain up to 15 subunits. Tabulation of the components of the three polymerases from various sources [70] reveals substantial interspecies variation in the M_r and numbers of the subunits. Thus polymerase III has been reported to have from 9 to 15 subunits depending on the species from which it is isolated. In general terms, however, each enzyme possesses two non-identical, large subunits of M_r 120 000–220 000 and up to 13 smaller subunits which, with the exception of an 80 000–90 000 subunit in polymerase III, have M_r of less than 50 000. Lewis and Burgess [78] have discussed the difficulty of ascertaining the precise number of subunits in the active enzymes. Some subunits may only be present in the purified enzyme at a molar ratio of less than one molecule per enzyme molecule. Several explanations are possible for such a lack of stoichiometry. The polypeptide concerned could be a contaminant or it could be a genuine subunit that is differentially lost during enzyme purification. If a subunit functioned as a factor, this could also explain low abundance. A factor like sigma in the *E. coli* enzyme, that is only required at certain points in the transcription cycle, would not necessarily occur in stoichiometric amounts on the purified enzyme.

Since it is not yet possible to reconstitute active enzyme from its subunits, the functional significance of any component is hard to ascertain but considerable progress is being made with the recently developed method of epitope tagging. In the first stage of this technique, a gene encoding a polymerase subunit is modified so that, during protein synthesis, an extra string of amino acids is inserted on the end of the polypeptide. The extra amino acids are chosen so as to minimize disruption of the protein and, most importantly, so that they are recognized by a well-characterized monoclonal antibody. The enzyme can then be purified by gentle immunoaffinity methods and the subunit composition compared with that derived after more traditional purification techniques.

Perhaps the most studied enzyme is RNA polymerase II which has been purified to near homogeneity from more than 20 different species. That of the yeast, *Saccharomyces cerevisiae*, has ten subunits (RPB1–RPB10) whether purified by conventional techniques or by immunoprecipitation with a monoclonal antibody against epitope-tagged subunit 3 [79]. Each subunit is encoded by a single copy gene but the enzyme appears to be composed of one polypeptide of each of RPB1, 2, 6, 8 and 10, two copies of each of RPB3, 5 and 9, and less than stoichiometric amounts of subunits RPB4 and 7.

The two largest subunits of the enzyme contain extensive regions of homology with the β and β'-subunits of *E. coli* RNA polymerase [80] and are related in size and sequence to the large subunits of RNA polymerases I and III [73]. The largest subunit of 220 kDa is the largest of any of the RNA polymerases and experiments involving cross-linking, together with susceptibility to antibodies and proteases, indicate that it is involved in binding to the DNA template [71]. It contains, in addition to the regions of homology with other enzymes, a carboxy-terminal domain (CTD) that forms an extension or tail to the core enzyme and consists of tandem repeats of a seven amino acid sequence: Tyr-Ser-Pro-Thr-Ser-Pro-Ser (reviewed in [81]). The tandem repeats are common to all known RNA polymerase II enzymes but the number of repeats varies from 17 in the malarial parasite to 52 in mammals. The extension is essential for enzyme function and may be unphosphorylated (RNA polymerase IIA) or extensively phosphorylated (RNA polymerase IIO). Phosphorylation causes a conformational change in the carboxyl domain [261] and recent data indicate that polymerase IIA forms the initiation complex but that phosphorylation of the *C*-terminal domain occurs before the initiation of transcription [82]. There is also evidence that transcription factors could interact with the domain [83].

The next largest subunit of RNA polymerase II, RPB2, appears to participate in substrate binding [85] and phosphodiester bond formation, thus showing functional homology with the β subunit of bacterial enzymes. The third largest subunit is also an essential component of the transcription apparatus [84], may be the equivalent of subunit RPC5 that is common to RNA polymerases I and III, and is probably the functional homologue of prokaryotic sigma factor [84]. At least three other subunits, RPB5, RPB6 and RPB8, are immunologically and biochemically indistinguishable

in RNA polymerases I, II and III [86]. It would thus appear that many of the enzymic components are substantially preserved between the three eukaryotic enzymes and indeed exhibit considerable similarities with prokaryotic RNA polymerase.

9.4 THE INITIATION OF EUKARYOTIC TRANSCRIPTION

The study of eukaryotic RNA polymerase activity has been much enhanced by the development of systems in which the enzymes correctly transcribe specific genes *in vitro*. The first of these was a transcription system for polymerase III [87] and led to the use of extracts of *Xenopus* oocytes as a major system to support transcription by this enzyme [88, 89]. Similar systems, employing extracts of tissue culture cells, were also described for the transcription of cloned eukaryotic genes and of viral genes by both polymerases II and III [90, 91]. The last to be developed was a cell-free extract that correctly initiated the transcription of pre-rRNA from cloned rDNA [92]. Such polymerase I-dependent systems are species-specific [93].

Transcription is also studied by the microinjection of cloned eukaryotic genes into *Xenopus* oocytes [94], by the stable introduction of genes into mammalian cells in culture [95] and by the insertion of cellular genes into an SV40 viral vector [96] or a plasmid containing the SV40 or other viral replication origins [97].

The use of such systems has shown that RNA synthesis by eukaryotic polymerases has much in common with that catalysed by the prokaryotic enzyme. Since there are three nuclear enzymes, however, one may expect differences between the three in promoter recognition as well as termination and control of transcription.

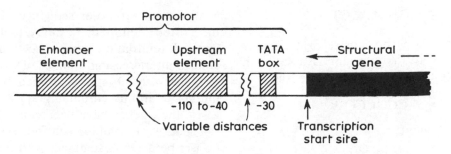

Fig. 9.6 A diagram of the promoter region for RNA polymerase II.

9.4.1 Initiation by RNA polymerase II

The promoter for RNA polymerase II consists of a number of sequence elements required for accurate and efficient initiation of transcription (reviewed in [71]). Two of these show similarities with the conserved sequences of prokaryotes. The first conserved element, identified when the 5'-regions of eukaryotic gene sequences were compared, was originally named the Goldberg/Hogness box after its discoverers [98]. It is now usually called the TATA box and is an AT-rich region with the consensus sequence:

$$5' \ TATA^A_TA^A_T \ 3'$$

It occurs approximately 30 bp upstream from the transcriptional start site in most eukaryotes (Fig. 9.6) but 60–120 nucleotides upstream of the start site of yeast genes. It has been called the selector sequence and its function is to fix the location of the start site. Mutations in the consensus sequence can profoundly affect correct initiation. Conversely, nucleotide deletions to the transcription start site region result in initiation at a new site, still approximately 30 bp downstream from the TATA box. Although this element is obviously very similar to the prokaryotic Pribnow box two differences are thus immediately apparent. First, the Pribnow box is 10 bp (one DNA helical turn) from the start site of prokaryotic transcription. Second, there is a degree of sequence conservation in the prokaryotic start site but much greater variability in eukaryotes. Transcriptional analysis and S1 nuclease mapping have shown that a functional TATA box precedes most genes transcribed by RNA polymerase II. There are, however, genes, such as the late SV40 genes and U1 snRNA genes, that contain equivalent sequences that can at best only be described as TATA-like [99, 100]. They may form another class of promoter. Other genes that lack TATA boxes are the so-called housekeeping genes that encode constitutively expressed proteins that are continuously expressed but at a low level [101]. The genes have a CG-rich promoter that is discussed in section 10.6.2 and their transcripts have multiple 5'-ends.

The second element in the RNA polymerase II promoter is the so-called upstream promoter element (UPE). It occurs in most protein-encoding genes and consists of one or more 8–12 nucleotide segments located a variable distance (−40 to −110) upstream of the transcription start site (Fig. 9.6). Whereas the function of the TATA box is to ensure accurate initiation of transcription, the UPE increases the rate of transcription. One common variant of the UPE is the so-called CAAT or CCAAT box which has the consensus sequence:

$$5' \ GG^T_CCAA^T_ACT \ 3'$$

and occurs 70–90 bp upstream from the start site. A further variant is a GC-rich sequence (GC box) with the consensus:

$$5' \ CCGCCC \ 3'$$

or its complement:

$$5' \ GGGCGG \ 3'$$

which can occur in one or more copies [102, 103] and may be present in addition to a CAAT box as the so-called −100 element [104].

More complex regulation of transcription is provided by enhancer elements (Fig. 9.6). These are short stretches of nucleotides which act in *cis* to increase the transcription of nearby genes. They function relatively independently of position, orientation and distance. Some show narrow tissue and temporal specificity whereas others permit expression in many cell types.

Although presented above as three separate elements, the distinction between promoters, upstream elements and enhancers is blurred. They share many properties and often overlap, with the same consensus sequence serving as both promoter and enhancer. It will be convenient here, however, to discuss the role of TATA boxes and UPEs in the initiation of transcription while leaving enhancers and their role in the regulation of transcription to be considered under the control of gene expression (chapter 10).

Most, probably all, of the above sequence elements are targets for proteins. Indeed, eukaryotic RNA polymerases are unable to recognize promoters without the aid of proteins. Often called transcription factors, these proteins are frequently present at very low concentrations but they are initially detected in cellular fractions required for transcription *in vitro* from a DNA template containing a promoter sequence. They are then often identified by DNA footprinting and gel retardation assays (section A.9.2) and, in an increasing number of cases, they have been substantially purified by recognition-site affinity chromatography. In some cases, their genes have been cloned.

The use of soluble, cell-free systems to transcribe a DNA template with a minimal TATA box promoter led to the identification of sets of transcription initiation factors from various species and cell types. Inevitably, this resulted in multiple names and acronyms that without doubt describe the same or closely related proteins. Table 9.1 presents the most generally accepted names that will be used for the remainder of this discussion and also lists the probable functional equivalents that have been described from other systems.

The pivotal factor in the recognition of the TATA box is most commonly called TFIID (transcription factor D for RNA polymerase II). It binds the TATA box independently, without the involvement of other factors or RNA polymerase ([111] and references therein). It is also apparently the factor that recognizes and binds to promoters with little TATA box homology [112]. TFIID has been described as a commitment factor as its binding is a prerequisite for the formation of the basal transcription apparatus (Fig. 9.7) that includes RNA polymerase II and several more factors: TFIIA, TFIIB, TFIIE (reviewed in [113]). Of these, TFIIA stimulates and facilitates the binding of TFIID but is not always required. TFIIE is multicomponent and has been separated into TFIIE and TFIIF of which the latter consists of two polypeptides of 30 kDa and 78 kDa. Impure TFIIF has helicase activity [109] but this is not present in pure FC which is probably the same protein [108]. More recently, a further factor, TFIIG, that is also an essential

Table 9.1 Transcription initiation factors for RNA polymerase II

Most common name	*Other names or probable functional equivalents in various eukaryotic systems*		*Function*	*Refs*
TFIID	TATA factor, BTF1, DB	B and D FD and FF	Formation of basal transcription complex	[71, 105, 107]
TFIIA	STF, AB	τ and ε	Stimulates TFIID binding but not always required	
TFIIB	BTF3, CBI, α	FB + FE	Stabilization of initiation complex.	
TFIIE	Multicomponent and including : TFIIE + [TFIIF (30 k + 78 k) RAP 30/74 FC 30 + 80] BTF2 $\beta\gamma$		May include DNA helicase activity [109]. The boxed components are probably the same two polypeptides.	[71, 106, 107 108, 110]

component of the transcription complex, has been identified [262].

TFIID is absolutely required for transcriptional initiation and its binding is stable through several rounds of transcription [114]. Both binding [111] and the activation of transcription [115] are dependent on a large *C*-terminal region of the protein that stretches from residue 63 to 240. Within this is a highly basic domain (residues 120–156) and a region (residues 197–240) that has sequence similarity with the −10-binding region of prokaryotic sigma factor [111, 116].

As well as being pivotal in the formation of the basal transcription factor, TFIID also plays a role in the modulation of transcription. A transcription factor SP1 interacts with the GC box of the UPE and in so doing causes a fivefold increase in transcription. Evidence collected by Pugh and Tjian [117] suggests that the complex of the GC box and SP1 causes stimulated transcription by interacting with TFIID promoter complex via a co-activator (Fig. 9.7). Similarly, the factor CTF activates transcription when it binds to the CAAT box and is again believed

to do so by interaction of the complex with TFIID [117]. There are multiple CAAT-binding proteins each of which binds to partially overlapping sets of genes [263] but whether they all interact with TFIID is unknown. Other, more specific transcription factors, are, however, thought to act through TFIID. These include the herpes virus activator protein VP16 [118], GAL 4 that regulates the genes encoding the galactose-metabolizing enzymes of yeast [119] and the mammalian activator protein ATF [264].

9.4.2 Initiation by RNA polymerase III

(a) *Internal promoters for RNA polymerase III*

The promoters for the tRNA and 5S rRNA genes do not lie in their 5′-flanking sequences but within the coding region of the genes themselves (reviewed in [73, 120, 121]). This unexpected finding resulted from the work of Brown and colleagues on the expression of *Xenopus* 5S RNA genes in

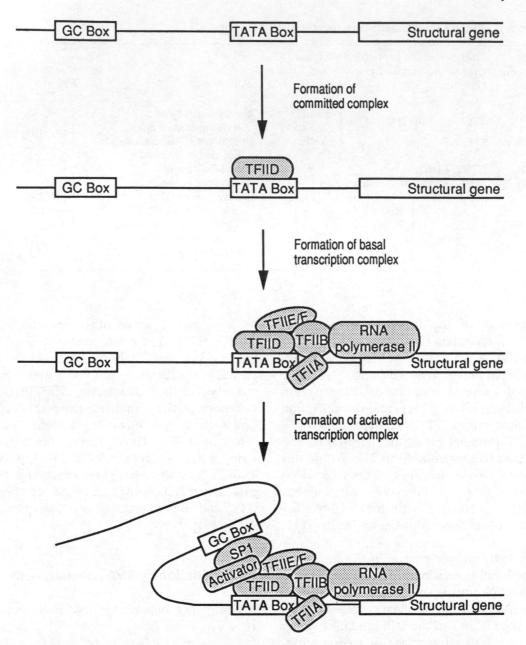

Fig. 9.7   Diagrammatic representation of initiation by RNA polymerase II.

oocytes and in systems for transcription *in vitro*. They found that the entire 5′-flanking sequence of the gene could be removed without the loss of 5S RNA tran-scription. Furthermore, when they deleted the 5′-end of the coding sequence, they found that 5S-sized RNA was still made from a new start site that corresponded to

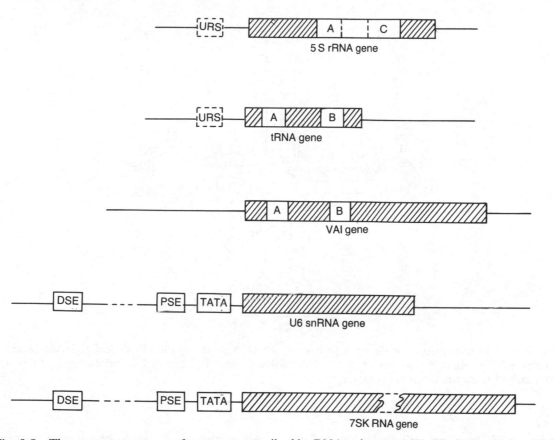

Fig. 9.8 The promoter structure for genes transcribed by RNA polymerase III. The boxes A, B, and C are the internal promoter elements of 5S RNA, tRNA and VAI genes. The dotted outline boxes, URS represent upstream regulatory elements that are present in the flanking sequences of some 5S and tRNA genes. TATA indicates the presence of RNA polymerase II-like TATA boxes. PSE and DSE are proximal and distal sequence elements of the U6 and 7SK genes.

where the 5′-end of the gene had been. Not until 50 coding nucleotides had been deleted was there a drop in the efficiency of transcription [122]. Subsequent experiments, in which the 3′-end of the gene was deleted [123], or in which extra nucleotide sequences were inserted into the gene [122], showed that the synthesis of 5S RNA was controlled by a promoter located between residues 50 and 83 of the *Xenopus* gene. The promoter is known as the *internal control region* or *ICR* (Fig. 9.8) and is required and sufficient for transcription. It is not, however, re-

cognized by RNA polymerase III [124] but by a transcription factor, known as TFIIIA [125], which specifically interacts with two factor-binding domains at the boundaries of the promoter region [126]. These domains have become known as Block A and Block C in reference to the two-block promoter of tRNA genes (see below).

TFIIIA was the first eukaryotic transcription factor to be discovered and is the archetype of a group of proteins that interact with DNA through so-called zinc fingers (section 10.5.2). The protein has nine of

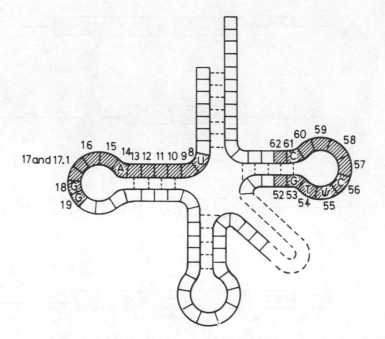

Fig. 9.9 The relationship of the split promoter of tRNA genes to the clover-leaf tRNA secondary structure. The shaded portions of the tRNA molecule and the labelled invariant nucleotides are transcribed from the internal promoter.

these Zn^{2+}-binding domains that jointly interact with approximately five helical turns of the ICR. The interaction is strongest with box C (base pairs 80–97) but also occurs with box A (bp 49–60) and weakly with some intervening regions. Most of the zinc fingers and the DNA binding activity are in the *N*-terminal portion of the polypeptide but this portion of the protein cannot by itself bring about formation of a stable transcription complex. Neither will it permit the cooperative binding that appears to occur between the tandemly arranged 5S RNA genes [127]. The binding of TFIIIA is the first stage in the formation of a 5S transcription complex and is followed by the sequential binding of TFIIIC and TFIIIB [128]. It is the complete complex that is recognized by RNA polymerase III [128].

Similar studies with tRNA genes have

shown that they too have an internal promoter (Fig. 9.8) but in this case it is clearly split into two blocks termed A and B [129]. Block A extends from nucleotides 8 to 19 and includes the portion of the gene that encodes the tRNA D loop and four invariant nucleotides. The B block runs from nucleotides 52 to 62 including those encoding the T-loop and five invariant nucleotides (Fig. 9.9). Thus, the invariant nature of some tRNA nucleotides appears to be important both in tRNA tertiary structure (section 12.3.1) and in the initiation of tRNA synthesis. Because of the length heterogeneity of the tRNA variable loop and the presence in some tRNA genes of inserts, the distance between the A and B blocks in the gene can vary between 25 and 95 bp.

Block B has no equivalent in the 5S gene intergenic promoter but block A is functionally equivalent to the 5S block A. Conversely, there is no equivalent to the C block in the promoter of tRNA genes which are not therefore targets for TFIIIA. Instead, the first step in the assembly of a transcription complex on tRNA genes occurs with the binding of TFIIIC to the B block of the promoter. The factor also interacts with the A block. Proteolysis of the yeast TFIIIC (also called Tau) shows that the 300 kDa protein has two functional domains one of which binds to B block and the other to A block [130]. Electron microscopy indicates that between the two domains the protein is very flexible allowing it to adopt a globular or dumbbell shape depending on the separation of the A and B blocks between the 25 and 95 bp minimum and maximum spacing[131]. There is evidence, in at least some systems, that TFIIIC is multicomponent and may include a polypeptide that specifically ·recognizes tRNA promoters [132]. However, the data are contradictory and await clarification (reviewed in [121]). As with the 5S promoter, the binding of TFIIIC is stabilized by TFIIIB so forming a pre-initiation complex that is recognized by RNA polymerase III.

TFIIIB is the protein that correctly positions RNA polymerase III at the start site for the initiation of transcription of both 5S and tRNA genes [133]. Kassavetis *et al.* demonstrated that, once TFIIIB is associated with the complexes, the factors TFIIIA and TFIIIC can be removed with heparin or high salt, leaving TFIIIB still attached. The partially stripped promoters still permit the accurate initiation of transcription by RNA polymerase III [133]. Thus, TFIIIA and TFIIIC function only as promoter recognition factors that sequester TFIIIB. Footprinting analyses, that show that TFIIIB protects 5'-flanking sequences of genes rather than the internal promoters, support these concepts [134].

(b) *Upstream promoters for RNA polymerase III*

Although the internal promoters of 5S and tRNA genes were unexpected, they nevertheless form a consistent picture whereby the sequential recognition of the promoter by transcription factors leads to the accurate positioning of RNA polymerase III. The adenovirus VA RNA gene, which is transcribed by RNA polymerase III into a RNA that regulates viral gene expression, supports the concept. Like tRNA genes, it has A and B block internal promoters (Fig. 9.8). However, it has recently become clear that this picture of RNA polymerase III transcription is too simplistic. Many 5S genes have been found to have 5'-sequences that dramatically influence the rate of their transcription. Furthermore, the elements involved are homologous in sequence and location with the TATA box of the RNA polymerase II promoter [135]. Other genes that are transcribed by RNA polymerase III blur the picture even more dramatically in that they have no internal promoters (reviewed in [136]). Thus, U6 snRNA and 7SK RNA genes have TATA boxes together with other upstream elements that are more usually associated with the enhancement of RNA polymerase II transcription (Fig. 9.8). These include proximal elements, that may function to position TFIIIB and thus RNA polymerase III [137], but are nevertheless homologous to the proximal elements of the snRNA genes transcribed by RNA polymerase II. In both *Xenopus* and plants it appears that the distance between the TATA box and the upstream element in snRNA genes is critical in determining whether they are recognized by RNA polymerase II or III [138, 270]. The promoters of U6 and 7SK genes also include distal elements that may be equivalent to enhancers (Fig. 9.8). In 7SK genes these include an essential CACCC motif that is similar to several eukaryotic regulatory elements [139]. Similarly, the distal element of U6 snRNA genes includes an octamer motif like that controlling the transcription of immunoglobulin and histone genes by RNA polymerase II [140]. Clearly these findings cloud the differences between the two polymerases and suggest that they evolved from a common ancestor.

(c) *TFIIIA may regulate the expression of 5S rRNA genes*

There is evidence that the developmental regulation of oocyte and somatic 5S RNA gene expression in *Xenopus* may be mediated through transcription factor TFIIIA. Oocytes express 5S RNA from the abundant oocyte 5S genes in the presence of high concentrations of TFIIIA whereas somatic cells have low concentrations of TFIIIA and transcribe 5S RNA from the low abundance family of somatic 5S RNA genes. Wolffe and Brown [141] have explained the differential regulation by differences in the stability of the TFIIIA–DNA complexes. They suggest that, because the somatic promoter–TFIIIA complexes are more stable, they are able to compete more effectively for the declining quantities of the factor that are present after the fertilization of the oocyte. However, Blanco *et al.* [142] have reported two different TFIIIA activities in this system. They claim that transcription of oocyte 5S RNA requires a 39 kDa form of the factor whereas a 42 kDa form activates the somatic genes and represses the oocyte genes.

9.4.3 Initiation by RNA polymerase I

The transcription of pre-rRNA represents approximately half of all cellular RNA synthesis and is regulated in response to changes in growth and metabolism (reviewed in [74, 265]). The high transcription rate may be presumed to reflect many features of the rDNA genes. There is a high gene dosage and, from yeasts to man, the rDNA is arranged in multiple tandem copies with the coding regions separated by intergenic spacers (section 8.4.3). The genes are concentrated in the nucleolus as is RNA polymerase I, the enzyme that apparently transcribes only pre-rRNA. Transcriptional regulation is focused in the intergenic spacer (formerly, but mistakenly, called the non-transcribed spacer). It contains at least three controlling elements; proximal promoters, upstream promoter elements and terminators. In many cases, upstream sequences also include elements with many of the properties of the enhancers of RNA polymerase II.

RNA polymerase I was the last of the eukaryotic nuclear RNA polymerases for which *in vitro* and *in vivo* transcription systems were developed [143], in part because, unlike polymerase II and III, this enzyme is very species-specific. Thus, mouse RNA polymerase recognizes its own promoter complex and that of closely related species like rat but it will not recognize that of human. Similarly, the human enzyme will recognize human and rhesus transcription systems but not that of mouse [144, 145]. Presumably, the singular specificity of the enzyme has permitted this evolutionary drift.

Notwithstanding species specificity, rDNA promoters of vertebrates share many common features (Fig. 9.10). In mammals, the core promoter element (CPE, also called the proximal promoter) overlaps the transcribed gene and spans from -31 to $+6$ with respect to the transcription start site. A second distinct but interacting domain, called the upstream control element (UCE)

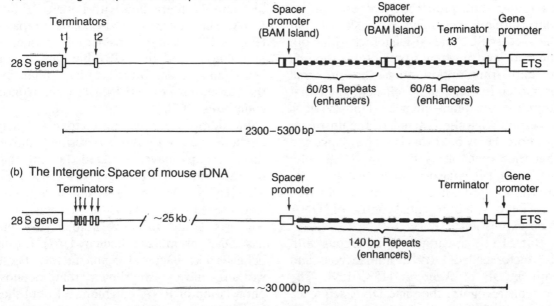

Fig. 9.10 The intergenic spacers of rDNA from (a) *Xenopus laevis* and (b) mouse. The black bar in the *Xenopus* spacer promoters indicates that enhancer elements occur within them. ETS is the external transcribed spacer (section 8.4.3).

or upstream promoter element (UPE) is located between −107 and −187 [146, 147]. The CPE is necessary for transcription and *in vitro* is sufficient for basal rates of rRNA synthesis. The UPE is required, however, for efficient transcription both *in vitro* and *in vivo* [148–150]. The intergenic spacers of some species contain a series of repetitive sequences upstream of the promoter that are, in effect, multiple duplications of promoter elements. Thus, in *Xenopus*, the core promoter again spans the transcription start site (−14 to +4) but in addition there are multiple promoter-like elements that are approximately 90% homologous to the promoter and are located in the so-called *Bam* islands (Fig. 9.10). *Drosophila* likewise has multiple internally repetitious promoters in its intergenic spacer [151, 152]. Transcription can initiate from within these intergenic promoters and the resultant

transcripts terminate upstream of the true gene promoters (reviewed in [265]); however the function of the transcripts, if any, is a matter of speculation.

In addition to the promoter-like elements, the *Xenopus* intergenic spacer contains repetitive short sequences, the 60/81 bp elements, that are clustered in tandem (Fig. 9.10) and which stimulate transcription when placed at a variable distance from the promoter and when placed in either orientation [153, 154]. They are thus analogous in their activity, but perhaps not in their mechanism, to the enhancers of RNA polymerase II (section 10.4). A 140 bp repeated element between the core and upstream promoters of mouse rDNA (Fig. 9.10) has also recently been shown to have an enhancer-like function [155] and enhancing elements have also been detected in rat and *Drosophila* [265].

At least two protein transcription factors are required for pre-rRNA synthesis. In rat, footprinting experiments have shown that the factors SL1 and SF1 interact with both the core and upstream promoters. Binding to and transcription from the proximal promoter is much more efficient than that from the upstream elements [156] so it is uncertain what the function of the upstream promoters may be. Moss [157] has suggested that they function as sinks for RNA polymerase but it is also possible that they are alternative promoters the use of which is associated with differential control of rDNA gene expression.

Rat SF1 is apparently homologous with the factor called UBF in human cells and with xUBF of *Xenopus* [158, 159]. The proteins recognize the same DNA sequence elements [160] and cannot therefore be responsible for species specificity of promoter recognition. Factor SL1, on the other hand, is a good candidate for the source of species specificity. Human SL1 (hSL1) will cause a mouse *in vitro* transcription system to recognize a human promoter [161]. Similarly, rat SL1 (rSL1) will reprogramme a human transcription system to recognize a rat promoter [156]. Bell *et al.* [162] have shown that mouse and human RNA polymerase I and UBF are functionally interchangeable but SL1 are distinct in both their binding and transcriptional activities. Thus, although mouse SL1 will selectively interact with DNA in the absence of UBF [162], human SL1 does not recognize DNA and associates with the promoter only by interaction with UBF [163]. A strong cooperative complex between the two proteins is critical to transcriptional activation [162, 164]. UBF has been purified from both human and *Xenopus* cells and shown to have multiple domains. The DNA binding domains exhibit the same specificity but the domains that interact with SL1 differ [165]. Thus, it

appears that species specificity resides in SL1 and its interaction with both UBF and DNA. The *Xenopus* 60/81 enhancer repeats (Fig. 9.10) exhibit considerable sequence homology with the UBF-binding region [266] and apparently act by enhancing the formation of UBF-SL1 transcription complexes [267].

In addition to detecting factors that may be homologous to those described above, three groups have detected factors that interact with RNA polymerase I [166–168]. That called TIF-1A by Grummt and colleagues, interacts with the enzyme and converts it into an active holoenzyme that has rDNA promoter specificity [167]. It thus behaves like bacterial sigma factor. These workers have also shown that dephosphorylation of RNA polymerase I abolishes its specificity for rDNA promoters. Another factor, designated TFIC by Thompson and coworkers [168, 169] is also tightly associated with RNA polymerase I and reduced levels of it have been implicated in the reduced transcription of rRNA in glucocorticoid-treated lymphosarcoma cells.

A further controlling influence on the initiation of transcription from tandemly arranged rRNA genes are the terminator sites of the previous gene which, in addition to their termination activity, are in some cases able to stimulate the adjacent gene promoter (reviewed in [170]). Thus, it is now known that, in *Xenopus*, RNA polymerase I transcribes right across the intergenic spacer to a terminator called t_3 that is approximately 60 bp upstream of the promoter for the next transcription unit (section 9.5.3). It has been suggested that it may act with the promoter in an interdependent complex bringing about the termination of one transcription unit and the initiation of the next. Mammalian RNA polymerase terminates after transcribing a much shorter distance into the intergenic spacer (section 9.5.3) but there are again extra terminator sequences close to the next promoter. They may well have two functions; the termination of transcripts that have initiated within the intergenic spacer and exerting a positive influence on the initiation of the next transcription unit. It is suggested that RNA polymerase is 'handed over' from the terminator to the promoter. However, recent experiments by Labhart and Reeder [171] have demonstrated that preventing the enzyme from reaching the *Xenopus* t_3 terminator had no effect on the density of polymerase molecules that collected at the 5'-ends of genes.

9.5 TERMINATION OF TRANSCRIPTION IN EUKARYOTES

Termination of transcription in eukaryotes is not well understood. Often it is difficult to even be sure of the relationship between observed 3'-termini of newly made RNA molecules and the real point at which transcription terminates. This is especially so with the transcripts of RNA polymerase II, the majority of which are post-transcriptionally modified by 3'-polyadenylation. The only method currently available to identify termination sites is run-on or run-off transcription (both names are confusingly used for the same process). By using purified nuclei for transcription *in vitro*, the synthesis of partially transcribed and still nascent RNA molecules is continued with the incorporation of radioactive nucleotides. After purification, the radioactive RNA is hybridized to fragments of DNA that correspond to the 3'-end and the 3'-flanking sequences of the mRNA encoding sequence. In this way, it has been demonstrated that many genes are transcribed well past their apparent 3'-termini and are then subject to processing to generate their 3'-termini

(processing is considered in sections 11.5, 11.6 and 11.7).

9.5.1 Termination by RNA polymerase II

(a) *Polyadenylated mRNA*

Use of the run-off transcription assay (section A.11) has demonstrated that RNA polymerase II characteristically terminates at heterogeneous sites up to several thousand base pairs beyond the polyadenylation site [172]. Thus, the termination of mouse β-globin occurs about 1 kb past the end of the gene [173] whereas mouse α-amylase transcripts terminate at multiple sites 2–4 kb past the 3'-terminus [174]. Proudfoot [175] has tabulated the run-off characterized termination of other genes and has reviewed termination by RNA polymerase II.

Section 11.5.1 describes how two sequence elements of polymerase II transcripts, the AAUAAA and a GC-rich sequence motifs, are specific recognition signals for the 3'-processing of pre-mRNA. It has become clear that these same elements are also involved in transcriptional termination. This was first indicated in a patient with α-thalassaemia in whom a base change in the α2-globin gene converted the AATAAA motif into AATAAG. This not only inactivated the processing of α2-globin pre-mRNA but also caused transcription to continue well past the normal termination sites [176]. Similarly, the AATAAA element of a mouse globin gene could be used in a gene construct to promote the termination of the adenovirus E1A gene but failed to do so if changed by point mutation [177]. Hybrid gene constructs in SV40 and polyoma have also been used to demonstrate that termination requires a functional 3'-processing site and that both the AATAAA and the GC-rich elements were required [178, 179]. Furthermore, the stronger the processing site the more efficient was the termination.

Two models have been put forward to account for the coupling between 3′-processing and termination. Darnell and coworkers suggest that a specific elongation factor could be released from RNA polymerase II during its passage over the processing signal sequences. This, they suggest would destabilize elongation and cause random termination [177]. The model most supported by the available evidence, however, is that the 3′-terminus is formed by the endonucleolytic cleavage of a transcript while it is still being synthesized. Cleavage is anticipated to occur at the processing site while RNA polymerase is still advancing a considerable distance downstream [172, 178]. After cleavage, the newly formed 3′-terminus becomes the site for polyadenylation (section 11.5.1) while the new 5′-terminus is not capped and is vulnerable to degradation by exonuclease. Proudfoot [175] has suggested that three factors may contribute to the eventual termination of transcription and release of the polymerase. These are the exonuclease activity, a RNA:DNA helicase that might be required to unwind the nascent RNA:DNA hybrids and pausing sites on the DNA required to slow the progress of RNA polymerase.

(b) *snRNA genes*

Termination of the transcription of the genes for snRNAs (snRNAs themselves are described in section 11.2) appears to occur at or close to the 3′-terminus of the mature snRNA [180, 181]. Three sequence elements are apparently required for this precise recognition event. The first of these is a preserved motif $GTTTN_{0-3}AAA$ $^{G}/_{A}NNAGA$ in the immediate 3′-flanking region of snRNA genes. The second is an internal sequence close to the 3′-end of all snRNA genes that is potentially able to form a stem–loop structure and the third element is the promoter of snRNA genes. The role of the promoter elements is not understood but snRNA termination signals do not work if combined in gene constructs with non-snRNA promoters. For example, RNA polymerase II initiating at an adenovirus late promoter, will transcribe through an snRNA termination sequence to terminate at a downstream polyadenylation site [182]. It thus seems likely that initiation at a snRNA promoter is accompanied by the incorporation into the initiation complex of a specific factor that then promotes disengagement of polymerase at the snRNA termination site.

(c) *Histone genes*

Histone genes illustrate a third type of termination by RNA polymerase II. As described in section 11.5.2, the 3′-processing of histone transcripts involves the recognition of a series of conserved elements by a number of factors that include snRNA U7. Termination of transcription occurs well downstream of these processing sites and, in contrast to polyadenylated pre-mRNA, the two processes can be independent of each other. Thus, the conserved processing signals of some histone genes can be removed without affecting termination and it has recently been shown that termination occurs in an A-rich region [183]. The motifs are close to but separate from the processing signals and, when placed in a globin gene intron, they cause premature termination of transcription. In contrast to these data however, recent studies with the mouse

H2A gene have shown that a processing site is required for termination [268].

9.5.2 Termination by RNA polymerase III

Termination of 5S RNA transcripts by RNA polymerase III occurs in a single consensus sequence that consists of a run of four or more T residues in the non-coding strand and is surrounded by GC-rich sequences [184]. Similar terminator sequences occur immediately or soon after the coding regions of almost all known tRNA genes, and mutations that create strings of four or more T residues within a gene cause premature termination [185, 186]. Deletion of the T clusters from the 3'-flanking regions causes readthrough beyond the normal 3'-terminus. Similar sequence motifs occur at the 3'-ends of other class III genes, however yeast RNA polymerase may require longer runs of Ts because the re-cognition sequence of most of its 5S RNA genes consist of 29 T residues.

Termination *in vitro* is stimulated by a highly conserved, 50 kDa nuclear phos-phoprotein called La. Sera from patients with autoimmune disease often carry anti-bodies to La and cell extracts depleted of the protein by the antisera lose the ability to transcribe class III genes [187, 188]. Purified La protein restores transcriptional activity to these extracts and the protein binds to the nascent polymerase III transcripts. Specifically, it binds to the U-rich 3'-ends of the RNAs that are created by transcription of the oligo(T) termination signal [189]. Gottleib and Steitz [190] have shown that La not only binds to the nascent RNA but facilitates termination. Furthermore, Bachmann *et al.* [191] have demonstrated that it is a nucleic acid-dependent ATPase that is capable of using phosphodiester

bond energy to melt RNA:DNA hybrids. It thus resembles the prokaryotic termination factor, rho (section 9.2.1).

9.5.3 Termination by RNA polymerase I

The ribosomal genes of *Xenopus* have three major sites that have been associated with RNA 3'-end formation and are designated t_1, t_2 and t_3 (Fig. 9.10). Their relationship with termination was first demonstrated by Labhart and Reeder [192] who showed that t_1 was at the 3'-end of 28S RNA but was not a terminus for transcription. Instead, transcription continued into the intergenic spacer, initially a further 235 bp to t_2. However, t_2 did not cause polymerase release either; rather it formed a boundary between relatively stable and very unstable transcript. Transcription continued from t_2 across the intergenic spacer to t_3 just 215 bp upstream of the initiation site for the next tandemly arranged rRNA gene and about 60 bp from its promoter. The transcript between t_2 and t_3 was rapidly degraded and was only detected by run-off transcription or electron microscopy. Thus, t_2 is a processing site and it defines the 3'-end of the largest relatively stable precursor, 40S pre-RNA. The true terminator, t_3, is a 12 bp element that is conserved between different am-phibia [193]. Within the 12 bp is a 7 bp sequence, GACTTGC, that is also present in t_2. In fact, t_2 can be converted into a t_3-like terminator by a single base change downstream of the GACTTGC box [194]. McStay and Reeder [195] have concluded that t_3 and the adjacent promoter of the next transcription unit act as one inter-dependent complex bringing about the termination of one transcription unit and the initiation of the next. They have also detected a DNA-binding protein that interacts with the t_3 sequence and is required for termination [196].

Termination elements have also been identified in mammalian rDNA, particularly that of the mouse where a repetitive 18 bp motif, known as the Sal box ($AGGTCGACCAG^A/_TT^T/_ANTCCG$) constitutes the termination signal (Fig. 9.10). RNA polymerase I specifically terminates 565 nucleotides downstream of the 3'-end of 28S RNA and 11 bp upstream of the first termination signal [197, 198]. The 3'-end of the transcript is then trimmed by 10 nucleotides to produce the 3'-terminus of the first relatively stable precursor, 45S RNA [199]. Termination again requires the binding of a specific factor (TTF-1) to the Sal box sequence [200]. The protein is obviously strongly conserved as it will function in species as divergent as yeast and mouse but it has no effect on termination by RNA polymerases II and III. However, it should be emphasized that, in contrast to the prokaryotic rho factor and to La protein in termination by RNA polymerase III, the factors so far identified for termination by RNA polymerase I interact with the DNA not the nascent RNA. Parallel studies of the termination of human rDNA transcription have identified a shorter, 11 bp, recognition motif (GGGTCGACCAG) that corresponds to the proximal part of the mouse terminator [201]. The protein that binds to this element exhibits different electrophoretic mobility but is otherwise very similar to the mouse protein [201].

As in *Xenopus*, mammalian rDNA intergenic spacers contain multiple terminator sequences including copies close to the initiation site for the next tandemly arranged transcription unit. These appear to have two functions. They terminate transcripts that have initiated in the intergenic spacer and they exert a positive effect on the initiation of the next rRNA gene. The possible importance of this in initiation is discussed in section 9.4.3.

9.6 DOES THE NUCLEOSKELETON PLAY A ROLE IN TRANSCRIPTION?

Studies with *in vitro* transcription systems and solubilized RNA polymerase argue that transcription is initiated, as described in the preceding sections of this chapter, by a diffusible enzyme binding to a promoter and then moving along the DNA. Cook [202] and others have pointed out that the evidence for this concept of transcription comes from extracts and nuclear fractions the preparation of which required the use of hypotonic salt concentrations. The aggregation of chromatin components precludes the use of isotonic or hypertonic solutions under most circumstances but preparations that do employ them suggest that active RNA polymerase is associated with the nucleoskeleton and is not freely diffusible (reviewed in [202]).

The structure and function of the nucleoskeleton is controversial and various structures that can be isolated (nuclear matrix, nuclear scaffold, nucleoid cages) may be derived from it to a greater or lesser extent. Thus, the nuclear matrix is the remnant that remains when nuclei are sequentially extracted with high salt concentrations (2 M NaCl) and non-ionic detergent (1% Triton) and are then treated with DNase and sometimes with RNase. When viewed by electron microscopy, such samples retain the overall outline of nuclei but are composed entirely of a network of thin proteinaceous fibres together with residual nucleolar structures, nuclear pores and connecting lamina.

Accumulating but conflicting data suggest that the nucleoskeleton may be intimately associated with both transcriptional and post-transcriptional processes [203]. DNA appears to be organized into a series of supercoiled loops anchored to the matrix and, when the loops are cleaved with endonucleases, the DNA sequences that remain attached to the matrix can be analysed. In this way the active ovalbumin [204, 205], conalbumin [204], globin [206, 207] and vitellogenin genes [208] have all been shown to be preferentially associated with the matrix. The data are not, however, universally accepted. Kirov *et al.* [207] have suggested that the apparent association of active genes with the matrix may be an artifact created by the association of active genes with proteins that are resistant to the high-salt methods of preparing the matrix. When DNase I is used in the absence of high salt no such association is observed. Mirkovitch *et al.* [209], also using a low-salt extraction method, have shown that histone gene clusters are attached to the matrix by an AT-rich region in the spacer between the clusters. They report, however, that the attachment is independent of transcription. Conversely, Cook and colleagues [210, 211] use gentle isotonic methods to make preparations of nucleoskeletons and find that RNA polymerase, nascent transcripts and the genes being transcribed are closely associated with it. Their method avoids the problems of aggregation by encapsulating cells in agarose microbeads. The beads can then be extracted with physiological buffers containing non-ionic detergent so that the cytoplasmic proteins and most RNA diffuses out to leave encapsulated chromatin surrounded by the nucleoskeleton. Transcriptional rates assayed in these preparations are twice those of conventional nuclei. Several groups have more precisely located matrix-associated regions (MARs) or scaffold-associated regions (SARs) and have shown that they coincide with the boundaries of DNase sensitive gene domains [212–214] or that they flank specific regulatory elements [215, 216]. Sippel and coworkers [217] have shown that when a reporter gene is flanked by a scaffold attachment site of the lysozyme gene, its expression in stably transfected cells is significantly elevated and is independent of chromosomal position.

Perhaps the most radical suggestion arising from these studies is a model of transcription in which it is suggested that the template moves past a polymerase molecule anchored to the nucleoskeleton. The skeleton is thus seen as the active site of transcription [202]. At best, the evidence to support the model is circumstantial but it does draw attention to uncertainties in the accepted concepts. The nuclear matrix has also been implicated as the site for the splicing of pre-mRNA and for the transport of mRNA to the cytoplasm (section 11.3.1).

Similar arguments to those discussed above centre on whether, in the mitotic chromosome, potentially active genes are associated with the so-called chromosomal scaffold (e.g. [218]).

9.7 THE TRANSCRIPTION OF MITOCHONDRIAL AND CHLOROPLAST GENES

The genomes of subcellular organelles are expressed by their own transcriptional machinery that includes specific RNA polymerases. Initially, the characterization of these systems lagged well behind those of nuclear transcription because of the low amounts of the enzymes and the lack of specific assays for initiation. However, the enzymes have now been purified from several sources and the triphosphates at the 5′-termini of their transcripts can be specifically labelled by the nuclear mRNA-capping enzyme, guanyltransferase.

9.7.1 Mitochondrial transcription

RNA polymerases have been purified and characterized from the mitochondria of man, *Xenopus* and yeast (reviewed in [219]). The amphibian and yeast enzymes consist of a core enzyme and a specificity factor that confers promoter specificity. The human enzyme is similar except that the core enzyme alone exhibits some promoter specificity but requires a factor called mitTF for efficient transcription. The human promoters consist of two components; a core promoter and an upstream region. The factor mitTF binds to the upstream region and in so doing is presumed to increase the efficiency with which the core enzyme interacts with the promoter [220, 221].

As described in section 8.5.1, vertebrate mitochondrial genomes are very compact and almost lack the intergenic spacers that would normally contain promoters. In fact, the whole genome is divergently transcribed from a single region that includes two independent promoters for leftward and rightward transcription (most genes are encoded on the so-called heavy DNA strand with just a few tRNA genes on the light strand). Furthermore, the promoters are closely associated with the primary origin of DNA replication and are within the D loop in which a short nascent DNA strand represents the resting stage of DNA replication (section 6.11.4). It has been suggested that vertebrate mitochondrial RNA polymerases serve a dual function in transcription and replication. Thus, DNA synthesis is seen as initiating with the synthesis of a primer RNA at one of the transcriptional promoters. The switch from RNA to DNA synthesis is then believed to occur within a 90 nucleotide region that encompasses three previously identified sequence blocks; CSB-I, CSB-II and CSB-III that are conserved in vertebrate mitochondrial DNAs [221].

Other vertebrate species (mouse and *Xenopus*) appear to have mitochondrial promoters of similar organization to that of man. In yeast, however, there are differences that reflect the less compact genome. All but one of the tRNA genes are on the same strand in the mitochondrial DNA of the yeast, *S. cerevisiae*, and some of the genes possess their own promoter whereas others are expressed from shared promoters. Each promoter consists of a block of 11 nucleotides with the sequence $^A/_T$TATATAAGTA. Transcription begins at the last A (position +1) of the sequence. The core enzyme reacts weakly with the core element or any other DNA and the 43 kDa specificity factor does not interact with DNA [222]. However, the two components co-operate together to form an initiation complex [223]. The 11 nucleotide promoter sequence also occurs in the DNA replication origin so it seems likely that in yeast, as in vertebrates, mitochondrial RNA polymerase may play a role in DNA synthesis [219, 221]. The yeast promoter does not appear to be strongly conserved in other fungi [224], and neither the nature nor the number of mitochondrial promoters is well characterized in higher plants [225].

9.7.2 Chloroplast transcription

Chloroplast gene arrangement and expression has much in common with that of prokaryotes (reviewed in [226-228]). Clusters of genes in some cases exhibit homology with gene sets in the cyanobacterial genome from which they are believed to have evolved. Most of the genes in these clusters are co-transcribed into polycistronic mRNAs from promoters that appear to be homologous with the prokaryotic −10 and −35 motifs. Many of the promoters can in fact be recognized by *E. coli* RNA polymerase. However, some chloroplast genes do not have typical prokaryotic promoters and there is evidence that chloroplasts may have more than one RNA polymerase. A polymerase that is isolated as a transcriptionally active DNA–protein complex, preferentially synthesizes chloroplast rRNA and may be a single polypeptide [229, 230]. Other preparations appear to be multi-subunit enzymes and may specifically transcribe the tRNA and mRNA genes of chloroplasts [231]. It is possible, however, that these apparent differences will prove to be artifacts of the procedures used and the impurity of the preparations [228].

9.8 TRANSCRIPTION OF DNA VIRUSES

9.8.1 Prokaryotic DNA viruses

Some bacteriophage redirect the RNA polymerase of their host to transcribe the phage genome (sections 10.1.6 and 10.3.2). Others, at least in part, employ their own enzymes (for a general review of these viral RNA polymerases, see [232]).

The RNA polymerase of bacteriophage T7 and its relatives (T3, SP6, gh-1) is encoded by one of the so-called early genes of the virus and is transcribed by host RNA polymerase early in the infective cycle (reviewed in [233]). The transcript is translated by the host's protein-synthesizing system and viral polymerase is then responsible for the transcription of the middle and late viral genes from class II and class III promoters, respectively. These pro-moters share a strongly conserved 23 bp consensus sequence, located at −17 to +6 with respect to the transcription start site and for which T7 polymerase has a stringent specificity [234]. Class III promoters have a second conserved motif in the region −22 to −18 [235].

The T7 enzyme consists of a single polypeptide of 98 000 kDa. Transcription is exclusively from promoters on the r stand of the viral DNA and occurs at a very fast rate of 200 nucleotides per second. It selectively transcribes DNA that is linked to the 23 bp promoter, binds to one face of the DNA [269] and all transcripts start with the sequence pppGGG [236]. Bacteriophage T3 RNA polymerase is closely related to that of T7 (82% amino acid identity). It is, however, specific for its own promoter motif which is also a 23 bp sequence but differs from that of T7 particularly in the region −10 to −12 [237].

Bacteriophage N4 does not rely on the host enzyme for early gene transcription but carries its own, very large (M_r 350 000) single polypeptide, RNA polymerase within the bacteriophage particle.

9.8.2 Eukaryotic DNA viruses

Many of the double-stranded DNA viruses that infect animal cells, including SV40, polyoma, papilloma, adenovirus and Epstein Barr virus (EBV), rely on host RNA polymerases to express their genetic programme. For instance, the genes of EBV virus are expressed from 24 promoters by RNA polymerase II [238] and from two promoters by RNA polymerase III [239]. The genomes of these viruses include enhancer elements and they encode regulatory proteins that allow them to hijack their host's transcriptional machinery; often ensuring that their genes are expressed in multi-tiered cascades. The regulatory mechanisms that control these processes in SV40 and adenovirus are discussed in sections 10.4.4 and 10.4.8, respectively.

Relatively few eukaryotic DNA viruses employ viral proteins for replication and transcription but the poxviruses are exceptions; presumably because they replicate in the cytoplasm of their host cells. They are large viruses. Thus, vaccinia virus (Copenhagen strain) has a genome of 191 636 bp that includes 198 open reading frames of at least 60 amino acids (vaccinia virus transcription is reviewed in [240] but many of the most recent findings described below are sited in a review of a symposium that was devoted to poxviruses [241]). The enzymes encoded in the vaccinia genome include those required for transcription (RNA polymerase, poly(A) polymerase and capping enzyme) and those needed for replication (DNA polymerase, DNA ligase, type I topoisomerase, thymidine kinase and thymidylate kinase).

When the virion is assembled, the enzymes required for early gene expression are encapsulated in the core. The uncoating process that follows infection liberates these proteins, transcription commences and results in the production of capped, polyadenylated mRNAs that are translated on host cell ribosomes [240]. Vaccinia RNA polymerase has been purified and characterized. It has a M_r of 425 000, comprises seven subunits, is resistant to α-amanitin and, *in vitro*, is dependent on Mn^{2+} ions for activity [242]. The two largest subunits show sequence homology with those of eukaryotic RNA polymerases and one small subunit exhibits homology with the elongation factor S-11 [241]. The polymerase does not of itself possess promoter specificity and recognition of the early gene promoters depends on vaccinia early transcription factor (vETF)

[243]. The factor consists of two subunits of 73 and 82 kDa of which the smaller has homology to DNA helicase whereas the larger has the leucine zipper and zinc-finger domains that are associated with many transcriptional factors (section 10.5). Footprinting experiments (section A.9.2) show that the factor interacts with the −13 to −28 A-rich promoter sequence, A_5TGA_8, that specifies the initiation of early genes. It also interacts with the region +7 to +10. When interacting with the factor the promoter assumes the conformation of bent DNA and the major groove widens in the region of the initiation site. ATP destabilizes the initiation-factor–polymerase complex and it has been suggested that the hydrolysis of ATP releases the complex from the promoter and allows the polymerase to initiate elongation (from [241]).

Termination of the early genes occurs at the sequence TTTTTNT. It requires another two-subunit protein which allows the transcriptional complex to recognize the U_5NU transcript of the termination motif. Termination is induced approximately 50 nucleotides further downstream. The termination factor also functions as a capping enzyme; its large subunit possesses triphosphatase and guanyl transferase activity whereas the small subunit possesses methyl transferase activity [244, 245].

Among the proteins synthesized from the expression of the early genes is an intermediate transcription factor (vITF). It confers specificity on the viral RNA polymerase for the intermediate gene promoters and these are expressed after the early proteins induce further uncoating of the viral DNA. Similarly, the genes transcribed in this intermediate class include those encoding three late gene transcription factors (vLTFs). They cause the polymerase to recognize the promoters of the so-called late genes [240, 241].

9.9 THE REPLICATION OF RNA VIRUSES BY RNA-DEPENDENT RNA POLYMERASE (REPLICASE)

9.9.1 RNA bacteriophages

The RNA bacteriophage R17, Qβ, MS2 and f2 have 3600–4500 nucleotide, single-stranded RNA genomes which function both as a mRNA from which viral proteins are translated, and as a template for a RNA-dependent RNA polymerase (replicase). Part of the replicase is translated from the infective (+) strand of the viral RNA which it then copies into (−) strands. These are then also copied to produce more (+) strands for packaging into progeny infective bacteriophage.

Replicase transcribes RNA from its 3'-end, initiating a new RNA chain with a 5'-GTP and continuing its synthesis in a 5' → 3' direction. It therefore moves along the viral genome in the opposite direction to the ribosomes that are reading it as a mRNA. *In vitro*, a RNA primer can replace GTP in initiating synthesis. The purification and properties of Qβ replicase have been reviewed [246] and the functional domains of the enzyme have been mapped [247]. The enzyme consists of four subunits of which subunit II is encoded by the viral genome whereas the others are normally part of the host's protein-synthesizing machinery. They include the ribosomal protein S1 and the elongation factors EF-Tu and EF-Ts (the normal function of these proteins is described in chapter 12). A hexameric protein called host factor is also required for Qβ replication. The role that proteins normally involved in protein synthesis could play in RNA transcription is of considerable interest, although incompletely understood. It is discussed in section 12.10.2. EF-Tu and EF-Ts are required for replication but are only loosely bound to the complex of S1

and II. They appear to have a stabilizing role whereas S1 and host factor probably function in the binding of replicase to Qβ RNA.

9.9.2 Eukaryotic RNA viruses

(a) *Picornaviruses (class IV)*

These have single-stranded RNA genomes which can function as mRNAs (+ strands). The most thoroughly studied example is poliovirus, the 7.5 kb genome of which is a polyadenylated mRNA linked at its 5'-terminus to a 22 amino acid peptide, VPg. The genome is translated by host ribosomes into a giant polyprotein. This is then cleaved by proteases into a number of proteins including the four viral capsid proteins, part of the replicase and a number of other products including VPg.

Polio virus RNA polymerase (replicase) has been purified from infected cells and from *E. coli* expressing the protein from recombinant plasmids. It is a 53 000 kDa protein that is unable to initiate RNA synthesis without either an oligonucleotide primer or factors produced by the host cell. The role of these factors in viral RNA replication is controversial [248] as is that of VPg ([249] and references therein).

There are two main models for the replication of the viral RNA (Fig. 9.11). In the first of these, VPg, or VPg linked to two uridine monophosphates in the form of VPg-pUpU, is postulated to prime RNA synthesis [250, 251]. Baltimore [252], suggests the uridylylated VPg hybridizes to the 3'-poly(A) tail of the (+) strand RNA and is then elongated by the replicase. The (−) strand so produced is copied by the replicase after VPg-pUpU has hybridized to its two 3'-A residues.

A second model proposes that a hairpin

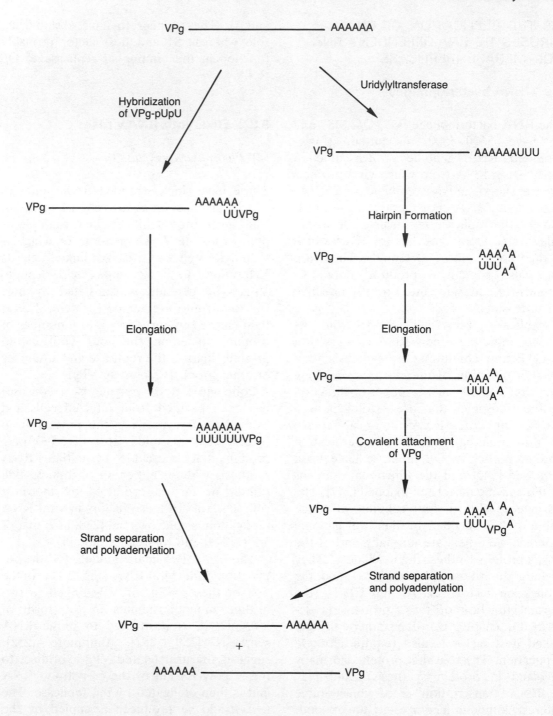

Fig. 9.11 Two models of polio virus replication.

forms at the 3'-end of the polio virion RNA, possibly as a direct result of uridylylation of the poly(A) terminus by terminal uridylyltransferase. The hairpin primes the elongation of the complementary (−) strand RNA. In this model (Fig. 9.11), VPg is thought to function in the trans-esterification required to break a phosphodiester bond within the hairpin. The template (+) strand and the product (−) strand then separate leaving a product with VPg linked to its 5'-terminus [249].

(b) *Rhabdoviruses (class V)*

Rhabdoviruses have single-stranded, (−) strand RNA genomes that cannot function as mRNAs. They must be copied into (+) strands before they can be expressed. To achieve this, the infective viruses contain an RNA-dependent RNA polymerase or replicase (also known as a transcriptase). The most thoroughly studied example is vesicular stomatitis virus (VSV), the single-stranded genome of which has an M_r of 4×10^6. It is tightly associated with 1258 copies of the nucleocapsid protein N and this complex, together with smaller quantities of the replicase (protein L [253, 254]) and the phosphoprotein NS, constitutes the transcribing ribonucleoprotein complex (RNP). The replicase synthesizes five distinct species of mRNA from the RNP complex *in vitro* and *in vivo*. These include the RNP proteins N, L and NS as well as the virion proteins M and G (in the virion, the RNP is coiled into a tight helix associated with the matrix protein M and the glycoprotein G projects through the outer lipid bilayer that is acquired from the host).

Much remains to be fully understood of the transcription and replication of VSV and other rhabdovirus (reviewed in [255]). RNA synthesis initiates at the 3'-end of the viral RNA with the synthesis of a leader RNA

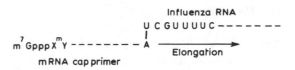

Fig. 9.12 The priming of influenza virus RNA synthesis.

but it is not clear whether subsequent transcription is continuous through the five genes or whether it results from multiple initiation events. The mRNAs synthesized *in vitro* are typical of those of eukaryotes in carrying a methylated 5'-cap and a 3'-polyadenylated tail. Virus-specific enzyme activities carry out these modifications but for the most part have not been identified with specific proteins. The replicase is also responsible for replication, namely the copying of complete (+) strands into (−) strands.

(c) *Myxovirus (class V)*

These viruses have a segmented, single-stranded RNA genome in seven or more distinct non-overlapping pieces and with negative polarity. The most-studied example is influenza. The virus encodes and packages its own replicating system. However, viral mRNA synthesis also requires cellular mRNA, the 5'-terminal portions of which are used as primers. The 5'-terminal cap (m⁷GpppXm – section 11.4) of cellular mRNA is cleaved by a cap-dependent viral endonuclease at a purine residue 10–13 nucleotides downstream from the cap [256]. The resulting fragments are weakly bonded to the viral RNA segments, it is presumed by a single base pair between a 3'-terminal A of the primer and a 3'-U of the viral RNA (Fig. 9.12). They then function as a primer for elongation by the viral transcriptase (reviewed in [257, 258]).

(d) *Reoviruses (class III)*

These are double-stranded RNA viruses of which human reovirus type 3 is a typical example (reviewed in [259]). The genome consists of ten segments of double-stranded RNA, each specifying a single polypeptide. Each is transcribed asymmetrically and conservatively by a virion-associated RNA polymerase into (+) RNA strands. These mRNAs are capped but contain no poly(A). However, they are functional mRNAs and are translated by host cell ribosomes. In viral replication the ten (+) strand mRNA molecules complex with some of the viral proteins. The (−) strand RNAs are synthesized on this complex.

(e) *Retroviruses (class VI)*

Retroviruses differ from other animal viruses in containing an RNA genome that replicates through a DNA intermediate that is stably inserted into the host chromosomes mutating and occasionally capturing cellular genes. The genome of the free viral particle consists of two identical single-stranded RNA chains wrapped around a core of viral protein. Each RNA chain is base paired with a host-specific tRNA molecule that functions as a primer for reverse transcriptase that copies the RNA into DNA. Reverse transcriptase and the replication of retroviruses are discussed in chapter 6 (sections 6.4.8 and 6.14, respectively).

The duplex DNA, that results from the replication, is stably integrated into the host chromosome (section 6.14). The encoded genes are then expressed by host cell enzymes under the direction of viral control elements. These typically include enhancers and promoters that are located mainly in the LTR and U3 regions of the viral DNA.

REFERENCES

1 Weiss, S. and Gladstone, L. (1959) *J. Am. Chem. Soc.*, **81**, 4118.
2 Chamberlin, M. J. (1976) in *RNA Polymerase* (eds R. Losick and M. Chamberlin), Cold Spring Harbor Laboratory, New York, p. 159.
3 Chamberlin, M. J. (1982) in *The Enzymes* (ed. P. Boyer), Academic Press, New York, vol. 15, p. 61.
4 von Hippel, P. H., Bear, D. G., Morgan, W. D. and McSwiggen, J. A. (1984) *Annu. Rev. Biochem.*, **53**, 389.
5 Miller, J. A., Serio, G. F., Howard, R. A., Bear, J. L., Evans, J. E. and Kimball, A. P. (1979) *Biochim. Biophys. Acta*, **579**, 291.
6 Burgess, R. R., Travers, A. A., Dunn, J. J. and Bautz, E. K. (1969) *Nature (London)*, **221**, 43.
7 Zillig, W., Palm, P. and Heil, A. (1976) in *RNA Polymerase* (eds R. Losick and M. Chamberlin), Cold Spring Harbor Laboratory, New York, p. 101.
8 Krakow, J. S., Rhodes, G. and Jovin, T. M. (1976) in *RNA Polymerase* (eds R. Losick and M. Chamberlin), Cold Spring Harbor Laboratory, New York, p. 127.
9 Scaife, J. (1976) in *RNA Polymerase* (eds R. Losick and M. Chamberlin), Cold Spring Harbor Laboratory, New York, p. 207.
10 McClure, W. R. (1985) *Annu. Rev. Biochem.*, **54**, 171.
11 Berg, O. G., Winter, R. B. and von Hippel, P. H. (1981) *Biochemistry*, **20**, 6929.
12 von Hippel, P. H., Bear, D. G., Morgan, W. D. and McSwiggen, J. A. (1984) *Annu. Rev. Biochem.*, **53**, 389.
13 Winter, R. B., Berg, O. G. and von Hippel, P. H. (1981) *Biochemistry*, **20**, 6961.
14 Park, C. S., Wu, F. Y-H. and Wu, C-W. (1982) *J. Biol. Chem.*, **257**, 6950.
15 Hannon, R., Richards, E. G. and Gould, H. J. (1986) *EMBO J.*, **5**, 3313.
16 Berg, O. G., Winter, R. B. and von Hippel, P. H. (1982) *Trends Biochem. Sci.*, **7**, 52.
17 Ricchetti, M., Metzger, W. and Heumann, H. (1988) *Proc. Natl. Acad. Sci. USA*, **85**, 4610.
18 Metzger, W., Schickor, P. and Heumann, H. (1989) *EMBO J.*, **8**, 2745.
19 Rosenberg, H. and Court, D. (1979) *Annu.*

Rev. Genet., **13**, 319.

20 Siebenlist, U., Simpson, R. B. and Gilbert, W. (1980) *Cell*, **20**, 269.

21 Hawley, D. K. and McClure, W. R. (1983) *Nucleic Acids Res.*, **11**, 2237.

22 Youderian, P., Bouvier, S. and Susskind, M. M. (1982) *Cell*, **30**, 843.

23 Brunner, M and Bujard, H. (1987) *EMBO J.*, **6**, 3139.

24 O'Neil, M. C. (1989) *J. Biol. Chem.*, **264**, 5522.

25 Auble, D. T. and deHaseth, P. L. (1988) *J. Mol. Biol.*, **202**, 471.

26 Smith, G. R. (1981) *Cell*, **24**, 599.

27 Waldberger, C., Gardella, T., Wong, R. and Suskind, M. (1990) *J. Mol. Biol.*, **215**, 267.

28 Liebig, H-D. and Rüger, W. (1989) *J. Mol. Biol.*, **208**, 517.

29 Buc, H. and McClure, W. R. (1985) *Biochemistry*, **24**, 2712.

30 Buc, H. (1987) in *Nucleic Acids and Mol. Biol.* (eds Eckstein, F. and Lilley, D.) Springer, Heidelberg, vol. 1, 186.

31 Amouyal, M. and Buc, H. (1987) *J. Mol. Biol.*, **195**, 795.

32 Buc, H., Amouyal, M., Buckle, M., Herbert, M., Kolb, A. *et al.* (1987) in *RNA Polymerase and the Regulation of Transcription* (eds W. S. Retznikof *et al.*), Elsevier, Amsterdam, p. 115.

33 Krakow, J. S., Rhodes, G. and Jovin, T. M. (1976) in *RNA Polymerase* (eds R. Losick and M. Chamberlin), Cold Spring Harbor Laboratory, New York, p. 127.

34 Krummel, B. and Chamberlin, M. J. (1989) *Biochemistry*, **28**, 7829.

35 Stackhouse, T. M., Telusnitsky, A. P. and Meares, C. F. (1989) *Biochemistry*, **28**, 7781.

36 Shimamoto, N., Kamigochi, T. and Utiyama, H. (1986) *J. Biol. Chem.*, **261**, 11859.

37 Liu, L. F. and Wang, J. C. (1987) *Proc. Natl. Acad. Sci. USA*, **84**, 7024.

38 Wu, H-Y., Shyy, S., Wang, J. C. and Liu, L. F. (1988) *Cell*, **53**, 433.

39 Giaever, G. N. and Wang, J. C. (1988) *Cell*, **55**, 849.

40 Tsao, Y-P., Wu, H-Y. and Liu, L. F. (1989) *Cell*, **56**, 111.

41 Chamberlin, M. J., Nierman, W. C., Wiggs, J. and Neff, N. (1979) *J. Biol. Chem.*, **254**, 10061.

42 Kassavetis, G. A. and Chamberlin, M. J. (1981) *J. Biol. Chem.*, **256**, 2777.

43 Gamper, H. B. and Hearst, J. E. (1982) *Cell*, **29**, 81.

44 Morgan, W. D., Bear, D. G. and von Hippel, P. H. (1983) *J. Biol. Chem.*, **258**, 9565.

45 Chamberiin, M., Kingston, R., Gilman, M., Wiggs, J. and dc Vera, A. (1983) *Methods Enzymol.*, **101**, 540.

46 Adhya, S. and Gottesman, M. (1978) *Annu. Rev. Biochem.*, **47**, 967.

47 Bear, D. G. and Peabody, D. S. (1988) *Trends. Biochem. Sci.*, **13**, 343.

48 Richardson, J. P. (1991) *Cell*, **64**, 1047.

49 Farnham, P. J. and Platt., T. (1981) *Nucleic Acids Res.*, **9**, 563.

50 Bertrand, K., Korn, L., Lee, F. and Yanofsky, C. (1977) *J. Mol. Biol.*, **117**, 227.

51 Martin. F. and Tinoco, I. (1980) *Nucleic Acids Res.*, **8**, 2295.

52 O'Hare, K. M. O. and Hayward, R. S. (1981) *Nucleic Acids Res.*, **9**, 4689.

53 Arndt, K. M. and Chamberlin, M. J. (1988) *J. Mol. Biol.*, **202**, 271.

54 Arndt, K. M. and Chamberlin, M. J. (1990) *J. Mol. Biol.*, **213**, 79.

55 Goliger, J. A., Yang, X., Guo, H-C. and Roberts, J. W. (1989) *J. Mol. Biol.*, **205**, 331.

56 Telisnitsky, A. and Chamberlin, M. J. (1989) *J. Mol. Biol.*, **205**, 315.

57 Roberts, J. W. (1969) *Nature (London)*, **224**, 1168.

58 Dombroski, A. J. and Platt, T. (1988) *Proc. Natl. Acad. Sci. USA*, **85**, 2538.

59 McSwiggen, J. A., Bear, D. G. and von Hippel, P. H. (1988) *J. Mol. Biol.*, **199**, 609.

60 Bear, D. G., Hicks, P. S., Escudero, K. W., Andrews, C. L., McSwiggen, J. A. *et al.* (1988) *J. Mol. Biol.*, **199**, 623.

61 Dombroski, A. J., LaDine, J., Cross, R. L. and Platt, T. (1988) *J. Biol. Chem.*, **263**, 18810.

62 Brennan, C. A., Dombroski, A. J. and Platt, T. (1987) *Cell*, **48**, 945.

63 Dombroski, A. J., Brennan, C. A., Spear, P. and Platt, T. (1988) *J. Biol. Chem.*, **263**, 18802.

64 Brennan, C. A., Steinmetz, E. J., Spear, P. and Platt, T. (1990) *J. Biol. Chem.*, **265**, 5440.

65 Morgan, W. D., Bear. D. G. and von

Hippel, P. H. (1983) *J. Biol. Chem.*, **258**, 9565.

66 Faus, I. and Richardson, J. P. (1990) *J. Mol. Biol.*, **212**, 53.

67 Morgan, W. D., Bear, D. G. and von Hippel, P. H. (1984) *J. Biol. Chem.*, **259**, 8664.

68 Morgan, W. D., Bear, D. G., Litchman, B. L. and von Hippel, P. H. (1985) *Nucleic Acids Res.*, **13**, 3739.

69 Albrechtsen, B., Squires, C. L., Li, S. and Squires, C. (1990) *J. Mol. Biol.*, **213**, 123.

70 Sentenac, A. (1985) *Crit. Rev. Biochem.*, **18**, 31.

71 Sawadogo, M. and Sentenac, A. (1990) *Annu. Rev. Biochem.*, **59**, 711.

72 Woychick, N. A. and Young, R. A. (1990) *Trends Bichem Sci.*, **15**, 347.

73 Geiduschek, E. P. and Tocchini-Valentini, G. P. (1988) *Annu. Rev. Biochem.*, **57**, 873.

74 Sollner-Webb, B. and Mougey, E. B. (1991) *Trends Biochem. Sci.*, **16**, 58.

75 Anderson, J. A., Juntz, G. P. P., Evans, H. H. and Swift, T. J. (1971) *Biochemistry*, **10**, 4368.

76 De Mercoyrol, L., Job, C. and Job, D. (1989) *Biochem. J.*, **258**, 165.

77 Himmelfarb, H. J., Simpson, E. M. and Friesen, J. D. (1987) *Mol. Cell. Biol.*, **7**, 2155.

78 Lewis, M. K. and Burgess, R. R. (1982) in *The Enzymes* (ed. P. Boyer), Academic Press, New York, vol. 15, p. 109.

79 Kolodziej, P. A., Woychick, N. A., Lao, S. M. and Young, R. A. (1990) *Mol. Cell. Biol.*, **10**, 1915.

80 Martin, C., Okamura, S. and Young, R. (1990) *Mol. Cell. Biol.*, **10**, 1908.

81 Corden, J. L. (1990) *Trends. Biochem. Sci.*, **15**, 383.

82 Laybourn, P. L. and Dahmus, M. E. (1990) *J. Biol. Chem.*, **265**, 13 165.

83 Allison, L. A. and Ingles, C. J. (1989) *Proc. Natl. Acad. Sci. USA*, **86**, 2794.

84 Kolodkiej, P. and Young, R. A. (1989) *Mol. Cell. Biol.*, **9**, 5387.

85 Riva, M., Christophe, C., Sentenac, A., Grachev, M. A., Mustaev, A. A. *et al.* (1990) *J. Biol. Chem.*, **265**, 16 498.

86 Woychick, N. A., Liao, S. M., Kolodziej, P. A. and Young, R. A. (1990) *Genes Devel.*, **4**, 313.

87 Wu, G-J. (1978) *Proc. Natl. Acad. Sci. USA*, **75**, 2175.

88 Weil, P. A., Segall, J., Harris, B., Ng, S. Y. and Roeder, R. G. (1979) *J. Biol. Chem.*, **254**, 6163.

89 Birkenmeier, E. H., Brown, D. D. and Jordan, E. (1978) *Cell*, **15**, 1077.

90 Weil, P. A., Luse, D. S., Segall, J. and Roeder, R. G. (1979) *Cell*, **18**, 469.

91 Wasylyk, B., Kedinger, C., Corden, J., Brison, O. and Chambon. P. (1980) *Nature (London)*, **285**, 367.

92 Grummt, I. (1981) *Proc. Natl. Acad. Sci. USA*, **78**, 727.

93 Grummt, I., Roth, E. and Paule, M. R. (1982) *Nature (London)*, **296**, 173.

94 Brown, D. P. and Gurdon, J. B. (1977) *Proc. Natl. Acad. Sci. USA*, **74**, 2064.

95 Pellicer, A. , Robins, D., Wold, B., Sweet, R., Jackson, J. *et al.* (1980) *Science*, **209**, 1414.

96 Hamer, D. and Leder, P. (1979) *Cell*, **21**, 697.

97 Mellon, P., Parker, V., Gluzrnan, Y. and Manniatis, T. (1981) *Cell*, **27**, 279.

98 Goldberg, M. (1979) Ph.D. thesis, Stanford University, California.

99 Brady, J., Radonovitch, M., Vodkin, M., Natarajan, V., Thoren, M. *et al.* (1982) *Cell*, **31**, 625.

100 Skuzeski, J. M., Lund, E., Murphy, J. T., Steinberg, T. H., Bargess, R. R. and Dahlberg, J. E. (1984) *J. Biol. Chem.*, **259**, 8345.

101 Dynan, W. S. (1988) *Trends Genet.*, **2**, 196.

102 McKnight, S. L. (1982) *Cell*, **31**, 355.

103 Everett, R. D., Baly, D. and Chambon, P. (1983) *Nucleic Acids Res.*, **11**, 2447.

104 Dierks, P., van Ooyen, A., Cochran, M. D., Dobkin, C., Reiser, J. and Weissmann, C. (1983) *Cell*, **32**, 695.

105 Conaway, J. W., Reines, D. and Conaway, R. C. (1990) *J. Biol. Chem.*, **265**, 7552.

106 Conaway, R. C. and Conaway, J. W. (1990) *J. Biol. Chem.*, **265**, 7559.

107 Kitajima, S., Kawaguchi, T., Yasukochi, Y. and Weissman, S. M. (1989) *Proc. Natl. Acad. Sci. USA*, **86**, 6106.

108 Kitajima, S., Tanaka, Y., Kawaguchi, T., Nagaoka, T., Sherman, M. *et al.* (1990) *Nucleic Acids Res.*, **18**, 4843.

109 Sopta, M., Burton, Z. F. and Greenblatt, J. (1989) *Nature (London)*, **341**, 410.

110 Flores, O., Ha, I. and Reinberg, D. (1990) *J. Biol. Chem*, **265**, 5629.

111 Horikoshi, M., Yamamoto, T., Ohkuma,

Y., Weil, P. A. and Roeder, R. G. (1990) *Cell*, **61**, 1171.

112 Hahn, S., Buratowski, S., Sharp, P. A. and Guarente, L. (1989) *Proc. Natl. Acad. Sci. USA*, **86**, 5718.

113 Lewin, B. (1990) *Cell*, **61**, 1161.

114 Van Dyke, M. W., Roeder, R. G. and Sawadogo, M. (1988) *Science*, **241**, 1335.

115 Hoey, T., Dynlacht, B. D., Peterson, M. G., Pugh, B. F. and Tjian, R. (1990) *Cell*, **61**, 1179.

116 Hoffmann, A., Sinn, E., Yamamoto, T., Wang, J., Roy, A. *et al.* (1990) *Nature (London)*, **346**, 387.

117 Pugh, B. F. and Tjian, R. (1990) *Cell*, **61**, 1187.

118 Stringer, K. F., Ingles, C. J. and Greenblatt, J. (1990) *Nature (London)*, **345**, 783.

119 Horikoshi, M., Carey, M. F., Kakidani, H. and Roeder, R. G. (1988) *Cell*, **54**, 665.

120 Sollner-Webb, B. (1988) *Cell*, **52**, 153.

121 Palmer, J. M. and Folk, W. M. (1990) *Trends Biochem. Sci.*, **15**, 300.

122 Sakonja, S., Bogenhagen, D. F. and Brown, D. D. (1980) *Cell*, **19**, 13.

123 Bogenhagen, D. F., Sakonju, S. and Brown, D. D. (1980) *Cell*, **19**, 27.

124 Cozzarelli, N. R., Gerrard, S. P., Schlissel, M., Brown, D. D. and Bogenhagen, D. F. (1983) *Cell*, **34**, 829.

125 Engelke, D. R., Ng, S-Y., Shastry, B. S. and Roeder, R. G. (1980) *Cell*, **19**, 717.

126 Bogenhagen, D. F. (1985) *J. Biol. Chem.*, **260**, 6466.

127 Windsor, W. T., Lee, T. C., Daly, T. J. and Wu, C. W. (1988) *J. Biol. Chem.*, **263**, 10272.

128 Bieker, J. J., Martin, P. L. and Roeder, R. G. (1985) *Cell*, **40**, 119.

129 Hall, B. D., Clarkson, S. G. and Toccini-Valentini, G. (1982) *Cell*, **29**, 3.

130 Gabrielsen, O. S., Marzouki, N., Ruet, A., Sentenac, A. and Frogageot, P. (1989) *J. Biol. Chem.*, **264**, 7505.

131 Schultz, P., Marzouki, N., Marck, C., Ruet, A., Oudet, P. *et al.* (1989) *EMBO J.*, **8**, 3815.

132 Johnson, D. L. and Wilson, S. L. (1989) *Mol. Cell. Biol.*, **9**, 2018.

133 Kassavetis, G. A., Braun, B. R., Nguyen, L. H. and Geiduschek, E. P. (1990) *Cell*, **60**, 235.

134 Huibregtse, J. M. and Engelke, D. R. (1989) *Mol. Cell. Biol.*, **9**, 3244.

135 Tyler, B. M. (1987) *J. Mol. Biol.*, **196**, 801.

136 Murphy, S., Moorefield, B. and Pieler, T. (1989) *Trends Genet.*, **5**, 122.

137 Moenne, A., Camier, S., Anderson, G., Margottin, F., Beggs, J. and Sentenac, A. (1990) *EMBO J.*, **9**, 271.

138 Waibel, F. and Filipowicz, W. (1990) *Nature (London)*, **346**, 199.

139 Kleinert, H., Bredow, S. and Benccke, B. L. (1990) *EMBO J.*, **9**, 711.

140 Murphy, S., Peirani, A., Scheidereit, C., Melli, M. and Roeder, R. G. (1989) *Cell*, **59**, 1071.

141 Wolffe, A. P. and Brown, D. D. (1988) *Science*, **241**, 1626.

142 Blanco, J., Milstein, L., Razik, M. A., Dilworth, S., Cote, C. and Gottesfeld, J. (1989) *Genes Devel.*, **3**, 1602.

143 Moss, T. (1982) *Cell*, **30**, 835.

144 Grummt, I., Roth, E. and Paule, M. (1982) *Nature (London)*, **296**, 173.

145 Onishi, T., Berglund, C. and Reeder, R. H. (1984) *Proc. Natl. Acad. Sci. USA*, **81**, 484.

146 Haltina, M. M., Smale, S. T. and Tjian, R. (1986) *Mol. Cell. Biol.*, **6**, 227.

147 Jones, M. H., Learned, R. M. and Tjian, R. (1988) *Proc. Natl. Acad Sci. USA.*, **85**, 669.

148 Cassidy, B., Hagland, R. and Rothblum, L. (1987) *Biochim. Biophys. Acta*, **909**, 133.

149 Yang-Yen, H-F. and Rothblum, L. (1988) *Mol. Cell. Biol.*, **8**, 3406.

150 Henderson, S. L. and Sollner-Webb, B. (1990) *Mol. Cell. Biol.*, **10**, 4970.

151 Miller, J. R., Hayward, D. C. and Glover, D. M. (1983) *Nucleic Acids Res.*, **11**, 11.

152 Simeone, A., La Volpe, A. and Boncinelli, E. (1985) *Nucleic Acids Res.*, **13**, 1189.

153 Labhart, P. and Reeder, R. H. (1984) *Cell*, **37**, 285.

154 Reeder, R. H. (1984) *Cell*, **38**, 349.

155 Pikaard, C. S., Pape, L. K., Henderson, S. L., Ryan, K., Paalman, M. H. *et al.* (1990) *Mol. Cell. Biol.*, **10**, 4816.

156 Smith, S. D., Oriahi, E., Yang-Yen, H-F., Xie, W., Chen, C. *et al.* (1990) *Nucleic Acids Res.*, **18**, 1677.

157 Moss, T. (1983) *Nature (London)*, **302**, 223.

158 Bell, S. P., Learned, R. M., Jantzen, H-M. and Tjian, R. (1988) *Science*, **241**, 1192.

159 Pikaard, C. S., Smith, S. D., Reeder, R. H. and Rothblum, L. (1990) *Mol. Cell. Biol.*, **10**, 3810.

160 Pikaard, C. S., McStay, B., Schultz, M. C., Bell, S. P. and Reeder, R. H. (1989) *Genes*

Devel., **3**, 1779.

161 Learned, R. M., Cordes, S. and Tjian, R. (1985) *Mol. Cell. Biol.*, **5**, 1358.

162 Bell, S. P., Jantzen, H-M. and Tjian, R. (1990) *Genes Devel.*, **4**, 943.

163 Learned, R. M., Learned, T. K., Haltiner, M. M. and Tjian, R. (1986) *Cell*, **45**, 847.

164 Bell, S. P., Learned, R. M., Jantzen, H-M. and Tjian, R. (1988) *Science*, **241**, 1192.

165 Bell, S. P., Pikaard, C. S., Reeder, R. H. and Tjian, R. (1989) *Cell*, **59**, 489.

166 Tower, J. and Solner-Webb, B. (1987) *Cell*, **50**, 873.

167 Schnapp, A., Pfleiderer, C., Rosenbauer, H. and Grummt, I. (1990) *EMBO J.*, **9**, 2857.

168 Mahajan, P. B. and Thompson, E. A. (1990) *J. Biol. Chem.*, **265**, 16225.

169 Mahajan, P. B., Gokal, P. B. and Thompson, E. A. (1990) *J. Biol. Chem.*, **265**, 16244.

170 Baker, S. M. and Platt, T. (1986) *Cell*, **47**, 839.

171 Labhart, P. and Reeder, R. H. (1989) *Proc. Natl. Acad. Sci. USA*, **86**, 3155.

172 Proudfoot, N. J. and Whitelaw, E. (1988) in *Frontiers in Molecular Biology*, IRL Press, Oxford, p. 97.

173 Citron, B., Falck-Pederson, E., Salditt-Georgieff, M. and Darnell, J. E. (1984) *Nucleic Acids Res.*, **12**, 8723.

174 Haggenbüchle, O., Welleur, P. K., Cribbs, D. L. and Schibler, U. (1984) *Cell*, **38**, 737.

175 Proudfoot, N. J. (1989) *Trends Biochem. Sci.*, **14**, 105.

176 Whitelaw, E. and Proudfoot, N. J. (1986) *EMBO J.*, **5**, 2915.

177 Logan, J., Falck-Pederson, E., Darnell, J. E. and Shenk, T. (1987) *Proc. Natl. Acad. Sci. USA*, **84**, 8306.

178 Connelly, S. and Manley, J. L. (1988) *Genes Devel.*, **2**, 440.

179 Lanoix, J. and Acheson, N. H. (1988) *EMBO J.*, **7**, 2515.

180 Kunkel, G. R. and Pedersen, T. (1985) *Mol. Cell. Biol.*, **5**, 2332.

181 Ciliberto, G., Dathan, N., Frank, R., Philipson, L. and Mattaj, I. W. (1986) *EMBO J.*, **5**, 2931.

182 Hernandez, N. and Weiner, A. M. (1986) *Cell*, **47**, 249.

183 Briggs, D., Jackson, D., Whitelaw, E. and Proudfoot, N. J. (1989) *Nucleic Acids Res.*, **17**, 8061.

184 Bogenhagen, D. F. and Brown, D. D. (1981) *Cell*, **24**, 261.

185 Koski, R., Clarkson, S. G., Kurjan, J., Hall, B. D. and Smith, M. (1980) *Cell*, **22**, 415.

186 Traboni, C., Ciliberto, G. and Cortese, R. (1984) *Cell*, **36**, 179.

187 Gottesfeld, J. M., Andrews, D. L. and Hoch, S. O. (1984) *Nucleic Acids Res.*, **12**, 3185.

188 Gottlieb, E. and Steitz, J. A. (1987) in *RNA Polymerase and the Regulation of Transcription* (eds W. S. Reznikoff *et al.*), Elsevier, New York, p. 465.

189 Stefano, J. E. (1984) *Cell*, **36**, 145.

190 Gottlieb, E. and Steitz, J. A. (1989) *EMBO J.*, **8**, 851.

191 Bachmann, M., Pfiefer, K., Schröder, H. C. and Müller, W. (1990) *Cell*, **60**, 85.

192 Labhart, P. and Reeder, R. H. (1986) *Cell*, **45**, 431.

193 Labhart, P and Reeder, R. H. (1987) *Mol. Cell. Biol.*, **7**, 1900.

194 Labhart, P. and Reeder, R. H. (1990) *Genes Devel.*, **4**, 269.

195 McStay, B. and Reeder, R. H. (1990) *Genes Devel.*, **4**, 1240.

196 McStay, B. and Reeder, R. H. (1990) *Mol. Cell. Biol.*, **10**, 2793.

197 Grummt, I., Maier, U., Öhrlein, A., Hassouna, N. and Bachellerie, J-P. (1985) *Cell*, **43**, 801.

198 Grummt, I., Rosenbauer, H., Niedermyer, I., Maier, U. and Örhlein, A. (1986) *Cell*, **45**, 837.

199 Kuhn, A. and Grummt, I. (1989) *Genes Devel.*, **3**, 224.

200 Bartsch, I., Schoneberg, C. and Grummt, I. (1988) *Mol. Cell. Biol.*, **8**, 3891.

201 Pfiederer, C., Smid, A., Bartsch, I. and Grummt, I. (1990) *Nucleic Acids Res.*, **18**, 4727.

202 Cook, P. R. (1989) *Eur. J. Biochem.*, **185**, 487.

203 Roberge, M., Duhmas, M. E. and Bradbury, E. M. (1988) *J. Mol. Biol.*, **201**, 545.

204 Robinson, S. I., Small, D., Idzerda. R., McKnight, G. S. and Vogelstein, B. (1983) *Nucleic Acids Res.*, **11**, 5113.

205 Ciejek, E. M., Tsai, M-J. and O'Malley, B. W. (1983) *Nature (London)*, **306**, 607.

206 Hentzen, P. C., Rho, J. H. and Bekhor, I. (1984) *Proc. Natl. Acad. Sci. USA*, **81**, 304.

207 Kirov, N., Djondjurov, L. and Roumen. T. (1984) *J. Mol. Biol.*, **180**, 601.

208 Jost, J-P. and Seldran, M. (1984) *EMBO J.*, **3**, 2005.

209 Mirkovitch, J., Mirault, M-E. and Laemmli, U. K. (1984) *Cell*, **39**, 223.

210 Jackson, D. A. and Cook, P. R. (1985) *EMBO J.*, **4**, 919.

211 Jackson, D. A., Yuan, J. and Cook, P. R. (1988) *J. Cell. Sci.*, **90**, 365.

212 Levy-Wilson, B. and Fortier, C. (1989) *J. Biol. Chem.*, **264**, 21 196.

213 Strätling, W. H., Dölle, A. and Sippel, A. E. (1986) *Biochemistry*, **25**, 495.

214 Phi Van, L. and Stätling, W. H. (1988) *EMBO J.*, **7**, 655.

215 Jarman, A. P. and Higgs, D. R. (1988) *EMBO J.*, **7**, 3337.

216 Cockerill, P. N., Yuen, M-H. and Garrard, W. T. (1987) *J. Biol. Chem.*, **262**, 5394.

217 Steif, A., Winter, D. M., Strätling, W. H. and Sippel, A. E. (1989) *Nature (London)*, **341**, 343.

218 Kuo, M. T. (1982) *J. Cell. Biol.*, **93**, 278.

219 Schinkel, A. H. and Tabak, H. F. (1989) *Trends Genet.*, **5**, 149.

220 Fisher, R. P., Topper, J. N. and Clayton, D. A. (1987) *Cell*, **50**, 247.

221 Chang, D. D., Fisher, R. P. and Clayton, D. A. (1987) *Biochim. Biophys. Acta*, **909**, 85.

222 Kelly, J. L. and Lehman, I. R. (1986) *J. Biol. Chem.*, **261**, 10 340.

223 Schinkel, A. H., Groot Koerkamp, M. and Tabak, H. F. (1988) *EMBO J.*, **7**, 3255.

224 Kubelik, A. R., Kennel, J. C., Akins, R. A. and Lambowitz, A. M. (1990) *J. Biol. Chem.*, **265**, 4515.

225 Levings, C. S. and Brown, G. G. (1989) *Cell*, **56**, 171.

226 Mullet, J. E. (1988) *Annu. Rev. Plant Physiol.*, **39**, 475.

227 Gruissem, W. (1989) in *The Biochemistry of Plants: a Comprehensive Treatise*, vol. 15, p. 151.

228 Gruissem, W. (1989) *Cell*, **56**, 161.

229 Reiss, T. and Link, G. (1985) *Eur. J. Biochem.*, **148**, 207.

230 Narita, J. O., Rushlow, K. E. and Hallick, R. B. (1985) *J. Biol. Chem.*, **260**, 11 194.

231 Orozco, E. M., Mullet, J. E. and Chua, N. H. (1985) *Nucleic Acids Res.*, **13**, 1283.

232 Ishihama, A. and Nagata, K. (1988) *Crit. Rev. Biochem.*, **23**, 27.

233 Studier, F. W. and Dunn, J. J. (1983) *Cold Spring Harbor Symp. Quant. Biol.*, **47**, 999.

234 Davanloo, P., Rosenberg, A. H., Dunn, J. J. and Studier, F. W. (1984) *Proc. Natl. Acad. Sci. USA*, **81**, 2035.

235 Gunderson, S. I., Chapman, K. A. and Burgess, R. R. (1987) *Biochemistry*, **26**, 1539.

236 Chamberlin, M. and Ryan, T. (1982) in *The Enzymes* (ed. P. Boyer), Academic Press, New York, vol. 15, p. 87.

237 Klement, J. F., Moorefield, M. B., Jorgenson, E., Brown, J. E., Risman, S. *et al.* (1990) *J. Mol. Biol.*, **215**, 21.

238 Buer, R., Bankier, A. T., Biggin, M. D., Deininger, P. L., Farell, P. J. *et al.* (1984) *Nature (London)*, **310**, 207.

239 Arrand, J. R. and Rymo, L. (1982) *J. Virol.*, **41**, 376.

240 Moss, B. (1990) *Annu. Rev. Biochem.*, **59**, 661.

241 Traktman, P. (1990) *Cell*, **62**, 621.

242 Nevins, J. R. and Joklik, W. K. (1977) *J. Biol. Chem.*, **252**, 6930.

243 Broyles, S. S., Yuen, L., Shuman, S. and Moss, B. (1988) *J. Biol. Chem.*, **263**, 10 754.

244 Shuman, S. (1990) *J. Biol Chem.*, **265**, 11 960.

245 Shuman, S and Morham, S. G. (1990) *J. Biol. Chem.*, **265**, 11 967.

246 Blumenthal, T. (1982) in *The Enzymes* (ed. P. D. Boyer), Academic Press, New York, vol. 15, p. 267.

247 Mills, D. R., Priano, C., DiMauro, P. and Binderow, B. D. (1989) *J. Mol. Biol.*, **205**, 751.

248 Paul, A. V., Yang, C. F., Jang, S-K., Kuhn, R. J., Tada, H. *et al.* (1987) *Cold Spring Harbor Symp. Quant. Biol.*, **52**, 343.

249 Tobin, G. J., Young, D. C. and Flanegan, J. B. (1989) *Cell*, **59**, 511.

250 Crawford, N. M. and Baltimore, D. (1983) *Proc. Natl. Acad. Sci. USA*, **80**, 7452.

251 Takeda, N., Yang, C. F., Kuhn, R. J. and Wimmer, E. (1987) *Virus Res.*, **8**, 193.

252 Baltimore, D. (1984) in *The Microbe 1984, Society for General Biology Symposium*, **36**, part 1 (eds B. W. J. Mahy and J. R. Pattison), Cambridge University Press, Cambridge, p. 109.

253 Ongradi, J., Cunningham, C. and Szilagyi, J. F. (1985) *J. Gen. Virol.*, **66**, 1011.

254 De, B. P. and Banerjee, A. K. (1985) *Biochem. Biophys. Res. Commun.*, **126**, 40.

255 Banerjee, A. K. (1987) *Miocrobiol. Rev.*, **51**, 66.

256 Plotch, S., Bouloy, M., Ulmanen, I. and Krug, R. M. (1981) *Cell*, **23**, 847.

257 Krug, R. M. (1981) *Curr. Top. Microbiol. Immunol.*, **93**, 125.

258 Krug, R. M. (1985) *Cell*, **41**, 651.

259 Wagner, R. R. (1975) in *Comprehensive Virology* (eds H. Fraenkel-Conrat and R. R. Wagner), Plenum Press, New York, vol. 4, p. 1.

260 Alifano, P., Rivellini, F., Limauro, D., Bruni, C. B. and Carlomagno, S. (1991) *Cell*, **64**, 553.

261 Zhang, J. and Corden, J. L. (1991) *J. Biol. Chem.*, **266**, 2297.

262 Sumimoto, H., Ohkuma, Y., Yamamoto, T., Horikoshi, M. and Roeder, R. G. (1990) *Proc. Natl. Acad. Sci. USA*, **87**, 9158.

263 Johnson, P. F. and McKnight, S. (1989)

264 Horikoshi, M., Hai, T., Lin, Y-S., Green, M. R. and Roeder, R. G. (1988) *Cell*, **54**, 1033.

265 Reeder, R. H. (1990) *Trends Genet.*, **6**, 390.

266 Pikaard, C. S., McStay, B., Schultz, M. C., Bell, S. P. and Reeder, R. H. (1989) *Genes Devel.*, **3**, 1779.

267 Reeder, R. H. (1989) *Curr. Opinion Cell Biol.*, **1**, 466.

268 Chodchoy, N., Pandey, R. N. and Marzluff, W. F. (1991) *Mol. Cell. Biol.*, **11**, 497.

269 Muller, D. K., Martin, C. T. and Coleman, J. E. (1989) *Biochemistry*, **28**, 3306.

270 Lescure, A., Carbon, P. and Krol, A. (1991) *Nucleic Acids Res.*, **19**, 435.

271 Platt, T. (1981) *Cell*, **24**, 10.

Annu. Rev. Biochem., **58**, 799.

10

Control of transcription

An organism must be able to modulate the expression of its genes. Some proteins, such as the enzymes of key metabolic pathways and the ribosomal proteins, will be required in large quantities. In an *E. coli* cell it has been calculated that there are 20 000 ribosomes, so it follows that there will need to be at least 20 000 copies of each of the constituent ribosomal proteins. Conversely, some proteins are normally present at very low concentrations but can be made in greater quantities when required. Thus, in *E. coli* growing on glucose, the enzyme β-galactosidase is normally present at approximately five molecules per cell but, under circumstances to be described in the following sections, it can be synthesized in a thousandfold greater quantities. Clearly, there must be mechanisms for ensuring and controlling the selective utilization of the genome and those mechanisms that operate at a transcriptional level are the subject of this chapter. Mechanisms that operate during RNA processing and protein synthesis are considered in chapters 11 and 12, respectively.

10.1 THE REGULATION OF PROKARYOTIC TRANSCRIPTIONAL INITIATION

The best understood transcriptional control systems are those of prokaryotes and their viruses. Bacteria must be flexible enough to employ many substrates for their metabolic requirements but must combine flexibility with economy by not making enzymes unless they are needed. Their answer to this challenge is to link the genes for functionally related proteins, such as the enzymes of a metabolic pathway, into clusters that are subjected to co-ordinated control and are known as operons. Typically, but far from exclusively, the operon is under the control of a single promoter from which RNA polymerase transcribes all of the constituent genes as a single *polycistronic* mRNA. The promoter is in turn controlled by a series of regulatory elements which respond to the bacterial environment and modulate the activity of RNA polymerase.

10.1.1 Induction of the *lac* operon – a negative control system

The *E. coli lac* operon is the classic example of the negative regulation of a bacterial operon by a *repressor protein*. It provides the mechanism by which the bacterial cell does not normally make the enzymes of lactose metabolism (except in very small quantities) but retains the capacity to do so should it find itself in an environment in which glucose is not available but lactose is present as an alternative. Under such conditions, the concentration of β-galactosidase, the enzyme which catalyses the splitting of lactose into galactose and glucose, rapidly increases from approximately five to several thousand molecules per cell. The enzyme is described as *inducible*, as are the functionally related lactose-metabolizing enzymes, β-galactoside permease and β-galactoside transacetylase.

Jacob and Monod [1] derived a series of mutant bacteria defective in the production of these enzymes. Some mutations affected the synthesis of one enzyme and were assumed to be in the gene for that enzyme. Others, known as *lacI* mutations, disturbed the regulation of all three genes such that, instead of being inducible in the presence of lactose or other β-galactosides, they were synthesized *constitutively*; that is they were made at the induced rate regardless of the presence or absence of an inducer. As a result of their studies [1], Jacob and Monod proposed the operon theory, a hypothesis which has since been proved by direct analysis. The features of the *lac* operon, in its currently understood form, are illustrated in Fig. 10.1 and are the subject of a number of reviews [2, 3]. They may be summarized as follows.

1. The *structural genes* for the three enzymes are contiguous on the DNA and are transcribed from a single promoter into a polycistronic mRNA.
2. The *I* region of the DNA is a regulator gene which is transcribed constitutively from a separate promoter. The resulting I mRNA is translated into a $38\,600\text{-}M_r$ polypeptide which aggregates into a tetrameric *repressor* protein and, in the absence of inducers, binds to the *operator*. Binding of the repressor protein to the operator blocks the transcription of the structural genes (Fig. 10.1(a)). The *lacI* mutations isolated by Jacob and Monod

[1] produced mutant repressors that were unable to bind to the operator or block transcription. Similar constitutive mutants may arise from mutations in the operator.
3. In the presence of lactose, a complex is formed between the inducer (a metabolite of lactose) and the repressor protein. The complex is unable to bind to the operator and transcription of the structural genes can thus proceed (Fig. 10.1(b)). The natural inducer is not lactose but allolactose formed from lactose by transglycosylation. In studying this system induction is maximized by using a non-metabolizable or *gratuitous inducer*, isopropyl thiogalactoside (IPTG).
4. The 3′-portion of the promoter (Fig. 10.2) is more or less typical of prokaryotic promoters (bacterial promoters are discussed in section 9.2.2 and the divergence of the *lac* promoter from that of the prokaryotic consensus sequence is reviewed by Reznikoff [3]). The 5′-portion of the promoter functions in catabolite repression, a further control mechanism that inhibits the induction of the *lac* operon when lactose and glucose are both available as substrates (catabolite repression is discussed in section 10.1.3).

The first bacterial repressor protein to be isolated was that of the *lac* operon [4]. In wild-type *E. coli* it represents approximately 0.001% of the total protein of the cell and its purification would have presented formidable problems had not Gilbert and Müller-Hill [4] increased its synthesis by genetic means. They first produced a strain of *E. coli* with a mutated promoter that overproduced the repressor tenfold. This mutated DNA they then incorporated into transducing phage that could occur in many copies per cell. The result was an overall thousandfold increase in the synthesis of the protein. Puri-

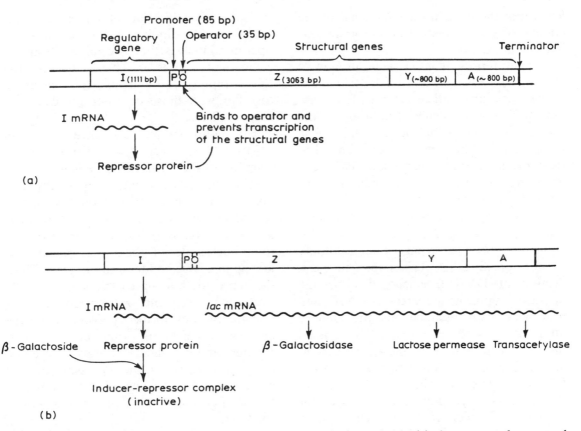

Fig. 10.1 A diagrammatic representation of the lactose (*lac*) operon in (a) the repressed state and (b) the induced state.

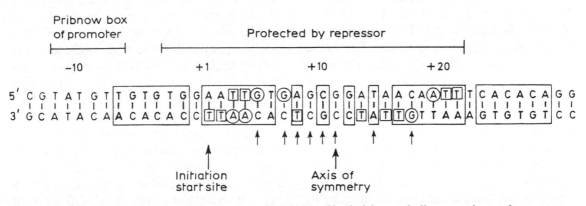

Fig. 10.2 The repressor binding region on *lac* DNA. Shaded boxes indicate regions of symmetry. Open squares enclose thymidine residues which, when replaced by 5-bromouracil and exposed to ultraviolet light, become cross-linked to the repressor protein. Open circles enclose residues which the repressor protects from modification by dimethyl sulphoxide. Short arrows identify the positions of point mutations that lead to constitutive expression of the structural genes.

fication of the protein then made use of the affinity of the repressor for the gratuitous inducer IPTG [4].

The *lac* repressor is an acidic protein of 360 amino acids with a M_r of 154 520 and it consists of four identical subunits. Each subunit binds one molecule of inducer but the protein interacts with the *lac* DNA as a tetramer [5]. When the inducer is bound to the repressor it causes an allosteric change in its configuration that greatly reduces its affinity for the operator (reviewed in [6]).

The binding of the repressor to the operator is very strong ($K_d = c.\ 10^{-13}$ M) and is about 4×10^6 times stronger than its affinity for non-specific DNA. This enabled the purification of DNA fragments carrying the operator region as a repressor–DNA complex which was retained on nitrocellulose filters whereas non-complexed DNA passed through. Isolation from these DNA fragments of sequences that were protected from DNase digestion by their association with the repressor yielded a 27-bp operator sequence of which 16 bases occurred as a hyphenated palindrome [7]. It has since been shown that the twofold symmetry extends beyond the protected region and involves 28 bp of a 35-bp sequence. The operator sequence is illustrated in Fig. 10.2 with the symmetrical sequences in boxes. The figure also indicates those nucleotides that have been shown to be important in repressor interaction. These include base pairs which if subject to mutation give rise to the constitutive phenotype. They also include nucleotides which the receptor protects from *in vitro* methylation by dimethyl sulphate, and thymidines which, when replaced by 5-bromouracil and exposed to ultraviolet light, can be cross-linked to the repressor. Experiments with partially digested repressor, mutant repressors and operators, analogy with other known repressors and most

recently, crystallization of intact tetramer, reveal that the binding of the protein to the operator involves the *N*-terminal 60 amino acids of the repressor. These data, and the DNA–repressor interaction, are considered together with those of other repressors in section 10.1.7.

The 35-bp operator sequence runs from −7 to +28 with respect to the transcription start site of the β-galactosidase gene. It thus overlaps the gene and butts on to the Pribnow box of the promoter. Evidence was presented in section 9.2.2 that RNA polymerase, when it binds to the promoter, covers up to 50 nucleotides on the 5′-side of the gene and up to 20 nucleotides into the gene. Thus, the repressor binding site at least partially overlaps the sequence recognized by RNA polymerase and, in theory, would deny the enzyme access to its promoter. However, Straney and Crothers [8] have presented data that suggest that the repressor does not block the initial binding of RNA polymerase. Rather it halts it in a pre-transcriptional complex.

As described above, the *lac* repressor protein is a tetramer. This is unusual as all other characterized repressor molecules are dimers of which each subunit binds to one half of the palindromic operator. However, in addition to the operator, there are two additional repressor binding sites in the *lac* operon that have been known as pseudo-operators (O_2 and O_3). Their affinities for the lac repressor protein are low and this, combined with their location (O_2 is 401 bp downstream and O_3 is 92 bp upstream), led to the belief that they played no significant role in repression. Recent data have shown, however, that they do contribute to the 1000-fold difference in gene expression between induction and full repression. The deletion of either O_2 or O_3 decreased repression 2–3-fold but, if both were deleted, repression decreased 50-fold

[9]. Despite the fact that the repressor binds DNA as a tetramer, only two subunits are required for operator recognition and Kania and Müller-Hill [10] have demonstrated that the protein can in fact simultaneously interact with two operators. It would appear that full repression requires such a cooperative interaction between the operators with the intervening DNA looping out to allow their coming together in a repression complex ([11] and references therein).

The speed with which the repressor protein is able to locate its operator, which represents less than one-millionth of the bacterial genome, is too fast to be accounted for by random association with and dissociation from the DNA. Rather it suggests that there must be some mechanism for the facilitated transfer of the protein to its regulatory site. Evidence has been presented that the protein first binds non-specifically to DNA and is then transferred to the operator either by sliding along the DNA or by direct intersegment transfer [12]. Similar evidence for the location of promoters by RNA polymerase was discussed in section 9.2.2.

10.1.2 Repression of the *trp* operon

Escherichia coli has the capacity to synthesize all of the amino acids required for protein synthesis. In many cases, however, the necessary enzymes are only made in the absence of an adequate exogenous source of the amino acid. Perhaps the most studied such system is the *trp* operon [13]. This consists of five structural genes which encode the five polypeptides of the three enzymes required to synthesize tryptophan from chorismate (chorismic acid is the common precursor for the synthesis of the three aromatic amino acids). The operon is again under the control of a single promoter, is transcribed into a polycistronic mRNA and is controlled by a repressor protein that is encoded in a regulator gene (Fig. 10.3). In this instance, however, the regulator gene (*trpR*) is remote from the rest of the operon. Furthermore, the repressor protein that it encodes is inactive and, in the absence of tryptophan, does not bind to the operator. When tryptophan is present, however, it binds to and activates the repressor which now interacts with the operator and blocks transcription. Tryptophan is thus a corepressor in that, whereas lactose decreases the affinity of the *lac* repressor for the *lac* operator, tryptophan greatly increases the affinity of the *trp* repressor for its operator [14]. Both systems are, however, examples of *negative control* because the active repressor switches off transcription. The *trp* operon is also controlled by attenuation (section 10.2.1).

The twofold axis of symmetry of the *lac* operator has proved to be a common theme and the *E. coli trp* operator has a 20-bp sequence of dyad symmetry that includes only one mismatch. This operator falls entirely within the promoter, and the region of symmetry includes the Pribnow box. The interaction of the *trp* repressor with its operator is discussed in greater detail in section 10.1.7.

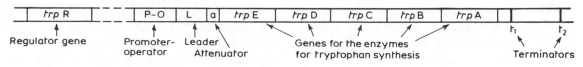

Fig. 10.3 A diagrammatic representation of the tryptophan (*trp*) operon.

10.1.3 Catabolite repression – a positive control system

When grown in the presence of both lactose and glucose, *E. coli* selectively metabolizes glucose. This is achieved by a control of the *lac* operon that is distinct from the negative control of the repressor protein. It is called *catabolite repression* and is controlled by the *catabolite activator protein* (CAP) (also known as the *cyclic AMP receptor protein*, CRP). CAP is a positive activator of the *lac* operon but is only active in the presence of cyclic AMP (cAMP). The dimeric protein interacts with cAMP and undergoes a conformational change [15]. This in turn enables the protein subunits to interact with the 5′-region (−72 to −52) of the *lac* promoter thereby facilitating the formation of an RNA polymerase initiation complex (reviewed in [16, 17]).

Catabolite repression depends on the ability of glucose to reduce the cellular levels of cAMP. In the absence of glucose, the levels of the nucleotide rise and result in the formation of CAP–cAMP which in turn activates transcription [16]. The control system is obviously of survival value to the bacteria as it is energetically desirable to metabolize glucose rather than lactose; especially if by so doing it becomes unnecessary to express the *lac* operon. Furthermore, catabolite repression is not restricted to the *lac* operon. The binding of the same CAP–cAMP complex is also necessary for the expression of other bacterial operons including those that enable the bacteria to metabolize a number of alternative carbohydrates that can replace glucose as a source of energy and carbon.

The DNA sequence to which the CAP–cAMP complex binds has been determined in a number of operons and, like those of operators, has been found to exhibit partial dyad symmetry. Furthermore, after an analysis of 18 such regions, Ebright *et al.* [18] derived a CAP-binding DNA consensus sequence of approximately 22 bp. They further demonstrated the importance of the interactions between specific base pairs within the consensus sequence and specific amino acid residues in the CAP protein. Thus, a mutant protein, in which a Glu at position 181 was replaced by Lys, Leu or Val, could be compensated for by an altered DNA sequence in which G:C base pairs at two positions in the recognition sequence were replaced by A:T base pairs [18, 19]. Interactions between CAP and DNA have also been investigated by hydroxyl radical footprinting; a technique that reveals each base of the DNA that is protected by a bound protein. The similarities in the mechanism by which CAP and repressor proteins interact with DNA is discussed in section 10.1.7.

Considerable evidence suggests that the CAP–cAMP complex stimulates initiation by first binding to the DNA and then interacting with RNA polymerase. Thus, it is known that the complex bends the *lac* promoter ([20] and references therein) and Buc [21] has proposed that this allows RNA polymerase to bind by direct protein–protein interaction. However, it is not clear that the protein always works in the same manner and notwithstanding the shared features of catabolite repression, there are sufficient differences between the target operons to make it difficult to envisage a common mode of action. Much of the uncertainty relates to variation in the location of the CAP-binding site. In the *lac* operon, the complex binds adjacent to the RNA polymerase binding site at −72 to −52 with respect to the transcription start site. In the galactose (*gal*) operon, it binds at −50 to −23 and appears to overlap the RNA polymerase binding site whereas in the arabinose (*araBAD*) operon the major CAP-binding site is much further away at −107 to −78. Gaston *et al.* [22] have pointed out that, on most operons, the binding sites for CAP–cAMP are located at one of three distances from the transcription start site. They have suggested that the binding to at least two of these need not preclude similar interactions between RNA polymerase and at least one of the CAP subunits. However, a model of the activation process will also have to explain the apparently variable manner in which initiation is stimulated. Thus, *in vitro* transcription assays employing the *lac* operon [23] indicate that CAP acts to stimulate the formation of a closed RNA polymerase–promoter complex [24]. Conversely, at the *gal P1* promoter, CAP accelerates the formation of the open complex without influencing the initial binding [25] whereas at the *malT* promoter, bound CAP has no effect on either closed or open promoter complex formation but stimulates the escape of RNA polymerase from the promoter complex into elongation [26]. (The above steps in initiation of transcription are considered in section 9.2.2.)

10.1.4 Dual control of the *araBAD* operon

The arabinose (*araBAD*) operon allows *E. coli* to use arabinose as an energy source. It is induced by arabinose and subject to catabolite repression when glucose is also present. Its three structural genes (*araB*, *araA* and *araD*) encode the enzymes required to convert arabinose into xylulose 5-phosphate, an intermediate in the pentose phosphate pathway. They are contiguous and are transcribed into a polycistronic mRNA. Two independent operons contain the *araE* and *araF* genes that encode a permease system involved in arabinose uptake into the bacterial cell.

The dual control of the *araBAD* operon is complex and may be summarized as follows.

1. The regulator gene *araC* encodes a regulator protein AraC that is involved in both positive and negative control of the *araBAD* genes. It achieves this by binding to three totally separate control regions of the operon; *araI*, *araO_1* and *araO_2*. Of these, *araI* is divisible into two regions *araI_1*, *araI_2* and *araO_2* is well separated from the other control regions and is located within the non-coding region of the *araC* gene (Fig. 10.4).

2. In section 10.1.1 it was seen that full repression of the *lac* operon required that the tetrameric repressor interacted with two operators that caused a looping out of the intervening DNA. Repression of the *araBAD* operon similarly requires that, in the absence of arabinose, the dimeric AraC protein binds to each of the control regions *araI_1* and *araO_2* and also to each other [27–29]. Two dimers thus interact and the intervening 211 bp of DNA forms the so-called *repression loop* that leaves the *araBAD* genes uninducible (Fig. 10.4(b)). AraC protein normally has a low affinity for *araO_2* but the ligand-free protein binds with high affinity to *araI_1* and also has potential to bind another dimer. This increases the effective concentration of dimer in the region of *araO_2* leading to cooperative binding and full repression [27]. Such a cooperative interaction is called the chelate effect. Gene constructs show that the control region of the operon has an 11.1 helical repeat and that maximal repression requires that the *araO_2* and *araI_1* regions are on the same face of the DNA. Thus, the insertion of five extra base pairs rotates one site halfway around the DNA double helix and greatly diminishes repression [30]. Constructs also show that if the separation of the two binding sites is greater than 500 bp, repression is considerably reduced but there does not ap-

(a) The *araBAD* operon

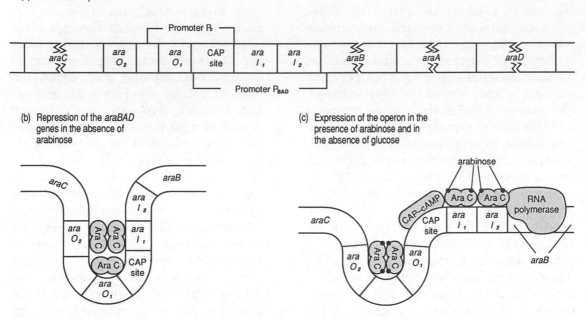

Fig. 10.4 The *araBAD* operon. A diagrammatic representation of the possible way in which two mutually exclusive cooperative interactions between AraC repressor protein and its binding sites may control gene expression. See text for details.

pear to be a lower limit to their separation for cooperativity [30].

3. Cooperative binding also gives rise to an alternative DNA loop that forms preferentially in the presence of arabinose and precludes the formation of the repression loop [29]. This second, *antirepression loop* results from the binding of AraC protein to a second high affinity site, *araO₁*, that again allows cooperative binding to *araO₂* (Fig. 10.4(c)).

4. Arabinose changes the conformation of AraC [31] and apparently changes its interaction with *araI*. In the absence of the inducer, the protein protects two turns of the DNA from modifying reagents (the region *araI₁*). In the presence of the sugar, occupancy extends into a region called *araI₂* (Fig. 10.4(c)) and the protein now

occupies four turns of the DNA that includes *araI₁* and *araI₂*. (It does not appear to be known whether this involves the binding of extra dimers as diagrammatically shown in Fig. 10.4(c) or whether conformational changes are responsible). The region *araI₂* overlaps the promoter P_{BAD} (Fig. 10.4(a)) and it is suggested that the binding of AraC protein produces a new conformation of the DNA that favours RNA polymerase activity [32].

5. When the concentration of cAMP is high, the cAMP–CAP protein complex binds between *araO₁* and *araI* and is presumed to stimulate transcription by influencing the effects of the AraC–arabinose complex on RNA polymerase.

6. Yet a further element in the control of the

araBAD operon is that AraC protein regulates its own synthesis [33]. Initial studies indicated that this occurred through its binding to *araO₁* which is within the promoter P_C for the *araC* gene (Fig. 10.4(a)). However, subsequent studies revealed that all three AraC binding sites are required for autoregulation [34]. Specifically, the low affinity binding site *araO₂* is essential for autoregulation, but the high affinity *araI₁* and *araO₁* sites are also required because they are necessary for cooperative binding with *araO₂*.

10.1.5 Other variations in the control of initiation at bacterial operons

(a) *The* bio *operon is an example of divergent transcription*

The biotin (*bio*) operon of *E. coli* contains five structural genes (*A*, *B*, *F*, *C* and *D*) and an unidentified open reading frame (ORS) that encode enzymes involved in biotin synthesis and which have an operator–promoter region located between genes *A* and *B*. Gene *A* and the ORS are transcribed leftwards from one promoter whereas the remaining four genes are transcribed rightwards from a second promoter on the other DNA strand. The operator between the two promoters appears to have a single binding site for the repressor–corepressor complex that negatively regulates the operon. It exhibits partial dyad symmetry and overlaps both promoters [35, 36]. Such a dual promoter system provides a mechanism for differential rates of expression of the different genes and is a common form of gene organization (divergent promoters are reviewed in [37]).

The most unusual feature of the *bio* operon is the nature of the repressor–corepressor complex. The repressor protein (BirA) is the same protein as biotin–protein ligase, the enzyme that catalyses the attachment of biotin coenzyme to various carboxylating/ decarboxylating enzymes. Furthermore, the corepressor is not biotin as might be expected from the resemblance between the regulation of the *trp* and *bio* operons. Instead it is biotoyl-AMP, the product of the first half of the reaction catalysed by biotin–protein ligase (the *bio* operon is reviewed in [38]).

(b) *Autogenic control of bacterial genes*

The synthesis of *E. coli* alanyl-tRNA synthetase is under autogenic control. The synthetase represses the transcription of its own gene by binding to an operator-like palindromic sequence that flanks the transcription start site of the gene [39]. High concentrations of the cognate amino acid enhance the repression by causing an even tighter association between the synthetase and the DNA.

(c) *The MetJ protein is an example of a repressor that controls multiple operons*

Several intermediates in the pathway to the synthesis of methionine are also metabolites for the synthesis of other amino acids. However, flux through the pathway is substantially controlled, at the level of gene expression, by the MetJ protein which acts as a repressor protein for at least six operators scattered through the *E. coli* genome [40]. The corepressor for MetJ is not methionine but another product of the pathway and the major intracellular methyl donor, *S*-adenosyl methionine [41].

Structural analyses [42, 43] suggest that MetJ has a unique mode of action. This is discussed in section 10.1.7.

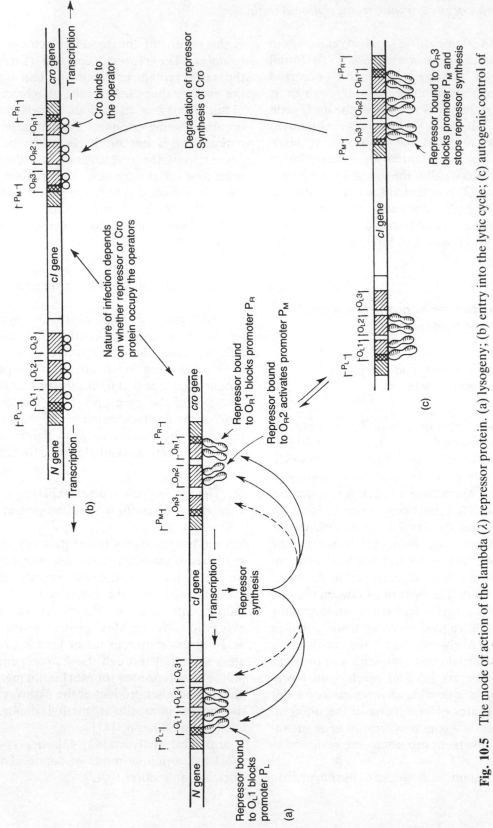

Fig. 10.5 The mode of action of the lambda (λ) repressor protein. (a) lysogeny; (b) entry into the lytic cycle; (c) autogenic control of repressor protein synthesis.

(a)

N gene

Transcription —

$\lceil P_L \rceil$ $\lceil O_L1 \rceil \lceil O_L2 \rceil \lceil O_L3 \rceil$

Repressor bound to O_L1 blocks promoter P_L

cl gene

Transcription

Repressor synthesis

$\lceil P_M \rceil$ $\lceil O_R3 \rceil \lceil O_R2 \rceil \lceil O_R1 \rceil$ $\lceil P_R \rceil$

cro gene

Repressor bound to O_R1 blocks promoter P_R

Repressor bound to O_R2 activates promoter P_M

(b)

N gene

— Transcription

$\lceil P_L \rceil$ $\lceil O_L1 \rceil \lceil O_L2 \rceil \lceil O_L3 \rceil$

Nature of infection depends on whether repressor or Cro protein occupy the operators

cl gene

$\lceil P_M \rceil$ $\lceil O_R3 \rceil \lceil O_R2 \rceil \lceil O_R1 \rceil$ $\lceil P_R \rceil$

cro gene

— Transcription

Cro binds to the operator

Degradation of repressor
Synthesis of Cro

(c)

$\lceil P_L \rceil$ $\lceil O_L1 \rceil \lceil O_L2 \rceil \lceil O_L3 \rceil$

cl gene

$\lceil P_M \rceil$ $\lceil O_R3 \rceil \lceil O_R2 \rceil \lceil O_R1 \rceil$ $\lceil P_R \rceil$

Repressor bound to O_R3 blocks promoter P_M and stops repressor synthesis

10.1.6 The repressor of the bacteriophage lambda

The repressor–operator interactions of the *E. coli* bacteriophage lambda (λ) have been studied in considerable detail (reviewed in [44–47]). This, and related viruses, can exhibit two different life-styles. In the lysogenic state, the viral DNA, known as a prophage, is integrated into the host genome and is replicated with it. Conversely, in the lytic cycle, the genes of the bacteriophage are transcribed by the host RNA polymerase leading ultimately to the production of many progeny bacteriophage that are released with the lysis of the bacterial cell (section 3.7.8).

(a) *Maintenance of the lysogenic state*

Maintenance of the lysogenic state of infection depends on a single protein, the lambda repressor. The way in which this protein suppresses the alternative lytic cycle of the phage is illustrated in Fig. 10.5 and can be summarized as follows.

1. The repressor polypeptide can be conceived as a dumbbell-shaped molecule in which *N*-terminal and *C*-terminal domains are joined by an intervening stalk. The active protein is a dimer of two of the 27 000-M_r polypeptides joined through their carboxy domains.
2. The viral DNA has two operators O_L and O_R each of which has three 17-bp repressor-binding sites designated O_L1, O_L2, O_L3 and O_R1, O_R2 and O_R3, respectively. The binding sites conform to a consensus sequence and show partial dyad symmetry. Each binding site will bind the *N*-terminal domains of a repressor dimer.
3. The binding sites O_L1 and O_R1 exhibit the greatest affinity for the repressor and they overlap with the leftward and right-

ward promoters P_L and P_R. The preferential binding of the repressor to these sites inhibits transcription from the promoters and in so doing stops the expression of the proteins required in the lytic cycle of viral infection.

4. The presence of repressor bound at O_L1 and O_R1 greatly stimulates the binding of further repressor dimers at O_L2 and O_R2 and the adjacent repressor molecules interact through their carboxyl domains (Fig. 10.5(a)). The presence of bound repressor at O_R2 stimulates the interaction of RNA polymerase with the promoter P_M and the transcription of the *cI* gene that encodes the repressor. The repressor thus exerts a negative influence on the promoters P_R and P_L but is a positive regulator of the promoter P_M and of its own synthesis.
5. The binding sites O_L3 and O_R3 have a relatively low affinity for repressor and normally remain unoccupied. If, however, large quantities of repressor accumulate, the site O_R3 will become occupied, so blocking the promoter P_M and shutting off repressor synthesis. The repressor thus exerts an *autogenic* control over its own synthesis (Fig. 10.5(c)). Repressor bound at O_R3 does not interact with repressor bound at O_R1 and O_R2.

(b) *Switching from lysogenic to lytic infection*

The lysogenic prophage retains the capacity to switch to lytic infection and this is best achieved in the laboratory by damaging the bacterial DNA with ultraviolet light or chemical mutagens. This stimulates a set of responses in the host which are known collectively as SOS functions and which include the activation of a protease activity in the recA protein that normally functions in genetic recombination (section 7.4.2). The activation results from the binding to the enzyme of single-stranded DNA generated by the damage to the host cell genome. Activated recA protein cleaves and inactivates the lambda repressor and thus derepresses the operator P_R (P_R is derepressed before P_L as the latter has the higher affinity for the protein). Transcription from P_R then leads to the production of the Cro protein, a dimer which like lambda repressor recognizes the three operator components O_R1, O_R2 and O_R3 (Fig. 10.5(b)). Its binding is, however, non-cooperative and its affinity is the reverse of that of the lambda repressor. Thus, at low concentrations, Cro binds preferentially to O_R3 so blocking the P_M promoter and inhibiting the replacement of the degraded repressor. At higher concentrations it also occupies O_R2 and O_R1 turning down the transcription of the early genes; a step which by then is a necessary part of the lytic cascade (section 10.2.2).

(c) *The establishment of the lysogenic state versus entry into the lytic cycle*

Establishment of lysogeny requires the synthesis of two proteins, the repressor and a DNA topoisomerase called integrase, that catalyses the integration of the prophage into the host DNA (section 7.5.1). It was seen in section 10.1.6(a), however, that the repressor activates and controls its own synthesis from the binding sites of the O_R operator. One might therefore question how sufficient repressor could be made to start the process. In fact the *cI* gene encoding the repressor is under the influence of a second promoter, P_E, which is functional during the establishment of lysogeny while the P_M promoter takes over for its maintenance. When lambda first infects a bacterium, no repressor is present so the operators O_R and O_L are unblocked and transcription will initiate from P_L and P_R. The first products are

the proteins Cro and N and, as will be seen in section 10.2.2, this permits the further transcription of the genes for the proteins cII and cIII. The combined effect of these two proteins is to stimulate the synthesis of repressor from its alternative promoter, P_E. They also stimulate the production of integrase. It can be seen that whether bacteriophage infection results in lytic or lysogenic infection is under delicate balance (Fig. 10.6). The early products of infection include the proteins Cro, cII and cIII one of which blocks the action of the repressor whereas the other two stimulate its synthesis. Whether the infection becomes lytic or lysogenic will depend on whether Cro or repressor occupies the viral operators. The control of the lytic cycle is considered in section 10.2.2.

10.1.7 The interaction of prokaryotic repressor and activator proteins with DNA

(a) *The helix-turn-helix motif*

Lambda repressor and Cro proteins, together with the equivalent proteins of the closely related phage 434, can be made in large quantities from genes inserted in recombinant plasmids and both have been subjected to sequencing and crystallographic studies. At first sight the two proteins are not very similar. The lambda Cro protein consists of only 66 amino acids whereas the repressor protein contains 236 residues. However, both proteins have at their DNA-binding *N*-termini a pair of α-helices separated by a β-turn. They are known as helix 2 and helix 3, are superimposable in space and contain conserved amino acids at key positions such as at the β-turn between the helices. Furthermore, hybrid proteins in which the phage repressor helix 3 is replaced

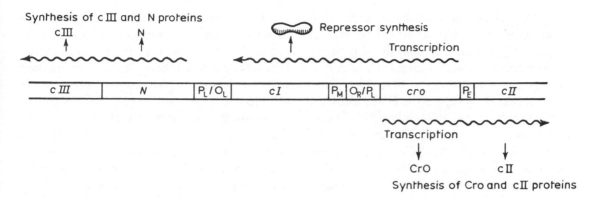

Fig. 10.6 The initial events following the infection of *E. coli* with bacteriophage lambda. Cro protein competes with the repressor for occupancy of the operators, O_R and O_L and whether the infection is lytic or lysogenic depends on the outcome. N is an antiterminator which permits the synthesis of cII and cIII proteins (section 10.2.2) which in turn stimulate repressor transcription from the promoter P_E.

by the helix 3 of Cro exhibit identical interactions with DNA. A wealth of biochemical and genetic evidence has accumulated suggesting that the α-helices are intimately involved in the interaction with the operator DNA and this is confirmed by structural studies. X-ray crystallographic analysis has revealed the structures of complexes between synthetic operators and the DNA binding domains of phage 434 repressor [48, 49], lambda repressor [50], Cro protein [51] as well as the *E. coli trp* repressor [52, 53]. In each case, the helix-turn-helix motif interfaces with the DNA such that helix 2 lies across the major grove helping to orient the protruding helix 3 (recognition helix) into the major groove. High resolution of the co-crystals reveals that the recognition helices make specific contacts with the bases of the major groove and this requires relaxation of the binding site that is accompanied by an overwinding of the operator DNA and a narrowing of the minor groove.

The lambda repressor also contains two *N*-terminal arms that wrap around the DNA and make contact with the central base pairs within the operator [54]. Rather unexpectedly, the two identical subunits of the lambda repressor bind to the partially symmetrical operator asymmetrically. Furthermore, this is built into the protein–DNA recognition as base substitutions that change the non-consensus half of the operator to the consensus sequence and decrease repressor binding [55]. As outlined in the previous section, the phage Cro and repressor proteins recognize six partially conserved operator elements with differing affinity. Base substitution experiments have shown that the proteins specifically recognize the slight sequence differences in their respective favoured binding sites and that they each cause the DNA to adopt different conformations [56, 57].

The helix-turn-helix motif is proving to be a common feature of prokaryotic DNA-binding proteins. The three-dimensional structure of *E. coli* CAP protein (section 10.1.3) has also been determined [58]. In contrast to the phage proteins it interacts with DNA through its *C*-terminus, whereas its larger *N*-terminus interacts with cyclic AMP. Nevertheless, the αD, αE and αF helices of the *C*-terminus can be superimposed on the α1, α2 and α3 helices of Cro. The superimposition between αE–αF and

$\alpha1-\alpha2$ is particularly striking and they appear to interact with DNA in the same way [18, 19]. Predictive analysis of protein sequence, coupled with structural and genetic analysis, strongly suggest that helix-turn-helix motifs are involved in the DNA–protein interface of many other proteins [59]. These include the *bio* repressor [60] and the *lac* repressor in which the helix-turn-helix is aligned in the opposite direction to that described above [61, 62]. The recent crystallization of the intact *lac* repressor tetramer complexed with inducer and with operator DNA [63] should lead to further understanding of this system.

(b) *Other forms of prokaryotic protein–DNA interaction*

Recent structural analyses of the three-dimensional crystal structures of the *E. coli* *met* repressor, with and without the corepressor (*S*-adenosylmethionine), reveal a different form of interaction with DNA. The protein is a dimer of intertwined monomers and has no helix-turn-helix motifs [42]. It has been proposed that it interacts with DNA either through a pair of symmetrically related α-helices or through a pair of β-strands [42]. The binding site for the *met* repressor, MetJ, is an eight base-pair consensus that is repeated from two to five times in different *met* operators. Evidence is accumulating that MetJ binds to these in tandem arrays [43].

The Arc and Mnt repressor proteins of bacteriophage P22 also appear to lack helix-turn-helix motifs. Evidence has been presented that they interact with DNA through their *N*-terminal regions [64].

(c) *How does a DNA-bound protein influence transcription?*

The importance of the helix-turn-helix interaction in influencing transcription has been

most dramatically demonstrated with the *trp* repressor (the *trp* operon is considered in section 10.1.2). In this dimeric protein the two sets of DNA-binding helices are located either side of a rigid protein core. In the absence of tryptophan they are folded inwards unable to interact with the operator. Thus, the repressor binds DNA with low affinity and no specificity. Tryptophan increases specific binding a thousandfold by promoting hydrophobic interactions that lock the helices into an active position in which they protrude into and recognize the major groove of the *trp* operator [53].

Repressors may sterically block a promoter or, as has been suggested for the *lac* repressor, may halt RNA polymerase in a pre-initiation complex [8]. Activators must in some way stimulate the initiation of transcription and most evidence indicates that they do so by positioning RNA polymerase on the DNA rather than by modifying the enzyme [65]. The *lac* activator protein, CAP, bends and distorts the DNA and it has been suggested that the free energy of bending may contribute to the ease with which the bound RNA polymerase can melt the DNA [66]. However, the apparent variability in the mode of action of CAP has already been discussed. Lambda repressor molecules, when bound to the two operator motifs O_R1 and O_R2, activate the promoter P_M (section 10.1.6). This results, at least in part, from a direct interaction between a negatively charged region of the repressor and a so-called activating surface of RNA polymerase (Fig. 10.7). This, in an unknown way, helps the enzyme to bind and to initiate transcription of the *cI* gene. The main function of the cooperativity in repressor binding appears to be that it makes the system very sensitive to small changes in repressor concentration. The cooperative interaction may also bend the DNA and certainly causes it to loop if the two operator elements are artificially

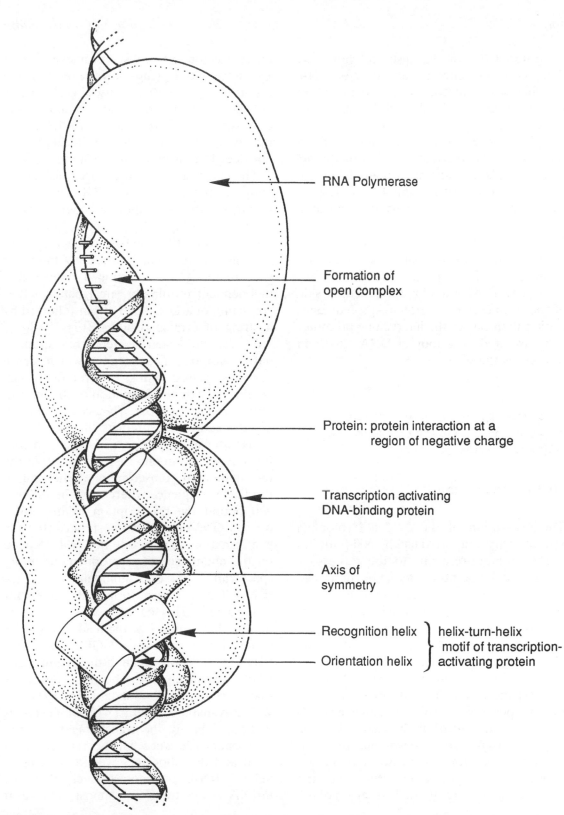

RNA Polymerase

Formation of
open complex

Protein: protein interaction at a
region of negative charge

Transcription activating
DNA-binding protein

Axis of
symmetry

Recognition helix ⎫ helix-turn-helix
            ⎬ motif of transcription-
Orientation helix ⎭ activating protein

Fig. 10.7 A diagrammatic representation of the formation of an open promoter complex under the influence of an activating protein.

separated [67]. In this instance, however, there is no evidence for an effect of DNA bending on transcriptional activation; in fact, transcription is abolished in the constructs with separated operator elements. Looping has nevertheless been implicated in repression. Its role in repression of the *lac* and *araBAD* operons has already been discussed (sections 10.1.1 and 10.1.4, respectively) and a similar mechanism is implicated in the repression of the *E. coli* galactose (*gal*) operon that encodes enzymes responsible for metabolizing galactose to glucose 1-phosphate. The repressor protein binds to both of two operators located on either side of the promoter [68]. Protein–protein association then causes the intervening promoter to be isolated on a loop of DNA ([69] and references therein).

10.2 THE REGULATION OF THE TERMINATION OF TRANSCRIPTION IN PROKARYOTES

10.2.1 Attenuation

The expression of the *E. coli* tryptophan operon varies approximately 600-fold but this is not entirely due to the repressor–corepressor–operator interaction described in section 10.1.2. Repression can reduce transcription 70-fold [70] whereas a controlled termination of transcription, known as *attenuation*, can reduce it by 8- to 10-fold [71] giving a combined effect of 600-fold. Attenuation is the subject of many reviews [72–74] and was discovered when mutants of the *trp* operon were isolated that gave rise to increased amounts of all the enzymes of the tryptophan-synthetic pathway and mapped, not at the operator or repressor, but at a site between the operator and *trpE*, the first structural gene (Fig. 10.3). This region of the

DNA was found to consist of a sequence of 162 nucleotides lying downstream of the promoter. It was thus transcribed by RNA polymerase and formed the 5′-end of the polycistronic mRNA known as the *leader sequence*. It is believed to be translated but it does not form a functional protein.

The leader has been sequenced, as has the entire tryptophan operon [75], and it was found that its transcript can undergo extensive base-pairing. The leader region numbered 1 in Fig. 10.8 can base-pair with region 2. This leaves region 3 free to base-pair with region 4 thus forming a typical rho-independent terminator site (section 9.2.4) consisting of a G:C-rich hairpin followed by a string of uridine residues (Fig. 10.8(c)). However, the leader sequence has an alternative base-pairing strategy in which region 2 is base-paired with 3, a structure which would preclude the formation of the terminator site (Fig. 10.8(b)). The accepted model of attenuation [76] predicts that the leader transcript can adopt either of these two conformations and in one case will result in the termination of transcription at the end of the leader sequence whereas in the other it will permit the transcription of the entire operon. The model also postulates that the conformation that is adopted will depend on the abundance of tryptophan, or, more specifically, on the abundance of Trp-tRNA^{Trp} (the charged form of the tRNA for tryptophan). This concept arose from the finding that region 1 of the leader transcript contains two successive UGG codons for tRNA^{Trp}, a codon which in the genome as a whole is relatively rare. The model envisages that, if tryptophan is abundant, Trp-tRNA^{Trp} will be available to interact with UGG codons without delaying the translating ribosome. The leader sequence will be rapidly translated and the ribosomes will follow closely behind RNA polymerase such that the mRNA will be translated almost as fast as it

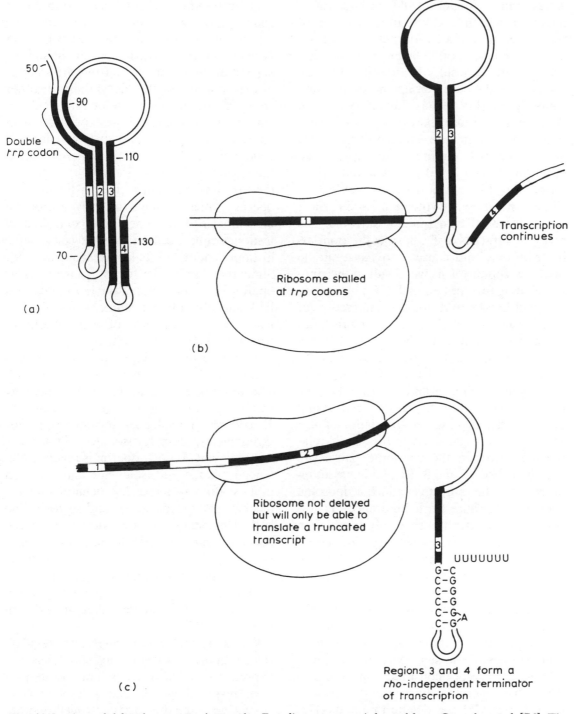

Fig. 10.8 A model for the attenuation of the *E. coli trp* operon. Adapted from Oxender *et al.* [76]. The shaded areas 1–4 indicate regions of the leader able to undergo extensive base-pairing. (a) base-pairing possibilities of the *trp* leader transcript; (b) base-pairing in low tryptophan concentrations; (c) base-pairing in high tryptophan concentrations.

is transcribed. Region 1 will have been translated and region 2 will be in the process of translation before a hybrid between regions 2 and 3 could form. Under these circumstances, there would be nothing to stop regions 3 and 4 base-pairing and terminating transcription (Fig. 10.8). Thus, if tryptophan is abundant, those polymerase molecules that manage to initiate transcription (i.e. are not blocked by a repressor occupying the operator – section 10.1.2) will only transcribe the 162 nucleotides of the leader. When, however, tryptophan is in short supply there will be a different outcome. The rarity of Trp-tRNA^{Trp} will slow the translating ribosome and cause it to pause at the dual Trp codons. This will result in the ribosome being further behind RNA polymerase and will create time for the interaction of leader regions 2 and 3. This in turn precludes the hybridization of regions 3 and 4 into a terminator and allows RNA polymerase to continue past the leader and into the structural genes (Fig. 10.8(b)). Clearly, the model depends on the close relationship between transcription and translation in prokaryotes. It also depends on the spatial relationship between the ribosome pause site and the regions 2 and 3. By making assumptions about the range over which a ribosome was likely to cause steric hindrance, Yanofsky [77] demonstrated that only tryptophan and, to a much lesser extent, arginine starvation could cause attenuation of the *trp* operon in *E. coli*. Only a ribosome stalled at these codons could allow the formation of the base-paired stem between regions 2 and 3. Considerable evidence has accumulated that supports the above model of attenuation. Studies of the sensitivity of the leader to ribonuclease are consistent with the predicted base-pairing [76] as are studies with base change [77] and deletion mutants ([78] and references therein) that modify the ability of the different regions to base-pair

with each other. The adjacent tryptophan codons in the *trp* leader are conserved in the *trp* operon of other bacterial species [77] but perhaps the best evidence for their importance in attenuation comes from the analysis of the operons for the biosynthesis of other amino acids. In every case investigated, the amino acids that control the expression of an operon have been found to have clustered codons in the leader of that operon. Often the clustering is more dramatic than that of the *trp* operon. Thus, the histidine (*his*) operon, which encodes the enzymes for the biosynthesis of histidine and is controlled by histidine concentration, has six consecutive histidine codons in its leader sequence [79]. Similarly, the leader of the leucine (*leu*) operon has four consecutive leucine codons [80] and the phenylalanine (*phe*) operon leader has a sequence of nine codons of which seven specify phenylalanine [81]. Even more convincing are those operons regulated by more than one amino acid. The threonine (*thr*) operon is regulated by threonine and isoleucine and a sequence of 13 codons in its leader includes eight for threonine and four for isoleucine [82]. Similarly, the *ilvGMEDA* operon is regulated by leucine, valine and isoleucine and a stretch of 17 codons in its leader includes four for leucine, five for isoleucine and six for valine [83]. Deficiency of any one of the three cognate aminoacyl-tRNAs relieves attenuation [84]. The leaders of some of these operons have been shown to be capable of forming mutually exclusive secondary structures similar to that described for the *trp* leader [85–87].

Charged tRNA molecules also regulate the transport of some amino acids. Thus, the expression of two operons that encode proteins involved in the transport of leucine, isoleucine and valine are controlled by transcriptional attenuation of a leader rich in leucine codons. In this case, however, a rho-dependent rather than a rho-independent terminator is involved [88, 89].

A variation of the attenuation of the tryptophan biosynthetic operon is seen in several species of *Bacillus* [90, 91]. The leader sequence of the mRNA still adopts alternative secondary structures that influence whether the transcript terminates. In this instance, however, a protein that is encoded in another, presumably tryptophan-controlled operon, dictates which alternative structure forms. Such attenuation-controlling proteins may best be regarded as examples of antiterminators (section 10.2.2).

The *pyrB1* and *pyrE* operons of *E. coli* exhibit another variant mechanism of attenuation. The operons encode enzymes involved in the biosynthesis of pyrimidine nucleotides and their expression is inversely correlated with the concentration of cellular UTP [92]. The transcript of the leader of the *pyrB1* operon can fold into two structures. The first of these is a hairpin followed by a series of uridine residues. However, the uridylate residues are sufficiently separated from the hairpin that the structure does not form a rho-independent terminator. Rather its template functions as a UTP-dependent pause site which, if UTP is in short supply, will delay the transcribing RNA polymerase molecules. The second structural element is downstream from the hairpin and is a functional rho-independent terminator. A model for the attenuation of this system, that is supported by elegant gene construct and site-directed mutagenesis studies [93–95], is as follows. At low UTP concentrations, RNA polymerase pauses at the hairpin and this allows the first translating ribosome to catch up with the enzyme. Transcription and translation subsequently remain tightly coupled such that, as the polymerase transcribes the downstream terminator, the closely coupled ribosome disrupts its secondary structure and prevents termina-

tion. If however, UTP concentrations are high, RNA polymerase will not be slowed at the hairpin and will proceed to synthesize and recognize the terminator without influence from the more distant ribosome. The translating ribosome must in fact be within 14–16 nucleotides of the terminator to disrupt termination [95].

Yet another mechanism of attenuation employs antisense RNA. The replication of the staphlococcal plasmid, pT181, is controlled by Rep C protein, the transcription of the mRNA of which is controlled by attenuation. A termination hairpin can form in the leader of the mRNA just upstream of the initiation codon. Its formation is promoted by association with antisense counter transcripts but in the absence of these an upstream sequence known as the pre-emptor pairs with one arm of the hairpin and prevents termination [96]. The control of gene expression by antisense RNA is considered in section 10.7.

All the variant mechanisms of attenuation depend on pausing and premature termination by RNA polymerase. It has recently been demonstrated that RNA polymerase II can also pause and terminate both *in vitro* and *in vivo* and evidence is growing that eukaryotic cells also employ attenuation to control transcription. As yet, however, little is known of the mechanisms involved.

10.2.2 Antiterminators of transcription

During lytic infection by bacteriophage lambda and related phage, the viral genes are expressed sequentially in what is known as the lytic cascade (Fig. 10.9). The primary control of this process is at the level of the termination of transcription which is inhibited by viral antiterminator proteins (reviewed in [97–99]). As described in section 10.1.6(c), when phage lambda first infects a

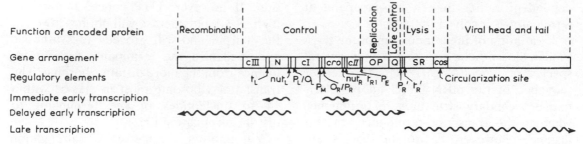

Fig. 10.9 A diagrammatic representation of the events in the lytic cycle of bacteriophage lambda. Note that, although it is illustrated as a linear molecule, the lambda DNA is circularized on infection by the annealing of its single strand *cos* ends and ligation by host cell DNA ligase (section 3.7.8). Some of the illustrated regulatory elements are discussed in section 10.1.6.

cell, two genes, *cro* and *N*, are expressed by host RNA polymerase transcribing in opposite directions from the promoters P_R and P_L (rightward and leftward promoters). These two genes are known as the *immediate early genes* and they encode the antirepressor protein Cro (the function of which is described in section 10.1.6(c)) and an antiterminator, the N protein. This protein permits the lytic cycle to continue with the expression of the next block of genes; the *delayed early genes*. It achieves this by allowing RNA polymerase to transcribe past rho-dependent terminators (t_L and t_R1) located at the 3′-ends of the *N* gene and *cro* gene, respectively (rho-dependent terminators are considered in section 9.2.4). The delayed early genes encode the cII and cIII proteins discussed in section 10.1.6(c). They also encode two proteins involved in the replication of the viral genome, a series of proteins that function in recombination and another antitermination factor, Q.

The protein Q unlocks the final block of bacteriophage genes, the so-called *late genes*, that encode the viral head and tail proteins as well as two proteins that bring about host cell lysis. They are transcribed from a separate promoter $P_{R'}$ from which RNA poly-

merase normally initiates RNA synthesis constitutively. However, in the absence of Q, the transcripts terminate at an adjacent rho-dependent terminator ($t_{R'}$) with the release of a 194-nucleotide RNA known as 6S RNA. The Q protein suppresses termination at $t_{R'}$ and allows RNA polymerase to transcribe through it and into the late genes. The lytic cycle is illustrated in Fig. 10.9.

The N gene encodes a $12\,000\,M_r$ protein that is a highly specific antiterminator for lambda genes. Related bacteriophage have equivalent proteins but they are not interchangeable. This specificity suggests that the protein recognizes unique sites on lambda DNA and such sites have been identified from mutations that abolish their function. They are known as N protein utilization sites (*nutL* and *nutR*) and were identified as 17-bp sequences located between P_L and t_L on the leftward strand and between P_R and t_R1 on the rightward strand of the lambda DNA (Fig. 10.9). They are identical for 16 of their 17 bp and contain a 5-bp region of inverted symmetry [100]:

A G C C C T G A A Pu A A G G G C A
T C G G G A C T T Py T T C C C G T

The functional significance of this recognition sequence, known as box B, was demonstrated when a cloned fragment that included *nutR* was found to cause termination-resistant transcription when placed between a non-lambda promoter and a termination signal [101]. The hairpin sequence within the box B is recognized by a conserved arginine-rich motif of the cognate N protein and these elements are also the site of species specificity [102]. The combination of N and *nut* will overcome any rho-dependent terminator and will also act at rho-independent termination sites. Because it will function with both sorts of terminators it seems likely that the N protein acts on an element common to them both, namely RNA polymerase.

E. coli mutations reveal that at least five host proteins are involved in antitermination by N protein. They are known as N utilization substances (Nus) and are designated NusA−E. *NusD* maps to the gene that encodes rho protein and *nusC* maps at the *rpoB* gene that encodes the β subunit of RNA polymerase. The *nusA* gene encodes a $69\,000\,M_r$ protein that binds to N protein [103]. It also binds with equimolar stoichiometry to RNA polymerase core enzymes [104] not, as previously suggested, by displacing sigma factor but after sigma is released [105]. In the absence of N protein, NusA induces transcriptional pausing and termination [106, 107] but in the presence of N-protein it forms a core antitermination complex that reads through some terminators [108]. To read through other terminators, the complex requires additional factors the function of which is unclear, although they may be necessary for stabilizing the core. Immunochemical studies [109] suggest that NusB and NusE are part of the complete antiterminator complex. The *nusB* gene product is a $15\,689\,M_r$ protein that directly stimulates transcriptional anti-

termination [110] and *nusE* encodes the ribosomal protein S10 that can function in *in vitro* antitermination systems either as the purified protein or as 30S ribosomal subunits [111]. A second element of the *nut* site may be a recognition sequence for one or more of the host factors [112]. It is known as box A, is highly conserved in various operons and has a consensus sequence:

$$5' \; {}^{C}_{T}\text{GCTCTT(T)A} \; 3'$$

Several groups have suggested that the antitermination complex actually recognizes and binds to RNA transcripts of *nut* rather than the DNA. For instance, N protein and some Nus factors are released from the transcription complex by treatment with ribonuclease [109] and mutations that prevent hairpin formation in the RNA transcript also prevent antitermination [113]. Furthermore, the finding by Barik *et al.* [114] that an anti-N protein antibody would precipitate RNA polymerase-containing transcription complexes but only after the enzyme had passed over an active *nut* site, is consistent with this interpretation. A currently favoured model thus suggests that N protein and Nus factors recognize and bind to the *nut* RNA. Subsequently or simultaneously, they also bind RNA polymerase to form a stable antitermination complex [109, 114].

The other lambda antiterminator, Q, is a $23\,000\,M_r$ protein that appears to depend on fewer host proteins. In an *in vitro* system it promotes readthrough of the $t_{R'}$ terminator with RNA polymerase and NusA protein as the only additional protein factors [115]. Grayhack *et al.* [116] have shown that, in the presence of NusA, Q protein binds to RNA polymerase and causes it to read through a pause site early in the λ late gene operon.

Sequences important for Q-dependent anti-termination (sometimes known as *qut* sequences) are close to the $P_{R'}$ promoter and cause RNA polymerase to pause at nucleotide +16. The paused complex is acted on by Q protein and this in some way modifies the enzyme so that it is able to read through downstream terminators [116].

Antitermination has been studied in considerably less detail in a number of other systems [117–119]. The elements involved in the antitermination of the *E. coli* rRNA operons are, however, known and they exhibit considerable similarities with those of phage lambda ([120] and references therein).

10.3 MODIFICATION OF PROKARYOTIC RNA POLYMERASE

10.3.1 Diversity in sigma factor

The specificity of transcription in bacteria is dependent on the association with RNA polymerase core enzyme of sigma factor; a subunit that allows the enzyme to recognize gene promoters but which leaves the ternary complex when transcripts are a few nucleotides in length (section 9.2.2). The mechanics of the system suggest a potential control mechanism in which different classes of promoter could be recognized by a series of promoter-specific sigma factors. Until recently, such control mechanisms were thought to be uncommon but the growing number of known examples is forcing a re-evaluation of their importance [121, 122]. In general, the use of alternative sigma factors is emerging as a means for the controlled transcription of groups of genes whose products contribute to a common physiological response or whose products are required at a precise time such as after infection or during sporulation. There fol-

lows a description of some of the best understood examples.

(a) *Sporulation and phage infection in Bacillus subtilis*

Sporulation in *Bacillus subtilis* results from a programmed, temporal expression of more than 50 different genes. The process is initiated in response to nutrient depletion, probably because of falling concentrations of intracellular GDP and GTP [123]. It is conventionally divided into seven, rather arbitrarily defined, morphological stages. Thus, the vegetative cell is classified as stage 0 and the release of the mature spore from the lysed mother cell is known as stage VII [124]. Genes, the mutation of which affect sporulation, are similarly classified. Thus, mutation of *spo0* genes prevents a cell from entering the sporulation pathway whereas *spoII* mutations are those which stop passage from stage II to stage III, i.e. they allow initiation and the formation of an asymmetric septum but they prevent forespore formation.

The temporal pattern of transcription results, at least in part, from the stage-specific appearance of different forms of RNA polymerase sigma factor which recognize different promoters and in so doing reprogramme the specificity of the enzyme (reviewed in [124–126]). Table 10.1 lists the known *B. subtilis* factors and the promoter consensus sequences of the genes that they recognize and regulate.

The principal sigma factor of *B. subtilis* vegetative cells was originally called sigma^{55} (σ^{55}) from an empirically determined molecular mass of 55 000 daltons. When, however, the protein was sequenced its 371 amino acids were found to have an actual M_r of 43 000 [127]. It was therefore renamed σ^{43} but, because similar ambiguity surrounds other factors, we here adopt the suggested

Table 10.1 The sigma factors of *B. subtilis* and its phage, SPO1

Sigma factor	Alternative notation	Notation of gene		Consensus Promoter Sequence	
				−35 region	−10 region
σ^A	σ^{55}, σ^{43}	*rpoD, sig A*	Housekeeping functions	TTGACA	TATAAT
σ^B	σ^{37}	*sig B*	Stationary phase functions	AGG–TT	GG–ATTG–T
σ^C	σ^{32}	*sig C*	Unknown	AAATC	TA–TG–TT–TA
σ^D	σ^{28}	*sig D*	Flagellar synthesis/chemotaxis	CTAAA	CCGATAT
σ^E	σ^{29}	*spoIIGB, sig E*	Sporulation genes	TT–AAA	CATATT
σ^F		*spoIIAC, sig F*	Sporulation genes		
σ^G		*spoIIIG, sig G*	Sporulation genes		
σ^H	σ^{30}	*spoOH, sig H*	Stationary phase and sporulation genes	RCAGGA	GAATT––T
σ^K		*spoIIIC/spoIVCB*	Sporulation genes		
σ^{gp28}		phage SP01 28	Phage middle genes	T–AGGAGA––A	TTT–TTT
$\sigma^{gp33-34}$		phage SP01 33, 34	Phage late genes	CGTTAGA	GATATT

nomenclature of Losick *et al.* [124] in which σ^{43} is designated σ^A (Table 10.1 compares the alternative notations that are in use). σ^A shares sequence homology with *E. coli* σ^{70} [128] and recognizes a promoter sequence similar to its −10 and −35 regions [129].

In addition to σ^A, vegetative cells of *B. subtilis* contain at least four minor sigma factors, σ^B, σ^C, σ^D and σ^H. Some of these factors become more abundant as the bacteria move out of logarithmic growth into stationary phase and, further into the programmed development, sporulation-essential factors are produced. These include σ^E and σ^F encoded in the stage II-specific genes and σ^G and σ^K that are encoded in stage III-specific genes. At least two of the sporulation-specific factors (σ^E and σ^K) are post-transcriptionally modified, thus providing a further potential level at which gene expression can be modified.

The function of the minor factors σ^B, σ^C and σ^D of vegetative cells is poorly understood but it appears that they are involved in non-sporulation events associated with entry into the stationary phase of growth. Such events include adaption to growth-limiting stress, synthesis of extracellular enzymes and development of chemotaxis and motility. The abundance of σ^B-containing RNA polymerase increases during late log phase and *in vitro* and *in vivo* it has been shown to direct the transcription of genes that are expressed only in the stationary phase [130, 131]. Another gene, which is expressed in both the growth and early-stationary phases, has promoters for both σ^A and σ^B, suggesting that expression of this gene is controlled by σ^A-associated RNA polymerase during rapid growth and by σ^B-associated RNA polymerase during stationary phase [132]. Sigma D has been implicated in the expression of genes associated with flagella synthesis and chemotaxis [133–135].

The production of the sporulation-specific

factor, σ^E is initiated at stage II (approximately 2 h after initiation) at a time when the forespore septum is produced but such that RNA polymerase associated with the factor is subsequently present in both the resulting cell compartments [136]. The gene that encodes σ^E can be transcribed by RNA polymerase associated with σ^A. However, the spacing of the promoter elements is unusual and other factors appear to be required to allow promoter recognition [137]. This may explain why several genes, including that encoding σ^H, must be transcribed before σ^E can be expressed. σ^E is translated as an inactive pre-protein and its activation requires the proteolytic removal of 29 amino acids. The protease that catalyses this maturation is also expressed from a developmentally regulated gene and its activity is first detected 1–2 h into the sporulation process. Clearly this provides further potential control of the temporal expression of σ^E-controlled genes.

σ^H recognizes the consensus promoter sequences RCAGGA and GAATTNNT for the −10 and −35 regions, respectively ([159] and references therein). (R represents an unspecified purine and N is any nucleotide.) It is present in both growing cells and during early sporulation and is involved both in reprogramming gene expression during the transition from growth to stationary phase and in the initiation of sporulation [138]. In common with other sporulation-associated genes, some of the genes transcribed by σ^H-containing RNA polymerase have complex promoters with recognition elements for two different sigma factors [139, 140]. In at least one case, this appears to correlate well with observed gene expression. Thus, the expression of the gene for fumarase is controlled from two promoters, P1 and P2. Of these, P1 is presumed to be recognized by σ^A and functions to provide the relatively low and stable basal concentration of fumarase found throughout growth. Conversely, the promoter P2 is recognized by σ^H and is responsible for the peak in enzyme levels 1–2 h after the end of exponential growth. P2 activity varies over 50-fold depending on growth rate, sporulation, carbon source etc. and factors additional to σ^H are involved in this regulation [140].

σ^F directs the transcription of genes that are expressed in the forespore chamber [446] and the other two sporulation-essential factors, σ^G and σ^K are thought to govern compartment-specific gene expression in the forespore and mother cell, respectively [141, 142].

The *B. subtilis* bacteriophage, SPO1, capitalizes on the flexible RNA polymerase conformation of the bacteria by regulating it for its own reproduction [121, 125, 126]. Three viral genes (genes 28, 33 and 34) encode sigma-like polypeptides that alter promoter recognition by the host cell polymerase (Table 10.1). As was seen with phage lambda (section 10.2.2), the genes of the bacteriophage can be divided into three transcriptional blocks: those expressed early in infection, those expressed in the middle of the lytic cycle and those expressed late in the cycle. The promoters for the phage early genes are recognized by the host polymerase with the σ^A subunit. The product of one of these early genes, gene 28, is a sigma-like, 26 000-M_r protein ($\sigma^{-gp^{28}}$) [143] that directs the host RNA polymerase to recognize and initiate transcription of the middle genes [144]. Two of the middle genes, genes 33 and 34, encode the sigma-like proteins, $\sigma^{-gp^{33}}$ and $\sigma^{-gp^{34}}$, which direct the host polymerase to transcribe the late genes [125].

(b) *The heat shock response in* E. coli

When cells are subjected to the stress of heat shock, there is an increase in the synthesis of a set of proteins known as the heat shock proteins. This response is universal, occurring in all organisms so far tested from bacteria to mammalian cells, although the mechanisms by which it is controlled vary. The function of the heat shock proteins is poorly understood but they are known to play roles in the assembly and disassembly of macromolecular complexes, intracellular transport and proteolysis.

When *E. coli* cells growing at 30°C are shifted to 42°C, the rate of synthesis of at least 17 proteins increases 5–20-fold, reaching a peak after 5–15 min. Thereafter, the rate of synthesis declines until a new rate, characteristic of the higher temperature, is attained. The genes for the 17 proteins are members of the so-called heat shock regulon [145] that is not recognized by the major *E. coli* RNA polymerase holoenzyme associated with σ^{70}. Instead, their synthesis depends on recognition of their promoters by a holoenzyme that is specific for heat shock genes [146], contains a 32 kDa sigma factor [147] and is known as $E\sigma^{32}$ or σ^{32}-holoenzyme. The factor interacts with a consensus promoter sequence CTTGAA and CCCCAT–AT at the −35 and −10 regions, respectively [148].

(c) *Expression of* σ^{54}-*dependent genes*

σ^{54} was first recognized in enteric bacteria as a regulatory factor required for the expression of glutamine synthetase and was later found to be required for the expression of other genes the products of which functioned in nitrogen assimilation [149]. Thus, in a wide range of organisms, it was found to be required for the expression of genes encoding proteins required for amino acid transport, amino acid degradation, and nitrogen fixation (reviewed in [149–151]). More recently, the factor has also been associated with the controlled expression of genes involved in a wider range of physiological functions including dicarboxylic acid transport, catabolism of toluene and xylene, formate degradation, oxidation of molecular hydrogen and the synthesis of flagella and pilin proteins [151].

All characterized promoters for the σ^{54}-holoenzyme (core RNA polymerase associated with σ^{54}) have a minimal conserved GC doublet at -11 to -14 with respect to the start site of transcription and a conserved GG doublet exactly 10 bp further upstream. It has become clear, however, that σ^{54} confers on RNA polymerase the ability to bind to these promoters but it remains as a closed complex [152, 153]. Formation of an open promoter complex and the initiation of transcription requires an additional activator protein and ATP. Furthermore, evidence is accumulating that the transcription at σ^{54}-dependent promoters is primarily controlled by a suite of these activator proteins.

A protein called NTRC is the activator that initiates transcription in response to low combined nitrogen levels. In the glutamine synthetase gene, it is specifically required for the formation of the open complex and does so by binding to a site upstream of the promoter that shares many of the properties of a eukaryotic enhancer. The protein is synthesized in an inactive form and its activity is positively and negatively controlled by phosphorylation and dephosphorylation (reviewed in [154]). A different protein, NIFA, activates the nitrogen fixation (*nif*) genes [155]. It is activated by low oxygen tension. Transcription of the genes encoding the proteins for toluene and xylene catabolism requires the activator protein XYLR which is apparently activated by binding the low-M_r substrates of the pathway [156, 157]. A fourth protein, DCTD is required for the activation of genes encoding dicarboxylic acid transport proteins. It is itself activated in the presence of dicarboxylic acids, probably by

phosphorylation by a partner protein [158].

It is clear then that each of the activator proteins allows the σ^{54}-holoenzyme to initiate transcription in response to a distinct physiological signal. The four proteins share regions of conserved sequence similarity but σ^{54} itself shares little similarity with other sigma factors.

10.3.2 Bacteriophage T4 modulation of host RNA polymerase

As with other bacteriophages, the development of the *E. coli* phage, T4, is characterized by the successive expression of several gene classes. All transcription is catalysed by the host RNA polymerase but the enzyme undergoes a series of phage-induced modifications that subverts the host enzyme to the temporal initiation of the different classes of T4 promoters (reviewed in [160]). At least three classes of viral gene, early, middle and late, are known and each has a distinct promoter type.

The early genes are at first transcribed by unmodified *E. coli* RNA polymerase from promoters indistinguishable from those of the host. However, immediately after infection, the polymerase undergoes its first modification carried out by the product of the T4 *alt* gene. This protein is a component of intact phage particles and is injected with the DNA. It brings about the addition of ADP-ribose to an arginine residue of one of the host enzyme α subunits. A few minutes later, a second ADP-ribosylation on the same residue (arginine 265) of the other α subunit is catalysed by the product of one of the early genes (*mod*). The function of the ADP-ribosylation of host RNA polymerase is not entirely clear but it is known to reduce the affinity of the enzyme for σ^{70} and to result in more efficient rho-dependent termination.

The promoters of at least some of the T4 middle genes include the so-called mot box (TA$^T/_A$GCTT) which occurs 13 bp upstream of the −10 element. Initiation from these genes requires the product of the viral *motA* gene and two further genes, *motB* and *motC*, may also be involved. The mode of action of these proteins is unknown but they arc not thought to function as alternative sigma factors. At this time, the protein products of two further phage genes, gp*rpbA* and gp*rpbB* (formerly called 15K and 10K proteins) associate with the host RNA polymerase. Again they are known to reduce its affinity for σ^{70} but their overall function is unclear.

The most dramatic change in the transcriptional specificity of host RNA polymerase occurs with the switch to the transcription of the late genes. Expression of these genes requires concurrent DNA synthesis and is stimulated by three T4-encoded DNA polymerase accessory proteins (gp44, gp45 and gp62) [161]. During T4 DNA replication, these proteins bind to the DNA at the primer–template junction and in transcription it appears that this binding site acts like a eukaryotic enhancer (enhancers are described in section 10.4.1). Indeed, Geiduschek and colleagues [161] have suggested that the viral DNA replication fork acts as a mobile enhancer for the transcription of late genes and that the three DNA polymerase accessory proteins are a part of the DNA replicating machinery when they activate transcription.

The promoters of the late genes are recognized by *E. coli* RNA polymerase core enzyme supplemented with the product of the T4 gene 55. This protein, gp55, is an alternative sigma factor and is sufficient for the recognition of the late promoter consensus sequence, TATAAAATA that is located a few bases upstream of the transcription start site. No −35 consensus element is required. Curiously, host sigma factor, σ^{70}, is dominant over gp55 in competing for RNA polymerase, and promoter specificity is generated by a further T4 gene product, gp33. Only RNA polymerase that has gp33 bound to it is subject to enhancement by the three DNA replication proteins [455].

10.4 THE CONTROL OF TRANSCRIPTION BY EUKARYOTIC RNA POLYMERASE II

The control of gene expression in eukaryotic cells is less well understood than that of prokaryotes. In addition to genes that can be turned on and off, such as those regulated by hormones, a eukaryotic cell will have many permanently shut-down genes. Thus, a liver cell will not express the genes encoding proteins found only in a brain cell or proteins made only during the differentiation of germ cells. Eukaryotic DNA exists in the nucleus as chromatin and it is assumed that the proteins of chromatin are intimately involved in the availability of genes for expression. The nucleus itself is a feature not found in prokaryotes and nuclear RNA processing and transport of RNA through the nucleoplasm and nuclear membrane are additional levels at which gene expression could be controlled. The control of gene expression is considered here whereas the control of processing is considered in chapter 11.

10.4.1 Promoter and enhancer organization

As in prokaryotes, the accurate and controlled transcription of eukaryotic mRNA-encoding genes requires a series of discrete DNA sequence elements that function as tar-

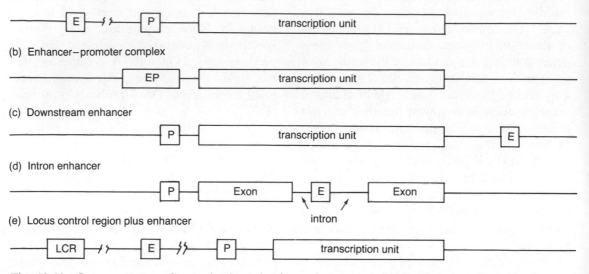

Fig. 10.10 Some patterns of organization of eukaryotic gene regulatory elements. Note that one or more of the above arrangements can be combined in genes with multiple enhancers. P, promoter; E, enhancer; EP, enhancer–promoter; LCR, locus control region.

gets for regulatory proteins. In eukaryotes, however, these elements may be dispersed over a large region of DNA and may extend 60 000 bp or more on either side of the start site of transcription. Operationally, the controlling elements are divided into promoters and enhancers (Figs 9.6 and 10.10). The TATA box of the promoter lies approximately 30 bp upstream of the transcriptional start site and is involved in fixing the position of initiation. This, together with commonly encountered upstream elements of the promoter (the CAAT box and GC-rich sequences), has been discussed in section 9.4.1.

Enhancer elements and their yeast equivalents, the upstream activator sequences (UAS), are regulatory in nature. They can stimulate transcription from a distance of tens of thousands of base pairs (the limit of operational effectiveness is not known), they operate from upstream or downstream positions (Fig. 10.10) and, in many cases,

their activity has been shown to be independent of their orientation in the DNA. Like promoters, they are organized in a modular pattern, i.e. they are composed of many different *transcriptional elements (TEs or enhansons)* each of about 10–20 bp and forming the binding site for one or more transcription regulating proteins. Often, there are multiple repeats of the same TE (e.g. Fig. 10.14) and the multiplicity is a requirement for maximal enhancer activity. Gene constructions with fewer repeats often function, but less well. There may also be several different TEs each providing binding sites for different proteins and allowing multiple regulatory pathways for the expression of the associated gene. Yeast upstream elements are generally considered to be less complex but the more they are systematically dissected the more complex they appear to be (section 10.4.3).

The distinction between promoters and enhancers is operational. The enhancer

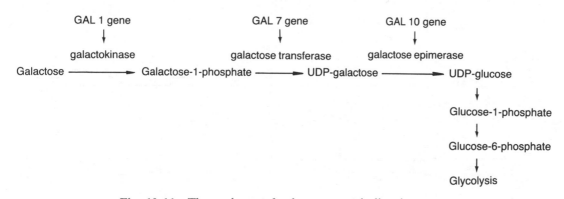

Fig. 10.11 The pathway of galactose metabolism in yeast.

must stimulate transcription but this is not necessarily a feature of promoters. Conversely, promoters must have one or more elements that direct and fix the position of initiation whereas accuracy is not a feature of enhancers. Nevertheless, promoters and enhancers can share the same sequence elements and they frequently overlap. Fig. 10.10 illustrates the possible arrangement of promoter and enhancer elements. (Enhancers are reviewed in [162–166].)

10.4.2 Enhancers are targets for transcription-regulating proteins

Most, probably all, of the transcriptional elements that comprise the modules of promoters and enhancers are binding sites for transcriptional factors (reviewed in [167–170]). Some of these, such as TFIIA, TFIIB, TFIID and TFIIE/F may be universally required in the formation of the transcriptional complex with RNA polymerase (section 9.4.1). Others, including the protein Sp1, that binds to the GC-rich box, may be common to many genes (section 9.4.1). Still other factors may be uniquely involved in the expression of genes in specific cell types or in signal-dependent regulation such as in response to heat shock, hormones or growth

factors. Different temporal or physiological signals may determine the relative abundance of factors or they may govern their post-transcriptional modification or activation to a conformation that can interact with their recognition elements. Furthermore, different factors, that may be positive or negative in their influence, may compete for the same or overlapping TEs whereas others may recognize other bound factors rather than elements of DNA. There are thus many known variations in control mechanisms and the selected examples that follow attempt to illustrate their diversity.

10.4.3 Yeast GAL gene organization superficially resembles a prokaryotic operon

One of the most studied of the gene regulatory systems of yeast is that encoding the enzymes of galactose utilization (reviewed in [171]). The three enzymes that allow cellular galactose to enter the glycolytic pathway (Fig. 10.11) are encoded in the GAL1, GAL7 and GAL10 genes that are clustered on chromosome II (Fig. 10.12). However, the analogy with prokaryotes immediately breaks down as they are not transcribed into a polycistronic mRNA. Each gene has a

(a)

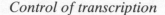

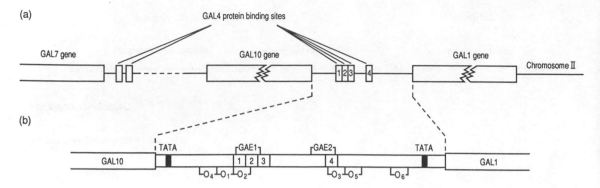

(b)

Fig. 10.12 (a) The organization of the GAL genes on chromosome II. (b) The GAL1–GAL10 divergent promoter. Sites 1–4 are GAL4 binding sites; O_1–O_6 are the so-called GAL operators; GAE_1 and GAE_2 are galactose-independent elements.

separate promoter. The genes for two further enzymes are located on chromosome XII. GAL2 encodes a permease that catalyses the transport of galactose into the cell and MEL1 is the gene for an *a*-galactosidase, an enzyme that allows the metabolism of galactose-containing disaccharides. Also encoded on separate chromosomes are two proteins that regulate the transcription of the structural genes. The GAL4 protein activates the transcription of the mRNAs for all five enzymes, whereas GAL80 protein binds to GAL4 and prevents its activity.

The most intensely studied aspect of GAL gene regulation is the transcription, from divergent promoters, of the GAL1 and GAL10 genes. In the absence of galactose, transcription is almost undetectable but, in the presence of the sugar, it can be induced 10 000-fold. Control of this extraordinary range of activity rests with a 365 bp upstream activating sequence (UAS_G) located between the two promoters (Fig. 10.12). It will be convenient to consider the various elements of this controlling region in turn.

(a) *GAL4 binding sites*

GAL4 is a 881 amino acid activator protein that binds to four 17 bp consensus elements

in the UAS_G. The consensus also occurs in the UAS of the other three structural genes and their transcription also requires the binding of GAL4. The protein binds as a dimer through its amino terminus [172] and the features of its DNA binding domain are discussed in section 10.5.2. The multiple binding sites for GAL4 allow a synergistic activation of gene expression. Thus, gene constructs reveal that the presence of two sites can elicit a rate of transcription that is more than twice that observed with a single site. Cooperativity does not apparently involve interaction of the GAL4 molecules with each other as for instance is observed with the lambda repressor (section 10.1.6). Rather synergism involves their simultaneous interaction with a common target molecule that has so far not been identified [173]. The probable value of four binding sites between the GAL1 and GAL10 genes is discussed below.

(b) *Inhibition by GAL80*

The regulatory proteins, GAL4 and GAL80 are produced constitutively and, in the absence of galactose, GAL4 does not activate transcription because GAL80 antagonizes its effects. It does so by recog-

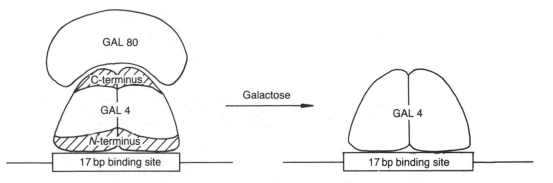

Fig. 10.13 The influence of galactose on the interaction of GAL4 and GAL80.

nizing and binding to the carboxyl terminus of GAL4 [174–176] thus creating a complex which, although still bound to DNA, does not activate transcription (Fig. 10.13). Galactose causes GAL80 to dissociate from the complex thereby allowing transcriptional activation.

(c) *Negative control elements*

The UAS_G also contains six negative control elements; the so-called GAL operators. They are located adjacent to or overlapping the GAL4 binding sites (Fig. 10.12) and in heterologous gene constructs, can repress a UAS placed upstream of them [177]. Their existence may, in part, explain why the control elements between GAL1 and GAL10 have four GAL4 binding sites whereas other GAL structural genes have two. It appears that the greater number is required to overcome the negative effects of the operators [178].

(d) *Galactose-independent activating elements*

Two other control elements have recently been identified in the UAS_G. Called GAE_1 and GAE_2, these elements, when incorporated into gene constructs, are able to activate transcription independently of galactose [179]. Their role is uncertain but it is clear that the above negative elements are necessary in order to inhibit their activity and to maintain galactose inducibility. It seems likely that the confusing diversity of overlapping elements in the UAS_G are required to permit the observed broad range of activity of the GAL1 and GAL10 genes [178] but it has also been suggested that the GAE sites are recognized by so far unidentified proteins and may be involved in physiological functions other than galactose catabolism [179].

(e) *Catabolite repression*

When glucose and galactose are available to the yeast cell, the GAL genes are repressed, apparently because GAL4 has a reduced affinity for its DNA binding sites. Under these conditions, the negative control elements are required to repress transcription [178].

10.4.4 Simian virus 40 has an enhancer/promoter complex

The SV40 enhancer was the first to be identified and remains one of the best charac-

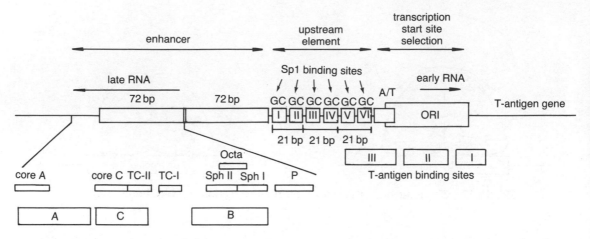

Fig. 10.14 A schematic diagram of the SV40 control region. Adapted from Nussinov [180].

terized. Structurally and functionally it is highly complex and forms a part of approximately 400 bp of DNA into which is compacted the regulatory elements for transcription of the early and late genes as well as those for DNA replication. The control of SV40 gene expression is reviewed in [163, 164, 166, 180].

The elements of the viral control region are best considered by working from the right to the left of Fig. 10.14.

(a) *The origin region*

This 70-bp domain includes within it the origin of DNA replication, the sites of early transcription initiation and three binding sites for T-antigen. T-antigen is the product of the first early gene and when it binds to the three sites, it inhibits early transcription and stimulates both late transcription and DNA replication.

(b) *AT-rich region*

This overlaps the origin region and is weakly homologous to the TATA box of eukaryotic cells. However, its deletion only reduces

transcription to about 20% and it appears to contribute little to overall promoter efficiency.

(c) *21-bp repeats*

Adjacent to the AT-rich region there are three 21-bp repeats, each of which contains two copies of the GC motif GGGCGG that is a component of the upstream element of many eukaryotic genes. As already described in section 9.4.1, the GC motif is the specific binding site for the transcription factor Sp1 and, in SV40, its binding stimulates the transcription of both early and late genes. The factor is thus able to function bidirectionally and this is confirmed by constructs in which the GC motif is inserted with opposite polarity. The complete deletion of the 21-bp repeats severely affects transcription but mutants with only two repeats are viable. Systematic mutagenesis has revealed that there is a gradient of importance in the six GC motifs with an order of binding affinities III, V, II, VI, I, IV. It is assumed that the order results from steric constraints and from variation in the sequences flanking the motifs.

(d) *72-bp enhancer repeats*

The enhancer has two levels of organization [181]. Three discrete functional domains, of 21, 22 and 15 bp that are called A, B and C, constitute the first level of organization [182]. In gene constructs, each of these elements is able to function as an enhancer in its own right; provided it is present as two or more copies [183]. Furthermore, each exhibits unique patterns of cell specificity [184, 185]. Examination of the sequences of the three domains reveals the second level of organization namely that they contain sequence motifs that are the binding sites for multiple transcription factors. Figure 10.14 illustrates the location of the motifs that include core A (also known as GTII), core C (also known as GTI), TCII, TCI, SphII (the doublet of Sph also forms an octamer binding site) and P. Each of these TEs or enhansons can be duplicated or interchanged in gene constructs that create new enhancer domains. Some, for instance core A, are apparently recognized by multiple transcription factors [186]. Where these species all occur in the same cell, they presumably compete but some are cell-specific. In contrast, Davidson *et al.* [187] have isolated a factor TEF-1 that specifically and cooperatively binds to the unrelated sequences of both GTII and Sph. Clearly, the SV40 enhancer presents multiple levels of organization that presumably allow a highly variable transcriptional response suited to its ability to infect many cell types.

10.4.5 Temporal and tissue-specific gene expression

(a) *Globin genes*

As outlined in section 8.4.2, the vertebrate globin genes are grouped into α-type and β- type clusters each containing different globin variants that are sequentially expressed through the embryonic, foetal and adult stages of development. Their expression is not only temporally controlled but is also highly tissue specific. Thus, the embryonic genes are expressed in the yolk sac, the foetal genes are active in the liver and the adult genes are expressed in the bone marrow. Control of these events is largely or totally transcriptional and the system has been a popular and profitable one for investigation.

The globin genes, particularly the β- and γ-globulin genes of various species, have been shown to have enhancer-like sequences just upstream, downstream and within their coding sequences (reviewed in [162]). Thus, the human β- and γ-globin genes can be expressed with the correct tissue and developmental specificity if they are transfected into mice with approximately 1 kb of 5′- and 3′-flanking sequence [188, 189]. Similarly, a 400–600 bp sequence 3′ to the chicken β-globin gene contains an enhancer that directs the tissue specific expression of the gene [190–192]. However, the extent of expression of these transfected genes is very low, independent of gene copy number and very sensitive to their site of insertion into the genome of the host. Up to 70% of inserted genes do not express, presumably because they are inserted into heterochromatic regions at transcriptionally inert sites that the controlling elements are unable to overcome (heterochromatin is discussed in section 3.3). Clearly, local sequences are insufficient for normal globin gene regulation.

Two sorts of observation demonstrated that regions far upstream of the globin genes are important for their controlled expression (reviewed in [193, 194]). First, rare forms of thalassaemia ($\gamma\delta\beta$-thalassaemia) arise from the deletion of large segments of upstream flanking sequence. These result in no ex-

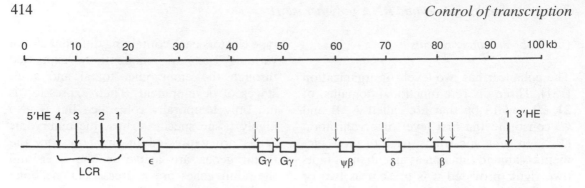

Fig. 10.15   Diagram of the human β-globin domain on chromosome 11. Solid arrows represent developmentally stable DNase I hypersensitive (HE) sites. Those clustered at the 5′-end of the domain form the locus control region (LCR). Discontinuous arrows indicate expression-related hypersensitive sites found specifically in embryonic, foetal or adult erythroid cells.

pression of β-type globins even though the genes themselves are often intact. Second, studies with DNase I indicate the existence of hypersensitive sites well upstream of the β-gene locus. In fact the whole gene cluster is sensitive to DNase digestion in erythroid as compared with non-erythroid cells but within the locus there are two sorts of hypersensitive site that are particularly sensitive to the enzyme. Expression-related hypersensitive sites, the appearance of which is specific to developmental stage, occur in the 5′-promoter regions of all the β-type genes (Fig. 10.15). Far more remote, however, are four sites that are erythroid specific but are present at all stages of development and are located 20 kb to the 5′-end of the embryonic ε-globin gene [195]. A further such site also marks the 3′-boundary of the so-called β-globin gene domain (Fig. 10.15). Direct demonstration of the regulatory role of the upstream hypersensitive (HS) sites came when the region containing them was directly ligated to globin genes and their expression monitored in transgenic mice and cultured erythroid cells. One example, presenting data on 74 transgenic mice, will serve to illustrate the nature of the data obtained in many laboratories [196]. Of 51 mice transfected with constructs in which HS sites were

linked to human β-globin genes, 50 expressed β-globin specifically in erythroid cells and to an extent comparable with that of endogenous mouse globin. Conversely, of 23 animals receiving the gene unlinked to HS regions, only seven expressed human globin and then only to 0.3% of that of the endogenous protein [196]. Similar studies showed that the HS regions would confer erythroid-specific expression on heterologous genes [197] and more recently segments of 0.5–1.0 kb, containing single HS sites have been shown to be active (reported in [194]). If the HS-containing region is deleted from the globin gene domain, specificity of expression is lost as is DNase hypersensitivity [198].

The region of DNA containing the four developmentally stable 5′-hypersensitive sites that is capable of conferring a high degree of position-independent expression on the globin genes, has received several names. These have included the domain control region (DCR) and the locus activating region (LAR). However, at the 1990 Haemoglobin Switching Meeting (reviewed in [194]), it was agreed that in future the region will be called the locus control region (LCR.) Thus, that described above for the β-globin genes is the β-LCR whereas the recently characterized equivalent for the α-globin genes is the α-LCR. The same meeting also standardized the notations for the numbering of HS sites and the naming of well characterized erythroid-specific transcriptional factors.

Analysis of minimum length HS segments, together with interspecies sequence comparisons, has led to the identification of three sequence elements that are consistently observed in the LCRs. They have much in common with those previously identified in globin promoters and enhancers. One element, TGAGTCA, is the binding site for the AP1 family of transcription factors that includes the products of the cellular oncogenes *jun* and *fos*. The family also includes an erythroid-specific factor called NF-E2 [199]. Two copies of the consensus occur in an 18-bp region within the LCR and, when they are linked to reporter genes, they function as a powerful enhancer [200]. Their deletion from the LCR cripples its activity so it is clearly required for its function. Nevertheless, it is not in itself sufficient for globin gene regulation as, if the tandem AP-1 sequence is linked to a β-globin gene, it does not cause high levels of expression in transgenic mice (reviewed in [194]).

A second sequence element, GATA, is recognized by an erythroid specific DNA-binding protein which has been cloned [201, 202] and is now called GATA-1 but was formerly known as Eryf1, NF-E1, GF-1, EF-1 or EF-γa. The GATA consensus sequence is present in the LCRs, enhancers and promoters of many erythroid-expressed genes [203] and GATA-1 is specifically required for the differentiation of erythroid cells [447]. In some of these, GATA elements even replace the TATA box and the protein can strongly activate transcription from minimal promoters containing a single GATA binding site (reported in [194]). A third sequence element, CACCC, is also commonly found in erythroid-specific promoters, enhancers and LCRs.

How then do these elements associated with erythroid-specific genes lead to the temporal control of globin gene expression? The β-LCR is seen primarily as a chromatin opening sequence that ensures that the globin domain is available for transcription in erythroid cells. It is envisaged that it is activated by specific protein factors early in erythroid cell development and is responsible for organizing the entire β-globin locus into an open chromatin domain that is stable throughout subsequent development. How this is achieved is unknown. One theory suggests that the LCR is a nuclear matrix attachment site and that the association exposes the domain to factors that change the conformation of the chromatin [204]. A second hypothesis is that the creation of the open domain is linked to DNA replication and that the LCR functions as an erythroid specific origin of DNA replication that allows the creation of a modified chromatin structure [205].

Once the open domain is established, the LCR is then thought to function as a powerful enhancer that governs the correct temporal regulation. In a theory first put forward by Choi and Engel [206], the embryonic, foetal and adult genes are envisaged as competing with the common LCR enhancer. Proteins produced in the yolk sac, foetal liver and adult bone marrow would influence the outcome of these competitive interactions. In the form of the model illustrated in Fig. 10.16 looping of the DNA would bring the LCR into contact with downstream promoters, the favoured interaction in any erythroid cell type being influenced by positive transcription factors and perhaps also by negatively acting 'silencing factors'. Considerable support for the model comes from recent studies in which gene switching has been observed in transgenic mice [207, 208].

(a) ε globin expression in embryonic yolk sac

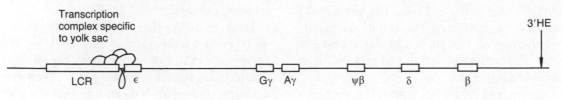

(b) γ globin expression in foetal liver

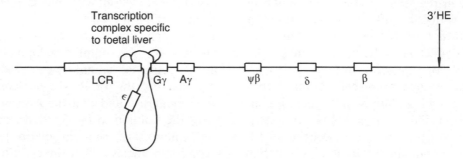

(c) β globin expression in adult bone marrow

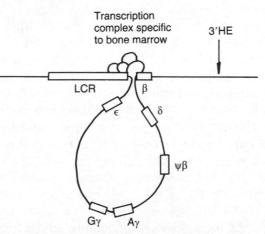

Fig. 10.16 A model for haemoglobin gene developmental control. Proteins specific to yolk sac, foetal liver or bone marrow regulate the competition among the globin genes for interaction with the LCR. Other proteins, not illustrated, may function to silence non-transcribed genes and preclude unwanted interactions.

When the γ- or β-globin genes were individually fused to the LCR, high tissue-specific expression was observed but temporal expression was lost. Either gene was expressed throughout development. When, however, the LCR was attached to both γ and β-genes in series, temporal expression was restored. There was a switch from foetal to adult glo-

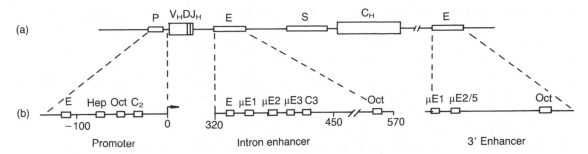

Fig. 10.17 (a) Diagram of the mouse immunoglobulin heavy chain. Boxed elements are the promoter (P), variable (V_H), diversity (D), joining (J), enhancer (E), switch (S) and constant (C_H) regions. (b) Protein binding sites of the promoter and enhancers. The boxed elements of the promoter and intron-enhancer are known protein binding sites; those of the 3'-enhancer are homologous sequences.

bin expression suggesting that the switch is controlled by a mutually exclusive interaction between the LCR and either the γ- or β-globin gene [207, 208].

(b) *Immunoglobulin genes*

Like the globin genes, the transcription of the immunoglobulin genes is specific to both cell type and developmental stage. The B-lymphoid-specific expression of IgH and IgK genes have been studied most extensively and the present discussion will concentrate on the control of the IgH genes (reviewed in [209–211]).

As previously described in section 7.5.7, the Ig genes are assembled during B cell development and, in the case of IgH genes, one of a large selection of variable (V) gene regions is joined to one of each of the diversity (D) and joining (J) regions. Later rearrangements, that occur at a switch (S) region lead to the attachment of the constant (C) region to the VDJ segment but an intron between the J and S regions is retained. Within this intron is an enhancer (Fig. 10.17). A second, recently described enhancer occurs 3' of the heavy chain domain in undifferentiated cells and remains 3' of the rearranged gene [212]. There are thus three known controlling ele-

ments in the IgH rearranged gene; a promoter 5' to the V region, the so-called intron enhancer and a 3'-enhancer (Fig. 10.17).

In contrast to the globin gene promoters, the IgH promoter contributes strongly to tissue specific gene expression. Thus, hybrid constructs, with the IgH promoter linked to reporter genes, are preferentially expressed in B lymphocytes[213, 214]. However, IgH genes with the promoter but lacking the intron enhancer are not normally expressed in B-lymphoid cells. The enhancer, which was the first to be discovered in eukaryotic cells, is crucial for their activation, is specifically active in B lymphoid cells and confers to linked genes a lymphocyte-specific developmental programme [215, 216]. The combination of enhancer and promoter has greater activity than when either of the individual elements is paired with a heterologous promoter or enhancer. Notwithstanding all these data, there is an apparent paradox in that a number of cell lines are known that still transcribe IgH despite the deletion of the intron enhancer [217, 218]. The recent discovery of the additional 3'-enhancer [212] may well explain these findings.

Footprinting experiments first demonstrated that both the enhancer and promoter regions have various sequence motifs that are the binding sites for various transcription factors. More recently, a number of the proteins involved have been purified and cloned. Fig. 10.17 illustrates the known protein-binding sites within the promoter and the intron enhancer. Other sites, for which binding *in vivo* has not been conclusively demonstrated, have nevertheless been implicated because deletion or mutation of the sites reduces or increases overall enhancer activity. Thus, they may yet prove to be elements for positive or negative regulation of gene expression. The illustrated sites on the 3'-enhancer have only so far been identified by sequence homology.

The octamer site ATGCAAAT is present in every known Ig promoter and enhancer and it binds a 60 kDa protein called Oct-2 (also called OTF−2, OTF−2b and NF-A2). Even though the activity of the IgH enhancer is specific to B-cells, this is the only known factor that is primarily found in lymphoid cells and, even here, a ubiquitous protein, Oct-1 (also called OTF-1, NF-A1, OBP100), also binds to the same sequence. In fact, there appears to be a family of octamer-binding transcription factors [219] and recent evidence indicates that additional B-cell specific components are required to interact with Oct-2 to facilitate its specificity [220]. The octamer cannot totally account for enhancer activity as its deletion reduces but does not abolish its activity [221]. Within the promoter however, the octamer cannot be deleted without abolishing activity [222] and, if the motif is placed upstream of a β-globin gene plus a TATA box, it is sufficient to confer lymphoid-specific expression [223].

The three enhancer motifs μE1, μE2 and μE3 (also known as B, C1 and C2) are required for full activity but again they are bound by ubiquitously occurring proteins and are also found in the control regions of unrelated genes [224–226]. Similarly, a ubiquitous factor μEMP-E binds to the site E in

both the enhancer and promoter. Clearly much remains to be understood of the protein−DNA interactions that give rise to cell-specific expression of IgH genes but it may be that the influence of negative factors are as important as positive controls [227].

10.4.6 Inducible enhancers

Inducible enhancers are those that respond to environmental signals such as hormones, growth factors, heat shock, heavy metals, light and viral infection [164].

(a) *Steroid hormones activate ligand-inducible transcription factors*

Long before the discovery of enhancers and transcriptional factors, it was known that steroid hormones influence the expression of specific genes in their target tissues. This was most graphically demonstrated as long ago as 1960 when Clever and Karlson showed that the insect hormone, ecdysone, induced a complex series of gene activations that could be seen microscopically as swellings known as puffs on the polytene chromosomes of the salivary glands [228, 229]. Other favoured systems for the investigation of steroid-induced gene expression have been: the induction of liver gluconeogenic enzymes by glucocorticoids, the production of egg white and egg yolk precursor proteins in the oviducts and liver, respectively, of birds and amphibians responding to oestrogens and progesterone, the control of mouse mammary tumour virus gene expression in response to various steroid hormones and the induction of a calcium-binding protein in intestinal cells in response to 1,25-dihydroxy vitamin D (reviewed in [230–232]).

In all these systems, those for other steroids and those responding to thyroxine and retinoic acid, the hormones easily dif-

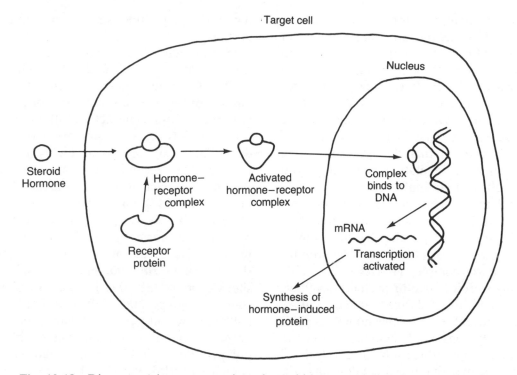

Fig. 10.18 Diagrammátic representation of steroid hormone-induced gene expression.

fuse across the membrane and into the target cell (Fig. 10.18). Once within the cell they bind to receptor proteins and this interaction is highly specific at two levels. First, receptors are only found in target cells. Thus, only a cell that responds to oestrogen would have an oestrogen receptor, in fact, it is the presence of the receptor that makes it a target cell. Second, the receptor proteins are highly specific for their cognate hormone. For instance, at physiological concentrations, the oestrogen receptor will not bind other structurally related steroids such as androgens and progesterone.

Early studies of steroid hormone action also demonstrated that the binding of the hormone in some way changed the receptor; activating it, allowing it to interact with the genome and to elicit the expression of specific genes (Fig. 10.18). It is now known

that receptor proteins are in fact ligand-inducible transcription factors that are inactive unless complexed with hormone. Their targets are well characterized sequence motifs, collectively known as hormone response elements. They are often present in multiple copies in the enhancers of the inducible genes and it is clear that they act synergistically, that is, the combined effects of two are greater than the sum of their individual effects ([233] and references therein). The glucocorticoid response element (GRE) is a 15-mer consensus sequence, GGTACANNNTGTTCT and the oestrogen response element (ERE) has the consensus AGGTCANNNTGTTCT. Both of these are palindromic and exhibit near or total rotational symmetry. Receptor proteins bind to them as dimers [234, 235]. Despite a prolonged search, no clear response element has been identified for the progesterone and androgen receptor–hormone complexes. However, both, together with mineralocorticoids, can regulate transcription via the GRE. It remains to be determined how these four hormone types can recognize the same response element yet each remain specific in the genes they induce. It may be that the receptor protein does much to regulate the specificity but some cells contain multiple receptor species and still maintain hormone specific transcriptional effects. Consensus sequences have also been defined for the thyroid hormone receptor complex (TRE) and the retinoic acid receptor complex (RARE).

Many steroid receptor proteins have been cloned and their derived amino acid sequences reveal considerable homology [236]. The highly conserved, cysteine-rich central domain is responsible for binding to DNA. It consists of two zinc finger subdomains (zinc finger domains are discussed in section 10.5.2). The first finger is responsible for specificity in that it differentiates between the different hormonal response elements [237–239]. The hormone-binding domain of receptor proteins is the moderately conserved *C*-terminal region of the protein [236]. Within it are conserved heptad repeats of hydrophobic amino acid residues that are not only involved in ligand-binding but are also essential for receptor dimerization [240]. The *N*-terminal regions of receptors are more variable both in size and sequence. They contain elements important in the *trans*-activation of gene expression but there appear to be multiple regions of the protein that are involved in activation and their multiplicity makes precise determination of their location difficult [241–244]. Other regions of the protein function in nuclear localization and the interaction with other proteins such as HSP 90 [245]. It is thus clear that receptors are complex multi-functional, multidomain transcriptional proteins.

Much remains to be discovered as to how transcription is activated. The classical model of steroid hormone action is that the binding of the ligand to the receptor protein is necessary for its interaction with the HRE. The model is certainly supported by some data [235, 246]. However, highly purified preparations of some receptors have been shown to interact with HREs independently of hormone [247]. One possible explanation of this is that most, probably all, hormone receptors are normally associated with other proteins. The best known of these is the heat shock protein HSP 90, but a second heat shock protein HSP 70 is also associated with some receptors and Toft and colleagues [248] have recently reported that unactivated progesterone receptor is associated with five polypeptides. It has been suggested that the complex of HSP 90 with the receptor prevents its translocation to the nuclei [249, 250] and/or binding to DNA; perhaps because the heat shock protein blocks dimerization [448]. Association of the hormone with the receptor is thought to cause dislocation of HSP-90 leaving the DNA binding domain free to interact with the HRE. Dislocation of the heat shock protein during purification may leave the receptor able to bind to DNA in the absence of the hormone. However, the hormone does more than just displace inhibitory proteins. Other data show that the binding of hormone to the receptor specifically accelerates DNA binding [251] and enhances the formation of a stable pretranscription complex [252].

Another ill-understood aspect of steroid hormone-induced transcription is whether, once positioned on the HRE, the hormone receptor complex requires additional transcription factors in order to initiate transcription. Schüle *et al.* [253] demonstrated that nuclear factor 1 (NF1) and other well-characterized factors could act in combination with a GRE but were only necessary when a GRE was weak in its enhancement of transcription. Even then, the NF1 binding site could be replaced with another GRE to equal effect. Nevertheless, Corthésy *et al.* [254] found that a binding site for a liver homologue of NF1 was part of the enhancer for the oestrogen-induced vitellogenin B1 gene and it was required for induction. A negative element in the same enhancer also bound a liver-specific protein and was responsible for the lack of promoter activity in the absence of oestrogen.

(b) *Cyclic AMP activates transcription factors by phosphorylation*

The cyclic AMP (cAMP) signalling pathway is the route by which the signals of many peptide hormones, growth factors and neurotransmitters are carried into the target cell. This second messenger system is well known as an activator of cAMP-dependent protein kinase which in turn brings about the phosphorylation of other cellular proteins. In this way, for instance, glucagon and adrenaline stimulate glycogen breakdown by activating glycogen phosphorylase and inhibiting glycogen synthetase.

It has been known for many years that cAMP can also activate the expression of a variety of eukaryotic genes. These include those encoding tyrosine aminotransferase [255], phospho*enol*pyruvate carboxykinase [256], enkephalin [257], α-chorionic gonadotropin [258] and somatotrophin [259]. Stimulation is mediated through a conserved DNA motif called the cAMP response element (CRE) with the consensus sequence TGACGTCA. Several transcription factors that bind to the CRE have been identified [260–262] and one of these, a 43 kDa protein called CREB or ATF, has been isolated and cloned [263, 264]. It is a phosphoprotein and

its phosphorylation at a specific serine is stimulated by cAMP [259]. It is thus likely that cAMP stimulates transcription by phosphorylating and thus activating specific CRE-binding transcription factors [259, 262]. The possibility that the expression of these genes might be modulated by the proto-oncoprotein Jun is discussed in section 10.4.8.

(c) *Plant genes can be activated by external signals like light and touch*

In addition to its role as an energy source in photosynthesis, light is essential for numerous developmental and metabolic processes in plant growth. It has been found to influence directly the expression of many genes of which the most extensively studied are those encoding the small subunit of ribose 1,5-biphosphate carboxylase-oxygenase and the proteins that bind chlorophylls *a* and *b* [265, 266]. Both are induced by light via the photoreceptor, phytochrome.

The enhancers of light-regulated genes are modular, contain a series of conserved motifs and bind multiple transcription factors (reviewed in [267]). It seems likely that these different factors interact with each other so providing a flexible regulatory response.

One of the most remarkable recently reported environmental controls of gene expression has been the discovery that the expression of plant calmodulin and calmodulin-regulated genes is induced by rain, wind and touch [268]. Calmodulin is ubiquitous in eukaryotes and, in animal cells, functions to mediate cellular responses to fluxes in calcium ion concentration. It is likely that it acts in a similar manner in plants and the implications of data on the regulation of calmodulin genes are that calcium ions are involved in the transduction of signals that enable plants to respond to external signals that, for instance, allow them

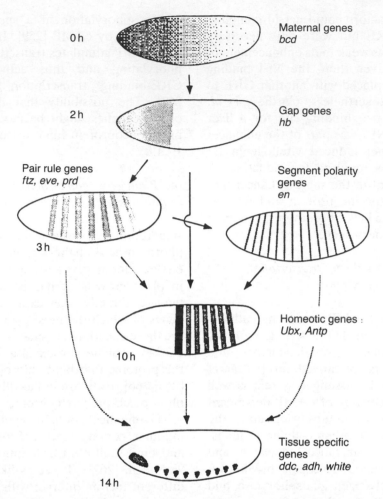

Fig. 10.19   The regulatory cascade of genes that control pattern formation in the *Drosophila* embryo. Examples of the pattern of expression of each class of gene are illustrated. Reproduced, with permission, from Biggin and Tjian [270].

to adopt a shorter, sturdier pattern of growth in a windy environment. Little is yet known of the regulatory processes at the level of the genes involved.

10.4.7 The control of embryonic development

In previous sections, selected examples have been chosen to examine what is known of

the control of cell- and tissue-specific gene expression. In embryonic development however, the patterns of gene expression, that will eventually establish the basic architecture of the animal, must be initiated before specialized cells emerge. In recent years, considerable progress has been made toward identifying the individual genes involved in this process, as well as where and when they are expressed. However, much still remains to be uncovered about the control of the processes involved.

(a) *Embryonic pattern formation in* Drosophila

Insect bodies are segmented and individual segments differentiate to generate particular structures such as wings, legs and antennae. The generation of segments in the fruit fly, *Drosophila*, has proved a popular system and its early development has been shown to be governed by a hierarchy of maternal and embryonic genes (reviewed in [269, 270]).

The first milestone in insect development is the formation of an anterior/posterior axis of which a major determinant is the product of the *bicoid (bcd)* gene. The gene is maternally expressed and the *bcd* mRNA is held, apparently by elements of the cytoskeleton, at the anterior pole of the egg. Similarly, the transcript of the *oskar* genes accumulate at the posterior pole of the egg. On fertilization, these RNAs are translated and their protein products diffuse away from the poles but not throughout the egg. Thus, the bcd protein comes to be distributed in a steep concentration gradient over the anterior half of the egg (Fig. 10.19) and the *oskar* gene products are similarly concentrated in the posterior half.

In an ill-understood manner, these maternal polarity gene products cause the transcription of the next set of genes in the developmental cascade (Fig. 10.19). These are collectively known as the gap genes and are expressed in the zygote when it is a syncitium (multinuclear cell), two nuclear divisions before cellularization (the embryos of *Drosophila* and other so called 'long germ-band insects', are syncitial until the blastoderm stage of development when they simultaneously establish their overall body plan). The best known gap genes are *hunchback* (hb), *Krüppel* (*Kr*) and *knirps* and they are expressed in broad overlapping zones that are established in part by the maternal polarity gradients and in part by regulatory interactions between themselves. The gene *Hb* is expressed in two broad domains at each end of the zygote whereas *Kr* is expressed in a single broad band in its middle. Thus, it appears that the maternal proteins are positive regulators of *hb* and negative regulators of *Kr*.

Although the zygotic gap genes are controlled by the maternal polarity genes, they in turn control the next class of genes in the hierarchical programme of gene expression (Fig. 10.19). These are the pair rule genes that were first discovered when Hafen *et al.* [271] reported the expression of the *fushi-tarazu (ftz)* gene as a series of seven mRNA stripes that encircled the embryo during its cellularization stage. There are eight known pair rule genes that derive their name from the correspondence between each stripe and interstripe with the ultimate segmental organization of the insect; that is there is one stripe per two segments. Other much studied examples of pair rule genes are *even-skipped (eve), hairy* and *runt*. The domains in which individual pair rule genes are expressed are at first wide but they become progressively restricted to form sharp bands (reviewed in [272]).

The pair-rule genes define the domain of expression of the segment polarity genes (*engrailed* and *wingless*) which are expressed at the end of the cellularization process as 14 narrow stripes each of which now represent what will ultimately become the segmental organization (Fig. 10.19). They are called parasegments and each contributes to the posterior half of one body segment and the anterior of the next.

The identity eventually adopted by each of the parasegments is dictated by the product of the homeotic genes. These were first identified as mutations that transformed one segment into another. Thus, the Bithorax mutation duplicates the thoracic segment carrying the wings to produce a fly with four true wings. Antennapedia mutants have

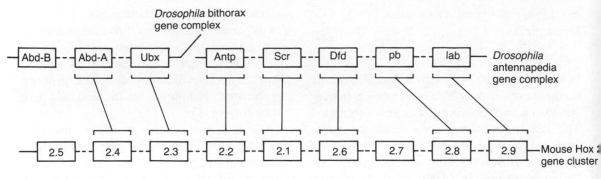

Fig. 10.20 The relationship between *Drosophila* and mouse homeobox genes. Those genes linked by brackets exhibit extensive sequence homology. In both species, the linear order on the chromosome relates to the anteroposterior axis of the body such that genes to the left of the figure are expressed at the posterior end of the embryo and, moving right along the clusters, the genes are expressed in progressively more anterior locations. In the insect at least, the encoded proteins define a cell's position along the body axis. Abd-B, abdominal B; Abd-A, abdominal A; ubx, ultrabithorax; Antp, antennapedia; scr, sex combs reduced; Dfd, deformed; pb, proboscapedia; lab, labial.

their antennae replaced by an extra pair of legs. Homeotic genes are first expressed in rather broad zones, specified at least in part by gap gene products. Expression is then modulated by interaction with pair-rule gene products and probably with each other to generate unique patterns of expression that specify parasegment identity. For instance, activity of the *ftz* pair-rule gene leads to the specific expression of the homeotic genes *antennapedia, sex combs reduced* and *ultrabithorax* in parasegments 2, 4 and 6, respectively [273]. This differential expression continues into embryonic development until segments are visible as morphological entities.

Many of the above proteins have been shown to be transcriptional factors that control the cascade of induction [270, 272, 274–276]. They without doubt also control the expression of other genes that encode proteins that are required to bring about the structural changes observed in the developing embryo. The homeotic genes have attracted particular attention as they have provided a tool for the study of the genes that control development in other species.

(b) *Homeobox genes are widely distributed*

The homeotic genes of *Drosophila* have given their name to a 183-bp, protein-encoding element called the homeobox that was first identified in *antennapedia* and *ultrabithorax* and in the pair rule gene *fushi tarazu* [277, 278]. The box encodes a 61 amino acid DNA-binding domain called the homeodomain, which NMR and X-ray diffraction has shown to be structurally related to the helix-turn-helix domain of prokaryotic DNA-binding proteins [279] (prokaryotic helix-turn-helix motifs are described in section 10.1.7). DNA encoding the homeobox element has been widely used to probe for similar genes in other species and has revealed that homeobox proteins have been strongly conserved through evolution. They occur widely in genes that commit cells to specific pathways of development. Thus, they form a part of the mating proteins of yeast and they occur in many genes that dictate the vertebrate body plan. The homeobox sequence itself is highly conserved; that of the Hox 2.2 protein of *Xenopus*, for instance, is closely related to the *Drosophila anten-*

napedia homeodomain. Similarity even extends to the arrangement of the genes such that both in insects and vertebrates they are found in similarly ordered clusters. The *Drosophila* segment polarity genes are clumped in two families; the antennapedia complex (*labial, proboscapedia, deformed, sex combs reduced* and *antennapedia*) and the bithorax complex (*ultrabithorax, abdominal A* and *abdominal B*). Vertebrates exhibit similar clustering and, in both cases, the genes are arranged in a precise order that reflects their anatomical location (Fig. 10.20). Genes expressed posteriorly in the animal are at the left of the cluster and those expressed anteriorly are to the right [280]. Sequence comparisons allow matching between the *Drosophila* and vertebrate genes [449] and most recently the linear relationship between the gene clusters has been established by gene swapping [450, 451]. For instance, the functional equivalence of the mouse Hox 2.2 gene with *Drosophila antennapedia* gene was demonstrated by placing the mouse gene in *Drosophila* constructs where it induced the same homeotic transformations [450].

(c) *The control of development*

Much then is known of the nature of the proteins involved in controlling the determination of a body plan. It has also become clear that many of the proteins discussed above, including the homeodomain proteins, are transcriptional factors [270, 272, 274–276] and considerable knowledge has accumulated of the complex interrelationships of the proteins with each other. In addition to controlling the hierarchical cascade of gene expression, it is known that many of the proteins of each class interact to control the expression of each other. Some also autoregulate their own expression. Other factors, such as retinoic acid have been implicated in

controlling the expression of some of the genes and it is also likely that the identified proteins control the expression of a host of other genes that have not been identified.

Much less is understood of the control of these events at a molecular level. Many investigations have been described *in vitro*, in cultured cells and in *Drosophila* in which P element transposons have been used to introduce engineered genes into chromosomes. As might be expected, given the complexity of the observed responses, it would appear that many of the genes have complex promoters that respond to many regulatory proteins. Somewhat surprisingly, however, no clear DNA consensus sequence to which homeobox proteins bind has yet emerged (reviewed in [274]). Recombinant homeobox proteins assayed *in vitro* appear to have relatively low sequence specific recognition capability but it remains to be ascertained whether they recognize the same sequences *in vivo* where they may be modified post-transcriptionally or by interaction with other proteins.

10.4.8 Malfunctional control of transcription can cause cancer – oncogenes

Oncogenes were first recognized in the cancer-causing acute transforming retroviruses, as the component of the viral genome that brought about cellular transformation (section 3.7.9). It was soon realized that these viral genes (v-*onc*) had cellular counterparts (c-*onc* or proto-oncogenes) that were slightly different in DNA sequence and did not normally cause cellular transformation. The concept thus arose that the protein product of oncogenes normally function to influence programmes of cell growth and differentiation. Their viral equivalents were seen as genes that were originally captured from host cells and had mutated such that vital controls, that modulated their growth stimulatory capacity, had been lost. Without such restraint the oncoprotein products of the genes caused uncontrolled growth – cancer. Support for the hypothesis came from the discovery that mutation of cellular proto-oncogenes in other ways, such as by chemical carcinogens, could also lead to cellular transformation and cancer as could the over-expression of the normal genes in cultured cells or transgenic animals.

In keeping with the concept that cancer results from a multi-step transformation process, it is notable that the cellular equivalents of oncoproteins are involved in many different cellular processes that are united only by their role in cell growth and development. Some are cytoplasmic or membrane proteins that influence various steps in the pathway by which a hormone or growth factor binds to a membrane receptor causing the transduction of an environmental signal into the cell through the activation of protein kinases [281]. Considcration of the oncoproteins that disrupt these processes are outwith the scope of this book and this discussion will concern only those oncogenes that directly affect transcriptional activity.

(a) The oncogenes jun and fos

The protein encoded by the oncogene v-*jun* is responsible for the initiation of fibrosarcomas in chicken by the avian sarcoma virus, SV17 [282]. The first evidence that its cellular equivalent, c-*jun*, functions as a transcription factor was the demonstration of homology between its amino acid sequence and that of the transcription factor GCN4 that binds to the UAS regions of several yeast genes and activates their transcription [283]. The homology centres on the DNA-binding carboxy-terminal region of the two proteins and this is sufficiently similar that it

can be exchanged between the two proteins without the loss of function [284].

A further indication of transcriptional function was the demonstration that both Jun and GCN4 proteins recognize and bind to the DNA consensus sequence, $TGA^C/_GTCA$ that had previously been identified as the binding site for the mammalian transcription factor AP1 [285]. The AP1 family of transcription factors is involved in the regulation of gene expression by various hormones, growth factors, tumour promoters, calcium ionophores and several other oncoproteins [286]. Among the the more recently recognized members of the AP1 family is the cellular equivalent of another oncoprotein, Fos (the oncogene v-*fos* is carried by the FBJ murine sarcoma virus and causes osteogenic sarcoma in mice). Indeed, it has become apparent that AP1 is in fact a dimer of Fos and Jun interacting with each other through their leucine zippers [287] and jointly recognizing the dyad symmetry of of the $TGA^C/_GTCA$ consensus DNA binding site (leucine zippers are discussed in section 10.5.3). Furthermore, the vertebrate genome contains at least three *jun*-related genes (c-*jun*, *jun*-B and *jun*-D) and several *fos*-related genes (c-*fos*, *fos-B*, *fra-1*, *fra-2*). Members of the Jun family can dimerize with each other or with any member of the Fos family. There are thus at least 18 pairing possibilities that give rise to the AP1 transcription factor family. This, in turn, may allow variety in the resultant stimulatory and inhibitory effects on gene transcription. It is known, for instance, that Jun B is less efficient in activating transcription than c-Jun and can inhibit transcriptional activation by c-Jun. After growth stimulation, virtually all Jun protein occurs as Jun–Fos heterodimers [288] and compared to homodimers, heterodimers are significantly enhanced in their ability to recognize the DNA binding site and in stimulating transcription [289]. Further variability is also built into the system. The expression of *jun* and *fos* genes is controlled both negatively and positively thus creating further possibilities for subtle control and AP1 is also known to be modified by post-transcriptional modification. Jun, for instance, can be phosphorylated at five sites and its activation has been associated with dephosphorylation [452]. Disruption of the delicate balance between members of the AP1 complex may induce malignant transformation and this may occur through mutation, as in the viral v-*fos* and v-*jun* genes, or it may result from the deregulated expression of the normal proteins (reviewed in [290–292]).

It has recently become apparent that Jun and Fos proteins may have a wider range of influences than the genes carrying the TGA^C/$_G$TCA recognition sequence. The cyclic AMP response element (CRE), that is recognized by the CREB transcription factors, has a very similar recognition sequence TGACGTCA (section 10.4.6b). Jun can dimerize, again through leucine zippers, with some of the CREB proteins and the resulting dimers are preferentially bound at the CRE motif [293]. In another example of apparent involvement of Jun and Fos at inducible enhancers, it appears that Fos can compete with the glucocorticoid receptor at some of its promoters [294]. The significance *in vivo* of these observations is unclear, as is the role of other proto-oncoproteins that are known to have a role in the regulation of transcription. Those known to be involved include Myc, Myb, Rel and Ets. Their possible roles are reviewed in [292].

(b) *Cancer can also be caused by DNA viruses*

A second set of transforming genes, that have no obvious homologies with the oncogenes of retrovirus, are those of the DNA tumour viruses. A well-studied example is the modulation of cellular gene expression by the products of the adenovirus early region gene-1A (E1A) (reviewed in [295]).

E1A protein comprises 289 amino acids. It is a potent activator of transcription of both viral and host cell genes and it is an oncoprotein. Three regions of the protein, designated 1, 2 and 3, are highly conserved among the adenovirus and their function has been reviewed by Moran and Mathews [296]. Region 3 is necessary and sufficient for transcriptional activation of viral early genes [297] but E1A is not a DNA-binding protein and transcriptional activation apparently requires interaction with other transcriptional factors ([298–300] and references therein).

Regions 1 and 2 of E1A protein are required for cellular transformation [297]. The process is ill understood but again appears to require interaction with other proteins. There is evidence that differential effects of E1A on different members of the AP1 transcription factor family may play a key role in the transformation process [301].

10.5 THE INTERACTION OF EUKARYOTIC TRANSCRIPTION FACTORS WITH DNA AND WITH EACH OTHER

The analysis of proteinaceous transcription factors and the genes that encode them has led to the identification of a number of DNA-binding motifs.

10.5.1 The homeodomain is a form of helix–turn–helix motif

The prokaryotic helix-turn-helix motif was discussed in section 10.1.7 and it was first suggested that the eukaryotic homeodomain (section 10.4.7) might adopt a similar structure by Ohlendorf and colleagues [302]. Subsequent sequence comparisons revealed little homology with the prokaryotic proteins [303] but the application of NMR analyses [304] and complete structural determinations [305] have shown that the predictions are correct. Indeed, comparison of the backbone of the motif of five prokaryotic proteins with that of Antennapedia protein shows that the structures can be superimposed (reviewed in [274, 279]).

Despite this apparently high degree of evolutionary conservation between prokaryotic and eukaryotic DNA-binding proteins, there are major differences in DNA binding. Prokaryotic helix-turn-helix proteins bind DNA as dimers whereas studies with *Drosophila* homeoproteins *in vitro* indicate that they bind as monomers; although the contacts do extend across two major grooves of the DNA and involve both the helix-turn-helix and adjacent residues [306, 307]. The amino acids of the domain that are involved in DNA recognition also differ. Alterations to homeodomain sequences indicate that the ninth amino acid of the recognition helix is critical for binding specificity whereas in prokaryotes it is the first and second amino acids that are most important [308, 309].

Homeodomains are often associated with a second 150–160 amino acid, DNA-binding domain called the POU domain [453] but the nature of the POU:DNA interaction is little understood.

10.5.2 The zinc-finger motif

The zinc-finger motif was first recognized in the *Xenopus laevis* RNA polymerase III transcription factor, TFIIIA (section 9.4.2). It had been previously noted that Zn was necessary for the transcription of 5S RNA genes [310] and that the primary sequence of TFIIIA included nine repeats of the primary amino acid sequence, $^{Tyr}/_{Phe}$-X-Cys-$X_{2,4}$-Cys-X_3-Phe-X_5-Leu-X_2-His-$X_{3,4}$-His-X_{2-6}, joined in tandem [311]. Klug and his colleagues [312] proposed that the repeating structure folded around a zinc atom such that it became coordinated with the invariant cysteine and histidine residues. The resulting structural domain has become known as a zinc finger (Fig. 10.21(a)). The validity of the zinc binding has been confirmed, most directly by X-ray absorption spectroscopy. Furthermore, DNA encoding the zinc finger has been employed as probe to isolate over a hundred further genes for proteins that contain one or more zinc fingers [313]. Two-dimensional NMR studies show, however, that the concept of a finger, implying a domain projecting vertically from the rest of the protein, is misleading [314, 315]. The finger consists of two anti-parallel β-sheet elements that contain the two cysteines and this is joined through a turn to an α-helix that includes the two histidine residues (Fig. 10.21(b)). The tandem repeats of this structure have been described as looking like sausage links that run parallel to the body of the protein [316]. By combining previous protection data, the structure of the zinc finger and their own use of hydroxy-radical footprinting, Klug and colleagues [317] have suggested that the fingers interact with DNA as illustrated in Fig. 10.21(c). The zinc fingers are seen as laying in the major grove; each finger spanning three nucleotides [454] with the linking sequences alternately located in the major grove or crossing the minor groove. The protein lies on one face of the DNA.

The steroid hormone receptors (section 10.4.6) and GAL4 protein (section 10.4.3) exhibit different forms of the 'zinc finger' that are not closely related to each other or to that of TFIIIA. The central DNA-binding portion of steroid receptor proteins contains nine conserved cysteine residues which

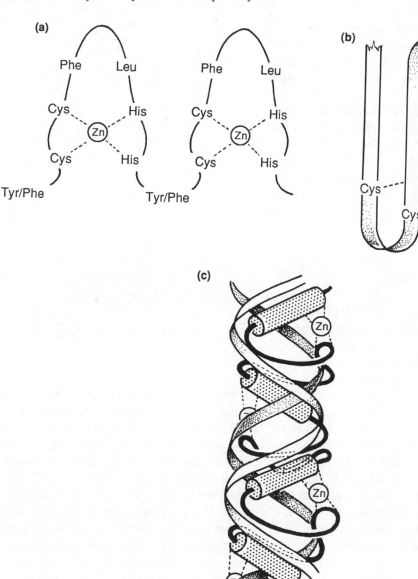

Fig. 10.21 The zinc finger motif. (a) the fingers as originally envisaged (based on Miller *et al.* [312]); (b) a structure based on NMR studies (from Berg [314]); (c) a suggested interaction with DNA (based on Churchill *et al.* [317].

mutagenesis experiments show are essential for DNA binding. X-ray atomic absorption spectroscopy demonstrates that eight of the nine residues are involved in coordination with zinc [318]. The bound metal ions are essential for DNA binding [318] and it appears likely that the DNA-binding domain includes two zinc finger-like structures in which two Zn atoms are each coordinated to the S atoms of four cysteines in the con-

served sequence $Cys-X_2-Cys-X_{13}-Cys-X_2-$
Cys. Evidence has been presented in section
10.4.6 that the two metal binding domains
play different roles in DNA binding and
receptor function.

The DNA-binding domain of the yeast
factor GAL 4 (section 10.4.3) include six invariant cysteine residues in the sequence Cys-
$X_2-Cys-X_6-Cys-X_6-Cys-X_2-Cys-X_6-Cys$. All
six cysteines are coordinated to zinc to form
a DNA binding motif that is not a zinc finger
but forms a $Zn(II)_2Cys_6$ binuclear cluster
[319].

10.5.3 The leucine zipper

In 1988, Landschulz *et al.* [320] proposed
that dimerization of the C/ERB transcription factor occurred by the interaction of
adjacent leucine-rich regions of the polypeptides. The leucines occur every seventh residue in this region of the primary sequence
thus ensuring that they are aligned on one
side and at every second turn of an α-helix
[321]. It was originally proposed that the
leucine side chains would interdigitate like
the teeth of a 'zipper' (US English for 'zip')
in a ladder of leucine–leucine interactions
(Fig. 10.22). Further investigation, however,
has revealed that this linear concept was
rather simplistic. In common with the α-
helices of many fibrous proteins, leucine zipper motifs interact as a coiled coil (left-hand
twisting of two a-helices around each other).
Rather than interdigitating, the leucines in
this coiled coil may interact with other hydrophobic amino acids on the other helix [321].

Leucine zippers are highly conserved in
the AP1 family of eukaryotic transcription
factors that includes C/ERP, Fos and Jun
(section 10.4.8) and the yeast protein GCN4.
The ability of the AP1 proteins to link
through their leucine zippers as heterodimers
as well as homodimers has been discussed in

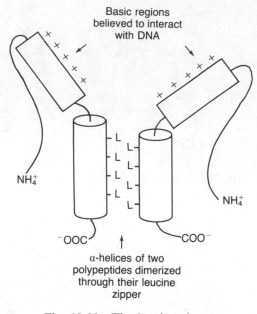

Fig. 10.22   The leucine zipper.

section 10.4.8 together with the potential
value of this flexibility in the control of gene
expression. Leucine zippers have also been
identified in many non-nuclear proteins
and appear to form the basis of a common
dimerization strategy that is not limited to
proteins involved in the control of transcription. However, transcription factors with the
motif also have a region of 30–40 residues
rich in positively charged amino acids that
precedes the zipper (Fig. 10.22). It was suggested that dimerization brings these basic
regions into a configuration that allows their
specific interaction with a DNA recognition
sequence and this concept has been confirmed in many laboratories by domain
swapping experiments. Circular dichroism
reveals that the basic region, as an isolated
peptide, or when attached to the leucine
zipper, is about 25% helical. On binding to
DNA, however, it assumes a stable helical
structure [322–324]. Two groups have proposed similar structures for the interaction of

leucine zipper proteins with DNA [322, 325]. They predict that the protein would have a Y-shaped conformation. The zipper would form the stem of the Y whereas the DNA-binding elements branch in opposite directions along the major groove of the DNA dyad-symmetrical recognition site.

10.5.4 The helix–loop–helix motif

A further motif that has recently been demonstrated in a number of transcription-regulating proteins is the so-called helix-loop-helix structure. The 60 amino acids that comprise this element have the potential to form two amphipathic helices in which hydrophobic amino acids are highly conserved [326, 327]. The proteins in which the motif occurs include: three factors that activate immunoglobulin gene enhancers [326, 328]; Myo D, a regulator of muscle-specific gene expression [326, 329]; Myc, (the product of the *myc* gene that can be oncogenic but normally functions in the control of growth) [326]; a protein called Id that inhibits gene activation by other helix-loop-helix proteins [330] and a series of *Drosophila* proteins that have cell-determination functions [331, 332]. The latter include proteins encoded in the *achaete* and *scute* genes that are involved in the establishment of clusters of cells that prime sensory organ development in the fly embryo and the product of the *emc* gene that antagonizes *achaete* and *scute* gene expression [331, 332].

Various helix-loop-helix proteins can associate as heterodimers and the binding of these to their respective DNA recognition sequences is stronger than that of homo-dimers [327]. There are thus obvious similarities between the interaction of these proteins and those that carry the leucine zipper (sections 10.4.8 and 10.5.3).

10.5.5 Transcriptional activation domains

The concept that some transcription factors might be expected to have at least two domains, one interacting with DNA and the other activating transcription, was first suggested by Ptashne [333] from work with the prokaryotic lambda repressor. His group noted that when the λ repressor protein binds to the operator $O_R 2$, it makes contact with RNA polymerase bound to the promoter P_M and thus activates transcription of the *cI* gene (section 10.1.6). The activating surface of the repressor protein consists in part of an amphipathic α-helix of which the exposed residues are predominately acidic (aspartic acid and glutamic acid). Mutations that reduce the negative charge also reduce activation. Transfer of the acidic region to the inhibitory Cro protein allows it to stimulate transcription [334].

The identification of activation domains in eukaryotic proteins has also revealed stretches of amino acids that can form amphipathic helices that are rich in acidic residues. Other than this feature, they have little in common with lambda repressor or with each other and, as in hormone receptor proteins (section 10.4.6), they may occur in several regions of the protein. The lack of obvious sequence homology does not, however, detract from the effectiveness of these sequences. Thus, the combination of the acidic domain of the yeast activation protein, GAL4 linked to a heterologous DNA binding domain, can activate the transcription of reporter genes fused to the cognate DNA recognition sequence [335]. Conversely, the DNA binding portion of GAL4 can be activated when fused to the acidic domain of other proteins [336] or a wide variety of acidic peptides ([337] and references therein).

It appears that in eukaryotes the acidic

domain interacts relatively non-specifically with the proteins of the transcription initiation complex and it has been found to stabilize the interaction between factor TFIID (section 9.4.1) and the TATA box element of the promoter [338, 339]. It may be that all eukaryotic activators of transcription by RNA polymerase II exert their effects by interacting with TFIID. This interaction might be direct and such factors would be universal in their effects. Alternatively, the interaction may be indirect via other factors bearing acidic activating sequences thus creating the opportunity for specific activation [340, 341].

A second form of activation sequence is the so-called glutamine-rich domain that has been associated with the activation region of the Sp1 group of transcription factors together with that of various *Drosophila* homeotic proteins (reviewed in [342]). Yet a third form contains 20–30% proline residues. In many cases these proline-rich activators are found on the same proteins as those with glutamine-rich activators (reviewed in [342]).

10.6 ACTIVE CHROMATIN

For nearly 20 years it has been observed that the chromatin of active or potentially active genes differs in a number of respects from that in inactive genes. Active chromatin is more sensitive to digestion by DNase I, it is relatively hypomethylated, exhibits changes in nucleosome phasing and may differ in the conformation, winding, bending and looping of the DNA. It has proved very difficult to draw these observed changes into a unified concept of the role of chromatin structure in the availability of genes for transcription and the consideration here of the features of active chromatin will inevitably be a list of

phenomena rather than casting light on the function of the events observed.

10.6.1 DNase hypersensitive sites in chromatin

Actively transcribed genes are usually found in regions of chromatin that are susceptible to nuclease digestion whereas inactive genes are often but not always located in resistant areas. Sensitivity to digestion often extends over several tens of kilobases of chromatin fibre and can be associated with domains of specific chromatin conformation (reviewed in [343]). Such domains may contain arrays of related genes such as those of the β-globin family (section 10.4.5). These moderately sensitive regions exhibit an approximately tenfold increase in susceptibility to DNase digestion but within them is a highly selective fraction of the genome known as hypersensitive sites that are typically two orders of magnitude more sensitive than bulk chromatin. Hypersensitive sites are nucleosome-free windows of chromatin and are typically found in DNA elements such as promoters and enhancers. Hypersensitive sites are thoroughly reviewed in [344] and those of the β-globin gene domain are considered in section 10.4.5.

In any given cell, DNase I hypersensitive sites fall into two categories; constitutive and inducible. Constitutive sites are found in promoters of genes that are available for transcription. They are thus found in the promoters of β-globin genes of erythroid cells but not in cells in which β-globin is not transcribed (section 10.4.5). The sites are present before the genes are transcribed and are independent of gene expression. Conversely, inducible sites appear before transcription and in response to an inducing agent such as a steroid hormone [345, 346], during differentiation [347] or in association with specific stages of the cell cycle [348]. They often persist long after the removal of the inducing agent and are not of themselves sufficient to ensure transcription.

The formation and maintenance of hypersensitive sites is poorly understood. Because they are free of nucleosomes, histones are absent from these regions but they are enriched in other proteins; including topoisomerases I and II and RNA polymerase. Other *trans*-acting proteins, that are presumed to play a role in site formation, have not, however, been identified. Other features of active chromatin, such as demethylation and DNA bending, may well play a part in the formation of hypersensitive sites. Thus Jost and colleagues [349], studying oestrogen induction of avian vitellogenin genes, have reported that the formation of an expression-linked hypersensitive site is linked to the demethylation of a specific mCpG sequence in the promoter. Furthermore, they have detected a protein that is bound to this region only when the cytosine is methylated. The precise interdependence of these events is nevertheless hard to unravel.

10.6.2 Demethylation of active chromatin

As described in section 4.6.1, methylcytosine is found in vertebrate DNA almost exclusively in the relatively rare sequence, CpG. This dinucleotide occurs at only 20–25% of its expected frequency in vertebrate DNA but, where it occurs, 60–90% of the cytosine residues are methylated in the 5-position. The apparent absence of methylation in some organisms, such as insects, suggests that it cannot play a key role in the control of gene expression. However, lack of the modification in some organisms does not mean that it is not important to others and it is clear that, in vertebrates at least, methyla-

tion does provide a mechanism for altering the local structure of DNA and for playing a part in the regulation of gene activity.

Many studies have described a correlation between hypomethylation of promoter CpG sequences and gene expression (reviewed in [350]). The first such observation was in 1978 when it was observed that several CpG dinucleotides in the rabbit β-globin gene were methylated in tissues where the gene was inactive but were unmethylated where it was expressed [351]. Many further examples also demonstrated an inverse correlation between methylation and gene activity [350]. In sperm, and presumably ova, genes are almost fully methylated. Most genes in most cells will remain in this repressed state throughout development and in adult tissues, but programmed demethylation is a part of the expression of specific genes in particular tissues. Thus, if methylated α-actin genes are introduced into fibroblasts they remain inactive and methylated. In myoblasts, however, in which the α-actin gene is normally expressed, the gene is demethylated and activated [352]. Similar tissue specific demethylation has been observed for immunoglobulin genes in B-lymphocytes [353] and in the progressive demethylation of liver-specific genes [354].

Most of these studies make use of isochizomeric restriction endonucleases that recognize the same nucleotide sequence but are differently affected by its methylation. Thus, *Msp* I cuts the sequence CCGG whether or not the second cytosine is methylated whereas *Hpa* II cuts the same sequence only if it is unmethylated. The disadvantage of this method is that only about 4% of methylcytosines occur in the tetranucleotide. Furthermore, the enzymes cannot distinguish between hemimethylated sites (DNA methylated on one strand) and those that are fully methylated. Jost and colleagues have described an alternative approach using genomic sequencing to identify methylcytosines and, when this technique is coupled to the polymerase chain reaction, it becomes very sensitive (section A.5.1) [355]. They have shown, for instance, that gene activation can be associated with partial demethylation to produce a hemimethylated site [356].

Another feature of DNA methylation, or the lack of it, was discovered by Bird *et al.* in 1985 [357]. They found an unusual clustering of non-methylated CpG-rich sequences in what have become known as CpG or HTF islands (reviewed in [358]). These regions can be several hundred base pairs long, constitute about 1% of vertebrate DNA (the HTF fraction) and contain CpG at close to the expected frequency as compared with the 20–25% of expected frequency in bulk chromatin. The regions surrounding CpG islands are heavily methylated but the islands themselves are unmethylated both in germ cells and somatic cells. There is then a paradox; a high percentage of CpG sequences are methylated but when the dinucleotide occurs at its greatest density it is unmethylated. CpG islands are closely, although not exclusively, associated with the 5′-ends of so-called housekeeping genes – genes that are constituitively expressed in all cells and encode universally required proteins. Thus, in a study of a large number of genes, CpG islands were associated with the 5′-ends of all housekeeping genes and also many tissue-specific genes [359]. They were also found at the 3′-end of some tissue-specific genes [359].

The transcription of genes with CpG islands is inhibited by their artificial methylation [360] and they are naturally methylated in the genes of inactivated X chromosomes [361–363]. Some viral genomes are CpG-rich and can be inactivated by methylation [364, 365]. Bird [358] has argued that CpG-rich islands function to distinguish regions of the genome that are always available

for transcription. Conversely tissue-specific genes are normally methylated but are demethylated as part of their activation. It is not clear, however, whether demethylation is a prerequisite or a consequence of the activation process.

What then is the role of methylation in the control of gene expression? Evidence is accumulating that tissue-specific expression may require demethylation in order that transcriptional factors can gain access to their recognition sequences. Thus, methylation of the cAMP-responsive enhancer, CRE, abolishes its interaction with its recognition factors [366, 367]. Optimal expression of the adenovirus major late promoter is abolished if the target sequence for the factor MLTF is methylated [369]. Conversely, the factor Sp1, that often associates with the promoters of housekeeping genes, will do so even if they are methylated. It is thought that in this case the factor may associate with the promoters to prevent their methylation [368].

A second possible control mechanism is that methylated CpGs could be a target for specific nuclear proteins. The resulting complex might then form a chromatin structure that denies access to transcription factors. A mammalian protein has recently been identified that specifically binds to methylated CpG dinucleotides [370] and this protects the sequence from digestion by DNase I [371].

10.6.3 The chromatin proteins of active genes

Nucleosomes consist of an octamer of the core histones (two molecules of each of H2A, H2B, H3 and H4) around which is wrapped 145 bp of DNA (section 3.3.2). The length of the linker DNA between nucleosomes is variable but a single molecule of histone H1 interacts with the octamer, and with 23 bp of linker effectively completing two turns of DNA around the complex. With such a commitment to the supercoiling of the DNA, it is pertinent to ask whether histones and other nucleosome-associated proteins can also play a role in uncoiling the DNA and in gene regulation.

(a) *Histone acetylation*

The *N*-termini of histones H1, H2A and H4 are irreversibly acetylated [372] but there is no suggestion that this plays any role in the control of gene expression. Conversely, the reversible acetylation of the ε-amino groups of internal lysines in the *N*-terminal domains of the core histones has been implicated both in transcriptional control and in DNA processing and replication (reviewed in [373]). In particular, high levels of acetylated H2B, H3 and H4 are strongly correlated with chromatin unfolding and elevated levels of transcription (e.g. [374–378]). The association of highly acetylated histones with actively transcribed genes does not, however, prove that modification is a prerequisite for transcriptional activation. Rather, it could be a consequence of transcription and arise from the associated reorganization of nucleosomes or the replacement of displaced histones.

(b) *Histone H1 interaction*

Several studies have indicated that histone H1 is non-randomly deposited in the nucleus [379] and depleted in active chromatin [380–382]. Weintraub [383] has presented evidence that it cross-links adjacent nucleosomes in inactive regions of the genome but not in active regions, where its loss may contribute to the more open structure and increased accessibility to nucleases and the transcription factors necessary for gene expression. Whether the observed depletion reflects a genuine loss of H1 from active genes or is a consequence of experimental manipulation is uncertain.

(c) *Other changes observed in the chromatin proteins of active genes*

Ubiquitin is a highly conserved 76-amino acid protein that is required for ATP-dependent non-lysosomal intracellular protein degradation [384]. It is also found conjugated to histones, principally histone H2A. Ubiquitinated histones are present at higher concentration in actively transcribed chromatin regions and Jentsch *et al.* [385] have speculated that the ubiquitination of histones by a protein called RAD6 induces chromatin remodelling.

Histone may also be phosphorylated, methylated and ADP-ribosylated and attempts have been made to link all these modifications to transcriptional activation [386, 387]. PolyADP-ribosylation of histones has been shown to inhibit RNA synthesis [388] and histone phosphorylation [389].

The non-histone proteins HMG 14 and HMG 17 are also strongly associated with active chromatin [390, 391]. They are responsible, at least in part, for the DNase I sensitivity of active chromatin [392, 393] and can even be immobilized on an affinity column and used to isolate actively transcribed chromatin [394]. They are differentially phosphorylated during the cell cycle [395].

10.6.4 Alterations to chromatin / DNA structure and transcriptional activation

(a) *Nucleosome phasing*

The possible influence that nucleosome array might have on the access of transcription factors to their DNA recognition sites is poorly understood. The nucleosome-free nature of DNase I hypersensitive site was discussed in section 10.6.1, but it is not yet clear whether their exclusion results from failure to form during DNA replication or whether they are displaced during transcriptional activation. Most investigation into the role of nucleosome positioning in gene expression has come from the study of inducible genes for which the location of nucleosomes can be precisely demonstrated and for which changes in positioning can be related to gene expression (reviewed in [396]). These systems include a number of yeast genes including that for acid phosphatase (PHO5) and the divergently transcribed genes of the galactose metabolizing enzymes, GAL1 and GAL10 (section 10.4.3). They also include the mouse mammary tumour virus (MMTV) long terminal repeat (LTR) which is a viral promoter regulated by glucocorticoids.

When yeast is grown in high phosphate media, the PHO5 gene is under the control of a weak upstream-activating sequence (UASp-1) and its TATA box is wrapped into a nucleosome [397]. In low phosphate, however, four precisely positioned nucleosomes, that bracketed UASp-1, are liberated so uncovering a stronger upstream element (UASp-2) as well as the TATA box.

In glucose-containing media, the yeast GAL1 and GAL10 are repressed (section 10.4.3) and are fully complexed with nucleosomes. In glycerol media, although they are still inactive, they appear to be primed and the nucleosomes are displaced from the shared promoter region. In this state, in the presence of galactose, the regulatory protein GAL4 can bind to its recognition sites and induce the genes. It apparently does this under the influence of a protein GRF-2 that may function to stop nucleosomes occupying the GAL4 binding sites [398].

The LTR promoter region of the tumour virus MMTV acquires six precisely posi-

tioned nucleosomes when introduced into mammalian cells [399, 400]. Reconstitution studies with pure histones and a fragment of the LTR demonstrated the assembly *in vitro* of two of the nucleosomes (called A and B) in the correct phasing pattern. Nuclear factor 1 (NF1), a component of the MMTV transcription complex was excluded from interacting with its target sequences in this fragment but the glucocorticoid–receptor complex was still able to bind to the GRE recognition site. During the subsequent hormone-induced transcriptional initiation, nucleosome B was displaced [401]. If these data accurately reflect the situation *in vivo*, it would appear that, in the absence of the hormone, the DNA sequences that are required to form a transcription complex are sequestered in chromatin structure. However, the interaction of the hormone–receptor complex with GRE causes the displacement of a nucleosome so uncovering the NF1 target sequence [401]. Similar results suggest a role for chromatin structure in the glucocorticoid-dependent activation of the rat tyrosine aminotransferase (TAT) gene [402].

(b) *Bending and looping of DNA*

Examples have already been cited in this chapter for which there is evidence that bending or looping of DNA can influence the transcription of both prokaryotic and eukaryotic genes (e.g. sections 10.1.3, 10.1.4, 10.1.5 and 10.4.5). Before transcription, the *E. coli* RNA polymerase is believed to be wrapped around up to 60 bp of the DNA upstream of the transcription start site [403]. This implies that sequences that are bent might enhance polymerase binding and the free energy of bending might contribute to the melting of DNA. Transcription factors, both in prokaryotes and eukaryotes, might promote transcription by inducing bending.

Thus, it is known that the eukaryotic 5S RNA promoter-binding factor, TFIIIA (section 9.4.2), bends DNA into a hairpin configuration [404].

One of the more dramatic illustrations that bending might be at least part of the function of some factors concerns the prokaryotic cAMP receptor protein, CAP (section 10.1.3). Again the protein complex is known to introduce a bend at its site of interaction with the DNA ([405] and references therein) but, if the binding site for CAP is replaced with an element of curved DNA, the requirement for the protein is overcome [406].

The possible importance of DNA bending is the subject of a recent review [407], as is the relationship between base sequence and curvature [408].

Looping of DNA permits regulatory elements and their associated proteins, that may be located well away from the transcription start site, to influence the initiation of transcription (e.g. section 10.4.5). This phenomenon of action at a distance has been detected directly in both prokaryotes and eukaryotes by electron microscopy [409, 410]. Alternatively it is detected as cooperativity by, for instance, measuring the effects of a protein bound at one site to influence the binding of another at a second site [411]. DNA looping is reviewed in [412, 413].

(c) *Active genes appear to be associated with the nuclear matrix*

The commonly accepted [414] although still controversial [415] model of chromatin organization is that DNA is organized into a series of supercoiled loops anchored to the nuclear matrix (section 3.3.5). The matrix was first defined as the remnant that remains when nuclei are sequentially extracted with high concentrations of NaCl (2 M) and detergent and are then treated with DNase [416].

Similar structures, believed to be of the same derivation but known as the nuclear scaffold, are isolated by extraction at low salt concentrations in the presence of diiodosalicylate [417] and a similar matrix-like material has been isolated by much more gentle techniques involving the encapsulation of nuclei in agarose followed by extraction with salts at physiological concentrations [418].

It is postulated that the attachment of the supercoiled DNA loops to the matrix plays a role in the control of gene expression. This arose from the finding that when the supercoiled loops of chromatin were digested with nuclease, the sequences that remained attached to the matrix were enriched in transcriptionally active genes [419–423]. More recently, regions of DNA that specifically attach to the matrix have been mapped to the boundaries of genes or gene domains and these elements have been called matrix attachment regions or MARs [424–426]. In some cases MARs have been shown to be close to but separate from enhancers [425, 427, 428]. Their importance was further supported by experiments that demonstrated that if MARs were placed either side of a reporter gene it enhanced position-independent expression of the gene in transfected cells [429]. This suggests that the MARs ensured that the reporter gene was inserted into the host genome in a conformation in which it was accessible for transcription.

The proteins presumed to be required to anchor DNA to the matrix are unknown but the consensus sequence for the binding of topoisomerase II is found in MARs. This suggests that matrix anchoring may influence the torsional stress in chromatin domains and may thus activate genes within the domain.

It must be emphasized that the above concepts are not universally accepted and Cook [415] in particular has marshalled evidence in support of a radical alternative. In this he proposes that it is RNA polymerase, not DNA, that is attached to the nuclear matrix and that transcription occurs as the template DNA moves past the anchored enzyme.

(d) *Z-DNA and gene transcription*

Perhaps the most extreme form of conformational change that might be associated with transcriptional control is a switch to the left-handed helical Z-DNA (the structure of Z-DNA has been considered in section 2.3.3). Most work with Z-DNA has concerned the sequence requirements and conditions required for its formation *in vitro*. The structure was first detected *in vivo* only in artificial DNA inserts consisting of 40 or more base pairs of alternating CG sequences [430]. The probability of finding such a long CG repeat in the controlling elements of a natural gene is low. However, Rahmouni and Wells [431] have recently described a sensitive detection system that has demonstrated that segments of as few as 12 bp of alternating CG residues can adopt the Z-DNA conformation *in vivo* and there is also the possibility that specific proteins could induce the conformational switch in less CG-rich sequences.

There is little convincing evidence that Z-DNA plays a role in gene expression. Furthermore, evaluation of the available evidence is not made easier by the doubts that have been raised over some of the methodology employed. Thus, the reliability of some of the antibodies purported to recognize Z-DNA has been questioned and it has recently been suggested that many of the so-called Z-DNA binding proteins are actually phospholipid-binding proteins [432]. Notwithstanding these doubts, some data do suggest that a conformational change to Z-DNA may be a transcriptional control mechanism. Thus, evidence has been presented that the DNA between two *Droso-*

phila transcription units switches from B to Z conformation when transcription is induced by the insect steroid hormone, ecdysterone [433].

10.7 REGULATION OF GENE EXPRESSION BY ANTISENSE RNA

An antisense RNA, that is an RNA with a sequence complementary to that of a mRNA, clearly has the potential to modulate gene expression. Since its discovery as a naturally occurring phenomenon in prokaryotes, it has become widely used as a tool to artificially manipulate the expression of specific genes both in prokaryotes and eukaryotes (reviewed in [434–437]).

One of the best characterized examples of control by antisense RNA *in vivo* is the expression of transposon Tn10 (transposable elements are described in section 7.7) which inserts at random in the *E. coli* genome [438]. Tn10 is 9300 bp in length with inverted repeats of an insertion sequence (IS10) at its ends. The right-hand IS10 sequence encodes a transposase protein that catalyses the insertion of the element into the bacterial chromosome. Transcription from one of the two transposase promoters produces the mRNA for the protein (RNA-IN). Transcription from the second promoter, on the other hand, begins from just inside the coding region, proceeds in an upstream direction and produces an RNA (RNA-OUT) that overlaps with the 5'-end of the transposase message (Fig. 10.23). This 35 bp region of complementarity includes the ribosome binding site of the transposase message and pairing between two RNAs prevents ribosome binding. The resultant inhibition of translation is believed to account for antisense control of IS10 *in vivo* [439].

Other examples in which prokaryotic genes are controlled by antisense sequences

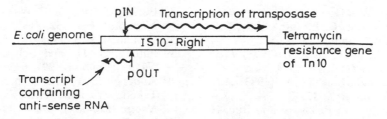

Fig. 10.23 The transcription of the insertion sequence IS10-Right. pIN is the promoter for the transcription of transposase. pOUT is a strong promoter that directs the synthesis of an RNA of which a 36 bp antisense sequence is complementary to the initial codons of the transposase mRNA.

include *E. coli ompC* and *ompE* genes that encode two outer membrane proteins of the bacteria [440], the initiation of the replication of the plasmid Col E1 (reviewed in [436]), and the C4 repressors that are responsible for the heteroimmunity of bacteriophages P1 and P7 [441].

Evidence for the control of eukaryotic genes by antisense sequences largely rests on the detection of RNA complementary to specific mRNA species within the relevant cells. Recent examples include the detection of antisense RNA to the dihydrofolate reductase–thymidylate synthase mRNA in the protozoan parasite, *Leishmania* [442]; sequences complementary to the mRNA encoding the P53 protein that is associated with the transformation of murine erythroleukaemia cells [443] and the detection of myelin basic protein antisense sequences in mouse [444].

In the most rigorously investigated systems, antisense RNA appears to function by hybridizing to the target RNA, thus masking the ribosome binding site and rendering the message functionally inactive. Artificial systems are also most efficient when antisense sequences include elements that are complementary to the ribosome binding site. However, other effects of antisense, including transcriptional attenuation and disruption of post-transcriptional processing, have not been ruled out and are supported by some data [445].

REFERENCES

1 Jacob, F. and Monod, J. (1961) *J. Mol. Biol.*, **3**, 318.
2 Miller, J. H. and Reznikoff, W. S. (1978) *The Operon*, Cold Spring Harbor Laboratory, New York.
3 Reznikoff, W. S. (1984) in *The Microbe* (part II) (eds D. P. Kelly and N. G. Carr), Cambridge University Press, Cambridge, p. 195.
4 Gilbert, W. and Müller-Hill, B. (1966) *Proc. Natl. Acad. Sci. USA*, **56**, 1891.
5 Riggs, A. D. and Bourgeois, S. (1968) *J. Mol. Biol.*, **34**, 361.
6 Beyreuther, K. (1978) in *The Operon* (eds J. H. Miller and W. S. Reznikoff), Cold Spring Harbor Laboratory, New York, p. 123.
7 Gilbert, W. and Maxam, A. (1973) *Proc. Natl. Acad. Sci. USA*, **70**, 3581.
8 Straney, S. B. and Crothers, D. M. (1987) *Cell*, **51**, 699.
9 Orhler, S., Eismann, E. R., Krämer, H. and Müller-Hill, B. (1990) *EMBO J.*, **9**, 973.
10 Kania, J. and Müller-Hill, B. (1977) *Eur. J. Biochem.*, **79**, 381.
11 Krämer, H., Amouyal, M., Nordheim, A. and Müller-Hill, B. (1988) *EMBO J.*, **7**, 547.
12 Winter, R. B., Berg, O. G. and von Hippel, P. H. (1981) *Biochemistry*, **20**, 6961.
13 Platt, T. (1978) in *The Operon* (eds J. H. Miller and W. S. Reznikoff), Cold Spring Harbor Laboratory, New York, p. 263.

14 Schevitz, R. W., Otwinowski, Z., Joachimiak, A., Lawson, C. L. and Sigler, P. B. (1985) *Nature (London)*, **317**, 782.

15 Krakow, J. S. and Pastan, I. (1973) *Proc. Natl. Acad. Sci. USA*, **70**, 2529.

16 Ullman, A. and Danchin, A. (1983) *Adv. Cyclic Nucleotide Res.*, **15**, 2.

17 de Crombrugge, B., Busby, S. and Buc, H. (1984) *Science*, **224**, 831.

18 Ebright, R., Cossart, P., Gicquel-Sanzey, B. and Beckwith, J. (1984) *Nature (London)*, **311**, 232.

19 Ebright, R., Cossart, P., Gicquel-Sanzey, B. and Beckwith, J. (1984) *Proc. Natl. Acad. Sci. USA*, **81**, 7275.

20 Travers, A. and Klug, A. (1987) *Nature (London)*, **327**, 280.

21 Buc, H. (1986) *Biochem. Soc. Trans.*, **14**, 196.

22 Gaston, K., Bell, A., Kolb, A., Buc, H. and Busby, S. (1990) *Cell*, **62**, 733.

23 McClure, W. R., Hawley, D. K. and Malan, T. P. (1982) in *Promoters – Structure and Function* (eds R. L. Rodriguez and M. J. Chamberlin), Praeger, New York, p. 111.

24 Malan, P. T., Kolb, A., Buc, H. and McClure, W. R. (1984) *J. Mol. Biol.*, **180**, 881.

25 Herbert, M., Kolb, A. and Buc, H. (1986) *Proc. Natl. Acad. Sci. USA*, **83**, 2807.

26 Menendez, M., Kolb, A. and Buc, H. (1987) *EMBO J.*, **6**, 4227.

27 Dunn, T., Hahn, S., Ogden, S. and Schleif, R. (1984) *Proc. Natl. Acad. Sci. USA*, **81**, 5017.

28 Martin, K., Huo, L. and Schleif, R. (1986) *Proc. Natl. Acad. Sci. USA*, **83**, 3654.

29 Huo, L., Martin, K. and Schleif, R. (1988) *Proc. Natl. Acad. Sci. USA*, **85**, 5444.

30 Lee, D-H and Schleif, R. F. (1989) *Proc. Natl. Acad. Sci. USA*, **86**, 476.

31 Wilcox, G. (1974) *J. Biol. Chem.*, **249**, 6892.

32 Lee, N., Francklyn, C. and Hamilton E. P. (1987) *Proc. Natl. Acad. Sci. USA*, **84**, 8814.

33 Lee, N., Gielow, W. O. and Wallace, R. G. (1981) *Proc. Natl. Acad. Sci. USA*, **78**, 752.

34 Hamilton, E. P. and Lee, N. (1988) *Proc. Natl. Acad. Sci. USA*, **85**, 1749.

35 Ketner, C. and Campbell, A. (1975) *J. Mol. Biol.*, **96**, 13.

36 Otsuka, A. and Abelson, J. (1978) *Nature (London)*, **276**, 689.

37 Beck, C. F. and Warren, R. (1988) *Microbiol. Rev.*, **52**, 318.

38 Cronan, J. E. (1989) *Cell*, **58**, 427.

39 Putney, S. D. and Schimmel, P. (1981) *Nature (London)*, **291**, 632.

40 Anderson, W. F., Ohlendorf, D. H., Takeda, Y. and Matthews, B. W. (1981) *Nature (London)*, **290**, 754.

41 Mondragon, A., Subbiah, S., Almo, S. C., Drottar, M. and Harrison, S. (1989) *J. Mol. Biol.*, **205**, 189.

42 Rafferty, J. B., Somers, W. S., Saint-Girons, I. and Phillips, S. (1989) *Nature (London)*, **341**, 705.

43 Phillips, S., Manfield, I., Parsons, I., Davidson, B. E., Rafferty, J. B. *et al.* (1989) *Nature (London)*, **341**, 711.

44 Ptashne, M., Jeffrey, A., Johnson, A. D., Maurer, R., Meyer, B. J. *et al.* (1980) *Cell*, **19**, 1.

45 Adhya, S. L., Garges, S. and Ward, D. F. (1981) *Prog. Nucleic Acids Res. Mol. Biol.*, **26**, 103.

46 Johnson, A. D., Poteete, A. R., Lauer, G., Sauer, R. T., Ackers, G. K. and Ptashne, M. (1981) *Nature (London)*, **294**, 217.

47 Ptashne, M. (1984) *Trends Biochem. Sci.*, **9**, 142.

48 Anderson, J. E., Ptashne, M. and Harrison, S. C. (1987) *Nature (London)*, **326**, 846.

49 Aggarwal, A. K., Rogers, D., Drottar, M. Ptashne, M. and Harrison, S. C. (1988) *Science*, **242**, 899.

50 Jordan, S. R. and Pabo, C. O. (1988) *Science*, **242**, 893.

51 Wolberger, C., Dong, Y., Ptashne, M and Harrison, S. C. (1988) *Nature (London)*, **335**, 789.

52 Otwinowski, R. W., Schevitz, R. W., Zhang, R-G., Lawson, C. L., Joachimiak, A. *et al.* (1988) *Nature (London)*, **335**, 321.

53 Zang, R-G., Joachimiak, A., Lawson, C. L., Schevitz, R. W., Otwinowski, R. W. *et al.* (1987) *Nature (London)*, **327**, 591.

54 Pabo, C. O., Krovatin, W., Jeffrey, A. and Sauer, R. T. (1982) *Nature (London)*, **298**, 441.

55 Sarai, A. and Takeda, Y. (1989) *Proc. Natl. Acad. Sci. USA*, **86**, 6513.

56 Koudelka, G. B., Harrison, S. C. and Ptashne, M. (1987) *Nature (London)*, **326**, 886.

57 Wolberger, C., Dong, Y., Ptashne, M. and

Harrison, S. C. (1988) *Nature (London)*, **335**, 789.

58 McKay, D. and Steitz, T. (1981) *Nature (London)*, **290**, 744.

59 Brennan, R. G. and Matthews, B. W. (1989) *J. Biol. Chem.*, **264**, 1903.

60 Buonchristiani, M. R., Howard, P. K. and Otsuka, A. J. (1986) *Gene*, **44**, 255.

61 Lehming, N., Sartorius, J., Niemöller, M., Genenger, G., Wilcken–Bergmann, B. *et al.* (1987) *EMBO J.*, **6**, 3145.

62 Lehming, N., Sartorius, J., Oehler, S., von Wilcken-Bergmann, B., Müller-Hill, B. *et al.* (1988) *Proc. Natl. Acad. Sci. USA*, **85**, 7947.

63 Pace, H. C., Lu, P. and Lewis, M. (1990) *Proc. Natl. Acad. Sci. USA*, **87**, 1870.

64 Knight, K. L., Bowie, J. U., Vershon, A. K., Kelly, R. D. and Sauer, R. T. (1989) *J. Biol. Chem.*, **264**, 3639.

65 Buc, H., Amouyal, M., Buckle, M., Herbert, M., Kolbe, A. *et al.* (1987) in *RNA Polymerase and the Regulation of Transcription* (eds Retznikof, W. S. *et al.*), Elsevier, Amsterdam, p. 115.

66 Lui-Johnson, H-N., Gartenberg, M. R. and Crothers, D. M. (1986) *Cell*, **47**, 995.

67 Griffith, J. Hochschild, A. and Ptashne, M. (1986) *Nature (London)*, **322**, 750.

68 Majumdar, A. and Adhya, S. (1984) *Proc. Natl. Acad. Sci. USA*, **81**, 6100.

69 Mandal, N., Su, W., Haber, R., Adhyer, S. and Echols, H. (1990) *Genes Devel.*, **4**, 410.

70 Jackson, E. and Yanofsky, C. (1972) *J. Mol. Biol.*, **69**, 307.

71 Bertrand, K. and Yanofsky, C. (1976) *J. Mol. Biol.*, **103**, 339.

72 Kolter, R. and Yanofsky, C. (1982) *Annu. Rev. Genet.*, **16**, 113.

73 Jensen, K. F., Bonekamp, F. and Poulsen, P. (1986) *Trends Biochem. Sci.*, **11**, 362.

74 Yanofsky, C. (1988) *J. Biol. Chem.*, **263**, 609.

75 Yanofsky, C., Platt, T, Crawford, I. P., Nichols, B. P., Christie, G. E. *et al.* (1981) *Nucleic Acids Res.*, **9**, 6647.

76 Oxender, D., Zarawski, G. and Yanofsky, C. (1979) *Proc. Natl. Acad. Sci. USA.*, **76**, 5524.

77 Yanofsky, C. (1981) *Nature (London)*, **289**, 751.

78 Kuroda, M. and Yanofsky, C. (1983) *Proc. Natl. Acad. Sci. USA*, **80**, 2206.

79 Barnes, W. M. (1978) *Proc. Natl. Acad. Sci. USA*, **75**, 4281.

80 Keller, E. B. and Calvo, J. M. (1979) *Proc. Natl. Acad. Sci. USA*, **76**, 6186.

81 Zurawski, G., Brown, K., Killingly, D. and Yanofsky, C. (1978) *Proc. Natl. Acad. Sci. USA*, **75**, 4271.

82 Gardner, J. F. (1979) *Proc. Natl. Acad. Sci. USA*, **76**, 1706.

83 Nargang, F. E., Subrahmanyam, C. S. and Umbarger, H. E. (1980) *Proc. Natl. Acad. Sci. USA*, **77**, 1823.

84 Lawther, R. P., Wek, R. C., Lopes, J. M., Periera, R., Taillon, B. E. *et al.* (1987) *Nucleic Acids Res.*, **15**, 2137.

85 Keller, E. B. and Calvo, J. M. (1979) *Proc. Natl. Acad. Sci. USA*, **76**, 6186.

86 Johnston, H. M., Barnes, W. M., Chumley, F. G., Bossi, L. and Roth, J. R. (1980) *Proc. Natl. Acad. Sci. USA*, **77**, 508.

87 Hauser, C. A. and Halfield, G. W. (1983) *Nucleic Acids Res.*, **11**, 127.

88 Quay, S. C. and Oxender, D. L. (1977) *J. Bacteriol.*, **139**, 1024.

89 Landick, R. (1984) in *Microbiology – 1984* (eds L. Leive and D. Schlessinger), American Society Microbiology, Washington. p. 71.

90 Kuroda, M. I., Shimotsu, H., Henner, D. J. and Yanofsky, C. (1986) *J. Bacteriol.*, **167**, 792.

91 Shimotsu, H., Kuroda, M. I., Yanofsky, C. and Henner, D. J. (1986) *J. Bacteriol.*, **166**, 461.

92 Schwartz, M. and Neuhard, J. (1975) *J. Bacteriol.*, **121**, 814.

93 Lynn, S. P., Burton, W. S. Donohue, T. J., Gould, R. M., Gumport, R. I. *et al.* (1987) *J. Mol. Biol.*, **194**, 59.

94 Roland, K. L., Powell, F. E. and Turnbough, C. E. (1985) *J. Bacteriol.*, **163**, 991.

95 Roland, K. L., Liu, C. and Turnbough, C. E. (1988) *Proc. Natl. Acad. Sci. USA*, **85**, 7149.

96 Novick, R. P., Iordanescu, S., Projan, S. J., Kornblum, J. and Edelman, I. (1989) *Cell*, **59**, 395.

97 Friedman, D. and Gottesman, M. (1983) in *Lambda II* (eds F. W. Stahl and R. A. Weisberg), Cold Spring Harbor Laboratory, New York, p. 21.

98 Friedman, D. I., Olsin, E. R., Georgopoulos, C., Tilly, K., Herskowitz, I. and

Banuell, F. (1984). *Microbiol. Rev.*, **48**, 299.

99 Roberts, J. W. (1988) *Cell*, **52**, 5.

100 Rosenberg, M., Court, D., Shimatake, H., Brady, C. and Wulff, D. L. (1978) *Nature (London)*, **272**, 414.

101 deCrombrugghe, B., Mudryj, M., Dihaura, R. and Gottesman, M. (1979) *Cell*, **18**, 1145.

102 Lazinski, D., Grzadzielska, E. and Das, A. (1989) *Cell*, **59**, 207.

103 Greenblatt, J. and Li, J. (1981) *J. Mol. Biol.*, **147**, 11.

104 Greenblatt, J. and Li, J. (1981) *Cell*, **24**, 421.

105 Shimamoto, N., Kamiguchi, T. and Utiyama, H. (1983) in *Microbiology – 1983* (ed. D. Schlessinger), American Society for Microbiology, Washington, p. 7.

106 Schmidt, M. C. and Chamberlin, M. J. (1987) *J. Mol. Biol.*, **195**, 809.

107 Faus, I., Chen, C-Y., Richardson, J. P. (1988) *J. Biol. Chem.*, **263**, 10 830.

108 Whalen, W., Ghosh, B. and Das, A. (1988) *Proc. Natl. Acad. Sci. USA*, **85**, 2494.

109 Horwitz, R. J., Li, J. and Greenblatt, J. (1987) *Cell*, **51**, 631.

110 Swindle, J., Zylicz, M., Georgopoulos, C., Li, J and Greenblatt, J. (1988) *J. Biol. Chem.*, **263**, 10229.

111 Das, A., Ghosh, B., Barik, S. and Wolska, K. (1985) *Proc. Natl. Acad. Sci. USA*, **82**, 4070.

112 Freidman, D. I. (1988) in *The Bacteriophages*, vol. 2 (ed. R. Calender) Plenum, New York, p. 263.

113 Szybalski, W., Brown, A. L., Hasan, N., Podhajska, A. J. and Somasekhar, G. (1987) in *RNA Polymerase and the Regulation of Transcription* (eds W. S. Retznikof *et al.*), Elsevier, Amsterdam, p. 381.

114 Barik, S., Ghosh, B., Whalen, W., Lazinski, D. and Das, A. (1987) *Cell*, **50**, 885.

115 Grayhack, E. J. and Roberts, J. W. (1982) *Cell*, **30**, 637.

116 Grayhack, E. J., Yang, X., Lau, L. and Roberts, J. W. (1985) *Cell*, **42**, 259.

117 Morgan, E. A. (1986) *J. Bacteriol.*, **168**, 1.

118 Mahadevan, S. and Wright, A. (1987) *Cell*, **50**, 485.

119 Stewart, V. and Yanofsky, C. (1986) *J. Bacteriol.*, **164**, 731.

120 Albrechtsen, B., Squires, C. L., Li, S. and Squires, C. (1990) *J. Mol. Biol.*, **213**, 123.

121 Doi, R. H. and Wang, L-F. (1986) *Microbiol. Rev.*, **50**, 227.

122 Helmann, J. D. and Chamberlin, M. J. (1988) *Annu. Rev. Biochem.*, **57**, 839.

123 Lopez, J. M., Dromerick, A. and Freese, E. (1981) *J. Bacteriol.*, **146**, 605.

124 Losick, R., Youngman, P. and Piggot, P. J. (1986) *Annu. Rev. Genet.*, **20**, 625.

125 Losick, R. and Pero, J. (1981) *Cell*, **25**, 582.

126 Losick, R. and Youngman, P. J. (1984) in *Microbiol. Development* (eds R. Losick and L. Shapiro), Cold Spring Harbor Laboratory, New York, p. 63.

127 Gitt, M. A., Wang, L. F. and Doi, R. H. (1985) *J. Biol. Chem.*, **260**, 7178.

128 Stragier, P., Parsot, C. and Bouvier, J. (1985) *FEBS Lett.*, **187**, 11.

129 Moran, C. P., Lang, N., LeGrice, S., Lee, G., Stephens, M. *et al.* (1982) *Mol. Gen. Genet.*, **186**, 339.

130 Igo, M., Lampe, M., Ray, C., Schafer, W., Moran, C. P. *et al.* (1987) *J. Bacteriol.*, **169**, 3464.

131 Igo, M. and Losick, R. (1986) *J. Mol. Biol.*, **191**, 615.

132 Wang, P-Z. and Doi, R. H. (1984) *J. Biol. Chem.*, **259**, 8619.

133 Gilman, M. Z., Wiggs, J. L. and Chamberlin, M. J. (1981) *Nucleic Acids Res.*, **9**, 5991.

134 Helmann, J. D. and Chamberlin, M. J. (1987) *Proc. Natl. Acad. Sci. USA*, **84**, 6422.

135 Helmann, J. D., Masiarz, F. R. and Chamberlin, M. J. (1988) *J. Bacteriol.*, **170**, 1560.

136 Carlson, H. C. and Haldenwang, W. G. (1989) *J. Bacteriol.*, **171**, 2216.

137 Kenney, T. J., York, K., Youngman, P. and Moran, C. P. (1989) *Proc. Natl. Acad. Sci. USA*, **86**, 9109.

138 Jaacks, K. J., Healy, J., Losick, R. and Grossman, A. D. (1989) *J. Bacteriol.*, **171**, 4121.

139 Zuber, P., Healy, J., Carter, H. L., Cutting, S., Moran, C. P. *et al.* (1989) *J. Mol. Biol.*, **206**, 605.

140 Price, V. A., Feavers, I. M. and Moir, A. (1989) *J. Bacteriol.*, **171**, 5933.

141 Kroos, L., Krunkel, B. and Losick, R. (1989) *Science*, **243**, 526.

142 Sun, D., Stragier, P. and Setlow, P. (1989) *Genes Devel.*, **3**, 141.

143 Costanzo, M. and Pero, J. (1983) *Proc. Natl. Acad. Sci. USA*, **80**, 1236.

144 Costanzo, M. and Pero, J. (1984) *J. Biol. Chem.*, **259**, 6681.

145 Neidhardt, F. C., VanBogelen, R. A. and

Vaughn, V. (1984) *Annu. Rev. Genet.*, **18**, 295.

146 Goff, S. A., Casson, L. P. and Goldberg, A. L. (1984) *Proc. Natl. Acad. Sci. USA*, **81**, 6647.

147 Straus, D. B., Walter, W. A. and Gross, C. A. (1987) *Nature (London)*, **329**, 348.

148 Cowing, D. W., Bardwell, J., Craig, E. A., Woolford, C., Hendrix, R. W. *et al.* (1985) *Proc. Natl. Acad. Sci. USA*, **82**, 2679.

149 Magasanik, B. (1982) *Annu. Rev. Genet.*, **16**, 135.

150 Haselkorn, R. (1986) *Annu. Rev. Microbiol.*, **40**, 527.

151 Kustu, S., Santero, E., Keener, J., Popham, D. and Weiss, D. (1989) *Microbiol. Rev.*, **53**, 367.

152 Popham, D., Szeto, D., Keener, J. and Kustu, S. (1989) *Science*, **243**, 629.

153 Sasse-Dwight, S. and Gralla, J. D. (1988) *Proc. Natl. Acad. Sci. USA*, **85**, 8934.

154 Keener, J., Wong, P., Popham, D., Wallis, J. and Kustu, S. (1987) in *RNA Polymerase and the Regulation of Transcription* (eds W. S. Retznikof *et al.*), Elsevier, Amsterdam, p. 159.

155 Gussin, G. N., Ronson, C. W. and Ausubel, F. M. (1986) *Annu. Rev. Genet.*, **20**, 567.

156 Ramos, J. L., Mermod, N. and Timmis, K. N. (1987) *Mol. Microbiol.*, **1**, 293.

157 Inouye, S., Nakazawa, A. and Nakazawa, T. (1987) *Proc. Natl. Acad. Sci. USA*, **84**, 5182.

158 Ronson, C. W., Astwood, P. M., Nixon, B. T. and Ausubel, F. M. (1987) *Nucleic Acids Res.*, **15**, 7921.

159 Daniels, D., Zuber, P. and Losick, R. (1990) *Proc. Natl. Acad. Sci. USA*, **87**, 8075.

160 Rabussay, D. (1983) in *Bacteriophage T4* (eds C. K. Matthews, E. M. Kulter, G. Mosig, and P. B. Berget), American Society of Microbiology, Washington, p. 167.

161 Williams, K. P., Müller, R., Rüger, W. and Geiduschek, E. P. (1989) *J. Bacteriol.*, **171**, 3579.

162 Müller, M. M., Gerster, T. and Schaffner, W. (1988) *Eur. J. Biochem.*, **176**, 485.

163 Wasylyk, B. (1988) *Biochim. Biophys. Acta*, **951**, 17.

164 Atchison, M. L. (1988) *Annu. Rev. Cell. Biol.*, **4**, 127.

165 Dynan, W. S. (1989) *Cell*, **58**, 1.

166 Wasylyk, B. (1988) *CRC Critical Rev. Biochem.*, **23**, 77.

167 Johnson, P. F. and McKnight, S. L. (1989) *Annu. Rev. Biochem.*, **58**, 799.

168 Maniatis, T., Goodbourn, S. and Fischer, J. A. (1987) *Science*, **236**, 1237.

169 Mitchell, P. J. and Tjian, R. (1989) *Science*, **245**, 371.

170 Ptashne, M. (1988) *Nature (London)*, **335**, 683.

171 Johnson, M. (1987) *Microbiol. Rev.*, **51**, 458.

172 Carey, M., Kakidani, H., Leatherwood, J., Mostashari, F. and Ptashne, M. (1989) *J. Mol. Biol.*, **209**, 423.

173 Carey, M., Lin, Y-S., Green, M. R. and Ptashne, M. (1990) *Nature (London)*, **345**, 361.

174 Ma, J. and Ptashne, M. (1987) *Cell*, **50**, 137.

175 Johnson, S. A., Salmeron, J. M. and Dincher, S. S. (1987) *Cell*, **50**, 143.

176 Lue, N. F., Chasman, D. I., Buchman, A. R. and Kornberg, R. D. (1987) *Mol. Cell. Biol.*, **7**, 3446.

177 West, R. W., Chen, S., Putz, H., Butler, G. and Banerjee, M. (1987) *Genes Devel.*, **1**, 1118.

178 Finley, R. L., Chen, S., Ma, J., Byrne, P. and West, R. W. (1990) *Mol. Cell. Biol.*, **10**, 5663.

179 Finley, R. L. and West, R. W. (1989) *Mol. Cell. Biol.*, **9**, 4282.

180 Nussinov, R. (1990) *Crit. Rev. Biochem.*, **25**, 185.

181 Ondeck, B., Gloss, L. and Herr, W. (1988) *Nature (London)*, **333**, 40.

182 Herr, W. and Clarke, J. (1986) *Cell*, **45**, 461.

183 Clarke, J. and Herr, W. (1987) *J. Virol.*, **61**, 3536.

184 Ondeck, B. A., Shepard, A. and Herr, W. (1987) *EMBO J.*, **6**, 1017.

185 Schirm, S., Jiricny, J. and Schaffner, W. (1987) *Genes Devel.*, **1**, 65.

186 Xiao, J. H., Davidson, I., Ferrandon, D., Rosales, R., Vigneron, M. *et al.* (1987) *EMBO J.*, **6**, 3005.

187 Davidson, I., Xiao, J. H., Rosales, R., Staub, A. and Chambon, P. (1988) *Cell*, **54**, 931.

188 Chada, K., Magram, J and Costantini, F. (1986) *Nature (London)*, **319**, 685.

189 Kollias, G., Wrighton, N., Hurst, J. and Grosveld, F. (1986) *Cell*, **46**, 89.

190 Hesse, J. E., Nickol, J. M., Lieber, M. R. and Felsenfeld, G. (1986) *Proc. Natl. Acad. Sci. USA*, **183**, 4312.

191 Kretsovali, A., Müller, F. W., Weber, F., Marcaud, L., Farache, G. *et al.* (1987) *Gene* **58**, 167.

192 Emerson, B. M., Nickol, J. M., Jackson, P. D. and Felsenfeld, G. (1987) *Proc. Natl. Acad. Sci. USA*, **84**, 4786.

193 Townes, T. M. and Behringer, R. R. (1990) *Trends Genet.*, **6**, 219.

194 Orkin, S. H. (1990) Cell, 63, 665.

195 Tuan, D., Solomon, D., Li, Q. and London, I. M. (1985) *Proc. Natl. Acad. Sci. USA*, **82**, 6384.

196 Ryan, T. M., Behringer, R. R., Martin, N. C., Townes, T. M., Palmiter, R. D. *et al.* (1989) *Genes Devel.*, **3**, 314.

197 Grosveld, F. van Assendelft, G. B., Greaves, D. R. and Kollias, B. (1987) *Cell*, **51**, 975.

198 Antoniou, M. and Grosveld, F. (1990) *Genes Devel.*, **4**, 1007.

199 Mignotte, V., Wall, L., deBoer, E. and Romeo, P. H. (1989) *Nucleic Acids Res.*, **17**, 37.

200 Ney, P. A., Sorrentino, B. P., Donagh, K. T. and Neinhuis, A. W. (1990) *Genes Devel.*, **4**, 993.

201 Tsai, S-F., Martin, D., Zon, L. I., D'Andrea, A. D., Wong, G. G. *et al.* (1989) *Nature (London)*, **339**, 446.

202 Evans, T. and Felsenfeld, G. (1989) *Cell*, **58**, 877.

203 Evans, T., Reitman, M. and Felsenfeld, G. (1988) *Proc. Natl. Acad. Sci. USA*, **85**, 5976.

204 Grosfeld, F., van Assendelft, B. G., Greaves, G. and Kollias, G. (1987) *Cell*, **51**, 975.

205 Smith, S. and Stillman, B. (1989) *Cell*, **58**, 15.

206 Choi, O. R. and Engel, J. D. (1988) *Cell*, **55**, 17.

207 Enver, T., Raich, N., Ebens, A. J., Papayannopoulu, T., Costantini, F. *et al.* (1990) *Nature (London)*, **344**, 309.

208 Behringer, R. R., Ryan, T. M., Palmiter, R. D., Brinster, R. L. and Townes, T. M. (1990) *Genes Devel.*, **4**, 380.

209 Calame, K. and Eaton, S. (1988) *Adv. Immunol.*, **43**, 235.

210 Sen, R. and Baltimore, D. (1989) in *Immunoglobulin Genes* (eds T. Honjo, F. Alt and T. Rabbits), Academic Press, New York, p. 97.

211 Calame, K. (1990) *Trends Genet.*, **5**, 395.

212 Pettersson, S., Cook, G. P., Brüggemann, M., Williams, G. T. and Neuberger, M. S. (1990) *Nature (London)*, **344**, 165.

213 Grosscheld, R. and Baltimore, D. (1985) *Cell*, **41**, 885.

214 Mason, J. O., Williams, G. T. and Neuberger, M. S. (1985) *Cell*, **41**, 479.

215 Adams, J. M., Harris, A. W., Pinkert, C. A., Corcoran, L. M., Alexander, W. S. *et al.* (1985) *Nature (London)*, **318**, 533.

216 Gerlinger, P., Le Meur, M., Irrmann, C., Renard, P., Wasylyk, C. *et al.* (1986) *Nucleic Acids Res.*, **14**, 6565.

217 Aguilera, R., Hope, T. J. and Sakano, H. (1985) *EMBO J.*, **4**, 3689.

218 Eckhardt, L. A. and Birshtein, B. K. (1985) *Mol. Cell. Biol.*, **5**, 856.

219 Schöler, H. R., Hazopoulis, A. K., Balling, R. Suzuki, N. and Grus, P. (1989) *EMBO J.*, **8**, 2543.

220 Pierani, A., Heguy, A., Fujii, H. and Roeder, R. G. (1990) *Mol. Cell. Biol.*, **10**, 6204.

221 Lenardo, M., Pierce, J. and Baltimore, D. (1987) *Science*, **236**, 1573.

222 Falkner, F. and Zachau, H. (1984) *Nature (London)*, **310**, 71.

223 Dreyfus, M., Doyen, N. and Rougeon, F. (1987) *EMBO J.*, **6**, 1685.

224 Weinberger, J., Baltimore, D. and Sharp, P. A. (1986) *Nature (London)*, **322**, 846.

225 Lenardo, M., Pierce, J. W. and Baltimore, D. (1987) *Science*, **236**, 1573.

226 Beckmann, H., Su, L-K. and Kadesch, T. (1990) *Genes Devel.*, **4**, 167.

227 Weinberger, J., Jap, P. and Sharp, P. (1988) *Mol. Cell. Biol.*, **8**, 988.

228 Clever, U. and Karlson, P. (1960) *Exp. Cell Res.*, **20**, 623.

229 Asburner, M. (1990) *Cell*, **61**, 1.

230 Beato, M. (1987) *Biochim. Biophys. Acta*, **910**, 95.

231 Yamamoto, K. R. (1985) *Annu. Rev. Genet.*, **19**, 209.

232 Beato, M. (1989) *Cell*, **56**, 335.

233 Martinez, E. and Wahli, W. (1989) *EMBO J.*, **8**, 3781.

234 Tsai, S. Y., Carlstedt-Duke, J., Weigel, N. L., Dahlman, K., Gustafsson, J-A. *et al.* (1988) *Cell*, **55**, 361.

235 Kumar, V. and Chambon, P. (1988) *Cell*, **55**, 145.

236 Evans, R. M. (1988) *Science*, **242**, 889.

237 Mader, S., Kumar, V., de Verneuil, H and

Chambon, P. (1989) *Nature (London)*, **338**, 271.

238 Danielsen, M., Hinck, L. and Ringold, G. M. (1989) *Cell*, **57**, 1131.

239 Umesono, K. and Evans, R. M. (1989) *Cell*, **57**, 1139.

240 Fawell, S. E., Lees, J. A., White, R. and Parker, M. G. (1990) *Cell*, **60**, 953.

241 Kumar, V., Green, S., Stack, G., Berry, M., Jin, J-R and Chambon, P. (1987) *Cell*, **51**, 941.

242 Godowski, P. J., Picard, D. and Yamamoto, K. (1988) *Science*, **241**, 812.

243 Webster, N., Green, S., Rui Jin, J. and Chambon, P. (1988) *Cell*, **54**, 199.

244 Dobson, A., Conneely, O. M., Beattie, W., Maxwell, B. L., Mak, P. *et al.* (1989) *J. Biol. Chem.*, **264**, 4207.

245 Pratt, W. B., Jolly, D. J., Pratt, D. V., Hollenberg, S. M., Giguere, V. *et al.* (1988) *J. Biol. Chem.*, **263**, 267.

246 Bagchi, M., Elliston, J. F., Tsai, S. Y., Edwards, D. P., Tsai, M-J. *et al.* (1988) *Mol. Endocrinol.*, **2**, 1221.

247 Willmann, T. and Beato, M. (1986) *Nature (London)*, **324**, 688.

248 Smith, D. F., Faber, L. E. and Toft, D. O. (1990) *J. Biol. Chem.*, **265**, 3996.

249 Groyer, A., Schweizer-Groyer, G., Cadepond, F., Mariller, M. and Baulieu, E. (1987) *Nature (London)*, **328**, 624.

250 Denis, M., Poellinger, L., Wikström, A. C. and Gustafsson, J. A. (1988) *Nature (London)*, **333**, 686.

251 Schauer, M., Chalepakis, G., Willmann, T. and Beato, M. (1989) *Proc. Natl. Acad. Sci. USA*, **86**, 1123.

252 Klein-Hitpass, L., Tsai, S. Y., Weigel, N. L., Allen, G. F., Reilly, D. *et al.* (1990) *Cell*, **60**, 247.

253 Schüle, R., Muller, M., Kaltschmidt, C. and Renkawitz, R. (1988) *Science*, **242**, 1418.

254 Corthésy, B., Cardinaux, J-R., Claret, F-X. and Wahli, W. (1989) *Mol. Cell. Biol.*, **9**, 5548.

255 Boshart, M., Eeih, H., Schmidt, A., Fournier, R. and Schütz, G. (1990) *Cell*, **61**, 905.

256 Wynshaw–Boris, A., Lugo, T. G., Short, J. M., Fournier, R. and Hanson, R. W. (1984) *J. Biol. Chem.*, **259**, 12161.

257 Comb, M., Birnberg, N. C., Seasholtz, A., Herbert, E. and Goodman, H. M. (1986) *Nature (London)*, **323**, 353.

258 Delegeane, A. M., Ferland, L. H. and Mellon, P. L. (1987) *Mol. Cell. Biol.*, **7**, 3994.

259 Gonzalez, G. A. and Montminy, M. R. (1989) *Cell*, **59**, 675.

260 Montminy, M. R. and Bilezikjian, L. M. (1987) *Nature (London)*, **328**, 175.

261 Andrisani, O. and Dixon, J. E. (1990) *J. Biol. Chem.*, **265**, 3212.

262 Merino, A., Buckbinder, L., Mermelstein, F. H. and Reinberg, D. (1989) *J. Biol. Chem.*, **264**, 21266.

263 Hoeffler, J. P., Meyer, T. E., Yun, Y., Jameson, L. J. and Habener, J. F. (1988) *Science*, **242**, 1430.

264 Gonzalez, G. A., Yamamoto, K. K., Fischer, W. H., Karr, D., Menzel, P. *et al.* (1989) *Nature (London)*, **337**, 749.

265 Tobin, E. M. and Silverthorne, J. (1985) *Annu. Rev. Plant Physiol.*, **36**, 569.

266 Dean, C., Pichersky, E. and Dunsmuir, P. (1989) *Annu. Rev. Plant Physiol.*, **40**, 415.

267 Gilmartin, P. M., Sarokin, L., Memelink, J. and Chua, N-H. (1990) *Plant Cell*, **2**, 369.

268 Braam, J. and Davis, R. W. (1990) *Cell*, **60**, 357.

269 Ingham, W. P. (1988) *Nature (London)*, **335**, 25.

270 Biggin, M. D. and Tjian, R. (1989) *Trends Genet.*, **5**, 377.

271 Hafen, E., Kuroiwa, A. and Gehring, W. (1984) *Cell*, **37**, 833.

272 Carroll, S. B. (1990) *Cell*, **60**, 9.

273 Ingham, P. W. and Martinez-Arias, A. (1986) *Nature (London)*, **324**, 592.

274 Hayashi, S. and Scott, M. P. (1990) *Cell*, **63**, 883.

275 Levine, M. and Hoey, T. (1988) *Cell*, **55**, 537.

276 Wright, C., Cho, K., Oliver, G. and DeRobertis, E. (1989) *Trends Biochem. Sci.*, **14**, 52.

277 McGinnis, W., Levine, M. S., Hafen, E., Kuroiwa, A. and Gehring, W. J. (1984) *Nature (London)*, **308**, 428.

278 Scott, M. P. and Weiner, A. J. (1984) *Proc. Natl. Acad. Sci. USA*, **81**, 4115.

279 Gehring, W. J., Müller, M., Affolter, M., Percival-Smith, A., Billeter, M. *et al.* (1990) *Trends Genet.*, **6**, 323.

280 Graham, A., Papalopulu, N., Lorimer, J., McVey, J., Tuddenham, E. *et al.* (1988)

Genes Devel., **2**, 1424.

281 Herrlich, P. and Ponta, H. (1989) *Trends Genet.*, **5**, 112.

282 Maki, Y., Bos, T. J., Davis, C., Starbuck, M. and Vogt, P. K. (1987) *Proc. Natl. Acad. Sci. USA*, **84**, 2848.

283 Vogt, P. K., Bos, T. J. and Doolittle, R. F. (1987) *Proc. Natl. Acad. Sci. USA*, **84**, 3316.

284 Struhl, K. (1988) *Nature* **322**, 649.

285 Bohmann, D., Bos, T. J., Admon, A., Nishimura, T., Vogt, P. K. *et al.* (1987) *Science*, **238**, 1386.

286 Wasylyk, C., Imler, J. L. and Wasylyk, B. (1988) *EMBO J.*, **8**, 2475.

287 Landschulz, W., Johnson, P. and McKnight, S. (1988) *Science*, **240**, 1759.

288 Rauscher, F. J., Cohen, D. R., Curran, T., Bos, T. J., Vogt, P. K. *et al.* (1988) *Science*, **240**, 1010.

289 Chiu, R., Boyle, W. J., Meek, J., Smeal, T., Hunter, T. *et al.* (1988) *Cell*, **54**, 541.

290 Vogt, P. K. and Bos, T. J. (1989) *Trends. Biochem. Sci.*, **14**, 172.

291 Herrlich, P. and Ponta, H. (1989) *Trends Genet.*, **5**, 112.

292 Gutman, A. and Wasylyk, B. (1991) *Trends Genet.*, **6**, 49.

293 MacGregor, P. F., Abate, C. and Curran, T. (1990) *Oncogene*, **5**, 451.

294 Diamond, M. I., Miner, J. N., Yoshinaga, S. K. and Yamamoto, K. (1990) *Science*, **249**, 1266.

295 Nevins, J. R. (1989) *Adv. Virus Res.*, **37**, 35.

296 Moran, E. and Mathews, M. B. (1987) *Cell*, **48**, 177.

297 Flint and Shenk (1990) *Annu. Rev. Genet.*, **23**, 141.

298 Liu, F. and Green, M. R. (1990) *Cell*, **61**, 1217.

299 Martin, K. J., Lillie, J. W. and Green, M. R. (1990) *Nature (London)*, **346**, 147.

300 Bagchi, S., Raychaudhuri, P. and Nevins, J. R. (1990) *Cell*, **62**, 659.

301 van Dam, H., Offringa, R., Meijer, I., Stein, B., Smits, A. M. *et al.* (1990) *Mol. Cell. Biol.*, **10**, 5857.

302 Ohlendorf, D. H., Anderson, W. F. and Matthews, B. W. (1983) *J. Mol. Evol.*, **19**, 109.

303 Brennan, R. G. and Matthews, B. W. (1989) *J. Biol. Chem.*, **264**, 1903.

304 Otting, G., Qian, Y-Q., Müller, M., Affolter, M., Gehring, W. J. *et al.* (1988) *EMBO J.*, **7**, 4305.

305 Qian, Y. Q., Billeter, M., Otting, G., Müller, M., Gehring, W. J. *et al.* (1989) *Cell*, **59**, 573.

306 Mann, R. S. and Hogness, D. S. (1990) *Cell*, **60**, 597.

307 Gibson, G., Schier, A., Lemotte, P. and Gehring, W. J. (1990) *Cell*, **62**, 1087.

308 Treisman, J., Gonczy, P., Vashishtha, M., Harris, E. and Desplan, C. (1989) *Cell*, **59**, 553.

309 Hanes, S. D. and Brent, R. (1989) *Cell*, **57**, 1275.

310 Hanas, J. S., Hazuda, D. J., Bogenhagen, D. F., Wu, F. and Wu, C. W. (1983) *J. Biol. Chem.*, **258**, 14120.

311 Ginsberg, A. M., King, B. O. and Roeder, R. G. (1984) *Cell*, **39**, 479.

312 Miller, J., McLachan, A. D. and Klug, A. (1985) *EMBO J.*, **4**, 1609.

313 Berg, J. M. (1990) *Annu. Rev. Biophys. Biophys. Chem.*, **19**, 405.

314 Berg, J. M. (1988) *Proc. Natl. Acad. Sci. USA*, **85**, 99.

315 Lee, M. S., Gimpert, G. P., Soman, K. V., Case, D. A. and Wright, P. E. (1989) *Science*, **245**, 635.

316 Berg, J. M. (1990) *J. Biol. Chem.*, **265**, 6513.

317 Churchill, M., Tullias, T. D. and Klug, A. (1990) *Proc. Natl. Acad. Sci. USA*, **87**, 5528.

318 Freedman, L. P., Luisi, B. F., Korszun, Z., Basavappa, R., Sigler, P. B. *et al.* (1988) *Nature (London)*, **334**, 543.

319 Pan, T. and Coleman, J. E. (1990) *Proc. Natl. Acad. Sci. USA*, **87**, 2077.

320 Landschulz, W. H., Johnson, P. F. and McKnight, S. (1988) *Science*, **240**, 1759.

321 O'Shea, E. K., Rutkowski, R. and Kim, P. S. (1989) *Science*, **243**, 538.

322 O'Neil, K. T., Hoess, R. H. and DeGrado, W. F. (1990) *Science*, **249**, 774.

323 Weiss, M. A., Ellenberger, T., Wobbe, C. R., Lee, J. P., Harrison, S. C. *et al.* (1990) *Nature (London)*, **347**, 575.

324 Patel, L., Abate, C. and Curran, T. (1990) *Nature (London)*, **347**, 572.

325 Vinson, C. R., Sigler, P. B. and McKnight, S. L. (1989) *Science*, **246**, 911.

326 Murre, C., McCaw, P. S. and Baltimore, D. (1989) *Cell*, **56**, 777.

327 Murre, C., McCaw, P. S., Vaessin, H., Caudy, M., Jan, L. Y. *et al.* (1989) *Cell*, **58**, 537.

328 Beckmann, H., Su, L-K. and Kadesch, T. (1990) *Genes Devel.*, **4**, 167.

329 Schäfer, B. W., Blakely, B. T., Darlington, G. J. and Blau, H. M. (1990) *Nature (London)*, **344**, 454.

330 Benezra, R., Davis, R. L., Lockshon, D., Turner, D. L. and Weintraub, H. (1990) *Cell*, **61**, 49.

331 Ellis, H. M., Spaan, D. R. and Posakony, J. W. (1990) *Cell*, **61**, 27.

332 Garrell, J. and Modolell, J. (1990) *Cell*, **61**, 39.

333 Ptashne, M. (1988) *Nature (London)*, **335**, 683.

334 Bushman, F. D. and Ptashne, M. (1988) *Cell*, **54**, 191.

335 Ma, J. and Ptashne, M. (1985) *Cell*, **51**, 113.

336 Berger, S. L., Cress, W. D., Cress, A., Triezenberg, S. J. and Guarente, L. (1990) *Cell*, **61**, 1199.

337 Himmelfarb, H. J., Pearlberg, J., Last, D. H. and Ptashne, M. (1990) *Cell*, **63**, 1299.

338 Horikoshi, M., Carey, M. F., Kakidani, H. and Roeder, R. G. (1988) *Cell*, **54**, 665.

339 Horikoshi, M., Hai, T., Green, M. R. and Roeder, R. G. (1988) *Cell*, **54**, 1033.

340 Stringer, K. F., Ingles, C. J. and Greenblatt, J. (1990) *Nature (London)*, **345**, 783.

341 Ptashne, M. and Gann, A. (1990) *Nature (London)*, **346**, 329.

342 Mitchell, P. J. and Tjian, R. (1989) *Science*, **245**, 371.

343 Goldman, M. A. (1988) *BioEssays*, **9**, 50.

344 Gross, D. S. and Garrard, W. T. (1988) *Annu. Rev. Biochem.*, **57**, 159.

345 Kaye, J. S., Pratt,-Kaye, S., Bellard, M., Dretzen, G., Bellard, F. *et al.* (1986) *EMBO J.*, **5**, 277.

346 Nitsch, D., Stewart, F., Boshart, M., Mestril, R., Weih, F. *et al.* (1990) *Mol. Cell. Biol.*, **10**, 3334.

347 Nahon, J. L., Venetianer, A. and Sala-Trapat, J. (1987) *Proc. Natl. Acad. Sci. USA*, **84**, 2135.

348 Chrysogelos, S., Riley, D. E. and Stein, J. (1985) *Proc. Natl. Acad. Sci. USA*, **82**, 7535.

349 Saluz, H. P., Feavers, I. M., Jiricny, J and Jost, J. P. (1988) *Proc. Natl. Acad. Sci. USA*, **85**, 6697.

350 Cedar, H. (1988) *Cell*, **53**, 3.

351 Waalwijk, C. and Flavell, R. A. (1978) *Nucleic Acids Res.*, **5**, 4631.

352 Yisraeli, J., Adelstein, R. S., Melloul, D., Nudel, U., Yaffe, D. *et al.* (1986) *Cell*, **46**, 409.

353 Kelley, D. E., Pollak, B. A., Atchison, M. L. and Perry, R. P. (1988) *Mol. Cell. Biol.*, **8**, 930.

354 Benvenisty, N., Mencher, D., Meyuhas, O., Razin, A. and Rashef, L. (1985) *Proc. Natl. Acad. Sci. USA*, **82**, 267.

355 Saluz, H. P. and Jost, J. P. (1989) *Proc. Natl. Acad. Sci. USA*, **86**, 2602.

356 Saluz, H. P., Jiricny, J. and Jost, J. P. (1986) *Proc. Natl. Acad. Sci. USA*, **83**, 7167.

357 Bird, A. P., Taggart, M. H., Frommer, M., Miller, O. J. and Macleod, D. (1985) *Cell*, **40**, 91.

358 Bird, A. P. (1986) *Nature (London)*, **321**, 209.

359 Gardiner-Garden, M. and Frommer, M. (1987) *J. Mol. Biol.*, **196**, 261.

360 Keshet, I., Yisraeli, J. and Cedar, H. (1985) *Proc. Natl. Acad. Sci. USA*, **82**, 2560.

361 Toniolo, D., D'Urso, M., Martini, G., Perisco, M., Tufano, V. *et al.* (1984) *EMBO J.*, **3**, 1987.

362 Yen, P. H., Patel, P., Chinault, A. C., Mohandas, T. and Shapiro, L. J. (1984) *Proc. Natl. Acad. Sci. USA*, **81**, 1759.

363 Wolf, S. F., Dintzis, S., Toniolo, D., Perisco, G., Lunnen, K. D. *et al.* (1984) *Nucleic Acids Res.*, **12**, 9333.

364 Kruczek, I. and Doerfler, W. (1983) *Proc. Natl. Acad. Sci. USA*, **80**, 7586.

365 Harbers, K., Schnieke, H., Stuhlmann, H., Jahner, D. and Jaenisch, R. (1981) *Proc. Natl. Acad. Sci. USA*, **78**, 7609.

366 Iguchi-Ariga, S. and Schaffner, W. (1989) *Genes Devel.*, **3**, 612.

367 Comb, M. and Goodman, H. M. (1990) *Nucleic Acids Res.*, **18**, 3975.

368 Höller, M., Westin, G., Jiricny, J. and Schaffner, W. (1988) *Genes Devel.*, **2**, 1127.

369 Watt, F. and Molloy, P. L. (1988) *Genes Devel.*, **2**, 1136.

370 Meehan, R. R., Lewis, J. D., McKay, S., Kleiner, E. L. and Bird, A. P. (1989) *Cell*, **58**, 499.

371 Antequera, F., Macleod, D. and Bird, A. P. (1989) *Cell*, **58**, 509.

372 Phillips, D. (1963) *Biochem. J.*, **87**, 258.

373 Csordas, A. (1990) *Biochem. J.*, **265**, 23.

374 Allegra, P., Sterner, R., Clayton, D. F. and Allfrey, V. G. (1987) *J. Mol. Biol.*, **196**, 379.

375 Hebbes, T. R., Thorne, A. W. and Crane-Robinson, C. (1988) *EMBO J.*, **7**, 1395.

376 Pfeffer, U., Ferrari, N., Tosetti, F. and Vidali, G. (1989) *J. Cell. Biol.*, **109**, 1007.

377 Zhang, D. and Nelson, D. (1988) *Biochem. J.*, **250**, 233.
378 Ip, Y. T., Jackson, V., Meier, J. and Chalkley, R. (1988) *J. Biol. Chem.*, **263**, 14044.
379 Leffak, M. and Trempe, J. P. (1985) *Nucleic Acids Res.*, **13**, 4853.
380 Pederson, T. (1978) *Int. Rev. Cytol.*, **55**, 1.
381 Wolffe, A. P. (1990) *New Biologist*, **2**, 211.
382 Gerrard, W. T. (1989) in *Tissue Specific Gene Expression*, (ed. R. Renkawitz), VCH, p. 1.
383 Weintraub, H. (1984) *Cell*, **38**, 17.
384 Finley, D. and Varshavsky, A. (1985) *Trends Biochem. Sci.*, **10**, 343.
385 Jentsch, S., McGrath, J. P. and Varshavsky, V. (1987) *Nature (London)*, **329**, 131.
386 Weisbrod, S. (1982) *Nature (London)*, **297**, 289.
387 Cartwright, I. C., Abmayr, S. M., Fleischmann, G., Lowenhaupt, K., Elgin, S. *et al.* (1982) *Crit. Rev. Biochem.*, **13**, 1.
388 Yu, F-L. (1985) *FEBS Lett.*, **190**, 109.
389 Ushiroyama, T., Tanigawa, Y., Tsuchiya, M., Matsuara, R., Ueki, M. *et al.* (1985) *Eur. J. Biochem.*, **151**, 173.
390 Levy, W. B. and Dixon, G. (1978) *Nucleic Acids Res.*, **5**, 4155.
391 Albanese, I. and Weintraub, H. (1980) *Nucleic Acids Res.*, **8**, 2790.
392 Weisbrod, S. and Weintraub, H. (1979) *Proc. Natl. Acad. Sci. USA*, **76**, 631.
393 Weisbrod, S., Groudine, M. and Weintraub, H. (1980) *Cell*, **19**, 289.
394 Weisbrod, S. and Weintraub, H. (1981) *Cell*, **23**, 391.
395 Bhorjee, J. S. (1981) *Proc. Natl. Acad. Sci. USA*, **78**, 6944.
396 Grunstein, M. (1990) *Trends Genet.*, **6**, 395.
397 Almer, A., Rudolph, H., Hinnen, A. and Horz, W. (1986) *EMBO J.*, **5**, 2689.
398 Chasman, D. I., Lue, N. F., Buchman, A. R., Lapointe, J. W., Lorch, Y. *et al.* (1990) *Genes Devel.*, **4**, 503.
399 Richard-Foy, H. and Hager, G. L. (1987) *EMBO J.*, **6**, 2321.
400 Archer, T. K., Cordingley, G., Marsaud, V., Richard-Foy, H. and Hager, G. L. (1989) in *Proceedings of the 2nd International CBT Symposium on the Steroid/Thyroid Receptor Family* (eds Gustafsson *et al.*), Birkhauser, Berlin, p. 221.
401 Archer, T. K., Cordingley, M. G., Wolford, R. G. and Hager, G. L. (1991) *Mol. Cell. Biol.*, **11**, 688.

402 Carr, K. D. and Richard-Foy, H. (1990) *Proc. Natl. Acad. Sci. USA*, **87**, 9300.
403 Amouyal, M. and Buc, H. (1987) *J. Mol. Biol.*, **195**, 795.
404 Bazett-Jones, D. P. and Brown, M. L. (1989) *Mol. Cell. Biochem.*, **9**, 336.
405 Gartenberg, M. R. and Crothers, D. M. (1988) *Nature (London)*, **333**, 824.
406 Bracco, L., Kotiarz, D., Kolb, A., Diekmann, S. and Buc, H. (1989) *EMBO J.*, **8**, 4289.
407 Travers, A. A. (1990) *Cell*, **60**, 177.
408 Hagerman, P. J. (1990) *Annu. Rev. Biochem.*, **59**, 755.
409 Krämer, H., Niemöller, M., Amouyal, M., Revel, B., Wicken-Bergmann, B. *et al.* (1987) *EMBO J.*, **6**, 1481.
410 Théveny, B., Bailly, A., Rauch, C., Rauch, M., Delain, E. *et al.* (1987) *Nature (London)*, **329**, 79.
411 Page, M. and Jencks, W. (1971) *Proc. Natl. Acad. Sci. USA*, **68**, 1971.
412 Schleif, R. (1987) *Nature (London)*, **327**, 369.
413 Schlief, R. (1988) *Science*, **240**, 127.
414 Pienta, K. J. and Coffey, D. S. (1984) *J. Cell Sci. Suppl.*, **1**, 123.
415 Cook, P. R. (1989) *Eur. J. Biochem.*, **185**, 487.
416 Verheijen, R., Van Venrooij, W. and Ramaekers, F. (1988) *J. Cell. Sci.*, **90**, 11.
417 Mirkovitch, J., Mirault, M-E. and Laemmli, U. K. (1984) *Cell*, **39**, 223.
418 Jackson, D. A. and Cook, P. R. (1985) *EMBO J.*, **4**, 913.
419 Robinson, S. I., Small, D., Idzerda, R., McKnight, G. S. and Vogelstein, B. (1983) *Nucleic Acids Res.*, **11**, 5113.
420 Ciejek, E. M., Tsai, M-J. and O'Malley, B. W. (1983) *Nature (London)*, **306**, 607.
421 Hentzen, P. C., Rho, J. H., Bekhor, I. (1984) *Proc. Natl. Acad. Sci. USA*, **81**, 304.
422 Kirov, N., Djondjurov, L. and Roumen, T. (1984) *J. Mol. Biol.*, **180**, 601.
423 Jost, J-P. and Seldran, M. (1984) *EMBO J.*, **3**, 2005.
424 Phi-Van, L. and Strätling, W. H. (1988) *EMBO J.*, **7**, 655.
425 Gasser, S. M. and Laemmli, U. K. (1986) *Cell*, **46**, 521.
426 Dijkwel, P. A. and Hamlin, J. L. (1988) *Mol. Cell. Biol.*, **8**, 5398.
427 Cockerill, P. N., Yeun, M-H. and Garrard, W. T. (1987) *J. Biol. Chem.*, **262**, 5394.

428 Cockerill, P. N. and Garrard, W. T. (1986) *Cell*, **44**, 273.

429 Stief, A., Winter, D. M., Strätling, W. H. and Sippel, A. E. (1989) *Nature (London)*, **341**, 343.

430 Herr, W. (1985) *Proc. Natl. Acad. Sci. USA*, **82**, 8009.

431 Rahmouni, A. R. and Wells, R. D. (1989) *Science*, **246**, 358.

432 Krishna, P., Kennedy, B. P., Waisman, D. M., van de Sande, J. H. and McGhee, J. D. (1990) *Proc. Natl. Acad. Sci. USA*, **87**, 1292.

433 Jimenez-Ruiz, A., Requena, J. M., Lopez, M. C. and Alonso, C. (1991) *Proc. Natl. Acad. Sci. USA*, **88**, 31.

434 Inouye, M. (1988) *Gene*, **72**, 25.

435 Simons, R. W. and Kleckner, N. (1988) *Annu. Rev. Genet.*, **22**, 567.

436 Takayama, K. M. and Inouye, M. (1990) *Crit. Rev. Biochem.*, **25**, 1990.

437 Mol, J., van der Krol, A. R., van Tunen, A. J. and van Blokland, R (1990) *FEBS Lett.*, **268**, 427.

438 Simmons, R. W. and Kleckner, N. (1983) *Cell*, **34**, 683.

439 Ma, S. and Simons, R. W. (1990) *EMBO J.*, **9**, 1267.

440 Mizuno, T., Chou, M-Y and Inouye, M. (1984) *Proc. Natl. Acad. Sci. USA*, **81**, 1966.

441 Citron, M. and Schuster, H. (1990) *Cell*, **62**, 591.

442 Kapler, G. M. and Beverly, S. M. (1989) *Mol. Cell. Biol.*, **9**, 3959.

443 Khochbin, S. and Lawrence, J. J. (1989) *EMBO J.*, **8**, 4107.

444 Tosic, M., Roach, A., De Rivaz, J. C., Dolivo, M. and Matthieu, J. M. (1990) *EMBO J.*, **9**, 401.

445 van der Krol, A. R., Mur, L. A., de Lange, P., Mol, J. and Stuitje, A. R. (1990) *Plant Mol. Biol.*, **14**, 457.

446 Schmidt, R., Margolis, P., Duncan, L., Coppolecchia, R., Moran, C. P. *et al.* (1990) *Proc. Natl. Acad. Sci. USA*, **87**, 9221.

447 Pevny, L. P., Simon, M. C., Robertson, E., Klein, W. H., Tsai, S-F. *et al.* (1991) *Nature (London)*, **349**, 257.

448 DeMarzo, A. M., Beck, C. A., Oñate, S. A. and Edwards, D. P. (1991) *Proc. Natl. Acad. Sci. USA*, **88**, 72.

449 Scott, M. P., Tamkun, J. W. and Hartzell, G. (1989) *Biochim. Biophys. Acta*, **989**, 25.

450 Malicki, J., Schughart, K. and McGinnis, W. (1990) *Cell*, **63**, 961.

451 McGinnis, N., Kiziora, M. A. and McGinnis, W. (1990) *Cell*, **63**, 969.

452 Boyle, W. J., Smeal, T., Defize, L., Angel, P., Woodgett, J. R. *et al.* (1991) *Cell*, **64**, 573.

453 Ruvkun, G. and Finney, M. (1991) *Cell*, **64**, 475.

454 Nardelli, J., Gibson, T. J., Vesque, C. and Charnay, P. (1991) *Nature (London)*, **349**, 175.

455 Herendeen, D. R., Williams, K. P., Kassavetis, G. A. and Geiduschek, E. P. (1990) *Science*, **248**, 573.

11

Processing of RNA transcripts and its control

11.1 THE NATURE OF GENE TRANSCRIPTS

11.1.1 Prokaryotic transcripts

Messenger RNA was discovered in bacteria. In 1953 Hershey [1] noticed that the appearance of metabolically active RNA followed the infection of *E. coli* with the bacteriophage T2. Subsequent analysis [2, 3] showed that the RNA was rapidly synthesized and had been formed from and would hybridize to the viral DNA. By prelabelling the host bacterial cells with ^{13}C- and ^{15}N-labelled growth medium and then infecting them in the presence of ^{12}C- and ^{14}N-labelled media, Brenner, Jacob and Meselson [4] were able to show that the newly made (isotopically light) viral mRNA was translated on pre-existing (isotopically heavy) cellular ribosomes. Furthermore, when they employed radioactive amino acids, they demonstrated that the translation product was viral protein. Rapidly labelled RNA with similar properties could also be detected in uninfected cells [5] but two other bacteriophages, α and SP8, were of importance in the demonstration that only one strand of the DNA serves as a template for mRNA syn-

thesis. The two DNA strands of these viruses are sufficiently different in density to allow their separation on caesium chloride density gradients and the mRNA formed in infected bacterial cells was shown to hybridize to only one DNA strand [6–8]. In the case of phage ϕX174, however, it was shown that both strands of the replicative form of the DNA were used as a template [8] thus establishing the principle that genes are not necessarily located on one strand of the genome.

In prokaryotes, transcription and translation are not physically separated by the barrier of a nuclear membrane and the two processes are so tightly coupled that ribosomes will attach to the 5'-end of the mRNA and commence translation before its 3'-synthesis is complete [9]. Since the rates of the two processes (considered per mRNA nucleotide) are similar (section 12.1) [10], the unit can be regarded as a transcription–translation complex in which RNA polymerase transcribing the gene is followed by a string of ribosomes translating the message (Fig. 11.1). Indeed the electron micrographs obtained by Miller *et al.* [9] allowed visualization of the complexes.

Many transcripts are of considerable

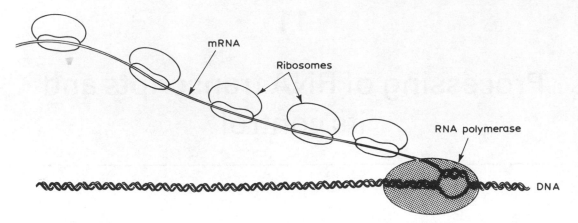

Fig. 11.1 Diagrammatic representation of a prokaryotic transcription–translation complex.

length because prokaryotic genes are often grouped in functional units known as operons in which, for instance, the genes encoding the enzymes of a particular metabolic pathway might occur consecutively on the DNA. Thus, the lactose or *lac* operon encodes the three enzymes of lactose metabolism and the *trp* operon contains the genes for the five polypeptides that form the three enzymes that catalyse tryptophan synthesis. (The value of this system in the control of gene expression is discussed in section 10.1.) Typically, an operon has a single 5′-promoter and the genes within it are transcribed by RNA polymerase into a single polycistronic mRNA. Such messages may encode up to 20 proteins.

Another feature of prokaryotic mRNA is a very short half-life, which averages 0.7 min [11] although it varies by a factor of at least 50 [12] and can depend on the growth rate [13]. This feature of prokaryotic mRNA, combined with its lack of the 3′-poly(A) tail that provides such a convenient 'handle' for the purification of the more stable eukaryotic mRNA, makes it very difficult to study. Some bacterial mRNAs, such as those for *E. coli* lipoprotein, *lac* operon and *gal* operon have been isolated but for the most

part they can only be detected by their translation products in cell-free protein synthesizing systems or by their hybridization to a radioactive complementary DNA (cDNA). Indeed, much of the control of bacterial gene expression rests on the fine tuning that mRNA instability allows. The mRNAs of bacteriophages, particularly the messenger/genome of the RNA phages R17, Qβ, MS2 and f2 are more stable than those of their host in part at least because they exhibit extensive secondary structure (section 12.9.1). This has made them an important source of prokaryotic mRNA. As is the case with many eukaryotic gene transcripts, bacterial and bacteriophage mRNAs are synthesized with additional sequences at their 5′- and 3′-ends. These sequences do not encode protein but the 5′ or leader sequence does include the Shine–Dalgarno sequence that is involved in the binding of mRNA to the ribosome (section 12.4.1). It also includes non-coding sequences that function as targets for site-specific endonucleases [14]. In at least some of the small number of prokaryotic genes known to contain introns, it appears that the intervening sequence is transcribed into a pre-mRNA and subsequently spliced out to produce the mature

RNA (section 11.3.2). Additional, inter-cistronic sequences of variable length (−1 to over 100 nucleotides in some bacteriophage) also occur between the coding regions of polycistronic mRNAs.

Usually, the translating ribosomes initiate protein synthesis separately at each coding region on the message. There is some evidence, however, that polycistronic messages may occasionally be processed into their individual protein encoding elements (reviewed in [15, 16]). The best evidence for this comes from studies with the polycistronic mRNA that encodes the five early bacteriophage T7 proteins. RNase III may process the transcript to monocistronic mRNAs [17]. It does not appear, however, that such processing is necessary for the expression of this [18] or any other polycistronic mRNA although mutations in the genes of the nucleases that are known to process bacterial rRNA and tRNA precursors do affect the efficiency with which some proteins are produced [19].

11.1.2 Eukaryotic transcripts

Three characteristics of eukaryotic mRNAs make them easier to study than those of prokaryotes. First, they are more stable and have half-lives ranging from 1 to 24 h [20, 21]. Second, they have 3′-polyadenylate tails which allow their purification by affinity chromatography on columns of oligo(dT)-cellulose or poly(U)-Sepharose (section A.2.2). Third, in some specialized cells, a high percentage of the total mRNA codes for one or a few abundant proteins. Thus 50% of the mRNA produced by the tubular gland cells of the magnum portion of the hen oviduct encodes the egg white protein ovalbumin. As a result this was one of the first mRNA species to be isolated [22]. Similar tissue or cell-specific abundance led to the

early isolation of globin mRNA from reticulocytes [23], silk fibroin mRNA from the silk glands of the larvae of the silk moth *Bombyx mori* [24], crystallin mRNA from chick lens [25] and immunoglobulin light chain mRNA from myeloma cells [26]. The studies with these and other abundant mRNA species led to the development of techniques whereby much less abundant species could be studied (Appendix).

It is now established that cytoplasmic mRNA of eukaryotic cells is the product of the maturation of longer precursors which form at least a part of the high-M_r nuclear RNA known as heterogeneous nuclear RNA (hnRNA). The early kinetic evidence for the relationship between hnRNA and mRNA was ambiguous. Most of a pulse of radio-active precursor incorporated into hnRNA turns over within the nucleus. This, combined with the difficulties encountered in conducting effective chase experiments and in adequately separating hnRNA from mRNA and precursor to rRNA, led to conflicting data. The relationship between the two types of molecule was strengthened with the finding that both possess 3′-polyadenylate tails and modified 5′-termini known as caps. The greatest problem with the precursor–product hypothesis, however, was the severalfold difference in size between the two classes of molecule. Heterogeneous nuclear RNA can have a sedimentation value of 100S indicating lengths up to 50 000 nucleotides [27], whereas even a large mRNA, such as that of the avian egg yolk precursor protein, vitellogenin (6700 nucleotides) [28], is very much smaller. Aggregation does contribute to the apparent large size of hnRNA molecules but, even under severely denaturing conditions, the average hnRNA molecule is still several times the length of the average mRNA [29].

Various models put forward to account for this paradox included the possibility that an

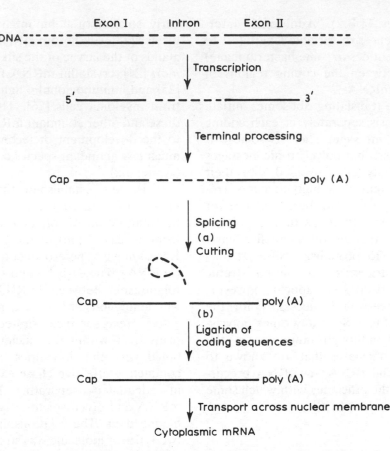

Fig. 11.2 A schematic representation of the processing of the transcript of a gene containing a single intron.

hnRNA might include more than one mRNA sequence or that the hnRNA might be cleaved and digested to conserve an mRNA sequence within it. The latter appeared to require the destruction of at least one protected end of the hnRNA and the synthesis, on the mRNA, of a new 5'-cap or 3'-polyadenylate. However, the alternative solution could be ruled out because eukaryotic mRNA is not normally polycistronic, although polygene transcripts are observed in trypanosomes (section 11.3.2) and it has recently been demonstrated that the entire chicken globin gene domain can be transcribed into a 17 kb polycistronic mRNA

[30]. A third possibility, that hnRNA might be processed by the removal of internal portions of the molecule, would have received scant attention before the discovery that genes are themselves discontinuous (section 8.2). However, it is precisely this solution that solves the precursor–product paradox; introns, as well as exons, are transcribed into pre mRNAs which make up at least a substantial proportion of hnRNA. Processing then involves the removal of intron transcripts as well as the addition of 5'-caps and 3'-poly(A) tails. Figure 11.2 is a diagrammatic representation of the processing of eukaryotic pre-mRNA. rRNA and tRNA

are also transcribed as precursor molecules
and each will be considered in turn in following sections.

11.2 snRNAs ARE INVOLVED IN MANY PROCESSING EVENTS

The members of a group of small nuclear
RNA molecules, collectively known as
snRNA or U snRNA, play a considerable
role in the processing of gene transcripts. In
vertebrate cells, six of these, U1–U6, are
present in nuclei at 10^5–10^6 copies per cell
whereas others, of which a further six have
been characterized, occur at a much lower
abundance (Table 11.1). Twenty to thirty
related species have been described from
yeast and plant cells [31, 32]. snRNAs are
very stable, have highly conserved primary
and secondary structures, unique modified
nucleotides form their 5'-caps and multiple
other base modifications tend to be clustered
at their 5'-ends. They occur as ribonucleoprotein particles (snRNP or Snurps) in
which one (U1, U2, U5) or two (U4/U6)
molecules of RNA are associated with at
least seven polypeptides that are common to
all snRNPs. The polypeptides are known as
B, B′, D, D′, E, F and G and have M_r of
29 000, 29 000, 16 000, 15 500, 12 000, 11 000
and 9000, respectively. In addition, some
snRNPs possess specific proteins that are
presumably uniquely associated with the
specific processing functions of that particle.
Thus, U1 snRNP has three additional polypeptides, U2 has two and U5 has three.
Table 11.1 lists some of the characteristics of
the snRNAs of vertebrate cells, compares
them with more abundant small RNA
species such as tRNA and 5S RNA and
specifies the sections of the subsequent text
in which their proposed function is discussed. (snRNA is reviewed in [33, 403,
404] and snRNP particles in [35], and a

compilation of snRNA sequences can be
found in [34].)

11.3 INTRON TRANSCRIPTS ARE REMOVED BY SPLICING

Most protein-encoding nuclear genes of
higher eukaryotes contain introns. They
have also been identified in some rRNA
and tRNA genes, in the genes of subcellular
organelles such as mitochondria, in some
prokaryotic genes such as thymidylate synthase [40] and two other genes of bacteriophage T4 and in the genomes of archaebacteria [41]. Furthermore, there would
appear to be at least four major mechanisms
and some minor variants by which the
transcripts of introns are removed. These
will be discussed in the following sections.

11.3.1 The splicing of pre-mRNA

(a) *Introns are transcribed into pre-mRNA*

The first demonstration that the precursors
of mRNA contain transcripts of introns was
provided by Tilghman *et al.* [42]. They
followed their initial observation that
the gene for β-globin contained at least
one intron with the demonstration that a
β-globin pre-mRNA contained a transcript
of the intron. Specifically, they showed that
when normal 10S globin mRNA hybridized
to its gene, part of the DNA did not hybridize but formed a loop of double-stranded
DNA (Fig. 11.3(a)) known as an R-loop.
The loop comprised the large intron of the
gene for which there was no corresponding
sequence in the mRNA (the small intron
was not detected by these techniques).
When, however, a 15S globin pre-mRNA
species was hybridized to the gene, no such
loop was observed because the precursor

Table 11.1 Small RNAs of vertebrate cells

Notation of RNA components	Length (nucleotides)	Abundance (copies/cell)	Major subcellular location	Transcribing RNA polymerase	Nature of 5'-terminus	References
U1	164	1×10^6	Nucleoplasm	II	2',2'-Dimethyl-7-methylguanosine	Section 11.3.1
U2	187	6×10^5	Nucleoplasm	II		Sections 11.3.1
U3	210–215	3×10^5	Nucleolus	II		Section 11.6.2
U4	145	2×10^5	Nucleoplasm	II		Section 11.3.1
U5	116	2×10^5	Nucleoplasm	II		Section 11.3.1
U6	106	4×10^5	Nucleoplasm	III	γ-Monomethyl phosphate	Section 11.3.1
U7	65	3×10^4	Nucleoplasm	II		Section 11.5.2
U8	136–141	2.5×10^4	Nucleolus	II		Section 11.6.2
U9	~130	10^3–10^4	Probably nucleoplasm	Presumed to be II	2',2'-Dimethyl-7-methylguanosine	[36]
U10	~65					[37, 38]
U11	131					[38]
U12	150					
U13	105		Nucleolus	III		Section 11.6.2
7S or L or 7SL	195–300	5×10^5	Cytoplasm	III	5'-Triphosphate	
7.1 7.2 } 7SM	260	1×10^5	Nucleolus	III		[34, 39]
7.3 or 7SK	290	0.5–1×10^5	Nucleolus	III		
	331		Nucleoplasm	III		
Y	83–112	10^5–10^6	Cytoplasm	III		
4.5S	90–99	3×10^5	Nucleoplasm	III	5'-Monophosphate	Section 12.6.1
5.8S	158	5×10^6	Cytoplasm	I	5'-Triphosphate	Section 12.6.1
5S	121	7×10^6	Cytoplasm	III	5'-Monophosphate	Section 12.3.1
tRNA	74–95	1×10^8	Cytoplasm	III		

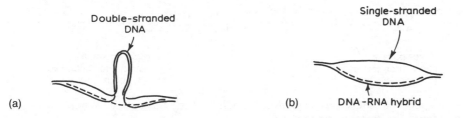

Fig. 11.3 A diagrammatic representation of (a) the hybridization of β-globin mRNA to its gene, (b) the hybridization of β-globin pre-mRNA to its gene.

contained a transcript of the intron and the molecule hybridized throughout its length (Fig. 11.3(b)). Tsai *et al.* [43] combined the technique of R-loop analysis with a kinetic approach in which finely minced oviduct was pulse-labelled to follow the processing of ovomucoid mRNA. They showed that all seven introns of this gene were transcribed into the largest pre-mRNA detected and that the seven intervening sequences were removed in a preferred but not obligatory order (Fig. 11.4). Similar processing events were soon observed in the transcripts of other genes including a 7.8 kb ovalbumin pre-mRNA [43], a 10.6 kb precursor to immunoglobulin light chain mRNA [44] and a large precursor to amphibian vitellogenin mRNA [45].

(b) *Signals for the removal of intron transcripts*

The obvious place to look for clues on the mechanism by which intron transcripts were removed from pre-mRNA was the nucleotide sequence of the intron–exon boundary. It was soon noticed that these so-called *splice sites* were strongly conserved and a number of authors published consensus sequences from which any known sequence deviated only marginally. Most recently, Shapiro and Senapathy [46] analysed 3700 splice sites and Jacob and Gallinaro [47] analysed 3294 5'-splice sites from a wide range of

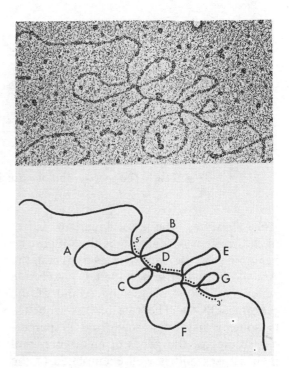

Fig. 11.4 Electron micrograph and line drawing of a hybrid molecule formed between a cloned avian ovomucoid gene and its mRNA. The R-loops labelled A–F are formed by the seven introns in the DNA. Reproduced from [43] with permission of the authors and publishers.

organisms. A consensus sequence derived from their data would give rise to the transcript illustrated in Fig. 11.5. The sequence immediately before the 3'-splice site is pyrimidine-rich and free of the dinucleotide

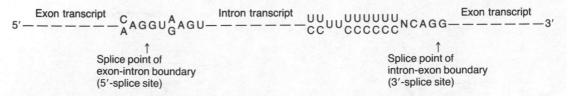

Fig. 11.5 Intron splice site consensus sequence. N indicates any nucleotide. Where two nucleotides are given, the upper is the more commonly used.

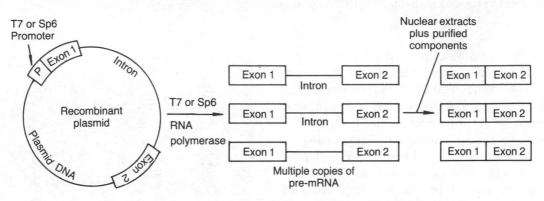

Fig. 11.6 The preparation of artificial pre-mRNA and its splicing *in vitro*.

AG. However, the most invariant aspect of the consensus sequence is the GU at the beginning of the intron transcript and the AG at its end (the so-called GT/AG rule when applied to the coding strand of the genomic DNA). The role of this sequence in splicing has been supported in numerous studies of the genes of eukaryotes and their viruses and is well exemplified in the β-thalassaemias; a group of hereditary anaemias in which the production of β-globin is either diminished (β^+ thalassaemia) or absent (β^0-thalassaemia). In one form of β^0-thalassaemia, lack of globin production was found to be due to a mutation in the GGT splice point at the exon–intron boundary of the large intron of the gene [48, 49]. As a result, processing of β-globin transcripts was impaired. In a β^+-thalassaemia, defective globin synthesis was shown to be associated with a single G → A mutation which created a sequence CTATTAG

within the large intron. The sequence closely resembled the normal intron–exon sequence CCGTTAG and competed with it as a splice point to the extent that 90% of transcripts were incorrectly spliced [50, 51].

From these and many similar studies it is clear that mutations in the GT and AG splice site sequences severely affect splicing and that mutations near the GT may also have an effect. Interference with normal splicing may lead to the utilization of previously unused cryptic splice sites. Thus, a mutation in, for instance, the 5′-splice site may result in abnormal splicing such that the 3′-exon is ligated to a new 5′-site which may show strong homologies with the normal splice site but would not normally be used. The existence of sequences, often within coding sequences, that resemble the splice site consensus sequences indicates that other factors must be involved in the selection of the correct intron boundaries.

(c) *The removal of intron transcripts from pre-mRNA of higher eukaryotes (splicing)*

The first clue to the mechanism by which intron–exon boundaries are recognized came with the observation that the 5′-terminal region of the most abundant snRNA, U1, exhibits considerable sequence complementarity with the conserved 5′-splice site boundary region. Further progress came with the development of *in vitro* splicing systems [52, 53]. These have been designed to employ synthetic pre-mRNAs that are transcribed from recombinant plasmids. The insert of such plasmids usually consists of an intron-containing gene fragment fused to the promoter for T7 or SP6 RNA polymerase. Synthetic pre-mRNAs, transcribed from the plasmids by the RNA polymerases, are incubated with various nuclear components and extracts to determine the sequence of events that lead to splicing (Fig. 11.6). Other methodological tools that have contributed to the elucidation of the splicing pathway include the use of antibodies that are produced in the serum of patients with systemic lupus erythematosus. Sm-type antibodies recognize antigenic determinants on U1, U2, U4, U5 and U6 snRNP whereas antibodies produced in other patients recognize the polypeptides that are specific to U1 and U2-containing particles [54, 55]. The unique trimethylated cap of snRNAs has also featured in the design of methods to study splicing. It has formed the basis for affinity chromatography and immunoprecipitation systems for the purification of snRNP [38]. More recently, non-denaturing polyacrylamide gels have been used to identify and fractionate splicing intermediates [56] and both affinity chromatography and gel filtration systems have been developed for their purification [57, 58].

The pathway of pre-mRNA maturation

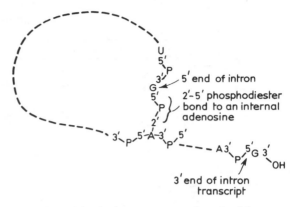

Fig. 11.7 The lariat structure of excised intron transcripts.

that has emerged from the use of these techniques is one in which splicing occurs in a two-step trans-esterification. The RNA is first cleaved at the 5′-splice site and a 2′–5′-phosphodiester bond is formed between the 5′-phosphorylated guanosine at the terminus of the cleaved intron and the 2′-OH of an A residue at a specific branch point within the intron. In the second step, the 3′-splice site is cleaved, the two exons are ligated together and the intron is released as a branched molecule known as a *lariat* (Figs 11.7 and 11.8). All of these events occur in ribonucleoprotein particles known as *spliceosomes*. The pathway of assembly and disassembly of the various complexes observed, the role of the various components and their precursor–product relationships have still to be clarified [59, 60]. However, a consensus view of the splicing pathway is shown in Fig. 11.9 and occurs in the following steps.

1. U1 SnRNP recognizes and binds to the 5′-splicing site by base pairing between the site and the 5′-terminal nucleotides of U1 RNA [47, 61] (Fig. 11.10). The binding is rapid, and does not require ATP. There is some evidence that U1 snRNP initially recognizes sequences at both

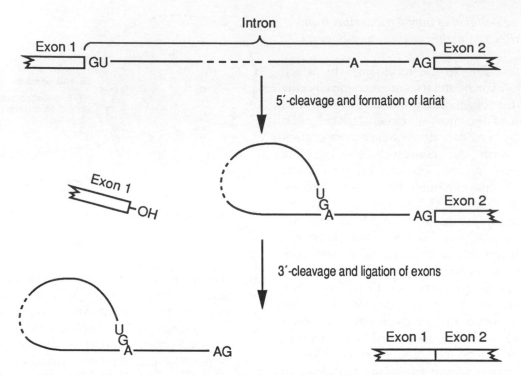

Fig. 11.8   The sequence of events in the splicing of exons in eukaryotic mRNA.

splice sites and that this must occur before the subsequent stages of spliceosome assembly [62].

2. U2 snRNP, in an ATP-dependent reaction, binds to the region of the intron that will eventually be the branch point of the lariat (Fig. 11.9). For U2 snRNP to bind, an auxiliary factor U2AF must first recognize and bind to the polypyrimidine/3′-splice site in an ATP-independent reaction [63]. Thus at both the 5′- and 3′-splice sites, the conserved elements are initially recognized by ATP-independent factors and this is followed by the ATP-dependent and highly stable binding of U2 snRNP in what may be the rate-limiting step in spliceosome assembly. Deletion of the 5′-splice site or depletion of U1 snRNP prevents the

binding of U2 snRNP again suggesting that the formation of the U1 snRNP/ 5′-splice site complex facilitates the subsequent steps [405]. There is evidence that at least one further auxiliary factor, called p62, is also required for the formation of the U2 snRNP/pre-mRNA complex [73].

3. In the next step a larger complex is formed, it is believed by the association of the above aggregate with a multi-snRNP particle containing U4, U5 and U6 snRNAs [56] (Fig. 11.9). As detected on non-denaturing gels these complexes no longer contain U1 snRNA [56] but 60S functional spliceosomes that do contain U1 can be isolated by gel filtration if splicing is blocked with EDTA [64]. It is not clear, therefore, whether U1 snRNP

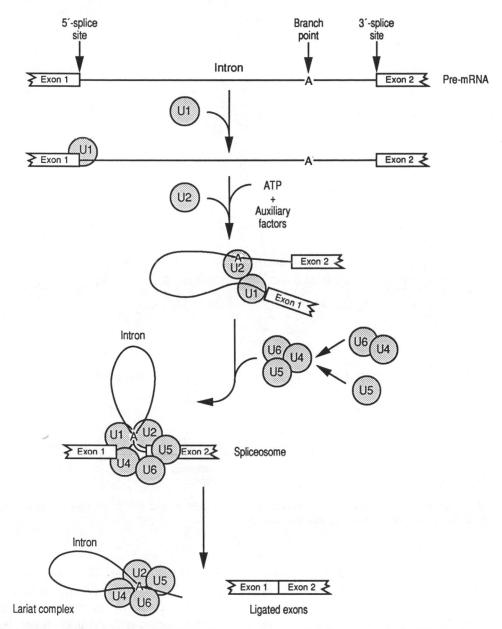

Fig. 11.9 Diagrammatic representation of the pathway of spliceosome assembly and pre-mRNA splicing.

naturally departs at this point or whether its association is too labile to survive non-denaturing gels.

4. The two-step splicing reaction described above takes place in association with and as a result of the formation of the spliceosome complex and leads to the efficient and accurate removal of the intron transcript. The 5′- and 3′-sites of cleavage are identified by their respec-

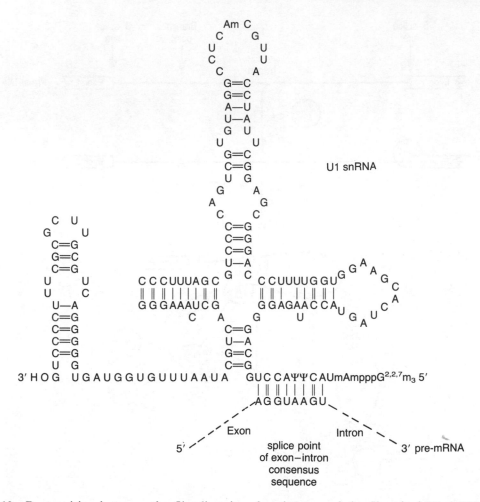

Fig. 11.10 Base pairing between the 5′-splice site of an intron and the 5′-end of U1 snRNA. The snRNA is drawn in its conserved secondary structure.

tive consensus sequences although the catalytic mechanisms of the process remain ill understood. The selection of the branch-site is also specific in that in almost all known introns only a single branch site is utilized. In yeast, the reason for specificity again resides in a conserved sequence, UACUAAC which base pairs with U2 snRNA (see following section). In higher eukaryotes, however, there is a weak branch site consensus with a base sequence, YNYURAY (A = branch-site

adenosine residue, N = any base, Y = pyrimidine and R = purine) in which only the branch-point adenosine residue is strictly conserved. At least three lines of evidence suggest that the selection of the branch site in higher eukaryotes depends on its proximity to the 3′-splice site. First, in most introns, the branch site occurs at an adenosine residue 18–37 nucleotides upstream of the 3′-splice site. Second, mutations of the branch site can lead to the use of cryptic sites the same distance from the splice site, and third, a few cases are known where there are no adenosine residues within 18–37 nucleotides of the 3′-splice site and U or C residues are used instead. Notwithstanding these observations, the selection of the branch site cannot be dependent on proximity to the 3′-splice site as introns have recently been described with branch points more than 150 nucleotides from the splice site. Smith *et al.* [65], studying such an intron in the rat α-tropomyosin gene, have concluded that, despite the weak sequence conservation, the branch site is specified by its sequence together with that of the neighbouring polypyrimidine tract. Furthermore, they demonstrated that the 3′-splice site itself was only recognized by reference to the branch point by a distance-independent scanning to the first downstream AG dinucleotide. Why the branch point sequence of higher eukaryotes should be so weakly conserved is unclear. It has been shown that the UACUAAC yeast branch point sequence is the most efficient sequence for mammalian splicing [66]. Furthermore, the complementary sequence is conserved in the U2 snRNA of higher eukaryotes. Presumably, the prior binding to the polypyrimidine tract by the proteins U2Af and p62, described above, precludes ambiguity and the flexibility of the higher eukaryotic system provides some evolutionary advantage. The secondary/tertiary structure of the branch point/3′-splice site region may also be important in its recognition [67, 406].

5. In the final step, splicing results in the ligation of the two exons and the release of the intron as a lariat-shaped RNA in a complex with U2, U5, U4 and U6 snRNPs (Fig. 11.9). Analyses on polyacrylamide gels appear to show that U4 snRNA is not a part of this post-splicing complex [56]. However, the use of oligonucleotide probes to affinity select complexes confirms that it is present [68]. It has, therefore, been proposed that during splicing the conformation of the U4/U6 snRNP complex changes, destabilizing U4 such that it can no longer remain bound to the complex during gel electrophoretic conditions [68].

(d) *The splicing pre-mRNA in fungi, higher plants and non-consensus vertebrate introns*

(i) *Fungi*. In contrast to higher eukaryotes, many yeast nuclear genes do not contain introns. Exceptions include the genes for actin, the mating-type proteins and some ribosomal proteins. Although the overall features of the splicing of these introns resemble those described above for higher eukaryotes, there are clearly defined differences that are sufficient to ensure that yeast cells are unable to splice the intervening sequences of vertebrate genes [69]. First, yeast appears to have a greater complexity of snRNA (*c.* 30 species) although the growing number of characterized vertebrate species may eventually nullify this difference. A genuine difference, however, is that the yeast RNAs are of low abundance (at most a few hundred copies per cell) and some, including snR7, snR14, snR19 and snR20, are much larger than the vertebrate species. Nevertheless, they appear to perform the same functions. Thus, the 1175 nucleotide snR20 is the yeast analogue of the 187 nucleotide U2 snRNA. It perhaps performs the functions of both U2 and U2Af in that it recognizes and binds to the conserved yeast branch point sequence, TACTAAC [70]. RNA snR14 is the yeast equivalent of U4 snRNA whereas snR6 is the counterpart of U6. The two species interact in the same way as U4/U6 [71]. The 569 nucleotide snR19 is the homologue of U1 snRNA [72] and also has extensive homology with snR20 (U2 equivalent) indicating that these two species may also interact by base pairing. snR7, which occurs in two forms, snR7S and snR7L, that differ in length by 35 nucleotides, is the equivalent of U5 snRNA [74]. Where homology is demonstrated, the yeast-specific notations are increasingly being dropped in favour of those of the vertebrate U-series snRNAs.

Analysis of yeast spliceosome formation reveals a step-wise assembly process similar to that of vertebrates (reviewed in [75]) but yeasts offer the advantage that molecular studies can be combined with genetics [76]. Thus, a set of temperature-sensitive mutations (*prp*2–*prp*11) affect splicing such that pre-mRNA precursors accumulate at non-permissive temperatures. This has allowed the identification of a series of proteins essential to the splicing process. Protein PRP8 associates with the U5 equivalent, snR7 and is a putative RNA helicase [77], PRP4 associates with the snR6/snR14 (U4/U6) complex [78] and PRP11 is also a part of spliceosomes [79]. The protein PRP2 appears to be an extrinsic factor acting after spliceosome assembly [80]. Recently a set of mutants, *prp*16–*prp*28 have been isolated, some at least of which affect late stages of splicing [81]. PRP16 protein is an RNA-dependent ATPase that interacts transiently with the spliceosome [407] and PRP22 is a RNA helicase-like protein required for mRNA release from the spliceosome [408].

Other fungi apparently employ a similar splicing mechanism to that of *Saccharomyces* [82, 83].

(ii) *Plants*. Plant introns exhibit similar consensus sequences to those of vertebrates except that the polypyrimidine tract is less pronounced [84]. Nevertheless, mammalian pre-RNAs are not properly processed in transgenic plants [85], although vertebrate systems will process plant pre-mRNA. The data of several groups indicate that intron recognition in plant cells resembles vertebrates in that branch site selection is relaxed but a polypyrimidine tract is not necessary and AU sequence elements are important in recognition [86, 87].

(iii) *Non-consensus vertebrate introns*. In their analysis of more than 3700 splice sites Shapiro and Senapathy [46] noted 13 non-consensus sites (nine 5'- and four 3'-site non-conformities). Nothing is known of how these transcripts are processed but the above authors speculated that, as most of the non-conforming sites were in immunoglobulin genes, they might be involved in the regulation of immunoglobulin gene expression.

(e) Trans-*splicing*

It has become apparent that splicing is not always intramolecular (in *cis*) but can occur between two RNA molecules (in *trans*) (reviewed in [88]).

The first evidence for *trans*-splicing came from studies with trypanosomes; organisms that display many biological novelties. All known mRNA species of *Trypanosoma brucei* have the same 39 nucleotides at their 5'-ends and other species have conserved leaders of similar length. These so-called *spliced leaders* (SL) or mini exons are not encoded in the genes from which the mRNAs are transcribed. Rather, they form the 5'-end of a 137 nucleotide SL RNA which is transcribed from a 1.35 kb sequence that is tandemly repeated 200 times in one or two trypanosomal chromosomes (reviewed in [89, 90]).

Addition of the SL sequence to the mRNA takes place by splicing 'in *trans*', i.e. the required components of the two molecules are spliced together. Furthermore, *trans*-splicing appears to be the rule for trypanosomes and their relatives (*Trypanosoma*, *Leishmania* and *Crithidia*) [91]. It also occurs in a minority of the mRNAs of the nematode *Caenorhabditis elegans*, in which two different 22-nucleotide spliced leaders (SL1 and SL2) are donated by leader RNAs of approximately 100 nucleotides [92]. A small number of *trans*-spliced mRNAs have been reported from chloroplasts [93, 94].

Most evidence suggests that *trans*-splicing occurs by a mechanism similar to *cis*-splicing. Thus, the sequence immediately adjacent to the spliced leader in SL RNA conforms to the 5′-consensus sequence for *cis*-splicing and the sequence at the joining site on the mRNA conforms to the 3′-consensus sequence. Second, *trans*-splicing proceeds via a linear Y-shaped branched molecule analogous to the lariat of *cis*-splicing [95]. Third, destruction of U2, U4 or U6 snRNA blocks *trans*-splicing in trypanosomes [96]. Steitz and colleagues [97] have suggested that SL RNAs have a dual function. The leader portion functions as an exon whereas the remainder functions as an snRNA. Thus, they suggest that the RNA forms an snRNP and that, just as U1snRNP base pairs with the 5′-splice site in *cis*-splicing, so SLsnRNP might activate its own 5′-splice site. This concept is supported by the fact that SL RNAs have an snRNA-like trimethyl guanosine cap, a conserved secondary structure [97], exist as ribonucleoprotein particles that are immunoprecipitable by Sn antibodies directed against mammalian snRNPs [98] and can substitute for U1 snRNA in mammalian splicing systems *in vitro* [409].

A further oddity of trypanosomal biology is that most protein-encoding genes occur in multiple copies that are organized in tandem arrays with each gene copy separated from the next by a few hundred nucleotides of non-coding sequence. It has been shown that these gene clusters are copied into polygene transcripts, presumably from a single 5′-promoter [99, 100]. It has been further suggested that these polygene transcripts are *trans*-spliced to yield monogenic mRNAs each with a SL RNA attached to its 5′-end. The *trans*-spliced mRNAs retain their trimethyl guanosine caps, so that in *Caenorhabditis*, where 10% of the mRNA species are *trans*-spliced, there will be two populations of mRNAs, those with the conventional monomethylated guanosine cap and those with the snRNA-like cap [101, 102].

(f) *The role of snRNP, hnRNP, splicing factors and the nuclear matrix in catalysing the splicing and transport of mRNA*

Many aspects of splicing remain poorly understood. Little is known of the roles of individual snRNPs in the catalytic processes. Neither is it known whether the proteins of the particles are the catalytic entities or whether, as many have suggested, the RNA components will prove to be further examples of catalytic RNA (sections 11.3.2, 11.3.3 and 11.3.4).

The proteins of snRNP are not the only ribonucleoproteins implicated in splicing. Pre-mRNA itself also exists as ribonucleoprotein complexes known as hnRNP [103] which form like beads on a string as transcripts elongate. The hnRNP C proteins are known to be required for pre-mRNA splicing in crude extracts [104, 105] but their function, and that of the other hnRNP proteins, remains to be determined.

In addition to the U2 auxiliary factors described above, a number of other protein factors have been described that are required for splicing ([410] and references therein) but again their precise functions are largely unidentified.

Also of uncertain function is the nuclear matrix, the fibrous nucleoskeleton that remains after nuclei have been extracted with high salt concentrations and digested with DNase. As described in section 9.6, transcriptionally active genes are associated with the matrix. Pre-mRNA is also associated with the matrix [106] and can be chased out when splicing is promoted [107]. This has led to suggestions that the matrix provides a support for pre-mRNA processing and perhaps controls its transport to the cytoplasm [59]. It has recently been demonstrated that the major components of nuclear splicing are located in discrete nucleoplasmic foci [109, 411] but the structural basis of this is so far unknown.

11.3.2 The self splicing of group I introns

(a) *The concept of self splicing*

As discussed in section 8.4.3, the extrachromosomal rDNA of some strains of the ciliated protozoan, *Tetrahymena thermophila* contain an intron. Cech *et al.* [108], studying the excision of this 413 nucleotide intervening sequence from the pre-rRNA transcript, demonstrated that splicing occurred by a fundamentally new mechanism that revolutionized concepts of biochemical mechanisms. They found that RNA can catalyse its own splicing. At least *in vitro*, the reaction required no enzyme or any other protein and no energy was consumed. The sole requirements were that the RNA should be in a solution of salts with Mg^{2+} ions and a guanosine cofactor that could be guanosine, GMP or GTP (reviewed in [110]).

The splicing of the insertion sequence, commonly known as a *ribozyme*, is illustrated in Fig. 11.11 and occurs in a series of steps as follows.

1. Guanosine binds at or close to the 5′-splice junction [111], apparently recog-

nizing and binding to a specific and highly conserved G:C base pair [112].
2. The RNA behaves like an enzyme in that the binding of guanosine promotes the cleavage of the phosphodiester bond at the 5′-splice junction and the transfer of the 5′-phosphate to the 3′-OH of guanosine [113].
3. The 3′-splice junction is cleaved in a similar way except that the 5′-phosphate created in the break is joined through a normal phosphodiester bond to the upstream exon so forming the spliced RNA.
4. The released insertion sequence circularizes in a cleavage/ligation reaction in which the linear molecule is cleaved 15 nucleotides from its 5′-end and the A residue to the right of the cleavage site is joined to the G residue at the 3′-end [114]. Thus, a circular and a short linear molecule are produced of which the circle opens and closes again with the loss of a further four nucleotides. This again reopens to produce a molecule known as L−19 IVS (linear minus 19 intervening sequence). It is assumed that the complex degradation route, together with mass action effects, forces forward a reaction that is otherwise fully reversible [115].

The self-splicing of the ribozyme is absolutely specific; the 5′- and 3′-splice sites being unambiguously selected from a precursor rRNA of more than 6000 nucleotides. This implies that the RNA must, in an enzyme-like manner, provide substrate binding sites for the guanine cofactor and for the 3′- and 5′-splice sites. It is not surprising, therefore, that ribozyme activity requires a specific folded structure and that activity is lost when this is denatured or disrupted by mutation.

Analysis of the important elements of the *Tetrahymena* IVS has been greatly aided by site-specific mutation and also by com-

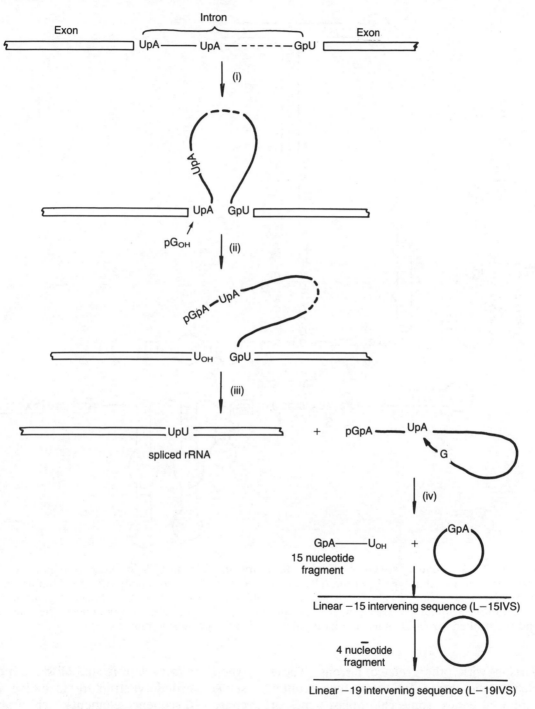

Fig. 11.11 The autocatalytic splicing of *Tetrahymena* pre-rRNA.

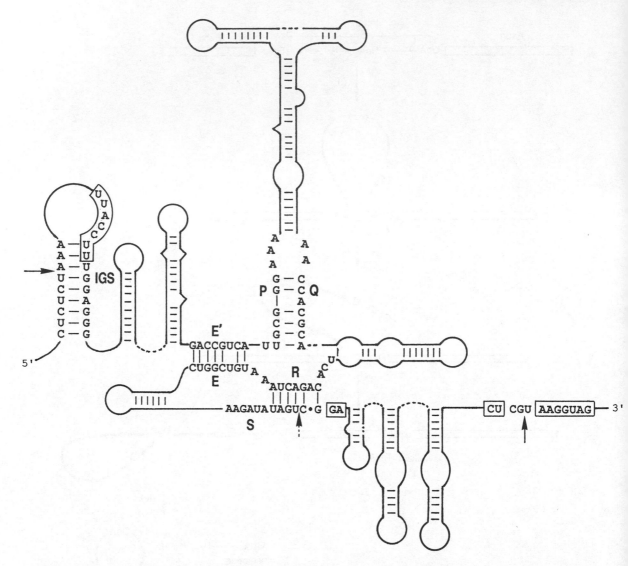

Fig. 11.12 A secondary structure model of the *Tetrahymena* intron. The base-paired regions, P, Q, R and S are conserved in type I introns. The elements E and E′ are not highly conserved but are always complementary. There is also evidence that complementarity between the boxed regions may be important in 3′-splice site selection (see text and Fig. 11.13). The solid arrows indicate the splice sites and the broken arrow identifies the base pair to which the guanosine binds.

parison with other related introns. These include the introns of many fungal mitochondrial genes, some chloroplast genes of higher plants, three introns in the genes of bacteriophage T4 and one in bacteriophage SPO1. Collectively, these are called *group I*,

type I or *class I introns* and all share a conserved secondary structure including four conserved sequence elements each of about ten nucleotides.

Figure 11.12 illustrates one of a number of similar models for the secondary structure

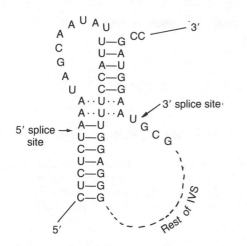

Fig. 11.13   A model by which the internal guide sequence (IGS) could align the two splice sites of the *Tetrahymena* IVS. Similar structures can be proposed for other group I introns.

of group I introns, here based on the IVS of *Tetrahymena*. The base-paired sequence elements P, Q, R and S are conserved in group I introns, although it should be noted that they have been given a confusing multiplicity of notations [109]. Two other sequence elements, E and E', that are widely separated on the IVS molecule, are not highly conserved in primary structure but are always complementary to each other. Mutation and compensatory mutation experiments have demonstrated that other elements of the base pairings in the proposed secondary structure are important for splicing [116]. One other sequence element, the internal guide sequence (IGS) is illustrated in Fig. 11.12 base-paired with the 5'-exon–intron boundary. In fact, it has the potential to interact with both the 5'- and the 3'-intron–exon boundaries (Fig. 11.13) but although there is clear cut evidence for its role in 5'-splicing (e.g. [117]) some evidence appears to preclude its participation in 3'-splicing [118]. Thus, in a popular model, the specificity of 5'-splicing is

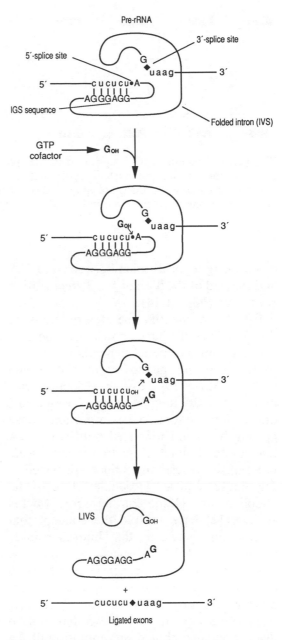

Fig. 11.14   A model for the splicing of the *Tetrahymena* group 1 intron. Adapted from Woodson and Cech [115]. ● denotes the 5'-splice site phosphate; ◆ the 3'-splice site phosphate that becomes the ligation junction phosphate.

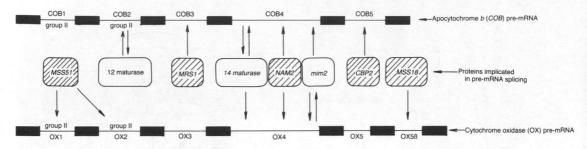

Fig. 11.15 Protein factors implicated in the splicing of mitochondrial pre-mRNAs in yeast. Proteins that are the product of nuclear genes are indicated by striped boxes whereas proteins that are mitochondrial gene products are indicated by open boxes. Black boxes are exons and the intervening lines are introns. Except where indicated the introns are group I introns. From Burke [126].

achieved by hybridization between the IGS and the end of the 5'-exon but 3'-recognition is unclear (Fig. 11.14).

Efforts to identify the determinants of the 3'-splice site have shown that the surrounding nucleotide sequence is important, particularly the guanosine at the 3'-end of the IVS [119]. Recent evidence also indicates that long distance base-pairing interactions may also be important. Base pairing has been identified between a GA nucleotide doublet close to the guanosine binding site (these nucleotides are boxed in Fig. 11.12) and a CU doublet close to the 3'-splice site [120]. Furthermore, Michel *et al.* [112] have revived the concept that interaction between the internal guide sequence and the 3'-splice site may be important. They have confirmed the findings of others that disruption of the putative pairing has little influence on 3'-splicing *in vitro* but have further shown that, if such disruption is combined with mutation of the above GA/CU pairing, splicing is greatly decreased. Compensatory mutations restored splicing. Suh and Waring [121] have also concluded that interaction between the 3'-exon and the IGS can play a role in selecting against the use of cryptic splice sites.

(b) *The use of group I introns as enzymes*

The excised *Tetrahymena* IVS RNA retains catalytic activity. A truncated version, L–19 IVS RNA that lacks the first 19 nucleotides and cannot therefore undergo intramolecular circularization, retains catalytic activity on other molecules. Thus, in an intermolecular version of 5'-splicing it carries out a sequence specific endonucleolytic cleavage comparable with that of a restriction endonuclease [122]:

$$5'\text{---CUCUCUA---}3'$$

$$G_{OH} \downarrow \quad L\text{--}19 \text{ IVS}$$

$$5'\text{---CUCUCU}_{OH} + GA\text{---}3'$$

When provided with oligo(C) as a substrate, L–19 IVS will exhibit poly(C) polymerase and phosphodiesterase properties [123] and if given a 3'-phosphorylated oligo(C) substrate, it acts as a phosphotransferase and an acid phosphatase [124]. As in normal 5'-splicing of group I introns, the specificity of these reactions depends on annealing the substrate to the internal guide sequence of the IVS. If this is replaced by a different template, the specificity of the reactions catalysed can be modified. It is anticipated that ribozymes derived in this way from this and other catalytic RNAs (section 11.3.3 and 11.3.4) could, in the future, provide important research and therapeutic tools for the selective interference with gene expression.

(c) *Proteins assist splicing of some, perhaps all, group I introns*

At least 12 of the more than 64 group I introns that have been characterized to date have been shown to be self-splicing *in vitro* [125]. However, some are not self-splicing *in vitro* and, of those that are, there is good reason to believe that their splicing *in vivo* is enhanced by proteins (reviewed in [126]). It seems likely that those introns that require proteins do so to assist in holding the RNA in the correct structure for self-splicing.

Some of the protein factors that are required for the efficient splicing of the introns in the mitochondrial genes of fungi are encoded in nuclear genes and translated in the cytoplasm. Others, known as maturases, are encoded in open reading frames within the introns of the mitochondrial pre-mRNAs and are translated on mitochondrial ribosomes [127]. In splicing, the maturase is believed to recognize short nucleotide sequences within the introns from which they are translated so that, as well as catalysing the removal of the intron, they participate in the destruction of their own message and are autoregulating. The system also has features that allow for the concerted control of more than one gene product. The maturase produced from the fourth intron of the apocytochrome *b* (COB) gene transcript is believed to recognize identical signal sequences in its own intron and an intron of the cytochrome oxidase (OX) gene transcript. Thus, it assists in the processing of both gene products [136, 145].

Figure 11.15 illustrates the protein factors implicated in splicing the COB and OX pre-mRNAs of *Saccharomyces cerevisiae*.

It should be mentioned in passing that most group I introns contain open reading frames capable of encoding proteins and that, in addition to maturases, a number are known to encode sequence specific endonucleases that control the mobility of group I introns (reviewed in [128, 129]). One of the nucleus-encoded splicing proteins of *Neurospora* is the product of the *cyt 18* gene. It appears to be involved in the splicing of a number of mitochondrial gene introns and intriguingly has been shown to be a tyrosyl-tRNA synthetase ([130] and references therein). Furthermore, a yeast protein that is involved in the activation of a maturase is a mitochondrial leucyl-tRNA synthetase [131]. The available data indicate that these enzymes are exerting a dual function; the splicing of pre-RNAs containing group I introns and the charging of mitochondrial tRNAs. This in turn drives speculation that part of the three-dimensional structure of group I introns might resemble that of tRNA (tRNA synthetases are considered in section 12.3.2).

11.3.3 The splicing of group II introns

The analysis of the introns of fungal mitochondria led to the realization that they could be arranged into at least two groups. The members of each group were related to each other by conserved sequence elements and a conserved secondary structure but the two groups were unrelated. Thus was born the designations group I and group II introns.

Michel *et al.* [132] have catalogued and compared 70 group II introns and their consensus secondary structure, which is one of a number of similar structures, is

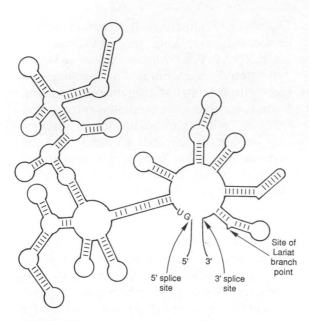

Fig. 11.16 A consensus secondary structure for group II introns. Based on Michel *et al*. [132].

illustrated in Fig. 11.16. As with group I inserts, at least some group II introns have the ability to self-splice *in vitro* although at a slow rate and under rather artificial conditions [133]. Splicing is reversible and, as in group I introns, it is probable that this makes it possible for the introns to transpose to a new location (reviewed in [134]).

Notwithstanding the above similarities, the splicing pathways for the two intron classes is very different. In fact, the excision of type II introns has more in common with that of nuclear genes (section 11.3.1). Thus, the first stage entails a 5'-cleavage and lariat formation and the second stage consists of 3'-cleavage and exon ligation. However, no guanosine nucleotide is involved in 5'-cleavage; rather it is initiated by an attack of the 2'-OH of a nucleotide within the intron on the 5'-splice junction [133, 135]. The similarity with pre-mRNA maturation has prompted suggestions that the splicing mechanism for nuclear introns could have evolved from self-splicing group II introns (reviewed in [427]).

11.3.4 Other catalytic RNAs and RNA-containing enzymes

The discovery of self-splicing has provided support for the concept, first raised by Crick in 1968 [137], that, in the early stages of the evolution of life, the first enzymes could have been made entirely of RNA. Further support for these ideas comes from the growing list of known catalytic RNA molecules which are considered here, despite being a diversion from the theme of splicing.

(a) *Ribonuclease P*

RNase P is a ribonucleoprotein enzyme that cleaves pre-tRNA molecules (section 11.7) to generate the 5'-termini of mature tRNA (reviewed in [138, 139]). In *E. coli*, the complex consists of two subunits of a 119 amino acid polypeptide (C5) and a 377 nucleotide RNA known as M1 RNA [140]. Protein-free M1 RNA exhibits catalytic activity in the presence of high concentrations of magnesium ions but the protein is essential for activity under physiological conditions [141]. This appears to be because the small, basic C5 protein provides a pool of counter ions that decreases the repulsion between the polyanionic RNAs that are the enzyme and substrate.

The sequence of M1 RNA is not strongly conserved in the equivalent enzymes of other eubacteria. Nevertheless, phylogenetic comparison of the RNA component of the RNase P of nine bacterial species belonging to two different phyla has been used to derive a substantially conserved secondary structure [142]. Furthermore, the deletion of non-conserved elements of the secondary structure produced a 263 nucleotide derivative of the 417 nucleotide M1 RNA that retained both catalytic specificity and efficiency [142].

Analysis of substrate recognition by

RNase P has suffered a number of false starts. However, the recent use of model substrates, such as partial tRNA molecules and derivatives of Turnip Yellow Mosaic Virus, indicate that recognition is dependent on the length and structure of the T stem and loop, the acceptor stem of the tRNA and the presence of a CCA terminus [143, 144].

Homologous RNase P enzymes that are also ribonucleoproteins have been purified from yeast, HeLa cells and their mitochondria. The RNA components of the enzymes are again required for enzymic activity but the isolated RNAs do not exhibit catalytic activity *in vitro* [138]. Indeed, nuclease inactivated yeast RNase P still binds to tRNA suggesting that, in this instance, the protein component of the enzyme might play a role in substrate binding [146].

(b) *Viroids, virusoids and satellite RNA*

Plant viroids are among the smallest infectious agents known. Their 300–400 nucleotide, single-stranded RNAs do not encode any proteins. They are, however, self-replicating and are transmitted as naked RNA from plant to plant through damaged cells. Virusoids and satellite RNAs are not self-replicating but rely on helper viruses for replication and encapsidation.

The RNAs of these agents can exist in circular or linear form and, during replication, the circular molecules are thought to act as templates for the production of oligomers that are then cleaved to monomers ([147] and references therein). In several satellite RNAs and one viroid, the cleavage of oligomers to monomers has been shown to occur by self-processing at unique sites and with the production of 2':3'-cyclic phosphate and 5'-hydroxyl termini [148, 149]. Comparison of a number of these self-cleaving RNAs has been employed to

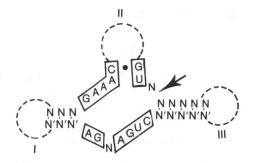

Fig. 11.17 The secondary structure of the hammerhead domain. The boxes enclose conserved nucleotides. N indicates an arbitrary nucleotide that is complementary to N'. The number of nucleotides in hairpin loops varies from 2 to more than 200. The arrow indicates the point of cleavage. From Uhlenbeck [150].

derive a consensus secondary structure for the self-cleaving domain [149]. In this, three double-stranded helices, arranged in the so-called *hammerhead structure* (Fig. 11.17), bring single-stranded regions of highly conserved sequence into close alignment. The hammerhead can also be assembled *in vitro* from two synthetic RNA fragments of which one strand is cleaved by the other [150].

(c) *Telomerase*

When DNA polymerase copies a DNA molecule, it requires not only a template to copy but a primer from which to initiate polymerization. Thus, if the 3'-end of a DNA strand is annealed to the primer it will not be copied by the polymerase and a linear DNA will tend to become shorter each time it is copied. In eukaryotic chromosomes, this problem is solved by telomeres; simple repeating units at the ends of the chromosomes, which at the 3'-ends have a generalized repeat sequence, T_mG_n. An enzyme, telomerase, capable of generating additional repeat units in compensation for loss during

(a) RNase P cleavage

(b) Viroid RNA autocatalytic cleavage

(c) Group I intron self-splicing

(d) Group II intron self-splicing, nuclear pre-mRNA splicing

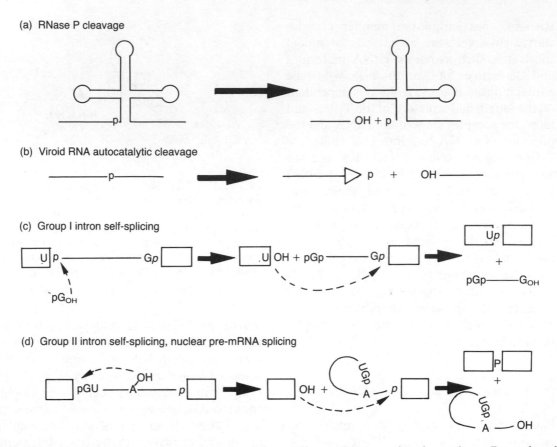

Fig. 11.18 A diagrammatic representation of four different RNA-catalysed reactions. Reproduced from Green [426].

replication, has been characterized from several ciliate protozoa and more recently from human cells [151].

Blackburn and colleagues [152, 153], working with *Tetrahymena*, first demonstrated that telomerase is a ribonucleoprotein and they suggested that the RNA component functioned as a template for the addition of the correct GGGGTT units to the chromosome end. Subsequently, they showed that the RNA of the telomerase contained the sequence 5'-CAACCCCAA-3' that is complementary to one and a half of the telomeric units. Furthermore, mutations in the CAACCCCAA sequence caused the

synthesis, *in vivo*, of modified telomeric sequences that were complementary to the mutated sequences [153]. Thus, the RNA of telomerase is not a ribozyme, in that the RNA is not itself catalytic; rather the RNA is an essential template in a catalytic ribonucleoprotein that functions as a specialized reverse transcriptase (reviewed in [154, 412]).

(d) *Less defined catalytic RNAs*

It is clear from the sections above that at least four types of catalytic RNA have been categorized – group I introns, group II

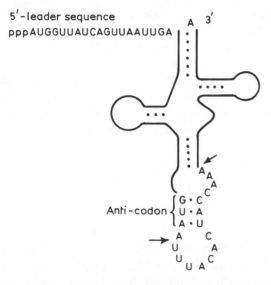

5'-leader sequence
pppAUGGUUAUCAGUUAAUUGA

Anti-codon

Fig. 11.19 A schematic illustration of a yeast 108 bp tRNA^{Tyr} precursor with a 10 nucleotide 5'-leader sequence and a 14 nucleotide intron that base pairs with the anticodon. The arrows indicate the points of cleavage and ligation.

introns, hammerheads and RNase P (Fig. 11.18). Additional mechanisms have probably been identified but await better characterization ([155] and references therein).

11.3.5 The splicing of pre-tRNA

The tRNA genes of eukaryotes [156–158] and archaebacteria [159] may contain introns and in yeast nuclei they are particularly common with approximately 40 of the 400 tRNA genes interrupted [159], affecting nine tRNA gene families [160]. Some tRNA species commonly appear to contain inserts. Thus, intron-containing tRNA^{Tyr} genes have been described for yeast, *Xenopus*, man and *Nicotiana* ([161] and references therein). The single intron of yeast genes may be from 14 to 60 nucleotides in length but in higher plants they are only 11–13 nucleotides long [162]. In all known

examples, the intron is inserted one nucleotide to the 3'-side of the anticodon (Fig. 11.19). Furthermore, it includes a sequence which base pairs with all or part of the anticodon and sometimes with a portion of the anticodon loop [159].

A yeast temperature sensitive mutant, *rna1*, accumulates the precursors of nine tRNAs when grown at the non-permissive temperature [160, 163]. These precursors have mature 5'- and 3'-termini but contain the intervening sequences. They can be correctly spliced *in vitro* in systems derived from yeast [164, 165], *Xenopus* germinal vesicle [166] and HeLa cells [167]. Of these, the HeLa cell system has become particularly widely used and will accurately splice tRNA precursors derived from many animal and plant sources ([161] and references therein). *In vitro* splicing by a wheat germ system is, however, much less flexible and will only splice plant pre-tRNAs [162].

In all systems, splicing can be separated into two stages [168]. In the first of these, the tRNA precursor is precisely cleaved at the intron boundaries thus generating two half tRNA molecules and a linear intervening sequence. In the second stage, the two half molecules are ligated together to produce the mature tRNA. The endonucleolytic cleavages are unusual in producing 5'-hydroxyl termini and 3'-phosphorylated termini; a feature that is common to degradative enzymes such as the RNases A and T1, but is not normally found in processing enzymes. The yeast endonuclease has been purified to homogeneity [169] and consists of three polypeptides of 31, 42 and 51 kDa. The single enzyme is able and sufficient to cleave all of the nine species of intron-containing pre-tRNA [170] apparently because it recognizes features that are common to all mature tRNAs, particularly the conserved nucleotides C56 and U8 [171, 172]. Presumably, the lack of

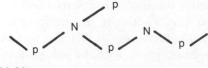

Fig. 11.20

flexibility in the wheat germ splicing system reflects intolerance by the plant endonuclease of the variability of intron length in animal cells. The yeast enzyme has two unusual features. It appears to be a membrane protein [173] and is presumably located in the nuclear membrane. Second, it produces a 2′:3′-cyclic phosphorylated terminus rather than a 3′-phosphate (Fig. 11.21). Clearly, the multimeric protein must contain catalytic sites for both 5′- and 3′-cleavage as well as a domain that ensures membrane insertion. It is to be hoped that recent progress towards the identification of the genes that encode the nuclease [169] will lead to their isolation and sequencing so that more may be understood of the structure/ function relationships of this enzyme.

Ligation has been studied in yeast [174], wheat germ [175, 176] and animal cell systems [167, 177] and again differs between plants and animals. In yeast, the 3′-half of the tRNA is phosphorylated and is then adenylated by an adenylated ligase. Condensation occurs between this activated 3′-half of the tRNA molecule and the 2′:3′-cyclic phosphate terminus of the 5′-half of the molecule. The phosphodiester linkage thus formed has the form illustrated in Fig. 11.20 with the 3′−5′ bond derived from the 5′-terminal phosphate. This reaction mechanism is similar to that of bacteriophage T4 ligase [178]. The 2′-phosphate is subsequently removed by phosphatase activity to generate spliced tRNA [179, 180]. The yeast tRNA ligase has been purified and is a 95.4 kDa single polypeptide with all the catalytic properties associated with liga-

tion. Thus, it exhibits 2′:3′-cyclic phosphodiesterase, 5′-polynucleotide kinase, adenylate synthase and tRNA ligase activities [181]. Wheat germ ligase has been partially purified and again the ligase and nucleotide kinase activities copurify [175]. The splicing pathway of yeast is diagrammatically illustrated in Fig. 11.21.

Animal cell ligation involves a different strategy [167, 177]. The 3′-half of the tRNA molecule is not phosphorylated and the phosphate of the phosphodiester bond is derived from the 5′-half of the molecule (Fig. 11.22). A HeLa cell ligase has been purified and is dependent on Mg^{2+}, ATP or dATP, and RNA substrates with 5′-hydroxyl and 2′:3′-cyclic phosphate termini [182].

11.3.6 A postscript on splicing

The preceding sections have discussed four major ways in which the transcripts of introns are removed together with variants such as fungal pre-mRNA splicing, *trans*-splicing and variations in the ligation mechanisms for pre-tRNA. Group I and group II introns are self-splicing, the former requiring a guanosine cofactor whereas in the latter the 2′-hydroxyl group at the branch site takes part in transesterification. The splicing of pre-mRNA has similarities with the self-splicing of group II introns. It produces a branched intermediate and may be evolutionarily related to self-splicing in that one or both splicing events may yet prove to be RNA catalysed by the snRNA components of the spliceosome. The splicing of pre-tRNA, on the other hand, has little in common with other mechanisms, rather, its enzymic reactions appear to be borrowed from elsewhere. Thus, yeast pre-tRNA ligase has much in common with bacteriophage T4 ligase.

That evolution should have produced

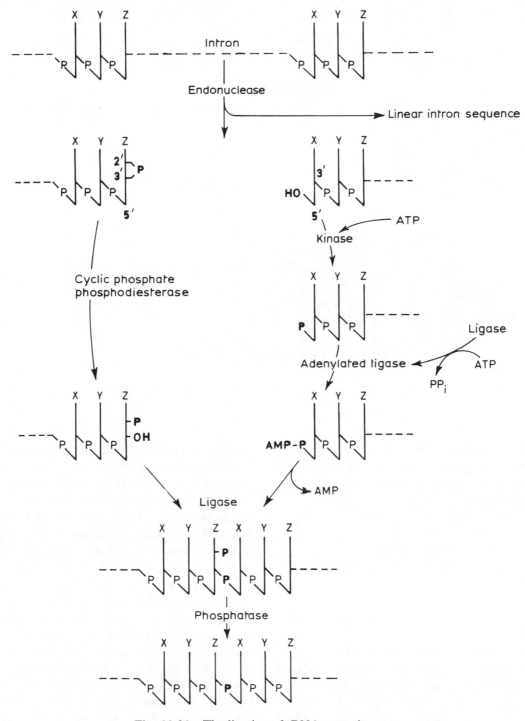

Fig. 11.21 The ligation of tRNA exons in yeast.

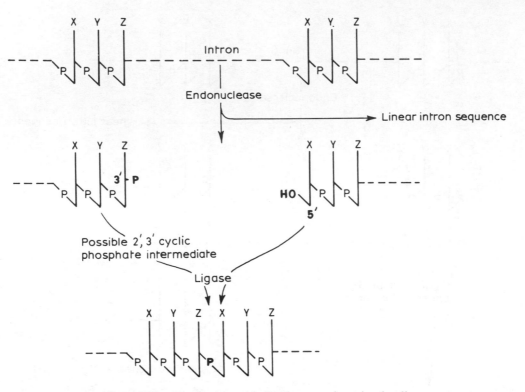

Fig. 11.22   The ligation of tRNA exons in animal cells.

such a plethora of splicing mechanisms perhaps emphasizes the advantage to the evolutionary process of having introns in the first place.

11.4 THE CAPPING AND 5'-PROCESSING OF PRE-mRNA AND snRNA

The mRNAs and pre-mRNAs of eukaryotes and most of their viruses are unique in having their 5'-termini blocked by the post-transcriptional addition of a methylated guanosine cap (reviewed in [183–185]). The guanosine is linked through a 5'–5' pyrophosphate bridge to the 5'-terminal nucleotide of the primary transcript and is then methylated at carbon-7 to form

cap 'zero'. Further modification may involve 2'-*O*-methylation of the ribose of the first or the first two nucleotides of the transcript thus forming 'cap one' and 'cap two', respectively (Fig. 11.23). The relative abundance of the different cap structure changes with evolutionary complexity. Yeast mRNAs have cap zero; those of slime moulds are mainly cap zero but 20% have cap one; messages of the brine shrimp and sea urchin are all terminated with cap one whereas mammals have high percentages of caps one and two [183]. The enzymatic reactions of cap formation were first elucidated with purified reovirus [186] and vaccinia virus [187]. More recently they have been characterized in cellular systems (reviewed in [188]) and are illustrated in Fig. 11.24.

Caps are added to the 5'-ends of incom-

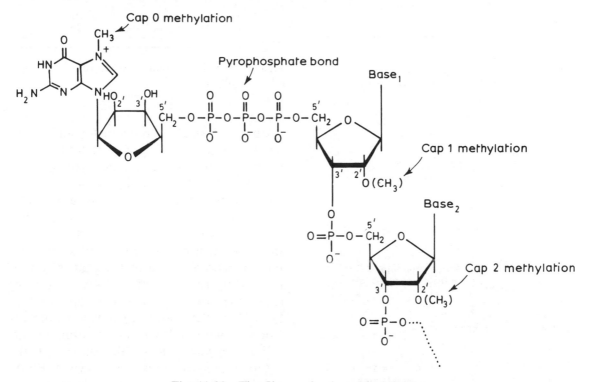

Fig. 11.23 The 5'-cap of eukaryotic mRNA.

plete transcripts [189] and, in those systems where sufficient data are available, the 5'-termini of both the mature mRNA and its precursors are identical and map at the same point in the gene sequence. Even where microheterogeneity exists in the initiation site of a gene [190, 191], this appears to be reflected in similar heterogeneity in the cytoplasmic mRNA and there is little if any 5'-trimming. In any case the addition of caps to trimmed transcripts seems unlikely as it would require different enzymatic reactions.

Although it has long been suggested that caps may function in the transport of mRNA from the nucleus, it is only recently that evidence to support the suggestion has been presented. Hamm and Mattaj [413] demonstrated that U1 snRNA synthesized by RNA polymerase II and thus processed by

the addition of a monomethylated cap, was exported out of the nucleus. Conversely, if it was synthesized by RNA polymerase III, it was given a different cap and remained in the nucleus [413]. The cap also appears to protect mRNA from nuclease attack, and has an important function in the attachment of the 40S ribosomal subunit to the mRNA (section 12.5.1). In an, as yet, unknown way, the cap is also involved in the splicing of pre-mRNA (section 11.3.1) in that messengers with caps are spliced more efficiently than those without [192, 193]. That this is not simply due to mRNA stabilization was elegantly demonstrated by Shimura and colleagues [194]. They prepared pre-mRNAs with the natural m^7GpppG cap and with an unnatural ApppG cap. The unnaturally capped precursor was as stable as the natural pre-mRNA but was not spliced as efficiently.

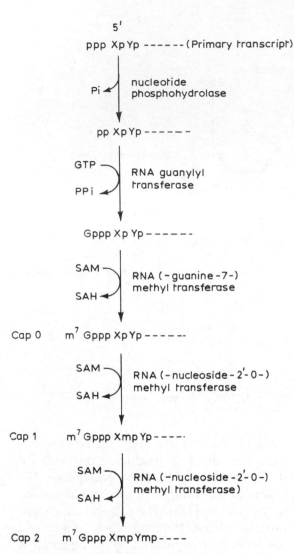

Fig. 11.24 The enzymic pathway for the biogenesis of 5'-terminal cap structures on eukaryotic pre-mRNA. SAM, *S*-adenosyl-1-methionine; SAH, *S*-adenosyl-1-homocysteine.

| 5' non-coding | Coding sequence | 3' non-coding | Poly (A) |

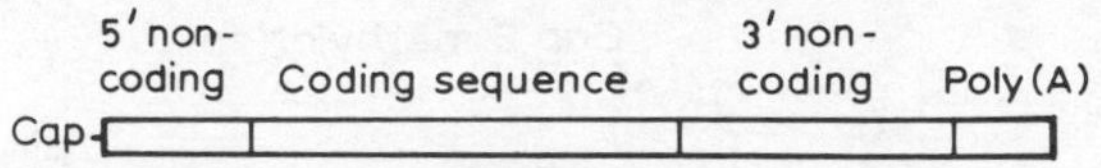

Fig. 11.25 Diagrammatic structure of a polyadenylated mRNA.

methylguanosine that is characteristic of and almost unique to this group of RNAs. The migration of snRNA to the cytoplasm is necessary for the formation of snRNP, the component proteins of which are made and stored in the cytoplasm. Mattaj and colleagues have recently demonstrated that the movement of the snRNP back into the nucleus requires the trimethylguanosine cap [195, 414]. The only other RNAs known to share the trimethylguanosine cap are the SL RNAs that appear to function both as snRNAs and as a leader in the formation of mRNAs by *trans*-splicing (section 11.3.1(e)).

Other cap structures are known. snRNA U6 differs from all other snRNAs in having a γ-monomethyl phosphate cap [196] and apparently does not leave the nucleus. A group of strongly expressed mRNAs of the cowpox virus, which replicates in the cytoplasm, have a 5–21 residue poly(A) 5'-terminus [197].

The trimethylguanosine cap in translation is recognized by the initiation factor eIF-4F complex, through its component eIF-4E (section 12.5.1). Nuclear cap-binding proteins have also been described and may function in nuclear stabilization, transport or splicing ([198] and references therein).

Between the cap and the translational initiation codon, AUG (section 12.2), there is a length of non-coding RNA known as the leader sequence (Fig. 11.25). It varies in length from three nucleotides in the immunoglobulin kappa chain [199] to 256 nucleotides for liver amylase [200] although most vertebrate leaders are between 20 and 100 nucleotides in length [201]. Leaders

As described in section 11.2, the snRNAs are also capped. Initially, cap formation follows the pre-mRNA model with the addition, through a pyrophosphate bond, of a 7-methylguanosine. After migration out of the nucleus, the cap is then further modified in the cytoplasm to a 2,2,7-tri-

contain no universally conserved sequence elements but within those of some gene families there do seem to be some conserved nucleotides and even some evidence of common secondary structure [185]. As is described in section 12.5.1, the sequence 'context' of the initiation codon conforms to a rather loose consensus [201].

11.5 POLYADENYLATION AND 3'-PROCESSING OF PRE-mRNA

11.5.1 Polyadenylated mRNA

In prokaryotes, the 3'-ends of most mRNAs are formed by transcription termination (section 9.4). Conversely, most mRNAs in eukaryotic cells have at their 3'-termini a homopolymer of adenosine residues. These stabilize the mRNA, stimulate its transcription and provide a means of regulating the manner and rate at which transcripts are made available for translation (section 12.9.3). Other possible roles, such as nuclear-cytoplasmic transport of mRNA are controversial ([226] and references therein).

The so-called poly(A) tail varies from approximately 20 to 300 residues in length. Mammalian cells exhibit a narrower size range distribution, around 250 nucleotides, whereas in lower eukaryotes the average size is nearer 100 residues.

The site of polyadenylation does not coincide with the transcript terminus. Indeed, termination by RNA polymerase II is ill-understood but clearly occurs hundreds or even thousands of nucleotides downstream of the polyadenylation site (section 9.5.1). Thus, polyadenylation is a two stage process (reviewed in [202–204, 415]) in which a 3'-terminus is first formed by endonucleolytic cleavage of the transcript. This occurs while it is still being synthesized and probably with the advancing RNA polymerase a considerable distance downstream. The cleavage generates a 3'-hydroxyl and a 5'-phosphate and in the second stage, adenosine residues are added one at a time to the former. The development of *in vitro* processing systems, in which synthetic RNAs that end in poly(A) addition sites are polyadenylated by nuclear extracts, has shown that the cleavage and polyadenylation can be separated [205]. Nevertheless, it is clear that *in vivo* the two events are mechanistically coupled and catalysed by a multiprotein complex. Thus, poly(A) polymerase is required for cleavage as well as for polyadenylation[206] and it is a part of a cleavage/polyadenylation complex [207].

The hexanucleotide AAUAAA is required for both cleavage and polyadenylation. It is very highly conserved in polyadenylated mRNAs and is typically located 15–25 residues upstream of the poly(A) addition site. One relatively common variant is the sequence AUUAAA which occurs in approximately one in ten mRNAs but other variants are very rare and are processed very inefficiently [204]. The importance of the AAUAAA hexanucleotide has been repeatedly demonstrated by experimental mutation ([202] and references therein) and is dramatically illustrated in forms of both α and β thalassaemias. In these, point mutations in the hexanucleotide result in very low amounts of mature mRNA in the patients' blood cells [208, 209]. Such mutations do not determine whether the AAUAAA sequence is required for cleavage or polyadenylation but experiments *in vitro*, in which the two reactions are studied independently, suggest that it is required for both. The hexanucleotide appears, in fact, to be the only sequence element required for polyadenylation. Substrates as short as 11 nucleotides will support polyadenylation *in vitro*, provided they contain the

AAUAAA element followed by eight or more nucleotides. The sequence of these downstream nucleotides is apparently unimportant [210, 211]. Conversely, the cleavage reaction requires the AAUAAA hexanucleotide and downstream elements. These are much less highly conserved although various consensus sequences have been put forward. They include the sequence YGUGUUYY which occurs approximately 30 nucleotides downstream [212] and the consensus sequence, CAYUG, which occurs upstream or downstream of the polyadenylation site [213, 214]. In truth, neither of these sequence elements is well conserved but it is clear that sequences downstream of the hexanucleotide are invariably U- or GU-rich [212]. Furthermore, although they lack either a consensus sequence or conserved secondary structure, these downstream elements are important [215]. The sequence AAUAAA occurs internally in many mRNA species and the reason that such potential cleavage sites are not recognized is apparently because they lack downstream GU- or U-rich elements [207].

Poly(A) polymerase, an enzyme catalysing the addition of poly(A) to RNA, was first described 30 years ago; long before the observation that mRNAs were polyadenylated. The enzyme was substantially purified by the mid-1970s, but until recently, it remained uncertain whether it was responsible for the polyadenylation of pre-mRNA. The purified enzyme would add poly(A) to unnatural substrates such as tRNA or even oligonucleotides and showed no requirement for the AAUAAA hexanucleotide [216]. It is now known that this non-specific activity reflected the high concentrations of enzyme or substrate employed. *In vitro* polyadenylation assays, employing the classical poly(A) polymerase and 1000-fold lower substrate concentrations, become

specific for AAUAAA-containing substrate but depend on the presence of a nuclear factor known as specificity factor or PF2 [206, 207, 217]. Indeed, at least three HeLa cell poly(A) polymerases, two nuclear and one cytoplasmic, exhibit similar characteristics [218]. It is the specificity factor that recognizes AAUAAA. It is presumed that it also makes contact with poly(A) polymerase and in so doing decreases its dissociation constant for AAUAAA-containing substrate. It is also presumed that the 15–25 residues between the hexanucleotide and the polyadenylation site reflect the distances between the AAUAAA-binding site and the catalytic site of the respective proteins.

The way in which specificity factor imparts specificity on an otherwise non-specific poly(A) polymerase is analogous to prokaryotic RNA polymerase for which promoter recognition depends on sigma factor. An increasing number of examples are known in which various prokaryotes and their phage employ sigma factors of differing specificity and, in so doing, control the expression of different genes. Clearly, a suite of specificity factors could similarly control the route and rate of pre-mRNA processing. The control of 3'-processing is further discussed in section 11.9.1.

The analogy between specificity factor and sigma factor can be taken further. Sigma imparts specificity on prokaryotic RNA polymerase but then dissociates from the initiation complex after the transcription of 8–10 nucleotides (section 9.2.2). Similarly, Sheets and Wickens [219] have presented evidence that polyadenylation occurs in two phases. The first phase is dependent on AAUAAA and specificity factor. In the second phase, which occurs after the addition of the tenth adenosine, further polymerization is independent of AAUAAA but requires the already made oligo(A) as a primer. It is suggested, but not proven, that

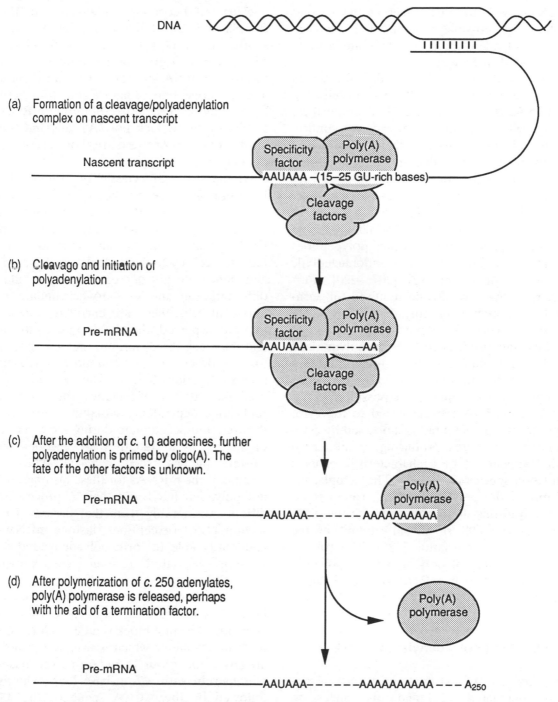

(a) Formation of a cleavage/polyadenylation complex on nascent transcript

Nascent transcript

Specificity factor / Poly(A) polymerase

AAUAAA –(15–25 GU-rich bases)

Cleavage factors

(b) Cleavage and initiation of polyadenylation

Pre-mRNA

Specificity factor / Poly(A) polymerase

AAUAAA – – – – – –AA

Cleavage factors

(c) After the addition of *c*. 10 adenosines, further polyadenylation is primed by oligo(A). The fate of the other factors is unknown.

Pre-mRNA

AAUAAA– – – – – AAAAAAAAAA

Poly(A) polymerase

(d) After polymerization of *c*. 250 adenylates, poly(A) polymerase is released, perhaps with the aid of a termination factor.

Poly(A) polymerase

Pre-mRNA

AAUAAA– – – – –AAAAAAAAAA – – – A_{250}

Fig. 11.26 The 3'-processing and polyadenylation of eukaryotic mRNA.

specificity factor disengages at this second phase. A possible scheme of 3′-cleavage and polyadenylation is diagrammatically illustrated in Fig. 11.26.

Specificity factor has a high M_r which may exceed 300000 and it may include the snRNA, U11 [207, 209, 221]. Total purification will be required, however, to confirm this and to investigate further its interaction with enzyme and substrate. It is nevertheless clear that both poly(A) and specificity factor form part of a massive cleavage/ polyadenylation complex that appears to have a M_r of 500000–1000000 [203] and includes at least three further polypeptides that are required for the endonucleolytic cleavage of pre-mRNA [207, 220]. Further factors may also be required for accurate termination of polyadenylation which occurs *in vitro* after the expected number of adenosine residues have been polymerized [219, 222]. Nuclear poly(A)-binding protein is a candidate for the activity but no evidence as yet exists to support this possibility.

Once mRNA is transported to the cytoplasm, the poly(A) tail is shortened by cytoplasmic nucleases producing a tail length that appears to be a characteristic of each mRNA species [223–225]. The cytoplasmic form of the poly(A)-binding protein is involved in this shortening and with the binding of the 60S ribosomal subunit in the initiation of translation [226]. The role of the shortening of poly(A) and of poly(A)-binding protein in mRNA stability is again controversial (section 12.9.3).

11.5.2 Non-polyadenylated mRNA

Up to 30% of cytoplasmic mRNA is not polyadenylated [227] and this appears to comprise two classes. Some poly(A)$^-$ mRNA is derived from mRNA that was polyadenylated for its transport out of the

nucleus but becomes deadenylated in the cytoplasm [223]. Evidence exists that the shortening of poly(A) tracts is associated with mRNA degradation (section 12.9.3) but some mRNA species, for instance actin mRNA, appear to remain stable with very little if any poly(A) tail [228]. Other mRNA species naturally lack poly(A) tails and the most prominent of these are those encoding the major class of histones.

Of the two main classes of histone, the least prevalent are polyadenylated and are presumably processed as described in section 11.5.1. They are known as replacement and tissue-specific variants, and are not subject to cell-cycle-dependent variation in abundance, do not decrease during cellular differentiation and tend to accumulate in quiescent cells. They are encoded in genes that are dispersed, they tend to have introns and they lack the 3′-termini that are shortly to be described for nonpolyadenylated species (section 8.4.1). The second and predominant class of histones, the so-called replication-dependent variants, are synthesized and accumulate during the S-phase of the cell cycle and they are encoded on clustered genes (section 8.4.1). The 3′-termini of the mRNAs for these histones are not polyadenylated and their 3′-processing differs considerably from that described in section 11.5.1 (exceptions, histone mRNAs apparently able to form polyadenylated or non-polyadenylated termini, are known [229, 230]).

The 3′-ends of replication-dependent histone genes have a series of conserved elements. The first block is a GC-rich region of dyad symmetry which is usually hyphenated by a run of four Ts and forms a six base-pair hairpin with a four-base loop. This is followed by the ACCA sequence the 3′-end of which corresponds to the 3′-end of the mature mRNA. A third, strongly conserved element occurs downstream into

Fig. 11.27 The 3'-consensus sequence of sea urchin histone genes.

the spacer between the clustered genes. In sea urchins, this is a highly conserved purine-rich sequence, CAAGAAAGA, which occurs about eight nucleotides downstream from the 3'-end of the message (Fig. 11.27). In vertebrates, the purine-rich consensus RAAAGAGCT is rather less highly conserved [231].

The termination of the transcription of histone genes appears to be at multiple sites and, in the case of sea urchin histone H2A, occurs heterogeneously approximately 100 nucleotides downstream from the mRNA 3'-terminus in an A-rich region [232, 233]. The correct 3'-end of the message is generated by endonucleolytic cleavage of these longer transcripts [234, 235] as a result of the recognition of the consensus sequence by at least three factors. One of these has as a component the rare snRNA, U7, the role of which was first indicated in sea urchins [236, 237]. It was shown that the 5'-end of the 57 nucleotide RNA had the capacity to base pair with six of the nine nucleotides of the CAAGAAAGA element [237]. Furthermore, U7 snRNA from sea urchin was able to compensate for the inefficiency of a *Xenopus* 3'-processing system [238] and, more conclusively, the processing of a defective sea urchin histone H3 gene was suppressed by introducing a compensatory change into U7 snRNA [239]. Subsequent analysis also demonstrated the role of U7 snRNA in the 3'-processing and endonucleolytic cleavage of mammalian histone mRNA [240–242].

It is becoming apparent that the 3'-sequence elements of replication-dependent histone mRNAs are not only important as recognition signals for processing but are also critically important in regulating the efficiency of processing [243], the cytoplasmic half-lives of the mRNAs [244] and in the control of cell cycle-dependent histone mRNA abundance. The 5'-end of mouse snRNA U7, which is complementary to the purine-rich consensus [241], is available to micrococcal nuclease whereas the remainder of the RNA is protected by the proteins of the snRNP particle [245]. However, this availability is cell cycle-dependent and it has recently been shown that the 5'-sequence of the RNA is occluded during the G0 stage of the cell cycle but exposed and free to interact with mRNA during the S-phase when histone mRNA synthesis is at its peak [246].

Several other factors are involved in the 3'-processing and may also participate in its control. A heat-labile component of the processing apparatus is present in relatively low amounts in G1-arrested cells that accumulate histone H4 pre-mRNA and in which mature H4 mRNAs are drastically reduced. Conversely, the factor is in excess in extracts of exponentially dividing cells in which H4 mRNA concentrations are two orders of magnitude higher [247]. The conserved hairpin-forming portion of the histone mRNA consensus sequence is not involved in the interaction with U7 snRNA [248] but it is a target element for a distinct, nuclease insensitive processing factor [242, 248] that increases the efficiency of 3'-cleavage [248].

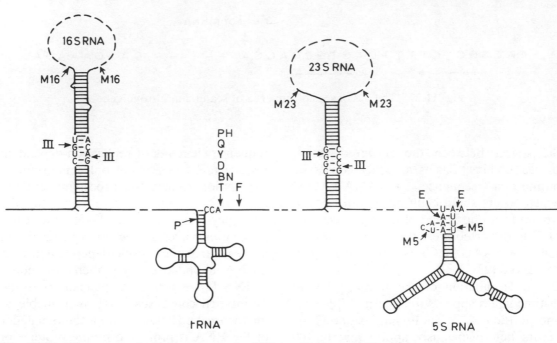

Fig. 11.28 The processing of prokaryotic rRNA, tRNA and 5S rRNA precursors. Arrows and associated symbols indicate sites cleaved by the following ribonucleases. III indicates sites cleaved by RNase III in the double-stranded regions flanking 16S and 23S RNA. P indicates the site of cleavage by RNase P to generate the 5′-end of mature tRNA. M16 and M23 are names that have been given to ill-characterized nucleases involved in the maturation of 16S and 23S RNA. F indicates the endonucleolytic cleavage by RNase F to the 3′-side of tRNA. PH, Q, Y, D, BN and T indicate the possible role of these nucleases in the 3′-trimming of tRNA. E indicates the initial endonucleolytic cleavage by *E. coli* ribonuclease E in the generation of 5S RNA. M5 indicates the final trimming of 5S RNA precursor in *B. subtilis* by RNase M5.

11.6 THE PROCESSING OF PRE-rRNA

11.6.1 The processing of prokaryotic pre-rRNA

The stoichiometric formation of 16S, 23S and 5S ribosomal RNA is ensured in eubacteria by their co-transcription from operons that encode all three genes (section 8.4.3). The primary transcript cannot normally be detected; presumably because it is efficiently processed by endonucleases while it is still being transcribed. The investigation of the maturation events involved has been made possible by the availability of processing deficient mutants that accumulate precur-

sors. The major steps involved are diagrammatically illustrated in Fig. 11.28.

(a) *The processing of the prokaryotic 16S and 23S transcripts*

The sequences flanking both the 16S and 23S gene transcripts form inverted repeats the two halves of which are, respectively, 1700 and 2900 nucleotides apart. These sequences interact to form base-paired stems [249, 250] at the base of giant loops that contain the complete sequences of 16S and 23S RNA and which, in the nascent transcript, are already in the form of a ribonucleoprotein. The base-paired stems form specific catalytic sites for the enzyme ribonuclease III (Fig. 11.28) and precursors containing them accumulate in mutants lacking the enzyme [251]. RNase III also cleaves the RNAs of several bacteriophage and there is growing evidence that it is important in limiting the half-lives of some mRNAs. How the enzyme selects its double-stranded cleavage sites is unclear. There is little sequence homology between its catalytic sites in pre-ribosomal RNA and those of other substrates. It will even act non-specifically on double-stranded RNA, cleaving it at intervals of approximately 15 nucleotides [252]. Nevertheless, it is clear that the major cleavage sites in pre-rRNA are very specific. Even a small perturbation strongly inhibits the enzyme [253] and there is recent evidence that specificity is determined by some common nucleotide features and by the distance from the end of a region of continuous base-pairing [254]. It is to be hoped that the recent cloning of the enzyme, its expression in recombinant cells and its purification to homogeneity [255] will lead to a clarification of the way in which it recognizes and cleaves its substrates.

RNase III produces staggered cuts in each stem of *E. coli* pre-ribosomal RNA (Fig. 11.28) and the products are precursor ribonucleoproteins containing RNAs that are slightly longer than the mature molecules. The nature of these precursors and the enzymes that catalyse their maturation is ill understood. Putative precursors and poorly characterized enzymes have been described but their relationship with the true processing reactions is uncertain (reviewed in [256]).

(b) *The processing of the prokaryotic 5S transcript*

At least three enzymatic steps are required for the generation of 5S rRNA from the transcripts of the *E. coli rrn* operons (Fig. 11.28). The first of these is the removal of the precursors to 23S rRNA from the growing nascent rRNA precursor chain. This is catalysed by RNase III as described above.

The second *E. coli* enzyme, RNase E has been purified as a $66\,000\,M_r$ protein but perhaps requires additional factors [257]. It generates a precursor of 5S RNA known as p5S which, compared with the mature molecule, has three extra nucleotides at both the 5′- and 3′-termini [258]. In a temperature sensitive, RNase E mutant, a 9S RNA accumulates which the added enzyme will convert into p5S [259]. Singh and Apirion [260] have analysed this molecule from the *rrnB* operon (which does not have a trailer tRNA gene after the 5S gene). It comprises the 5S rRNA with 85 additional 5′-nucleotides that extend to the RNase III cleavage site near the 3′-end of 23S RNA. It also has extra 3′-nucleotides that extend to the terminator of the operon. The precursor would contain tRNA transcripts were it from those operons containing distal tRNA genes and if it was not processed by RNase P (section 11.7). The catalytic site for RNase E occurs in an extension of the double-stranded stem, which is formed by the base pairing of the 5′- and 3′-ends of the 5S RNA sequence and is known as a molecular stalk [261]. Processing apparently occurs on the ribosome after the binding of the three 5S-associated ribosomal proteins (section 12.6.1) [262].

The third processing step in *E. coli* is catalysed by uncharacterized enzyme activities that complete the processing of 5S RNA by trimming the residual 5′- and 3′-tails. Conversely, in *B. subtilis*, a single enzyme, RNase M5, catalyses an endonucleolytic cleavage of pre-5S RNA to yield both the 3′- and 5′-termini of mature 5S RNA [263]. One of the two subunits of RNase M5 is

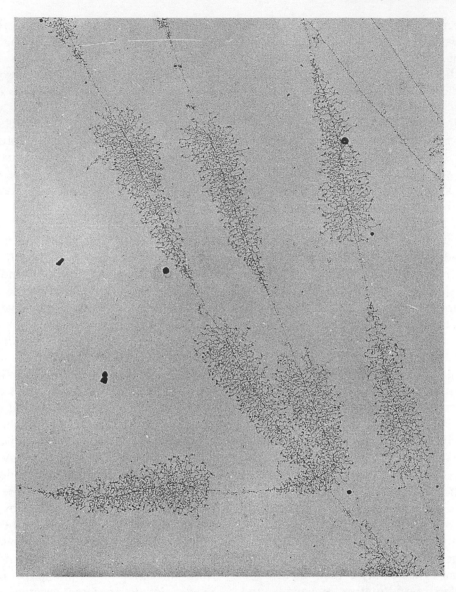

Fig. 11.29 An electron micrograph of actively transcribed RNA genes of the amphibian *Triturus*. The gradient of fibrils radiating from each gene are transcripts of pre-rRNA in successive stages of completion. They are the product of approximately 100 molecules of RNA polymerase I that are simultaneously transcribing each gene. The DNA fibre between each gene is the intergenic spacer. Reproduced with permission from Miller and Beatty [265], copyright 1969, AAAS.

the *B. subtilis* ribosomal protein BL16 (equivalent to *E. coli* ribosomal protein L18 – section 12.6.1). It is thought that the ribosomal protein binds to the pre-5S RNA and that this ensures that it adopts the correct conformation for the binding of the catalytic, 'alpha' subunit of the enzyme [264]. Processing then occurs with the simul-

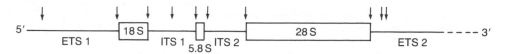

Fig. 11.30 The processing of vertebrate 47S pre-rRNA. The arrows indicate identified cleavage points.

taneous cleavage of both strands of the base-paired stem of the precursor [263].

(c) *The processing of tRNA in the pre-rRNA transcript*

Spacer and trailer tRNA sequences in the transcripts of *rrn* operons are removed and processed by a series of RNases as described in section 11.7.

11.6.2 The processing of eukaryotic pre-rRNA

(a) *The processing of eukaryotic 18S, 5.8S and 28S transcripts*

The transcription of rRNA, together with most processing and assembly into ribosomal particles, occurs in the nucleolus (the transcribing enzyme, RNA polymerase I, its structure, initiation and termination have been considered in chapter 9). Electron micrographs of oocyte rDNA engaged in rRNA transcription reveal the rDNA fibre forming the 'trunk' of a 'Christmas-tree-like' structure of which the 'bole of the trunk' is the non-transcribed spacer which separates the rRNA gene units [265]. The 'branches' on the transcription unit are represented by about 100 nascent transcripts which increase in length along the length of the rDNA fibre axis and have at their base a transcribing molecule of RNA polymerase (Fig. 11.29). The longest branches are complete transcripts of the transcription unit but they are only approximately one-twelfth its length

because they are already partially wrapped into ribonucleoproteins.

RNA polymerase I produces a single long pre-rRNA molecule that contains the 18S, 5.8S and 28S rRNAs joined to each other by transcripts of the transcribed spacers (Fig. 11.30). Where, as in yeast, the rDNA repeat unit includes the gene for 5S RNA, this is separately transcribed from the opposite DNA strand by RNA polymerase III ([266] and references therein). The initial transcript is subject to multiple ribonucleolytic cleavages which produces the mature ribosomal species via a succession of intermediate pre-rRNA species (reviewed in [267]). The more stable of these precursors have been identified on sucrose density gradients, in polyacrylamide gels, by hybridization and by sequencing. Historically, like the mature rRNAs, they have been named according to their sedimentation constants but as the precise lengths and cleavage points are identified, this notation is becoming increasingly irrelevant. Fig. 11.30 indicates the identified cleavage sites in the maturation of the mammalian transcript and Fig. 11.31 presents a generalized scheme for the detected processing intermediates in mammals.

As described in section 9.5.3, it has become apparent in recent years that the so-called non-transcribed spacer separating the transcription units in the clustered genes are at least partially transcribed. In *Xenopus laevis*, RNA polymerase I transcribes across most of the intergenic spacer and terminates at a region known as T3 just 215 bp upstream of the next initiation site. However, the

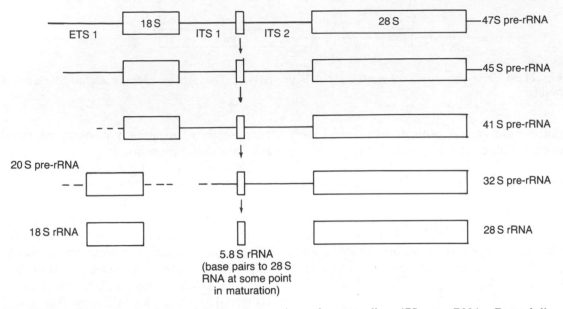

Fig. 11.31 Generalized scheme for the maturation of mammalian 47S pre-rRNA. Dotted lines indicate uncertainty in the cleavage points and in the sequence of maturation events.

3′-end of the largest relatively stable rRNA precursor (7860 nucleotides) corresponds to a processing site T2. The transcript between T3 and T2 is very unstable and is only detected in nuclear run-off experiments and by electron microscopy [416, 417]. Both T2 and T3 contain a conserved seven-nucleotide sequence GACTTGC and both are required for processing to T2 [418]. Mouse RNA polymerase terminates 565 nucleotides downstream of the 3′-end of the 28S RNA gene and the transcript is then trimmed by 10 nucleotides to produce the 3′-end of the first relatively stable precursor [419]. This first observable, 14 000 nucleotide short-lived species, known in mammals as 47S pre-rRNA [268], is processed to the first relatively stable and easily detected intermediate called 45S pre-rRNA. In mouse, this processing step results from the cleavage of the transcript approximately 650 nucleotides from its 5′-end and 4 kb upstream of the 18S sequence. Analogous cleavages occur at

residue *c.* 414 in human and at position 790 in rat 47S pre-rRNA [269, 270]. This step in the maturation pathway is the best understood because it can be reproduced *in vitro* [269, 272] and it has been shown to occur in a large ribonucleoprotein complex [420]. Three nucleolar snRNP particles containing the snRNAs, U3, U8 and U13, appear to be involved in pre-rRNA processing in mammalian cells [273] and the most abundant of these, U3 snRNP, is specifically involved in the first maturation step [274]. It interacts with and can be cross-linked to a region just downstream of the cleavage site. Thus, it interacts with nucleotides 438–695 of human pre-rRNA where cleavage is at position *c.* 414 [275]. Similarly, it cross-links to nucleotides in the region 767–1149 of the rat transcript where cleavage is at nucleotide 790 [271]. The role of U3 snRNP in the cleavage is so far unknown but the particle contains at least six proteins of which one, fibrillarin, gives rise to an autoimmune response in certain scleroderma patients [276].

A 51-kDa endonuclease, that has been purified from the nucleoli of ascites tumour cells, has also been implicated in the initial processing of 47S pre-rRNA [277] and has also recently been shown to catalyse the generation *in vitro* of the 3′-terminus of 18S rRNA. The purified enzyme cleaved pre–18S RNA at three specific sites; the site of the mature 3′-terminus, a second site 55 nucleotides downstream and a third site 35 nucleotides upstream [278]. The first two sites correspond with rat pre-rRNA cleavage sites identified in hamster cells after they had been transfected with mouse rDNA [279].

Other processing sites are known with less precision. As indicated in Fig. 11.30, Bowman *et al.* [280] suggested that the 5′-end of 18S RNA was generated by a single endonucleolytic cleavage of 45S pre-rRNA. However, in a processing system *in vitro*, the maturation of 18S RNA appeared to involve an endonucleolytic cleavage followed by exonucleolytic trimming [281]. Similarly, the generation of the 5′-end of 5.8S rRNA occurs by cleavage at both the mature terminus and at a second position 5–7 nucleotides upstream [280]. The 5′-terminus of 28S rRNA may also be formed by an initial cleavage within the ITS 2 followed by trimming to generate the mature terminus [280]. The removal of the ITS 2 transcript from 32S pre-rRNA to generate 28S and 5.8S rRNAs is not followed by their separation. The two species remain hydrogen-bonded.

The most studied lower eukaryote is the yeast, *Saccharomyces cerevisiae*, in which a number of mutants have assisted in the identification of the processing pathway (reviewed in [266]). RNA polymerase I transcribes across the intergenic region to a terminator T2 positioned at +210 with respect to the 3′-terminus of the mature 26S rRNA. This terminus is rapidly degraded

to a processing site T1 [421]. The endonucleolytic cleavage of a further seven nucleotides then creates the 3′-end of the first easily detectable 37S pre-RNA [282]. The 20S pre-rRNA intermediate has the 5′-end of mature 18S rRNA but an extra 207 nucleotides at the 3′-end derived from ITS 1 [283]. Its further processing to 18S rRNA takes place in the cytoplasm [284]. 5.8S rRNA may also be generated via a 7S intermediate.

At least nine snRNPs (snR3, 4, 5, 8, 9, 10, 17, 128 and 190) have been reported to be localized in the yeast nucleolus and one of these snR17 is apparently the analogue of snRNP U3 of higher eukaryotes [285, 286]. It is essential for cell growth; as are some of the others but their role in processing is unknown.

(b) *The processing of eukaryotic 5S transcripts*

Xing and Worcell [287] have identified a 3′-exonuclease that processes the 130 and 142 base precursors of the major *Xenopus* oocyte 5S RNA to the 120 nucleotide mature molecule. The same nuclease completely degrades transcripts of the pseudogene (section 8.4.4) but is blocked from doing the same to the 5S transcript by the complementarity between the 5′- and 3′-ends of the molecule.

11.6.3 Post-transcriptional base modification of rRNA

The analyses of the late 1950s established that rRNA contains methylated nucleotides [288] and it has since become apparent that it also contains pseudouridine [289] and, in eukaryotes, both N^4-acetylcytosine [290] and the hypermodified nucleotide 3-(3-amino-carboxypropyl)-1-methyl pseudouridine [291]. Some modified bases are strongly conserved and presumably have important functions. Thus, as is described in section 12.6.2, a conserved sequence, $m_2^6Am_2^6A$, found at the 3'-ends of both prokaryotic and eukaryotic small subunit rRNAs seems to have a role in the decoding phase of protein synthesis [292]. Conversely, some modifications are not shared between prokaryotes and eukaryotes. 5-methyl-cytosine is found only in prokaryotic rRNA [290], whereas the much larger methylated nucleotide content of eukaryotic rRNA is largely attributable to nucleotides that carry 2'-*O*-methyl groups on the ribose. Of 70 methyl groups on HeLa cell 28S rRNA, 65 are 2'-*O*-ribose methylations and the remainder are on bases. Similarly, there are 40 2'-*O*-methyl riboses on 18S rRNA and six base methylations [293]. These methylations are strongly conserved between eukaryotic species [294] and their overall number appears to show a one to one relationship with the number of pseudouridine residues [295]. All of the 2'-*O*-methylations occur on the rRNA encoding region of 45S rRNA [293] but pseudouridine occurs on conserved and non-conserved sequences [289]. Both of these modifications occur in the nucleus but base methylation occurs in the cytoplasm [293].

Analysis of the 18S rRNA of *Xenopus laevis* and man by Maden and coworkers established the location of all of the modified nucleotides both within the primary sequence [296, 297] and, for methyl groups, within the consensus secondary structure of the molecule [298]. In general, they occur in non-helical regions, at hair-pin loop ends or at helix boundaries and imperfections. The base methylations are all in the 3'-domain of the molecule and one cluster of 2'-*O*-ribose methylation occurs in an area of complicated secondary structure in the 5'-third of the molecule.

11.6.4 Precursor ribonucleoproteins in the formation of ribosomes

In prokaryotes, many of the ribosomal proteins attach sequentially to the nascent rRNA during its transcription and each of the pre-rRNA species occurs as a ribonucleoprotein particle [300]. King *et al.* [256] have reviewed the growing evidence that the association with proteins is essential to at least some of the processing steps.

The assembly of new ribosomes in eukaryotic cells occurs predominantly in the nucleolus and, as in prokaryotes, the precursor rRNA species do not occur as free RNA but as ribonucleoprotein particles. Thus, the mammalian 45S pre-rRNA transcript and the 32S processing intermediate (Fig. 11.28) occur as 80S and 55S particles, respectively (reviewed in [301]). A 40S nucleolar precursor to the 40S ribosomal subunit can also be isolated [302]. Similar maturation pathways have been followed in other organisms. The 37S initial transcript of yeast occurs in a 90S pre-ribosome which matures to 40S and 60S mature ribosomal subunits via 66S and 43S precursor particles [303].

In lower eukaryotes, as in bacteria, there are only one or two copies of each ribosomal protein gene [304–306] but in mammals there are multiple copy genes. Those of mouse vary from 7 to 20 per protein species and they are scattered over more than one chromosome [307, 308]. The amphibian, *Xenopus*, appears to have an intermediate repetitiveness with 2–5 gene copies per protein [309].

A number of groups have demonstrated the sequential addition of ribosomal proteins during the formation of ribosomes. Todorov *et al.* [302] showed that the mammalian 80S pre-ribosome contained most of the small subunit proteins. Only two, S3 and S21, are added during the formation of the 40S nucleolar precursor and four others, S2, S19, S26 and S29, are added later; presumably in the cytoplasm. Similar sequential addition of the large subunit proteins has also been observed [310–312], and recent evidence suggests that ubiquitin may be involved in either chaperoning certain ribosomal proteins to the site of ribosome biogenesis or in the assembly process *per se* (section 12.10.2) [299]. The nucleolar pre-ribosomal particles contain some proteins that are not present in the mature ribosome [310, 311].

11.7 THE PROCESSING OF PRE-tRNA

11.7.1 The excision from precursors: 5'- and 3'-trimming

(a) *Prokaryotic pre-tRNA processing*

As has been seen in previous sections, prokaryotic pre-tRNA can be a transcript of a single gene or of gene clusters that may also contain mRNA or rRNA sequences (section 9.4.1). The maturation process has been the subject of a number of reviews [313–315].

A key enzyme in the maturation process is RNase P. This is a complex ribonucleoprotein consisting of a polypeptide of approximately $18\,000\,M_r$ and a RNA of 375–377 nucleotides known as M1 RNA [316, 317]. The M1 RNA is the catalytic portion of the enzyme and has already been discussed as an example of catalysis by RNA in section 11.3.4. Endonucleolytic catalysis by RNase P generates the mature 5'-terminus of the tRNA (Fig. 11.28). In monomeric precursors, this might be the primary cleavage. Thus, it specifically removes the 41 nucleotide sequence upstream of the tRNA gene of the *E. coli* tRNA^{Tyr} transcript [318]. Multimeric precursors, however, may first

be cleaved further to the 5'-side of the gene by other ribonucleases such as RNase III (section 11.6.1).

The cleavage events that generate the 3'-end of the mature prokaryotic tRNA appear to be less precise and are not well understood. An initial endonucleolytic cleavage of precursors generates a 3'-end that still contains a few extra nucleotides. These are then removed by exonucleases (Fig. 11.28). Ribonuclease F [319] is a possible contender for the endonucleolytic function although some of the properties of this enzyme are those associated with degradative activities [319, 320]. Of the possible exoribonuclease activities, Deutscher [315] has reviewed the evidence implicating *E. coli* ribonucleases Q, Y, D and BN, and to these possibilities must be added the more recently characterized RNase T [321]. Although these enzymes can all with varying efficiencies generate the correct 3'-terminus of tRNA, mutant strains devoid of their activity can still accurately trim their tRNA. Even a multiple mutant deficient in RNases II, BN, T and D is still able to synthesize mature tRNA [322]. More recently, a further candidate enzyme, RNase PH has been isolated that favours tRNA as a substrate, trims it accurately and, unusually, requires phosphate for exonucleolytic activity [323]. It remains to be seen whether its credentials as a candidate for the true 3'-trimming enzyme will survive further experimentation.

(b) *Eukaryotic pre-tRNA processing*

Although they are all monocistronic, the primary transcripts of tRNA genes in eukaryotic cells also have extra sequences at their 5'- and 3'-ends. In some instances they also contain insertion sequences that are spliced out during processing (section 11.3.5). The lack of mutants in all but the lowest eukaryotes has slowed an under-

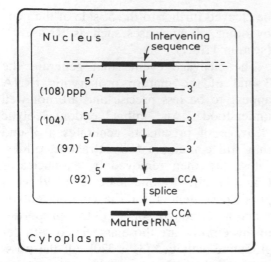

Fig. 11.32 A schematic diagram of the processing steps in the maturation of a 108 bp precursor to yeast tRNA[Tyr]. The numbers in brackets indicate the number of nucleotides in each of the intermediates. The 5'-leader is removed in three stages, the last of which is accompanied by excision of the 3'-trailer and the addition of the CCA 3'-terminus. Nucleotide modifications occur between several of the maturation steps [324].

standing of the trimming reactions involved in processing but this has been partially offset by the use of cloned genes in *in vitro* processing systems. Such a system is the nucleus of *Xenopus* oocytes (sections A.9.3 and A.11) and Melton *et al.* [324] first made use of this to study the expression of a yeast tRNA gene microinjected directly into the nucleus. They were able to detect mature tRNA, its precursors and the subcellular location of processing (Fig. 11.32). Subsequent studies have employed a wide range of pre-tRNA species and have investigated the processing activities in cells derived from species as diverse as *Drosophila* and man (e.g. [325] and references therein). It has also been shown that splicing does not necessarily follow trimming but can precede it [422].

The excision of the 5'-leader sequence and the generation of the mature 5'-terminus is catalysed by RNase P. This ribonucleoprotein has been best characterized in *E. coli* and other eubacteria where it has been shown that the RNA component is the catalytic entity (section 11.3.4). In contrast to the bacterial enzymes, the RNA subunits of eukaryotic nuclear and mitochondrial RNase P enzymes are not catalytic in assays so far devised *in vitro* [326–329] but Forster and Altman [330] have shown that they can be folded into cage-shaped structures very similar to a large conserved domain of bacterial M1 RNA. They suggest that the RNA component of eukaryotic enzymes conserves the hydrolytic capacity but lacks a substrate-binding component that is presumably provided by the protein portion of the enzyme[330]. The RNase P activities of chloroplasts and the archaebacterium, *Sulfolobus solfataricus*, are not affected by treatment with micrococcal nuclease, suggesting that there is no RNA component or that it is protected from nuclease action [331, 332].

At the 3'-end of tRNA-containing transcripts, the processing reactions of eukaryotic cells must precisely remove the extra sequence before the addition of the CCA terminus (section 11.7.1(c)). Activities detected in *Xenopus* oocyte nuclei cleave the entire 3'-extension of a *Bombyx mori* pre-tRNA[Ala] as a single 22 nucleotide fragment [333, 334] and a partially purified human activity cleaves the 8–12 nucleotide trailer of a pre-tRNA[Met] [335]. The latter enzyme required a mature 5'-end to the tRNA precursor suggesting that processing followed an obligatory order of 5'-maturation followed by 3'-maturation. A similar processing order has been described in other systems, including mouse tRNA[Gly] [325] but mouse tRNA[His] was shown to be processed

Table 11.2 Base modification in tRNA

Modified base	tRNA nucleotide modified*	Enzyme	Ref.
7-Methylguanosine	46	m^7G, tRNA methyltransferase	[344]
2-Methylguanosine	6, 26	m^2G, tRNA methyltransferase	[345]
1-Methyladenine	1, 4, 22, 58	m^1A, tRNA methyltransferase	[346]
5-Methylcytidine	Various	m^5C, tRNA methyltransferase	[347]
Ribosylthymine	54	rT, tRNA methyltransferase	[348]
5-Methylaminomethyl-2-thiouridine	34	mnm^5s^2U tRNA methyltransferase	[349]
Pseudouridine	Various	Pseudouridine synthase	[351]
4-Thiouridine	8	s^4U tRNA sulphurtransferase	[352]
Queuosine	34	tRNA guanine ribosyltransferase	} Section 4.8
Inosine	34	tRNA hypoxanthine ribosyltransferase	

* The numbering of the modified nucleotides is based on the numbering system derived from yeast phenylalanine tRNA. It indicates the usual location of the modified nucleotide and does not imply that the position is so modified in all tRNAs.

in the reverse order with 3′-processing preceding that of the 5′-terminus [325].

The processing of chloroplast pre-tRNA resembles that of eukaryotic nuclear processing rather than that of prokaryotes. Thus, the 3′-terminus is generated by a single endonucleolytic cleavage at the 3′-end of the encoded tRNA and the CCA terminus is added post-transcriptionally [331].

(c) Synthesis of the CCA 3′-terminus

In prokaryotes, the 3′-terminal trinucleotide sequence CCA, that forms the amino acceptor stem of the mature molecule, is usually though not invariably encoded within the gene [336, 337]. Conversely, no known eukaryotic genes encode the CCA terminus, and this includes those of mitochondria and chloroplasts. The terminus is added post-transcriptionally by tRNA nucleotidyltransferase EC 2.7.7.25 [338], a highly specific enzyme which, in the presence of ATP, CTP and tRNA lacking the 3′-terminal sequence, will catalyse the reaction:

$$tRNA + ATP + 2\,CTP \rightarrow$$
$$tRNA\text{-}C\text{-}C\text{-}A + 3\,PP_i$$

The enzyme has been identified in a wide variety of organisms, subcellular particles, bacteria and viruses [338]. It is present in *E. coli* and required for its normal growth despite the fact that all known tRNA genes in this organism encode the CCA terminus. The explanation for this is that tRNA nucleotidyltransferase is also required for the repair of damaged tRNA molecules which arise by end turnover, mainly of the 3′-AMP residue, when tRNA molecules are not fully charged [339].

11.7.2 Post-transcriptional base modification of tRNA

More than 50 different modified nucleotides have been found in tRNA [340, 341] and all of these arise by post-transcriptional base modification or base transposition. In many cases modification occurs by a single enzymic catalysis such as a methylation.

Others, such as the formation of wybutosine and 2-methylthio-6-isopentenyladenine, require a sequence of enzymic steps [342]. Much remains to be learned both of the enzyme activities catalysing the modification and of the function of the modified nucleotides. Both formation [342, 343] and function [341] have, however, been the subject of reviews and Table 11.2 lists some of the commonest modifications and the enzymes involved in their formation. It should be emphasized, however, that the listed enzymes represent activities that may be shared by multiple enzymes. Thus, Samuelsson *et al.*, used transcription *in vitro* by T7 RNA polymerase to prepare a completely unmodified tRNA substrate [350] and have employed this to show that pseudouridines at positions 13, 32 and 55 in the tRNA are each synthesized by different enzymes [351]. Similarly, different enzymes catalyse the synthesis of the same modified nucleotide in tRNA and rRNA. One unusual form of modification, by which inosine and queuosine are inserted into the 'wobble' position of the anticodon of a number of tRNA species, is a direct base replacement. This is discussed in chapter 4 (section 4.8).

The base-pairing patterns of modified nucleotides in the 'wobble' position of tRNAs are considered in section 12.2.2, and the role of queuosine in suppression is discussed in section 12.9.6.

11.8 RNA EDITING

A pillar of molecular biology is the so-called 'central dogma'; namely that DNA encodes and is copied into RNA that in turn encodes and is the template for the synthesis of protein. The dogma has survived considerable erosion of its foundations, including the discovery that part of the process can be driven backwards by reverse transcriptase,

the finding that the coding sequence of DNA could be interrupted by introns and the concept that non-coding regions of mRNA could be altered by capping and polyadenylation. Recently, however, a major implication of the dogma, that all the coding nucleotides of a mRNA are encoded in the DNA template, has been seriously challenged by the discovery of RNA editing.

Editing can be seen as the counterpart of splicing. Although the latter removes non-coding regions of transcripts, editing alters the sequence and sense of coding regions (reviewed in [353–355, 423]).

11.8.1 Uridine insertion and deletion

The first example of RNA editing was discovered by Benne *et al.* [356] in the mitochondria of the kinetoplastid flagellate, *Trypanosoma brucei*. The mitochondrial DNA of this and related organisms is unique in comprising 25–50 maxicircles and 1000–10 000 minicircles. Maxicircles are equivalent to conventional mitochondrial DNA in that they encode mitochondrial RNA and some proteins but the 22 kb circles are catenated into a network. Minicircles are only 1 kb in length and until recently had no known function. The discovery of RNA editing came from the realization that the nucleotide sequences of a number of mRNAs from trypanosome species differed from that of the maxicircle genes from which they were transcribed [356–358]. It was initially considered likely that the genes themselves were edited before transcription but all attempts to find evidence in support of this failed (reviewed in [353]). Instead it was found that the incomplete genes, now known as cryptogenes [359], were transcribed into unedited pre-mRNA and that editing gave rise to mature message [356].

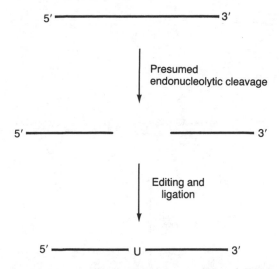

Fig. 11.33 A diagrammatic overview of RNA editing (uridine insertion).

Initially, the observed examples of editing involved the insertion of a few extra uridines into the unedited transcript (Fig. 11.33). Thus, a total of four uridines were inserted into one region of the mRNA encoding cytochrome oxidase subunit II and it appeared that the process might involve a relatively simple sequence recognition event by an editing enzyme. Such simplistic ideas were soon dispelled by the discovery that over 55% of the nucleotide sequence of cytochrome oxidase subunit III mRNA was created by editing (Fig. 11.34).

The discovery, by Blum *et al.* [360], of a class of RNAs known as guide RNA (gRNA) has led to a more plausible model for RNA editing. These small RNAs are encoded in both the maxicircle and minicircle DNA [360, 361] and their 5'-ends exhibit extensive sequence complementarity

with the region that is 3' to the editing sites of cryptogene transcripts (provided that U:G base pairing, already known from RNA secondary structure, is permitted). In the guide RNA model (Fig. 11.35), hybridization of these complementary regions forms what has been called a 3'-anchor. At up to 16 base pairs in length, this anchor is rather short but may be stabilized by a 5'-anchor involving the recently discovered non-encoded 3'-oligo(U) tails of the gRNAs [362].

Two versions of the guide model for the deletion and insertion of uridines are illustrated in Fig. 11.35. After the base pairing between the guide RNA and the cryptogene transcript, the earlier model envisages a series of enzyme activities – an endonuclease, a 3'-terminal uridylate transferase and an RNA ligase – catalyse the removal or addition of uridines as specified by the sequence of the gRNA. The guide is not acting as a conventional template because a G residue in the guide specifies the addition of U rather than C. Nevertheless, editing is still occurring by the formation of complementary base pairs. It has been proposed that the three enzymes act coordinately as part of an 'editosome' that moves in a 3' → 5' direction along the base-paired RNAs, editing as it goes. It should be emphasized, however, that, although all of the above enzyme activities have been detected in mitochondria, they have not yet been specifically associated with editing. Furthermore, it has not yet been demonstrated unambiguously that the gRNAs participate in editing. There are in any case good reasons to believe that the model is

AuGuAuuuGuGuGuGuAAuuuuAuuGGuGuuuuUUUAGUUGuuGAuuA GuuAAuuuGu

Fig. 11.34 Part of the mature mRNA sequence of *Trypanosoma brucei* cytochrome oxidase subunit III. Upper case letters identify genomically encoded nucleotides. Lower case letters indicate uridines inserted by editing and the arrow indicates the position of a uridine removed by editing.

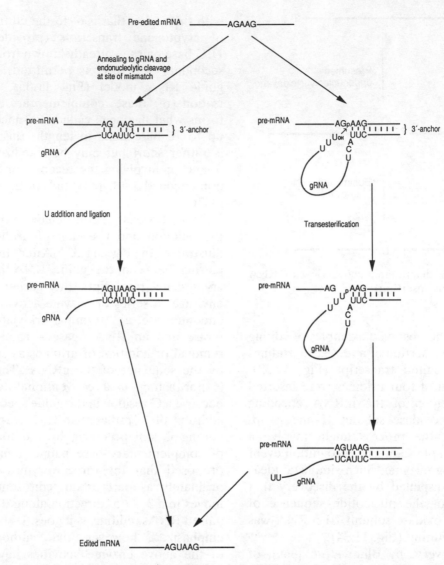

Fig. 11.35 Two models for the role of guide RNAs in the insertion of a single uridine into pre-mRNA. In the left hand series of reactions editing depends on the nucleolytic cleavage of the precursor, followed by addition of uridine and ligation. The right-hand series of reactions depicts a more recent alternative model in which the uridine at the 3'-terminus of the gRNA is inserted into the precursor by transesterification. Phosphates (p) are shown only for those phosphodiester bonds involved in transesterification.

simplistic and will require revision. Partially edited RNAs have been detected that have correct and incorrect additions and deletions in the editing domain [363]. This suggests either that correct editing is a multistage process or that mis-editing can occur; perhaps as a result of using the wrong gRNA.

A very recent alternative model of editing suggests that the U residues are inserted from the 3'-oligo(U) tail of the gRNA [424]. Base pairing between the gRNA and the region downstream of the editing site is still envisaged (Fig. 11.35) but the 3'-end of the guide is seen as looping round so that its terminal uridine can be inserted into the pre-mRNA by transesterification. The idea is attractive as the insertion or deletion of a uridine could then be analogous to the insertion or removal of an intron, and the process could be catalysed by a single enzyme activity, mitochondrial uridylyl-transferase, rather than requiring the multienzyme complex of earlier models. However, the verification of any model must await the development of systems that perform editing *in vitro*.

How such an apparently cumbersome and wasteful process as editing has evolved has been the subject of much conjecture [353] but is far from understood.

11.8.2 Other forms of mRNA editing

A number of other modifications that alter the coding potential of mRNA have been described in various organisms.

In the slime mould, *Physarum poly-cephalum*, the α-subunit of mitochondrial ATP synthetase contains multiple frameshift deletions that are repaired by the addition of single cytidines into the primary transcript to create an accurate reading frame [364]. It is not known whether guide RNAs are involved.

Site-specific conversions of C to U residues have been described in mammalian apolipoprotein B mRNA [365, 366] and various plant mRNAs [367–369]. In the former, the modification creates a stop codon whereas in the latter it changes the nature of the amino acid that is inserted into the protein. The mechanism of the modification is unknown but there is no reason to suppose that it is not a simple enzymic deamination as in similar tRNA base modifications.

The addition of guanosine residues occurs in specific internal sites in paramyxoviral mRNA. The mechanism of this is uncertain but some evidence suggests that it involves RNA polymerase 'stuttering' when transcribing a stretch of template homopolymer. Thus, the modification may be transcriptional rather than post-transcriptional [370, 371].

11.9 CONTROL OF THE PROCESSING OF RNA

11.9.1 Control of the splicing and polyadenylation of mRNA

The splicing and polyadenylation of pre-mRNA depends on the precise identification of splice sites and polyadenylation sites, respectively. Many cases are known, however, where a single pre-mRNA transcript is processed into several different mRNAs of differing function. Such differential processing may be cell-, tissue- or development-specific. It is known for instance to control the production of foetal or adult forms of myosin [372], the synthesis of membrane bound and secreted forms of IgM heavy chain [373–375], the cellular and plasma forms of fibronectin [376], prohormones that include or exclude various neuro-peptides [377, 378], sex determination in *Drosophila* [379, 380] and multiple isoforms of more that 50 different known proteins. Despite considerable investigation much remains to be learned of the mechanisms whereby this differential processing is controlled.

Smith *et al.* [381], in a recent review of alternative splicing in the control of gene expression, has classified the variant forms of processing into (a) spliced or retained introns, (b) alternative donor and acceptor splice sites, (c) alternative promoters and polyadenylation sites, (d) mutually exclusive exons and (e) cassette exons. It is not clear that these variants employ different regulatory processes but it will be convenient to consider examples of each.

(a) *Spliced or retained introns*

An intron may be removed or retained and its retention may result in the addition of extra amino acids to a protein or may totally disrupt protein synthesis by, for instance, inserting a premature stop codon. A well-studied example is the *Drosophila* P element the transposition of which is dependent on a transposase that is only expressed in germ-line cells (section 7.7.3) [382]. In somatic cells, the third intron of the P element pre-mRNA is retained. The resulting 66 kDa protein lacks transposase activity and functions as a negative regulator of transposition. Conversely, in germ-line tissues, the third intron is spliced normally, resulting in a 87 kDa transposase protein. Siebel and Rio [383] have described a 97 kDa protein that specifically interacts with the 5′-regions of the third intron and inhibit its splicing.

(b) *Alternative 5′-donor and 3′-acceptor splice sites*

The transcript of the SV40 early gene is a model of the use of alternative splice sites. It can be processed to produce the mRNA encoding large T antigen, or that of small t antigen by the utilization of alternative 5′-splice sites but a common 3′-splice site. Thus, what is part of an intron in the splicing of small t antigen mRNA becomes part of an exon during processing to produce large T antigen (Fig. 11.36). The polypyrimidine element of the 3′-splice site is involved in the selection [384]. Mutations that increase

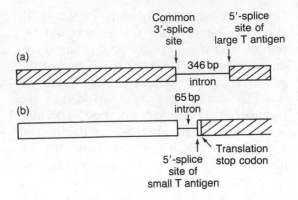

Fig. 11.36 The splicing of the pre-mRNA for (a) large T antigen and (b) small t antigen. Boxes are exons and the intervening lines are introns. Shaded boxes represent portions of the mRNAs that are translated into protein whereas open boxes represent portions that are left untranslated after termination at the indicated stop codon.

the pyrimidine content result in an increased production of large T antigen whereas those that increase purine content have enhanced amounts of small t [384]. *Trans*-acting factors have also been implicated in the differential splicing [385].

(c) *Alternative promoters and polyadenylation/cleavage sites*

The use of alternative promoters may dictate a variant splicing pattern. This is well demonstrated in the alternative processing of chicken myosin light chain [386] in which whether the mRNA 5′-end comprises exons 1 and 4 or exons 2 and 3 depends on which of two promoters is used during transcription (Fig. 11.37).

Many examples are known of the use of variant 3′-exons coupled with alternative polyadenylation sites. However, it is not always clear if the alternative cleavage/polyadenylation dictates the splicing pattern or whether the splicing pattern is the critical

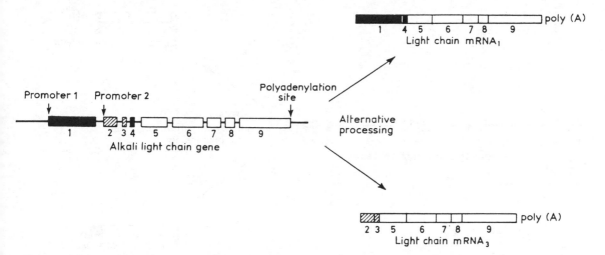

Fig. 11.37 A schematic representation of the alternative processing of myosin alkali light chain gene transcripts. Boxes 1–9 represent exons and the intervening lines are introns. Shaded exons are unique to light chain mRNA$_1$ which is transcribed from promoter 1. Hatched exons are unique to light chain mRNA$_3$ which is transcribed from promoter 2.

event controlling the 3′-processing. One example where it is established that the polyadenylation site rather that splice site selection governs the process, is the regulated production of IgM heavy chain (Fig. 11.38). Alternative processing during β lymphocyte differentiation produces the mRNAs encoding either the membrane-bound μ_m protein or the secreted μ_s protein [373–375]. Altered forms of the transcription unit have been employed to demonstrate that spacing between the two polyadenylation sites affects their usage independently of the presence or absence of splice sites [387].

Calcitonin and calcitonin gene-related peptide (CGRP) are encoded in a single gene of six exons. Calcitonin is the primary mRNA species observed in thyroid C cells and is produced by splicing together the first four exons combined with polyadenylation at the 3′-end of the fourth exon. CGRP mRNA is the major species observed in specific neurons of the central and peripheral nervous system. It results from the splicing of exons 1, 2, 3, 5 and 6 and the use of an alternative polyadenylation site at the 3′-end of exon 6 [388]. The possibility that control of the process was transcriptional was excluded when it was demonstrated that, in both cases, transcription was terminated 1 kb downstream from the second polyadenylation site [389]. By the introduction of fusion genes, Rosenfeld and colleagues [390] showed that most cell types had the capability to process precursor to calcitonin mRNA whereas only the neuronal cells that normally do so, together with heart cells, could produce CGRP transcripts. They suggested that neuronal cells possess specific regulatory components able to repress the production of calcitonin mRNA with the result that the alternative splicing programme was employed by default. More recently, the group has identified a specific splice acceptor site near the calcitonin 3′-splice junction that is associated with the inhibition of calcitonin mRNA processing in

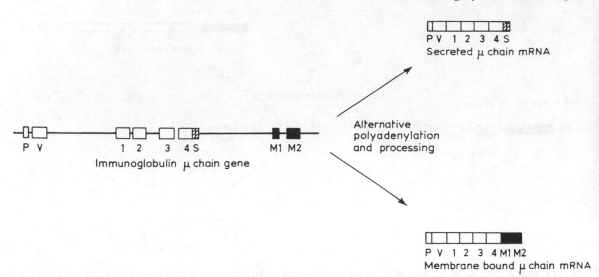

Fig. 11.38 A schematic representation of the alternative processing of transcripts of the immunoglobulin μ chain gene. Boxes represent exons and the intervening lines, introns. P refers to the signal peptide exon and V to the rearranged variable gene. Open boxes 1–4 are shared exons. The hatched boxes, S, encode the 3′-terminus of secreted μ chain mRNA and the shaded boxes M1 and M2 are unique to membrane-bound μ chain mRNA.

CGRP-producing cells [391]. Furthermore, they have cloned a snRNP-associated protein, Sm51, that is expressed in neuronal tissues and which may play a role in the differential processing [392]. The possible role that different specificity factors might play in the selection of alternative polyadenylation sites was discussed in section 11.5.1.

The differential use of two different 3′-splice sites is also employed in the alternative splicing of kininogen pre-mRNA and, in this case, repetitive sequences complementary to U1 snRNP are thought to govern the selection process [393].

(d) *Mutually exclusive exons*

In this variant of alternative splicing, one of a pair of exons is always spliced into an mRNA but never both.

The rat β-tropomyosin gene contains 11 exons and gives rise to two isoforms by differential splicing of the transcript. Exons 1–5, 8 and 9 are common to both mature mRNAs but exons 6 and 11 are only incorporated into the mRNA of fibroblast tropomyosin 1 whereas exons 7 and 10 are used in that of skeletal muscle β-tropomyosin. Unusually, far upstream lariat branch sites, as well as an associated polypyrimidine region, are implicated in the choice of exon 7 whereas the use of exon 6 is seen as the default pattern [394, 395]. There is also evidence of long-range interactions in the selection process as neither exon 6 or 7 can be spliced to exon 5 until the downstream splicing to exon 8 has occurred [396].

(e) *Cassette exons*

Cassette exons can be included or excluded from the mature mRNA independently of others and usually with the maintenance of the reading frame. Where multiple cassette exons occur in a transcript, various patterns of exclusion or inclusion can give rise to a high degree of mRNA diversity.

The gene encoding leukocyte-common antigen spans 130 kb and includes 33 exons. Various isoforms of the complete transcript are generated by the variable use of exons 4, 5 and 6. Extreme variants of the resultant proteins are the 180 kDa isoform that lacks all three cassette exons and a 220 kDa polypeptide that includes all three. At least six of the eight possible isoforms have been detected [397]. Streuli and Saito have detected three *cis*-elements in exon 4 that are required for differential splicing and have presented evidence that *trans*-acting factors govern cell-specific isoform generation [397].

The inclusion or exclusion of exon 5 in the processed mRNA of cardiac troponin T also depends on the recognition of elements within the exon [398] and *trans*-acting factors are also involved in the much more complex cell specific splicing of muscle troponin T [399].

(f) *Sex determination in* Drosophila *is controlled by a cascade of controlled splicing*

The above analysis of various alternative splicing strategies presents examples of the growing evidence that *cis*-sequence elements in the pre-mRNA and *trans*-acting protein factors control alternative processing. No mention has yet been made of the cascade of regulated splicing that controls sex determination in *Drosophila*. It is this system however, which employs several of the above variants and which can be subjected to genetic analysis, that perhaps provides the most promising system for a more complete understanding of the control elements and the way they work (reviewed in [379, 380]).

Sex determination in *Drosophila* depends on the ratio of X chromosomes to autosomes (the X:A ratio). Thus, flies with one X chromosome and two sets of autosomes are male whereas flies with equal numbers of X chromosomes and sets of autosomes are female. A one to one X:A ratio activates the so-called sex-lethal locus (*Sxl*) which both positively controls its own expression and controls three other groups of genes. These are the genes involved in (a) the sexual differentiation of somatic cells, (b) the sexual differentiation of germ-line cells and (c) dosage compensation – a mechanism that ensures, through hypertranscription, that the single X chromosome of males is transcribed into as much RNA as the two X chromosomes of females.

Through most of the life of the *Drosophila*, the *Sxl* gene produces female-specific or male-specific spliced transcripts. These have identical 5'-ends but differ in the presence or absence of a small male-specific exon that inserts an in-frame stop codon into the transcript [400] (Fig. 11.39). The protein encoded in the mature *Sxl* mRNA that is produced in the female includes an 80 amino acid RNA-binding domain. The same motif is also found in a number of other RNA-binding proteins, including those of snRNP and hnRNP, and it perhaps provides the key to how it controls both its own processing and that of further proteins in the regulatory cascade.

The next gene product involved in the control of the sexual differentiation of somatic cells is that of the *tra* gene. This also only functions in females because only in females and only under the influence of the *Sxl* gene product, is it spliced to produce a functional mRNA. In males default splicing causes the inclusion in the mRNA of exon 2 that again encodes an early, in-frame stop codon that terminates translation (Fig. 11.39). The *tra* gene product is of key importance in controlling sex determination in somatic cells because it controls the

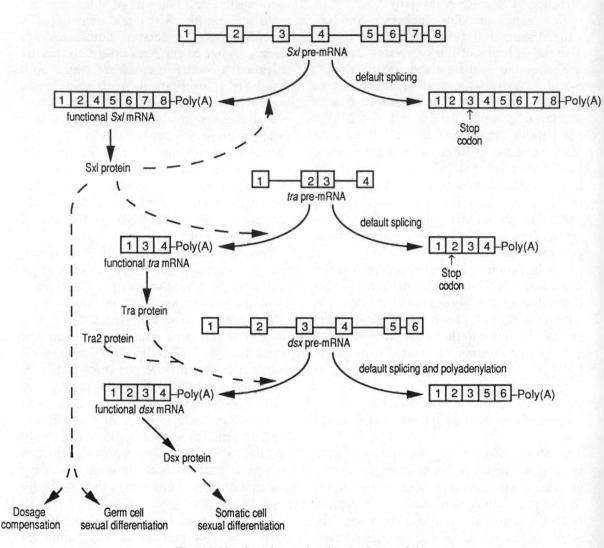

Fig. 11.39 Sex determination in *Drosophila*.

splicing and polyadenylation route of a third pre-mRNA – the product of the *dsx* gene. It does this with the assistance of the product of the *tra*-2 gene; another protein with the 80 amino acid RNA binding domain (Fig. 11.39) . The female *dsx* mRNA includes the first four exons of the six exon gene and its protein product represses male development. The default mRNA, produced in males without the influence of *tra* and *tra*-2 proteins, includes exons 1, 2, 3, 5 and 6, and uses an alternative polyadenylation site at the 3′-end of intron 6. Its protein product represses female development.

11.9.2 Control of the transport of RNA from the nucleus to the cytoplasm

The transport of RNA to the cytoplasm is little understood. It is not known for instance whether transcription and/or processing are coupled to transport although possible connections between splicing and transport have been reviewed by Chang and Sharp [425]. Almost nothing is known of the nature of the transport mechanism or what features of RNA molecules are recognized in their transport. However, Schroder *et al.* [401] have reviewed the quite considerable literature suggesting that both maturation and transport of mRNA occurs in association with the nuclear matrix, and de la Peña and Zasloff [402] have suggested that sequences within the promoter activate mRNA transport. The possible role of the mRNA 5′-cap in transport was discussed in section 11.4.

Even less is known about any possible control of RNA transport than is understood of transport itself. It is known, however, that considerably more RNA is transcribed than enters the cytoplasm.

REFERENCES

1 Hershey, A. D. (1953) *J. Gen. Physiol.*, **37**, 1.

2 Volkin, E. and Astrachan, L. (1956) *Virology*, **2**, 149.

3 Spiegelman, S., Hall, B. D. and Storck, R. (1961) *Proc. Natl. Acad. Sci. USA*, **47**, 1135.

4 Brenner, S., Jacob, F. and Meselson, M. (1961) *Nature (London)*, **190**, 576.

5 Gros, F., Hiatt, H., Gilbert, W., Kurland, C. G., Risebrough, R. W. and Watson, J. D. (1961) *Nature (London)*, **190**, 581.

6 Marmar, J., Greenspan, C. M., Policek, E., Kahan, F. M., Levine, J. and Mandel, M. (1963) *Cold Spring Harbor Symp. Quant. Biol.*, **28**, 191.

7 Tocchini-Valentini, G. P., Slodolsky, M., Aurisicchio, A., Sarnat, M., Fraziosi, F., Weiss, S. B. and Geiduschek, E. P. (1963) *Proc. Natl. Acad. Sci. USA*, **50**, 935.

8 Hayashi, M., Hayashi, M. N. and Spiegelman, S. (1963) *Science*, **140**, 1313.

9 Miller, O. L., Hamkalo, B. A. and Thomas, C. A. (1970) *Science*, **169**, 392.

10 Morse, D. E., Baker, R. F. and Yanofski, C. (1968) *Proc. Natl. Acad. Sci. USA*, **60**, 1428.

11 Baracchini, E. and Bremer, H. (1987) *Anal. Biochem.*, **167**, 245.

12 Nilsson, G., Belasco, J. G., Cohen, S. N. and von Gabain, A. (1984) *Nature (London)*, **312**, 75.

13 Melefors, O. and von Gabain, A. (1988) *Cell*, **52**, 893.

14 Brawerman, G. (1989) *Cell*, **57**, 9.

15 Gegenheimer, P. and Apirion, D. (1981) *Microbiol. Rev.*, **45**, 502.

16 Allman, S., Guerrier-Takada, C., Frankfort, H. M. and Robertson, H. D. (1982) in *Nucleases* (eds S. M. Linn and R. J. Roberts), Cold Spring Harbor Laboratory, New York, p. 243.

17 Dunn, J. J. and Studier, F. W. (1973) *Proc. Natl. Acad. Sci. USA*, **70**, 3296.

18 Dunn, J. J. and Studier, F. W. (1974) in *Processing of RNA* (ed. J. J. Dunn), *Brookhaven Symp. Biol.* vol. 26, Brookhaven National Laboratory, New York, p. 267.

19 Gitelman, D. R. and Apirion, D. (1980) *Biochem. Biophys. Res. Commun.*, **96**, 1063.

20 Singer, R. H. and Penman, S. (1973) *J. Mol. Biol.*, **78**, 321.

21 Spradling, A., Hui, J. and Penman, S. (1975) *Cell*, **4**, 131.

22 Rosenfeld, G., Comstock, J. P., Means, A. R. and O'Malley, B. W. (1972) *Biochem. Biophys. Res. Commun.*, **47**, 387.

23 Williamson, R., Morrison, M., Lanyon, G., Eason, R. and Paul, J. (1971) *Biochemistry*, **10**, 3014.

24 Suzuki, Y. and Brown, D. D. (1972) *J. Mol. Biol.*, **63**, 409.

25 Williamson, R., Clayton, R. and Truman, D. E. S. (1972) *Biochem. Biophys. Res. Commun.*, **46**, 1936.

26 Swan, D., Aviv, H. and Leder, P. (1972) *Proc. Natl. Acad. Sci. USA*, **69**, 1967.

27 Greenberg, J. R. and Perry, R. P. (1971) *J. Cell Biol.*, **50**, 774.

28 Wilks, A., Cato, A. C., Cozens, P. J., Mattaj, I. W. and Jost, J-P. (1981) *Gene* **16**, 249.

29 Federoff, N., Welleuer, P. K. and Wall, R. (1977) *Cell*, **10**, 597.

30 Broders, F., Zahraoui, A. and Scherrer, K. (1990) *Proc. Natl. Acad Sci. USA*, **87**, 503.

31 Riedel, N., Wise, J. Swerdlow, H. Mak, A. and Guthrie, C. (1986) *Proc. Natl. Acad. Sci. USA*, **83**, 8097.

32 Tollervey, D. (1987) *J. Mol. Biol.*, **196**, 355.

33 Birnstiel, M. L. (1988) *Structure and Function of Major and Minor Small Nuclear Ribonucleoprotein Particles*, Springer, Heidelberg.

34 Reddy, R. (1988) Nucleic Acids Res., **16**, r71.

35 Parry, H. D., Scherly, D. and Mattaj, I. W. (1989) *Trends Biochem. Sci.*, **14**, 15.

36 Reddy, R., Henning, D. and Busch, H. (1985) *J. Biol. Chem.*, **260**, 10930.

37 Kramer, A. (1987) *Proc. Natl. Acad. Sci. USA*, **85**, 8408.

38 Montzka, K. A. and Steitz, J. A. (1988) *Proc. Natl. Acad. Sci. USA*, **85**, 8885.

39 Busch, H., Reddy, R., Rothblum, L. and Choi, Y. C. (1982) *Annu. Rev. Biochem.*, **51**, 617.

40 Chu, F. K., Maley, G. F. and Belfort, M. (1984) *Proc. Natl. Acad. Sci. USA*, **81**, 3049.

41 Kaine, B. P., Gupta, R. and Woese, C. R. (1983) *Proc. Natl. Acad. Sci. USA*, **80**, 3309.

42 Tilghman, S. M., Curtis, P. J., Tiemeier, D. C., Leder, P. and Weissmann, C. (1978) *Proc. Natl. Acad. Sci. USA*, **75**, 1309.

43 Tsai, M., Ting, A., Nordstrom, J., Zimmer, W. and O'Malley, B. W. (1980) *Cell*, **22**, 219.

44 Herbert, M. G. and Wall, R. J. (1979) *J. Mol. Biol.*, **135**, 879.

45 Ryffel, G. U., Wyler, T., Muellenes, D. and Weber, R. (1980) *Cell*, **19**, 53.

46 Shapiro, M. B. and Senapathy, P. (1987) *Nucleic Acids Res.*, **15**, 7155.

47 Jacob, M. and Gallinaro, H. (1989) *Nucleic Acids Res.*, **17**, 2159.

48 Baird, M., Driscoll, C., Schreiner, H., Sciarratta, G. V., Sansone, G. *et al.* (1981) *Proc. Natl. Acad. Sci. USA*, **78**, 4218.

49 Treisman, R., Proudfoot, N. J., Shander, M. and Maniatis, T. (1982) *Cell*, **29**, 903.

50 Weatherall, D. J. and Clegg, J. B. (1982) *Cell*, **29**, 7.

51 Mount, S. and Steitz, J. (1983) *Nature (London)*, **303**, 380.

52 Goldenberg, C. J. and Hauser, S. D. (1983) *Nucleic Acids Res.*, **78**, 5430.

53 Padgett, R. A., Hardy, S. F. and Sharp, P. A. (1983) *Proc. Natl, Acad. Sci. USA*, **80**, 5230.

54 Mimori, T., Hinterberger, M., Pettersson, I. and Steitz, J. A. (1984) *J. Biol. Chem.*, **259**, 560.

55 Pettersson, I., Hinterberger, M., Mimori, T., Gottlieb, E. and Steitz, J. A. (1984) *J. Biol. Chem.*, **259**, 5907.

56 Konarska, M. M. and Skarp, P. A. (1987) *Cell*, **49**, 763.

57 Reed, R., Griffith, J. and Maniatis, T. (1988) *Cell*, **53**, 949.

58 Grabowski, P. J. and Sharp, P. A. (1986) *Science*, **233**, 1294.

59 Sharp, P. A. (1987) *Science*, **235**, 766.

60 Mattaj, I. W. (1991) *Curr. Opinion Cell Biol.*, **2**, 528.

61 Zhuang, Y. and Weiner, A. (1986) *Cell*, **46**, 827.

62 Ruby, S. W. and Abelson, J. (1988) *Science*, **242**, 1028.

63 Zamore, P. D. and Greem, M. R. (1989) *Proc. Natl. Acad. Sci. USA*, **86**, 9243.

64 Abmayr, S. M., Reed, R. and Maniatis, T. (1988) *Proc. Natl. Acad. Sci. USA*, **85**, 7216.

65 Smith, C., Porro, E. B., Patton., J. G. and Nadal-Ginard, B. (1989) *Nature*, **342**, 243.

66 Zhuang, Y., Goldstein, A. M. and Weiner, A. M. (1989) *Proc. Natl. Acad. Sci. USA*, **86**, 2752.

67 Blencowe, B. J., Sproat, B. S., Ryder, U., Barabino, S. and Lammond, A. I. (1989) *Cell*, **59**, 531.

68 Hall, K. B., Green, M. R. and Redfield, A. G. (1988) *Proc. Natl. Acad. Sci. USA*, **85**, 704.

69 Langford, C., Nellen, W., Niessing, J. and Gallwitz, D. (1983) *Proc. Natl. Acad. Sci. USA*, **80**, 1496.

70 Parker, R., Siliciano, P. G. and Guthrie, C. (1987) *Cell*, **49**, 229.

71 Siliciano, P. G., Brow, D. A., Rolha, H. and Guthrie, C. (1987) *Cell*, **50**, 585.

72 Kretzner, L., Rymond, B. C. and Rosbash, M. (1987) *Cell*, **50**, 593.

73 Garcia-Blanco, M., Jamison, S. F. and Sharp, P. A. (1989) *Genes Devel.*, **3**, 1874.

74 Petterson, B. and Guthrie, C. (1987) *Cell*, **49**, 613.

75 Vijayraghavan, U. and Abelson, J. (1989) *Nucleic Acids Mol. Biol.*, **3**, 197.

76 Warner, J. R. (1987) *Genes Devel.*, **1**, 1.

77 Jamieson, D. J., Rahe, B., Pringle, J. and

Beggs, J. D. (1991) *Nature (London)*, **349**, 715.

78 Banroques, J. and Abelson, J. (1989) *Mol. Cell. Biol.*, **9**, 3710.

79 Vijayraghavan, U. and Abelson, J. (1990) *Mol. Cell. Biol.*, **10**, 324.

80 Lin, R-J., Lustig, A. R. and Abelson, J. (1987) *Genes Devel.*, **1**, 7.

81 Vijayraghavan, U., Company, M. and Abelson, J. (1989) *Genes Devel.*, **3**, 1206.

82 Boel, E., Hansen, M. T., Hjort, I., Hoegh, I. and Fiil, N. P. (1984) *EMBO J.*, **3**, 1581.

83 Mertins, P. and Gallwitz, D. (1987) *EMBO J.*, **6**, 1757.

84 Brown, J. W. S. (1986) *Nucleic Acids Res.*, **14**, 9549.

85 Barta, A., Sommergruber, K., Thompson, D., Hartmuth, K., Matzke, M. A. et al. (1986) *Plant Mol. Biol.*, **6**, 347.

86 Wiebauer, K., Herrero, J-J and Filipowicz, W. (1988) *Mol. Cell. Biol.*, **8**, 2042.

87 Goodall, G. J. and Filopowicz, W. (1990) *Cell*, **58**, 473.

88 Agablan, N. (1990) *Cell*, **61**, 1157,

89 Boothroyd, J. C. (1985) *Annu. Rev. Microbiol.*, **39**, 475.

90 Borst, P. (1986) *Annu. Rev. Biochem.*, **55**, 701.

91 Laird, P. W. (1989) *Tends Genet.*, **5**, 204.

92 Huang, X-Y and Hirsh, D. (1989) *Proc. Natl. Acad. Sci. USA*, **86**, 8640.

93 Koller, B., Fromm, H., Gulun, E. and Edelman, M. (1987) *Cell*, **48**, 111.

94 Choquet, Y., Goldschmidt-Clermont, M., Girard-Bascou, J., Kück, U., Bennoun, P. et al. (1988) *Cell*, **52**, 903.

95 Sutton, R. E. and Boothroyd, J. C. (1988) *EMBO J.*, **7**, 1431.

96 Tschudi, C. and Ullu, E. (1990) *Cell*, **61**, 459.

97 Bruzik, J. P., Van Doren, K., Hirsh, D. and Steitz, J. A. (1988) *Nature (London)*, **335**, 559.

98 Van Doren, K. and Hirsh, D. (1988) *Nature (London)*, **335**, 556.

99 Tschudi, C. and Ullu, E. (1988) *EMBO J.*, **7**, 455.

100 Muhich, M. L. and Boothroyd, J. C. (1988) *Mol. Cell. Biol.*, **8**, 3837.

101 Liou, R-F. and Blumenthal, T. (1990) *Mol. Cell. Biol.*, **10**, 1764.

102 Van Doren, K. and Hirsh, D. (1990) *Mol. Cell. Biol.*, **10**, 1769.

103 Dreyfuss, G., Swanson, M. S. and Pinol-Roma, S. (1988) *Trends Biochem. Sci.*, **13**, 86.

104 Choi, Y. D., Grabowski, P. J., Sharp, P. A. and Dreyfuss, G. (1986) *Science*, **231**, 1534.

105 Swanson, M. S. and Dreyfuss, G. (1988) *EMBO J.*, 7, 3519.

106 Ciejek, E. M., Nordstrom, J. L., Tsai, M-J. and O'Malley, B. W. (1982) *Biochemistry*, **21**, 4945.

107 Zeitlin, S., Parent, A., Silverstein, S. and Efstratiadis, A. (1987) *Mol. Cell. Biol.*, **7**, 111.

108 Cech, T. R., Zaug, A. J. and Grabowski, P. J. (1981) *Cell*, **27**, 487.

109 Zamore, P. D. and Green, M. R. (1991) *EMBO J.*, **10**, 207.

110 Cech, T. R. (1990) *Annu. Rev. Biochem.*, **59**, 543.

111 Bass, B. L. and Cech, T. R. (1984) *Nature (London)*, **308**, 820.

112 Michel, F., Hanna, M., Green, R., Bartel, D. P. and Szostak, J. W. (1989) *Nature (London)*, **342**, 391.

113 Zaug, A. J. and Cech, T. R. (1982) *Nucleic Acids Res.*, **10**, 2823.

114 Zaug, A. J., Grabowski, P. J. and Cech, T. R. (1983) *Nature (London)*, **301**, 578.

115 Woodson, S. A. and Cech, T. R. (1989) *Cell*, **57**, 335.

116 Williamson, C. L. Desai, N. M. and Burke, J. M. (1989) *Nucleic Acids Res.*, **17**, 675.

117 Barfod, E. T. and Cech, T. R. (1989) *Mol. Cell. Biol.*, **9**, 3657.

118 Been, M. and Cech, T. R. (1985) *Nucleic Acids Res.*, **13**, 8389.

119 Price, J. V. and Cech, T. R. (1988) *Genes Devel.*, **2**, 1439.

120 Michel, F., Netter, P., Xu, M. and Shub, D. (1990) *Genes Devel.*, **4**, 777.

121 Suh, E. R. and Waring, R. B. (1990) *Mol. Cell. Biol.*, **10**, 2960.

122 Zaug, A. J., Been, M. D. and Cech, T. R. (1986) *Nature (London)*, **324**, 429.

123 Zaug, A. J. and Cech, T. R. (1986) *Science*, **231**, 470.

124 Zaug, A. J. and Cech, T. R. (1986) *Biochemistry*, **25**, 4478.

125 Cech, T. R. (1988) *Gene*, **73**, 258.

126 Burke, J. M. (1988) *Gene*, **73**, 273.

127 Lazowska, J., Jacq, C. and Slonimski, P. P. (1980) *Cell*, **22**, 333.

128 Perlman, P. S. and Butow, R. A. (1989) Science, 246, 1106.

129 Dujon, B., Belfort, M., Butow, R. A., Jacq, C., Lemieux, C. *et al.* (1989) *Gene*, **82**, 115.

130 Akins, R. A. and Lambowitz, A. M. (1987) *Cell*, **50**, 331.

131 Herbert, C. J., Labouesse, M., Dujardin, G. and Slonimski, P. P. (1988) *EMBO J.*, **7**, 473.

132 Michel, F., Umesono, K. and Ozeki, H. (1989) *Gene*, **82**, 5.

133 Peebles, C. L., Perlman, P. S., Mecklenberg, K. L., Petrillo, M. L., Tabor, J. H. *et al.* (1986) *Cell*, **44**, 213.

134 Grivell, L. A. (1990) *Nature (London)*, **344**, 110.

135 Van der Veen, R., Arnberg, A. C., van der Horst, G., Boren, L., Tabak, H. F. *et al.* (1986) *Cell*, **44**, 225.

136 De La Salle, H., Jacq, C. and Slonimski, P. P. (1982) *Cell*, **28**, 721.

137 Crick, F. (1968) *J. Mol. Biol.*, **38**, 367.

138 Altman, S. (1989) *Adv. Enzymol.*, **62**, 1.

139 Pace, N. R. and Smith, D. (1990) *J. Biol. Chem.*, **265**, 3587.

140 Kole, R. and Altman, S. (1979) *Proc. Natl. Acad. Sci. USA*, **76**, 3795.

141 Guerrier-Takada, C., Gardiner, K., March, T., Pace, N. and Altman, S. (1983) *Cell*, **35**, 849.

142 Waugh, D. S., Green, C. J. and Pace, N. R. (1989) *Science*, **244**, 1569.

143 McClain, W. H., Guerrier-Takada, C. and Altman, S. (1987) *Science*, **238**, 527.

144 Guerrier-Takada, C., van Belkum, A., Pleij, C. and Altman, S. (1988) *Cell*, **53**, 267.

145 Netter, P., Jacq, C., Carignani, G. and Slonimski, P. P. (1982) *Cell*, **28**, 733.

146 Nichols, M., Söll, D. and Willis, I. (1988) *Proc. Natl. Acad Sci. USA*, **85**, 1379.

147 Collmer, C. W., Hadidi, A. and Kaper, J. M. (1985) *Proc. Natl. Acad. Sci. USA*, **82**, 3110.

148 Prody, G. A., Bakos, J. T., Buzayan, J. M., Schneider, I. R. and Bruening, G. (1986) *Science*, **231**, 1577.

149 Hutchins, C. J., Rathjen, P. D., Forster, A. C. and Symons, R. H. (1986) *Nucleic Acids Res.*, **14**, 3627.

150 Uhlenbeck, O. C. (1987) *Nature (London)*, **328**, 596.

151 Morin, G. B. (1989) *Cell*, **59**, 521.

152 Greider, C. W. and Blackburn, E. H. (1987) *Cell*, **51**, 887.

153 Yu, G-L., Bradley, J. D., Attardi, L. D. and Blackburn E. H. (1990) *Nature (London)*, **344**, 126.

154 Boeke, J. D. (1990) *Cell*, **61**, 193.

155 Saville, B. J. and Collins, R. A. (1990) *Cell*, **61**, 685.

156 Ogden, R. C., Knapp, G., Peebles, C. L., Johnson, J. and Abelson, J. (1981) *Trends Biochem. Sci.*, **6**, 154.

157 Müller, F. and Clarkson, S. G. (1980) *Cell*, **19**, 345.

158 Robinson, R. R. and Davidson, N. (1981) *Cell*, **23**, 251.

159 Kaine, B. P., Gupta, R. and Woese, C. R. (1983) *Proc. Natl. Acad. Sci. USA*, **80**, 3309.

160 Ogden, R. C., Lee, M-C. and Knapp, G. (1984) *Nucleic Acids Res.*, **12**, 9367.

161 van Tol, H., Stange, N., Gross, H. J. and Beier, H. (1987) *EMBO J.*, **6**, 35.

162 Stange, N., Gross, H. J. and Beier, H. (1988) *EMBO J.*, **7**, 3823.

163 Hopper, A. K., Banks, F. and Evangelidis, V. (1978) *Cell*, **14**, 211.

164 O'Farrell, P. Z., Cordell, B., Valezuela, P., Rutter, W. J. and Goodman, H. M. (1978) *Nature (London)*, **274**, 438.

165 Knapp, G., Ogden, R. C., Peebles, C. L. and Abelson, J. (1979) *Cell*, **18**, 37.

166 Ogden, R. C., Beckmann, J. S., Abelson, J., Kang, H. S., Söll, D., Schmidt, O. *et al.* (1979) *Cell*, **17**, 399.

167 Filipowicz, W. and Shatkin, A. J. (1983) *Cell*, **32**, 547.

168 Peebles, C. L. Ogden, R. C., Knapp, G. and Abelson, J. (1979) *Cell*, **18**, 27.

169 Rauhut, R., Green, P. R. and Abelson, J. (1990) *EMBO J.*, **9**, 1245.

170 Peebles, C. L., Gegenheimer, P. and Abelson, J. (1983) *Cell*, **32**, 525.

171 Mattoccia, E., Baldi, I. M., Gandini-Attardi, D., Ciafré, S. and Tocchini-Valentini, G. P. (1988) *Cell*, **55**, 731.

172 Reyes, V. M. and Abelson, J. (1988) *Cell*, **55**, 719.

173 Green, P. R. and Abelson, J. (1990) *Methods Enzymol.*, **181**, 471.

174 Greer, C. Peebles, C. L., Gegenheimer, P. and Abelson, J. (1983) *Cell*, **32**, 537.

175 Pick, L. and Hurwitz, J. (1986) *J. Biol. Chem.*, **261**, 6684.

176 Pick, L., Furneux, H. and Hurwitz, J. (1986) *J. Biol. Chem.*, **261**, 6694.

177 Filipowicz, W., Konarska, M., Gross, H. J.

and Shatkin, A. J. (1983) *Nucleic Acids Res.*, **11**, 1405.

178 Uhlenbeck, O. C. and Gumport, R. I. (1981) in *The Enzymes* (ed. P. Boyer), Academic Press, New York, vol. 15, p. 31.

179 Greer, C. L., Peebles, C. L., Gegenheimer, P. and Abelson, J. (1983) *Cell*, **32**, 537.

180 McCraith, S. M. and Phizicky, E. M. (1990) *Mol. Cell. Biol.*, **10**, 1049.

181 Phizicky, E. M., Schwartz, R. C. and Abelson, J. (1986) *J. Biol. Chem.*, **261**, 2978.

182 Perkins, K. K., Furneaux, H. and Hurwitz, J. (1985) *Proc. Natl. Acad. Sci. USA*, **82**, 684.

183 Shatkin, A. J. (1976) *Cell*, **9**, 645.

184 Shatkin, A. J., Darzynkiewicz, W., Furuichi, Y., Kroath, H., Morgan, M. A. *et al.* (1982) *Biochem. Soc. Symp.*, **47**, 129.

185 Baralle, F. E. (1983) *Int. Rev. Cytol.*, **81**, 71.

186 Faruichi, Y., Mulhukrishnan, S., Tomasz, J. and Shatkin, A. J. (1976) *J. Biol. Chem.*, **251**, 5043.

187 Moss, B., Gershowitz, A., Wei, C-M. and Boone, R. (1976) *Virology*, **72**, 341.

188 Mizumoto, K. and Kaziro, Y. (1987) *Prog. Nucleic Acid Res. Mol. Biol.*, **34**, 1.

189 Babich, A., Nevins, J. R. and Darnell, J. E. (1980) *Nature (London)*, **287**, 246.

190 Lai, E. C., Roop, D. R., Tsai, M. J., Woo, S. L. and O'Malley, B. W. (1982) *Nucleic Acids Res.*, **10**, 5553.

191 Gertlinger, P., Krust, A., LeMeur, M., Perrin, F., Cochell, M., Gannon, F., Dupret, D. and Chambon, P. (1982) *J. Mol. Biol.*, **162**, 345.

192 Konarska, M. M., Padgett, R. A. and Sharp, P. A. (1984) *Cell*, **38**, 731.

193 Edery, I. and Sonenberg, N. (1985) *Proc. Natl. Acad. Sci. USA*, **82**, 7590.

194 Inoue, K., Ohno, M., Sakamoto, H. and Shimura, Y. (1989) *Genes Devel.*, **3**, 1472.

195 Hamm, J., Darzynkiewicz, E., Tahara, S. and Mattaj, I. W. (1990) *Cell*, **62**, 569.

196 Singh, R. and Reddy, R. (1989) *Proc. Natl. Acad. Sci. USA*, **86**, 8280.

197 Patel, D. D. and Pickup, D. J. (1987) *EMBO J.*, **6**, 3787.

198 Ohno, M., Kataoka, N. and Shimura, Y. (1990) *Nucleic Acids Res.*, **18**, 6989.

199 Hamlyn, P. H., Gait, M. J. and Milstein, C. (1981) *Nucleic Acids Res.*, **9**, 4485.

200 Hagenbüchle, P., Tosi, M., Schibler, U., Bovey, R., Welleur, P. K. and Young, R. A. (1981) *Nature (London)*, **289**, 643.

201 Kozak, M. (1987) *Nucleic Acids Res.*, **15**, 8125.

202 Manley, J. L. (1988) *Biochim. Biophys. Acta*, **950**, 1.

203 Manley, J. L., Proudfoot, N. J. and Platt, T. (1989) *Genes Devel.*, **3**, 2218.

204 Wickens, M. (1990) *Trends Biochem. Sci.*, **15**, 1990.

205 Zarkower, D., Stephenson, P., Sheets, N. and Wickens, M. (1987) *Mol. Cell. Biol.*, **7**, 1518.

206 Christofori, G. and Keller, W. (1988) *Mol. Cell. Biol.*, **9**, 193.

207 Gilmartin, G. M. and Nevins, G. R. (1989) *Genes Devel.*, **3**, 2180.

208 Higgs, D. R., Goodburn, S., Lamb, J., Clegg, J. B., Weatherall, D. J. and Proudfoot, N. J. (1983) *Nature (London)*, **306**, 398.

209 Orkin, S. H., Cheng, T-C., Autonarkis, S. E. and Kazazain, H. H. (1985) *EMBO J.*, **4**, 453.

210 Wigley, P. L., Sheets, M. D., Zarkower, D. A., Whitmer, M. E. and Wickens, M. (1990) *Mol. Cell. Biol.*, **10**, 1705.

211 Conway, L. J. and Wickens, M. (1988) *EMBO J.*, **6**, 4177.

212 Mclauchlan, J., Gaffney, D., Whitton, J. L. and Clements, J. B. (1985) *Nucleic Acids Res.*, **13**, 1347.

213 Berget, S. M. (1984) *Nature (London)*, **309**, 179.

214 Benoist, C., O'Hare, K., Breathnach, R. and Chambon, P. (1980) *Nucleic Acids Res.*, **8**, 127.

215 Levitt, N., Briggs, D., Gil, A. and Proudfoot, N. J. (1989) *Genes Devel.*, **3**, 1019.

216 Edmonds, M. (1989) *Methods Enzymol.*, **181**, 161.

217 Bardwell, V. J., Zarkower, D. A., Edmonds, M. and Wickens, M. (1990) *Mol. Cell. Biol.*, **10**, 846.

218 Lisa, L. C., Takagaki, Y. and Manley, J. L. (1989) *Mol. Cell. Biol.*, **9**, 4229.

219 Sheets, M. D. and Wickens, M. (1989) *Genes Devel.*, **3**, 1401.

220 Takagaki, Y., Ryner, L. C. and Manley, J. L. (1989) *Genes Devel.*, **3**, 1711.

221 Christofori, G. and Keller, W. (1988) *Cell*, **54**, 875.

222 Osborne, B. I. and Guarente, L. (1989) *Proc. Natl. Acad. Sci. USA*, **86**, 4097.

223 Krowczynska, A., Yenofsky, R and Brawerman, G. (1985) *J. Mol. Biol.*, **181**, 231.

224 Muschel, R., Khoury, G. and Reid, L. M. (1986) *Mol. Cell. Biol.*, **6**, 337.

225 Carrazana, E. J., Pasieka, K. B. and Majoub, J. A. (1988) *Mol. Cell. Biol.*, **8**, 2267.

226 Sachs, A. B. and Davis, R. W. (1990) *Cell*, **58**, 857.

227 Katinakis, P. K., Slater, A. and Burdon, R. H. (1980) *FEBS Lett.*, **116**, 1.

228 Brawerman, G. (1981) *Crit. Rev. Biochem.*, **10**, 1.

229 Kirsh, A. L., Groudine, M. and Challoner, P. B. (1989) *Genes Devel.*, **3**, 2172.

230 Challoner, P. B., Moss, S. B. and Groudine, M. (1989) *Mol. Cell. Biol.*, **9**, 902.

231 Birnstiel, M. L., Busslinger, M. and Strub, K. (1985) *Cell*, **41**, 801.

232 Birchmeier, C., Schümperli, D., Sconzo, G. and Birnstiel, M. L. (1984) *Proc. Natl. Acad. Sci. USA.*, **81**, 1057.

233 Briggs, D., Jackson, D., Whitelaw, E. and Proudfoot, N. J. (1989) *Nucleic Acids Res.*, **17**, 8061.

234 Krieg, P. A. and Melton, D. A. (1984) *Nature (London)*, **308**, 203.

235 Price, D. H. and Parker, C. S. (1984) *Cell*, **38**, 423.

236 Galli, G., Holfteller, H., Stannenberg, H. G. and Birnstiel, M. L. (1983) *Cell*, **34**, 823.

237 Strub, K., Galli, G., Busslinger, M. and Birnstiel, M. L. (1984) *EMBO J.*, **3**, 2801.

238 Strub, K. and Birnstiel, M. L. (1986) *EMBO J.*, **5**, 1675.

239 Schaufele, F., Gilmartin, G. M. Bannwarth, W. and Birnstiel, M. L. (1988) *Science*, **238**, 1682.

240 Soldati, D. and Schümperli, D. (1988) *Mol. Cell. Biol.*, **8**, 1518.

241 Cotten, M., Gick, O., Vasserot, A., Schaffner, G. and Birnstiel, M. L. (1988) *EMBO J.*, **7**, 801.

242 Mowry, K. L. and Steitz, J. A. (1988) *Science*, **238**, 1682.

243 Liu, T-J., Levine, B. J., Skoultchi, A. I. and Marzluff, W. F. (1989) *Mol. Cell. Biol.*, **9**, 3499.

244 Capasso, O., Bleecker, G. C. and Heintz, N. (1987) *EMBO J.*, **6**, 1825.

245 Gilmartin, J. M., Schaufele, F., Schaffner, G. and Birnstiel, M. L. (1988) *Mol. Cell. Biol.*, **8**, 1076.

246 Hoffmann, I. and Birnstiel, M. L. (1990) *Nature (London)*, **346**, 665.

247 Lüscher, B. and Schümperli, D. (1987) *EMBO J.*, **6**, 1721.

248 Vasserot, A. P., Schaufele, F. J. and Birnstiel, M. L. (1989) *Proc. Natl. Acad. Sci. USA*, **86**, 4345.

249 Young, R. A. and Steitz, J. A. (1978) *Proc. Natl. Acad. Sci. USA*, **75**, 3593.

250 Bram, R. J., Young, R. A. and Steitz, J. A. (1980) *Cell*, **19**, 393.

251 King, T. C., Sirdeshmukh, R. and Schlessinger, D. (1984) *Proc. Natl. Acad. Sci. USA*, **81**, 185.

252 Schweitz, H. and Ebel, J-P. (1971) *Biochimie*, **53**, 585.

253 Stark, M., Gourse, R. L., Jemiolo, D. K. and Dahlberg, A. E. (1984) *J. Mol. Biol.*, **182**, 205.

254 Krinke, L. and Wulff, D. L. (1990) *Nucleic Acids Res.*, **18**, 4809.

255 March, P. E. and Gonzalez, M. A. (1990) *Nucleic Acids Res.*, **18**, 3293.

256 King, T. C., Sirdeskmukh, R. and Schlessinger, D. (1986) *Microbiol. Rev.*, **50**, 428.

257 Roy, M. K. and Apirion, D. (1983) *Biochim. Biophys. Acta*, **747**, 200.

258 Roy, M. K., Singh, B., Roy, B. K. and Apirion, D. (1983) *Eur. J. Biochem.*, **131**, 119.

259 Ghora, B. K. and Apirion, D. (1978) *Cell*, **15**, 1055.

260 Singh, B. and Apirion, D. (1982) *Biochim. Biophys. Acta*, **698**, 252.

261 Pieler, T. and Erdmann, V. A. (1982) *Proc. Natl. Acad. Sci. USA*, **79**, 4599.

262 Feunteun, J., Jordan, B. K. and Monier, R. (1972) *J. Mol. Biol.*, **70**, 465.

263 Meyhack, B., Pace, B. and Pace, N. R. (1977) *Biochemistry*, **16**, 5009.

264 Stahl, D. A., Pace, B., Marsh, T. and Pace, N. R. (1984) *J. Biol. Chem.*, **259**, 11448.

265 Miller, O. L. and Beatty, B. R. (1969) *Science*, **164**, 955.

266 Warner, J. R. (1989) *Microbiol. Rev.*, **53**, 256.

267 Hadjiolov, A. A. (1985) in *The Nucleolus and Ribosome Biogenesis*, Springer, New York, p. 54.

268 Tiollais, P., Galibert, F. and Boiron, M.

(1971) *Proc. Natl. Acad. Sci. USA*, **68**, 1117.

269 Kass, S., Craig, N. and Sollner-Webb, B. (1987) *Mol. Cell. Biol.*, **7**, 2891.

270 Bourbon, H., Michot, B., Hassouna, N., Feliu, J. and Bachellerie, J-P. (1988) *DNA*, **7**, 181.

271 Stroke, I. L. and Weiner, A. M. (1989) *J. Mol. Biol.*, **210**, 497.

272 Miller, K. G. and Sollner-Webb, B. (1981) *Cell*, **27**, 165.

273 Tyc, K. and Steitz, J. A. (1989) *EMBO J.*, **8**, 3113.

274 Kass, S., Tyc, K., Steitz, J. A. and Sollner-Webb, B. (1990) *Cell*, **60**, 897.

275 Maser, R. L. and Calvet, J. P. (1989) *Proc. Natl. Acad Sci. USA*, **86**, 6523.

276 Ochs, R. L., Lischwe, M. A., Spohn, W. H. and Busch, H. (1985) *Biol. Cell.*, **54**, 123.

277 Shumard, C. M. and Eichler, D. C. (1988) *J. Biol. Chem.*, **263**, 19346.

278 Shumard, C. M., Torres, C. and Eichler, D. C. (1990) *Mol. Cell. Biol.*, **10**, 3868.

279 Raziuddin, R., Little, D., Labella, T. and Schlessinger, D. (1989) *Mol. Cell. Biol.*, **9**, 1667.

280 Bowman, H., Goldman, W. E., Goldberg, G. I., Herbert, M. B. and Schlessinger, D. (1983) *Mol. Cell. Biol.*, **3**, 1501.

281 Hannon, G. L., Maroney, P. A., Branch, A., Benenfield, B. J., Robertson, H. D. and Nilsen, T. W. (1989) *Mol. Cell. Biol.*, **9**, 4422.

282 Veldman, G. M., Klootwijk, J., de Jonge, P., Leer, R. J. and Planta, R. J. (1980) *Nucleic Acids Res.*, **9**, 6935.

283 De Jonge, P., Klootwikj, J. and Planta, R. J. (1977) *Eur. J. Biochem.*, **72**, 361.

284 Udem, S. A. and Warner, J. R. (1973) *J. Biol. Chem.*, **248**, 1412.

285 Tollervey, D. (1987) *EMBO J.*, **6**, 4169.

286 Zagorsky, J., Tollervey, D. and Fournier, M. J. (1988) *Mol. Cell. Biol.*, **8**, 3282.

287 Xing, Y.Y. and Worcell, A. (1989) *Genes Devel.*, **3**, 1008.

288 Smith, J. D. and Dunn, D. B. (1959) *Biochim. Biophys. Acta*, **31**, 573.

289 Jeanteur, P., Amaldi, F. and Attardi, G. (1968) *J. Mol. Biol.*, **33**, 757.

290 Thomas, G., Gordon, J. and Rogg, H. (1978) *J. Biol. Chem.*, **253**, 1101.

291 Brand, R. C., Klootwijk, J., Planta, R. J. and Maden, B. E. H. (1978) *Biochem. J.*, **169**, 71.

292 Helser, T. L., Davies, J. E. and Dahlberg, J. A. (1971) *Nature (London) New Biol.*, **233**, 12.

293 Maden, B. E. H. and Salim, M. (1974) *J. Mol. Biol.*, **88**, 133.

294 Khan, S. N., Salim, M. and Maden, B. E. H. (1978) *Biochem. J.*, **169**, 531.

295 Hughes, D. G. and Maden, B. E. H. (1978) *Biochem. J.*, **171**, 781.

296 Salim, M. and Maden, B. E. H. (1980) *Nucleic Acids Res.*, **8**, 2871.

297 Maden, B. E. H. (1986) *J. Mol. Biol.*, **189**, 681.

298 Almadja, J., Brimacombe, R. and Maden, B. E. H. (1984) *Nucleic Acids Res.*, **12**, 2649.

299 Finley, D., Bartel, B. and Varshavky, A. (1989) *Nature (London)*, **338**, 394.

300 Dunn, J. J. (1982) in *The Enzymes* (ed. P. D. Boyer), Academic Press, New York. vol. 15, p. 485.

301 Maden, B. E. H. (1971) *Prog. Biophys. Mol. Biol.*, **22**, 127.

302 Todorov, I. T., Noll, F. and Hadjiolov, A. A. (1983) *Eur. J. Biochem.*, **131**, 271.

303 Planta, R. J. and Mager, W. H. (1982) in *The Cell Nucleus* vol. XII (eds H. Busch and L. Rothblum), Academic Press, New York, p. 213.

304 Nomura, M., Morgan, E. A. and Jaskunas, S. P. (1977) *Annu. Rev. Genet.*, **11**, 297.

305 Woolford, J. C., Hereford, L. M. and Rosbash, M. (1979) *Cell*, **18**, 1247.

306 Fried, H. M., Pearson, N. J., Kim, C. H. and Warner, J. R. (1981) *J. Biol. Chem.*, **256**, 10176.

307 Monk, R. J., Meyuhas, P. and Perry, R. P. (1981) *Cell*, **24**, 301.

308 D'Eustachio, P., Meyuhas, O., Ruddle, F. and Perry, R. O. (1981) *Cell*, **24**, 307.

309 Bozzoni, I., Beccari, E., Xun Luo, Z., Amaldi, F., Pierandrei-Amaldi, L., Campioni, N. et al. (1981) *Nucleic Acids Res.*, **9**, 1069.

310 Auger-Buendia, M. A. and Longuet, M. (1978) *Eur. J. Biochem.*, **85**, 105.

311 Fujisawa, T., Imai, K., Tanaka, Y. and Ogata, K. (1979) *J. Biochem. (Tokyo)*, **85**, 277.

312 Lastick, S. M. (1980) Eur. J. Biochem., **113**, 175.

313 Gegenheimer, P. and Apirion, D. (1981) *Microbiol. Rev.*, **45**, 502.

314 Altman, S. (1981) *Cell*, **23**, 3.

315 Deutscher, M. P. (1984) *CRC Crit. Rev. Biochem.*, **17**, 45.
316 Stark, B. C., Kole, R., Bowman, E. J. and Altman, S. (1978) *Proc. Natl. Acad. Sci. USA*, **75**, 3717.
317 Guthrie, C. and Atchison, R. (1980) in *tRNA, Biological Aspects* (eds D. Söll, J. Abelson and P. Schimmel), Cold Spring Harbor Press, New York, p. 83.
318 Robertson, H. D., Altman, S. and Smith, J. D. (1972) *J. Biol. Chem.*, **247**, 5243.
319 Watson, N. and Apirion, D. (1981) *Biochem. Biophys. Res. Commun.*, **103**, 543.
320 Gurevitz, M., Watson, N. and Apirion, D. (1982) *Eur. J. Biochem.*, **124**, 553.
321 Deutscher, M. P., Marlor, C. W. and Zaniewski, R. (1984) *Proc. Natl. Acad. Sci. USA*, **81**, 4290.
322 Zaniewski, R., Pelkaitis, E. and Deutscher, M. P. (1984) *J. Biol. Chem.*, **259**, 11651.
323 Deutscher, M. P., Marshall, G. T. and Cudny, H. (1988) *Proc. Natl. Acad. Sci. USA*, **85**, 4710.
324 Melton, D. A., DeRobertis, E. M. and Cortese, R. (1980) *Nature (London)*, **284**, 143.
325 Rooney, R. J. and Harding, J. D. (1986) *Nucleic Acids Res.*, **14**, 4849.
326 Nichols, M., Söll, D. and Willis, I. (1988) *Proc. Natl. Acad. Sci. USA*, **85**, 1379.
327 Bartkiewicz, M., Gold, H. and Altman, S. (1989) *Genes Devel.*, **3**, 488.
328 Lee, J-Y. and Engelke, D. R. (1989) *Mol. Cell. Biol.*, **9**, 2536.
329 Morales, M. J., Wise, C. A., Hollingsworth, M. J. and Martin, N. C. (1989) *Nucleic Acids Res.*, **17**, 6865.
330 Forster, A. C. and Altman, S. (1990) *Cell*, **62**, 407.
331 Wang, M. J., Davis, M. W. and Gegenheimer, P. (1988) *EMBO J.*, **7**, 1567.
332 Darr, S. C., Pace, B. and Pace, N. R. (1990) *J. Biol. Chem.*, **265**, 12927.
333 Garber, R. L. and Gage, L. P. (1979) *Cell*, **18**, 817.
334 Hagenbüchle, O., Larsen, D., Hall, G. I. and Sprague, K. U. (1979) *Cell*, **18**, 1217.
335 Castaño, J. G., Tobian, J. A. and Zasloff, M. (1985) *J. Biol. Chem.*, **260**, 9002.
336 Green, C. J. and Vold, B. S. (1983) *Nucleic Acids Res.*, **11**, 5763.
337 Wawrousek, E. F. and Hansen, J. N. (1983) *J. Biol. Chem.*, **258**, 291.
338 Deutscher, M. P. (1982) in *The Enzymes* (ed. P. Boyer), Academic Press, New York, vol. 15, p. 183.
339 Deutscher, M. P., Lin, J. C. and Evans, J. A. (1977) *J. Mol. Biol.*, **117**, 1081.
340 Sprinzl, M. and Gauss, D. H. (1982) *Nucleic Acids Res.*, **10**, r1.
341 Björk, G. R., Ericson, J. U., Gustafsson, C., Hagervall, T. G., Jönsson, Y. H. *et al.* (1987) *Annu. Rev. Biochem.*, **56**, 263.
342 Kline, L. K. and Söll, D. (1982) in *The Enzymes* (ed. P. Boyer), Academic Press, New York, vol. 15, p. 567.
343 Söll, D. and Kline, L. K. (1982) in *The Enzymes* (ed. P. Boyer), Academic Press, New York, vol. 15, p. 557.
344 Aschhoff, H. J., Elton, H., Arnold, H. H., Mahal, G., Kersten, W. and Kersten, H. (1976) *Nucleic Acids Res.*, **3**, 3109.
345 Izzo, P. and Gantt, R. (1977) *Biochemistry*, **16**, 3576.
346 Glick, J. M. and Leboy, P. S. (1977) *J. Biol. Chem.*, **252**, 4790.
347 Keith, J. M., Winters, E. M. and Moss, B. (1980) *J. Biol. Chem.*, **255**, 4636.
348 Delk, A. S., Nagle, D. P. and Rabinowitz, J. C. (1980) *J. Biol. Chem.*, **255**, 4387.
349 Taya, Y. and Nishimura, S. (1973) *Biophys. Biochem. Res. Commun.*, **51**, 1062.
350 Samuelsson, T., Borén, T., Johansen, T-I. and Lustig, F. (1988) *J. Biol. Chem.*, **263**, 13692.
351 Samuelsson, T. and Olsson, M. (1990) *J. Biol. Chem.*, **265**, 8782.
352 Abrell, J. W., Kaufman, E. E. and Lipsell, M. N. (1971) *J. Biol. Chem.*, **246**, 294.
353 Bene, R. (1990) *Trends Genet.*, **6**, 177.
354 Stuart, K., Feagin, J. E. and Abraham, J. M. (1989) *Gene*, **82**, 155.
355 Weiner, A. M. and Maizels, N. (1990) *Cell*, **61**, 917.
356 Bene, R., van den Burg, J., Brakenhoff, J., Sloof, P., van Bloom, J. R. and Tromp, M. C. (1986) *Cell*, **46**, 819.
357 Feagin, J. E., Jasmer, D. P. and Stuart, K. (1987) *Cell*, **49**, 337.
358 Shaw, J. M., Feagin, J. E., Stuart, K. and Simpson, L. (1988) *Cell*, **53**, 401.
359 Simpson, L. and Shaw, J. (1989) *Cell*, **57**, 355.
360 Blum, B., Bakalara, N. and Simpson, L. (1990) *Cell*, **60**, 189.

361 Sturm, N. R. and Simpson, L. (1990) *Cell*, **61**, 879.

362 Blum, B. and Simpson, L. (1990) *Cell*, **62**, 391.

363 Decker, C. J. and Sollner-Webb, B. (1990) *Cell*, **61**, 1001.

364 Mahendran, R., Spottswood, M. R. and Miller, D. L. (1990) *Nature (London)*, **349**, 434.

365 Powell, L. M., Wallis, S. C., Pease, R. J., Edwards, Y. H., Knott, T. J. *et al.* (1987) *Cell*, **50**, 831.

366 Chen, S-H., Li, X., Liao, W., Wu, J. H. and Chan, L. (1990) *J. Biol. Chem.*, **265**, 6811.

367 Gaulberto, J. M., Lamattina, L., Bonnard, G., Weil, J.-H. and Grienenberger, J-M. (1989) *Nature (London)*, **341**, 660.

368 Covello, P. S. and Gray, M. W. (1989) *Nature (London)*, **341**, 662.

369 Schuster, W., Wissenger, B., Unsted, M. and Brennicke, A. (1990) *EMBO J.*, **9**, 263.

370 Thomas, S. M., Lamb, R. A. and Paterson, R. G. (1988) *Cell*, **54**, 891.

371 Cattaneo, R., Kaelin, K., Baczko, K. and Billeter, M. A. (1989) *Cell*, **56**, 759.

372 Nabeshima, Y., Fuji-Kuriyama, Y., Muramatsu, M. and Ogata, K. (1984) *Nature (London)*, **308**, 333.

373 Alt, F. W., Bothwell, A. L., Knapp, M., Siden, E., Mather, E. *et al.* (1980) *Cell*, **20**, 293.

374 Early, P., Rodgers, J., Davis, M., Calame, K., Bond, M. *et al.* (1980) *Cell*, **20**, 313.

375 Nelson, K. J., Hairnovich, J. and Perry, R. P. (1983) *Mol. Cell. Biol.*, **3**, 1317.

376 Schwartzbauer, J. E., Tamkun, J. W., Lemischka, I. R. and Hynes, R. O. (1983) *Cell*, **35**, 421.

377 Kitamura, N., Takogaki, Y., Furuto, S., Tanaka, T., Nawa, H. et al. (1983) *Nature (London)*, **305**, 545.

378 Nawa, H., Kolani, H. and Nakanishi, S. (1984) *Nature (London)*, **312**, 7299.

379 Hodgkin, J. (1989) *Cell*, **56**, 905.

380 Baker, B. S. (1989) *Nature (London)*, **340**, 521.

381 Smith, C., Patton, J. G. and Nadal-Ginard, B. (1989) *Annu. Rev. Genet.*, **23**, 527.

382 Laski, F. A., Rio, D. C. and Rubin G. M. (1986) *Cell*, **44**, 7.

383 Siebel, C. W. and Rio, D. C. (1990) *Science*, **248**, 1200.

384 Fu, X-Y., Ge, H. and Manley, J. L. (1988) *EMBO J.*, **7**, 809.

385 Ge, H. and Manley, J. L. (1990) *Cell*, **62**, 25.

386 Nabeshima, Y-I., Fuji-Kuriyama, Y., Muramatsu, M. and Ogata, K. (1984) *Nature (London)*, **308**, 333.

387 Galli, G., Guise, J., Tucker, P. W. and Nevins, J. R. (1988) *Proc. Natl. Acad. Sci. USA*, **85**, 2439.

388 Rosenfeld, M. G., Mermod, J. J., Amara, S. G., Swanson, L.W. and Sawchenko, P. E. (1983) *Nature (London)*, **304**, 129.

389 Amura, S. G., Evans, R. M. and Rosenfeld, M. G. (1984) *Mol. Cell. Biol.*, **4**, 2151.

390 Crenshaw, E. B., Russo, A. F., Swanson, L. W. and Rosenfeld, M. G. (1987) *Cell*, **49**, 389.

391 Emeson, R. B., Hedjran, F., Yeakley, J. M., Guise, J. W. and Rosenfeld, M. G. (1989) *Nature (London)*, **341**, 76.

392 Li, S., Klein, E. S., Russo, A. F., Simmonds, D. M. and Rosenfeld, M. G. (1989) *Proc. Natl. Acad. Sci. USA*, **86**, 9778.

393 Kakizuka, A., Ingi, T., Murai, T. and Nakanishi, S. (1990) *J. Biol. Chem.*, **265**, 10102.

394 Helfman, D. M. and Ricci, W. M. (1989) *Nucleic Acids Res.*, **17**, 5633.

395 Helfman, D. M., Roscigno, R. F., Mulligan, G. J., Finn, L. A. and Weber, K. S. (1990) *Genes Devel.*, **4**, 98.

396 Helfman, D. M., Ricci, W. M. and Finn, L. A. (1988) *Genes Devel.*, **2**, 1627.

397 Streuli, M and Saito, H. (1989) *EMBO J.*, **8**, 787.

398 Cooper, T. A. and Ordahl, C. P. (1989) *Nucleic Acids Res.*, **17**, 7905.

399 Breitbart, R. E. and Nadal-Ginard, B. (1987) *Cell*, **49**, 793.

400 Bell, L. R., Maine, E. M., Schedl, P. and Cline, T. W. (1988) *Cell*, **55**, 1037.

401 Schroeder, H. C., Bachmann, M., Diehl-Seifert, B. and Müller, W. (1987) *Prog. Nucleic Acid Res. Mol. Biol.*, **34**, 89.

402 De la Peña, P. and Zasloff, M. (1987) *Cell*, **50**, 613.

403 Lührmann, R., Kastner, B. and Bach, M. (1990) *Biochim. Biophys. Acta*, **1087**, 265.

404 Zieve, G. W. and Sauterer, R. A. (1990) *Crit. Rev. Biochem.*, **25**, 1990.

405 Barabino, S., Blencowe, B. J., Ryder, U.,

Sproat, B. S. and Lamond, A. (1990) *Cell*, **63**, 293.

406 Watakabe, A., Inoue, K., Sakamoto, H. and Shimuru, Y. (1989) *Nucleic Acids Res.*, **17**, 8159.

407 Schwer, B. and Guthrie, C. (1991) *Nature (London)*, **349**, 494.

408 Company, M., Arenas, J. and Abelson, J. (1991) *Nature (London)*, **349**, 487.

409 Bruzik, B. J. and Stietz, J. A. (1990) *Cell*, **62**, 889.

410 Utans, U. and Krämer, A. (1990) *EMBO J.*, **9**, 4119.

411 Carmo-Fonseca, M., Tollervey, D., Pepperkok, R., Barbarino, S., Merdes, A. *et al.* (1991) *EMBO J.*, **10**, 195.

412 Blackburn, E. H. (1990) *J. Biol. Chem.*, **265**, 5919.

413 Hamm, J. and Mattaj, I. W. (1990) *Cell*, **63**, 109.

414 Lamond, A. I. (1990) *Trends Biochem. Sci.*, **15**, 451.

415 Proudfoot, N. (1991) *Cell*, **64**, 671.

416 Labhart, P. and Reeder, R. H. (1990) *Mol. Cell. Biol.*, **7**, 1900.

417 DeWinter, R. and Moss, T. (1986) *Nucleic Acids Res.*, **14**, 6041.

418 Labhart, P. and Reeder, R. H. (1987) *Cell*, **50**, 51.

419 Kuhn, A. and Krummt, I. (1989) *Genes Devel.*, **3**, 224.

420 Kass, S. and Sollner-Webb, B. (1990) *Mol. Cell. Biol.*, **10**, 4920.

421 Van der Sande, C., Kulkens, T., Kramer, A. B., de Wijs, I. J., van Heerikhuizen, H. *et al.* (1989) *Nucleic Acids Res.*, **17**, 9127.

422 O'Conner, J. P. and Peebles, C. L. (1991) *Mol. Cell. Biol.*, **11**, 425.

423 Stuart, K. (1991) *Trends Biochem. Sci.*, **16**, 68.

424 Blum, B., Stum, N. R., Simpson, A. M. and Simpson, L. (1991) *Cell*, **65**, 543.

425 Chang, D. D. and Sharp, P. A. (1990) *Science*, **249**, 614.

426 Green, M. R. (1988) *Nature (London)*, **336**, 716.

427 Jacquier, A. (1990) *Trends Biochem. Sci.*, **15**, 351.

12

The translation of mRNA: protein synthesis

12.1 AN OVERVIEW OF PROTEIN BIOSYNTHESIS

In the preceding chapters the reader has already encountered the concept that the mRNA (messenger RNA) is an intermediary in the expression of that portion of the genetic information in the DNA that encodes proteins. The present chapter presents a detailed consideration of the process of *translation* of the mRNA. Although this will be prefaced with a summary of the main features of translation, the reader is directed to suitable textbooks of biochemistry (e.g. [1]) for a more elementary account of this topic.

In essence, protein biosynthesis involves translating the information of the sequence of the four different nucleotides of DNA or RNA into a protein sequence with twenty different possible amino acid units. The genetic code provides the conceptual basis of this in the relationship of single amino acids to groups of three nucleotides in the mRNA (triplet codons); whereas tRNA (transfer RNA) provides its physical basis through possessing an anticodon complementary to a mRNA codon, and a specific covalently attached amino acid. Protein biosynthesis involves the successive reading of the codons of the mRNA by the aminoacyl-tRNAs in an ordered manner, and the linking of the amino acids to form a polypeptide chain. This is a complex process and takes place on an elaborate organelle, the ribosome. The direction of growth of the polypeptide chain is from the N-terminus to the C-terminus [2], and the direction of reading of the mRNA is $5' \rightarrow 3'$ [3, 4]. Figure 12.1 provides a schematic summary of the interactions of mRNA, tRNA and ribosome. It represents a stage in protein biosynthesis just before the aminoacyl ester bond of the peptidyl-tRNA is broken and the polypeptide chain transferred to the α-amino group of the aminoacyl-tRNA. Also represented in Fig. 12.1 are two sites on the ribosome to which tRNA can bind, the A-site and the P-site, and the integral ribosomal enzymic activity that catalyses the formation of peptide bonds, *peptidyltransferase*.

The overall length of the mRNA and the rate of initiation are usually such that a second ribosome can attach to the mRNA before the first one has completed its polypeptide chain. In fact, several ribosomes are normally found on a given molecule of mRNA, translating different parts of it

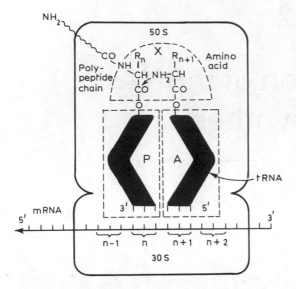

Fig. 12.1 Diagrammatic representation of a prokaryotic ribosome. Two tRNA molecules are bound to the ribosome in response to the mRNA codons designated n and $n + 1$. The tRNA bearing the growing polypeptide chain is occupying the peptidyl site (rectangular area, marked P), and the tRNA bearing an amino acid is occupying the aminoacyl site (rectangular area, marked A). The peptidyltransferase centre, where the peptide bond formation is catalysed, is represented by the semicircular area, marked X. Note the exaggeration of the amino acid (1 Å − 0.1 nm) relative to the tRNA (75 Å − 7.5 nm), and the misrepresentation of the shape of the ribosomal subunits (cf. Fig. 12.21).

simultaneously, and such groups of ribosomes are termed *polyribosomes or polysomes* (Fig. 12.2). The size of the polysomes increases with the size of the mRNA: polysomes synthesizing haemoglobin β-chains ($M_r = c.$ 16000) contain four to five ribosomes [5], whereas those synthesizing myosin heavy chains ($M_r = c.$ 200000) contain about 50–60 ribosomes [6].

The rates of protein synthesis in prokaryotes and eukaryotes appear to be quite similar: values of 15 amino acids s^{-1} for β-galactosidase in *E. coli* [7], and seven amino acids s^{-1} for globin chains in rabbit reticulocytes [8] have been reported. These are, however, much lower than the rates estimated for DNA or RNA synthesis (800 and 50 nucleotides s^{-1}, respectively, in prokaryotes).

It should be mentioned that the synthesis of certain small bacterial peptides occurs in a manner not dependent on mRNA and ribosomes. The reader interested in this subject is directed elsewhere [9].

12.2 THE GENETIC CODE

12.2.1 The standard genetic code

The elucidation of the genetic code represented one of the major breakthroughs in modern biology. Here we shall merely describe the features of the code, as accounts of the history of the code are to be found in a number of reviews (e.g. [10–12]).

The genetic code is a *triplet* code with individual amino acids represented in the mRNA by code words (*codons*) of three nucleotides. Although certain codons also specify initiation and termination signals, the code is uninterrupted, with no 'commas' between codons, and these follow one another in succession and do not overlap. The code was at one time thought to be universal, i.e. each triplet codon had the same meaning, regardless of the species. This assumption derived from comparison between *E. coli* and higher mammals, where initially it was shown that the tRNAs recognized the same codon triplets *in vitro* [13], a result subsequently corroborated by comparison of protein and nucleic acid sequences. It is now known that in certain organisms and in the mitochondria of eukaryotes the genetic code differs from that first established in *E. coli* (section 12.2.3). However, the latter code is so wide-

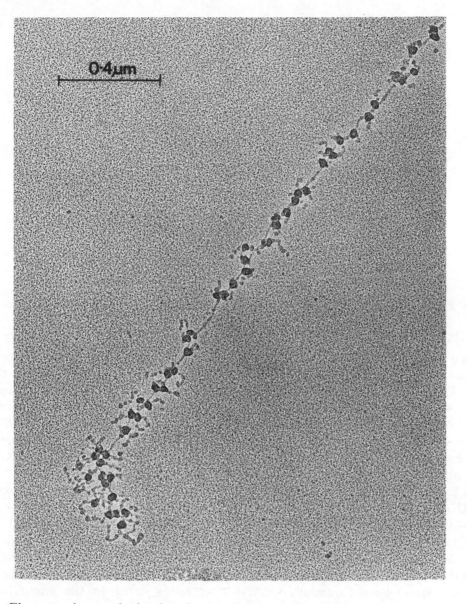

Fig. 12.2 Electron micrograph showing the translation of silk fibroin mRNA on polysomes. The extended fibrous fibroin molecules can be seen emerging from the ribosomes (dark irregular particles). The length of the fibroin molecules increases from the top right to the bottom left of the frame, indicating that this is the $5' \rightarrow 3'$ direction along the mRNA (courtesy of Dr Steven L. McKnight and Dr Oscar L. Miller, Jr).

spread (it is found in the vast majority of prokaryotes and eukaryotes, animals and plants) that we shall refer to it as the *standard* genetic code. It is presented in Fig. 12.3.

It can be seen from Fig. 12.3 that 20 different amino acids are specified by the genetic code. Other amino acids are found in proteins, but almost all of these are

Second letter

		U		C		A		G		
First letter (5')	U	UUU UUC	} Phe	UCU UCC	} Ser	UAU UAC	} Tyr	UGU UGC	} Cys	U C	Third letter (3')
		UUA UUG	} Leu	UCA UCG		UAA UAG	Stop Stop	UGA UGG	Stop Trp	A G	
	C	CUU CUC	} Leu	CCU CCC	} Pro	CAU CAC	} His	CGU CGC	} Arg	U C	
		CUA CUG		CCA CCG		CAA CAG	} Gln	CGA CGG		A G	
	A	AUU AUC	} Ile	ACU ACC	} Thr	AAU AAC	} Asn	AGU AGC	} Ser	U C	
		AUA AUG	Ile Met	ACA ACG		AAA AAG	} Lys	AGA AGG	} Arg	A G	
	G	GUU GUC	} Val	GCU GCC	} Ala	GAU GAC	} Asp	GGU GGC	} Gly	U C	
		GUA GUG		GCA GCG		GAA GAG	} Glu	GGA GGG		A G	

Fig. 12.3   The standard genetic code. Termination codons are indicated as 'Stop'.

generated by post-translational enzymic modification of these 20 genetically-defined amino acids. The one known exception is the rare amino acid, selanocysteine, which is encoded by UGA, otherwise a termination codon (see below). As this is not a complete reassignment of the meaning of the codon, but an alternative translational possibility, thought to depend on the broader context of the mRNA, we shall consider it in section 12.9.6 with other examples of 'suppression' of termination codons.

It is also apparent from Fig. 12.3 that three of the codons (UAA, UAG and UGA, designated 'Stop') do not normally specify an amino acid, but are all signals for the termination of the polypeptide chain. For historical reasons relating to the type of suppressor mutations (see below) that characterized them, UAG and UAA are also referred to as 'amber' and 'ochre', respectively; and UGA is sometimes called 'opal' or 'umber'. Although the termination process does *not* involve tRNA (section 12.4.3), certain mutant tRNAs have been found that can recognize individual termin-

ation codons. These tRNAs occur in the 'suppressor' strains of *E. coli*, so called for their ability to suppress particular classes of 'nonsense' mutants. Nucleotide sequence analysis of such a mutant of a tRNA[Tyr] from an amber suppressor strain showed that its anticodon is changed from 3'-AUG-5' to 3'-AUC-5'. Thus, it is able to insert tyrosine into a polypeptide chain in response to the termination codon UAG, rather than to the tyrosine codon UAC [14]. The reason that such suppressor mutants are viable, and do not exhibit premature termination of the bulk of normal proteins, is because the mutations occur in the minor, otherwise redundant, representatives of certain pairs of isoaccepting tRNAs.

Although inspection of Fig. 12.3 does not reveal a codon, the sole role of which is to specify the start of a polypeptide chain, the codon AUG fulfills this role as well as that of encoding methionine residues in the body of the polypeptide chain. In *E. coli* the initiation of protein synthesis involves the AUG codon being decoded by a unique species, *N*-formylmethionyl-tRNA (fMet-tRNA) [15]. The tRNA that inserts the initiating fMet into polypeptide chains, $tRNA_f^{Met}$, has a different nucleotide sequence from the tRNA that inserts Met internally: $tRNA_m^{Met}$ [16, 17]. Both species accept methionine, but only the Met-$tRNA_f$ can then be formylated by a *transformylase* enzyme that has N^{10}-formyltetrahydrofolic acid as a cofactor [18]. Although formylation of Met-$tRNA_f^{Met}$ is normally an absolute requirement for initiation in *E. coli* and many other bacteria, the specific recognition of the initiator tRNA in the initiation process (sections 12.4.1 and 12.5.1) is most certainly influenced by the structure of the tRNA itself. Thus, in eukaryotes the methionyl residue of the initiator tRNA is not formylated, and the initiator tRNA of at least one bacterium can compensate for lack of formylation by undergoing a base substitution [19].

In bacteria AUG, although the main initiation codon, is not the sole one. The usage of GUG and certain other much rarer codons for initiation is discussed in section 12.4.1; but at this point it is important to stress that these minor initiation codons, like AUG, are recognized by fMet-tRNA.

Although fMet or Met initiate polypeptide chains in prokaryotes and eukaryotes, respectively, these amino acids are not necessarily found at the *N*-terminus of the mature protein. In bacteria the formyl group is removed by a deformylase [20] and the methionine group is frequently removed by an aminopeptidase [21], the action of which is governed by the nature of the penultimate residue in the polypeptide chain [22].

12.2.2 The degeneracy of the genetic code

Since many of the 20 amino acids are encoded by more than one triplet (Fig. 12.3), the code is said to be degenerate.

Triplets coding for the same amino acid are not distributed at random, but are grouped together so that they generally share the same 5' and middle base (although there are two separate groups of codons in the cases of Ser and Leu). This has the consequence that mutations producing a change in the base at the 3'-position of the codon often have no effect on the amino acid specified. Furthermore, the different amino acids are segregated to a considerable extent on the basis of chemical similarity (hydrophobicity, hydrophilicity, acidity and basicity). Thus a mutation in the 5'-base of any of the six leucine codons would give a codon specifying another hydrophobic amino acid. Such a change might not impair the function of a particular globular protein if the altered amino acid merely performed a structural role in the hydrophobic core of this protein. It has therefore been argued that the specific arrangement of codons in the genetic code serves to reduce the potentially harmful effect of mutations.

It might have been expected that each degenerate codon would require its own tRNA with a corresponding anticodon. Although there are some such discrete *isoaccepting* tRNAs, recognizing the same amino acid, their number is less than the 61 required if all codon–anticodon interactions involved three standard Watson–Crick base pairs. This situation is viable because there is a degree of latitude or *wobble* in the complementary base-pairing between the base in the 3'-position of the codon and that in the 5'-position of the anticodon. Rules governing the extent of this latitude were proposed by Crick in his *wobble hypothesis* [23]. He considered the stereochemistry of possible non-standard base-pairing that might allow the two pairs of 3'-codon bases that are grouped together in degenerate codons, U and C, and A and G, to be recognized by a common 5'-anticodon base. He found that relatively small movements from the

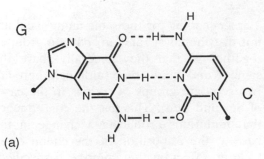

(a)

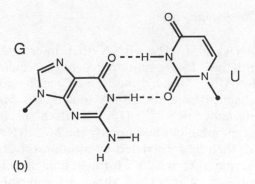

(b)

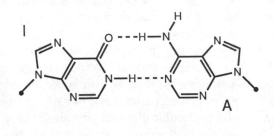

(c)

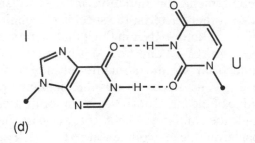

(d)

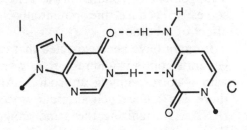

(e)

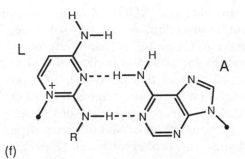

(f)

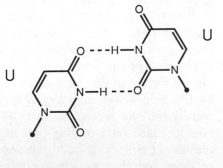

(g)

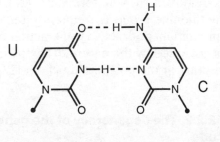

(h)

Fig. 12.4 Some proposed patterns of hydrogen bonding involving nucleosides in the 'wobble' position of the anticodon of tRNA. In each case the 5′-nucleoside of the tRNA anticodon is shown on the left, and that of the 3′-nucleoside of the mRNA codon on the right, with the ribose represented as ●. (a) and (b) 'wobble' for guanosine; (c), (d) and (e) 'wobble' for inosine; (f) proposed base-pairing for lysidine, $R = -CH_2CH_2CH_2CH_2CH(NH_3^+)COO^-$; (g) and (h) possible interactions of uridine with pyrimidines in mitochondrial tRNAs.

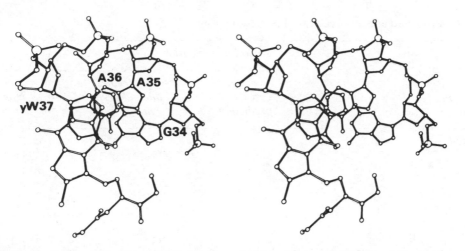

Fig. 12.5 Stereoscopic view of the anticodon bases 34–36 of yeast tRNA^{Phe} and the 3′-hypermodified base yW37 as viewed from the interior of the molecule. It can be seen that the stacking between residues yW37, A36 and A35 is greater than that between A35 and the 'wobble' base G34. (From [49], with permission.)

standard positions might allow the 5′-anticodon base, G, to pair with *both* pyrimidines, U and C, in the 3′-position of the codon; and allow the 5′-anticodon base, U, to pair with *both* purines, A or G, in the 3′-position of the codon. Moreover it had already been found that several tRNA anticodons contained the nucleoside inosine (which pairs only with C in a double helix), and he suggested that this might be able to pair with A, U or C in the 3′-position of the mRNA codon. The hydrogen bonding patterns of these wobble interactions are shown in Fig. 12.4. The elucidation of the three-dimensional structure of tRNA (section 12.3) helps explain how this wobble occurs. It is possible to accommodate a small conformational change into the tertiary structure of the anticodon of yeast tRNA^{Phe}, such that the 5′-anticodon wobble base Gm (the hydrogen bonding pattern of which is identical to G) could move into an alternative position that would allow pairing with U rather than C. The reason that such a change is confined to the 5′-anticodon base

seems to lie in the fact that the base immediately 3′ to the anticodon, which is always a purine (Fig. 12.7) and frequently heavily modified, has strong base-stacking interactions with the two adjacent bases of the anticodon (Fig. 12.5). This serves to anchor these bases in a helical conformation that only allows standard Watson–Crick base pairs between the anticodon and the two first codon positions.

Although the basic concept of the wobble hypothesis has been validated, it is important to realize that not all its detailed predictions have turned out to be correct. Furthermore, one must consider the rules for the base-pairing of important modified 5′-nucleosides of tRNA which were unknown at the time of the formulation of the wobble hypothesis. Table 12.1 presents a comparison of Crick's proposals with the observed pattern of codon–anticodon interaction in prokaryotes and the cytoplasm of eukaryotes, on the one hand, and mitochondria (which show a unique pattern of interaction), on the other hand. The wobble predictions that

Table 12.1 Predicted and observed 'wobble' base-pairing

5'-Anticodon nucleoside	3'-Codon nucleoside		
	Crick's prediction	Prokaryotes, and eukaryotic cytoplasm	Non-plant mitochondria
Unmodified			
C	G	G	G*
G	C, U	C, U	C, U
A	U	n.f.[†]	n.f.
U	A, G	n.f.[‡]	A, U, G, C
Modified			
(C) m^5C, Cm, ac^4C	–	G	n.f.
L	–	A	n.f.
(G) Gm, Q	–	C	C, U[§]
(A) I	U, C, A	U, C, A	n.f.
(U) mcm^5U, mcm^5s^2U, mnm^5s^2U	–	A	n.f.
mo^5U, cmo^5U, $mcmo^5U$	–	A, G, U	n.f.
$cmnm^5U$	–	n.f.	A, U

–: No prediction was made for the modified 5'-nucleoside (unknown at that time).
n.f.: The 5'-nucleoside is not found in tRNAs of that category.
* Except for that occurring in $tRNA_m^{Met}$, which decodes both A and G.
[†] Except for isolated examples in which the prediction in probably incorrect (see text).
[‡] Except for isolated examples in which all four 3'-nucleosides are decoded (see text).
[§] With Q in the 5'-anticodon position of tRNA.

have been substantiated are the ability of the 5'-anticodon nucleoside G to recognize 3'-codon nucleosides U and C (but not A or G), and the 5'-anticodon nucleoside I to recognize 3'-codon nucleosides U, C and A (but not G). Furthermore, with one exception (considered below), 5'-anticodon nucleoside C only recognizes 3'-codon nucleoside G. The problems arise with the wobble predictions for the recognition properties of 5'-anticodon nucleosides A and U, neither of which occur in an unmodified form in the vast majority of prokaryotic or eukaryotic cytoplasmic tRNAs.

The almost complete absence of A from the 5'-position of anticodons has been rationalized on the basis that it is more economical to use G, as all codons with U in

the wobble position are in the same family as a codon with C in that position. Likewise in three- or four-codon families, an A in the initial transcript is replaced by an I in the mature tRNA, which can decode three codons. In the rare cases in which A does occur in the wobble position of the anticodon (a tRNA[Thr] of mycoplasma and a tRNA[Arg] of yeast mitochondria are examples) the pattern of base-pairing is unclear, however. For example, the tRNA[Arg] of yeast mitochondria with anticodon 3'-GCA-5' would seem to have to decode the Arg codons in the 5'-CGN-3' family, all of which are represented in the yeast mitochondrial genome. Thus, a suspicion existed that there was some more fundamental reason for the rarity of an A in the wobble position of tRNA anticodons. This has been confirmed by experiments in which a 3'-CCA-5' anticodon was engineered into an *E. coli* $tRNA^{Gly}$, which did not then show the predicted discrimination against glycine codons other than 5'-GGU-3' [24]. It therefore appears likely that the reason that A is avoided in the wobble position of anticodons is that, contrary to prediction, it would wobble.

The situation with U and the 5'-position of anticodons is somewhat similar. In prokaryotes and the cytoplasm of eukaryotes U in the anticodon position of tRNAs is almost always modified. The reason for this would seem to be that the wobble that could occur with unmodified U would exceed the limitation predicted by the hypothesis. In two-codon amino acid families this would cause violation of the genetic code; in four-codon families the extended wobble is generally not efficient enough to be utilized. The basis for this viewpoint is as follows. Unmodified U does occur in mitochondria (and also in mycoplasma [25]), where it can decode all four 3'-codon bases. In mitochondria this has the important consequence that fewer tRNAs are needed to decode four-codon families (section 12.7.2). When the 3'-CCC-5' anticodon of the aforementioned *E. coli* $tRNA^{Gly}$ was mutated to 3'-CCU-5', the unmodified wobble U was able to decode Gly codon GGA, but was much less efficient than the mycoplasma $tRNA^{Gly}$ with the same anticodon in translating the codons GGG, GGU or GGC [26]. Thus, other structural features of the tRNA besides the anticodon are required to allow unmodified U to decode effectively all four bases in the third position of the codon. Even the modified derivatives of U employed by the majority of prokaryotes and the eukaryotic cytoplasm do not decode according to the wobble rules: mcm^5s^2U, mcm^5U and mnm^5s^2U decode only A;

whereas mo^5U and cmo^5U decode all three of A, G and U (these abbreviations are defined at the front of the book). Structural explanations for the extension or restriction of wobble recognition with some of these modified U residues have been proposed in terms of the adoption of the flexible C3'-endo or the rigid C2'-endo conformation (Fig. 2.16) of the ribose [27, 28]. It is interesting, however, that mitochondria have evolved tRNAs with a derivative of U, $cmnm^5U$, that allow them to decode codon pairs, NMA and NMG, according to the wobble rules, hence minimizing their complement of tRNAs [29]. Because this modified base has not been found in the wobble position of non-mitochondrial tRNAs one suspects that there are other features in the structure of the mitochondrial tRNAs that contribute to the pattern of decoding.

Similar extra-anticodon structural features must be invoked to explain the ability of the C in the 3'-UAC-5' anticodon of yeast mitochondrial $tRNA_m^{Met}$ to recognize both AUA and AUG [30]. One possible structure for the C:A base-pair that is presumed to form is that found in crystals of an artificial deoxyoligonucleotide [31].

The essence of the wobble hypothesis is the possibility of alternative base-pairing interactions involving the 5'-base of the anticodon. It is evident from the foregoing discussion that a sound structural basis exists for the standard wobble base-pairings of unmodified and modified anticodon bases. What, however, of the decoding of four-codon families described above for the unmodified U found in some mitochondrial and mycoplasma tRNAs? One possibility is that the unique structure of these tRNAs allow a relaxed pattern of base-pairing in which, by rotation of the C4'–C5' and P–O bonds, U:U and U:C become close enough together to form satisfactory hydrogen bonds (Fig. 12.4) [32]. However, an alternative possibility is that there is no inter-action in the third position in this situation, and that decoding is by the 'two-out-of-three' mechanism proposed by Lagerkvist [33]. This 'two-out-of-three' hypothesis was actually developed before the situation in mitochondrial tRNAs became apparent, and proposes that in certain circumstances a single tRNA that can recognize the first two codons of a four-codon family has the potential to decode the whole of such a family. The four-codon families for which such 'two-out-of-three' recognition was pro-posed are UCN, CUN, CCN, CGN, ACN, GUN, GCN and GGN (Fig. 12.3). All of these involve at least one strong G:C base pair in the first two positions. Some experi-mental evidence to support this hypothesis was obtained from the successful translation of phage MS2 RNA in a cell-free system from *E. coli* in which the individual tRNAs that can normally translate particular codons were omitted. These could be replaced by tRNAs that would be predicted to translate them according to the 'two-out-of-three' hypothesis, but not according to the 'wobble' hypothesis [34]. The efficiency of this pro-cess was such that it would be unable to compete with the normal tRNA that can make a 'three-out-of-three' interaction, and would, therefore, be most unlikely to par-ticipate in contemporary codon–anticodon interactions in prokaryotes (or eukaryotic cytoplasm). Nevertheless it cannot be ex-cluded that in the peculiar environment of the mitochondrion similar interactions might be involved in the decoding of four-codon families by anticodons containing unmodified U in the wobble position.

12.2.3 Alternative genetic codes

There are two types of divergence from the standard genetic code. Outside mitochon-dria the most widespread of these is the reassignment of one or two of the three termination codons to an amino acid. Thus, in some ciliated protozoa and algae UAA and UAG code for glutamine rather than acting as chain-termination codons [35], whereas in several species of the prokaryote, *Mycoplasma*, these codons are chain ter-minators but UGA codes for tryptophan [36]. In *Euplotes octocarinatus*, another ciliate, UGA codes for cysteine, with UAA functioning as a termination codon and UAG apparently not being employed in any capacity [310]. Species of tRNA with appro-priate anticodons exist in these species to decode these erstwhile termination codons.

How can these changes be accounted for in evolutionary terms? The first stage is thought to have been the loss of the par-ticular termination codon or codons from the repertoire in use. This could have been facilitated by a small coding capacity in the case of mitochondria, and by pressure against usage of G in the extremely AT-rich genome of *Mycoplasma*. Next, the for-mation of a new tRNA or a mutation in the anticodon of a tRNA would have been required to allow recognition of the codon as an amino acid when it re-emerged in the mRNA. For example, the mutation of the tryptophan anticodon from 3'-ACC-5' to 3'-ACU-5' (with appropriate post-trans-criptional base modification) would have allowed it to recognize the codon UGA as well as UGG (Table 12.1). Such a change would have had to be accompanied by a loss of, or alteration in specificity of, the termin-ation factor that had previously recognized the codon. In *Mycoplasma* the loss of RF-2 (section 12.4.3) could have accomplished this.

The second type of divergence from the standard genetic code involves changes in the assignment of codons specifying amino acids. Most of the examples are encountered

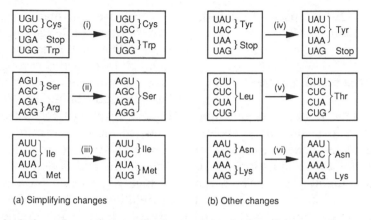

(a) Simplifying changes   (b) Other changes

Fig. 12.6 Some deviations from the standard genetic code found in non-plant mitochondria. The figure shows the changes in relation to the other members of the four codon family involved. The changes indicated are found in: (i) all non-plant mitochondria examined; (ii) echinoderms, molluscs, nematodes and platyhelminths; (iii) vertebrates, arthropods, molluscs, nematodes and yeasts; (iv) platyhelminths; (v) yeasts; (vi) echinoderms and platyhelminths. Details of other changes can be found in [37].

in non-plant mitochondria, which will be the focus of consideration, but a non-mitochondrial example is known. In the fungus, *Candida cylindracea*, CUG encodes serine rather than leucine [35]. Some of the changes found in non-plant mitochondrial genetic codes [37] are illustrated in Fig. 12.6. It should be stressed that there is no single mitochondrial genetic code, implying that different mitochondrial genomes have undergone separate evolution. Indeed, plant mitochondria (and chloroplasts) employ the standard genetic code. Examination of the nature of the changes suggest that although some appear quite gratuitous (e.g. Leu to Thr for CUN in *S. cerevisiae*), others can be regarded as simplifying the code into four-member or two-member families (e.g. change from Arg for AGR). This facilitates the decoding of the mitochondrial genome by a smaller number of tRNAs according to the special rules already described in section 12.2.2.

Such changes in the value of individual codons from one amino acid to another can also be envisaged in terms of temporary abandonment of use (all the examples are degenerate, i.e. other codons for the same amino acid remain) followed by reassignment. In the case of a gross change, such as that in which all the codons in yeast mitochondria of the type CUN changed from Leu to Thr, all that would be required would be for a mutation (outwith the anticodon) in the single tRNA that decodes this family to cause a change in aminoacylation specificity (section 12.3.2). In the case of the conversion of the AAA codon from Lys to Asn in echinoderm mitochondria, mutation of the Asn anticodon from 3'-UUG-5' (which can only decode AAC or AAU) to 3'-UUI-5' could have accomplished this if accompanied by the loss of the tRNA^{Lys} uniquely decoding AAA. In other cases there has been no change in the anticodon, but a change in the decoding capacity of the tRNA, presumably as a result of mutation elsewhere in the molecule. Thus, the Met-tRNA anticodon 3'-UAC-5' normally decodes only AUG, but in yeast mitochondria it has acquired the capacity to decode AUA as well (section 12.2.2).

12.2.4 Differential codon usage

Because the genetic code is degenerate, there exist, in principle, alternative choices between 'synonymous' codons for most of the amino acids, for the termination signal, and even for the initiation signal. The usage of alternative initiation (section 12.4.1) and termination codons (section 12.4.3) is discussed later: the present section considers the pattern of choice of amino acid codons and its implications.

The usage of such synonymous amino acid codons is definitely non-random, and differs between prokaryotes and eukaryotes, viruses and their hosts. In some cases, e.g. certain eukaryotic viruses and mycoplasma, codon choice seems to be determined by the pressure to a particular extreme base composition. However, in bacteria (and to a certain extent in yeast) there is evidence that the reason for non-random codon usage is that it can influence the rate of translation. In *E. coli* it was observed that certain codons that were rarely used in the genes for highly expressed proteins (e.g. ribosomal proteins) were found at a significantly greater frequency in the genes for weakly expressed proteins (e.g. the *lac* and other repressors) [38]. Circumstances where such translational, rather than transcriptional, regulation is important are nicely illustrated by the case of DNA primase. The gene for this protein is in the same transcription unit as those for ribosomal protein S21 and the σ factor of RNA polymerase, proteins that are expressed under similar circumstances but approximately 800 and 50 times more strongly, respectively. The difference in codon usage between this gene and its cotranscribed neighbours is most striking [39].

There are two main ideas as to how the pattern of codon usage might regulate translation. The first suggestion, for which the indirect evidence is strongest, is that the speed of translation of synonymous codons varies with the abundance of the corresponding isoaccepting tRNAs (section 12.2.2). A study of the iso-accepting tRNAs in *E. coli* showed a strong correlation between tRNA abundance and codon choice in the genes of highly expressed proteins [40]. A similar correlation between codon usage in highly expressed genes and the relative abundance of iso-accepting tRNAs was found in yeast [41] and for the fibroin mRNA of the silkworm, *B. mori* [42]. Nevertheless, in *E. coli* there is also a difference in usage of codons of the type NMU and NMC, which cannot be explained in terms of abundance of iso-accepting tRNAs as they are both decoded by a tRNA with G or I in the 'wobble' position (Table 12.1). It was observed in such pairs that the usage correlated well with the predicted codon–anticodon interaction energy, those with either maximum (e.g. CGC) or minimum (e.g. AUU) energies being less frequent in the genes for highly expressed proteins than those (e.g. CGU or AUC) with intermediate energies [38]. It was argued that intermediate codon–anticodon interaction energies would produce more efficient decoding by allowing the optimal balance between binding and release of tRNA. Although this latter idea has not found universal acceptance, it is striking that a similar pattern of preference in NMY codons has been reported in yeast [41].

Such correlations between codon usage and gene expression, although highly suggestive, cannot be taken as proof of a causal effect of codon usage on translation. However, strong experimental support has been obtained in both bacteria and yeast. One elegant experiment involved site-directed mutagenesis of

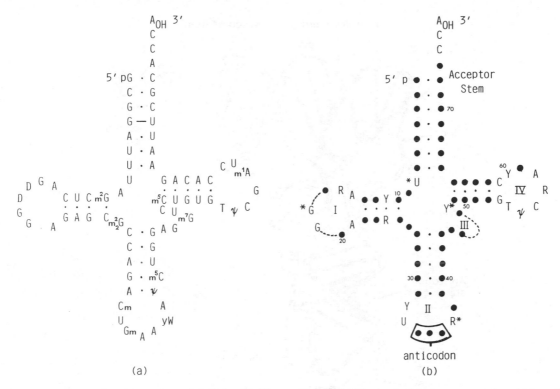

Fig. 12.7 Secondary structure of tRNA. (a) Yeast tRNA[Phe]; (b) generalized structure in which only the invariant or semi-invariant bases are named, the others being represented by filled circles. An asterisk indicates that a base may be modified. Hydrogen bonds in the helical stems are indicated by dots, except for the G:U pair in (a) which is indicated by a line for emphasis. The dotted lines indicate a variable number of nucleotides. R and Y stand for purine and pyrimidine, respectively, other symbols being defined at the beginning of the book. An explanation for the numbering of the variable loops can be found in [45].

rare Leu codons to more highly used codons in the attenuating leader of the leucine operon of *Salmonella typhimurium* (cf. section 10.2.1), and this was found to have a profound effect on the attenuation [43]. In yeast the highly expressed phosphoglycerate kinase gene was mutated so that up to 39% of major codons were replaced by minor ones, and this was found to cause a dramatic decline in the extent of expression [44].

In concluding this discussion of the usage of amino acid codons it should be mentioned that in the small genomes of mitochondria, some codons are completely unrepresented in the repertoire of mRNAs. As mentioned

in section 12.2.3, this is one of the prerequisites for the evolution of variant genetic codes.

12.3 THE STRUCTURE AND AMINOACYLATION OF tRNA

12.3.1 The structure of tRNA [27, 45]

The key role of transfer RNA in decoding the genetic information has already been mentioned. To reiterate, each tRNA can covalently attach one specific amino acid to its 3'-end, and each possesses a sequence of

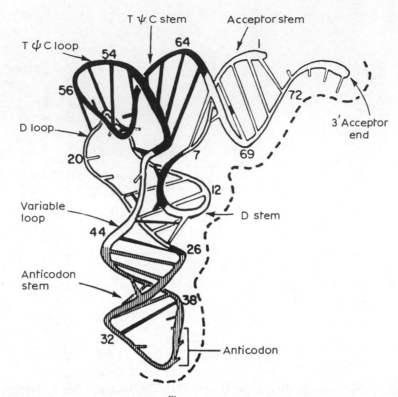

Fig. 12.8 Tertiary structure of yeast tRNA^Phc. The sugar–phosphate backbone is shown as a coiled tube, the numbers referring to the nucleotide residues starting from the 5'-end. Hydrogen bonding interactions between bases are shown as cross-rungs, tertiary interactions being shaded solid black. Bases not involved in hydrogen bonding are shown as shortened rods attached to the backbone (from [49], with permission).

three bases, the anticodon, complementary to and able to interact by hydrogen-bonded base pairing with a mRNA codon for this, so-called, cognate amino acid.

The tRNAs range in length from 73 to 94 nucleotides and are characterized by a relatively large proportion of modified or non-standard nucleosides (sections 2.1.2 and 11.7.2). Almost all these primary structures (mitochondrial tRNAs are the exception – section 12.7.2) allow themselves to be arranged into a common secondary structure, specific and generalized examples of which are shown in Fig. 12.7(a) and (b), respectively. The common features of these structures are as follows.

1. A stem, the acceptor or aminoacyl stem, containing the 5' and 3'-extremities of the molecule. It consists of a helix of seven pairs of bases generally making Watson–Crick base pairs (i.e. A:U or G:C) but occasionally (e.g. Fig. 12.7(a)) with a G:U base pair (cf. Fig. 12.4), together with an unpaired sequence of four bases at the end of the 3'-strand of the stem. The last three bases are always CCA and it is to the 3'-terminal adenosine residue that the amino acid is attached.

2. An arm, the dihydrouridine (D) arm, comprising a stem of three or four base pairs together with the D loop (loop I) of eight to eleven nucleotides, some of

which are invariant. Although loop I generally contains one or more dihydrouridine residues (hence one of its common names), several examples are known in which this base is absent.

3. An arm, the anticodon arm, comprising a helical stem of five base pairs together with the anticodon loop (loop II) of seven nucleotides. It is worth pointing out that the base 5′ to the centrally placed anticodon is always U, and that two other bases in the loop are semi-invariant (i.e. restricted to being purines in one case and pyrimidines in the other). The 5′-base of the anticodon, the base in the 'wobble' position, is frequently a modified or non-standard base.

4. An extra arm (III) of extreme variability, ranging from four to 21 nucleotides. This may be either a helical stem together with a loop of three or four nucleotides (13–21 nucleotides in all), or merely a loop of three to five nucleotides.

5. An arm, the TΨC arm, comprising a helical stem of five base pairs and a loop (loop IV) of seven nucleotides.

To date, X-ray crystal structures have been determined for four tRNAs uncomplexed to any protein [27, 45–48]. The structure of yeast tRNA^Phe, for which greatest resolution has been obtained, is shown in Fig. 12.8. One way of regarding it is as having an L-shape, with two angularly oriented domains: one comprising the TΨC stem and acceptor stem with the 3′-terminal CCA at its furthest extremity; and the other comprising the D-stem, the variable loop (III) and the anticodon arm, with the anticodon loop (II) at its furthest extremity. The manner in which the individual helical stems augment one another to form these two domains of extensive base stacking is perhaps the most fundamentally important feature of the structure. The corner made by the

junction of the two domains comprises the D-loop (I) and the TΨC-loop (IV). The dimensions of the domains are similar: approximately 60 Å (6 nm) from the TΨC-loop at the corner to either anticodon or 3′-acceptor end; whereas the distance between these extremities is approximately 75 Å (7.5 nm).

The inner surface of the two domains (broken line in Fig. 12.8) rather belies the description, L-shaped, that applies to the outer surface. It comprises a more or less planar region that in two dimensions can be

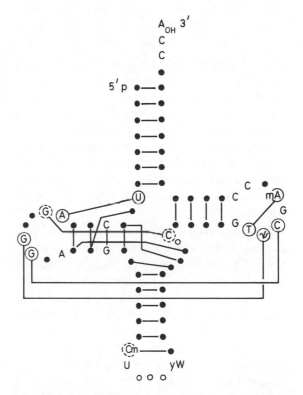

Fig. 12.9 Relationship of tertiary hydrogen bonding interactions to the cloverleaf secondary structure of yeast tRNA^Phe. The generalized convention of Fig. 12.7(b) has been applied to the structure in Fig. 12.7(a) to highlight the involvement of invariant or semi-invariant bases in the tertiary interactions (extended lines). (After [49], with permission.)

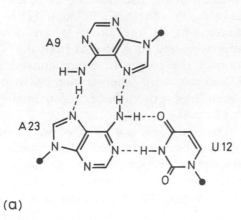
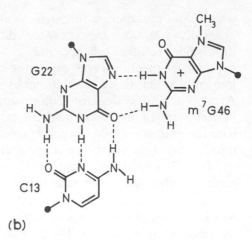

(a)   (b)

Fig. 12.10 Examples of hydrogen-bonding in yeast tRNA^{Phe} that involve three bases. The numbering is as in Fig. 12.7(b).

regarded as the top of a T, the base of which (at about nucleotide 56) is equivalent to the angle of the L. This inner surface contains many of the variable bases of the D-stem and variable loop (III), which, furthermore, are not involved in any of the tertiary interactions discussed below. They are thus available for interaction with other molecules, particularly the aminoacyl-tRNA synthetase (section 12.3.2).

The tertiary structure of yeast tRNA^{Phe} is maintained by a large number of hydrophobic stacking interactions in the augmented helices, together with additional specific hydrogen-bonding interactions between nucleotide residues that are often widely separated in the secondary structure cloverleaf (Fig. 12.9). Many of these interactions are between invariant or semi-variant bases, suggesting a rationale for their restricted variance and a generality for the yeast tRNA^{Phe} tertiary structure. These hydrogen-bonding interactions are in no way confined to Watson–Crick base pairs, but include a variety of non-standard interactions, some involving three bases (Fig. 12.10). Indeed, some of the interactions involve the sugar-

phosphate backbone. Although these hydrogen bonds may at first sight seem esoteric, their occurrence in tRNAs merely reflects the fact that the bases involved are not confined to the relative spatial orientations which they are forced to adopt in a standard RNA A-helix. The potential for the 2'-OH of the ribose to participate in hydrogen bonding may be the basis of the ability of RNA molecules to form such non-helical structures, not possible in the case of DNA.

The general base-stacked two-domain structure stabilized by tertiary interactions described above for yeast tRNA^{Phe} also obtains in yeast tRNA^{Asp} [48], and in yeast and *E. coli* tRNA$_f^{Met}$ (section 12.2.1), suggesting that it is a general feature of tRNAs. (There are certain differences in the conformation of the anticodon loop in the initiator tRNAs which are discussed in section 12.4.1.) There is also evidence indicating that the crystal structure corresponds to the structure adopted in solution [25]. However, as described in the next section, changes in this structure may occur on interaction with proteins.

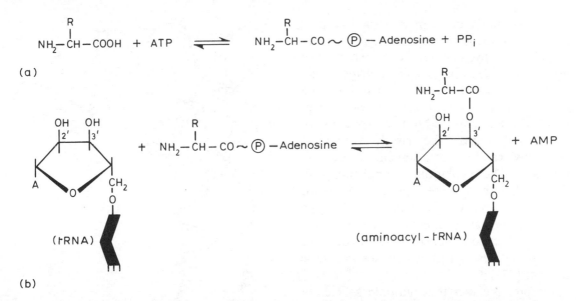

Fig. 12.11 The activation of amino acids and their attachment to tRNA. The reaction occurs in two stages, (a) and (b), both of which are catalysed by the same enzyme, and to which the intermediate is bound (see text). The symbol ~ represents a bond with a relatively high standard free energy of hydrolysis.

12.3.2 The aminoacylation of tRNA

Before a tRNA molecule can act as an adaptor by interacting with its corresponding anticodon in the decoding process, it must first be 'charged' with its cognate amino acid. The enzymes responsible for this process are called *aminoacyl-tRNA synthetases* (*amino acid-tRNA ligases* EC 6.1.1.) and catalyse the reaction illustrated in Fig. 12.11.

The reaction occurs in two stages [50, 51], in the first of which the amino acid is activated by ATP to form an aminoacyl adenylate:

$$\text{ATP} + \text{amino acid}_1 + \text{E}_1 \rightleftharpoons$$
$$(\text{amino acid}_1\text{-AMP})\text{E}_1 + \text{PP}_i$$

The aminoacyl-adenylate complex then reacts with the terminal adenosine moiety of the appropriate tRNA to form an aminoacyl-tRNA:

$$(\text{amino acid}_1\text{-AMP})\text{E}_1 + \text{tRNA}_1 \rightleftharpoons$$
$$\text{tRNA}_1\text{-amino acid}_1 + \text{AMP} + \text{E}_1$$

Although the amino acid is located at the 3'-OH of the ribose of the terminal adenosine moiety of tRNA during peptide bond formation, the initial point of attachment can be the 2'-OH, the 3'-OH, or either, depending on the amino acid [52]. After attachment, rapid migration between the two positions is possible [53]. The aminoacyl ester linkage has a relatively high standard free energy of hydrolysis, derived from the ATP hydrolysed during its activation. This is important as it provides the necessary energy for the subsequent peptide-bond formation to occur [54].

On the basis of primary structure it is possible to divide the aminoacyl-tRNA synthetases into two distinct classes, which have no discernible relationship between one another. Furthermore, it is observed that the class I synthetases generally catalyse attachment of the amino acid to the 2'-position of the ribose, whereas the class II synthetases generally catalyse attachment to the 3'-position. The evolutionary implication of this — that there were seperate origins for the two classes of synthetase — although startling, has been supported by the finding that the three-dimensional structures of representatives of the two classes are quite distinct [321].

Despite the fact that multiple species of tRNA (isoaccepting tRNAs) exist for a single amino acid (section 12.2.2), there appears to be only one aminoacyl-tRNA synthetase for each amino acid. Even the different initiating and elongating methionyl-tRNAs are recognized by the same enzyme. A fundamental question that arises is what features of these isoaccepting tRNAs are recognized by individual aminoacyl-tRNA synthetases to ensure that no misacylation occurs. Chemical and genetic manipulation of tRNAs has now led to a clearer general understanding of this, and the determination of the X-ray crystal structure of tRNA–synthetase complexes has provided detailed information.

In general, the studies with artificial tRNA constructs [55] have identified two main regions on tRNAs that act as determinants for recognition by synthetases: the anticodon loop and the acceptor stem. Which of these regions provides the major determinant varies from tRNA to tRNA, and it has been suggested that this may reflect the structural classes of aminoacyl-tRNA synthetase, mentioned above [56]. Although not all tRNAs show such clear-cut determinants, the cases of tRNA^{Ala} and tRNA^{Met} will be used to

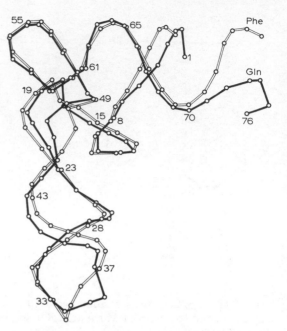

Fig. 12.12 Comparison of the conformation of free yeast tRNA^{Phe} with that of *E. coli* Gln-tRNA^{Gln} in a complex with glutaminyl-tRNA synthetase. The solid lines and open lines represent the path of the backbones of tRNA^{Gln} and tRNA^{Phe}, respectively (After [57], with permission.)

illustrate this point. A G:U base-pair unique to position 3:70 of the anticodon loop of tRNA^{Ala} determines recognition by the alanyl-tRNA synthetase. Mutation of this causes loss of Ala-acceptor activity, and mutation of the corresponding C:G base-pair of tRNA^{Cys} to G:U conveys Ala-acceptor activity to this latter tRNA. In the case of tRNA^{Met}, conversion of the cytosine of the 3'-UAC-5' anticodon to its modified derivative, lysidine (found in tRNA^{Ile}), conveys Ile-acceptor activity on the tRNA; whereas conversion of the 3'-CAU-5' anticodon of the tRNA^{Val} to 3'-UAC-5' conveys upon it Met-acceptor activity.

The first tRNA–synthetase complex to have its structure determined was that for *E. coli* tRNA^{Gln}, a class I enzyme [57]. Both

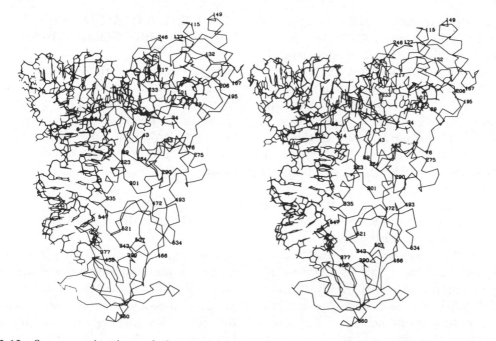

Fig. 12.13 Stereoscopic view of the complex between *E. coli* glutaminyl-tRNA synthetase and Gln-tRNA. (From [57], with permission.)

the central U of the anticodon of this tRNA and its acceptor stem are determinants of synthetase recognition. The striking feature of the X-ray crystal structure is the distortion of both these features of the tRNA compared with tRNA^{Phe} (unfortunately the structure of tRNA^{Gln} alone has not yet been determined) and the interaction of these regions of the tRNA with the synthetase (Figs 12.12 and 12.13). In the complex the base pairing of the acceptor stem is actually broken, explaining why only a weak G:U pair (or a non-pairing C:A) can be tolerated in the 1:72 position. In addition, the observation that this distorted structure is stabilized by a hydrogen bond to the phosphate backbone made by the 2-amino group of the unpaired residue G73 explains the requirement for a guanine residue in this position of the acceptor stem. It is perhaps a little surprising that the major role of the determining residues of the acceptor stem in this

tRNA does not lie in direct interaction with the synthetase, the most striking feature of which in this region are amino acid residues that disrupt the helix. However, it should be stressed that there are specific interactions between the enzyme and the tRNA in these regions. For example, the carboxyl group of Asp^{235} of the synthetase makes hydrogen bonding interactions with the the 2-amino group of the guanine of the G3:C70 base-pair, and the importance of this is indicated by the fact that mutations that convert this Asp to Asn or Gly cause misacylation [58]. Furthermore, the anticodon loop, which is also involved in specifying the identity of the tRNA, makes multiple specific interactions with amino acid residues in the protein [322].

The three-dimensional structure of a complex of a class II synthetase with its tRNA has also been determined, that of tRNA^{Asp} [323]. This complex also shows interaction between the synthetase and the acceptor stem and anticodon loop of the tRNA. Although in this case there is no separation of the acceptor stem of tRNA^{Asp} (this is not surprising as most of the identity determinants are in the anticodon loop) the distortion of the anticodon loop found in tRNA^{Gln} is also observed, and in this case there is a structure for tRNA^{Asp} alone (rather than that of tRNA^{Phe}) as a basis for direct comparison.

The recognition of the correct amino acid by the aminoacyl-tRNA synthetase is equally as important for the fidelity of translation as the recognition of the correct tRNA, just described. This is because if an incorrect amino acid is enzymically attached to the tRNA it will be misincorporated into protein on the basis of the codon–anticodon interaction. (This was shown in the classic experiment in which Cys-tRNA^{Cys} was reduced with Raney nickel to give Ala-tRNA^{Cys}, which then incorporated alanine in response to codons for cysteine in a cell-free system [59]). There are several pairs of amino acids differing in structure by no more than a single methyl group, and this poses a real problem in discrimination for the synthetases. For example, it was observed that valine bound appreciably to Ile-tRNA synthetase. There is good evidence to support the existence of a 'proofreading' or 'editing' mechanism in those synthetases for which there are inappropriate isosteric or smaller amino acids with which the tRNA may be mischarged. This involves a hydrolytic site on the enzyme, close to, but distinct from, the acylation site, to discharge such inappropriate aminoacyl-tRNAs [60]. The accuracy of recognition at the hydrolytic site can, of course, be no greater than that at the initial acylation site; but by requiring the amino acid to be recognized twice, an error frequency of (e.g.) 1 in 10^2 would be reduced to 1 in 10^4.

12.4 THE EVENTS ON THE BACTERIAL RIBOSOME [61–63]

Bacterial protein synthesis will be described using the traditional two-site model of the ribosome (cf. Fig. 12.1) as a convenient visual aid. There are, however, several different proposals for additional sites on the ribosome which will be discussed subsequently.

12.4.1 Chain initiation [64–66]

In polypeptide chain initiation fMet-tRNA is bound to the initiation codon of the mRNA on the 30S ribosomal subunit and the resulting 30S initiation complex then reacts with the 50S ribosomal subunit to give a 70S initiation complex. This process requires GTP and the initiation factors, IF-1, IF-2 and IF-3, and in polycistronic mRNAs can occur independently at several different initiation sites.

The exact sequence of events in initiation and the precise roles of all the factors is not entirely certain, but that thought most likely is presented in Fig. 12.14. The initiating 30S subunit most probably has bound to it IF-3 and IF-1 (Fig. 12.14(a)) which are involved in generating free ribosomal subunits after polypeptide chain termination (section 12.4.3). Their role in initiation is distinct from this latter as they are required for the formation of a 30S initiation complex (Fig. 12.14(b)) even when 50S subunits are not present. IF-3 (M_r = 21 000) is primarily involved in binding mRNA to the ribosome. Although it is needed for translation of natural mRNAs, there is no absolute requirement for IF-3 in either the AUG-dependent ribosome binding of fMet-tRNA or the translation of artificial polynucleotides such as AUGA$_n$. This suggests that, either directly or indirectly, IF-3 facilitates the

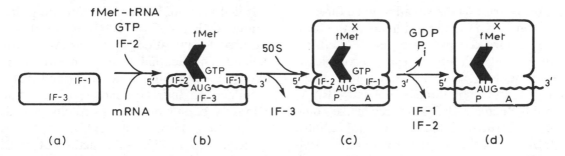

Fig. 12.14 A schematic diagram of prokaryotic polypeptide chain initiation. The ribosome and tRNA are represented as in Fig. 12.1.

16S rRNA	3′	HOAUUCCUCCACUAG.......
		⊢—7±2—⊣
Phage φX174 *A*	5′....	AAUCUUGGAGGCUUUUUU**AUG**..
Phage MS2 coat protein	5′....	UCAACCGGAGUUUGAAGC**AUG**..
Phage λ *cro*	5′....	AUGUACUAAGGAGUUGUA**UG**...
galE	5′....	CCUAAUGGAGCGAAUU**AUG**....
β-lactamase	5′....	AUUGAAAAAGGAAGAGU**AUG**...
Ribosomal protein S12	5′....	AAACCAGGAGCUAUUUA**AUG**...
RNA polymerase β	5′....	GCGAGCUGAGGAACCCU**AUG**...
lacI	5′....	CUUCAGGGUGGUGAAU**GUG**....

Fig. 12.15 Some ribosome binding sites of *E. coli* and bacteriophage mRNAs. The key polypyrimidine region in the 16S rRNA and the bases complementary to this are shown with background shading, and the initiation codons are indicated in bold.

recognition of the untranslated 'leader' sequences that precede the initiating AUGs of natural mRNAs. Such bacterial 'leader' sequences contain short polypurine regions complementary to a polypyrimidine sequence at the 3′-end of the 16S rRNA (Fig. 12.15); and it was suggested by Shine and Dalgarno [67] that base-pairing between these regions is the means by which bacterial ribosomes select the correct AUG codon for initiation. There is now overwhelming evidence (reviewed in [68]) that this hypothesis is correct, although other factors, such as secondary structure, may prevent every potential 'Shine and Dalgarno' sequence being used for initiation. Three known exceptions in which initiation codons lack such a polypurine region in their 'leader' sequence

nicely prove the rule, as these occur in mRNAs coding for extremely weakly expressed proteins: the cI repressor protein of phage lambda, and the DNA primase (the particular codon usage pattern of which was mentioned in section 12.2.4) and *trp* repressor of *E. coli*. The polypurine region is generally found at quite a specific distance 5′ to the initiation codon (Fig. 12.15), as would be expected if its function is to bring the initiation codon into the P-site of the ribosome.

The primary role of IF-2 ($M_r = 97\,000$) is to bind fMet-tRNA to the ribosome in a reaction that requires GTP and is stimulated by the other initiation factors, especially IF-1. The fMet-tRNA most probably binds to the ribosome after the mRNA has done so, in contrast to the situation in eukaryotes (section 12.5.1), and most likely in the form of a ternary complex between IF-2, fMet-tRNA and GTP, although such a complex is experimentally much less stable than those formed by evolutionarily related tRNA-binding factors. IF-2 must recognize some specific structural feature of the initiator tRNA as it will not interact with aminoacyl-tRNAs, not even Met-tRNA. One such feature may be the unique unpaired end of the acceptor stem, C1:A72, as mutation of this to a C:G base-pair allowed the tRNA to function in elongation [69]. Another, possibly relevant, structural difference of the initiator tRNA from elongator tRNAs is seen in the X-ray crystal structure, where the anticodon loop has an external-facing rather than internal-facing disposition [47], a feature also conserved in eukaryotic initiator tRNA [46].

Once a 30S initiation complex, containing mRNA and fMet-tRNA, has been formed (Fig. 12.14(b)) the 50S ribosomal subunit can associate with it (Fig. 12.14(c)) causing the release of IF-3. The non-hydrolysable analogue of GTP, 5′-guanylylmethylene diphosphonate, has been used to show that hydrolysis of GTP is not required for this step, but GTP hydrolysis is required for the fMet-tRNA to become available for reaction with puromycin. This reactivity with puromycin is used to define occupancy of the P-site, as puromycin is an analogue of aminoacyl-tRNA, and hence binds to the A site. After hydrolysis of GTP, IF-2 and (probably) IF-1 are released. As GTP hydrolysis does not cause relative movement of mRNA and the ribosome, fMet-tRNA must be bound directly at the P-site [70]. The role of the GTP hydrolysis, which is discussed in more detail below (section 12.4.2), cannot, therefore, be the provision of energy for movement of the fMet-tRNA from A-site to P-site.

It will be evident from the foregoing that although IF-1 ($M_r = 8000$) is absolutely required for initiation, its role is not clearly defined. The fact that it cycles on and off the ribosome during initiation established that it is indeed an initiation factor, rather than a loosely bound ribosomal protein such as S1 (section 12.6). Although IF-1 seems especially to facilitate the action of IF-2, it is clear that it does not have a function analogous to that of EF-Ts in elongation (section 12.4.2).

Lower-M_r subspecies of IF-2 and IF-3 exist, lacking portions of the N-termini of the larger species. The smaller form of IF-2 (IF-2β), rather than being a proteolytic fragment of the larger form (IF-2α), is the product of a second initiation on the mRNA [71]. These subspecies appear to be functionally equivalent to their parents.

The crucial role of the Shine and Dalgarno sequence in the selection of the initiation codon helps explain the fact, mentioned in section 12.2.1, that this latter is not always AUG. The most common variant, GUG, is used at about 3–4% the frequency of AUG. There are, in addition, some examples of UUG [68] and one of AUU [72] functioning as initiation codons in natural bacterial mRNAs. The other initiation codons are translationally less efficient than AUG: relative activities of AUG > GUG > UUG have been demonstrated [73]. However, the occurrence of GUG and UUG codons in highly expressed mRNAs confirms that this effect is not large and may be compensated for by a strong Shine and Dalgarno sequence. In the case of the AUA initiation codon, which is only known to occur in the mRNA for IF-3, the low intrinsic activity of this appears to be important for autoregulation by the cellular concentration of IF-3 itself. It has been suggested that the translation of the mRNA for IF-3 is independent of IF-3, and that a low intrinsic initiation activity because

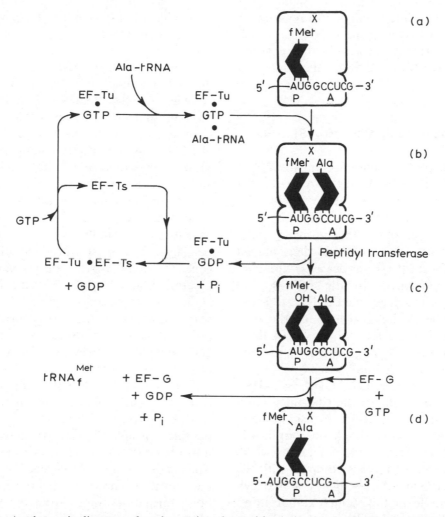

Fig. 12.16 A schematic diagram of prokaryotic polypeptide chain elongation. For convenience the 70S initiation complex of Fig. 12.14(d) has been taken as the starting point, (a), although the scheme applies equally for 70S ribosomes bearing peptidyl-tRNA in the P-site (e.g. (d)). Likewise, the designation of the mRNA triplet in the A-site as coding for Ala, and the third triplet for Ser are purely arbitrary.

of the AUA codon will mean that the mRNA cannot compete for ribosomal subunits unless a low concentration of IF-3 prevents other mRNAs from doing so [74]. Consistent with this idea, mutation of the AUA codon to AUG causes a 40-fold increase in translation and abolishes the autoregulation [75].

Some comment is necessary on the fact

that the recognition of the alternative initiation codons, GUG and UUG, involves 'wobble' at the first, rather than the third, position of the codon. It is noteworthy that, in contrast to elongator tRNAs (section 12.2.2), the base 3' to the anticodon in $tRNA_f^{Met}$ is unmodified. Hence the base-stacking that facilitates fidelity in the first two codon positions of elongator tRNAs is decreased. It is interesting in this regard that in eukaryotic cytoplasmic $tRNA_f^{Met}$, which does not normally show such first position 'wobble' (section 12.5.1), this base *is* modified.

12.4.2 Chain elongation [76–79]

In polypeptide chain elongation an aminoacyl-tRNA binds to the A-site of the ribosome and reacts with the peptidyl-tRNA (Fig. 12.16) or fMet-tRNA (Fig. 12.14) in the P-site, accepting the growing polypeptide chain. The tRNA is then moved across to the P-site (translocated), with concomitant movement of the mRNA and expulsion of the deacylated tRNA, in order to make the A-site available for the next aminoacyl-tRNA. Elongation requires three soluble factors, EF-Tu, EF-Ts, EF-G, and the hydrolysis of two molecules of GTP.

Elongation factor EF-Tu ($M_r = 43\,000$ [126]) is responsible for the ribosomal binding of the aminoacyl-tRNA corresponding to the mRNA codon in the A-site (arbitrarily designated as Ala in Fig. 12.16), before which it forms a soluble ternary complex with the tRNA and GTP. All elongator aminoacyl-tRNAs will form this complex, but fMet-tRNA will not [80]. The non-hydrolysable analogue of GTP, 5'-guanylylmethylene diphosphonate, will allow the aminoacyl-tRNA to bind to the 70S ribosome, but the GTP must be hydrolysed before peptide-bond formation can occur. The GTP hydrolysis is not required for the peptidyl-transferase reaction itself and its possible role is discussed below. The EF-Tu and GDP are released from the ribosome as a complex. In this form the EF-Tu cannot react with GTP or aminoacyl-tRNA, and it is the function of EF-Ts ($M_r = 30\,000$) to displace GDP from the EF-Tu.GDP complex. This results in the formation of an EF-Tu.EF-Ts complex, from which the EF-Tu.GTP complex can be regenerated (Fig. 12.16).

Analysis of its primary structure shows that EF-Tu, like IF-2 and EF-G (below), belongs to the family of GTP-binding proteins that include the signal-transducing G-proteins [81]. The X-ray crystal structure of EF-Tu has been determined [82, 83], but, as yet the structure of the complex with aminoacyl-tRNA has not been solved, so that at present only the position at which GTP binds is known. There are two, apparently functionally equivalent forms of EF-Tu in *E. coli*, the products of separate genes (*tufA* and *tufB*), differing only in their *C*-terminal amino acids. EF-Tu is extremely abundant, constituting some 5% of total bacterial cell protein, and occurring in approximately sixfold excess over ribosomes and other elongation factors. The significance of this is not known.

The aminoacyl-tRNA bound in the A-site (Fig. 12.16(b)) can now be linked to the carboxyl group of the fMet or nascent peptide, through the catalytic activity of the intrinsic *peptidyltransferase* centre of the 50S ribosomal subunit. As already mentioned, the thermodynamic free energy for peptide bond formation comes from the hydrolysis of the 'energy-rich' acyl-ester bond of the aminoacyl-tRNA, which in its turn is derived from the ATP hydrolysed during amino-acylation. The fact that GTP and supernatant factors are not required for the trans-peptidation was perhaps most convincingly confirmed when it was discovered that, in the presence of ethanol (about 50%), a 3'-hexanucleotide fragment of fMet-tRNA could react with puromycin on the isolated 50S ribosomal subunit [84]. Extension of this 'fragment reaction' to even smaller oligo-nucleotide fragments has shown that CCA-fMet is the smallest species that can occupy the P-site of the peptidyltransferase, and that puromycin can be replaced by CA-Gly at the A-site.

The translocation of the peptidyl-tRNA from the A-site to the P-site requires the elongation factor EF-G ($M_r = 77\,000$) and GTP. This reaction has been shown to allow movement of the peptidyl end of the tRNA so that it becomes reactive towards puromycin, movement of the mRNA relative to the ribosome, and ejection of the deacylated-tRNA from the P-site. The reaction requires hydrolysis of the GTP, the non-hydrolysable analogue being inactive although it will allow EF-G to bind to ribosomes. The molecular mechanism underlying this process is perhaps the most intriguing and the least understood aspect of protein biosynthesis. It is clearly possible that a large part of the structural complexity of the ribosome, including even the division of the ribosome into subunits, may be a consequence of the need for this specific and concerted movement of macromolecules.

After translocation (Fig. 12.16(d)), one cycle of elongation has been completed (cf. Fig. 12.16(a)). The vacant A-site now contains a new mRNA codon, to which a corresponding aminoacyl-tRNA can bind, starting another round of elongation.

The function of GTP hydrolysis in the elongation reactions was unclear for many years. The earliest ideas were influenced by the biochemical precedents of ATP hydrolysis and the GTP-utilizing phosph*enol*pyruvate carboxykinase, and focused on the provision of energy for movement of molecules: the translocation of peptidyl-tRNA to the P-site in the case of EF-G, and the 'accommodation' of the aminoacyl-tRNA into the A-site in the case of EF-Tu. However, as mentioned above, it is now clear that the GTP-utilizing protein-synthesis factors are members of a larger family of GTPases which includes the G-proteins, in which the function of GTP hydrolysis is

generally to terminate the action of the protein by causing its dissociation from its target [81]. Against this context the idea has emerged of a unitary role for GTP hydrolysis in protein biosynthesis to expel the factors (including IF-2) from the ribosome after they have fulfilled their functions. The presence of the factor on the ribosome can be regarded as preventing the next reaction from occurring, and the role of EF-G envisaged in this model is to prime the ribosome for subsequent translocation which does *not*, in itself, require GTP. In the case of EF-Tu, the time taken for the GTP to be hydrolysed after the ternary complex has bound to the ribosome is important in the kinetic 'proofreading' that ensures accuracy of the codon–anticodon interaction (section 12.6.3). However, assertions that the function of GTP hydrolysis is to provide the energy for this proofreading are, in our view, misleading.

As already stated, there are models of the ribosome containing additional sites, and these have implications for models of the elongation cycle. Lake [85] proposed that before occupying the A-site proper, the aminoacyl-tRNA entered through a recognition or R-site, which would allow the anticodon to be proofread for correspondence to the codon. Nierhaus (reviewed in [86]), on the other hand, has proposed a model of the ribosome in which the deacylated tRNA is first transferred to an exit or E-site (i.e. after stage (c) in Fig. 12.16), before being expelled from the ribosome. More recently Moazed and Noller [87] obtained evidence for the employment of an additional site during translocation; although one rather different from that proposed by Nierhaus. Chemical footprinting indicated that the translocation occurred in two discrete steps involving first the acceptor end and then the anticodon end of the tRNA. To explain their results they proposed an extra E-site, solely

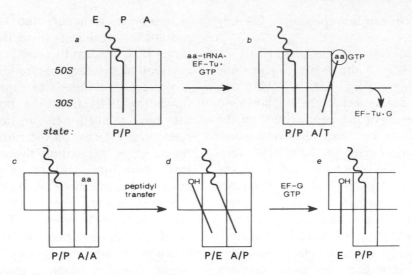

Fig. 12.17 Model for the movement of tRNA during translocation involving hybrid occupancy of A- or P-sites on the 30S subunit, and A-, P- or E-sites on the 50S subunit of the ribosome. In state A/T, no site on the 50S subunit is occupied. (From [87], with permission.)

on the 50S ribosomal subunit, with hybrid occupancy by the tRNA of various 50S and 30S sites occurring during the elongation cycle (Fig. 12.17). It is interesting that there is structural similarity between the RNA in the proposed E-site and that of part of the RNA component of *E. coli* ribonuclease P (section 11.3.4(a)) involved in interaction with the 3'-end of precursor tRNAs [319].

12.4.3 Chain termination [88–90]

In polypeptide chain termination (Fig. 12.18), the ester linkage of the peptidyl-tRNA is hydrolysed in response to one of three termination codons (section 12.2.1) in a reaction involving two of the three release factors, RF-1, RF-2 and RF-3. The deacylated tRNA and the mRNA are expelled from the ribosome in the presence of release factor, RRF, and EF-G, liberating free ribosomal subunits. The subunits will associate to form 70S ribosomes unless pre-

vented from doing so by IF-3 and IF-1 (section 12.4.1).

In contrast to the other 61 codons, the three specific terminators are not read by tRNAs. This was shown using RNA from an amber mutant of bacteriophage R17 in which the first six codons of the coat protein are followed by a stop codon. Only the six appropriate aminoacyl-tRNAs and supernatant proteins were required for release of the hexapeptide. This same system was subsequently used with purified elongation factors and release factors to show that release of the peptide required the peptidyl-tRNA to be at the P-site of the ribosome [91]. To study the factor requirements for termination at all three codons, an assay was developed in which the termination codons could direct the release of fMet-tRNA, previously bound to ribosomes in the presence of the triplet AUG. This led to the resolution of two release factors, RF-1 ($M_r = 36\,000$) and RF-2 ($M_r = 38\,000$), of different codon specificities:

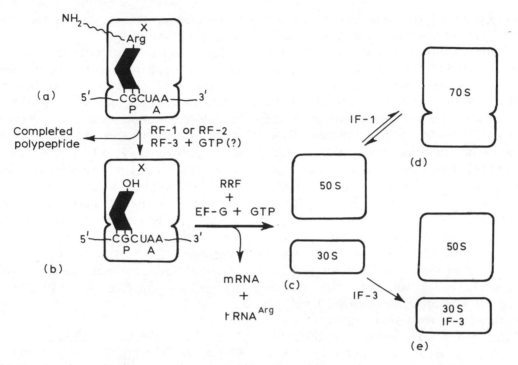

Fig. 12.18 A schematic diagram of prokaryotic polypeptide chain termination. The amino acid designation in the P-site is purely arbitrary. Other possible termination codons in the A-site are UAG and UGA (see text).

RF-1 for UAA or UAG

RF-2 for UAA or UGA

The third release factor, RF-3 (M_r = 46 000), is not codon-specific and has no release activity in the absence of the other factors. It enhances the release of polypeptide promoted by the other factors and seems to stimulate both binding and release of these latter from the ribosome. Its activity is increased by GTP, but a requirement for GTP hydrolysis during termination in prokaryotes is not firmly established as it has been reported that GDP can replace GTP in the reaction *in vitro*. In this respect it would be useful to know the primary structure of RF-3 to see whether it is related to those of the known GTPases discussed in section 12.4.2.

There are quite strong grounds for thinking that the actual hydrolysis of the peptidyl ester linkage is catalysed by the peptidyltransferase centre of the ribosome, the reaction specificity of which has been modified by the binding of the release factors. This was suggested by the finding that the peptidyltransferase would catalyse the formation of an ester link to fMet-tRNA or its hexanucleotide fragment (section 12.4.2) if certain alcohols were presented to the ribosome instead of aminoacyl-tRNA. If the hydroxyl groups of an alcohol could replace the α-amino group of an aminoacyl-tRNA as a reactive nucleophile, it seemed possible that the hydroxyl group of water might do likewise. This suggestion was supported by the fact that a number of antibiotics (e.g. sparsomycin and chloramphenicol) and ionic conditions known to inhibit the peptidyltransferase reaction were also found to inhibit the termination reaction *in vitro*.

More recently a new perspective has been generated on the termination process in the form of a possible role for ribosomal RNA. A ribosomal mutation that specifically suppressed UGA termination codons was found to be the deletion of a cytosine (C_{1054}) from a highly conserved region of the 16S rRNA, and it was suggested that this mutation might prevent codon-specific base-pairing to a proximal 3′-ACU-5′ sequence [92]. There is, as yet, no direct evidence for such an interaction, nor is there any indirect evidence for analogous interactions involving the other two termination codons.

After the release of the peptide, the mRNA and deacylated tRNA are still attached to the ribosome (Fig. 12.18(b)) and must be removed before subunits can be regenerated for another round of protein synthesis. This requires GTP, EF-G and ribosome release factor, RRF ($M_r = 18\,000$). Although it might be expected that the primary role of EF-G in this process would be the expulsion of deacylated tRNA, there appear to be no data bearing on this question.

The released ribosomes can be in the form of 70S particles or 30S and 50S subunits. The supply of isolated subunits for reinitiation is controlled by IF-3, acting as an *anti-association factor*, preventing the 50S subunit from associating with the 30S.IF-3 complex [93]. Inactive 70S ribosomes do accumulate in cells, especially when inhibition of initiation results in a relative excess of 30S subunits over IF-3. To regenerate subunits when conditions improve there must be an equilibrium between 70S ribosomes and ribosomal subunits. Initiation factor IF-1 is thought to accelerate the interconversion of

these species, without altering the position of its equilibrium [94].

Although all three termination codons are found to occur in mRNAs, the distribution of these is not random. The codon UAG is highly disfavoured, and in a sample of *c*. 800 *E. coli* termination codons the relative frequencies of occurrence, UAA:UAG:UGA, were approximately 10:4:1 [95, 96]. Furthermore, the context of termination codons is also not random, with a tendency for a following U being the most obvious feature of this. These features appear to provide protection against readthrough by natural suppressor tRNAs (section 12.9.6), rather than contributing to the recognition of the codons by termination factors [97, 98].

12.5 THE EVENTS ON THE EUKARYOTIC RIBOSOME [99, 100]

Eukaryotic ribosomes catalyse essentially the same process as prokaryotic ribosomes. Although the details of eukaryotic protein synthesis are less well understood, it is clear that the differences from prokaryotic protein synthesis are relatively minor for elongation and termination, but much greater for initiation. The following discussion relates to the nucleo-cytoplasmic protein-synthesizing system, the protein-synthesizing systems of mitochondria and chloroplasts being dealt with in section 12.7.

12.5.1 Chain initiation [101, 102]

One way in which eukaryotic initiation differs from that in prokaryotes is in the initiating amino acid: instead of fMet, the initiating amino acid is Met. Nevertheless, there is a specific species of methionine tRNA for initiation, distinct from that used in elongation. The eukaryotic initiator tRNA will be referred to here as tRNA$_f^{Met}$ (and the elongator as tRNA$_m^{Met}$), although it is sometimes designated tRNA$_i^{Met}$. This nomenclature emphasizes the unity in structure between the eukaryotic and prokaryotic initiator tRNAs: despite the fact that there is no transformylase in the eukaryotic cytoplasm, the eukaryotic Met-tRNA$_f$ can be formylated *in vitro* by *E. coli* transformylase [103].

The more fundamental difference in eukaryotic initiation is the mode of selection of the initiating AUG, and it is clear that it is this aspect of initiation, although still incompletely understood, that accounts for a plethora of eukaryotic initiation factors, far exceeding the number in prokaryotes. Instead of the 'Shine and Dalgarno' interaction (eukaryotic 18S rRNA lacks the key CCUCC sequence involved in this), allowing independent internal initiations on a polycistronic mRNA, in eukaryotic initiation there is a different mechanism: the Kozak 'scanning mechanism' [68, 104, 105]. The 40S ribosomal subunit binds to the 5'-end of a monocistronic mRNA and 'scans' along this until it encounters an appropriate (usually the first) AUG initiation codon, when attachment of the 60S ribosomal subunit can occur. In contrast to the situation in prokaryotes, the initiator tRNA binds to the small ribosomal subunit before this attaches to the mRNA [106, 107], and experiments in which the 3'-UAC-5' anticodon of the tRNA$_f^{Met}$ was mutated to 3'-UCC-5' have demonstrated that the anticodon is involved in the scanning for the appropriate AUG codon [108].

A model for the mechanism of eukaryotic initiation is presented in Fig. 12.19. The first stage is the formation of a (stable) ternary complex between Met-tRNA$_f^{Met}$, eIF-2 and GTP. Although eIF-2 appears functionally analogous to prokaryotic IF-2, the eukaryotic factor differs from the latter in comprising three subunits (α, β and γ). The ternary complex then binds to a 40S ribosomal subunit bearing eIF-3 and eIF-4C, giving a 43S preinitiation complex. One of the roles of eIF-3 is as an anti-association factor, generating free 40S subunits; but it should be stressed that eIF-3 is an extremely complicated species, consisting of at least seven antigenically discrete polypeptide chains with an aggregate M_r of approximately 700 000 [109].

The next step is a most distinctive feature of eukaryotic initiation, the binding of the 43S preinitiation complex to the 5'-end of the mRNA, involving melting of mRNA secondary structure and recognition of the 5'-cap structure. Four factors are involved at this stage: the three factors eIF-4A, eIF-4E, and p220 ($M_r = 220 000$), which together make up a complex known as eIF-4F, and a fourth factor, eIF-4B. The factor responsible for promoting unwinding of the secondary structure of mRNA is eIF-4A, which is an ATP-dependent helicase. (At least in mouse, there are two distinct functional genes for eIF-4A, but the significance of this is unclear [110].) The recognition of the 5'-cap structure is the property eIF-4F. Although all three components of the factor are required for this, purification on 'cap' affinity columns has shown that it is eIF-4E ($M_r = 24 000$) that is the actual cap-binding protein. The factor eIF-4E must play a pivotal role in the regulation of eukaryotic protein biosynthesis, for transfected fibroblasts over-expressing this factor exhibit malignant transformation [111]. The precise role of eIF-4B at this step of initiation has yet to be defined.

The 40S initiation complex, having bound to the mRNA, scans for the correct AUG codon, when the 60S subunit is incorporated, forming the 80S complex. Factor eIF-5 appears to be necessary to promote GTP hydrolysis and the release of the other

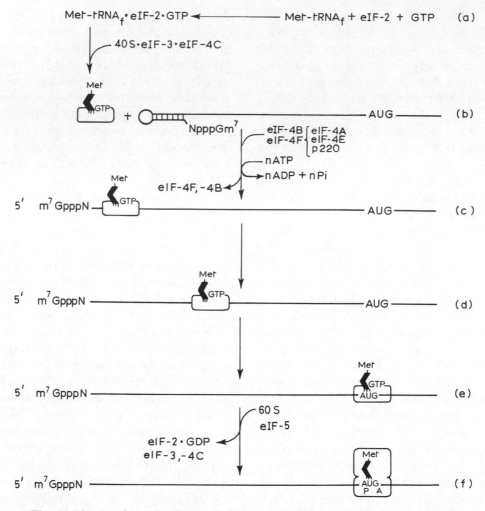

Fig. 12.19 A schematic diagram of eukaryotic polypeptide chain initiation.

factors before this can occur. The eIF-2 released at this stage is complexed to GDP which must be displaced by a factor, eIF-2B (also called eRF and GEF), before it can continue to function (Fig. 12.35). This reaction, which is analogous to that between EF-Tu and EF-Ts (Fig. 12.16), is subject to regulation and is discussed in more detail in section 12.9.5. Another factor, termed eIF-4D, has been implicated in the formation of the first peptide bond, but its role has not been clearly defined and it is not shown in Fig. 12.19. This protein is interesting, however, as it contains a unique modified derivative of lysine, hypusine, which is essential for its biological activity [112].

The key tenets of the original 'scanning model' of Kozak [104], presented above, are attachment of the ribosome to the 5'-end of the mRNA and initiation at the first AUG codon; and the corollary of these is that

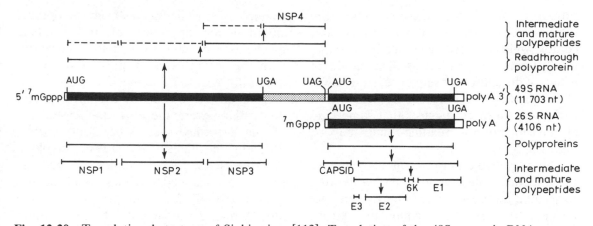

Fig. 12.20 Translational strategy of Sinbis virus [113]. Translation of the 49S genomic RNA occurs only from the first AUG, a second internal initiation site being recognized only on the subgenomic 26S RNA, where it has now become the first AUG. The initial transcripts are polyproteins which are proteolytically processed to give the mature non-structural and structural proteins indicated. Suppression (section 12.9.6) of the first termination codon produces a larger polyprotein from the 49S RNA, the processing of which generates NSP4 (and perhaps more of the other non-structural proteins: indicated by broken lines).

there should be no polycistronic mRNAs with internal initiation, in contrast to the situation in prokaryotes. Certainly, in the vast majority of cases this holds true, and it is striking how many RNA viruses (e.g. [113]) resort to strategies such as the production of polyproteins or the generation of nested mRNA subspecies that allow then to be functionally monocistronic despite having structurally polycistronic genomes (Fig. 12.20). It is true that there are some minor deviations from the original predictions, but these can be accommodated by slight refinement of the scanning model [105]. However, there are also some rare major violations of the predictions that can only be explained by the existence of an alternative mechanism of eukaryotic initiation [114]. We shall consider each of these situations in turn.

In approximately 700 vertebrate mRNAs analysed, the most 5'-terminal AUG is the initiation codon in 90–95% of cases. It was realized that the initiation codons actually used have a similar 'context', the consensus being:

$$GCCGCCRCCAUGG$$

and it was therefore proposed that this context somehow influenced the selection of the initiation codon, with AUG codons in weak contexts being bypassed. Not all the consensus nucleotides exert equal influence, but the presence of either a purine at position '−3' or a G at position '+4' to the start of the AUG seem to be particularly important. The majority of those 5–10% of mRNAs that did not initiate at the first AUG were found to initiate at the first AUG occurring in a strong context. Moreover, some rare cases where two consecutive AUG codons are used as alternative initiation codons are explicable in terms of this context effect if it is assumed that a proportion of 43S initiation complexes ignore the first AUG because of a relatively poor context which, nevertheless, does not deviate totally from the consensus. The mode of action of the consensus sequence is unknown, although the occurrence of purines with a periodicity of three suggests that some kind of 'phasing' of the reading-frame may be occurring.

There are two other refinements that have to be made to the 'scanning' model in order to account for cases where initiation does not occur at the first AUG codon in a strong context. One is when the first AUG occurs within 10 nucleotides of the 'cap', in which case it is assumed to be inaccessible to the 43S initiation complex. The second is where the AUG codon from which the protein initiates is preceded by a short open reading-frame. It appears that the first open reading-frame is translated, and that, after termination, continued scanning and re-initiation can occur [115]. The explanation of this phenomenon, which is restricted to short open reading-frames culminating in a termination codon, is unclear. Such preceding short open reading-frames have the effect of decreasing translation of the authentic initiation reading-frame, and Nature appears to have used this as a device to repress the translation of certain proteins. The most studied example of this is the mRNA for the GCN4 regulatory protein of yeast, where four such short open reading-frames are found preceding the initiation codon [116].

There is ample experimental evidence that mutation of AUG initiation codons markedly decreases translation (reviewed in [105]), so that the first reports of initiation at non-AUG codons generated considerable surprise. These, the best cellular examples of which involve CUG codons, appear to be for very weakly expressed proteins [117–119], and therefore seem to represent a further example of how deviations from the optimal situation for operation of the scanning mechanism are used to regulate translation.

In the instances cited above there is no question that the basic mechanism of 'scanning' from the 5'-end of the mRNA still holds. However, there is one case, that of the picornaviruses, where an entirely different mechanism has to be invoked. Picornaviruses, such as polio, are peculiar in having uncapped mRNAs which do not require eIF-4F for their translation; and indeed this is fundamental to the manner in which they subvert the protein synthetic machinery of the cell (section 12.9.5). Although this does not itself preclude scanning from the 5'-end of the mRNA, this appeared unlikely because of the long 5'-leader with multiple AUG codons, some in strong contexts, preceding the actual initiation codon of the polyprotein that polio encodes. There now seems no doubt that there is internal initiation in this case, as it has been shown that a 450-nucleotide segment of this leader, not including the extreme 5'-end, can allow translation of the erstwhile silent second member of laboratory-generated bicistronic mRNAs when inserted in front of the latter [114]. The secondary structure of the internal entry site provided by this 450 nucleotide segment seems to be important, and there is some evidence that this may be recognized by a cellular protein [120], but otherwise the details of the mechanism of such internal initiation are obscure. It is clear, however, that a completely distinct mechanism exists for the complex and fundamental process of initiation of protein synthesis, raising the question whether this could have evolved in picornaviruses. It is perhaps easier to envisage the virus acquiring a mechanism of initiation evolved by a few host mRNAs to enable them to operate the type of translational regulation described in section

12.9.5, and the existence of a functional internal initiation sequence has recently been demonstrated for one host mRNA [121].

12.5.2 Chain elongation [122]

The eukaryotic factors required for binding aminoacyl-tRNA to the ribosome during the elongation cycle are analogous to EF-Tu and EF-Ts of prokaryotes. EF-1α, which is structurally related to EF-Tu [81], forms a ternary complex with aminoacyl-tRNA and GTP; and a second activity, EF-1$\beta\gamma$ (or perhaps EF-1$\beta\gamma\delta$), functionally analogous to the monomeric EF-Ts, promotes release from EF-1α of GDP, and exchange for GTP. Like prokaryotic EF-Tu, eukaryotic EF-1α is an extremely abundant cellular protein [123], and is encoded by two separate genes [124].

The elongation factor involved in the translocation reaction in eukaryotes, EF-2, is structurally and functionally related to EF-G in prokaryotes. One point of interest is the specific inactivation of EF-2 (but not EF-G) by bacterial diphtheria toxin [125], which transfers ADP-ribose from NAD^+ to an unusual modified histidine residue [126] in EF-2. Although there has been a report of an enzyme with an analogous activity to that of diphtheria toxin in eukaryotic cells [127], it is still uncertain whether ADP-ribosylation of EF-2 represents a normal mechanism of cellular regulation.

Eukaryotic, like prokaryotic, ribosomes possess an intrinsic peptidyltransferase activity, which has also been studied using the 'fragment' reaction (section 12.4.2) [128]. Although the eukaryotic peptidyltransferase is inhibited by certain antibiotics (e.g. sparsomycin) that inhibit prokaryotic peptidyltransferase, it is resistant to the action of others (e.g. chloramphenicol).

One surprising feature of eukaryotic elongation is a species difference: yeast and some fungi have a third elongation factor, EF-3, not required by a range of other eukaryotes [122]. Although yeast EF-3 has been purified and its gene cloned, its precise function is as yet unclear.

12.5.3 Chain termination

In eukaryotic termination there is a single factor, RF, to recognize all three termination codons, UAA, UAG and UGA, and hence this is functionally equivalent to both RF-1 and RF-2 of prokaryotes [129]. A second eukaryotic release factor, equivalent to bacterial RF-3, has also been reported and warrants further study in view of the fact that the eukaryotic termination reaction shows a clear requirement for GTP hydrolysis *in vitro*. The amino acid sequence of eukaryotic RF shows no similarity to those of prokaryotic RF-1 and RF-2 (which are quite similar to one another [130]), although, intriguingly, it contains a region homologous to one in the tryptophanyl-tRNA synthetases [131].

Although one imagines that there is a eukaryotic factor analogous to RRF, none has so far been described. Eukaryotic ribosomes are liberated from polysomes as subunits, and a pool of inactive 80S monomers is present in eukaryotic cells [132]. In contrast to prokaryotes there is both an anti-association activity, eIF-3 (section 12.5.1), which binds to 40S subunits, and an 80S subunit dissociation factor, eIF-6 [101].

The usage of the three termination codons in eukaryotes is much more random than in prokaryotes (section 12.4.3), although UAG is still the codon least frequently used. In vertebrates the relative occurrence of UGA:UAA:UGA is approximately 2.5:1.7:1 [133].

12.6 THE RIBOSOME

The focus of this section will be *E. coli*, the ribosomes of which have been the subject of extensive study [134–138]. Further information about eukaryotic ribosomes can be found elsewhere [139, 140].

12.6.1 The structure of the ribosome

(a) *Overall features*

The overall size and shape of ribosomes have been analysed by several different techniques. At the most basic level was the physical separation of the ribosomes into subunits (by removal of the Mg^{2+} that holds them together) and the determination of their sedimentation coefficients in the ultracentrifuge: 30S and 50S in the case of the 70S bacterial ribosome, and 40S and 60S in the case of the 80S eukaryotic ribosome of mammals. (The sizes of the ribosomes of other eukaryotes may differ from this.) Initial small-angle X-ray scattering studies provided values for the overall dimensions of the ribosomes and their subunits, but their gross features have been deduced from electron microscopy of individual negatively stained particles. Although inherent problems in this latter method limit the detail that can be obtained, useful models have been derived from such work, and the one presented in Fig. 12.21 is generally used as a reference for studies on the organization of the ribosomal components [141]. Its most obvious features are a 'cleft' and 'platform' on the 30S subunit, and an elongated 'stalk' protruding to one side of the 'central protuberance' in the 50S subunit.

Refinements have been made in the methodology for interpreting such electron micrographs [142], but the way to more substantial advances would seem to lie through

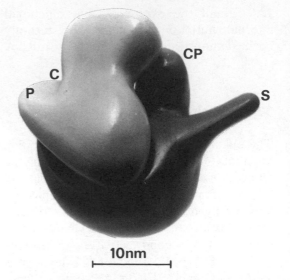

Fig. 12.21   Model of the 70S ribosome of *E. coli* based on electron-microscopic studies by Lake [141]. The 30S subunit is light, and the 50S subunit is dark. The cleft (C) and platform (P) of the 30S subunit, and the stalk (S) and central protuberance of the 50S subunit, referred to in the text, are indicated.

diffraction methods. Three-dimensional crystals of ribosomes which diffract X-rays have been obtained, but these have not yet yielded sufficient resolution to be useful. However, significant progress has been made with three-dimensional image reconstruction of the electron diffraction patterns obtained with two-dimensional crystalline sheets of ribosomes [143]. One striking feature revealed by this technique is a 'tunnel' through the centre of the 50S ribosomal subunit (Fig. 12.22), and a similar feature has been observed in independent studies of eukaryotic 60S ribosomal subunits [144]. This tunnel is large enough to accommodate a nascent peptide of the length (25–40 amino acid residues) that the ribosome is known to protect against proteolytic digestion [145], and it is most likely that this feature represents an exit channel for the nascent peptide.

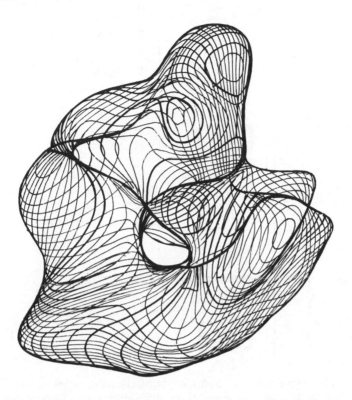

Fig. 12.22 Computer graphic representation of a reconstructed model of the 50S ribosomal subunit of *Bacillus stearothermophilus* obtained from two-dimensional sheets, negatively stained with gold thioglucose (courtesy of Dr A. Yonath).

What of the physical location of other species interacting with the ribosome? The 'length' of the tRNA (75 Å − 7.5 nm) is such that it could be accommodated in the space between the two ribosomal subunits; but this is purely speculation. Immune electron microscopy (see below) suggests that puromycin, and hence the acceptor stem of aminoacyl-tRNA, may interact with the 50S subunit near the base of the 'central protuberance' [146]. Related experiments suggest that the anticodon of the tRNA interacts with the mRNA at the 'platform' of the 30S subunit [147]. It has long been known from nuclease protection experiments that 35−50 nucleotides of mRNA (i.e. 12−17 codons) are in contact with the ribosome. A more recent sophisticated analysis revealed two internal cleavage sites that could represent 'bends' in the mRNA, and have been interpreted in terms of the mRNA 'looping' around the ribosome to re-emerge near its point of entry [148].

(b) *Components*

The composition of the ribosome is approximately 60% RNA and 40% protein. The smaller ribosomal subunit contains a single species of RNA (16S or 18S RNA for 30S and 40S subunits, respectively) together with ribosomal proteins; the larger ribosomal subunit contains a major species of RNA (23S or 28S RNA for 50S and 60S subunits, respectively), an additional small 5S RNA, together with a number of proteins exceeding that in the small subunit (Fig. 12.23). In eukaryotes the portion of the primary

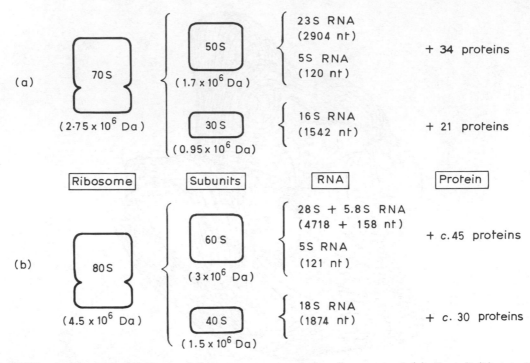

Fig. 12.23 A schematic diagram of the components of the ribosomes of (a) *E. coli* (b) rat. The molecular masses of the particles are the mean of physical determinations. The chemical values for the M_r of 70S, 50S and 30S ribosomal particles from *E. coli*, based solely on RNA and protein content, are 2.3×10^6, 1.45×10^6 and 0.85×10^6. The discrepancy between the two sets of values can be accounted for by the presence of metal ions and spermidine.

transcript that gives rise to the major species of rRNA in the 60S subunit undergoes further nucleolytic cleavage after it has started adopting its secondary structure (section 8.4.3). This results in a 5.8S RNA species hydrogen-bonded to the 28S RNA in the case of mammals. Similar 5'-post-transcriptional processing gives rise to a 2S RNA species in *Drosophila* and a 4.5S RNA species in plant chloroplasts (sections 9.4.2, 9.4.3 and 9.7.2). As discussed further below, from a functional point of view these RNAs are better considered to be integral parts of the larger rRNA species.

The ribosomal RNAs contain a small number of specific modified (largely methylated) nucleotides (section 11.6.3)

[149, 150]. In the case of *E. coli*, 10 base methylations have been identified in both 16S and 23S rRNA; and the 23S rRNA also contains three pseudouridine residues and a few ribothymidine residues and ribose methylations. In mammals there are 46 and 71 methylated groups on the 18S and 28S RNA, respectively, most of which involve the 2'-*O* of the ribose moiety. They are predominantly clustered in the 5'-half of 18S rRNA and the 3'-half of 28S rRNA [151]. Eukaryotic rRNAs also contain pseudouridine residues: approximately 37 and 60 for 18S and 28S rRNA, respectively. Neither eukaryotic nor prokaryotic 5S rRNA contains modified nucleotides.

The ribosomes of eukaryotes, prokaryotes,

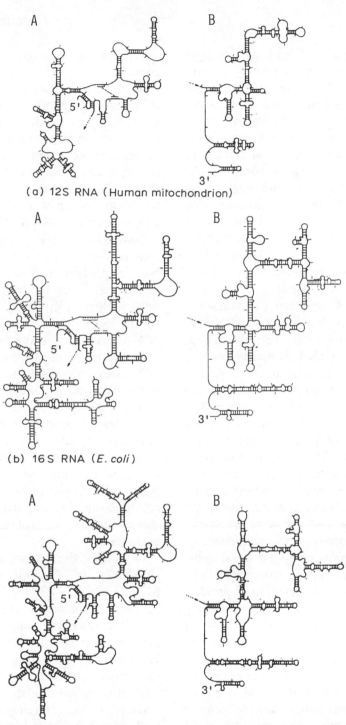

(a) 12S RNA (Human mitochondrion)

(b) 16S RNA (E. coli)

(c) 18S RNA (X. laevis)

Fig. 12.24 Comparison of the secondary structures of the smaller ribosomal RNA from (a) a human mitochondrion, (b) a prokaryote and (c) a eukaryote. Each structure has been separated into two parts for clarity of presentation. These early proposals have undergone subsequent detailed refinement, but without in any way invalidating the comparative approach. (After [152], with permission.)

mitochondria and chloroplasts catalyse similar reactions. It seems reasonable, therefore, to assume that their RNAs serve similar functions, whatever these might be. In comparing rRNAs that diverge markedly in size (from *c*. 600 to 2000 nucleotides for small subunit rRNAs, and from *c*. 1200 to 4700 nucleotides for large subunit rRNAs [149]) attention has therefore centred on common structural features that may be crucial to such functions. The most evident similarity lies in the secondary structure, in which a common core is present which becomes progressively more elaborated as the rRNA increases in size (Fig. 12.24). It is such comparison of secondary structures that makes it apparent that the 5.8S rRNA of mammalian ribosomes, when hydrogen-bonded to the 28S rRNA, is equivalent to a structural feature present in *E. coli* 23S rRNA [153]. At corresponding regions in the secondary structures of rRNAs from different species (predominantly in single-stranded regions) it is possible to identify a small number of relatively short, highly conserved, regions of primary structure. The significance of these will be discussed in section 12.6.2.

The initial models of a common secondary structure for the rRNAs were primarily based on comparisons of the sequences of rRNAs of different phylogenetic origins. Evidence was obtained from situations where there is poor conservation of primary structure in a feature of proposed secondary structure, which is maintained because of compensatory base changes (e.g. an A → G change in one strand of the RNA is compensated for by a U → C change in the complementary base of the other strand). Experimental confirmation and refinement of the models was then necessary, and was of especial importance where there was either extreme conservation or divergence of primary structure. One approach was to identify regions of single-stranded RNA, which can be done using certain nucleases, chemical reagents and oligonucleotide probes. A particularly extensive study used chemical probes for all four bases, and identified the resulting modified exposed bases by their property of terminating reverse transcription, primed by a battery of oligonucleotides complementary to rRNA [154]. The other, more laborious approach, was identification of double-stranded regions, which involved RNA–RNA cross-linking followed by isolation and identification of base-paired fragments. Such cross-links have been important in building models for the tertiary structure of the rRNAs [155, 156], one of which is shown in Fig. 12.25. Models for the secondary and tertiary structure of 5S rRNA have also been proposed [157].

One feature of the hairpin loops in the rRNA secondary structure (section 2.6) is worth reiterating. This is the frequency with which the loops are found to consist of the tetranucleotides, GNRA or UNCG. Analysis of these has revealed particular structures in which there is a base-pair between the first and fourth nucleotide (G:A and U:G, respectively) [317]. The resulting stability imparted to the loops by these sequences explains their frequent occurrence.

Originally the ribosomal proteins of the ribosomes of *E. coli* were enumerated S1–S21 and L1–L34 for the small and large subunits, respectively. It is still thought that the 30S ribosomal subunit has these 21 distinct proteins, one copy per 30S subunit. However, there has been subsequent revision of the status of the proteins of the 50S subunit. It emerged that the species designated L8 was, in fact, a complex of L7/L12 and L10, and that L26 is identical to S20, there being, on average, 0.2 copies of L26 and 0.8 copies of S20 per 70S ribosome [158]. Furthermore, two additional proteins, A and B, were detected on the ribosome, and as

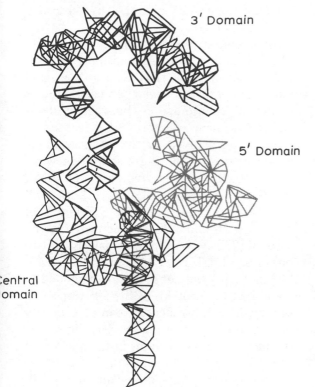

3′ Domain

5′ Domain

Central
domain

Fig. 12.25 Model proposed by Noller and co-workers for the folding of *E. coli* 16S rRNA. The 3′-domain is at the top of the figure and the 5′-domain at the right. (From [155], with permission.)

plete amino acid sequences of all the ribosomal proteins of *E. coli* have been determined. The structures of a couple of ribosomal proteins have been determined by X-ray crystallography and, on the basis of these, possible regions for interaction with rRNA have been proposed [160, 161].

Amino acid sequences are at present available for approximately half the eukaryotic ribosomal proteins, but so far only eight of these have been shown to be statistically significantly related to the sequences of proteins of *E. coli* (to S3, S5, S10, S14, S16, S17, S19 and L6 of the latter). In addition, there are grounds for thinking that certain ribosomal proteins of eukaryotes are functionally homologous to L7/L12 of *E. coli*. These include the size of these proteins, their acidic nature and their occurrence in multiple copies. There is also an indirect primary structure link through mutual relatedness to a corresponding archaebacterial (section 12.7.1) ribosomal protein [162]. Two forms of this acidic ribosomal protein, with similar but distinct primary structures, occur in eukaryotes. A third related, but larger, protein may be functionally equivalent to L10 (section 12.6.2) [163]. All these eukaryotic acidic proteins are phosphorylated to various extents; a modification found in another eukaryotic ribosomal protein (section 12.9.4) but not in prokaryotic ribosomal proteins [164].

(c) *Organization*

We now turn to the question of the organization of RNA and ribosomal proteins in the ribosome. The relative positions of ribosomal proteins to one another have been determined by cross-linking with cleavable bifunctional reagents [165] and by neutron scattering [166]. In the latter, extremely powerful, technique, ribosomes are reconstituted from their components, but with

these are chemically basic and have genes that map in clusters of other ribosomal proteins (section 12.9.2) they have now been redesignated as ribosomal proteins L35 and L36 [159]. Thus it is best regarded that there are 34 distinct proteins on the 50S subunit. All but two of these proteins are completely dissimilar and are present as single copies. The exceptions are L7 and L12, L7 being the *N*-acetylated form of L12. These two proteins together are present in a total of four copies per 50S subunit. Apart from proteins S1, S6, L7 and L12, the ribosomal proteins are all chemically basic, with values of M_r in the range 9000–35 000. The com-

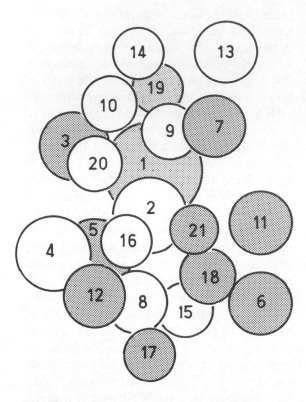

Fig. 12.26 The relative positions of the proteins of the 30S ribosomal subunit of *E. coli* derived from the neutron scattering studies of Moore and co-workers. (From [166], with permission.)

two of their proteins replaced by ones of a neutron density that contrasts with that of the rest of the particle. This is achieved by growing *E. coli* in media with appropriately different proportions of D_2O. The results of such measurements for the proteins of the 30S ribosomal subunit are shown in Fig. 12.26. A more indirect method is 'immune electron-microscopy', although this has the advantage that it relates the positions of the proteins to the gross features of the ribosomes. This method involves electron microscopic examination of ribosomal particles cross-linked by specific antibodies to individual proteins [141, 167]. It is re-assuring that there is broad agreement

between the results obtained by these different techniques.

To integrate models such as that in Fig. 12.26 with the three-dimensional models for the structure of rRNA both direct and indirect approaches have been used. Studies of the direct interactions between proteins and RNA have shown that a subset of proteins is involved in the primary interaction with RNA during assembly *in vitro*, and probably also *in vivo*. For 16S RNA and 5S RNA these proteins are fairly well defined as S4, S7, S8, S15, S17 and S20; and L5, L18 and L25, respectively. In the case of 23S RNA an original list of 10 proteins (L1, L2, L3, L4, L6, L13, L16, L20, L23 and L24) was subsequently extended to about half the proteins of the 50S ribosomal subunit. This may reflect a greater complexity of structure, but probably also illustrates the fact that most proteins in the assembled ribosome are in contact with RNA. Thus, using chemical cross-linking reagents [156], it was possible to demonstrate interaction with RNA for proteins other than those involved in the assembly of ribosomes [134, 166]. This approach has provided more precise information than obtained previously by nuclease-protection experiments. The indirect approach to determining the relative positions of RNA and protein has been to compare their electron microscopic locations on the ribosome. Initially electron-microscopic visualization was only possible for regions of rRNA against which antisera could be raised, such as the 5'- and 3'-ends and certain methylated bases. However, after the secondary structure of rRNA had been established it became possible to use biotinylated oligonucleotides complementary to single-stranded regions, and to visualize these after cross-linking with avidin [168].

On the basis of all these different types of information, models of the relative positions

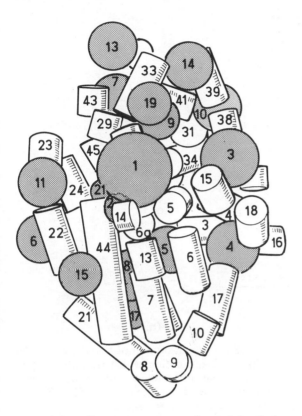

Fig. 12.27 Computer-generated model of the 30S ribosomal subunit of *E. coli* proposed by Brimacombe and co-workers. Helical regions of the rRNA are represented as cylinders and the proteins as spheres, as in Fig. 12.26. (From [156], with permission.)

of RNA and protein in the 30S subunit have been proposed [155, 156], one of which is shown in Fig. 12.27.

12.6.2 Ribosomal structure–function relationships [169, 170]

What is the relationship between the structure of the ribosome already described and its function in catalysing protein biosynthesis? It is evident that some of the components – proteins and portions of the RNA – play only structural roles; moreover bacterial ribosomes with certain individual

proteins deleted have been found to be viable [171]. Although it is not yet possible to describe the operation of the ribosome in terms of its constituent parts, it has been possible to identify components of importance to different aspects of ribosomal function. These are present in different domains on the ribosome, and we shall concentrate our discussion on three of these, with emphasis on RNA, rather than on proteins. This is because the experimental evidence is increasingly in line with the evolutionary speculation [172, 173] that RNA has the primary functional role in ribosomes.

(a) *The elongation factor domains*

There is a domain on the 50S subunit involved in the action of EF-G and its intrinsic GTPase activity. It is associated with the 'stalk' of the subunit (Fig. 12.21), which is made up of the elongated L7/L12 tetramer, at the base of which is ribosomal protein L10, and in the vicinity of which is ribosomal protein L11. Proteins L7/L12 are essential for the binding and GTPase activity of EF-G, and L11, although not functionally indispensable, is labelled by affinity analogues of GDP, and has an rRNA binding site (nucleotides 1052–1112) that overlaps the region (1055–1081) to which EF-G can be cross-linked (Fig. 12.28(a)). Strong evidence that this region of 23S RNA is functionally involved in the action of EF-G comes from studies involving the antibiotic thiostrepton, a specific inhibitor of the action of this factor. The natural producer of thiostrepton, *Streptomyces azureus*, unlike *E. coli*, is methylated at nucleotide A_{1067}. Mutation of this same nucleotide from A to a pyrimidine, either under antibiotic selection or by site-directed mutagenesis, conveys resistance to thiostrepton. Furthermore, this region of rRNA (in Domain II of Noller [155]) contains highly conserved portions of primary

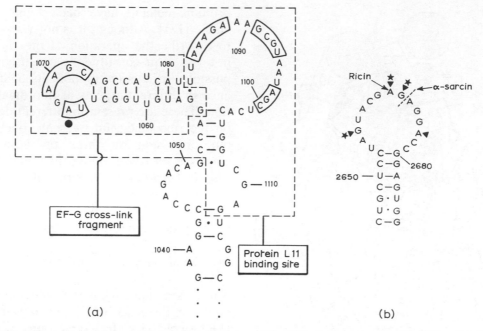

Fig. 12.28 The proposed elongation factor domains of the 50S ribosomal subunit of *E. coli* (after [169] and [155], with permission). (a) Part of Domain II of Noller: ● nucleotide which is methylated in *Streptomyces azureus* and confers resistance to thiostrepton; (b) part of Domain II of Noller: ★, base protected from chemical modification by EF-Tu; ▲, base protected from chemical modification by EF-G. The boxed areas are highly conserved in eubacterial and organelle ribosomes.

structure, consistent with a functional role.

There is a body of evidence suggesting that EF-G and EF-Tu interact with the ribosome in the same region. However, it appears that EF-Tu does not bind to the region just described, consistent with the fact that its action is not inhibited by thiostrepton. It has been shown by RNA 'footprinting' with chemical probes that there is a distinct region of the ribosome where there is overlap between the binding of EF-G and EF-Tu, and this involves the conserved loop in the region of nucleotide 2600 (in Domain VI of Noller [155]). Moreover this loop (Fig. 12.28(b)) is known to be functionally important because it is the target of the cytotoxins α-sarcin (which cleaves between G_{2661} and A_{2662}) and ricin (which removes A_{2660}). One possibility is

that these nucleotides are important for triggering the GTPase activities of the two elongation factors to similar effect. In the case of EF-G this would also need to involve the other domain, even though the two regions of 23S rRNA are far apart in both primary and secondary structure.

In focusing on the domains on the ribosome associated with the action of EF-G, one should not neglect the dynamic aspects of the translocation of peptidyl-tRNA from the A- to the P-site. This may involve changes in the conformation of the ribosome as a result of events on the ribosome. It is possible that those changes of conformation that have been detected (by distal changes in protected nucleotides) may involve 'switches' between alternative rRNA secondary structures.

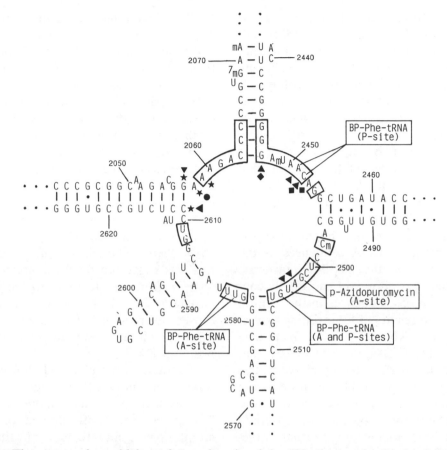

Fig. 12.29 The proposed peptidyltransferase domain of the 50S ribosomal subunit of *E. coli* (after [174], with permission). ★, Site of base change conferring resistance to erythromycin; ▲, site of base change conferring resistance to chloramphenicol; ◆, site of base change conferring resistance to anisomycin; ● nucleotide which is methylated in *Streptomyces erythraeus* and confers resistance to erythromycin; BP-Phe-tRNA, 3-(4'-benzoylphenyl)propionyl-Phe-tRNA. The boxed nucleotides are highly conserved between different organisms.

(b) *The peptidyltransferase domain*

The peptidyltransferase catalytic activity is the most fundamental functional feature of the ribosome, and, before the discovery of ribozymes (section 11.3), it was assumed that it must reside in an individual ribosomal protein. The major tools in probing this activity have been the drugs that specifically inhibit it; but, although several proteins (especially L2, L16 and L27) were implicated in the peptidyltransferase centre by affinity labelling with antibiotic derivatives, none of them satisfied the criteria for catalytic activity. Although, likewise, it has not been possible to demonstrate catalytic activity associated solely with the rRNA, attention became focused on the latter because it was discovered to be the site of point mutations that confer certain mitochondrial ribosomes with resistance to drugs such as chloramphenicol or erythromycin. These mutations fall in a non-hydrogen-bonded loop (in Domain V of Noller [155]) that forms a junction from which several hairpin stems emanate (Fig. 12.29), and many nucleotides in this junction are conserved. Furthermore, affinity-labelling and 'foot-printing' experiments with aminoacyl- or peptidyl-tRNAs or their derivatives allow assignment of parts of the A- and P-sites to conserved nucleotides in this central loop [155, 174, 175]. The key role of the 3'-terminal CCA of tRNAs in the peptidyl-transferase 'fragment' reaction (section 12.4.2) suggests that this might base-pair to complementary RNA sequences at the A-and P-sites of the peptidyltransferase centre. There are, however, no totally conserved 3'-UGG-5' sequences in the 'central loop', and the results of footprinting experiments with tRNAs lacking increasing portions of the CCA sequence suggest that any nucleotides that base-pair with this region would be physically separated from one another in the primary structure and require to be brought together in the tertiary structure [176]. Models for the tertiary structure of 23S rRNA are starting to emerge (e.g. [318]) and should help clarify this question.

(c) *The decoding domain*

This is the domain on the 30S subunit where the codon–anticodon interaction occurs, and this is close to the region of the 'Shine and Dalgarno' interaction between the mRNA and the 3'-end of the 16S RNA (section 12.4.1), which may be modulated by ribosomal protein S21. Although the 'Shine and Dalgarno' interaction is specific to pro-karyotes (section 12.4.1), the decoding domain has features that are conserved in prokaryotes and eukaryotes, especially the non-base-paired region from nucleotides 1392–1407 of the 16S RNA of *E. coli* (Fig.

12.30). This region contains a nucleotide (C_{1400}) that can be cross-linked to the modified base, cmo^5U, at the wobble position of $tRNA^{Val}$ when this is bound to the P-site of *E. coli* or yeast ribosomes [177]. Furthermore, resistance to certain antibiotics that cause misreading was associated with methylation or mutation at positions 1405, 1408 and 1409 [178, 179]. Site-directed mutagenesis suggests that C_{1400} is important for ribosome assembly, but that the adjacent base (G_{1401}) is necessary for function [180].

Also in this general area is the highly conserved sequence $m_2^6Am_2^6A$ (nucleotides 1518–1519). The functional importance of this was suggested by the characteristics of an *E. coli* mutant resistant to the antibiotic kasugamycin, which inhibits initiation and causes misreading. The ribosomal RNA from this mutant was unmethylated at these positions [181].

It should be emphasized, however, that not all antibiotics that cause misreading interact with this domain in the 30S subunit. The most thoroughly studied of these is streptomycin, mutations conferring resistance to which occur in both the 900 region (C_{912}) and the 530 region (A_{523} and C_{525}) of 16S RNA, the latter involving a pseudoknot [320]. The role of these latter regions of 16S rRNA may relate to the participation of other species in the control of ribosomal accuracy, as is described in the next section.

12.6.3 Ribosomal optimization of translational accuracy [182]

We have already seen how a mechanism exists to minimize misreading of the genetic message as a result of mischarging of tRNA (section 12.3.2). Misreading can also occur as the result of incorrect codon–anticodon recognition, and reference has been made in section 12.6.2(c), above, to the effects of certain antibiotics on this. Although some mu-

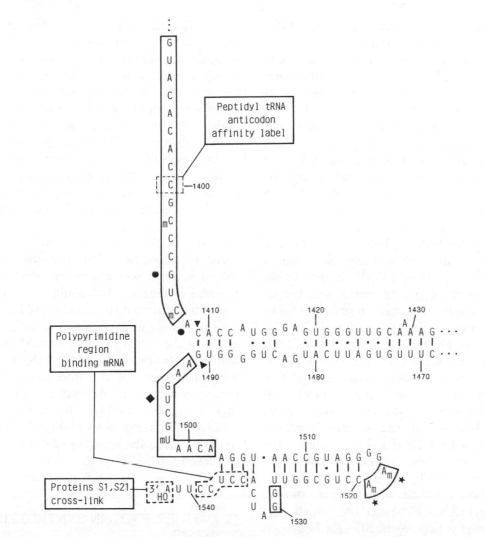

Fig. 12.30 The proposed decoding domain of the 30S ribosomal subunit of *E. coli*. The RNA secondary structure is from [156] and diverges from that of [155] with respect to the hydrogen bonding of the polypyrimidine region at the 3'-end. ★, Nucleotides $m_2^6A_{1518}$ and $m_2^6A_{1519}$, the demethylation of which confers resistance to kasugamycin; ▲, site of base change conferring resistance to paromomycin; ◆, site of base change conferring resistance to hygromycin B; ● nucleotides which are methylated (as m^7G_{1405} and m^1A_{1408}) in organisms producing certain aminoglycoside antibiotics, and confer them with resistance to these. The boxed nucleotides are those not involved in secondary structure that are highly conserved between different organisms.

tations conveying resistance to streptomycin have now been shown to involve RNA, the mutants most pertinent to the question of the accuracy of decoding involve ribosomal proteins. These led to the realization that ribosomes with different error frequencies are possible. Thus, many streptomycin-resistant mutants have alterations in ribosomal protein S12 that confer greater accuracy to the ribosomes; whereas revertants from streptomycin-dependence (also involving S12) have mutations in ribosomal protein S4 that result in increased misreading [183]. Indeed, a role for ribosomal protein S12 in modulating translational accuracy is indicated by the fact that ribosomes lacking S12 are more accurate than those of the wild-type [184].

The accuracy of decoding found *in vivo* is difficult to reconcile with the energy difference between cognate and non-cognate codon–anticodon interactions. Although recognition may involve more than the anticodon of the tRNA (there is a suppressor tRNA with an unaltered anticodon, but a base change in loop I [185]) it seems that the accuracy of the ribosome is achieved by the multiplicative effect of several steps of 'kinetic' proofreading (cf. section 12.3.2), as proposed by Hopfield [186] and Ninio [187]. There is evidence that the first step of proofreading (some workers prefer to call this stage 'recognition', and reserve 'proofreading' for the subsequent checking step) occurs in the binding of the ternary complex, aminoacyl-tRNA.EF-Tu.GTP, when the complex either binds to the ribosome, with hydrolysis of GTP, or dissociates; whereas the second proofreading is of the bound aminoacyl-tRNA.EF-Tu.GDP, which can either form a peptide bond, with liberation of EF-Tu.GDP, or dissociate from the ribosome [188]. In effect, the recognition of the cognate aminoacyl-tRNA by the ribosome is a question of whether the aminoacyl-tRNA dissociates from the ribosome before the GTP is hydrolysed or the EF-Tu.GDP is released: the time taken for GTP hydrolysis provides a 'kinetic yardstick' for proofreading. It has been suggested that the hydrolysis of GTP during the action of EF-G is also involved in proofreading, but, although there are some data that are consistent with this view, it cannot be regarded as proven [182].

The general view of the nature of ribosomal accuracy described above made it possible to resolve an apparent paradox. This was that, if it is possible to select for ribosomes that have increased accuracy, why has Nature not already done this? The most reasonable answer (and the one that is thought to be correct) is that perfect accuracy is only possible if translation is infinitely slow, and hence the actual error frequency is the best compromise between accuracy and speed. However, studies *in vitro* showed that ribosomes of different accuracy exhibited no differences in their maximum speed of translation. This paradox was resolved when it was realized that what was relevant to kinetic proofreading was not the maximum rate that ribosomes could achieve with saturating ternary complex, but the rate at which the ternary complex could interact with the ribosome. When the values of K_m were measured for the interaction of the ternary complex with ribosomes it was found that the K_m increased (and hence the rate of cellular protein synthesis would be expected to decrease) as the accuracy of the ribosome increased [182, 188].

12.7 OTHER PROTEIN-SYNTHESIZING SYSTEMS

12.7.1 Archaebacteria [189, 190]

In the discussion of protein-synthesizing systems so far, a distinction has been made between the nucleo-cytoplasmic system of eukaryotes and the system found in those prokaryotes represented by the commonly studied bacteria (e.g. *E. coli*) and their viruses. There is good reason, however, to think that the prokaryotes comprise not one, but two, kingdoms. These are the *eubacteria*, into which kingdom the majority of bacteria fall, and the *archaebacteria*, a kingdom containing methanogens, extreme halophiles and certain thermoacidophiles. It is appropriate to discuss briefly the protein-synthesizing system of archaebacteria; not only because this shows certain differences from eubacteria and eukaryotes, but because analysis of the archaebacterial translation apparatus has provided the strongest evidence that the archaebacteria do in fact constitute a separate kingdom.

Archaebacteria have certain structural characteristics that distinguish them from eubacteria. Their cell walls (where these are found) lack peptidoglycan, and their lipids are unique in consisting of branch-chain fatty acids, and being linked to glycerol by an ether, rather than an ester, link. This, in itself, would hardly be sufficient to establish them as a separate kingdom. The evidence that does so comes from phylogenetic studies on ribosomal RNA. Woese and Fox [191] showed that the sequence divergence between the smaller (16S or 18S) rRNAs of eukaryotes and eubacteria clearly distinguished these two kingdoms. (This work was actually done by analysis of oligonucleotides before complete sequences were available.) When the rRNAs of the different archaebacteria were examined it was clear that they were related to one another, but were no more closely related to eubacteria than they were to eukaryotes.

The protein-synthesizing systems of different archaebacteria are not all identical, but, as a generalization, they can be said to show some similarities to those of eubacteria, some to those of eukaryotes, and to have some unique features. Although archaebacterial ribosomes have a sediment coefficient of 70S, rather than 80S, they have an additional morphological feature (a 'bill') also present in eukaryotic ribosomes. Most striking, however, is the sensitivity of archaebacterial ribosomes to antibiotics. Thus, certain archaebacterial ribosomes are insensitive to some antibiotics (chloramphenicol and kanamycin) that had previously been regarded as specific for prokaryotic ribosomes, and sensitive to at least one inhibitor (anisomycin) that had been regarded as specific for eukaryotic ribosomes. The position of archaebacteria in relation to eukaryotes and eubacteria is also apparent from the sequences of their ribosomal proteins [192, 193]. Although the genes for these are organized in clusters similar to those in *E. coli*, only some of the genes in these clusters encode proteins that are related to eubacterial ribosomal proteins. Some are related to eukaryotic ribosomal proteins, and others appear to be unique (although a relationship to eukaryotic proteins not yet characterized cannot be excluded at present). Two proteins of particular interest are the equivalents of *E. coli* L10 and L7/L12, the components of the 'stalk' on the large ribosomal subunit important for translocase function (section 12.6.2(a)). Although it is difficult to see a relationship between the eukaryotic and eubacterial proteins here, the similarity of the archaebacterial proteins to both eubacterial and eukaryotic proteins establishes this and argues for a functionally conserved role.

The sequences of the elongation factors from archaebacteria bear much greater similarity to those of eukaryotes than they do to those of *E. coli*. This is reflected in the fact that, as in eukaryotes, the translocase from archaebacteria has the modified histidine residue that can be ADP-ribosylated by diphtheria toxin, and an EF-Tu/EF-1α equivalent that is insensitive to the antibiotics, pulvomycin and kinomycin. Another apparent difference from eubacteria is the lack of formylation of the methionyl residue on the initiator tRNA, although further characterization of the initiation process is lacking at present.

12.7.2 Mitochondria [194–196]

The majority of mitochondrial proteins are encoded by mRNAs transcribed from nuclear genes, and these mRNAs are translated on 80S ribosomes in the cytosol, the resulting proteins being transported into the mitochondrion (section 12.8). The mitochondrion has its own genome (section 3.4.1) and its own machinery for transcription (section 8.5) and translation. Although the size and coding capacity of mitochondria vary quite widely between species (from *c.* 17 kb in man, to *c.* 2500 kb in some plants) most mitochondria encode their own rRNAs and tRNAs, together with a variable subset of the proteins of the electron-transfer and oxidative phosphorylation systems. The amount of coding DNA in the larger mitochondrial genomes is little greater than that in the smaller ones, as much of the extra DNA in the former is simple-sequence DNA (chapter 3).

It is generally thought that the precursors of mitochondria were aerobic bacteria that were engulfed by, and evolved in a symbiotic relationship with, an anaerobic host cell. Comparisons of rRNA sequences indicate that mitochondria are most closely related to purple photosynthetic bacteria [197]. This relationship between eubacteria and mitochondria is seen in certain functional aspects of the translational machinery, although there are unique features to the mitochondrial translational apparatus [198]. Mitochondrial ribosomes, other than those of plants, differ from the ribosomes of prokaryotes and eukaryotes in lacking the otherwise essential 5S rRNA, vestiges of which can be seen in the rRNA of the large subunit, which may subserve its function. Mitochondrial rRNAs show fewer secondary modifications (methylations, pseudouridine substitutions) than other rRNAs. As already mentioned (cf. Fig. 12.29), the sensitivity of mitochondrial ribosomes to antibiotics shows an extensive similarity to that of eubacteria. The mitochondrial ribosomal proteins (encoded in the nucleus) are not well characterized, but appear to exceed those of bacteria in number. The mitochondrial elongation factor responsible for binding aminoacyl-tRNA is functionally interchangeable with *E. coli* EF-Tu, but the translocase is not interchangeable with EF-G. Mitochondria, like eubacteria, use fMet-tRNA to initiate protein synthesis [199], and there is some evidence that the mitochondrial initiation factors resemble those of prokaryotes [198]. However, the smaller rRNAs from mitochondria lack the bacterial polypyrimidine 3′-sequence; and in any case there is no opportunity for 'Shine and Dalgarno'-type base-pairing to mammalian mitochondrial mRNAs that start directly, or almost directly, with AUG. The unique post-transcriptional editing of protist mitochondrial mRNAs has been described in section 11.8.

Perhaps the most striking features of the mitochondrial translational system (again, with the exception of higher plants) are the deviations from the standard genetic code (section 12.2.3) and the unique manner in which the codons are read by their tRNAs. This seems to be a consequence of the restricted set of mitochondrial tRNAs. The actual number varies between species (mammals have 22, yeast 25), but the number is less than the 32 required to read the standard code (Fig. 12.3) according to the wobble hypothesis (Table 12.1). In the case of mammalian mitochondria (although not in those with larger genomes) there are apparently no separate genes for $tRNA_f^{Met}$ and $tRNA_m^{Met}$, even though both Met-tRNA and fMet-tRNA are found. It is assumed that the single gene gives rise to both tRNA species through differential post-transcriptional modification. As already described, these restricted sets of tRNAs are able to decode the mitochondrial genome by a different set of wobble rules (Table 12.1). However, there are some protists (*Tetrahymena pyriformis*, *Trypanosoma brucei*, *Chlamydomonas reinhardtii*) in which the number of tRNAs encoded by the mitochondrial genome is too small even for this. In these cases it must be assumed that the tRNAs are encoded in the nucleus and imported from the cytoplasm.

The mitochondrial tRNAs do not conform to the general structural pattern described in section 12.3.1 and summarized in Fig. 12.7, and it is possible that these deviations are important for their unique decoding properties. In particular the mitochondrial tRNAs violate some or all of the following general features previously enumerated: the conservation of the GTΨCRA sequence in loop IV, the constant seven-nucleotide length of loop IV, the pattern of conserved bases in loop I. In individual cases even more striking violations are observed: the mammalian mitochondrial tRNA^{Ser} completely lacks loop I and its stem, and in several tRNAs from *Caenorhabditis elegans* loop IV and its stem have been replaced by a loop of variable size. Nevertheless, these tRNAs may adopt a conformation similar to that of yeast tRNA^{Phe} (Fig. 12.8) [200].

12.7.3 Chloroplasts [201–203]

The protein-synthetic apparatus of chloroplasts resembles that of bacteria even more clearly than does that of mitochondria. Indeed, rRNA sequence comparison (cf. section 12.7.1) shows that chloroplasts are specifically related to cyanobacteria, with which they also share similarities in the structure of their chlorophylls and carotenoids [197]. The genomes of chloroplasts (120–200 kb) contain much more genetic infor-

mation than those of mitochondria, and the apparent lack of a corresponding intense pressure for space has left them with a full complement of normal-sized rRNAs (including 5S rRNA). Although the standard genetic code is employed in chloroplasts, there are not quite enough tRNAs (31 in tobacco) to decode the mRNAs according to the standard pattern of observed wobble shown in Table 12.1. Unmodified Us are employed in certain anticodons to allow reading of all members of certain four-codon families [37, 202].

The chloroplast genome encodes other components of its protein-synthetic apparatus, including an initiation factor (IF-1), an elongation factor (EF-Tu, although this is absent in some cases) and somewhat less than half the ribosomal proteins. These latter show an average of 44% identity with clearly analogous proteins of *E. coli*, although there is one example of a chloroplast ribosomal protein with no eubacterial homologue [203]. The polypyrimidine sequence capable of 'Shine and Dalgarno' base-pairing is present at the 3'-end of chloroplast 16S RNA, but some chloroplast mRNAs seem to lack canonical Shine and Dalgarno sequences. If initiation codon selection involves a mechanism similar to that in eubacteria it may perhaps operate in a modified form.

12.8 THE CONTROL OF THE CELLULAR LOCATION OF THE PRODUCTS OF TRANSLATION [204]

Proteins synthesized on the cytoplasmic ribosomes of eukaryotic cells can be either retained in the cytoplasm, transferred to subcellular organelles, inserted into the plasma membrane, or secreted. Although, of course, prokaryotes do not have subcellular organelles, the other possibilities mentioned are also available to them, together with sequestration in the periplasmic space. The molecular mechanisms governing these possible fates of proteins involve recognition of features in their amino acid sequences – *targetting signals*. There are two main classes of such signals: signals that direct the translocation of proteins across a membrane separating two compartments, and signals that cause the retention of a protein in a particular cellular compartment. Some of the signals consist of a conserved amino acid sequence, but more frequently the targetting sequence can only be described in terms of its general chemical nature and position within the protein.

12.8.1 Translocation across the endoplasmic reticulum or the bacterial inner membrane [204–209]

Much work has centred on the execution of the signal for the translocation of proteins across the membrane of the endoplasmic reticulum of eukaryotes, or across the inner membrane of bacteria. This pathway is taken by secreted proteins, which have been the main focus of study, but it is also the common first step in the processes leading to compartmentation of other proteins (see below). Ribosomes synthesizing proteins destined for secretion (unlike the majority of ribosomes synthesizing proteins to be retained intracellularly) are located on the membranes of the rough endoplasmic reticulum of eukaryotes. The proteins are extruded through the membrane as they are synthesized and pass from the cisternae of the endoplasmic reticulum, via the Golgi apparatus, to secretory vacuoles. The chemical basis for the initial membrane translocation, and that occurring in prokaryotes, lies in a largely hydrophobic peptide, the 'signal peptide', at or near the

N-terminus of the nascent protein [210]. In most cases the signal peptide is an extension of the *N*-terminus of the mature protein, approximately 15–30 amino acids in length, and is later removed by proteolytic cleavage. (In some cases the signal peptide is internal, and in some cases it is not cleaved.) The core of this signal peptide is predominantly hydrophobic, but there are usually basic or neutral residues at the extreme *N*-terminus.

It seems reasonable to suppose that the hydrophobic nature of the signal peptide is necessary for it to cross the membrane. However, in order for this hydrophobic region to interact with the membrane it must remain exposed, and this means that the nascent protein must not be allowed to adopt its native conformation before the interaction can occur. It is now known that there are two mechanisms to achieve this in prokaryotes and lower eukaryotes (such as yeast), one of which is similar to the single well-studied mechanism that operates in higher eukaryotes. In higher eukaryotes the signal peptide interacts with an 11S signal recognition particle (SRP), which contains six polypeptide chains and a 7S RNA species, 260 nucleotides in length, to which the *Alu* family of repeated DNA sequences is related (Fig. 7.36, section 7.7.4(c)). This interaction, which involves the 54 kDa subunit of SRP (SRP54), may temporarily arrest protein synthesis, but, in any case, maintains the exposure of the signal peptide until the SRP interacts with a specific receptor ('SRP receptor' or 'docking protein') on the inner face of the membrane. This allows the ribosome to attach to the membrane, after which the SRP and its receptor are released, translation is resumed, and the nascent peptide is translocated through a specific channel [311] into the intracisternal space. The signal peptide is, in most cases, cleaved by a peptidase as it emerges on the luminal side of the endoplasmic reticulum (Fig. 12.31).

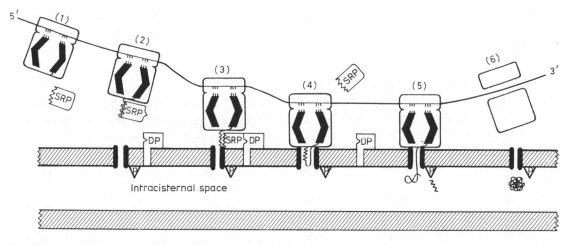

Fig. 12.31 Model for the segregation of secretory proteins into the lumen of the endoplasmic reticulum. SRP, signal recognition particle; DP, docking protein (signal recognition particle receptor); P signal peptidase. For further details see text.

There is evidence that GTP (and presumably its hydrolysis) is involved in the release of the SRP and its receptor [211], and SRP54 and the SRP receptor contain conserved motifs found in other GTPases [208]. It would be consistent with the precedents set by many other GTPases (including the elongation factors of protein synthesis) if the role of the GTP hydrolysis were to disengage these proteins after they had fulfilled their function. By analogy with EF-Tu, there have also been speculations that the time taken for GTP hydrolysis to occur may act as a 'kinetic standard' to allow dissociation of the *N*-terminal peptides of erroneously bound proteins, not destined for translocation across the membrane.

The SRP of prokaryotes escaped detection for a considerable time, even though it was known that bacterial signal peptides were functionally interchangeable with those of eukaryotes. The overall structure of bacterial SRP is still unclear, but it contains a functional counterpart of 7S RNA in the essential and abundant bacterial 4.5S RNA (the function of which had remained elusive for

many years), and a structural counterpart to SRP54 in the product of the *ffh* gene. Furthermore, a structural counterpart to the eukaryotic SRP receptor exists in the product of the *ftsY* gene [209]. One of the reasons that the SRP escaped detection for so long in bacteria was that the genetic approach that was applied to the problem only revealed components of the second mechanism for preventing protein folding (see below). However, it did lead to the identification of integral membrane proteins necessary for the process of translocation, and thus common to both mechanisms [212]. There is evidence that two of these, the products of genes *secE* (*prlG*) and *secY* (*prlA*), constitute the actual translocase, but the functions of the other two, the products of *secD* and *secF*, remain to be elucidated. Much less is currently known about the membrane translocase proteins in eukaryotes, and there may be some differences expected here, because in prokaryotes, but not in eukaryotes, the interaction of the negatively charged inner surface of the bacterial cell membrane with a positively charged residue near the *N*-terminus of the signal peptide is absolutely required for attachment to occur [213]. It also appears that the signal peptidase in eukaryotes is more complex than the general signal peptidase in prokaryotes, although their specificity is similar [214]. This cleavage specificity cannot be characterized simply, but involves recognition of an amino acid with a small sidechain in a relatively hydrophilic environment [210].

The second bacterial mechanism to prevent protein folding before translocation (also found in yeast) involves proteins termed 'molecular chaperones' or 'chaperonins'. Some of these (the GroE protein product of the *groEL*, *groES* genes, and the product of *dnaK*) are heat-shock proteins, which have a more general role in binding partially denatured proteins. However, one, the product of the *secB* gene, is specific for the secretory pathway, and transfers the nascent peptide to SecA, a peripheral membrane protein with ATPase activity [212]. The proteins protected from folding by this mechanism, unlike those protected by the mechanism involving SRP, pass through the membrane post-translationally, rather than co-translationally. The relative roles of the two pathways to the membrane still remains to be resolved. It seems likely that the one involving SRP is the more ancient, and it has been suggested that this pathway is still absolutely required for certain proteins (yet to be identified), although other proteins have evolved the ability to use the second pathway [215]. If this is so, it should be possible to identify features of the signal peptide that confer the ability to use the second pathway.

Finally it should be mentioned that a small minority of secretory proteins do not appear to possess a signal peptide of the type required by the mechanism described above,

and it has been suggested that an alternative mechanism must operate in these cases [216].

12.8.2 Fate of non-secretory proteins after translocation across the membrane of the endoplasmic reticulum

Proteins destined for secretion are thought to pass by default from the cisternae of the endoplasmic reticulum to the secretory vacuoles. However, proteins destined for other locations have additional signals to direct them to these or retain them in them.

(a) *Retention as membrane proteins [217]*

Certain plasma membrane proteins do not complete passage through the membrane of the endoplasmic reticulum, but remain embedded in it. The eventual fusion of membrane derived from the endoplasmic reticulum with the plasma membrane results in the internally facing *N*-termini of such proteins becoming exposed to the external surface of the plasma membrane. What stops such membrane proteins being secreted is a hydrophobic sequence at or near the *C*-terminus, the 'stop transfer' sequence, that anchors the protein into the membrane. This is especially well illustrated in the case of the membrane-bound and secreted forms of immunoglobulin heavy chain (Fig. 11.38), which differ only in approximately 20 amino acid residues at their *C*-termini [218]; and this *C*-terminal segment has been shown to convey membrane location when genetically 'grafted' onto a protein that is normally secreted [219].

It should be stressed that there are other orientations of proteins in the plasma membrane: external orientation of the *C*-terminus or multiple loops back and forth through the membrane. However, proteins with such orientations are inserted into the membrane in specialized manners involving internal signal sequences, for a description of which the reader is directed elsewhere [217, 220].

(b) *Retention in the lumen of the endoplasmic reticulum [221]*

The lumen of the endoplasmic reticulum contains a number of resident soluble proteins (e.g. protein disulphide isomerase, prolyl-4-hydroxylase and others that interact with newly synthesized proteins). These are prevented from being secreted by a specific *C*-terminal sequence, which in animal cells is KDEL, and in yeast is generally HDEL. It is thought that this sequence is bound by a membrane receptor in a salvage compartment, on route to the Golgi, and this receptor recycles the proteins to the endoplasmic reticulum.

(c) *Transfer to the lysosomes [222]*

Proteins such as hydrolases that are resident in the lysosomes of mammalian cells follow the secretory pathway as far as the Golgi. In the *cis* Golgi an unknown feature of these proteins is recognized by *N*-acetylglucosaminyl-1-phosphotransferase, which modifies *N*-linked oligosaccharides on the proteins, and further modification by other enzymes generates manose 6-phosphate groups. These latter constitute the signal that directs the proteins, bound by specific receptors, to the lysosomes.

A different mechanism to the above operates in yeast, and, although the details have not yet been completely elucidated, this involves the action of a protein kinase [312].

12.8.3 Translocation into the mitochondrion and direction to submitochondrial locations [223–225]

Although mitochondria encode some of their own proteins (section 12.7.2) the vast majority are encoded in the nucleus and synthesized on free cytoplasmic ribosomes. Irrespective of their ultimate submitochondrial location, most of these bear a targetting signal that directs them post-translationally to a translocation system which leads through the outer and inner mitochondrial membranes (at a point of contact between the two) into the mitochondrial matrix. Proteins of the outer and inner mitochondrial membrane have additional 'stop transfer' signals that arrest their transfer before they reach the matrix. Proteins of the intermembrane space are re-exported to that compartment on reaching the matrix.

The mitochondrial targetting sequence takes the form of an *N*-terminal extension of the mature protein, but differs from the signal peptide described in 12.8.1 in being more variable in length (*c.* 12 to > 70 residues) and being much more polar in nature. An important property of the targetting sequence is a potential for forming a positively charged amphiphilic α-helix. Soluble cytoplasmic chaperonins (including a family of 70-kDa heat-shock proteins, cytoplasmic hsp70) prevent the mitochondrial protein from folding completely after translation. The protein is then transferred from the cytoplasmic hsp70 to a receptor on the outer mitochondrial membrane, with the concomitant hydrolysis of ATP. Receptor components have been identified [224, 225], but as yeast bearing mutants in these are viable it is assumed that there are alternative receptor systems. The receptor is thought to transfer the protein to the actual transfer pore, one component of which would appear to be an essential integral outer-membrane protein (ISP42 in yeast, MOM38 in *Neurospora*). An electric potential across the inner membrane is required for transfer of the protein to the mitochondrial matrix, although this is of opposite polarity to that required for periplasmic sequestration in bacteria. Within the matrix the protein first interacts with one of the essential components of the translocation machinery, mitochondrial hsp70. It is then transferred to a second essential matrix chaperonin, mitochondrial hsp60 (a protein structurally related to GroEL of bacteria), with hydrolysis of ATP. There is evidence that hsp60 is required for correct folding of matrix proteins. The final step for matrix proteins is the cleavage of the signal peptide by a protease (MAS protease), which has two non-identical subunits.

It appears that the mechanism for the re-export of proteins to the intermembrane space is basically similar to that employed in bacterial periplasmic sequestration: the relationship of the mitochondrial matrix to the intermembrane space is similar to that of the bacterial cytoplasm to the periplasmic space, and the directionality of the membrane polarity is, of course, the same. In fact, the presumed evolutionary pre-existence of this translocation system allows one to rationalize the initial transfer of proteins to the matrix before transfer to the intermembrane space. A second, hydrophobic, portion of the signal sequence is required for this latter process, and this is finally removed by a protease, perhaps inner-membrane protease I, which is structurally related to bacterial signal peptidase.

Before concluding this section it should be stressed that the general mechanism described above does not apply to all mitochondrial proteins, some of which have specialized individual means of translocation. One example of such an exception

is cytochrome *c*, which does not have an *N*-terminal extension and is inserted directly into the intermembrane space, rather than first passing through the matrix.

12.8.4 Translocation across the chloroplast membrane [226, 227]

The problem of translocation to different compartments of the chloroplast has not received the same intense study accorded to the mitochondrion, but the picture emerging is not dissimilar. Initial targetting to the chloroplast is conveyed by an *N*-terminal extension somewhat resembling the mitochondrial one, although with the difference that it is not predicted to form an amphiphilic α-helix. This signal delivers the protein through the double membrane of the chloroplast envelope into the stroma. There is evidence for a requirement for ATP for import into the stroma, but not a membrane potential, and a stromal signal peptidase has been described. From the stroma certain proteins need to be transported further into the lumen of the thylakoids, and this process formally resembles that of mitochondrial transport from the matrix to the intermembrane space. It, too, requires a second, hydrophobic, signal sequence, and, indeed, certain proteins synthesized in the stroma on chloroplast ribosomes have just this single targetting signal.

12.8.5 Translocation to microbodies [228]

Microbodies, which include peroxisomes, glyoxisomes and glycosomes, are structurally related organelles which share the possession of a β-oxidation system. Microbody proteins are synthesized on cytoplasmic ribosomes and their transport is directed by targetting signals that do not undergo subsequent cleavage. In the case of some of these proteins this targetting signal is a *C*-terminal tripeptide sequence, SKL. However, other proteins appear to carry a different, but as yet undefined, internal targetting signal.

12.8.6 Translocation across the nuclear membrane [229, 230, 313]

At present two methods of post-translational translocation of proteins across the nuclear membrane have been identified. Small proteins pass through the nuclear pores by simple diffusion, and are then concentrated in the nucleus by specific binding to other nuclear proteins. Large proteins, however, require a targetting signal. The targetting signal, which may be as short as a pentapeptide, does not consist of a precise sequence, but contains three or more basic amino acids (lysine being more frequent than arginine). An example of such a sequence is PKKKRKVE of SV40 large T antigen, which can cause nuclear translocation when artificially introduced into proteins normally resident in the cytoplasm. The nuclear targetting sequence does not appear to have to be located at any particular position in the protein, provided that it is exposed. The translocation appears to be a two-step process: initial rapid binding to nuclear pores, probably involving an acidic receptor region, followed by slower ATP-dependent transfer through the pores.

12.9 THE REGULATION OF TRANSLATION

The regulation of protein synthesis by altering the rate of transcription of mRNA has been described in chapter 10. In eukaryotes

the sites of synthesis and usage of mRNA are physically separated, and this, and the relative stability of most eukaryotic mRNAs, provide scope for the translational control of protein synthesis. Moreover, even in prokaryotes, examples of translational regulation are to be found. The following account of translational control classifies these examples primarily in terms of the mechanisms involved. The post-transcriptional editing of mRNAs (section 11.8), and regulation of translation by antisense RNAs (section 10.7) and by codon usage (section 12.2.4) have already been described.

12.9.1 mRNA secondary structure

(a) *RNA bacteriophages [231, 232]*

The influence of mRNA secondary structure on translation differs in prokaryotes and eukaryotes because of their different modes of initiation. In the small bacteriophages, such as MS2 and Qβ, there is a striking example of the role of the secondary structure of their mRNA in the differential expression of different cistrons of a polycistronic mRNA. These single-stranded RNA phages direct the translation of four proteins from their structural RNA, and these include the coat protein and the viral subunit of the replicase. During infection, or translation *in vitro*, there is much greater synthesis of coat protein than of replicase; and it was shown that destruction of the secondary structure of RNA by mild formaldehyde treatment or by heating redressed the rate of synthesis of replicase to that of coat protein. The proposed secondary structure for MS2 RNA [233] shows the initiation site of the replicase subunit hydrogen-bonded to part of the coat protein cistron, the initiation site of the latter being well exposed (Fig. 12.32). Thus, it appears that initiation of replicase trans-

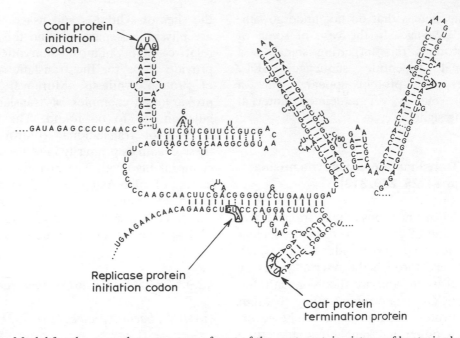

Fig. 12.32 Model for the secondary structure of part of the coat protein cistron of bacteriophage MS2. The positions of the coat protein initiation and termination codons are indicated, as well as the initiation codon for the replicase. The numbers 6, 50 and 70 indicate codons which have been mutated in the related bacteriophage f2 to produce premature termination (see text). (After [233], with permission.)

lation is restricted because it is only accessible when the coat protein cistron is being translated (which would temporally disrupt this hydrogen bonding). Experimental support for this model came from a study of chain-termination mutations in the coat protein gene. An amber mutation at the codon specifying amino acid position 6 of the coat protein severely repressed expression of the cistron for the replicase subunit, but this effect was abolished if the phage RNA was treated with formaldehyde. In contrast, amber mutations at the codons specifying the 50th, 54th and 70th amino acids did not have this effect. It can be seen from inspection of Fig. 12.32 that translation of only the first six codons of the coat protein cistron would not break the putative hydrogen bonds to the replicase subunit initiation site, but that translation to the

50th or subsequent mutant termination codons would. It now appears that the expression of a fourth, minor, protein of MS2 (the lysis protein) is also constrained by the secondary structure (and not by a requirement for frameshifting, as originally proposed).

(b) *Eukaryotic mRNAs*

Although initiation of eukaryotic protein synthesis involves the RNA helicase action of eIF-4A (section 12.5.1), there is experimental evidence that the translational efficiency of mRNAs decreases if artificial stem–loop structures are inserted in the 5′-untranslated region [234]. It is possible that a difference in secondary structure between different mRNAs could affect their translation if eIF-4A were limiting. Examples might be the α- and β-globin mRNAs, where a difference in translation was eliminated by addition of eIF-4A and eIF-4B [235], and reovirus mRNAs, which are translated to different extents during infection [236]. It has recently been suggested that such secondary structure may be a reason for the translational repression of maternal mRNAs of *Xenopus* oocytes; and evidence has been obtained that at fertilization, when translation of maternal mRNAs occurs, oocytes suddenly become capable of translating artificial mRNAs with extensive secondary structure [237]. In this respect it is interesting that an inhibitor of eIF-4F has been reported in unfertilized sea urchin eggs, which also contain translationally repressed maternal mRNA [238].

12.9.2 RNA–protein interaction

Another aspect of the role of mRNA secondary structure in translation control is that it can provide specific sites for the binding of regulatory proteins, and there are several well studied examples of this in both prokaryotes and eukaryotes.

(a) Bacteriophage coat protein

In the case of the translation of the cistrons of bacteriophage RNAs, discussed above, it is thought that the decline in synthesis of the replicase at later times of infection is caused by the binding of the coat protein (as the concentration of this increases) to a region of secondary structure containing the ribosome-binding site [239].

(b) Autogenous regulation of bacterial ribosomal protein synthesis [240, 241]

A more intensively studied prokaryotic example, and one involving cellular rather than phage mRNAs, is the autogenous control of the synthesis of bacterial ribosomal proteins. The syntheses of the ribosomal proteins of *E. coli* are closely co-ordinated with one another and with that of rRNA. There are approximately 25 transcriptional units (also referred to as operons) for the 54 ribosomal protein genes, some containing several ribosomal protein genes, whereas others contain only one. Genes for other proteins involved in macromolecular synthesis are also found in these units, some of which are shown in Fig. 12.33. When transducing phages or plasmids were used to increase the copy number of the ribosomal protein genes, without a concomitant increase in the number of rRNA genes, it was found that there was no increase in synthesis of the corresponding ribosomal proteins. This indicated that some form of feedback repression was occurring, and it transpired that a single protein in each transcription unit was responsible for repressing the synthesis of the other members of the unit *in vivo*, as indicated in Fig. 12.33. Experiments *in vitro* (which show consistent, if not identical, results to those *in vivo*) demonstrated that the target of the regulation was the translation of the ribosomal protein mRNA, and that the repressor proteins also inhibited their own translation.

A possible mechanism for the regulation was suggested by the fact that many of the proteins responsible for the feedback inhibition (e.g. S4, S7 and S8) had been shown to bind to rRNA early in the assembly of ribosomes (section 12.6.1). Thus, it was proposed that these proteins were also able to recognize a site on the polycistronic mRNA that resembled their rRNA-binding site, and the binding to which would block translation. Binding to this site would be of lower affinity than that to rRNA so as not to interfere with assembly of ribosomes, but would occur when the ribosomal protein was

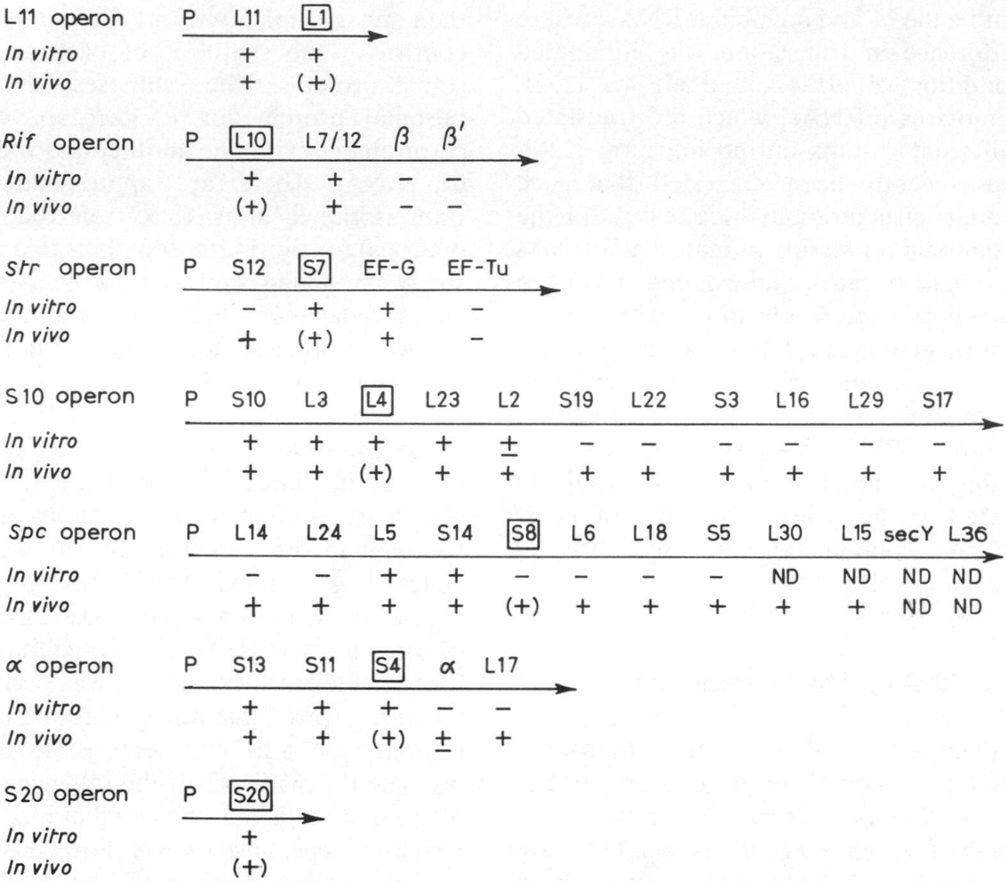

L11 operon   P   L11   [L1] →

In vitro	+	+
In vivo	+	(+)

Rif operon   P   [L10]   L7/12   β   β′ →

In vitro	+	+	−	−
In vivo	(+)	+	−	−

Str operon   P   S12   [S7]   EF-G   EF-Tu →

In vitro	−	+	+	−
In vivo	+	(+)	+	−

S10 operon   P   S10   L3   [L4]   L23   L2   S19   L22   S3   L16   L29   S17 →

In vitro	+	+	+	+	±	−	−	−	−	−	−
In vivo	+	+	(+)	+	+	+	+	+	+	+	+

Spc operon   P   L14   L24   L5   S14   [S8]   L6   L18   S5   L30   L15   secY   L36 →

In vitro	−	−	+	+	−	−	−	−	ND	ND	ND	ND
In vivo	+	+	+	+	(+)	+	+	+	+	+	ND	ND

α operon   P   S13   S11   [S4]   α   L17 →

In vitro	+	+	+	−	−
In vivo	+	+	(+)	±	+

S20 operon   P   [S20] →

In vitro	+
In vivo	(+)

Fig. 12.33 Autogenous regulation of ribosomal protein synthesis in *E. coli*. Each operon is indicated by an arrow, the direction of which is that of its transcription. The promoters are indicated by P and the individual genes by the names of their protein products. Regulatory ribosomal proteins are boxed, and the effects of these on the synthesis of proteins in the same operon *in vivo* and *in vitro* are indicated (+, inhibition; −, no effect; (+), presumed to inhibit *in vivo*; ND, not determined). The autogenous regulation of the synthesis of proteins L14, L24 and S12 operates by retroregulation, involving nucleolytic cleavage of the mRNA [324]. The non-ribosomal proteins in the operons are α, β, and β′, subunits of RNA polymerase, and secY, a component of the protein export machinery. (After [240], with permission.)

produced in excess of the available rRNA. As the cotranscribed 16S and 23S rRNAs would regulate the uptake of all rRNA-binding proteins in the same way, the feedback of the excess of these would explain how the translation of the polycistronic mRNAs from different transcription units was also co-ordinately regulated. Extensive mutagenic studies have established that the target of the translational feedback is generally a single site, located near the translational start of the first gene of the transcriptional unit (the *str* and *spc* operons are exceptional: Fig. 12.33). The binding sites on rRNA of certain ribosomal proteins have been determined (section 12.6.1) and, in the case of S8, L1 and L10, similar secondary structures can be predicted for their regulatory binding sites on the mRNA [241]. A more complex structure, a pseudoknot (Fig. 2.27), has been demonstrated as the recognition site for ribosomal protein S4 in the region of the S13 mRNA initiation codon, and it is assumed, although not yet established, that the S4 binding site on rRNA has a similar tertiary structure. Indeed, in this case, the deduction of the structure of the mRNA has provided an impetus to studies on the more complex tertiary structure of rRNA.

This proposed mechanism of translational regulation (analogues of which appear to operate on the mRNAs for a range of other RNA-binding proteins) raises the question of how the binding of a ribosomal protein to a single site on a polycistronic mRNA can prevent translation starting at separate initiation sites on the same molecule, and why certain members of the transcription units (e.g. EF-Tu and the β and β' subunits of RNA polymerase) are not subject to this feedback regulation (Fig. 12.33). The most likely explanation is that the initiation sites of all but the first mRNA are masked by secondary structure and only become accessible when the first mRNA is translated (cf. the coat protein and replicase genes of the RNA phages – section 12.9.1). This would explain those proteins that escape feedback regulation (EF-Tu etc.) in terms of exposed translational initiation regions, and explain the lack of repression *in vitro* for certain distal mRNA cistrons in terms of exposure through nuclease attack *in vitro*.

(c) *Ferritin mRNA [242]*

The best characterized example of the regulation of a eukaryotic mRNA by interaction with a protein is that of ferritin mRNA. Ferritin is an iron-storage protein, the translation of the mRNA for which increases when the concentration of iron increases. A secondary structure element in the 5'-untranslated region of the mRNA has been demonstrated by mutagenic experiments to be sufficient and necessary for the regulation; and this element has been termed the iron-responsive element (IRE). A 90-kDa protein that specifically binds this element *in vitro* has been identified, and this represses translation of the mRNA, which may also lead to its degradation. It is assumed that *in vivo* the binding of this protein to the IRE is regulated by the concentration of available iron. There is suggestive evidence that this regulation may be mediated by the iron-porphyrin, haemin, which may oxidize free sulphydryl groups on the protein [243, 244]. This may be a general mechanism for iron-binding proteins, as an IRE has also been identified in the 5'-untranslated region of the mRNA for the transferrin receptor [245].

12.9.3 Degradation of mRNA and modulation of its half-life [246, 247]

The destruction of a mRNA will obviously prevent its translation. Although the induction of ribonuclease is used as a regulatory mechanism in certain circumstances, the structural features of individual mRNAs can influence their half-lives, and it is this aspect that will be considered first.

(a) *The role of the poly(A) tail of the mRNA [248, 249]*

There is a large body of evidence indicating that the 3'-poly(A) segment of mRNAs influences their degradation, with the loss of this segment or its reduction to a certain minimum size leaving the mRNA vulnerable to exonuclease action. Thus, the poly(A) sequences of cytoplasmic mRNAs become progressively shorter with time; histone mRNAs, which lack poly(A), have shorter half-lives than most eukaryotic mRNAs; artificially deadenylated mRNAs are translated less efficiently than normal mRNAs *in vivo* and *in vitro*; and histone mRNAs have been stabilized by artificially polyadenylating them. Although initial polyadenylation occurs in the nucleus, it is possible for the length of the poly(A) segment of mRNAs to be extended in the cytoplasm [250]. This occurs during oogenesis and embryogenesis in certain maternal mRNAs that contain U-rich regions in their 3'-untranslated regions. This is thought to selectively increase the translation of these mRNAs.

Poly(A)-containing mRNA decays exponentially, and this implies that the susceptibility of such mRNA to degradation does not increase with decreasing size of poly(A). Rather, there is evidence that when the size of the poly(A) segment falls below a certain threshold value the mRNA suddenly becomes extremely susceptible to degradation. An approach to understanding the mechanism of this effect has come through the characterization of a specific poly(A)-binding protein containing a four-times tandemly repeated RNA-binding domain [251]. It is the association of this protein with the 3'-poly(A) segment that is thought to protect it from nuclease attack, and the critical poly(A) size below which rapid degradation of mRNA occurs is thought to be the minimum size for binding of the protein.

(b) *The AU-rich instability sequence [252]*

Apart from the cases in which the extent of polyadenylation is increased in the cytoplasm, the 3'-poly(A) segment would appear to be a common determinant of the half-life of eukaryotic mRNAs. Nevertheless, there must be some structural feature of different mRNAs to explain the differences in their half-lives. In the case of some very short-lived mRNAs, e.g. that for the granulocyte–monocyte colony stimulating factor, this has been shown to be an AU-rich component of the 3'-untranslated region of the mRNA. Several other short-lived mRNAs (including the cellular oncogenes, *fos* and *myc*) have a similar sequence, and transfer of this to a stable mRNA, such as globin, dramatically decreases its half-life.

At present nothing is known of the way in which this AU-rich sequence conveys instability on the mRNA. One suggestion is that it competes with the 3'-poly(A) segment for the poly(A)-binding protein [251], although a different protein that specifically binds AU-rich sequences has been described [253]. A further intriguing aspect of the instability of these mRNAs is that their half-lives can be dramatically increased by various mitogenic stimuli [252], and in view of the potential oncogenic nature of the products of some of these mRNAs there is considerable interest in elucidating the mechanism of this regulation.

(c) *Interferon and the degradation of viral RNA [254–257]*

Animal cells infected with viruses produce interferon, a family of glycoproteins which act on adjacent uninfected cells, enabling them to resist infection by inhibiting the replicative cycle of the virus. Interferons are species-specific but are effective against a wide spectrum of animal viruses. (They also have inhibitory effects on cell growth which may be of importance in uninfected cells.) Although in this section we are only concerned with the ability of interferon to inhibit translation of viral mRNAs, it appears probable that the transcription is also inhibited. The common signal for the induction of

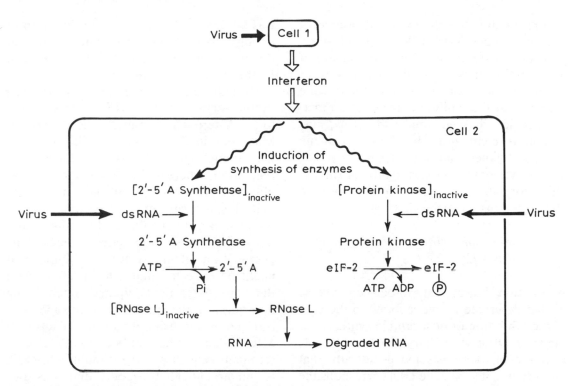

Fig. 12.34 Pathways by which double-stranded RNA and interferon, resulting from infection of cells by viruses, may cause the inhibition of viral protein synthesis. 2′–5′A is an abbreviation for $ppp(A2'p5'A2'p5'A)_n$. For other details, see text.

interferon by different viruses is thought to be double-stranded RNA, which is also the signal that activates the gene-products of interferon that counter the translation of viral mRNA in the 'protected' cell. One of these gene-products is a protein kinase, and this is discussed in section 12.9.4. The other is an enzyme called 2–5A synthetase, which catalyses the synthesis of the tri-nucleotide, $pppA2'p5'A2'p5'A$ (Fig. 12.34). This nucleotide, in turn, activates the re-latively non-specific ribonuclease L (section 4.3.1(e)).

Although it can be seen how this nuclease could prevent the translation of viral mRNA, it is not immediately obvious how host mRNA could escape degradation. Nor is the interferon-induced protein kinase any more

specific in its mode of action (section 12.9.4). Nevertheless, there is a considerable body of evidence that both these mechanisms can be important for the translational inhibition *in vivo*, their relative importance depending on the virus. In the case of the action of 2–5A synthetase, the importance of which appears restricted to certain picornaviruses, it has been suggested that the 2′-5′-oligo-adenylate required for the activation of the ribonuclease is rapidly degraded as it dif-fuses away from its site of synthesis, the double-stranded RNA [254]. It is argued that its action would be largely restricted to this viral double-stranded RNA, thought to consist of replicative intermediates in the case of these RNA viruses. (In the case of DNA viruses, overlapping mRNA transcripts are thought to be the source of double-stranded RNA.) However, it is more difficult to extend this idea to the action of the protein kinase. An alternative way of resolving the problem is possible if one accepts that the inhibition need not, in fact, be specific. The cell, unlike the virus, may be able to compensate for the effect of the nuclease by increasing its synthesis of RNA, and by resuming protein synthesis on de-phosphorylation of the cellular target protein after the virus has been destroyed.

(d) *Herpes simplex virus and the degradation of host mRNA [258, 259]*

A situation that appears to be the converse of that described above is found in the case of the inhibition of host protein synthesis by the large DNA virus, herpes simplex type-1. It has been demonstrated genetically that the product of viral gene UL41 is responsible for the early phase of the shut-off of host protein synthesis. However, if this non-essential gene is deleted there is an increase in the half-life of virus mRNA as well as host mRNA. Thus, it would appear that the ribo-nuclease responsible for the degradation (there are some indications that this is a cellular enzyme, rather than the product of gene UL41 itself) is not intrinsically specific for host mRNA. In this case it may be the concomitant selective shut-off of host trans-cription that produces an apparent specific effect on translation.

(e) *Autoregulation of the stability of tubulin mRNAs [260]*

The co-ordinated regulation of the α- and β-tubulins (the principal subunits of the microtubules) involves the regulation of the degradation of their mRNAs by a mech-anism that is, to date, unique to them. It was demonstrated that the synthesis of tubulin was inhibited when there was a build up of unpolymerized tubulin monomers, and that this involved degradation of their mRNAs. In the case of the mRNA for β-tubulin, it was found that this regulation was conveyed by the section of the mRNA encoding the four *N*-terminal amino acids, MREI. Mutations in this region that did not alter the coding potential did not destroy the regulation, whereas the opposite result was obtained with those that did. It appears that, in some way yet to be determined, the interaction of unpolymerized tubulin mono-mers with this peptide activates nucleolytic degradation of the mRNA. Although it might be expected that other proteins with the same *N*-terminal sequence might bring into question the specificity of this mechan-ism, no other example of an *N*-terminal MREI sequence has yet been found. The *N*-terminal sequence of α-tubulin, MREC, is similar enough to expect that it is also recognized by unpolymerized tubulin, al-though this has not yet been demonstrated.

12.9.4 Protein phosphorylation

(a) *Phosphorylation of eIF-2 [261–263]*

Initial studies of the phosphorylation of eIF-2, performed in reticulocyte lysates, will be summarized before considering the more general but similar phenomenon in cells responding to interferon. Reticulocyte lysates provide a very active protein-synthesizing system, but if haem is not added there is a rapid cessation of protein syn-thesis. (Haem is rapidly converted in aqueous solution into an oxidized form, haemin, which is what is actually used in such studies.) Although one can rationalize this effect in terms of the co-ordination of the synthesis of the predominant reticulocyte protein, globin, with the availability of its prosthetic

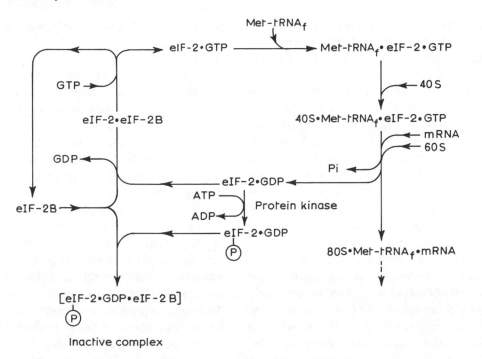

Fig. 12.35 A scheme for the role of factors eIF–2 and eIF–2B in the initiation of eukaryotic protein synthesis, and the effect of phosphorylation of the α-subunit of eIF2. The point at which the phosphorylation of eIF-2 has been indicated is purely arbitrary.

group, haem, it must be emphasized that, in fact, the synthesis of all reticulocyte proteins is similarly affected. It was demonstrated that the step of protein synthesis affected in reticulocytes deprived of haem was the binding of Met-tRNA$_f$ to the 40S ribosomal subunit, the reaction that requires eIF-2 (Fig. 12.19). An inhibitor (known as HCR, haem-controlled repressor; or HRI, haem-regulated inhibitor) was purified from the lysates and shown to be a cytoplasmic protein, pre-existing in an inactive form in normal cells. It was eventually found to be a specific protein kinase that phosphorylates the α-subunit of eIF-2. This has the effect of preventing the liberation of GDP from the eIF-2 released from the ribosome, although eIF-2B still forms a complex with the released species (Fig. 12.35). Although the formation of the complex is readily reversible, phos-

phorylated eIF-2 has a much greater affinity for eIF-2B than the unphosphorylated factor; in effect sequestering the exchange protein. Phosphorylation of only approximately 20% of eIF-2 can cause complete inhibition of initiation because there are only 20–25% as many molecules of eIF-2B as eIF-2.

It was also found in reticulocytes that low concentrations of double-stranded RNA could also provoke the phosphorylation of eIF-2, but that a different protein kinase (although of the same substrate-specificity) was activated. This double-stranded RNA-activated protein kinase is also found in nucleated cells, although, depending on the cell type, its endogenous concentration may be quite low. However, when cells are treated with interferon there is an induction of the synthesis of this protein kinase [256]. Attempts have been made to explain why low concentrations, but not high concentrations, of double-stranded RNA are able to activate this enzyme. It is known that activation of the protein kinase by double-stranded RNA is accompanied by autophosphorylation of the protein kinase, and it has been suggested that this is an intermolecular reaction that is essential for the activation. If this is so, then at high concentrations the ratio of double-stranded RNA to protein kinase would be such that any single molecule of RNA would be unlikely to bind the two molecules of protein kinase required for intermolecular autophosphorylation [264].

The ability of the double-stranded RNA-dependent protein kinase to inhibit viral protein synthesis *in vivo* has been clearly shown for adenovirus [265]. Adenovirus synthesizes a small RNA, VAI RNA, which is required to allow translation of late mRNAs, and this acts by binding to the protein kinase and inhibiting its activity. Other viruses also have mechanisms to inactivate this kinase [265]. In HIV-1 (human immunodeficiency virus type 1), it is possible that the stem–loop structure of the Tat-responsive region RNA acts in a similar way to adenovirus VAI RNA [314]. In contrast, vaccinia virus encodes a truncated homologue of eIF-2α lacking the target serine residue phosphorylated by the double-stranded RNA-dependent protein kinase. This has been shown to abrogate the effect of interferon, presumably by binding to the protein kinase as a pseudo-substrate inhibitor [315]. In the case of influenza virus the precise details of the anti-interferon mechanism are not yet known, but it is interesting that this involves a cellular, rather than viral, protein, which would suggest a role in normal metabolism [316]. Certainly, there is a substantial body of evidence that uninfected cells can regulate their protein synthesis through the double-stranded RNA-dependent eIF-2 protein kinase [266].

(b) *The phosphorylation of eukaryotic EF-2*

Although the dogma of translational regulation insists that control can only be exerted at initiation, there is now clear evidence that the eukaryotic translational translocase, EF-2, is the target of regulatory phosphorylation. Thus, it has been shown that the major (perhaps sole) substrate for Ca^{2+}/ calmodulin-dependent protein kinase III among cellular proteins *in vitro* is EF-2 [267], that this phosphorylation inactivates the factor [268], and that there are changes in the extent of phosphorylation of EF-2 in certain situations *in vivo*. It is rather curious, however, that several of these situations are typified by an increase, rather than a decrease, in protein biosynthesis (e.g. stimulation of quiescent cells by insulin [269]); and it must be remarked that there is no obvious rationale for the control of protein synthesis by changes in the cellular concentration of calcium ions.

(c) *The phosphorylation of other components of the protein synthetic apparatus*

Other eukaryotic initiation factors besides eIF-2 have been shown to be phosphorylated *in vivo* [266]. Most strongly implicated in regulation is the phosphorylation of eIF-4E, which in certain circumstances shows a correlation with protein synthetic activity. However, as yet, there has been no demonstration of a direct effect of phosphorylation on the activity of the factor *in vitro*.

Certain eukaryotic ribosomal proteins are also phosphorylated *in vivo* [270, 271]. On the 60S subunit, the acidic proteins related to *E. coli* L7/L12 (section 12.6.1) are phosphorylated, and there is evidence that this is a prerequisite for their incorporation into the subunit, an event that appears to occur in the cytoplasm for these proteins [272, 273].

On the 40S subunit, the basic ribosomal protein, designated (eukaryotic) S6, can accept up to five phosphoryl residues per molecule *in vivo*. The phosphorylation of this protein is greatest in rapidly growing cells and there is much interest in the possible role of this phosphorylation in the transduction of extracellular signals for growth inside the cell. Two ribosomal protein S6 kinases have been characterized [274, 275], one with two different protein kinase domains, but the function of the phosphorylation of ribosomal protein S6 is still uncertain. As it is clear that the phosphorylation is not an obligatory requirement for ribosome function, some more subtle regulatory role is indicated. One intriguing possibility is of an interaction with the system controlling the half-life of those unstable mRNAs that are induced following a mitotic stimulus (section 12.9.3).

Finally it should be mentioned that certain eukaryotic aminoacyl-tRNA synthetases are also phosphorylated, although the regulatory significance of this is as yet unclear [276].

12.9.5 Initiation factor proteolysis [264, 277]

The unorthodox mode of initiation adopted by poliovirus mRNA, which lacks the 5'-'cap' structure, has been described in section 12.5.1. Its lack of requirement for a functional eIF-4F complex is the basis of its inhibition of the translation of capped host mRNAs, which involves proteolytic cleavage of p220, the 220-kDa component of eIF-4F. The poliovirus mRNA, like that of Sinbis virus (Fig. 12.20), is translated into a polyprotein which is cleaved by two virally coded proteases. One of these, protease 2A, has been shown to be necessary for the proteolysis of p220, but is not itself directly

responsible for the degradation of the factor. It appears that an, as yet uncharacterized, cellular protease is the enzyme responsible, and this raises the possibility that there exist circumstances where this proteolysis is used as a cellular mechanism for the regulation of protein synthesis.

12.9.6 Suppression and frameshifting [278, 279]

The methods of regulation of protein synthesis discussed so far operate within the framework of rigid adherence to the genetic code. However, in both eukaryotes and prokaryotes there are cases, predominantly although not exclusively involving RNA viruses, where regulation involves the disregard of a termination codon (suppression) or its avoidance by change of reading frame (frameshifting).

(a) *Suppression [278–281]*

Suppression of termination codons is used by certain RNA viruses to effect the synthesis of an alternative extended version of a particular protein, containing additional functional potential. An example of this has already been encountered in the case of the alphavirus, Sinbis virus (Fig. 12.20); and bacteriophage Qβ, mentioned in section 12.9.1, synthesizes minor amounts of a fourth protein, thought to be involved in release of progeny phage particles from the bacterium, as a result of readthrough of the coat protein termination codon. Other examples include several plant viruses, such as tobacco mosaic virus, and certain retroviruses, such as Moloney murine leukaemia virus, where a *gag–pol* 'fusion protein' is produced by the relatively efficient suppression of the *gag* termination codon. In the case of the alphaviruses and retroviruses, there is proteolytic processing of the fusion protein to generate one or more separate functional polypeptides.

There are two requirements for such specific suppression of termination codons: there must be some distinguishing feature in the 'context' of the codon, and there must be a tRNA (a suppressor tRNA) to insert an amino acid in response to this. As regards a context effect, it has long been known to bacterial geneticists that amber mutations (i.e. artificially induced UAG stop codons) are particularly inclined to be 'leaky' (i.e. to be subject to readthrough suppression) but that the degree of 'leakiness' varied with the position of the mutation. It has been shown in experiments in bacteria that UAG codons followed by a purine, especially A, are well suppressed, whereas those followed by a pyrimidine are weakly suppressed; and it is precisely a UAG codon followed by an A that is suppressed in the readthrough of the $Q\beta$ coat protein termination codon. This, incidentally, allows one to rationalize in terms of efficient termination the high frequency with which U follows natural bacterial stop codons, and the low frequency with which UAG occurs (section 12.4.3). The tRNA responsible for the suppression in $Q\beta$ has been shown to be a normal *E. coli* tRNA^{Trp} with a 3'-ACC-5' anticodon, which would predict a C:A mismatch in the middle position of the codon. Although an attempt was made to explain the context effect of the A, 3' to the suppressed termination codon, in terms of base-pairing with the (universal) U, 5' to the anticodon, this has now been excluded, and at present the molecular basis of the abnormal codon–anticodon interaction is unknown [282].

In eukaryotic viruses there is no consistent pattern to the suppressed termination codon and the following base: in Sinbis virus it is a UGA codon followed by a C, in Moloney murine leukaemia virus a UAG followed by a G, and in tobacco mosaic virus a UAG followed by a C. Furthermore, in the case of Moloney murine leukaemia virus, mutation of the UAG termination codon to UAA did not abolish its suppression. Thus, it seems likely that a wider context is important, as has been found for frameshifting (below), although there are few other data available at present that bear on this question. Instead, the focus of much work in this area has been identifying the tRNAs responsible for the suppression *in vivo*.

A normal mouse tRNA^{Gln} has been implicated as the suppressor of the UAG termination codon of Moloney murine leukaemia virus, and this has an anticodon 3'-GUC-5', implying a G:U interaction at the first position of the codon [283]. In the case of tobacco mosaic virus, the UAG suppressor in tobacco leaves has been shown to be a normal tRNA^{Tyr} with an anticodon 3'-AΨG-5', implying abnormal G:G hydrogen bonding in the 'wobble' position. It is interesting that another natural tRNA^{Tyr} does not act as a suppressor. This has the 5'-anticodon G replaced by queuosine (Q), a modified form of G, which shows the same pattern of base-pairing with C in the wobble position. One imagines that the conformation adopted by the termination codon must be different from that usually adopted in order to allow such abnormal decoding to occur. Other potential natural suppressors identified include two calf liver tRNA^{Leu} species with anticodons 3'-GAC-5' and 3'-AAC-5', both of which are able to suppress UAG in tobacco mosaic virus RNA *in vitro*.

It is convenient in this section to describe one case in which there is suppression of certain cellular UGA codons, even though it does not have any regulatory significance. Rather, it is the means by which the genetic code is extended to a twenty first amino acid, *selanocysteine*, an analogue of cysteine in which selenium replaces the sulphur atom

[284, 309]. This amino acid is rare, but is found in *E. coli* formate dehydrogenase and mammalian glutathione peroxidase. In both eukaryotes [285] and prokaryotes [286] a specific tRNA, tRNA[Ser]Sec, has been identified as responsible. In each case the primary transcript has an anticodon of sequence 3′-ACU-5′, although the U in the 'wobble' position undergoes a variety of modifications in different organisms. The tRNAs have other unusual structural features that presumably ensure that they only recognize UGA codons in the specific context of the mRNAs for proteins containing selanocysteine. In the case of formate dehydrogenase this context includes 39 nucleotides 3′, and 9 nucleotides 5′, to the codon [287], and a common feature of the context in different cases is a stem–loop structure immediately 3′ to the UGA codon. It has been shown in *E. coli* that the tRNA[Ser]Sec is first charged with serine by the normal seryl-tRNA synthetase, and the serine then undergoes enzymic modification to selenocysteine [288]. A special elongation factor, structurally related to EF-Tu (section 12.4.2), is required to bind this tRNA to the ribosome [289].

(b) *Frameshifting [278, 279]*

The synthesis of extended fusion-proteins by frameshifting is encountered in retroviruses, such as Rous sarcoma virus, mouse mammary tumour virus and human immunodeficiency virus (HIV); in coronaviruses, such as the avian infectious bronchitis virus (IBV); in hepadnaviruses, such as mouse hepatitis virus; and in certain transposons such as Ty (section 7.7.3). However, there are also two examples known in which frameshifting is necessary for the synthesis of normal cellular proteins. Frameshifting in the translation of the mRNA encoded by the *E. coli dnaX* gene is responsible for the out-of-frame

termination that generates the 47-kDa γ subunit of DNA polymerase in addition to the 71-kDa τ subunit [290]; and frameshifting is necessary in the normal translation of the *E. coli* mRNA for RF-2 [291]. Although mutant tRNAs have been described that can cause abnormal frameshifting, the natural frameshifting involves normal tRNAs, and the focus of research has been on defining the mRNA context responsible.

The frameshifting in all the above cases is to the '−1' reading frame, with the exception of Ty and RF-2, where it is to the '+1' reading frame. In the case of shifting to the '−1' frame, a common feature of the mRNA in the region immediately preceding the position of shift is a sequence containing tandemly repeated nucleotides. This has been shown to be necessary for the frameshifting, and is thought to facilitate slippage of the (normal) tRNA. The most striking examples of such 'slippery sequences' are A AAA<u>AAC</u> and U UUU <u>UUA</u> in certain of the retroviruses (the codon recognized by the tRNA involved in the frameshift is underlined), although G GGA <u>AAC</u> is also encountered. It is not merely the tandem repetition that is important, as individual nucleotides are also essential in particular cases. However, such slippery sequences, though necessary, are not sufficient for frameshifting. A second feature is required that is thought to cause the ribosome to pause for a sufficient length of time to allow the frameshifting to occur. In some cases this takes the form of 3′-secondary structure: in many retroviruses (although not in HIV) a stem–loop, and in IBV a pseudoknot (Fig. 2.27). In Ty (where the slippery sequence, CUU <u>AGG</u> C, does not conform to the pattern of tandem repeats) the pausing is thought to be caused by a preceding AGG codon, which is decoded by a rare tRNA[Arg] (cf. section 12.2.4). In the case of RF-2 two features are involved: a 5′-Shine and Dalgarno sequence

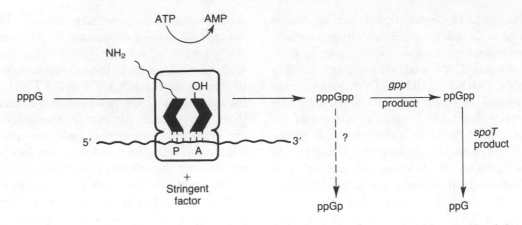

Fig. 12.36 Pathways for the synthesis and degradation of guanine nucleotides involved in the 'stringent response'.

and a UGA termination codon immediately 3′ to the CUU at which the shift occurs. The mechanism by which the Shine and Dalgarno sequence can facilitate frameshifting in this case is obscure, although pausing at the termination codon is easier to envisage. Indeed, the rationale for the frameshifting here is thought to be to regulate the synthesis of RF-2 through control of the ratio of frameshifting to termination. Termination would be expected to vary inversely with the cellular concentration of RF-2, which is required to read the UGA codon. It is curious, however, that there is no comparable regulation in the case of RF-1.

Finally, we should mention some cases where the frameshifting does not involve slippage of the tRNA and ribosome, but 'hopping' over considerable distances. An example of this is gene *60* of bacteriophage T4, where 50 nucleotides are bypassed with almost 100% efficiency. No rationalization of this apparent translational perversity has yet been proposed, but some requirements for its occurrence have been defined, and these include matching sequences at either side of the region bypassed, and a particular region of nascent peptide [292].

12.10 OTHER FUNCTIONS OF THE PROTEIN SYNTHETIC APPARATUS

12.10.1 Ribosomes and the stringent reaction

In addition to their role in protein synthesis, the ribosomes of bacteria such as *E. coli* are able to synthesize guanine nucleotides which act as 'alarmones' to integrate the regulation of RNA and protein synthesis. This phenomenon (which is not found in eukaryotes) occurs when amino acids are limiting, and is known as the 'stringent response' [293, 294]. It results in the selective inhibition of the synthesis of rRNA and tRNA, but not of the bulk of mRNA; hence avoiding wasteful synthesis of more ribosomes and tRNA, and conserving the amino acids provided by protein turnover for the synthesis of essential proteins.

Certain mutations preventing the stringent response map to the *relA* locus. This encodes the 'stringent factor', which is required for the ribosome to respond to the presence of a deacylated tRNA cognate to the codon in the A-site with the synthesis of guanosine 5′-

triphosphate 3′-diphosphate (pppGpp) in an 'idling' reaction (Fig. 12.36). There is further non-ribosomal conversion of pppGpp to yield what is probably the more physiologically important 'alarmone', guanosine 5′-diphosphate 3′-diphosphate (ppGpp); and other guanine nucleotides are generated during the degradation of these so-called 'magic spots'.

The concentration of ppGpp generally correlates well with the inhibition of stable RNA synthesis, but there has been a long controversy regarding the effects of this nucleotide *in vitro* [294, 295]. It is a striking reflection of the shift of scientific interest (and funding) from prokaryotic to eukaryotic systems that, over 20 years after its discovery, the mechanism of action of ppGpp is still unknown.

12.10.2 Other functions of individual ribosomal proteins and elongation factors

In prokaryotes, there are situations in which certain ribosomal proteins and elongation factors are recruited for other roles. It was mentioned in section 12.9.1 that the RNA phages encode a replicase enzyme; however this also requires three host subunits in order to replicate the RNA 'minus' strand. These are EF-Tu, EF-Ts, and ribosomal protein S1. This may merely be a case of the bacteriophage adapting for its own advantage the RNA-binding properties of EF-Tu and ribosomal protein S1. In the case of EF-Tu, it has been suggested that this may involve specific binding to putative tRNA-like features at the 3′-end, although the RNA phages do not show the strong resemblance to tRNA seen in certain plant viruses (section 12.10.3). For ribosomal protein S1 there is the fact that it is able to inhibit the synthesis of the bacteriophage coat protein, and it has been suggested that the presence of this protein in the replicase might either be connected with the temporal lagging of coat protein synthesis behind replicase synthesis, or to play a role in preventing collision between replicase and ribosomes on the RNA [231, 232].

An additional function is suggested in normal cells for the particular eukaryotic ribosomal protein that is termed S27a in mammals and S37 in yeast. In all eukaryotes examined this protein is generated by post-translational processing of a larger species which has *N*-terminal sequences specifying the protein ubiquitin [296]. Ubiquitin is perhaps best known for its role in targetting proteins for degradation [297], but in certain circumstances can apparently have the opposite effect. There is evidence to suggest that the ubiquitin portion remains with the ribosomal protein until it is incorporated into the ribosome [298], and it has been found that a yeast mutant with the ubiquitin tail removed from the ribosomal protein is defective in assembling the 40S ribosomal subunit [299]. It is therefore possible that the ubiquitin sequence acts as a 'chaperonin' for this protein or, a more provoking thought, plays a role in the assembly of the 40S subunit as a whole.

12.10.3 Other functions of aminoacyl-tRNA synthetases

The ribosome is not the only component of the protein synthetic machinery with the capacity to synthesize a nucleotide in an alternative to the reaction normally catalysed. Under abnormal conditions such as heat shock or oxidative stress, the aminoacyl adenylate produced in the first step of the aminoacyl-tRNA synthetase reaction can react with ATP to give diadenosine $5′,5′′′-P^1$, P^4-tetraphosphate, AppppA [300]:

(amino acid-pA)E + pppA

↓

E + amino acid + AppppA

Other related nucleotides have also been reported to be synthesized by this reaction, which can occur in both prokaryotes and eukaryotes. The function of the nucleotide is still in dispute, however. It has been suggested that AppppA may act as an 'alarmone' to induce the heat shock response [300], but this idea is not supported by experiments with *E. coli* mutants in the *apaH* gene, which specifies the hydrolase that degrades the nucleotide [301].

More recently it has been found that specific aminoacyl-tRNA synthetases are components of the intron-splicing machinery of fungal mitochondria [302]. It is possible to produce a mutant Tyr-tRNA synthetase that has lost its ability to charge tRNA with Tyr, but still functions in the splicing reaction. However, this does not exclude a role in RNA recognition of those parts of the enzyme already evolved to recognize features of tRNA^{Tyr}, especially as we have already seen that the acceptor stem, with its 5'- and 3'-termini, is one of the features that are recognized by certain tRNA-synthetases (section 12.3.2). The possibility that such recognition is a very ancient feature of life is discussed in the final section of this chapter.

12.10.4 Other functions of tRNAs and tRNA-like structures

In addition to their role on the ribosome, specific tRNAs are important for several other cellular processes. Some of these appear quite natural extensions of the major contemporary role of tRNAs. Thus, certain aminoacyl-tRNAs act as donors of amino acids in the synthesis of the peptide components of the peptidylglycans of bacterial cell walls [303]. A more recently described related function [304] appears to be in the post-translational tagging of the *N*-terminus of proteins with chemically basic amino acids to prepare them for proteolysis by the ubiquitin system [297].

Somewhat more esoteric is the involvement of tRNA^{Glu} in the synthesis of δ-aminolevulinate from glutamate in plants, where it is an important precursor of chlorophyll [305]. For its NADPH-linked reduction to glutamate 1-semialdehyde, glutamate must be in the form of Glu-tRNA. As there seems no obvious chemical advantage in this it has been suggested that the use of the tRNA may be a mechanism for co-ordinating chloroplast protein synthesis and chlorophyll synthesis.

The above examples all involve the amino acid attached to the tRNA. However, there are two virus-related examples involving tRNA structures which, although apparently quite different, may be 'molecular fossils' of a role of this structure that preceded protein biosynthesis. The first example is the use of particular cellular tRNAs by different retroviruses to prime reverse transcription, where a specific sequence on the tRNA hybridizes to the viral 5'-LTR [306], as has already been mentioned in section 6.14. At first sight the second example is quite dissimilar from the first. It is the role of the tRNA-like structures that are found at the 3'-ends of the genomic RNAs of certain plant viruses, such as tobacco mosaic virus [307]. These can be aminoacylated by specific amino acids (His in the virus mentioned), but for many years no plausible function could be suggested for them. An attractive hypothesis has now been proposed by Weiner and Maizels [308], in which it is suggested that the tRNA structures might serve a telomeric function (section 6.16.4). This would be through their recognition by the tRNA nucleotidyl transferase, which adds CCA to the 3′-end of tRNA transcripts (and the virus RNAs) – a process formally similar to the addition to telomeres of the polymeric C_mA_n sequences by telomerases. The point of convergence between these two examples is Weiner and Maizels' further speculation that tRNAs may have evolved from the 3′-terminal structures of primitive RNA genomes in the 'RNA world' believed to have existed before proteins (and protein synthesis) emerged; and which the proteins that subsequently took over the major part of the replicative function would have had to be able to recognize. In this case the use of extra-genomic tRNA by retroviruses might be more than just fortuitous use of a cellular RNA for its base-complementarity to their genome, but reflect an ancient ability of the replicative enzymes of the virus to recognize the tertiary structure of the tRNA molecule. Indeed, as the original replicative enzymes in an RNA world would have been RNA, a specific base-pairing interaction would also have been a feature of their interaction with the 3′-end of the genome.

REFERENCES

1 Stryer, L. A. (1989) *Biochemistry* (3rd edn), Freeman, New York.
2 Dintzis, H. M. (1961) *Proc. Natl. Acad. Sci. USA*, **47**, 247.
3 Thach, R. E., Cecere, M. A., Sundrararajan, T. A. and Doty, P. (1965) *Proc. Natl. Acad. Sci. USA*, **54**, 1167.
4 Terzaghi, E., Okada, Y., Streisinger, G., Emrich, J., Inouye, M. and Tsugita, A. (1966) *Proc. Natl. Acad. Sci. USA*, **56**, 500.
5 Lodish, H. F. and Jacobsen, M. (1972) *J. Biol. Chem.*, **247**, 3622.
6 Heywood, S. M., Dowben, R. M. and Rich, A. (1967) *Proc. Natl. Acad. Sci. USA*, **57**, 1002.
7 Lacroute, F. and Stent, G. (1968) *J. Mol. Biol.*, **35**, 165.
8 Knopf, P. M. and Lamfrom, H. (1965) *Biochim. Biophys. Acta*, **95**, 398.
9 Kleinkauf, H. and von Döhren, H. (1983) *Trends Biochem. Sci.*, **8**, 281.
10 Crick, F. H. C. (1967) *Proc. R. Soc. London, Ser. B*, **167**, 331.
11 Woese, C. R. (1967) *The Genetic Code. The Molecular Basis for Genetic Expression*, Harper and Row, New York.
12 Jukes, T. H. (1977) in *Comprehensive Biochemistry* (eds M. Florkin and E. H. Stotz), Elsevier, Amsterdam, vol. 24, p. 235.
13 Marshall, R. E., Caskey, C. T. and Nirenberg, M. (1967) *Science*, **155**, 820.
14 Goodman, H. M., Abelson, J., Landy, A., Brenner, S. and Smith, J. D. (1968) *Nature (London)*, **217**, 1019.
15 Marcker, K. A. and Sanger, F. (1964) *J. Mol. Biol.*, **8**, 835.
16 Dube, S. K., Marcker, K. A., Clark, B. F. C. and Cory, S. (1968) *Nature (London)*, **218**, 232.
17 Cory, S., Marcker, K. A., Dube, S. K. and Clark, B. F. C. (1968) *Nature (London)*, **220**, 1039.
18 Marcker, K. A., Clark, B. F. C. and Anderson, J. S. (1966) *Cold Spring Harbor Symp. Quant. Biol.*, **31**, 279.
19 Delk, A. S. and Rabinowitz, J. C. (1974) *Nature (London)*, **252**, 106.
20 Adams, J. M. (1968) *J. Mol. Biol.*, **34**, 571.
21 Miller, C. M., Strauch, K. L., Kukral, A. M., Miller, J. L., Wingfield, P. T. *et al.* (1987) *Proc. Natl. Acad. Sci. USA*, **84**, 2718.
22 Sherman, F., Stewart, J. W. and Tsunasawa, S. (1985) *BioEssays*, **3**, 27.
23 Crick, F. H. C. (1966) *J. Mol. Biol.*, **19**, 548.
24 Borén, T., Elias, P., Samuelsson, T., Claesson, C., Barciszewska, M. *et al.* (1992) *J. Biol. Chem.*
25 Offengand, J. (1982) in *Protein Biosynthesis in Eukaryotes* (ed. R. Pérez-Bercoff), Plenum, New York, p. 1.
26 Lustig, F., Borén, T., Guindy, Y., Elias, P., Samuelsson, T. *et al.* (1989) *Proc. Natl. Acad. Sci. USA*, **86**, 6873.
27 Hillen, W., Egert, E., Lindner, H. J. and Gassen, H. G. (1978) *FEBS Lett.*, **94**, 361.
28 Yokoyama, S., Watanabe, T., Marao, K., Ishikura, H., Yamaizumi, Z. *et al.* (1985) *Proc. Natl. Acad. Sci. USA*, **82**, 4905.
29 Martin, R. P., Sibler, A.-P., Gehrke, C. W., Kuo, K., Edmonds, C. G. *et al.*

(1990) *Biochemistry*, **29**, 956.

30 Sibler, A. P., Dirheimer, G. and Martin, R. P. (1985) *Nucleic Acids Res.*, **13**, 1341.

31 Hunter, W. N., Brown, T. and Kennard, O. (1987) *Nucleic Acids Res.*, **15**, 6589.

32 Grosjean H. J., de Henau S. and Crothers, D. M. (1978) *Proc. Natl. Acad. Sci. USA*, **75**, 610.

33 Lagerkvist, U. (1978) *Proc. Natl. Acad. Sci. USA*, **75**, 1759.

34 Samuelsson, T., Axberg, T., Boren, T. and Lagerkvist, U. (1983) *J. Biol. Chem.*, **258**, 13 178.

35 Caron, F. (1990) *Experientia*, **46**, 1106.

36 Osawa, S., Muto, A., Ohama, T., Andachi, Y., Tanaka, R. *et al.* (1990) *Experientia*, **46**, 1097.

37 Jukes, T. H. and Osawa, S. (1990) *Experientia*, **46**, 1117.

38 Grosjean, H. and Fiers, W. (1982) *Gene*, **18**, 199.

39 Konigsberg, W. and Godson, G. N. (1983) *Proc. Natl. Acad. Sci. USA*, **80**, 687.

40 Ikemura, T. (1981) *J. Mol. Biol.*, **146**, 1.

41 Bennetzen, J. L. and Hall, B. D. (1982) *J. Biol. Chem.*, **257**, 3026.

42 Garel, J.-P. (1976) *Nature (London)*, **260**, 805.

43 Carter, P. W., Bartkus, J. M. and Calvo, J. M. (1986) *Proc. Natl. Acad. Sci. USA*, **83**, 8127.

44 Hoekma, A., Kastelein, R. A., Vasser, M. and de Boer, H. A. (1987) *Mol. Cell. Biol.*, **7**, 2914.

45 Goddard, J. P. (1977) *Prog. Biophys. Mol. Biol.*, **32**, 233.

46 Schevitz, R. W., Podjarny, A. D., Krishnamachari, N., Hughes, J. J. and Sigler, P. B. (1979) *Nature (London)*, **278**, 188.

47 Woo, N. H., Roe, B. A. and Rich, A. (1980) *Nature (London)*, **286**, 346.

48 Westhof, E., Dumas, P. and Moras, D. (1985) *J. Mol. Biol.*, **184**, 119.

49 Quigley, G. J. and Rich, A. (1976) *Science*, **194**, 796.

50 Hoagland, M. B., Stephenson, M. L., Scott, J. F., Hecht, L. I. and Zamecnik, P. C. (1958) *J. Biol. Chem.*, **231**, 241.

51 Lagerkvist, U., Rymo, L. and Waldenström, J. (1966) *J. Biol. Chem.*, **241**, 5391.

52 Julius, D. J., Fraser, T. H. and Rich, A. (1979) *Biochemistry*, **18**, 604.

53 Griffin, B. E., Jarman, M., Reese, C. B., Sulston, J. E. and Trentham, D. R. (1966) *Biochemistry*, **5**, 3638.

54 Zachau, H. G. and Feldman, H. (1965) *Prog. Nucleic Acid Res. Mol. Biol.*, **4**, 217.

55 Yarus, M. (1988) *Cell*, **55**, 739.

56 Schimmel, P. (1991) *Trends Biochem. Sci.*, **16**, 1.

57 Rould, M. A., Perona, J. J., Söll, D. and Steitz, T. A. (1989) *Science*, **246**, 1135.

58 Perona, J. J., Swanson, R. N., Rould, M. A., Steitz, T. A. and Söll, D. (1989) *Science*, **246**, 1152.

59 Chapeville, F., Lipmann, F., von Ehrenstein, G., Weisblum, B., Ray, W. J. *et al.* (1962) *Proc. Natl. Acad. Sci. USA*, **48**, 1086.

60 Fersht, A. R. (1979) in *Transfer RNA: Structure, Properties and Recognition* (eds P. R. Schimmel, D. Söll and J. N. Abelson), Cold Spring Harbor Laboratory Monograph Series, p. 247.

61 Lengyel, P. (1974) in *Ribosomes* (eds M. Nomura, A. Tissières and P. Lengyel), Cold Spring Harbor Laboratory Monograph Series, p. 13.

62 Weissbach, H. (1979) in *Ribosomes: Structure, Function and Genetics* (eds G. Chambliss, G. R. Craven, J. Davies, K. Davis, L. Kahan and M. Nomura), University Park Press, Baltimore, p. 377.

63 Clark, B. F. C. and Petersen, H. U. (1984) *Gene Expression. The Translational Step and its Control*, Munskgaard, Copenhagen.

64 Grunberg-Manago, M. (1980) in *Ribosomes: Structure, Function and Genetics* (eds G. Chambliss, G. R. Craven, J. Davies, K. Davis, L. Kahan and M. Nomura), University Park Press, Baltimore, p. 445.

65 Maitra, U., Stringer, E. A. and Chaudhuri, A. (1982) *Annu. Rev. Biochem.*, **51**, 869.

66 Gualerzi, C. O. and Pon, C. L. (1990) *Biochemistry*, **29**, 5881.

67 Shine, J. and Dalgarno, L. (1974) *Proc. Natl. Acad. Sci. USA*, **71**, 1342.

68 Kozak, M. (1983) Microbiol. Rev., **47**, 1.

69 Seong, B. L. and RajBhandary, U. L. (1987) *Proc. Natl. Acad. Sci. USA*, **84**, 8859.

70 Kuechler, E. (1971) *Nature (London) New Biol.*, **234**, 216.

71 Morel-Deville, F., Vachon, G., Sacerdot, C., Cozzone, A., Grunberg-Manago, M.

et al. (1990) *Eur. J. Biochem.*, **188**, 605.

72 Sacerdot, C., Fayat, G., Dessen, P., Springer, M., Plumbridge, J. A. *et al.* (1982) *EMBO J.*, **1**, 311.

73 Reddy, P., Peterkofsky, A. and McKenny, K. (1985) *Proc. Natl. Acad. Sci. USA*, **82**, 5656.

74 Gold, L., Stormo, G. and Saunders, R. (1984) *Proc. Natl. Acad. Sci. USA*, **81**, 7061.

75 Butler, J. S., Springer, M. and Grunberg-Manago, M. (1987) *Proc. Natl. Acad. Sci. USA*, **84**, 4022.

76 Harris, R. J. and Pestka, S. (1977) in *Molecular Mechanisms of Protein Biosynthesis* (eds H. Weissbach and S. Pestka), Academic Press, New York, p. 413.

77 Miller, D. L. and Weissbach, H. (1977) in *Molecular Mechanisms for Protein Biosynthesis* (eds H. Weissbach and S. Pestka), Academic Press, New York, p. 324.

78 Bermek, E. (1978) *Prog. Nucleic Acid Res. Mol. Biol.*, **21**, 63.

79 Clark, B. F. C. (1983) in *DNA Makes RNA Makes Protein* (eds T. Hunt, S. Prentis and J. Tooze), Elsevier Biomedical, Amsterdam, p. 213.

80 Ono, Y., Skoultchi, A., Klein, A. and Lengyel, P. (1968) *Nature (London)*, **220**, 1304.

81 Bourne, H. R., Sanders, D. A. and McCormick, F. (1991) *Nature (London)*, **349**, 117.

82 La Cour, T. F. M., Nyborg, J., Thirup, S. and Clark, B. F. C. (1985) *EMBO J.*, **4**, 2385.

83 Jurnak, F. (1985) *Science*, **230**, 32.

84 Monro, R. E. (1967) *J. Mol. Biol.*, **26**, 147.

85 Lake, J. A. (1977) *Proc. Natl. Acad. Sci. USA*, **74**, 1903.

86 Nierhaus, K. H. (1990) *Biochemistry*, **29**, 4997.

87 Moazed, D. and Noller, H. F. (1989) *Nature (London)*, **342**, 142.

88 Caskey, C. T. (1977) in *Molecular Mechanisms of Protein Biosynthesis* (eds H. Weissbach and S. Pestka), Academic Press, New York, p. 443.

89 Tate, W. P. (1984) in *Peptide and Protein Reviews* (ed. M. T. W. Hearn), Marcel Dekker, New York, vol. 2, p. 173.

90 Tate, W. P., Brown, C. P. and Kastner, B. (1990) in *The Ribosome. Structure, Fuction*

and Evolution (eds W. E. Hill, A. Dahlberg, R. A. Garrett, P. B. Moore, D. Schlessinger and J. R. Warner), American Society for Microbiology, Washington, p. 393.

91 Capecchi, M. R. and Klein, H. A. (1969) *Cold Spring Harbor Symp. Quant. Biol.*, **34**, 469.

92 Mugola, E. J., Huazi, K. A., Göringer, A. E. and Dahlberg, A. E. (1988) *Proc. Natl. Acad. Sci. USA*, **85**, 4162.

93 Kaempfer, R. (1974) in *Ribosomes* (eds M. Nomura, A. Tissières and P. Lengyel), Cold Spring Harbor Monograph Series, p. 679.

94 Naaktgeboren, N., Roobol, K. and Voorma, H. O. (1977) *Eur. J. Biochem.*, **72**, 49.

95 Kohli, J. and Grosjean, H. (1981) *Mol. Gen. Genet.*, **182**, 430.

96 Brown, C. M., Stockwell, P. A., Trotman, C. N. A. and Tate W. P. (1990) *Nucleic Acids Res.*, **18**, 2079.

97 Miller, J. H. and Albertini, A. M. (1983) *J. Mol. Biol.*, **164**, 59.

98 Bossi, L. (1983) *J. Mol. Biol.*, **164**, 73.

99 Pérez-Bercoff, R. (1982) *Protein Biosynthesis in Eukaryotes*, Plenum, New York.

100 Moldave, K. (1985) *Annu. Rev. Biochem.*, **54**, 1109.

101 Pain, V. M. (1986) *Biochem. J.*, **235**, 625.

102 Rhoads, R. E. (1988) *Trends Biochem. Sci.*, **13**, 52.

103 Caskey, C. T., Redfield, B. and Weissbach, H. (1967) *Arch. Biochem. Biophys.*, **120**, 119.

104 Kozak, M. (1978) *Cell*, **15**, 1109.

105 Kozak, M. (1989) *J. Cell Biol.*, **108**, 229.

106 Darnborough, C., Legon, S., Hunt, T. and Jackson, R. (1973) *J. Mol. Biol.*, **76**, 379.

107 Schreier, M. H. and Staehelin, T. (1973) *Nature (London) New Biol.*, **242**, 35.

108 Cigan, A. M., Feng, L. and Donahue, T. F. (1988) *Science*, **242**, 93.

109 Milburn, S. C., Duncan, R. F. and Hershey, J. W. B. (1990) *Arch. Biochem. Biophys.*, **276**, 6.

110 Nielsen, P. J. and Trachsel, H. (1988) *EMBO J.*, **7**, 2097.

111 Lazaris-Karatzas, A., Montine, K. S. and Sonenberg, N. (1990) *Nature (London)*, **345**, 544.

112 Smit-McBride, Z., Schnier, J., Kaufman, R. J. and Hershey, J. W. B. (1989) *J. Biol. Chem.*, **264**, 18527.

113 Strauss, E. G., Rice, C. M. and Strauss, J. H. (1984) *Virology*, **133**, 92.

114 Jackson, R. J., Howell, M. T. and Kaminski, A. (1990) *Trends Biochem. Sci.*, **15**, 477.

115 Hunt, T. (1985) *Nature (London)*, **316**, 580.

116 Hinnebusch, A. G. (1988) *Trends Genet.*, **4**, 169.

117 Hann, S. R., King, M. W., Bentley, D. L., Anderson, C. W. and Eisenman, R. N. (1988) *Cell*, **52**, 185.

118 Florkiewicz, R. Z. and Sommer, A. (1989) *Proc. Natl. Acad. Sci. USA*, **86**, 3978.

119 Bernards, A. and de la Monte, S. M. (1990) *EMBO J.*, **9**, 2279.

120 Meerovitch, K., Pelletier, J. and Sonenberg, N. (1989) *Genes Devel.*, **3**, 1026.

121 Macejak, D. G. and Sarnow, P. (1991) *Nature (London)*, **353**, 90.

122 Riis, B., Rattan, S. I. S., Clark, B. F. C. and Merrick, W. C. (1990) *Trends Biochem. Sci.*, **15**, 420.

123 Slobin, L. I. (1980) *Eur. J. Biochem.*, **110**, 555.

124 Nagata, S., Nagashima, K., Tsunetsugu-Yokota, Y., Fujimura, K., Miyazaki, M. *et al.* (1984) *EMBO J.*, **3**, 1825.

125 Collier, R. J. (1975) *Bacteriol. Rev.*, **39**, 54.

126 Van Ness, B. G., Howard, J. B. and Bodley, J. W. (1980) *J. Biol. Chem.*, **255**, 10 710.

127 Lee, H. and Iglewski, W. J. (1984) *Proc. Natl. Acad. Sci. USA*, **81**, 2703.

128 Neth, R., Monro, R. E., Heller, G., Battaner, E. and Vázquez, D. (1970) *FEBS Lett.*, **6**, 198.

129 Goldstein, J. L., Beaudet, A. L. and Caskey, C. T. (1970) *Proc. Natl. Acad. Sci. USA*, **67**, 99.

130 Craigen, W. J., Cook, R. G., Tate, W. P. and Caskey, C. T. (1985) *Proc. Natl. Acad. Sci. USA*, **82**, 3616.

131 Lee, C. C., Craigen, W. J., Muzny, D. M., Harlow, E. and Caskey, C. T. (1990) *Proc. Natl. Acad. Sci. USA*, **87**, 3508.

132 Kaempfer, R. (1969) *Nature (London)*, **222**, 950.

133 Wada, K., Aota, S., Tsuchiya, R., Ishibashi, F., Gojobori, T. *et al.* (1990) *Nucleic Acids Res.*, **18**, 2367.

134 Nomura, M., Tissières, A. and Lengyel, P. (1974) *Ribosomes*, Cold Spring Harbor Monograph Series, New York.

135 Chambliss, G., Craven, G. R., Davies, J., Davis, K., Kahan, L. and Nomura, M. (1980) *Ribosomes: Structure, Function and Genetics*, University Park Press, Baltimore.

136 Hardesty, B. and Kramer, G. (1986) *Structure, Fuction and Genetics of Ribosomes*, Springer, New York.

137 Hill, W. E., Dahlberg, A., Garrett, R. A., Moore, P. B., Schlessinger, D. and Warner, J. R. (1990) *The Ribosome. Structure, Function and Evolution*. American Society for Microbiology, Washington.

138 Moore, P. B. (1988) *Nature (London)*, **331**, 223.

139 Bielka, H. (1982) *The Eukaryotic Ribosome*, Springer, Berlin.

140 Wool, I. G. (1990) in *The Ribosome. Structure, Function and Evolution* (eds W. E. Hill, A. Dahlberg, R. A. Garrett, P. B. Moore, D. Schlessinger and J. R. Warner), American Society for Microbiology, Washington, p. 203.

141 Lake, J. (1985) *Annu. Rev. Biochem.*, **54**, 507.

142 Frank, J., Verschoor, A., Wagenknecht, T., Radermacher, M. and Carazo, J.-M. (1988) *Trends Biochem. Sci.*, **13**, 123.

143 Yonath, A. and Wittmann, H. G. (1989) *Trends Biochem. Sci.*, **14**, 329.

144 Milligan, R. A. and Unwin, P. N. T. (1986) *Nature (London)*, **319**, 693.

145 Rich, A. (1974) in *Ribosomes* (eds M. Nomura, A. Tissières and P. Lengyel), Cold Spring Harbor Monograph Series, p. 871.

146 Olsen, H. McK., Grant, P. G., Cooperman, B. S. and Glitz, D. G. (1982) *J. Biol. Chem.*, **257**, 2649.

147 Olsen, H. McK., Lasater, L. S., Cann, P. A. and Glitz, D. G. (1988) *J. Biol. Chem.*, **263**, 15 196.

148 Kang, C. and Cantor, C. R. (1985) *J. Mol. Biol.*, **181**, 241.

149 Noller, H. F. (1984) *Annu. Rev. Biochem.*, **53**, 119.

150 Maden, B. E. H., Khan, M. S. N., Hughes, D. G. and Goddard, J. P. (1977) *Biochem. Soc. Symp.*, **42**, 165.

151 Maden, B. E. H. (1980) *Nature (London)*, **288**, 293.

152 Brimacombe, R. (1984) *Trends Biochem. Sci.*, **9**, 273.

153 Walker, T. A. and Pace, N. R. (1983) *Cell*, **33**, 320.

154 Moazed, D., Stern, S. and Noller, H. F. (1986) *J. Mol. Biol.*, **187**, 399.

155 Noller, H. F., Moazed, D., Stern, S., Powers, T., Allen, P. N. *et al.* (1990) in *The Ribosome. Structure, Function and Evolution* (eds W. E. Hill, A. Dahlberg, R. A. Garrett, P. B. Moore, D. Schlessinger and J. R. Warner), American Society for Microbiology, Washington, p. 72.

156 Brimacombe, R., Greuer, B., Mitchell, P., Osswald, M., Rinke-Appel, J. *et al.* (1990) in *The Ribosome. Structure, Function and Evolution* (eds W. E. Hill, A. Dahlberg, R. A. Garrett, P. B. Moore, D. Schlessinger and J. R. Warner), American Society for Microbiology, Washington, p. 93.

157 Westhof, E., Romby, P., Romaniuk, P. J., Ebel, J. P., Ehresmann, C. *et al.* (1989) *J. Mol. Biol.*, **207**, 417.

158 Wittmann-Liebold, B. (1986) in *Structure, Function and Genetics of Ribosomes* (eds B. Hardesty and G. Kramer), Springer, New York, p. 326.

159 Wada, A. and Sako, T. (1987) *J. Biochem. (Tokyo)*, **101**, 817.

160 Wilson, K. S., Appelt, K., Badger, J., Tanaka, I. and White, S. W. (1986) *Proc. Natl. Acad. Sci. USA*, **83**, 7251.

161 Leijonmarck, M. and Liljas, A. (1987) *J. Mol. Biol.*, **195**, 555.

162 Lin, A., Wittmann-Liebold, B., McNally, J. and Wool, I. G. (1982) *J. Biol. Chem.*, **257**, 9189.

163 Rich, B. E. and Steitz, J. A. (1987) *Mol. Cell. Biol.*, **7**, 4065.

164 Leader, D. P. (1980) in *Molecular Aspects of Cellular Regulation* (ed. P. Cohen), Elsevier/North-Holland, Amsterdam, vol. 1, p. 203.

165 Traut, R. R., Lambert, J. M., Boileau, G. and Kenny, J. W. (1980) in *Ribosomes. Structure, Function and Genetics* (eds G. Chambliss, G. R. Craven, J. Davies, K. Davis, L. Kahan and M. Nomura), University Park Press, Baltimore, p. 89.

166 Moore, P. B. and Capel, M. (1988) *Annu. Rev. Biophys. Biophys. Chem.*, **17**, 349.

167 Stöffler-Meilicke, M. and Stöffler, G. (1990) in *The Ribosome. Structure, Function and Evolution* (ed. W. E. Hill, A. Dahlberg, R. A. Garrett, P. B. Moore, D. Schlessinger and J. R. Warner), American Society for Microbiology, Washington, p. 123.

168 Oates, M. I., Scheinman, A., Atha, T., Schankweiler, G. and Lake, J. A. (1990) in *The Ribosome. Structure, Function and Evolution* (eds W. E. Hill, A. Dahlberg, R. A. Garrett, P. B. Moore, D. Schlessinger and J. R. Warner), American Society for Microbiology, Washington, p. 180.

169 Garrett, R. (1983) *Trends Biochem. Sci.*, **8**, 189.

170 Dahlberg, A. E. (1989) *Cell*, **57**, 525.

171 Dabbs, E. R., Hasenbank, R., Kastner, B., Rak, K-H., Wartusch, B. *et al.* (1983) *Mol. Gen. Genet.*, **192**, 301.

172 Crick, F. H. C. (1968) *J. Mol. Biol.*, **38**, 367.

173 Darnell, J. E. and Doolittle, W. F. (1986) *Proc. Natl. Acad. Sci. USA*, **83**, 1271.

174 Steiner, G., Keuchler, E. and Barta, A. (1988) *EMBO J.*, **7**, 3949.

175 Hall, C. C., Johnson, D. and Cooperman, B. (1988) *Biochemistry*, **27**, 3983.

176 Moazed, D. and Noller, H. F. (1989) *Cell*, **57**, 585.

177 Ehresmann, C., Ehresmann, B., Millon, R., Ebel, J.-P., Nurse, K. and Ofengand, J. (1984) *Biochemistry*, **23**, 429.

178 Li, M., Tzagoloff, A., Underbrink-Lyon, K. and Martin, N. C. (1982) *J. Biol. Chem.*, **257**, 5921.

179 Beauclerk, A. A. D. and Cundliffe, E. (1987) *J. Mol. Biol.*, **193**, 661.

180 Denman, R., Nègre, D., Cunningham, P. R., Nurse, K. *et al.* (1989) *Biochemistry*, **28**, 1012.

181 Helser, T. L., Davies, J. E. and Dahlberg, J. E. (1971) *Nature (London) New Biol.*, **233**, 12.

182 Kurland, C. G., Jörgensen, F., Richter, A., Ehrenberg, M., Bilgin, N. *et al.* (1990) in *The Ribosome. Structure, Function and Evolution* (eds W. E. Hill, A. Dahlberg, R. A. Garrett, P. B. Moore, D. Schlessinger and J. R. Warner), American Society for Microbiology, Washington, p. 513.

183 Gorini, L. (1974) in *Ribosomes* (eds M. Nomura, A. Tissières and P. Lengyel), Cold Spring Harbor Monograph Series, p. 791.

184 Ozaki, M., Mizushima, S. and Nomura, M. (1974) *Nature (London)*, **222**, 333.

185 Hirsch, D. (1971) *J. Mol. Biol.*, **58**, 439.

186 Hopfield, J. J. (1974) *Proc. Natl. Acad. Sci. USA*, **71**, 4135.

187 Ninio, J. (1975) *Biochimie*, **57**, 587.

188 Thompson, R. C. (1988) *Trends Biochem Sci.*, **13**, 91.

189 Woese, C. R. and Wolfe, R. S. (1985) *The Bacteria*, vol. 8, *Archaebacteria*, Academic Press, New York.

190 Kandler, O. and Zillig, W. (1986) *Archaebacteria '85*, Gustav Fischer, New York.

191 Woese, C. R. and Fox, G. E. (1977) *Proc. Natl. Acad. Sci. USA*, **74**, 5088.

192 Wittmann-Liebold, B., Köpke, A. K. E., Arndt, E., Krömer, W., Hatakeyama, T. *et al.* (1990) in *The Ribosome. Structure, Function and Evolution* (eds W. E. Hill, A. Dahlberg, R. A. Garrett, P. B. Moore, D. Schlessinger and J. R. Warner), American Society for Microbiology, Washington, p. 598.

193 Mathieson, A. T., Auer, J., Ramirez, C. and Böck, A. (1990) in *The Ribosome. Structure, Function and Evolution* (eds W. E. Hill, A. Dahlberg, R. A. Garrett, P. B. Moore, D. Schlessinger and J. R. Warner), American Society for Microbiology, Washington, p. 617.

194 Borst, P., Grivell, L. A. and Groot, G. S. P. (1984) *Trends Biochem. Sci.*, **9**, 48.

195 Attardi, G. (1985) *Int. Rev. Cytol.*, **93**, 93.

196 Chomyn, A. and Attardi, G. (1987) *Curr. Top. Bioeng.*, **15**, 295.

197 Gray, M. W. (1989) *Trends Genet.*, **5**, 294.

198 O'Brien, T. W., Denslow, N. D., Anders, J. C. and Courtney, B. C. (1990) *Biochim. Biophys. Acta*, **1050**, 174.

199 Smith, A. E. and Marcker, K. A. (1968) *J. Mol. Biol.*, **38**, 241.

200 De Bruijn, M. H. L. and Klug, A. (1983) *EMBO J.*, **2**, 1309.

201 Whitfeld, P. R. and Bottomley, W. (1983) *Annu. Rev. Plant Physiol.*, **34**, 279.

202 Sugiura, M. (1989) *Annu. Rev. Cell Biol.*, **5**, 51.

203 Subramanian, A. P., Smooker, P. M. (1990) in *The Ribosome. Structure, Function and Evolution* (eds W. E. Hill, A. Dahlberg, R. A. Garrett, P. B. Moore, D. Schlessinger and J. R. Warner), American Society for Microbiology, Washington, p. 655.

204 Das, R. C. and Robbins, P. W. (1988) *Protein Transfer and Organelle Biogenesis*, Academic Press, San Diego.

205 Pugsley, A. P. (1989) *Protein Targeting*, Academic Press, San Diego.

206 Walter, P., Gilmore, R. and Blobel, G. (1984) *Cell*, **38**, 5.

207 Verner, K. and Schatz, G. (1988) *Science*, **241**, 1307.

208 Rothman, J. E. (1989) *Nature (London)*, **340**, 433.

209 Rapoport, T. A. (1991) *Nature (London)*, **349**, 107.

210 Gierasch, L. M. (1989) *Biochemistry*, **28**, 923.

211 Connolly, T. and Gilmore, R. (1989) *Cell*, **57**, 599.

212 Bieker, K. L. and Silhavy, T. J. (1990) *Trends Genet.*, **6**, 329.

213 Inouye, S., Soberon, X., Franceschini, T., Nakarnura, K., Itakura, K. *et al.* (1982) *Proc. Natl. Acad. Sci. USA*, **79**, 3438.

214 Schelness, G. S. and Blobel, G. (1990) *J. Biol. Chem.*, **265**, 9512.

215 Poritz, M. A., Bernstein, H. D., Straub, K., Zopf, D., Wilhelm, H. and Walter, P. (1990) *Science*, **250**, 1111.

216 Muesch, A., Hartmann, E., Rohde, K., Rubartelli, A., Sitia, R. *et al.* (1990) *Trends Biochem. Sci.*, **15**, 86.

217 Wickner, W. T. and Lodish, H. F. (1985) *Science*, **230**, 400.

218 Rogers, J., Early, P., Carter, C., Calaine, K., Bond, M., Hood, L. *et al.* (1980) *Cell*, **20**, 303.

219 Yost, C. S., Hedgpeth, J. and Lingappa, V. R. (1983) *Cell*, **34**, 759.

220 Singer, S. J., Maher, P. A. and Yaffe, M. P. (1987) *Proc. Natl. Acad. Sci. USA*, **84**, 1960.

221 Pelham, H. R. B. (1990) *Trends Biochem. Sci.*, **15**, 483.

222 Pfeffer, S. R. and Rothman, J. E. (1987) *Annu. Rev. Biochem.*, **56**, 829.

223 Douglas, M. G., McCammon, M. T. and Vassarotti, A. (1986) *Microbiol. Rev.*, **50**, 166.

224 Hartl, F.-Z. and Neupert, W. (1990) *Science*, **247**, 930.

225 Baker, K. P. and Schatz, G. (1991) *Nature (London)*, **349**, 205.

226 Keegstra, K. and Bauele, C. (1988) *BioEssays*, **9**, 15.

227 Smeekens, S., Weisbeek, P. and Robinson, C. (1990) *Trends Biochem. Sci.*, **15**, 73.

228 Borst, P. (1989) *Biochim. Biophys. Acta*, **1008**, 1.

229 Dingwall, C. and Laskey, R. A. (1986) *Annu. Rev. Cell Biol.*, **2**, 367.

230 Roberts, B. (1989) *Biochim. Biophys. Acta*, **1008**, 263.

231 Hindley, J. (1973) *Prog. Biophys. Mol. Biol.*, **26**, 269.

232 Zinder, N. D. (1975) *RNA Phages*, Cold Spring Harbor Monograph Series.

233 Min Jou, W., Haegeman, G., Ysebaert, M. and Fiers, W. (1972) *Nature (London)*, **237**, 82.

234 Kozak, M. (1988) *Mol. Cell. Biol.*, **8**, 2737.

235 Kabat, D. and Chappell, M. R. (1977) *J. Biol. Chem.*, **252**, 2684.

236 Lawson, T. G., Cladaras, M. H., Ray, B. K., Lee, K. A., Abrahamson, R. D. et al. (1988) *J. Biol. Chem.*, **263**, 7266.

237 Fu, L., Ye, R., Browder, L. W. and Johnston, R. N. (1991) *Science*, **251**, 807.

238 Huang, W.-I., Hansen, L. J., Merrick, W. C. and Jagus, R. (1987) *Proc. Natl. Acad. Sci. USA*, **84**, 6359.

239 Romaniuk, P. J., Lowary, P., Wu, H.-N, Stormo, G. and Uhlenbeck, O. C. (1987) *Biochemistry*, **26**, 1563.

240 Nomura, M., Gourse, R. and Baughman, G. (1984) *Annu. Rev. Biochem.*, **53**, 75.

241 Draper, D. E. (1989) *Trends Biochem. Sci.*, **15**, 73.

242 Starzyk, R. M. (1988) *Trends Biochem. Sci.*, **13**, 119.

243 Haile, D. J., Hentze, M. W., Rouault, T. A., Harford, J. B. and Klausner, R. D. (1989) *Mol. Cell. Biol.*, **9**, 5055.

244 Lin, J.-J., Daniels-McQueen, S., Patino, M. M., Gaffield, L., Walden, W. E. et al. (1990) *Science*, **247**, 74.

245 Müllner, E. W. and Kühn, L. C. (1988) *Cell*, **53**, 815.

246 Raghow, R. (1987) *Trends Biochem. Sci.*, **12**, 358.

247 Cleveland, D. W. and Yen, T. J. (1989) *New Biologist*, **1**, 121.

248 Brawerman, G. (1981) *Crit. Rev. Biochem.*, **10**, 1.

249 Jackson, R. J. and Standart, N. (1990) *Cell*, **62**, 15.

250 Wickens, M. (1990) *Trends Biochem. Sci.*, **15**, 277.

251 Bernstein, P. and Ross, J. (1989) *Trends Biochem. Sci.*, **14**, 373.

252 Shaw, G. and Kamen, R. (1986) *Cell*, **46**, 659.

253 Malter, J. S. (1989) *Science*, **246**, 664.

254 Baglioni, C. (1979) *Cell*, **17**, 255.

255 Lengyel, P. (1982) in *Protein Biosynthesis in Eukaryotes* (ed. R. Pérez-Bercoff), Plenum, New York, p. 459.

256 Pestka, S., Langer, J. A., Zoon, K. C. and Samuel, C. E. (1987) *Annu. Rev. Biochem.*, **56**, 727.

257 Samuel, C. E. (1991) *Virology*, **183**, 1.

258 Kwong, A. D. and Frenkel, N. (1987) *Proc. Natl. Acad. Sci. USA*, **84**, 1926.

259 Kwong, A. D. and Frenkel, N. (1989) *J. Virol.*, **63**, 4834.

260 Cleveland, D. W. (1988) *Trends Biochem. Sci.*, **13**, 339.

261 Farrell, P. J., Balkow, K., Hunt, T. and Jackson, R. J. (1977) *Cell*, **11**, 187.

262 Ochoa, S. (1983) *Arch. Biochem. Biophys.*, **223**, 325.

263 Voorma, H. O., Goumans, H., Amesz, H. and Benne, R. (1983) *Curr. Topics Cell Regul.*, **22**, 51.

264 Schneider, R. J. and Shenk, T. (1987) *Annu. Rev. Biochem.*, **56**, 317.

265 Sonenberg, N. (1990) *New Biologist*, **2**, 402.

266 Hershey, J. W. B. (1989) *J. Biol. Chem.*, **264**, 20 823.

267 Nairn, A. C. and Palfrey, H. C. (1987) *J. Biol. Chem.*, **262**, 17 299.

268 Ryazanov, A. G., Shestakova, E. A. and Natapov, P. G. (1988) *Nature (London)*, **334**, 170.

269 Levenson, R. M. and Blackshear, P. J. (1989) *J. Biol. Chem.*, **264**, 19 984.

270 Leader, D. P. (1980) in *Molecular Aspects of Cellular Regulation*, vol. 1 (ed. P. Cohen), Elsevier/North Holland, Amsterdam, p. 203.

271 Kozma, S. C., Ferrari, S. and Thomas, G. (1989) *Cell Signal.*, **1**, 219.

272 MacConnell, W. P. and Kaplan, N. O. (1982) *J. Biol. Chem.*, **257**, 5359.

273 Vidales, F. J., Robles, M. T. S. and Balleta, J. P. G. (1984) *Biochemistry*, **23**, 390.

274 Jones, S. W., Erikson, E., Blenis, J., Maller, J. L. and Erikson, R. L. (1988) *Proc. Natl. Acad. Sci. USA*, **85**, 3377.

275 Kozma, S. C., Ferrari, S., Bassand, P., Siegmann, M., Totty, N. and Thomas, G. (1990) *Proc. Natl. Acad. Sci. USA*, **87**, 7365.

276 Clemens, M. (1990) *Trends Biochem. Sci.*, **15**, 172.

277 Sonnenberg, N. (1990) *Curr. Top. Microbiol. Immunol.*, **161**, 23.

278 Valle, R. P. C. and Morch, M.-D. (1988) *FEBS Lett.*, **235**, 1.
279 Atkins, J. F., Weiss, R. B. and Gestland, R. F. (1990) *Cell*, **62**, 413.
280 Murgola, E. F. (1985) *Annu. Rev. Genet.*, **19**, 57.
281 Eggertson, G. and Söll, D. (1988) *Microbiol. Rev.*, **52**, 354.
282 Ayer, D. and Yarus, M. (1986) *Science*, **231**, 393.
283 Kuchino, Y., Beier, H., Akita, N. and Nishimura, S. (1987) *Proc. Natl. Acad. Sci. USA*, **84**, 2668.
284 Chambers, I. and Harrison, P. R. (1987) *Trends Biochem. Sci.*, **12**, 255.
285 Lee, B. J., Rajagopalan, M, Kim, Y. S., You, K. H., Jacobson, B. *et al.* (1990) *Mol. Cell. Biol.*, **10**, 1940.
286 Schön, A., Böck, A., Orr, G., Sprinzl, M. and Söll, D. (1989) *Nucleic Acids Res.*, **18**, 1989.
287 Zinoni, F., Heider, J. and Böck, A. (1990) *Proc. Natl. Acad. Sci. USA*, **87**, 4660.
288 Leinfelder, W., Forchhammer, K., Veprek, B., Zehelein, E. and Böck, A. (1990) *Proc. Natl. Acad. Sci. USA*, **87**, 543.
289 Forchhammer, K., Leinfelder, W. and Böck, A. (1969) *Nature (London)*, **342**, 453.
290 Tsuchihashi, Z. and Kornberg, A. (1990) *Proc. Natl. Acad. Sci. USA*, **87**, 2516.
291 Craigen, W. J. and Caskey, T. (1987) *Cell*, **50**, 1.
292 Weiss, R. B., Huang, W. M. and Dunn, D. M. (1990) *Cell*, **62**, 117.
293 Edlin, G. and Broda, P. (1968) *Bacteriol. Rev.*, **32**, 206.
294 Gallant, J. A. (1979) *Annu. Rev. Genet.*, **13**, 393.
295 Lamond, A. I. and Travers, A. A. (1985) *Cell*, **40**, 319.
296 Warner, J. R. (1989) *Nature (London)*, **338**, 379.
297 Finley, D. and Varshavsky, A. (1985) *Trends Biochem. Sci.*, **10**, 343.
298 Müller-Taubenberger, A., Graak, H.-R., Grohmann, L., Schleicher, M. and Gerisch, G. (1989) *J. Biol. Chem.*, **264**, 5319.
299 Finley, D., Bartel, B. and Varshavsky, A. (1990) in *The Ribosome. Structure, Function and Evolution* (eds W. E. Hill, A. Dahlberg, R. A. Garrett, P. B. Moore, D. Schlessinger and J. R. Warner), American Society for Microbiology, Washington, p. 636.
300 Varshavsky, A. (1983) *Cell*, **34**, 711.
301 Farr, S. B., Arnosti, D. N., Chamberlin, M. J. and Ames, B. N. (1989) *Proc. Natl. Acad. Sci. USA*, **86**, 5010.
302 Benne, R. (1988) *Trends Genet.*, **4**, 181.
303 Soffer, R. L. (1974) *Adv. Enzymol.*, **40**, 91.
304 Ferber, S. and Ciechanover, A. (1987) *Nature (London)*, **326**, 808.
305 Kannangara, C. G., Gough, S. P., Bruyant, P., Hoober, J. K. *et al.* (1988) *Trends Biochem. Sci.*, **13**, 139.
306 Litvak, S. and Arya, A. (1982) *Trends Biochem. Sci.*, **7**, 361.
307 Haenni, A.-L., Joshi, S. and Chapeville, F. (1982) *Prog. Nucleic Acid Res.*, **27**, 85.
308 Weiner, A. M. and Maizels, N. (1987) *Proc. Natl. Acad. Sci. USA*, **84**, 7383.
309 Burk, R. F. (1991) *FASEB J.*, **5**, 2274.
310 Meyer, F., Schmidt, H. J., Plümper, E., Haslik, A., Mersmann, G. *et al.* (1991) *Proc. Natl. Acad. Sci. USA*, **88**, 3758.
311 Lingappa, V. R. (1991) *Cell*, **65**, 527.
312 Herman, P. K., Stark, J. H., DeModena, J. A. and Emr, S. D. (1991) *Cell*, **64**, 425.
313 Silver, P. A. (1991) *Cell*, **64**, 489.
314 Gunnery, S., Rice, A. P., Robertson, H. and Mathews, M. B. (1990) *Proc. Natl. Acad. Sci. USA*, **87**, 8687.
315 Beattie, E., Tartaglia, J. and Paoletti, E. (1991) *Virology*, **183**, 419.
316 Lee, T. G., Tomita, J., Hovanessian, A. G. and Katze, M. G. (1990) *Proc. Natl. Acad. Sci. USA*, **87**, 6208.
317 Heus, H. A. and Pardi, A. (1991) *Science*, **253**, 191.
318 Mitchell, P., Osswald, M., Schueler, D. and Brimacombe, R. (1990) *Nucleic Acids Res.*, **18**, 4325.
319 Altman, S. (1990) *J. Biol. Chem.*, **265**, 20053.
320 Powers, T. and Noller, H. F. (1991) *EMBO J.*, **10**, 2203.
321 Leberman, R., Härtlein, M. and Cusack, S. (1991) *Biochim. Biophys. Acta*, **1089**, 287.
322 Rould, M. A., Perona, J. J. and Steitz, T. A. (1991) *Nature (London)*, **352**, 213.
323 Ruff, M., Krishnaswamy, S., Boeglin, M., Poterszman, A., Mitschler, A. *et al.* (1991) *Science*, **252**, 1683.
324 Mattheakis, L., Vu, L., Sor, F. and Nomura, M. (1989) *Proc. Natl. Acad. Sci. USA*, **86**, 448.

Appendix: methods of studying nucleic acids

This appendix describes some of the most important general methods used in the study of nucleic acids not considered elsewhere in this book. The emphasis is on the principles underlying the methods, and no attempt is made to provide laboratory protocols, which are now available in abundance, and to which reference will be made. It is inevitable that some of the methods dealt with here will be superseded in the lifetime of this edition. The reader wishing information about current methodologies is therefore advised to consult the most recent texts. However, at the time of writing the cited references are recommended for more detailed information on principles [1–5] and practice [6–10].

A.1 OCCURRENCE AND CHEMICAL ANALYSIS

A.1.1 Chemical determination of nucleic acids in tissues

The usual approach to the determination of nucleic acids in tissues involves fractionation to remove lipids and proteins etc., followed by assay on the basis of either phosphorus, ribose or deoxyribose, or purine and pyrimidine. The tissue is treated with cold dilute trichloroacetic or perchloric acid to precipitate nucleic acid and protein, and lipid is removed by extraction with organic solvent. In the procedure of Schneider [11] total nucleic acid (DNA + RNA) can be separated from protein by hydrolysis in hot acid, the liberated products being soluble (section 2.7.3). These latter are then assayed for ribose and deoxyribose. In the procedure of Schmidt and Thannhauser [12] the DNA and RNA are separated by alkaline hydrolysis of the latter followed by acidification (Fig. A.1).

Analysis of nucleic acid phosphorus in such fractions usually employs a colour reaction involving the formation of a phosphomolybdate complex [13, 14]. Following depurination (section 2.7.2), deoxyribose can be determined by a colour reaction with diphenylamine (section 2.7.1) [15]. This reaction is specific for DNA. In contrast, the orcinol colour reaction employed for depurinated RNA also occurs to a lesser extent with DNA [16]. The spectral properties of the purine and pyrimidine bases of such hydrolysed nucleic acids may also be used as a basis of nucleic acid assay. These

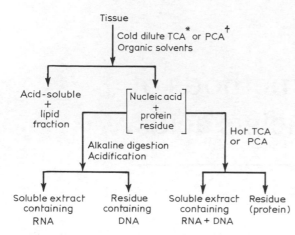

Fig. A.1 Extraction and fractionation of nucleic acids from tissues. *TCA – trichloroacetic acid, †PCA – perchloric acid.

(section 2.7.4). Protein also absorbs at 260 nm, although to a much lesser extent, dependent on the content of the aromatic amino acids phenylalanine, tryptophan and tyrosine. Because the latter two of these have an absorption maximum at 280 nm it is possible to obtain an approximate estimation of total nucleic acid and protein in a mixture by employing the following formulae based on [17]:

$$\text{Nucleic acid} = 0.064\,A_{260} - 0.031\,A_{280} \text{ (mg/ml)}$$

$$\text{Protein (mg/ml)} = 1.45\,A_{280} - 0.74\,A_{260}$$

Typical values for the nucleic acid and protein contents of some rat tissues are given in Table A.1. These may be expressed per gram tissue wet weight, but it is often useful to express them in terms of tissue protein, which may be easily estimated

bases have absorption maxima in the region of 260 nm (Fig. A.2). The spectral properties of double-stranded and single-stranded DNA have already been discussed (section 2.4.1), as has the effect of pH thereon

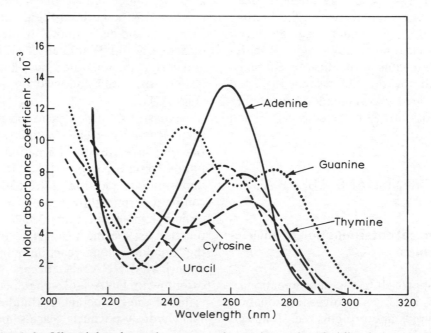

Fig. A.2 Ultraviolet absorption spectra for purine and pyrimidine bases at pH7.

Table A.1   Typical values for the concentrations of nucleic acids in different rat tissues

Tissue	RNA (mg/g protein)	DNA (mg/g protein)	RNA/DNA
Brain	40	15	2.7
Kidney	43	18	2.4
Liver	54	12	4.5
Skeletal muscle	62	13	4.8
Spleen	58	58	1.0
Thymus	30	93	0.3

Table A.2   Molar percentages of bases in RNAs from various sources

Type		Adenine	Guanine	Cytosine	Uracil
Ribosomal*					
E. coli	16S rRNA	25.2	31.6	22.9	20.3
	23S rRNA	26.2	31.4	22.0	20.4
	5S rRNA	19.2	34.2	30.0	16.7
Rat	18S rRNA	22.5	29.2	26.5	21.8
	28S rRNA	16.6	35.5	31.6	16.2
	5S rRNA	18.3	32.5	27.5	21.7
Messenger†					
E. coli	Outer-membrane lipoprotein	28.3	24.2	22.7	24.8
Rabbit	β-globin	23.6	27.2	23.0	26.2
Rat	α-actin	21.7	26.4	30.6	21.3
Viral*					
Bacteriophage MS2		23.4	26.0	26.1	24.5
Poliomyelitis virus		29.6	23.0	23.3	24.1
Rous sarcoma virus		23.8	28.8	25.3	22.1
Tobacco mosaic virus		29.1	24.2	19.1	27.6

* Ignoring base modifications.
† Ignoring 3′poly(A) tails.

[18, 19]. Values of haploid DNA content for various organisms are given in Table 3.1.

A.1.2 Analysis of base composition and nearest-neighbour frequency

The molar proportions of bases in nucleic acids are determined by hydrolysis, separation and spectral quantification, as described for DNA in section 2.2.2. Typical values for DNA are given in Table 2.3 and the variation in percentage GC content in different genomes can be found in Table 2.6. The principles of determining the base composition of RNA molecules are similar to those for DNA, and values for some types of RNA (actually based on sequence analysis) are given in Table A.2. Transfer RNA is not included because of the large proportion of modified bases (sections 2.1.2 and 11.7.2). In the case of rRNA (section 11.6.3) the modified bases have been ignored in Table A.2. It may be remarked that, although it was stated in chapter 12 that rRNA is GC-rich, this need only be true in relation to the overall base composition of the genome, rather than in absolute terms. The examples of mRNAs in Table A.2 are rather arbitrary, and dramatic variations in the base compositions of different mRNAs are seen in the same organism.

The determination of nearest-neighbour frequencies (i.e. the relative frequency of occurrence of the sixteen different dinucleotides, NpM) in DNA, although to a considerable extent superseded by nucleotide sequence determination (section A.5), still has applications. It is described in section 6.4.3.

A.1.3 Estimation of the molecular weight of DNA

Methods of estimating the size of small DNA molecules (up to 10 kb) using *electrophoresis* on acrylamide and agarose gels are described in section A.2.1. Such methods can be accurately calibrated using DNA molecules of known sequence. DNA molecules of up to even 1000 kb in length can be analysed by pulsed field gel electrophoresis (section A.2.1).

Of the more classical approaches, the best absolute methods of determining the molecular weight of DNA are light scattering on low-angle instruments [20–22], equilibrium analytical ultracentrifugation in gradients of caesium chloride [23], and viscoelastic relaxation [24]. Most other methods have an empirical basis and rely for their calibration on the few light-scattering experiments that have been performed. These more empirical methods include the measurement of intrinsic viscosity [25] and

sedimentation rate [25], electron microscopy [26], and radioautography [27]. Generally a combination of two or more methods has been used [28].

Although measurement of the *sedimentation rate* can be used to estimate the molecular weight of an RNA molecule (section A.2.2), this method is not readily applicable to DNA molecules because of their extremely large size and asymmetric shape. The theory behind the calculations depends on a knowledge of diffusion coefficients but these are equally difficult to obtain either by sedimentation analysis or optical mixing spectroscopy [29–31]. However, empirical formulae relating sedimentation coefficient ($s_{20,w}^{o}$) and molecular weight (M) have been derived by comparing the sedimentation rates of molecules of known size. For example, the equation for linear, double-stranded DNA is

$$s_{20,\ w}^{o} = 2.8 + 0.00834\ M^{0.479}\ [28, 32]$$

On sedimentation of DNA to *equilibrium* in density gradients (usually of CsCl – section A.2.1) the width of the band of DNA is inversely proportional to the square root of the molecular weight. As with light scattering (see below) the result is dependent on the concentration of the DNA and a series of values must be obtained and extrapolated to zero concentration [33]. This is the most widely used, fundamental method of estimating the size of DNA molecules and is applicable to molecules of up to 10^8 daltons (150 kb).

Light scattering was widely used in the 1950s to measure molecular weights of DNA up to 5000 kb. This technique, together with conventional analytical velocity centrifugation to determine sedimentation coefficients, should lead to reliable absolute determination of molecular weight [34]. But only since the introduction of the laser and the production of machines to enable study of scattering at low angles and changes in the Doppler shift in the wavelength of the scattered laser light has the method been extended to molecules of larger size [29–32]. However, at low angles the scattering by dust particles becomes a major problem and with large molecules different regions of the same molecule serve as scattering centres.

The above methods all suffer from the disadvantage that they are badly affected by any heterogeneity in the size of the DNA molecules. The problem is greatest when the intention is to prepare whole DNA molecules since high-molecular-weight DNA is very susceptible to shearing forces and to the action of contaminating nucleases. Thus all molecular-weight determinations of an unknown DNA must be carried out in nuclease-free conditions (achieved by heat-sterilization), without pipetting or any other manipulation which puts shear stress on the DNA. It is extremely difficult to prepare high molecular-weight DNA intact from chromosomes and these methods tend to reflect the size of the smaller molecules.

Attempts to relate molecular weight to *viscosity* involve subjecting the DNA solution to shear forces which tend to break long DNA molecules, and must be used with caution. However, Zimm developed a low-shear viscometer consisting of a slowly rotating tube (the Cartesian diver) immersed in a DNA solution under pressure, and a relationship has been established between intrinsic viscosity (η) and molecular weight (M) of linear duplex DNA similar to that between sedimentation coefficient and molecular weight:

$$0.665 \log M = 2.863 + \log (\eta + 5) \, [35]$$

A more important observation is that, when the Zimm viscometer is used to measure the size of very large DNA molecules, the Cartesian diver rotates in the reverse direction for a while after the power

has been switched off. This *viscoelastic* effect is caused by the stretched DNA molecules relaxing to the unstressed configuration. The theory derived by Zimm to explain this effect has been used to measure the size of the DNA in *Drosophila* chromosomes (each of which contains one DNA molecule [24]) showing the method to be applicable to DNA molecules of molecular weight up to 79×10^9 (over 10^5 kb). The method has the major advantage that it measures the size of the largest DNA molecules in the solution [36].

Two 'visual' methods of measuring the size of DNA molecules involve electron microscopy and autoradiography. With *electron microscopy* the Kleinschmidt technique involves spreading the DNA on grids coated with polylysine or polystyrene-4-vinylpyridine [37]. A drop of DNA solution in 1 M ammonium acetate containing 0.1 mg/ml denatured cytochrome c is applied to the surface of a solution of 0.2 M ammonium acetate. A film of cytochrome c spreads across the surface, binding to the DNA which is disentangled and spread out. A grid is touched to the surface and the sample is dehydrated in ethanol. It may be stained with uranyl acetate or by evaporation of platinum on to the surface. In the latter process the grid is rotated so as to cause the metal to pile up against all surfaces of the DNA, which stands out from the grid [30].

Double-stranded DNA is readily visualized by this technique but single-stranded DNA (or RNA) molecules require the presence of formamide to prevent their collapse [38]. Even so they appear as rather wispy threads and may be stretched out to different extents. The incorporation of proteins which bind strongly to single-stranded DNA (e.g. bacteriophage T4 gene 32 protein: section 6.5.2) causes these regions to assume a profile thicker even than the native DNA regions and they now have a regular mass per unit length. The presence, on the same grid, of marker DNA molecules of known length (e.g. SV40 DNA in the relaxed circular form) allows the size of an unknown DNA molecule to be found by measuring the relative contour lengths on electron micrographs. This method can readily be used with DNA molecules of up to about 10^3 kb, but larger molecules cannot be accommodated in a single frame.

In denaturation mapping (section 6.8.1) double-stranded DNA is partially denatured (usually by alkali treatment) and the un-paired bases reacted with formaldehyde to prevent their reannealing. The duplex DNA molecule now exhibits single-stranded bubbles in the (A + T)-rich regions to give a characteristic map.

When DNA is labelled *in vivo* with [^{3}H]thymidine and the cells lysed on a microscope slide to release the intact DNA molecules their position can be visualized by *radioautography*. The radioactive DNA on the slide is covered with a layer of photographic film which is activated by the β-particles produced by the decay of the ^{3}H to produce black grains. As with electron micrographs, the size of the DNA can be obtained by measuring the contour length of the radioautographic image. This method uses light microscopy and so the field size is much larger than with electron microscopy. Cairns applied it to the study of the *E. coli* chromosome, which is 3000 kb long [39].

A.2 ISOLATION AND SEPARATION OF NUCLEIC ACIDS

A.2.1 Isolation and separation of DNA

The main problems to be overcome in isolating high-molecular-weight chromosomal DNA from bacterial [40] or eukaryotic cells [41] are the removal of protein (especially

deoxyribonucleases) and RNA, and the avoidance of mechanical shearing. For eukaryotic cells a combination of pronase or protease K, a detergent and extraction with phenol can be used directly to disrupt the cells and remove protein. For the cells of Gram-negative bacteria such as *E. coli* gentle lysis can be effected using lysozyme. Phenol extraction is used widely in the isolation of both DNA and RNA, which are retained in the aqueous phase whereas denatured protein collects at the interface between this and the phenol phase, which contains the lipids. RNA may be removed from the DNA by using pure pancreatic ribonuclease, or by isopycnic ultracentrifugation in a gradient of CsCl, which is, in any case, a useful step for further purification.

In *isopycnic ultracentrifugation* a density gradient is established which encompasses the buoyant densities of the molecules to be separated. Molecules will assume a position on the gradient corresponding to their own buoyant density, after which no further separation can occur. In this method an equilibrium is established, and the method is sometimes referred to as 'equilibrium ultracentrifugation' to distinguish it from the rate-zonal method (section A.2.2). As already mentioned (section 2.4.3), the buoyant density of double-stranded DNA varies with the mole fraction of (G + C). In general, however, RNA has a much higher buoyant density (about 1.9 g/ml) than double-stranded DNA (about 1.7 g/ml), which, in turn, is higher than that of protein (about 1.3 g/ml). Single-stranded DNA is of slightly higher buoyant density than double-stranded DNA (e.g. for a double-stranded DNA of buoyant density 1.703 g/ml, the value for the single-stranded form is 1.717 g/ml).

Isopycnic ultracentrifugation is also used to separate small bacterial plasmids from chromosomal DNA. The basis of separation in this case is not a difference in GC contents (these are, in general, similar) but the fact that the chromosomal DNA is linear (normal methods do not extract the large circular bacterial chromosome intact) whereas the plasmids are circular. This difference is exploited by the addition of saturating amounts of the intercalating fluorescent dye, ethidium bromide. The intercalation requires that the DNA strands be forced apart, with concomitant decrease in buoyant density and partial unwinding of the double helix. This latter process is hindered in the closed-circular plasmid molecules with the result that they bind less ethidium bromide and have a higher buoyant density than the linear chromosomal (and open-circular and linearized plasmid) molecules (Fig. A.3). Before attempting to separate plasmid and chromosomal DNA it is advisable to enrich the preparation in plasmid DNA. The most frequently employed method for achieving this in preparative work involves the use of an alkaline pH during lysis and extraction [42]. This denatures the DNA, causing strand-separation in the case of linear molecules. On neutralization the latter will not renature, whereas the closed-circular plasmid DNA will and can be separated from the insoluble denatured DNA. For small-scale isolation of plasmid DNA, a boiling step is usually used in place of the treatment with alkali [43].

Ethidium/CsCl gradients are also used in the purification of small animal viruses (e.g SV40) which contain circular genomes. In this case the initial purification involves removal of most of the high-molecular-weight cellular DNA as a precipitate adsorbed to sodium dodecyl sulphate, leaving the virus DNA in the so-called Hirt extract [44].

Although most preparations of chromo-

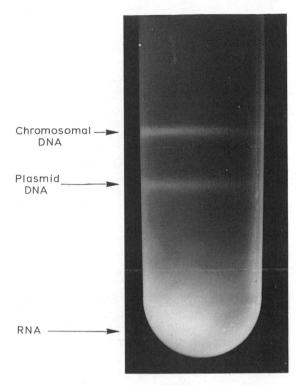

Chromosomal DNA

Plasmid DNA

RNA

Fig. A.3 Separation of closed-circular DNA of plasmid pBR322 from *E. coli* chromosomal DNA by isopycnic centrifugation in a CsCl density gradient in the presence of ethidium bromide. The band marked 'chromosomal DNA' may also contain nicked plasmid DNA molecules.

somal DNA consist of sheared fragments, these do not in general separate on isopycnic ultracentrifugation because the fragmentation is random and the base compositions of individual fragments are generally similar. Exceptions do occur in the case of regions of repetitive DNA in eukaryotes (sections 3.2.2 and 7.7) which may be generated as quite homogeneous species in relatively large amounts if the fragments are of an appropriate size. Should the base composition of such a repeated DNA be markedly different from the average base composition of the chromosomal DNA, resolution from the bulk of the latter will occur. Such DNA

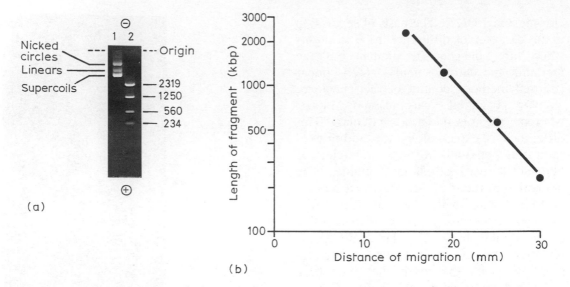

(a)

(b)

Fig. A.4 Agarose gel electrophoresis of DNA. (a) Separation of: 1, different forms of DNA of plasmid pBR322; 2, fragments of DNA (lengths indicated in kb) derived from plasmid pBR322 by double-digestion with restriction endonuclease *Bam*H I and *Bgl* I; (b) Plot of length of DNA fragment (log scale) against distance of migration (linear scale) of data from (a)2, illustrating linear relationship.

was originally given the descriptive name 'satellite DNA' (section 3.2.2(b)), a designation which is sometimes applied to all repetitive DNA, regardless of base-composition or function. Such loose terminology is to be discouraged. It is also possible to isolate the tandemly repeated ribosomal RNA and histone genes of certain organisms by this method [45].

Small-scale *column chromatography* has been used for more rapid separation of plasmid and chromosomal DNA. The chromedia employed include anion-exchange resins, gel permeation resins and finely divided glass, and proprietary columns are available in which the separation can be further accelerated by centrifugal elution.

Separation of DNA by isopycnic ultra-centrifugation or chromatography is largely confined to the different chromosomal species that one might find in a given cell. Precise fragments of such DNAs can be generated by restriction endonucleases (section A.6) or by the polymerase chain reaction (section A.7), and the separation of such fragments is a common requirement in recombinant DNA manipulations. Most frequently this is achieved by horizontal *agarose gel electrophoresis* [46]. This separates different linear DNA molecules according to size (the mobility is inversely proportional to the $\log_{10}$ of the molecular weight), as the charge per unit length of different molecules is identical and the larger molecules will be retarded by the gel matrix (Fig. A.4). The DNA is 'visualized' by use of ethidium bromide. The percentage of agarose can be increased to separate smaller-sized DNA fragments, but for very small species *polyacrylamide gel electrophoresis* (which is often used preparatively for fragments up to 1 kb) must be used [47]. This method is also employed in rapid nucleic acid sequencing methods (section A.5), in which case 8 M urea is included to ensure denaturation.

As agarose gel electrophoresis is frequently used to analyse preparations of

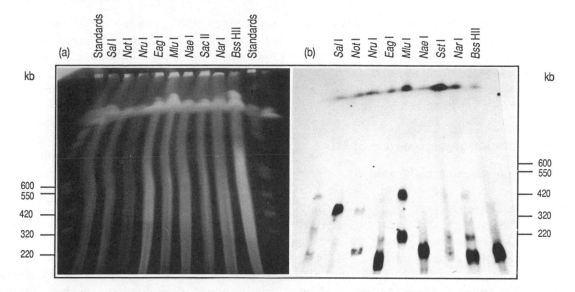

Fig. A.5 Pulsed field gel electrophoresis of fragments of human leukocyte DNA. The restriction endonucleases indicated, which cleave vertebrate DNA infrequently, were used to generate large fragments which were subjected to 62 s pulses at 290 V for 39 h in 0.8% agarose using a hexagonal arrangement of electrodes. (a) DNA stained with ethidium bromide, (b) autoradiograph after hybridization to a radioactive probe derived from a highly polymorphic region of DNA located downstream of the human α-globin gene complex. (Illustration kindly provided by Dr A. M. Frischauf.)

plasmid DNA which may be slightly nicked, it is worth mentioning that the closed-circular supercoiled form of the plasmid migrates more rapidly than the less compact relaxed-circular form produced by single-stranded nicks. Any linearized plasmid molecules generally migrate at an intermediate position between these two, a result that is perhaps a little unexpected (Fig. A.4). Agarose gel electrophoresis in the presence of varying amounts of ethidium bromide can be used to determine the number of super-helical turns in a closed-circular DNA, as the interchelating ethidium bromide progressively reduces the number of super-helical turns, with a consequent effect on the rate of electrophoretic migration [48].

Very large molecules of DNA (greater than about 100 kb) cannot be separated by conventional agarose gel electrophoresis in an electric field of constant direction. The

introduction of the technique of *pulsed field gel electrophoresis* [49, 50], in overcoming this limitation, has made it possible to separate DNA molecules of about 1000 kb and above (Fig. A.5). In this technique the electric field is alternated at intervals between the forward direction and one at an angle to this (it may even be directly opposite – there are many variations). Although the basis of the separation can be envisaged simplistically in terms of different rates of reorientation of the DNA molecules as the direction of the field changes, there is still debate regarding a rigorous physical explanation [51]. The resolving power of this method is such that it was necessary to devise new methods of preparing DNA that avoided shear breakage. The technique usually employed for this purpose is to embed the live cells in molten agarose, allowing it to solidify into blocks on cooling, after which the DNA can be treated *in situ* by allowing reagents to diffuse into (and out of) the blocks. Pulsed field gel electrophoresis can be performed on the gel blocks, and in this way the separation of intact yeast chromosomes has been achieved [49].

A.2.2 Isolation and separation of RNA

In principle the methods used to extract RNA from cells are generally similar to those for DNA. Degradation of RNA by ribonuclease is a greater threat than that of deoxyribonuclease degradation of DNA, especially in the case of mRNA, but there are a variety of methods to counter this. The major separation problem is posed by the large number of different RNA species in eukaryotes, and to study a particular species it is necessary to separate it from the others.

One preliminary that may be useful in particular cases is *subcellular fractionation* [52]. Thus, the nuclei may be removed by low-speed centrifugation ($700\,g$ for 5 min) after gentle disruption of the cells in the presence of sucrose (to preserve the osmolarity), the mitochondria may be removed by further centrifugation ($10000\,g$ for 10 min), and the ribosomes separated from soluble RNA species such as tRNA by ultracentrifugation (e.g. $100000\,g$ for 90 min). It may of course be necessary to purify further the subcellular fraction in which one is interested, and methods for nuclei [53], nucleoli [54], mitochondria [55], and ribosomes [56] are available. It is worth mentioning that the most appropriate method for isolating an organelle, such as the nucleus, for subsequent RNA extraction (where purity is of greatest importance) may be quite different from the method which will yield the most biologically active material [57].

One of the most widely used methods of separating RNA molecules is *rate-zonal ultracentrifugation* employing *sucrose density gradients* [58]. In this method the separation of RNA is based primarily on size (in principle, molecular shape can also have an influence, but with RNAs differences in shape make no significant contribution to the separation). It should be emphasized that the density range of sucrose solutions does not extend to that of RNA species and, in contrast to isopycnic ultracentrifugation involving CsCl gradients (section A.2.1), no equilibrium is reached. The species to be separated are applied in a narrow zone at the top of the sucrose gradient, the main purpose of which is to prevent diffusion of the zones of individual species during separation (Fig. A.6). Isopycnic ultracentrifugation in gradients of dense sucrose (often employing a discontinuous or 'step gradient') may be used to separate 'free' ribosomes from those bound to the membrane of the endoplasmic reticulum (section 12.8.1) [59].

Rate-zonal ultracentrifugation in sucrose density gradients is also used to separate different size classes of polyribosomes (section 12.1) and to separate the large and small subunits of ribosomes after their dissociation (section 12.6).

Rate-zonal ultracentrifugation is clearly inappropriate for the separation of different species of tRNA, which are broadly similar in size. Methods involving separation on the basis of charge and hydrophobicity differences (e.g. chromatography on BD-cellulose and RPC5) may be employed for this purpose [60].

Although *mRNA* may be isolated from purified polysomes, direct extraction from tissues is more frequently employed to minimize degradation by ribonuclease. The use of high concentrations of the chaotropic salt, guanidinium thiocyanate [61], is perhaps more widespread than the methods

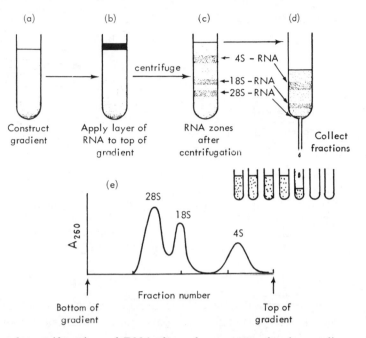

Fig. A.6 Rate zonal centrifugation of RNA through a sucrose density gradient. A sucrose density gradient is constructed in a centrifuge tube (a) and the RNA solution applied on top (b). During ultracentrifugation the main components of the RNA separate into zones, primarily on the basis of molecular weight (c). These zones may be recovered by puncturing the bottom of the tube and collecting different fractions in separate tubes (d). The separated RNA may be visualized and quantified by measurement of the absorbance at 260 nm (e). Steps (d) and (e) may be conveniently combined by pumping the gradient through the flow-cell of a recording spectrophotometer.

involving phenol and detergents, mentioned above for DNA. Another alternative is guanidine hydrochloride in combination with organic solvents [7]. Inhibitors of ribonuclease, such as placental ribonuclease inhibitor protein [62], vanadyl ribonucleoside complexes [63], or macaloid [64], are sometimes used in preparing mRNA. The major problem to be overcome in the isolation of mRNA is purification from other species of RNA, which are approximately 20 times more abundant. Affinity chromatography using oligo(dT)-cellulose [65] or poly(U)-Sepharose [66], materials that bind the poly(A) tails of mRNAs, is the basis of the separation of mRNA from other species of RNA. The minimum size of poly(A) tail required for binding to poly(U)-Sepharose is smaller than that for oligo(dT)-cellulose (about 6–10 residues, compared with 15 residues), and although the latter material is in more widespread use, the former has applications to mRNAs having very short poly(A) tails. A mixture of different mRNAs can be separated to a certain extent on the basis of size, either by sucrose density gradient centrifugation or, on a smaller scale, by agarose gel electrophoresis – broadly as for DNA but generally under denaturing conditions (section A.2.1). Both agarose gel electrophoresis and polyacrylamide gel electrophoresis have been applied to other types of RNA [67, 68]. The isolation of individual mRNA species is usually achieved by recombinant DNA techniques (section A.8).

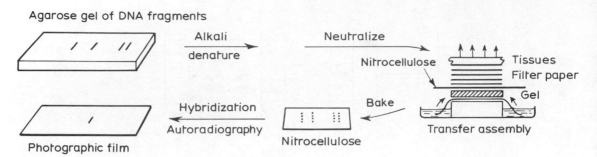

Fig. A.7 Southern blotting. Fragments of DNA are fractionated by electrophoresis on agarose, the DNA denatured in alkali and, after neutralization, transferred (blotted) from the gel to nitrocellulose membrane by capillary action at relatively high ionic strength in the transfer assembly shown. The transferred DNA is baked onto the nitrocellulose, after which it can be hybridized in solution with a radioactive probe and the hybridizing bands visualized by autoradiography. In the hypothetical example illustrated only one of the original four bands has hybridized to the probe.

Nucleic acids may be precipitated with ethanol, 2.5 volumes in the presence of 0.3 M sodium acetate, being frequently employed. This is useful for purification and concentration purposes.

A.3 HYBRIDIZATION OF NUCLEIC ACIDS

The denaturation and renaturation of homoduplex DNA molecules has been dealt with in section 2.4, where the effects of base composition, temperature and ionic strength were mentioned. The application of DNA renaturation to determining the copy number of different portions of eukaryotic genomes (C_0t value analysis) has been described in section 3.2.2. An important technique in molecular biology is the formation of heteroduplexes between different DNA molecules or between DNA and RNA. The application of electron microscopic visualization of such heteroduplexes formed in solution to the determination of the positions of introns (*R-loop mapping*) has already been described (section 8.2), and the electron microscopic analysis of

heteroduplexes can be most useful in a variety of other situations. However, for most routine work heteroduplexes are detected by the radioactivity of one of their components. (Both radioactive and non-radioactive labelled nucleic acid 'probes' may be used (section A.4), but in the following we refer only to radioactively labelled probes for economy of expression).

Originally, in order to identify which component of DNA fractionated by agarose gel electrophoresis hybridized to a particular radioactively labelled probe one had to perform time-consuming and inconvenient multiple hybridizations in solution. To overcome this problem Southern [69] devised a method of transferring the DNA from the fragile gel to a solid nitrocellulose membrane by means of simple capillary action (Fig. A.7), although this can be performed more rapidly with the aid of a vacuum [70]. The DNA must be denatured with alkali before transfer, and immobilized on the membrane afterwards. This latter can be achieved by baking at 80°C (in a vacuum oven to prevent the nitrocellulose igniting) or, in the case of the subsequently introduced nylon membranes, by ultraviolet

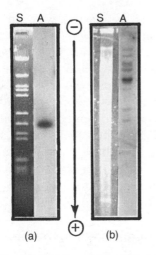

(a) (b)

Fig. A.8 Example of results of Southern blotting experiment. (a) Cloned DNA from a mouse/bacteriophage lambda recombinant (cf. Fig. A.24), or (b) chromosomal DNA from mouse liver, were digested with (different) restriction endonucleases, subjected to electrophoresis on agarose gels and Southern blotting performed as in Fig. A.7. Hybridization was to a ^{32}P-labelled mouse actin cDNA clone (cf. Fig. A.21). S: ethidium bromide stained gels photographed under illumination with ultraviolet light. A: autoradiographs of the nitrocellulose.

light. This method of transfer is commonly called *Southern blotting*, after its originator. The membrane is usually hybridized with the radioactive DNA probe in a minimum volume of solution in a sealed polythene bag at 68°C and conditions of relatively high ionic strength that promote heteroduplex formation. (If 50% formamide is included the hybridization is performed at 42°C.) The membrane is washed under suitable conditions (see below), dried, and subjected to autoradiography. This method is most easily applied to detecting which restriction fragment of a larger piece of cloned DNA contains sequences homologous to a particular probe (Fig. A.8(a)). However, the sensitivity of the method is such that it can detect individual fragments in the continuum

produced by restriction enzyme digestion of total genomic DNA (Fig. A.8(b)), and in fact the method was originally applied to uncloned genomic and subgenomic DNAs. Such genomic Southern blotting has become a routine tool in screening for human genetic disorders by the detection of restriction fragment length polymorphisms (section 3.1.4(d)) [71].

The conditions of washing the membranes depend on the degree of homology between the two members of the heteroduplex. (The homology need not be 100% as hybridization may be between a variety of related but non-identical sequences.) When the two members of the heteroduplex are identical, the washing is usually performed at 12°C below the melting temperature (T_m), which in a solution 0.2 M in Na$^+$ is related to the mole percentage of (G + C) by the equation:

$$T_m = 69.3 + 0.41 (G + C) \text{ [72]}$$

The effect of ionic strength (μ) on the melting temperature is given by the equation [73]:

$$T_{m2} - T_{m1} = 18.5 \log_{10} \frac{\mu_2}{\mu_1}$$

Thus, to allow for the fact that the T_m of duplex DNA decreases by 1°C per 1–1.5% mismatch [74, 75] the 'stringency' of the hybridization can be decreased by lowering the temperature of the wash and/or by increasing the ionic strength of the washing buffer.

Southern's method for the transfer of DNA from gels to a solid membrane has been extended to RNA, where it has acquired the rather illogical name of *Northern blotting*. (There is a third point on the compass, 'Western blotting', the electrophoretic transfer of proteins from polyacrylamide gels to nitrocellulose.) The methodology of Northern blotting is different from that of Southern blotting because RNA does not bind to nitrocellulose paper under conditions in which DNA does, and is hydrolysed by alkali. To overcome this problem, diazobenzyloxymethyl (DBM)-cellulose paper, which binds RNA (and DNA), was introduced [76]. However, the use of this has declined since it was discovered that RNA is, in fact, immobilized on nitrocellulose membranes if it has first been subjected to electrophoresis under denaturing conditions. This may be achieved by including formaldehyde [77], methyl mercuric hydroxide [78], or glyoxal and dimethyl sulphoxide [79] in the agarose gel.

Hybridization of immobilized nucleic acids on membrane supports is not restricted to material transferred from agarose gels. It can be used for detecting recombinant DNA in bacterial colonies or bacteriophage plaques after transfer from petri dishes (section A.8) or for multiple samples of total cellular RNA applied directly to the nitrocellulose (the so-called *dot-blot* technique).

A.4 METHODS OF LABELLING NUCLEIC ACIDS

There are many circumstances in which the detection of DNA is only possible if the DNA is labelled in some way. Furthermore, certain techniques require DNA labelled specifically at one end. By far the most common method of achieving such labelling is with radioactivity (usually ^{32}P; less commonly ^{35}S or ^{3}H), and for economy of expression the discussion below will be in terms of radioactively labelled probes. However, *non-radioactive labelling* methods are increasingly being used. Most frequently biotin-11-dUTP [80] is used to label the

DNA, which is reacted with an antibiotin antibody and then with a second antibody coupled to an enzyme such as horseradish peroxidase or alkaline phosphatase. Originally the enzyme was used to catalyse a colour reaction to allow visualization of the DNA, but more recently detection has been improved by using chemiluminescent substrates for alkaline phosphatase [81], or by coupling the peroxidase reaction to a photochemical reaction.

A.4.1 General labelling methods

When the objective of labelling a nucleic acid is merely that of allowing its detection, methods that cause the incorporation of radioactivity throughout the molecule are suitable, as well as the specific end-labelling methods described in section A.4.2. Historically, RNA and DNA were made radioactive from ^{3}H-, ^{14}C- or ^{32}P-labelled precursors *in vivo*, but methods involving the incorporation of radioactive label *in vitro* are more convenient for recombinant DNA work and produce material of higher specific activity.

One highly efficient method of labelling double-stranded DNA *in vitro* is by '*nick translation*' (Fig. A.9) using [α-^{32}P]dNTPs [82]. In this method nicks are introduced into the DNA with DNase and the 5'-phosphate ends produced can serve as substrates for the 5' → 3' exonucleolytic activity of *E. coli* DNA polymerase I, which at the same time repairs the gaps by addition of the [α-^{32}P]dNTPs to the 3'-OH end of the 'nick' (section 6.4.2). In equally widespread use for labelling double-stranded DNA to high specific activity is the method of '*random-priming*' [83, 84]. A random mixture of oligonucleotides (6–12 mers) is used to prime the synthesis of copies of the DNA template, which must be denatured

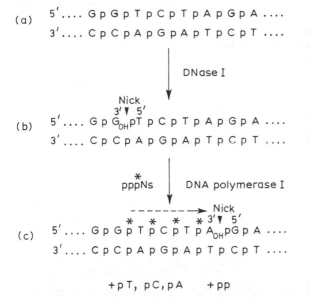

Fig. A.9 Nick translation. Double-stranded DNA (a) is treated with a low concentration of pancreatic DNase producing occasional nicks with 3′-OH groups (b) on which DNA polymerase can start polymerization. The use of [α-³²P]dNTPs produces a radioactive phosphodiester bond p̃ and the 5′ → 3′ exonuclease action of the enzyme successively cleaves further non-radioactive phosphodiester bonds which are replaced by radioactive ones (c). In the course of this replacement the position of the nick undergoes vectorial translation, as indicated.

for this purpose. The Klenow fragment of DNA polymerase I is used to avoid destruction of the primers and products by the 5′ → 3′ exonucleolytic activity of the intact enzyme, and [α-³²P]dNTPs are again used as labelled precursors.

For certain purposes it is necessary to prepare single-stranded labelled probes. One method of doing this [85] is by cloning the DNA into the single-stranded phage vector M13, and copying it *in vitro* using an oligonucleotide primer and [α-³²P]dNTPs (cf. section A.5.1). An alternative method of producing single-stranded probes is to clone into a vector that puts it under the control of a bacteriophage promoter [86], and generate radioactively labelled RNA copies of the DNA by transcription *in vitro* with the bacteriophage RNA polymerase and [α-³²P]NTPs. Systems based on bacteriophages SP6, T7 and T3 have been used for this purpose, as their RNA polymerases are highly specific for their own promoters. An advantage of generating single-stranded labelled RNA, rather than DNA, probes is that they are easier to remove from the DNA template, which has the potential to interfere in subsequent hybridizations.

Small oligonucleotides used as probes are usually made radioactive by end-labelling (section A.4.2), but procedures have been devised for producing uniformly labelled probes in situations where a higher specific activity is required [87].

It is possible to label mRNA directly *in vitro* by certain end-labelling methods (section A.4.2) but a copying method is generally preferred because it produces a more stable product of higher specific activity. This involves the use of [α-³²P]dNTPs and reverse transcriptase to produce a ³²P-labelled single-stranded cDNA copy (section A.8.2).

The labelling of tRNA *in vitro* is important in sequence determination in cases where insufficient material is available for spectral analysis and it is not possible to label with ³²P *in vivo*. As well as enzymic end-labelling methods (see below) chemical labelling using [³H]borohydride has been employed [88, 89].

A.4.2 End-labelling methods

End-labelling does not produce DNA probes of such a high specific activity as do methods of uniform labelling; but, nevertheless, this is necessary for certain specific applications. These include se-

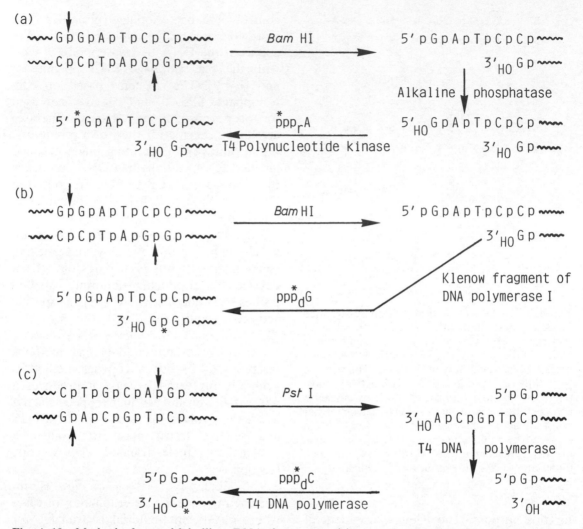

Fig. A.10 Methods for end-labelling DNA fragments. (a) use of polynucleotide kinase and [γ-^{32}P]ATP for 5′-labelling of cohesive ends with 5′-overhang produced by restriction endonuclease such as *Bam*H I; (b) use of Klenow fragment of DNA polymerase I and [α-^{32}P]dNTPs for 3′-labelling of cohesive ends with 5′-overhang produced by restriction endonuclease such as *Bam*H I. In this case any of the [α-^{32}P]dNTPs could have been used for labelling if the other non-radioactive dNTPs had been included to allow complete fill-in; (c) use of phage T4 DNA polymerase and [α-^{32}P]dNTPs for 3′-labelling of cohesive ends with 3′-overhang produced by restriction endonuclease such as *Pst* I. An excess of the other three non-radioactive dNTPs prevents the exonuclease activity proceeding further. Methods (a) and (c) may also be applied to blunt ends produced by restriction endonucleases such as *Sma* I. N.B. In each case the other end of the fragment (not illustrated) will also be labelled if it is similar, and a second restriction endonuclease digestion or strand-separation will be required to obtain a fragment of DNA labelled at one end only.

quence analysis by the method of Maxam and Gilbert (section A.5.1), mapping of RNA transcripts (section A.9.1), and 'DNA footprinting' (section A.9.2). It is also the most usual way of labelling oligonucleotide probes.

The DNA to be labelled is generated by cleavage with a restriction endonuclease (section A.6) and the method of labelling depends on whether flush ends, or 'sticky' ends (in which there is a 5'-PO_4 or 3'-OH overhang) are produced. Ends with a 5'-PO_4 overhang are easily labelled by either of two methods (Fig. A.10(a) and (b)), in one of which the 5'-PO_4 is removed by alkaline phosphatase and replaced by the γ-phosphate of [γ-^{32}P]ATP in a reaction catalysed by bacteriophage-T4 polynucleotide kinase (section 4.7.1) [90]. In the other the Klenow fragment of *E. coli* DNA polymerase I (section 6.4.3) is usually used to fill the gap in the strand with the recessed 3'-OH end, using an appropriate [α-^{32}P]dNTP, although bacteriophage T4 DNA polymerase may also be used (Fig. A.10(b)). (The 5' $\rightarrow$ 3' exonuclease of the complete *E. coli* enzyme would result in degradation of the template for this reaction.) Thus, these two methods complement each other in labelling different strands.

The polynucleotide kinase method may be adapted for labelling the 5'-PO_4 of flush ends by effecting local denaturation of the latter, but cannot be applied efficiently where there is a 3'-OH overhang. The 3' $\rightarrow$ 5' exonuclease and 5' $\rightarrow$ 3' polymerase activities of the Klenow fragment of DNA polymerase I can be used to label flush ends or 3'-OH protruding ends by a replacement synthesis method (Fig. A.10(c)), although bacteriophage-T4 DNA polymerase (which has a more powerful 3' $\rightarrow$ 5' exonuclease activity and also no 5' $\rightarrow$ 3' exonuclease) is more effective [91].

Several methods for end-labelling are available when the nucleic acid is only to be used as a probe. With DNA, extension of 3'-overhanging ends using terminal transferase and [α-^{32}P]dNTP [92] is an alternative to the use of DNA polymerase. For mRNA, extension of the 3'-end with *E. coli* poly(A) polymerase can be used for labelling with [α-^{32}P]ATP, or [α-^{32}P]cordycepin (3'-deoxy ATP) if the addition of only a single nucleotide is required [93]. An alternative to covalent incorporation as a method of end-labelling, useful in certain restricted circumstances, is hybridization of a labelled fragment to suitable single-stranded regions. In the case of eukaryotic mRNAs, [^{3}H]poly(U) can be hybridized to the poly(A) tails [94].

A.5 DETERMINATION OF NUCLEIC ACID SEQUENCES

A.5.1 Determination of DNA sequences

There are two rapid methods in use for sequencing large fragments of DNA. Although the method of Maxam and Gilbert [95]) has been largely replaced by that of Sanger [96] for general purposes, we shall describe it here as it is still important in particular situations. A description of a third method (the 'Wandering Spot' or 'Mobility Shift' method), which still finds application to small oligonucleotides, can be found in [97]. Both the method of Maxam and Gilbert and that of Sanger involve the generation of a 'ladder' of fragments of different sizes, but with one common end. The basis of these methods is to generate specific sets of radioactively labelled fragments, each set terminating at a particular base (in the ideal case), so that by using high-resolution polyacrylamide gels [98, 99] fragments differing by only a single

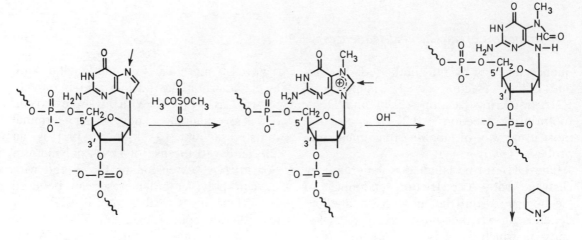

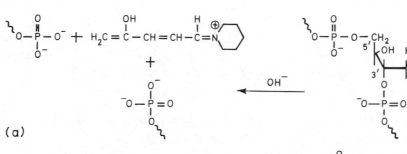

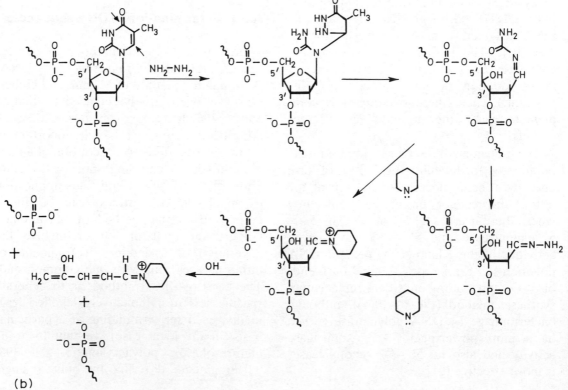

(a)

(b)

Fig. A.11 Base-specific cleavage reactions used in the sequencing method of Maxam and Gilbert. (a) Dimethyl sulphate reaction for guanine residues; (b) Hydrazine reaction for pyrimidine residues (thymine illustrated). For further details see text and reference [95], from which this figure is adapted, with permission.

nucleotide may be resolved and the sequence can be deduced. In both cases cloned DNA is normally used, although the method of Maxam and Gilbert can be applied to uncloned genomic DNA.

As sequences accumulate, it becomes necessary to use computer programs to handle and analyse them. Pioneers in developing such programs were Staden in Cambridge [100], and a group at the University of Wisconsin in the USA [101].

(a) *The method of Maxam and Gilbert [95]*

The essence of this method is the chemical fragmentation of either single- or double-stranded DNA by base-specific reactions. The reactions most commonly used are those absolutely specific for guanine residues or for cytosine residues, and those that are specific only for purines or pyrimidines. (There are some other, less frequently used, reactions [102, 103].) Cleavage at guanine residues is effected by methylation with dimethyl sulphate at the N7 position, leading to instability of the glycosidic linkage which is then hydrolysed by piperidine, followed by β-elimination of both phosphates from the sugar (Fig. A.11(a)). Purine nucleotide linkages are hydrolysed with acid (section 2.7.2), again followed by piperidine treatment. Pyrimidine residues are hydrolysed by hydrazine (Fig. A.11(b)), a reaction which, in the case of thymine, can be inhibited by 2 M NaCl, thus allowing specific cleavage at cytosine residues.

Conditions of chemical cleavage are generally adjusted to try to obtain a single scission per DNA molecule. Even so, each scission would produce two fragments. In order to visualize on polyacrylamide gels only fragments of increasing length emanating from one end of the DNA, a single end of the DNA must be radioactively labelled using one of the methods described

in section A.4.2. Such methods, in fact, generally label both ends of a piece of duplex DNA, and fragments with a single-labelled end are generated by restriction endonuclease digestion of this labelled DNA followed by separation on polyacrylamide gels. Less commonly, single end-labelled molecules are obtained by strand separation of the DNA. An example of a Maxam–Gilbert sequencing gel and its interpretation is given in Fig. A.12.

Because the advantage of the Sanger method of sequencing depends on cloning DNA into appropriate vectors and copying it *in vitro*, it is in circumstances where this is either impossible or undesirable that the method of Maxam and Gilbert finds its most frequent use today. Such circumstances include DNA that is unstable when cloned, and DNA with regions of secondary structure (generally arising from a high (G + C)-content) that prevent the DNA polymerase used in the Sanger method from completing its copying. However, the main current application of the method is for sequencing uncloned genomic DNA to study either its methylation state or its interaction with proteins. The latter application (genomic footprinting) is described in section. A.9.2. The study of DNA methylation by the method of Maxam and Gilbert is based on the fact that, under the conditions normally used, cleavage by hydrazine does not occur at 5-methylcytosine residues, causing a gap in the sequencing ladder [104]. It is necessary to analyse the methylation state of the genomic DNA *in situ*, as this is altered after cloning into prokaryotic vectors.

Genomic sequencing was devised by Church and Gilbert [105], and the method has subsequently undergone refinement [106]. The genomic DNA is digested with an appropriate restriction endonuclease to generate a fragment containing the region of interest with suitable end points for

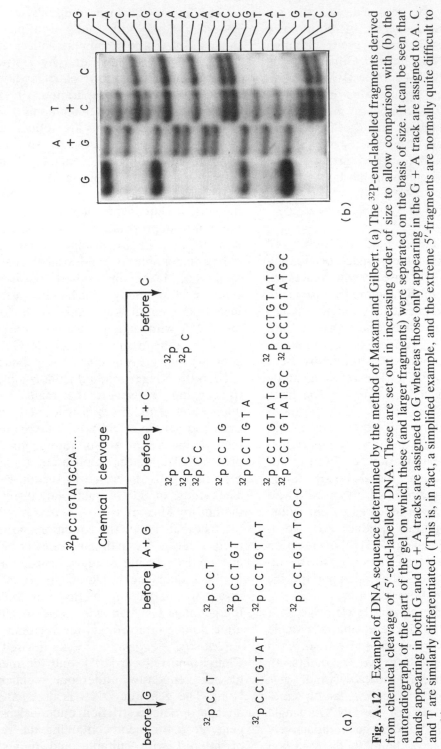

Fig. A.12 Example of DNA sequence determined by the method of Maxam and Gilbert. (a) The ^{32}P-end-labelled fragments derived from chemical cleavage of 5'-end-labelled DNA. These are set out in increasing order of size to allow comparison with (b) the autoradiograph of the part of the gel on which these (and larger fragments) were separated on the basis of size. It can be seen that bands appearing in both G and G + A tracks are assigned to G whereas those only appearing in the G + A track are assigned to A. C and T are similarly differentiated. (This is, in fact, a simplified example, and the extreme 5'-fragments are normally quite difficult to read.)

sequencing, and the base-specific reactions are performed. Because the restriction fragment of interest cannot be purified from all the other fragments generated, it is not possible to end-label it covalently as for conventional sequencing. Instead, the ends of individual strands of the restriction fragment are labelled after the chemical cleavages have been performed. Initially this was done indirectly by hybridization to suitable single-stranded radioactively labelled probes following gel electrophoresis and transfer to a nitrocellulose membrane. (This, of course, presupposes some prior knowledge of the sequence from analysis of cloned DNA.) Such indirect end-labelling had a number of disadvantages [106], but the most severe was the weakness of the radioactive 'signal' obtained. However, more recently the polymerase chain reaction (section A.7), primed unidirectionally by an oligonucleotide hybridized to the common end of the ladder of fragments, has been used to generate more highly labelled copies of the fragments, which can be subjected to electrophoresis and autoradiography in the normal manner [107].

(b) *The Sanger dideoxy method [98]*

The essence of this method is the primed synthesis of partial copies of the DNA to be sequenced, with random base-specific premature termination of the copying producing ladders of different length fragments terminating in each of the four different bases. The copying is catalysed by the Klenow fragment of *E. coli* DNA polymerase I, which requires a primer and that the template (the DNA to be copied) be in the single-stranded form. (The Klenow fragment, lacking the $5' \rightarrow 3'$ exonuclease, is used to prevent attack on the 5'-end of the primer.) The copies are made radioactive by inclusion of an $[\alpha\text{-}^{32}P]dNTP$ or a

$[\alpha\text{-}^{35}S]dNTP$ in the reactions, and base-specific chain termination is effected by the addition of the appropriate 2',3'-dideoxynucleoside triphosphate (ddNTP), which, lacking a 3'-OH, cannot be extended further. An example of a Sanger sequencing gel and its interpretation is given in Fig. A.13.

Initially, the widespread application of this method was prevented by the requirement for a single-stranded template and the need for a separate oligonucleotide primer for each piece of DNA to be sequenced. However, these limitations were overcome when the single-stranded bacteriophage, M13, was adapted for use as a sequencing vector [108, 109]. The DNA to be sequenced is subcloned into one of a variety of adjacent sites in the double-stranded replicative form of the bacteriophage vector, and can be sequenced in the easily prepared single-stranded DNA using oligonucleotide primers complementary to regions of the bacteriophage DNA flanking the cloning sites (Fig. A.14).

This development led to a dramatic increase in the use of the Sanger method, for which a number of refinements have been made. These include the replacement of the Klenow fragment by chemically modified bacteriophage T7 DNA polymerase to extend the size range of the copies generated [110], or by the thermostable DNA polymerase from *Thermus aquaticus* (*Taq* DNA polymerase) to allow the copying reaction to be performed at a temperature at which regions of DNA secondary structure no longer cause premature termination. The effect of DNA secondary structure causing anomalous migration during gel electrophoresis (so-called 'compressions') can be eliminated or decreased by replacing dGTP by 7-deaza-2'-dGTP. Sanger sequencing using M13 vectors always entails starting the copying from a similar position

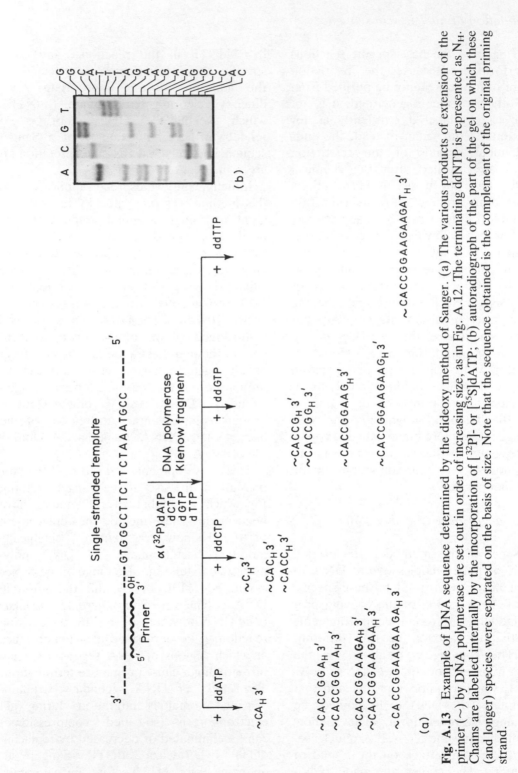

Fig. A.13 Example of DNA sequence determined by the dideoxy method of Sanger. (a) The various products of extension of the primer (~) by DNA polymerase are set out in order of increasing size, as in Fig. A.12. The terminating ddNTP is represented as N_H. Chains are labelled internally by the incorporation of [^{32}P]- or [^{35}S]dATP; (b) autoradiograph of the part of the gel on which these (and longer) species were separated on the basis of size. Note that the sequence obtained is the complement of the original priming strand.

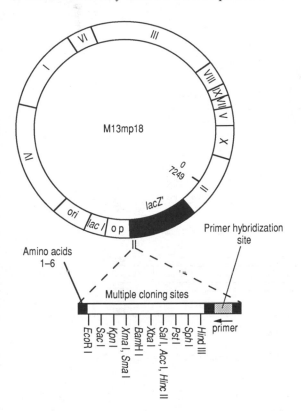

Fig. A.14 The bacteriophage vector M13mp18. The double-stranded replicative form of this single-stranded DNA phage is shown with authentic genes I–X indicated. The multiple cloning sequence interrupts the *E. coli lacZ'* (truncated β-galactosidase) gene, and the position of hybridization of one of the primers that may be used for sequencing by the Sanger dideoxy method is shown (cf. Fig. A.13).

in the multiple-cloning site near to the universal primer. Therefore, different fragments of the sequence of interest must be subcloned into the vector to obtain their sequences. This process has been accelerated by a technique in which successive deletions of increasing length generate subclones in which different portions of a long sequence of interest are brought into the proximity of the multiple-cloning site [111] (Fig. A.15). Special vectors are available with suitable restriction endonuclease sites

to facilitate this process, although there is also a more recent variant of the method in which the need for subcloning is actually bypassed [112].

A.5.2 Determination of RNA sequences

Before the advent of rapid DNA sequencing methods, considerable effort was expended on developing techniques for sequencing RNA. The availability of partially or totally base-specific endonucleases allowed employment of approaches formally rather similar to those used for sequencing proteins, but requiring different methods for separating and analysing the oligoribonucleotides. Although the sequencing of a DNA copy must now be the method of choice for sequencing any RNA, this approach may not be applicable to small RNAs or oligonucleotides, and in any case does not identify modified nucleotides. In these cases (especially for tRNAs) direct methods are still employed [113].

Rapid sequencing methods involving base-specific cleavage and separation of fragments on polyacrylamide gels have been introduced for 5'-end-labelled RNA [114, 115]. The cleavage is effected enzymically with the following ribonucleases (section 4.3):

Ribonuclease T1   : cleaves 3' of G

Ribonuclease U2   : cleaves 3' of A

Ribonuclease Phy M   : cleaves 3' of A and U

B. cereus ribonuclease : cleaves 3' of U and C

In other respects the method and the interpretation of the results are formally similar to that of Maxam and Gilbert, described in section A.5.1(a).

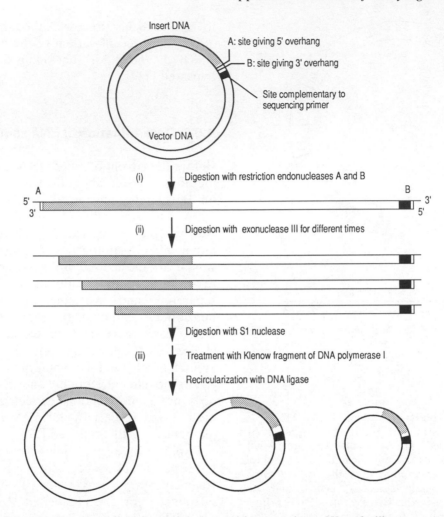

Fig. A.15   Generation of unidirectional deletions with exonuclease III to facilitate sequencing by the Sanger dideoxy method. The DNA to be sequenced is cloned into the double-stranded form of a vector in which there are two unique restriction endonuclease sites separating it from the site to which the sequencing primer hybridizes. (i) Digestion at the site proximal to the insert (A) generates a 5′-overhang, whereas digestion at the site proximal to the primer site (B) generates a 3′-overhang. (ii) Digestion with the exonuclease (which is specific for 3′-OH ends in double-stranded DNA) only occurs at site A, causing deletions the size of which depend on the time of digestion. (iii) The single-stranded regions are removed with S1 nuclease, any remaining 5′-overhangs 'filled' using the Klenow fragment of *E. coli* DNA polymerase I, and the DNA religated to bring the new end-points of the deleted insert in the proximity of the primer site. After transformation of *E. coli* and preparation of templates, sequencing can be performed from these new end points, which correspond to internal positions in the original insert. For further details see [111].

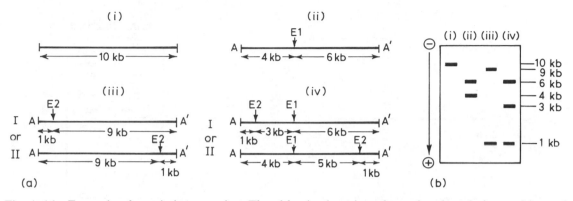

Fig. A.16 Example of restriction mapping. The objective here is to determine the relative positions of the recognition sites for two restriction endonucleases, E1 and E2, along a hypothetical 10 kb fragment of DNA. (a) If the 10 kb fragment (i) is cleaved by E1 to give 4 kb and 6 kb fragments the two ends of the DNA can be distinguished as A and A' (ii). This then defines two possible positions of cleavage (I and II) of a second enzyme E2, which generates 1 kb and 9 kb fragments (iii). The two alternatives I and II may be distinguished by the different sized fragments they predict in the case of double-digestion with E1 and E2 (iv); (b) diagrams of the results of agarose gel electrophoresis of (i) − (iv). The result in lane (iv) is only consistent with alternative I, and hence (iv)/I represents an (albeit limited) restriction map of the 10 kb fragment.

A.6 RESTRICTION MAPPING OF DNA

When studying specific regions of genomic or cloned DNA it is necessary to have some sort of 'map' to differentiate one area from another. There are two different types of map. A *genetic map* can be constructed (in suitable organisms) in which the positions of functional genes contributing to particular phenotypes can be interrelated by studying the frequency of genetic recombination between normal and mutated alleles. A *physical map*, on the other hand, can be constructed in the absence of genetic data or knowledge of any function the DNA might have, and uses purely physical techniques. These include the formation of hetero-duplexes between standard DNA molecules and the DNA under investigation, and restriction mapping, the subject of this section.

A restriction map indicates the positions along a piece of DNA at which there are recognition sites for particular type II restriction endonucleases (section 4.5.3). The method in its simplest form involves digesting the DNA separately and in combination with different restriction endonucleases, separating the resulting fragments by agarose (or sometimes polyacrylamide) gel electrophoresis (section A.2.1), visualizing these under ultraviolet light after staining with ethidium bromide, and estimating their sizes in relation to standards [116]. If there are not too many recognition sites for the enzymes it is possible to deduce the position of these, as is illustrated in Fig. A.16.

It can be appreciated that for enzymes with many recognition sites it is both difficult and tedious to construct a restriction map by this procedure. In the case of large pieces of DNA (e.g. those cloned in bacteriophage lambda or cosmid vectors – section A.8.4) most restriction endonucleases will have many recognition sites, so that it is difficult to circumvent this problem by choice of enzyme. In any case it is often useful to have a restriction map for enzymes that cleave a piece of DNA at many positions. An alter-

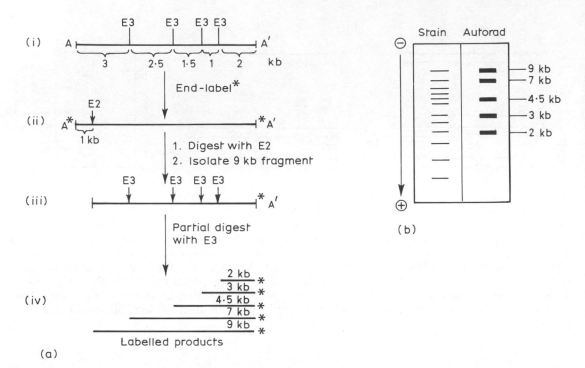

Fig. A.17 Example of partial digestion restriction mapping. (a)(i) The objective here is to determine the relative positions of the four recognition sites for the restriction endonuclease E3 on the hypothetical DNA fragment of Fig. A.16. This is end-labelled (ii), digested with enzyme E2, which is known to cleave close to one end, and the larger (9 kb) fragment isolated (iii). This is subjected to partial digestion with E3 giving a mixture of end-labelled fragments together with non-labelled fragments not illustrated (iv); (b) diagram of the results of electrophoresis of the partial digestion products. The distances of the E3 recognition sites from end A′ can be read off the autoradiography ladder using molecular size markers (cf. Fig. A.4(b)). This allows the restriction map of (a)(i) to be deduced.

native approach was devised involving end-labelling of the DNA and *partial* digestion with *single* restriction endonucleases [117]. In essence the strategy is similar to that for Maxam and Gilbert DNA sequencing using a single base-specific reaction, except that analysis of the larger fragments usually employs the standard agarose gels. Although the pattern of stained fragments is complex, the autoradiograph of the dried gel reveals a ladder of fragments, the increasing sizes of which represent increasing distances from the labelled end. This is illustrated schematically in Fig. A.17. A modification of this

strategy, for application to DNA cloned into bacteriophage or cosmid vectors, replaces end-labelling by hybridization of a specific ^{32}P-labelled oligonucleotide to either the right or left cohesive end of the molecule [118]. This avoids the need for secondary cleavage and gel fractionation to produce a fragment of DNA labelled at one end.

A.7 THE POLYMERASE CHAIN REACTION

The standard method for purifying individual genes or mRNAs is by cloning in

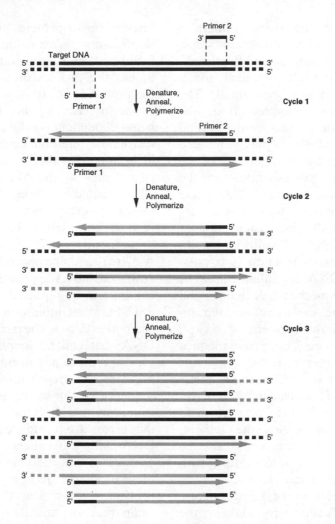

Fig. A.18 The polymerase chain reaction. The figure shows three cycles of polymerization. For details see text.

bacteria, as described in section A.8 below. However, it is now possible to use an enzymic reaction, the *polymerase chain reaction (PCR)*, to amplify minute quantities of specific regions of DNA without the intermediary of bacteria [119]. The polymerase chain reaction is not to be viewed as a substitute for cloning, rather it is an extremely powerful technique that can be used in concert with cloning methods to study and manipulate nucleic acids.

The principle of the method is illustrated in Fig. A.18. Two oligonucleotides are chosen to encompass the region of DNA to be amplified. Each must be complementary to a different strand in such a way that their 3'-ends are directed towards one another. The double-stranded template DNA is denatured and then allowed to anneal to the oligonucleotides. These are then used to prime the copying of the DNA template by bacterial DNA polymerase *in vitro*, producing multiple copies, but with random end-points. For clarity, Fig. A.18 shows extension from just one of the primers in this first reaction, and in subsequent steps. The DNA is again denatured and the daughter molecules allowed to reanneal and a second copying reaction performed. The daughter molecules generated from the first reaction will be primed by the second oligonucleotide, and the copies generated will all terminate at the common end of the daughter molecules corresponding to the first oligonucleotide primer. Successive reactions with alternate primers will generate a geometrically increasing number of molecules of DNA of uniform length. Using 25–45 cycles of reaction it is possible to amplify DNA a million times or more. Initially the method was tedious and costly because the denaturation step inactivated the Klenow fragment of *E. coli* DNA polymerase at each cycle, and the products were very heterogeneous. However, the introduction of the thermostable DNA polymerase from *T. aquaticus* allowed repeated denaturation steps to be performed at 94°C without loss of enzyme activity, and the elongation reaction could also be performed at a higher temperature with resulting greater specificity [120].

The outstanding feature of the polymerase chain reaction is its sensitivity, allowing the detection of a single mRNA molecule per cell. Thus, the most spectacular applications of the method (reviewed in [121]) have been of a general nature: the analysis of DNA from mummified museum specimens in the field of evolutionary biology, from single hair roots for forensic purposes, and from trace amounts of viruses in experimental medicine. The method has likewise found application in the diagnosis of genetic diseases and in determining the sex of fertilized embryos. However, the application of the method to the study of nucleic acids has, in its way, produced equally remarkable results. The cloning of cDNAs cor-

responding to proteins for which sequence information is available has been greatly facilitated and accelerated (section A.8), as has the process of mutagenesis *in vitro* (section A.9.4). It is now much easier to excise precise regions of DNA, without being dependent on the presence of suitable restriction endonuclease sites; and the inclusion of suitable restriction endonuclease sites at the 5'-ends of the primers allows equally precise insertion into, for example, expression vectors (cf. section A.9.3). The method has also been applied to genomic DNA sequencing (section A.5.1(a)) and chromosome walking (section A.8), and novel applications of the method continue to be reported.

The current limitation of the polymerase chain reaction is the size of the region of DNA that can be amplified (above about 1 kb, the efficiency declines); and its major problem is cross-contamination: a consequence of the exquisite sensitivity that is its major strength. One should also mention that errors are introduced into the amplified DNA with a frequency that requires the exercise of considerable caution. However, these problems are being actively addressed [214], and it is clear that the polymerase chain reaction and its various adaptations will continue to exert a profound influence on developments in biology.

A.8 CLONING DNA

A.8.1 The principles

The study of individual mRNAs and of particular regions of genomic DNA was hampered historically by two problems. The first was the difficulty of devising chemical methods to isolate one particular nucleic acid species from, perhaps, tens of thousands of chemically similar distinct species. The second was that, even if a chemical separation method had been found, the amount of nucleic acid isolated would in most cases have been too small to allow subsequent study. The cloning of DNA employs a *biological*, rather than a *chemical*, strategy to overcome both these problems.

This biological strategy can be illustrated by first considering how one can study the properties of a single mutant bacterium, present in a population of a million similar but non-mutant ('wild-type') bacteria. To do this one spreads all the bacteria out on agar plates so thinly that each is separated from its neighbour. Each bacterium gives rise to a visible colony, which is, in fact, composed of identical bacteria derived from the division of the original one. Such an assembly of genetically identical individuals is called a *clone*. In the case of the clone derived from the mutant bacterium, one could also say that one had cloned the mutant DNA. Even though this mutant DNA originally existed as a single chromosome in minute amount, one could obtain large quantities of the DNA from a bulk liquid culture seeded by a single colony. The basis of the purification and isolation of an mRNA or a chromosomal DNA segment using a biological strategy is to introduce them into individual bacteria so that they may be cloned in an analogous way to the hypothetical mutant chromosomal DNA considered above. For this reason the process is often referred to as *molecular cloning*.

There are several problems to be overcome before this strategy can be realized in practice. These are enumerated below, and the next section considers different ways in which they are tackled.

1. The foreign DNA must be presented as part of a molecule that can replicate along with the bacterial (or other host) chromosome. Such vehicles for introducing the DNA are called *vectors*, and the chimeric DNA produced by the insertion of the DNA to be cloned is a *DNA recombinant*. The two main types of vector are bacterial *plasmids* and *bacteriophages* (section A.8.2). (In theory one could actually integrate the foreign DNA into the bacterial chromosome, but in practice it is generally preferable to use extrachromosomal vectors.)

2. The DNA (or a DNA copy of the mRNA) must be inserted into the vector in a manner which allows its subsequent removal. In all cases this is achieved by the use of *restriction endonucleases*.

3. The vector must be introduced into the bacterium. In the case of plasmid vectors this process is called *transformation*, and requires the cell wall of the bacteria to be made permeable to the plasmid. In the case of bacteriophage lambda this is achieved through the normal process by which the phage infects the bacteria – referred to as *transfection* in this context.

4. One must be able to distinguish bacteria containing the vector from those which do not. This is relatively simple in the case of bacteriophage, which usually produce visible lysis of the bacteria. In the case of plasmids it is achieved by incorporating a gene for antibiotic resistance into the vector.

5. Bacteria containing recombinant DNAs must be distinguished from those containing non-recombinant vector. This may be achieved by a variety of chemical or genetic stratagems (*selection*), or by more tedious physical analysis.

6. In many cases it is also necessary to be able to identify which one (if any) of thousands of clones contains the particular DNA of interest (*screening*). The methods used in screening will be considered specifically in relation to cDNA and genomic cloning (section A.8.3).

The discussion of molecular cloning above has assumed that *E. coli* will be used as the biological *host* for the foreign DNA. This is because, whatever the source of the DNA, it is easier to perform initial cloning in bacteria. It may subsequently be of interest to introduce the cloned DNA into other types of cells, e.g. those of yeast, higher animals, or plants. This is discussed in section A.9.3 in relation to higher animals and a specific use of yeast is described in section A.8.3. However, the reader is directed elsewhere for further details of cloning in other systems [3, 5, 122, 123].

A.8.2 The major cloning vectors

As already stated, the cloning of a piece of foreign DNA in *E. coli* requires its introduction into a suitable vector. Here we outline the way in which the two main types of cloning vector, plasmids and bacteriophages, are used, considering separately bacteriophage lambda and bacteriophage M13.

(a) *Plasmid vectors*

Plasmids are self-replicating double-stranded circular DNA molecules found in certain bacteria (section 3.6). Especially useful are those that occur in multiple copies per cell, but natural plasmids had to be modified extensively before they could serve as efficient cloning vehicles. The most well known of the plasmid vectors is probably pBR322 [124], but this has been superseded by more sophisticated derivatives, and it is one of these, pUC18 [125], that has been chosen to illustrate the use of plasmids as vectors (Fig. A.19). The essential features of pUC18 are an origin of replication that allows the plasmid to replicate as the bacteria divide and multiply, unique restriction

endonuclease sites into which foreign DNA may be inserted without disabling the plasmid, a gene (Ap^R in Fig. A.19, although its correct genetic designation is *bla*) for resistance to the antibiotic ampicillin permitting the selection of bacteria which have taken up the plasmid, and a fragment of the *lac* operon (the *lac* operator and promoter regulatory regions and the *N*-terminal portion – α-fragment – of the β-galactosidase gene) which are included to allow one to distinguish between recombinants and non-recombinants.

In the simplest case a fragment of foreign DNA generated from a larger molecule by digestion with a restriction endonuclease (e.g. *Eco*R I) is ligated with DNA ligase (section 6.5.1) to pUC18 DNA that has been digested with the same enzyme, generating the chimeric molecule shown in Fig. A.19(b). To enable the plasmid DNA to enter the *E. coli* cells (*transformation*) these are generally pretreated with calcium chloride which, together with a short exposure to elevated temperature, appears to act by altering the structure of the cell wall [126]. Transformed cells are selected by plating the bacteria on agar containing the antibiotic, ampicillin (a derivative of penicillin), which will prevent the growth of cells that do not contain any plasmid. It is possible to use a chemical strategy to select for recombinants: pretreating the linearized plasmid (but not the foreign DNA) with alkaline phosphatase will prevent re-ligation of plasmid to itself. However, in many cases some of the transformed bacteria will contain plasmid with no insert. Colonies of such bacteria may be identified because the cloning region of the vector is at the end of the *lacZ'* gene, specifying the α-fragment of β-galactosidase. Normally, if the plasmid *lacZ'* gene is induced with IPTG (isopropylthio-β-galactoside) in an *E. coli* strain with a β-galactosidase gene lacking the

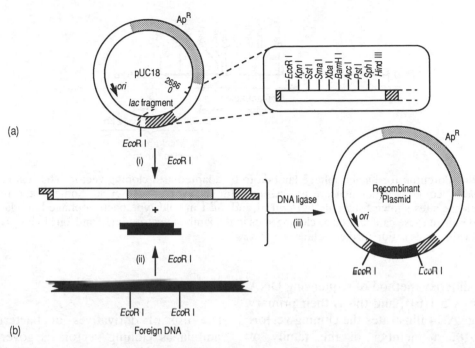

Fig. A.19 Use of the plasmid vector pUC18. (a) The gene for ampicillin resistance (Ap^R), the origin of replication (*ori*), and the *lac*· fragment containing the *E. coli lacZ'* (truncated β-galactosidase) gene interrupted by the multiple cloning sequence (cf. Fig. A.14) of pUC18 are shown; (b) to clone into e.g. the *Eco*R I site in the multiple cloning region, the vector DNA is digested with this enzyme (i), as is the foreign DNA (ii), and these are ligated together (iii), among the products being the desired recombinant, shown, in which the *lacZ'* gene has been inactivated by insertion of the foreign DNA.

N-terminal portion (actually carried on the F′ episome), complementation of the two peptides will occur giving enzymically active β-galactosidase (the inclusion of the multiple cloning sites does not alter the reading frame or result in inactivation of the β-galactosidase). The β-galactosidase can be detected by the blue colour produced when 'X-gal' (5-bromo-4-chloro-3-indolyl-β-D-galactoside) is present as a substrate. However, in most cases the insertion of foreign DNA into the multiple cloning sites at the extreme *N*-terminal region of β-galactosidase will cause inactivation (so-called 'insertional inactivation'). Thus, if the transformed bacteria are plated on agar containing IPTG and X-gal (as well as ampicillin) the white colonies will generally

contain plasmid with inserts, whereas the blue colonies generally will not.

Plasmid vectors are small enough that they are relatively easy to construct with specific unique restriction endonuclease sites. They were originally much employed for cDNA cloning, but more recently this use has declined dramatically. Currently they are used for subcloning larger pieces of cloned DNA, and as the basis of more sophisticated vectors, especially those used for the expression of the genetic information in the DNA (section A.9.3).

(b) *Bacteriophage M13*

Vectors based on bacteriophage M13 have already been mentioned in relation to the

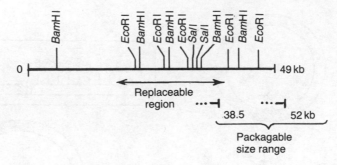

Fig. A.20 Potential for bacteriophage lambda to be adapted as a cloning vector. The region marked 'replaceable' contains genes that are not essential for the lytic growth cycle. Some of the restriction endonuclease sites shown for *Eco*R I, *Bam*H I and *Sal* I in wild-type bacteriophage lambda must be removed before these can be used as cloning sites in the lambda vectors λgt10 and λgt11 (Fig. A.22) and EMBL3 (Fig. A.24). For further details see text.

Sanger dideoxy method of sequencing DNA (section A.5.1(b)), and this is their primary use. Fig. A.14 illustrates the cloning vector, M13mp18, a member of the family of M13mp vectors, for which the *lac* fragment of the pUC vectors was initially engineered [127]. Although this is technically a single-stranded phage vector, the double-stranded replicative form illustrated is formally analogous to a plasmid, and the method of cloning, transformation, and detection of recombinants is analogous to that described for pUC18, above. The main difference from cloning in pUC18 is that there is no gene for antibiotic resistance: identification of cells that have taken up the vector is provided by the visible formation of 'plaques' (in fact areas of more slowly growing bacteria, as this bacteriophage does not lyse cells).

As already mentioned, the clustering of the restriction endonuclease sites together in M13mp18 not only allows identification of recombinants by insertional inactivation of β-galactosidase, but also allows a single oligonucleotide to be used to prime copying of the single-stranded form of the recombinant, irrespective of the nature of the insert being sequenced.

(c) *Bacteriophage lambda*

The use of derivatives of bacteriophage lambda as cloning vectors is governed by somewhat different considerations than the use of plasmids. In plasmids there is no restriction on the size of the DNA that may be inserted (although larger inserts often transform less efficiently and cause slower growth), but in bacteriophage lambda there is a size restriction of 38.5–52 kb for the DNA in order for it to be 'packageable' into phage heads. The size of wild-type bacteriophage lambda is about 49 kb, which would not give much scope for its use as a vector, were it not that a continuous portion of the phage genome of about 16 kb is not required for lytic growth (Fig. A.20). Thus, lambda vectors can be designed to allow replacement of this region of DNA by foreign DNA (*replacement* vectors), or they can be truncated in the dispensable region to a size at which they are still packageable, but which allows the insertion of an appreciable amount of foreign DNA (*insertion* vectors). Examples of these types of vector are described in sections A.8.3 and A.8.4.

The use of bacteriophage lambda as a cloning vector involves the same initial use of restriction endonucleases as for plasmids. However, to introduce the recombinants into the cell they are first 'packaged' into phage particles *in vitro*, using an extract from infected cells containing the necessary enzymes [128], and the bacteria are then infected (transfected) with the packaged recombinant. Clones of bacteriophage are identified as plaques: clear areas of the bacterial 'lawn' where the infection has produced lysis. There is no single common way of distinguishing recombinant and non-recombinant phage, and indeed the most useful vectors are those in which some method is available to prevent infection by the non-recombinants. Examples are mentioned later.

One feature of using cloning vectors based on bacteriophage lambda is that packaging *in vitro* followed by transfection results in a higher efficiency of transformation (number of transformants per mole of DNA) than does the transformation of plasmids by the calcium chloride procedure.

A.8.3 cDNA cloning

For cloning *cDNA* – DNA copies of mRNA – it is now usual to employ one of two insertion vectors based on bacteriophage lambda, λgt10 or λgt11, depending on what screening procedure one intends to adopt. We shall first describe how double-stranded DNA for cloning is prepared from poly(A)-containing (poly(A)$^+$) mRNA (Fig. A.21). The poly(A)$^+$ mRNA is usually purified on an affinity column of oligo(dT)-cellulose (section A.2.2) and reverse-transcribed using retroviral reverse transcriptase (section 6.4.8) primed by oligo(dT) hybridized to the poly(A) tail of the mRNA. This results in a single-stranded cDNA, complementary to the mRNA. For a number of years the usual procedure for synthesiz-

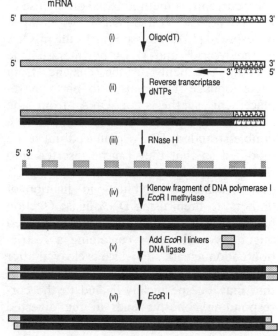

Fig. A.21 Preparation of cDNA for insertion into cloning vectors. Poly(A)$^+$ mRNA is primed with oligo(dT) (i) and a cDNA copy synthesized (ii). After RNase H treatment to nick the mRNA strand of the hybrid (iii), the Klenow fragment of DNA polymerase I is used to fill the gaps and replace the RNA by DNA, and the DNA is methylated with *Eco*R I methylase (iv). *Eco*R I linkers are then ligated to the ends of the double-stranded cDNA (v) and cleaved with *Eco*R I (vi) ready for insertion into the *Eco*R I site of an appropriate vector (Fig. A.22).

ing the second strand relied on the self-priming that occurs when (after removing the mRNA) the single-stranded cDNA folds back on itself. However, the S1 nuclease used to destroy the loop had the disadvantage that it destroyed the extreme 5′ end of the cDNA. For this reason the use of S1 nuclease has been superseded by methods that allow production of full-length cDNA copies of mRNAs. Although a variety of solutions to the problem of priming the second strand have been employed [129,

130], currently a method based on the use of *E. coli* RNase H (section 4.3.4) is most widespread [131]. RNase H nicks the mRNA in the hybrid, and DNA polymerase (the Klenow fragment or the enzyme from bacteriophage T4) is used to perform replacement synthesis of DNA from the 3′-OH groups generated. The ends of such double-stranded DNA produced must be modified so that they can be inserted into the *Eco*R I cloning sites in λgt10 and λgt11. This is achieved by 'blunt-end' ligation of 'linkers' to them using DNA ligase (section 6.5.1). Such linkers are double-stranded oligodeoxynucleotides containing a restriction endonuclease site which is cleaved after the ligation. This generates 'cohesive ends' that may be specifically ligated to the corresponding cohesive ends in the digested vector. (This is more efficient than blunt-end ligation directly into a blunt-ended site. Blunt-end ligation is much less efficient *per se* than ligation of fragments with cohesive ends. However, the high molar concentration of the small linkers can drive this reaction in a manner impossible with the larger molecules.) The use of *Eco*R I linkers raises the problem of preventing cleavage of any *Eco*R I sites in the cDNA. This is overcome by prior methylation of the cDNA with *Eco*R I methylase.

The choice between λgt10 and λgt11 [132] for cDNA cloning relates to the problem, not so far discussed, of identifying the cDNA corresponding to the mRNA of interest among the large number of different clones (the cDNA 'library') that will be generated (*screening* the library). In general, such identification depends on particular information about the protein encoded by the mRNA. Although many different screening procedures have been used, the two vectors under consideration rely on there being either partial amino acid sequence information for the protein (in the

case of λgt10) or on there being antibodies available against the protein (in the case of λgt11).

With *λgt10* (Fig. A.22(a)) the strategy is to screen the plaques with a ^{32}P-labelled *oligonucleotide probe* corresponding to the nucleotide sequence predicted by a part of the amino acid sequence. Because of the degeneracy of the genetic code (section 12.2.2) such nucleotide sequences cannot be predicted with certainty. It is necessary to select sequences containing amino acids of minimal codon degeneracy, to consider what bias in codon usage might exist in the organism from which the mRNA is derived (section 12.2.4), and to employ enough different oligonucleotides of sufficient length to allow for mistakes. Such probing with oligonucleotides was initially by the technique of plaque hybridization [133], a variant of the earlier technique of colony hybridization [134], and is illustrated in Fig. A.23. It involves transferring duplicate copies of the DNA in the plaques from a 'master' agar plate to nitrocellulose membranes (replica plating) and then immobilizing it, as in Southern blotting (section A.3). It is the lower background of bacterial DNA in plaques, compared with colonies, that is one of the reasons that lambda vectors are now preferred to plasmid vectors for cDNA cloning. Nevertheless plaque hybridization with oligonucleotides can be tedious and result in false 'positives', so that since the advent of the polymerase chain reaction it has become usual to screen a cDNA library (or single-stranded uncloned cDNA) by PCR using two oligonucleotides, clone the amplified DNA, check its sequence, and use the cloned DNA to probe the library by plaque hybridization for a full-length cDNA clone.

The specific advantage of using λgt10 for cDNA cloning is that it allows discrimination against non-recombinants. This is by virtue

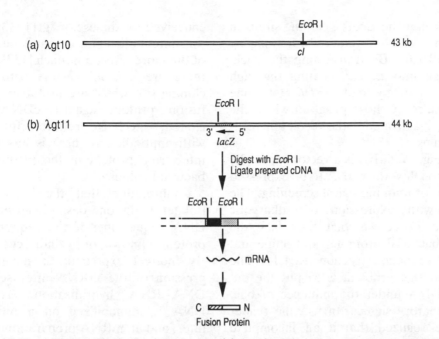

Fig. A.22 λgt10 and λgt11 cDNA cloning vectors. (a) λgt10, showing the *Eco*R I cloning site in the *cI* gene. (b) λgt11, showing the *Eco*R I cloning site in the *lacZ* gene (enlarged detail below the vector) used to generate a β-galactosidase fusion protein. For details see text.

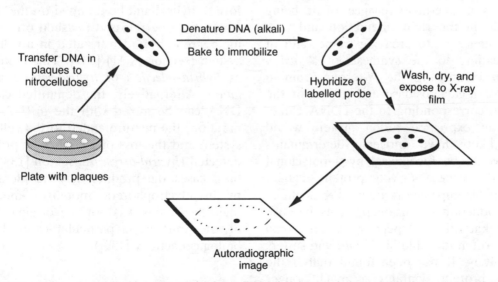

Fig. A.23 Plaque hybridization. Bacteriophage DNA in plaques is immobilized on nitrocellulose and hybridization to a suitable ^{32}P-labelled probe (see text) allows identification of bacteriophage carrying inserts of interest. In the diagram the number of plaques has been greatly decreased for clarity.

of the fact that the *Eco*R I cloning site is in the *cI* gene, which is required for lysogeny (section 10.1.6). By transfecting the packaged phage into *E. coli* bearing the high frequency lysogeny mutant *Hfl*A150, one ensures that only those phage in which the *cI* gene is inactivated – the recombinants – form plaques.

λgt11 (Fig. A.22(b)) is a vector especially designed to allow the cDNA to be expressed as protein for immunological screening. The problem with expression of eukaryotic cDNAs in *E. coli* is that they lack the requisite bacterial promoter and Shine and Dalgarno sequence (section 12.4.1). The solution to this problem is to put the eukaryotic DNA under the influence of bacterial regulatory signals. In fact the reader may have realized that if an incomplete eukaryotic cDNA clone (i.e. one lacking sequences corresponding to the 5′-untranslated portion of the mRNA) is cloned into the pUC or M13 vectors (section A.8.2), there is a one-in-six chance of it being inserted in the right orientation and with maintenance of the reading frame, and of this leading to the synthesis of a *fusion protein* comprising the *N*-terminal amino acids of *β*-galactosidase and a part of the protein corresponding to the cDNA clone. Such an expressed fusion protein would most likely have antigenic determinants that would be recognized by a polyclonal antibody to the eukaryotic protein. In fact, pUC or M13mp vectors are not very suitable for production of fusion proteins because the eukaryotic polypeptide is for some reason often unstable in *E. coli*, and suffers proteolysis. It has been found that if the fusion protein contains a much longer *N*-terminal portion of bacterial protein it is generally much more stable, and for this reason in *λ*gt11 the foreign DNA is cloned near the *C*-terminus of the *β*-galactosidase gene. Originally a lysogenic strategy was

conceived for the use of *λ*gt11 [135], but this was unreliable and was abandoned in favour of the more direct approach [132]. Although there are plasmid vectors with a similar cloning site which are useful for producing fusion protein against cDNAs already cloned, *λ*gt11 is preferred for screening with antibodies as there is less potentially interfering protein in the plaques than in bacterial colonies.

Confirmation that the clone one has isolated is the one desired can be either by DNA sequencing, if the sequence of the protein is known, or by 'indirect expression'. By indirect expression is meant the expression of the mRNA after selection by DNA–RNA hybridization. The cloned DNA is immobilized on a nitrocellulose filter, and an mRNA preparation (consisting of a mixture of species) from an appropriate tissue or cell is passed through the filter [136]. Only the mRNA complementary to the particular immobilized DNA clone will form a hybrid and be retained on the filter. It can then be released (washed off at low ionic strength) and translated in a cell-free system (section A.11) [137]. This is known as *hybrid-selection* or *hybrid-release translation*. Alternatively, the denatured cloned DNA can be mixed with the mRNA preparation, the mixture added to the cell-free system and the loss of a particular product detected (*hybrid-arrest translation* [138]). In both cases the product may be identified by its electrophoretic mobility, immunologically (if this was not the original basis of selection) or, in particular cases, by its biological activity [139]).

A.8.4 Genomic cloning

Other vectors than those described above are used when the objective is to prepare a eukaryotic genomic library; i.e. a collection

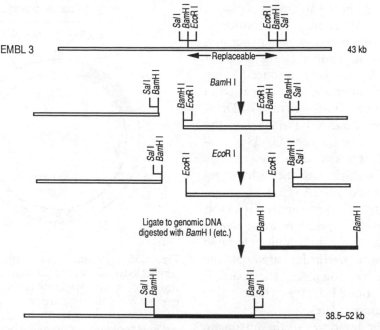

Fig. A.24 Bacteriophage lambda vector EMBL3 for cloning genomic DNA. The figure shows the cloning of a *Bam*H I fragment of genomic DNA between the *Bam*H I sites of the vector, with digestion by *Eco*R I being used to prevent religation of the central fragment of the vector. If genomic DNA digested with *Sau*3A I rather than *Bam*H I is used (see text) the recombinant will generally lack the *Bam*H I sites indicated. However, the insert can be recovered using *Sal* I. Only the desired product of ligation is illustrated.

of clones in which the whole DNA of the genome of a particular eukaryotic organism or cell is represented. The main reason for this is that the large size of eukaryotic genomes requires vectors with the largest capacity for foreign DNA. For example, to be 99% certain that a library of 15 kb inserts is representative of the 3×10^6 kb mouse genome, 10^6 clones are required. Plasmid vectors have the capacity for large inserts, but the efficiency of transformation of calcium-treated cells by the resulting larger DNA is low. Lambda insertion vectors can be transfected efficiently into cells, but, although sometimes used in genomic cloning (e.g. Charon 16A [140]), do not have optimal capacity. It is the *lambda replace-*

ment vectors, that can accept 15–20 kb of DNA that are preferred for genomic cloning, and a representative vector, EMBL 3, will be considered to illustrate their use.

EMBL 3 [141], is shown in Fig. A.24. This has been engineered so that restriction endonuclease digestion with *Bam*H I, *Eco*R I or *Sal* I cuts out a non-essential piece of DNA of 14 kb, which can then be replaced by foreign DNA (up to about 23 kb) with compatible cohesive ends. To prevent religation of the lambda 'centre' when using replacement vectors, the lambda 'arms' can either be first purified, or use made of the fact that bacteria that are lysogens for bacteriophage P2 do not support growth of phages which contain the *red* and *gamma* genes which reside in the fragment replaced in the recombinant. It may be mentioned that the compatible cohesive ends generated in the fragments of genomic DNA need not be (and generally are not) produced using the same restriction endonuclease as used in cleaving the vector. Thus, cleavage of DNA with the enzyme *Sau*3A I (which has the 4-base recognition sequence 'GATC) produces identical cohesive ends to cleavage with *Bam*H I (which has the 6-base recognition sequence G'GATCC). If high-molecular-weight chromosomal DNA is subjected to partial digestion with *Sau*3A I one is more likely to obtain clonable fragments of any particular area of the genome than by complete (or partial) digestion with *Bam*H I, for which there are fewer recognition sites.

Because of the large size of the introns in many eukaryotic genes, a 20 kb insert in a particular lambda genomic clone may not contain the whole of a given gene of interest. Vectors have therefore been developed which, although more difficult to handle, will accept even larger inserts. One such class of vector is the *cosmid*, a simple representative of which, pHC79 [142], is shown in Fig. A.25. Cosmids are based on bacteriophage lambda but retain only the cohesive ends (*cos*) of the bacteriophage, two of which are sufficient to allow packaging *in vitro*. This may allow insertion of up to 45 kb of foreign DNA, but the packaged recombinant DNA is no longer infective after it penetrates the host cell. For this reason the cosmid vector has an origin of replication and antibiotic resistance gene that allow it to behave like a plasmid, and selection of transformed colonies can be made on antibiotic plates. More recently vectors based on *bacteriophage P1* have been developed that can accept up 100 kb of DNA [143]. These vectors, the genetics of which are too complex to discuss here, can

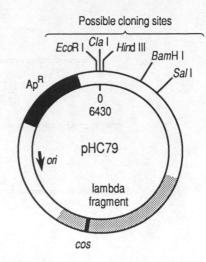

Fig. A.25 Cosmid vector pHC79. This is a simple cosmid vector containing an ampicillin resistance gene, origin of replication and unique cloning sites derived from the plasmid pBR322 (the precursor of pUC18 – Fig. A.19), together with a portion of circularized bacteriophage lambda DNA containing the *cos* site recognized by the lambda packaging system. The cleaved vector is ligated to large restriction fragments of genomic DNA that allow the formation of linear assemblies containing an insert flanked by two molecules of linearized cosmid to provide the two *cos* sites required for packaging *in vitro*.

be considered as being formerly analogous to cosmids, although they take advantage of a natural plasmid stage in the life cycle of bacteriophage P1.

To encompass a length of genome greater than the capacity of the vector containing the genomic clones it is necessary to perform the operation known as *chromosome walking*, in which a fragment of unique-sequence DNA near the extremity of the clone is used as a probe to re-screen the genomic library for an overlapping clone containing the adjoining area of the genome. With the start of efforts to determine the sequence of complete genomes it has become necessary to construct overlapping sets of clones covering such distances that even the vectors just mentioned are in-

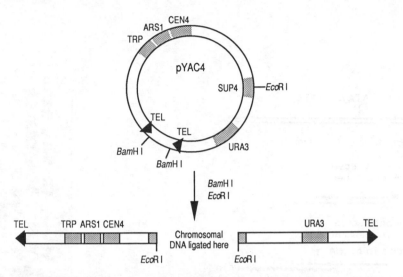

Fig. A.26 Yeast artificial chromosome vector pYAC4. The plasmid form, maintained in *E. coli*, is digested with *Bam*H I and *Eco*R I generating two vector arms, each with a yeast telomere (TEL) at one end. Large fragments of chromosomal DNA produced by partial digestion with *Eco*R I can be ligated between these, producing artificial chromosomes which can be maintained in yeast. ARS1 and CEN4 are a yeast replication origin and centromere, respectively, and TRP, SUP4 and URA3 yeast selectable markers. For further details see text.

adequate. A major breakthrough in this respect has been the development of *yeast artificial chromosomes* (YACs) [144, 145], which, using a eukaryotic microorganism as host, also allow the cloning of certain eukaryotic sequences that are unstable in a bacterial environment. The essential features for the replication of a yeast chromosome are the centromere, the telomeres and an origin of replication; and these (*CEN4*, *TEL*, and *ARS1*, respectively) are the essential constituents of pYAC4 (Fig. A.26). Markers must be included to select for transformants after introduction of the DNA (by permeablization with calcium chloride and ethylene glycol) into yeast spheroplasts (yeast lacking their cell wall). In pYAC4 these are provided by the *TRP1* and *URA3* genes, which allow growth on medium lacking tryptophan and uracil in a yeast mutant in which these genes are non-functional. The *Eco*R I cloning site of

pYAC4 is in the *SUP4* gene, encoding a suppressor tRNA (section 12.9.6) which is inactivated by the insertion of foreign DNA. Recombinants can be detected on the basis of the colour of the colonies when a host with an *ade2* ochre mutation is used. Because the YACs exist in the same environment as natural chromosomes, the recombinants could, in principle, be as large as the latter. In practice, the cloning of inserts of 1000 kb (1 Mb) into these vectors has been achieved.

A.9 ANALYSIS AND MANIPULATION OF CLONED DNA

There are many different manipulations that an investigator might wish to perform on a cloned nucleic acid. This section is restricted to describing the principles underlying the most common of these.

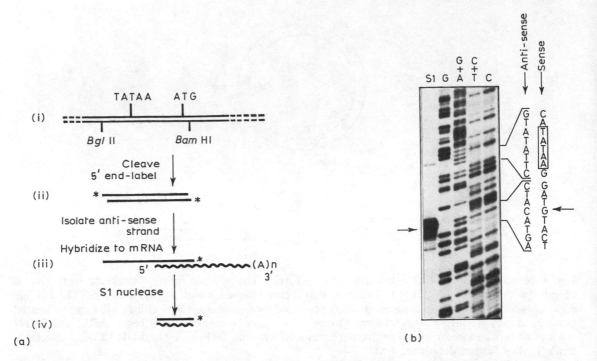

Fig. A.27 Example of nuclease S1 mapping to determine a transcriptional start point. (a) A cloned fragment of genomic DNA hybridizing to a particular mRNA has been sequenced and the likely initiation codon and TATAA box identified (i). A restriction fragment that is likely to contain the transcriptional start point is isolated from this region and labelled at the 5′-end (ii), the anti-sense strand is isolated and hybridized to mRNA (iii), and the single-stranded regions of the DNA (and mRNA) not involved in the hybrid is digested with S1 nuclease leaving the hybrid (iv); (b) the size of the protected DNA in (iv) is determined on a polyacrylamide sequencing gel. In this case Maxam and Gilbert sequencing reactions have been performed on the digested and undigested portion of (iii) and have been subjected to parallel electrophoresis to allow exact definition of the transcriptional start point in the sequence (the G indicated by the arrow). The position of the TATAA box is also indicated on the sequencing gel. (Data relating to the mRNA for the small subunit of herpes simplex virus 1 ribonucleotide reductase, kindly provided by Dr Barklie Clements.)

A.9.1 Mapping of RNA transcripts

One general problem related to cloned genomic DNA encoding a protein is to determine precisely which parts of the DNA are represented in the final processed RNA transcript. The main technique for determining the positions of intron splice sites and the points at which transcription of mRNA starts and finishes is *nuclease S1 mapping* [146]. The principle of this method is to form a DNA:RNA hybrid between the mRNA encoded by the cloned DNA and the complementary single strand of the latter. This is possible by judicious choice of hybridization temperature, because of the greater stability of the DNA:RNA hybrids compared with the original DNA:DNA hybrid. The endonuclease S1 (section 4.2.1) is used to digest the non-hybridized, single-stranded regions of the DNA, leaving the regions of DNA complementary to the mRNA intact. The size of the undigested DNA can be determined by electrophoresis

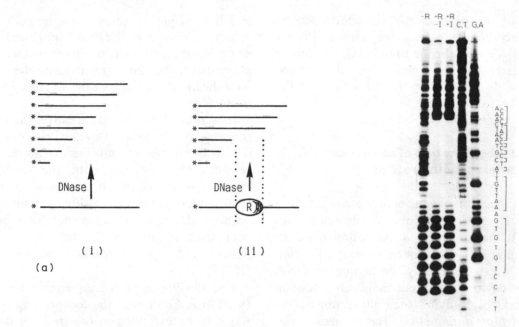

Fig. A.28 Example of DNase footprinting. (a) Partial digestion of a piece of end-labelled DNA to which a protein (R) is bound (ii) results in the absence of the end-labelled fragments having a cleavage point in this region, although these are found in the digest (i) of the unprotected DNA; (b) the results of separation of the products of such an experiment on a polyacrylamide sequencing gel. The example is of the *lac* operator DNA and the *lac* repressor protein (R). I is IPTG (0.3 M), which does not prevent the binding of a mutant repressor used in this study. C,T, and A,G represent the results of a Maxam–Gilbert C + T and A + G reactions on the undigested end-labelled fragment. Adapted from [149], with permission.

on alkaline agarose gels (to hydrolyse the RNA), and different fragments are identified by hybridizing to appropriate ^{32}P-labelled probes. For a precise determination of the transcription boundaries an appropriate ^{32}P-end-labelled restriction fragment of the cloned DNA is used in the hybridization, and this can subsequently be subjected to electrophoresis on a polyacrylamide sequencing gel alongside suitable standards of the restriction fragment in question (Fig. A.27).

In determining the position of introns it can be helpful to employ exonuclease VII [147] (section 4.5.2) in addition to the

nuclease S1. Exonuclease VII will digest single-stranded ends protruding from a mRNA/DNA hybrid but, unlike the endonuclease S1, cannot digest a single-stranded intron looped out from such a hybrid.

For determination of the 5′-ends of mRNAs, nuclease S1 mapping is usually used in combination with *primer extension* [148]. This latter technique involves a small restriction fragment or oligonucleotide known to lie near to, but 3′ of, the 5′-end of the mRNA; and this is used to prime copying of the mRNA with reverse transcriptase. If the primer is ^{32}P-labelled at its 5′-end or the DNA labelled during synthesis, the size of the product can be determined by polyacrylamide gel electrophoresis. Primer extension can also be used in cDNA cloning to 'extend' cDNA clones already isolated which do not reflect the full extent of the mRNA to its 5'-end.

A.9.2 Identification of regions of DNA that interact with proteins

The regulation of gene expression is achieved through the interaction of proteins with DNA; hence when a particular piece of genomic DNA has been cloned it is often of interest to identify the regions of DNA involved in such interactions. One general method of studying such interactions is by *DNA footprinting* [149]. This involves forming a complex between the protein and a fragment of the double-stranded DNA, radioactively end-labelled (section A.4.2) at one end only, as for Maxam–Gilbert sequencing (section A.5.1(a)). It is then subjected to digestion with bovine DNase I (section 4.5.1) and the region of DNA in contact with the protein is protected by the latter from digestion. If complete digestion occurred the protected fragment would need to be isolated and analysed. However, if the digestion is performed under conditions that on average result in a single endonucleolytic cut per molecule (cf. the use of similar conditions for chemical cleavage) an easier analysis is possible. This involves subjecting the DNA to electrophoresis in the type of denaturing polyacrylamide gel used for sequencing. If no protein is present during the DNase digestion a ladder of end-labelled oligonucleotides encompassing all possible sizes is produced on autoradiography. However, when the DNA–protein complex is subjected to DNase treatment no scission occurs in the protected region and no end-labelled oligonucleotides terminating in this region will be generated. This appears

as a hole or gap on the sequencing gel (Fig. A.28). Unprotected DNA subjected to sequencing reactions is analysed by electrophoresis on the same gel to allow the easy identification of the region of the DNA protected.

It is possible to use alternative cleavage reagents to DNase I which allow one to obtain more precise information about the residues that interact with the protein. For example, chemical methylation and 5-bromodeoxyuridine substitution with subsequent cleavage by ultraviolet light have been used to address the role of specific purines and thymine residues, respectively [150].

The development of sequencing *in vivo* (section A.5) allowed the footprinting technique to be extended to the study of interactions between uncloned DNA and protein *in vivo* [106, 151]. One approach has been to irradiate nuclei with ultraviolet light, producing pyrimidine dimers (section 7.2.5). The resulting saturation of the 5,6-double bond allows ring opening through reduction with sodium borohydride [152]. Alternatively, it is possible to apply the use of dimethyl sulphate for the methylation of purines to intact systems such as *E. coli* cells [153] and mammalian nuclei [154].

The objective of the footprinting technique is to identify the regions of DNA involved in interaction with proteins. In a few cases this has been followed by X-ray crystallography of complexes formed between purified proteins and chemically synthesized oligonucleotides corresponding to their binding sites (e.g. cover illustration [155]). However, it is often the case that the identity of the protein components of the individual interactions are unknown, and in this situation the technique of *band-retardation* [156] (also known as band-shift, gel-retardation and mobility-shift) may be undertaken to aid the identification and purification of such proteins. The basis of

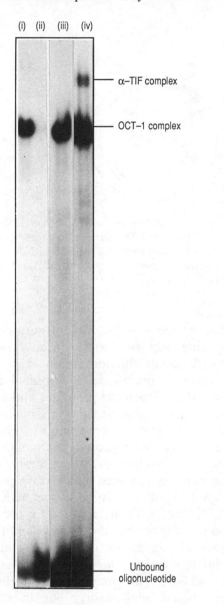

(i) (ii) (iii) (iv)

— α–TIF complex

— OCT–1 complex

— Unbound
oligonucleotide

this technique is the lower mobility during gel electrophoresis of a protein–DNA complex compared with the DNA alone. If one optimizes conditions to avoid dissociation of the complex, one is able to see a diminution in the intensity of the free DNA and the appearance of a slower-moving species (Fig. A.29). The technique can be applied to quite small DNA fragments and synthetic oligonucleotides, and so can also be used as a complement to footprinting in studying regions of DNA that bind proteins.

A.9.3 Expression

After having cloned a cDNA or gene encoding a particular protein one may well wish to use the clone to express the protein. However, there are quite different objectives one might have in seeking to do this, and the precise objective will determine the appropriate method to adopt. In this section we shall consider three possible objectives and corresponding expression systems. The objective may be to obtain large amounts of protein, in which case one is most likely to attempt to over-express a cloned cDNA from a bacterial or yeast plasmid or from a eukaryotic viral vector; one may merely wish to detect some product, or be interested primarily in the non-coding parts of a eukaryotic genomic

Fig. A.29 Band retardation assay of complexes of protein transcription factors with regulatory DNA sequences. The figure shows an autoradiograph obtained from a non-denaturing polyacrylamide gel to which had been applied a ^{32}P-labelled oligonucleotide (a 29-mer containing the promoter of the herpes simplex virus 1 immediate-early gene α-0) incubated with various protein fractions: (i) pure OCT-1 transcription factor (section 10.5), (ii) pure OCT-1 transcription factor preincubated with an anti-serum against it, (iii) nuclear extracts from uninfected HeLa cells, (iv) nuclear extracts from HeLa cells infected with herpes simplex virus. It can be seen that a complex of similar mobility to that formed with purified OCT-1 (i) is formed with nuclear extract from both uninfected (iii) and infected (iv) cells, and that an additional complex (with a virus transcriptional regulatory protein known as α-TIF) is formed with nuclear extract from infected cells only. Lanes (i) and (ii) are from a separate experiment to that of lanes (iii) and (iv), which have been exposed for a longer time to allow visualization of the α-TIF complex. The data for this figure were kindly provided by Drs Frances Purves and Bernard Roizman.

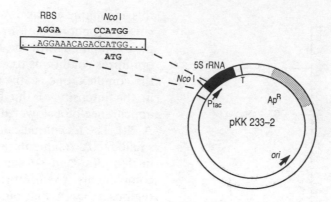

Fig. A.30 Bacterial plasmid expression vector pKK-233-2. An exploded view is shown of the *Nco* I cloning site containing an ATG initiation codon 3′ to a Shine and Dalgarno ribosome binding site (RBS). Transcription is driven by the 'tac' promoter (Ptac) and terminated by the 5S rRNA transcriptional terminator (T). The vector also contains the ampicillin resistance gene (Ap^R) and origin of replication (*ori*) of pUC18 (Fig. A.19). For further details see text.

clone that regulate its transcription, in which case *transient expression* may be most appropriate; or one may wish to study the expression of a eukaryotic gene in a cellular context, in which case it is necessary to attempt to *integrate* it into the host genome.

(a) *Over-expression from vectors*

The expression of a bacterial gene from a bacterial vector is reasonably straightforward, but the expression of the protein encoded by a eukaryotic cDNA or gene from a bacterial vector presents more problems. These relate to the fact that bacteria do not contain the splicing system necessary to process correctly genes that contain introns, and that the signals for the start of transcription and translation are quite different in eukaryotes and prokaryotes (chapters 9 and 12). Because of the ease with which bacteria can be handled and the large scale on which they can be grown, special bacterial vectors have been designed for expression of eukaryotic proteins. These are designed for cDNA

clones (there is no way to surmount the splicing problem), and typically contain a translational initiation codon at or near a cloning site preceded at the optimal distance by a strong bacterial promoter and a strong Shine and Dalgarno ribosome-binding site. In some vectors the position of cloning will result in a fusion protein containing a non-authentic *N*-terminus, which must either be tolerated or removed by a specific protease. However, other vectors such as pKK 233–2 [157] (Fig. A.30) are adapted for the precise positioning within the context of the prokaryotic signals of cDNAs in which the initiation codon is contained in a *Nco* I site, **CCATGG**. This may at first sight seem of limited applicability, but in fact this sequence conforms well to the Kozak consensus for initiation of eukaryotic translation (section 12.5.1) and many eukaryotic initiation codons either lie within such *Nco* I sites or can easily be made to do so by site-directed mutagenesis (section A.9.4). The vector pKK 233–2 contains quite a strong *E. coli* promoter, but it has been found that even greater expression can be obtained using vectors such as the pET series [158], in which the DNA is cloned into a site controlled by a promoter for bacteriophage T7 and is then expressed in a strain of bacteria which can synthesize the highly specific T7 RNA polymerase.

Although some eukaryotic proteins can be expressed from such vectors (which are also useful for over-expression of those bacterial proteins which are normally only weakly expressed), in many cases little protein is obtained. The major reason for this seems to be that many eukaryotic proteins are unstable in a bacterial environment, as already mentioned in relation to the design of λgt11 (section A.8.3). Indeed, if the objective in expressing a protein is to raise antibodies, it is better to generate fusion proteins using λgt11 or conceptually similar, and perhaps more easily handled, plasmid vectors such as those of the pUEX series [159]. However, if the objective is to obtain authentic functional protein, eukaryotic vectors may be the only alternative. One possibility is to use yeast vectors (e.g. [160]), but for mammalian proteins yeast still may not be a suitable environment. It is for this reason that certain viruses of higher eukaryotes have been employed as vectors for over-expression of eukaryotic proteins. One virus that has been much exploited is the *baculovirus, Autographa californica* mononuclear polyhedrosis virus, in insect cells (*Spodoptera frigiperda*; Sf9) [161]. This virus has an extremely abundant late gene-product, polyhedrin, that can constitute up to 50% of the total cellular protein. Because this protein is not essential for production of infectious extracellular virus, foreign genes can be cloned in front of the polyhedrin promoter in place of the polyhedrin gene and over-expressed. In this, and other vectors based on large eukaryotic viruses, it is necessary to introduce the foreign DNA into the virus vector by genetic recom-

bination with the foreign DNA cloned into a suitable plasmid; and there must be some means of selecting for recombinants. In the baculovirus vector it is necessary to isolate non-viable cells (as indicated by staining) that do not contain polyhedrin protein inclusion bodies.

The baculovirus system is technically rather demanding, and other virus vectors have been developed that are easier to handle. The greatest promise seems to be offered by sophisticated vectors [162] based on *vaccinia* virus, a lytic virus that infects a variety of mammalian cells. One vector based on this virus, Vac/Op/T7, contains the gene for bacteriophage T7 RNA polymerase, the expression of which is inducible (constitutive expression is detrimental to this virus) and under the indirect control of the strong vaccinia late promoter P11. (It is the *lacI* gene that is under the direct control of the vaccinia promoter, and the T7 polymerase gene is under the control of the *lac* operator, and hence can be induced by IPTG.) The foreign gene of interest is cloned into a site where it is under the control of a strong bacteriophage T7 promoter, and hence can be expressed at high levels. Indeed, much more mRNA is made than can be 'capped', and for this reason the cloning site of the vector is preceded by the 5′-leader sequence of encephalomyocarditis (EMC) virus, which, like polio, allows cap-independent translation (sections 12.5.1 and 12.9.5). The cDNA to be expressed is initially cloned into a plasmid, optimally into an *Nco* I site (as for pKK 233–2, above), and this is flanked by sequences for the vaccinia thymidine kinase gene (*tk*), to allow recombination and selection for *tk*⁻ virus after introducing the DNA into cells using the calcium phosphate technique (A.9.3(b)). The selection is done by using *tk*⁻ cells (the *tk* gene is not essential for either cells or virus in culture) and growing on bromodeoxyuridine (BrUdr), which can only be incorporated into DNA by tk^+ virus in which thymidine kinase converts BrUdr to BrdUMP. Ultraviolet light will fragment the DNA of those viruses in which thymine has been replaced by bromouracil (cf. section A.9.2), killing them and hence allowing selection of recombinant tk^- virus.

(b) *Transient expression in eukaryotic cells*

For some experimental purposes it is sufficient to obtain low level expression of a eukaryotic gene. One way of obtaining transient transcription of a cloned gene is *microinjection* into the nucleus of frog oocytes [163]. It is more usual to inject mRNA, rather than DNA, and this can be generated by transcription *in vitro* of DNA cloned into a vector that puts it under the control of a promoter for bacteriophage T7 [164]. Less specialized methods of introducing DNA into cells are also available, however. Most widespread is the technique of *calcium phosphate co-precipitation* of DNA onto tissue culture cells [165], which take up the precipitate, apparently by a process of phagocytosis. An alternative is *electroporation*, in which the cell membrane is made permeable by a brief pulse of high-voltage electricity [166].

The amount of DNA introduced into cells by calcium phosphate co-precipitation or electroporation is normally too small to produce detectable amounts of protein *per se*. It is for this reason the DNA is usually cloned into a vector containing a eukaryotic origin of replication, so that after 24 or 48 h sufficient DNA is generated to give detectable quantities of transcripts and their products. The most frequently used of such vectors are based on the small circular DNA virus, *SV40*; although it is important to stress that such vectors are

unable to form virus particles, and they are referred to as *viral mini-replicons*. Nor can they be maintained permanently in cells in the way in which bacterial plasmids can, as the massive replication makes the cells unviable.

For vectors based on SV40 to retain the ability of the viral DNA to replicate they require the presence of an origin of replication and the large T antigen, although the latter is usually provided by the use of a transformed COS cell line [167]. The vectors also contain bacterial plasmid sequences for cloning and propagation, and hence may be referred to as *shuttle vectors* (they 'shuttle'

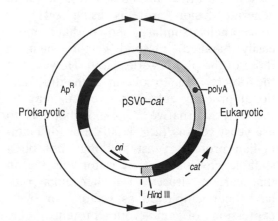

Fig. A.31 The CAT vector pSV0-*cat* [169]. The prokaryotic portion of the vector contains the ampicillin resistance gene (Ap^R) and origin of replication (*ori*) of pUC18 (Fig. A.19) for propagation in *E. coli*. The eukaryotic portion derives predominantly from the early region of SV40 into which the *E. coli* chloramphenicol transacetylase gene (*cat* – strictly speaking *cml*) has been inserted in an orientation that allows it to use the polyadenylation/processing site indicated (polyA). The SV40 promoters have been removed in this vector (and also the SV40 origin of replication) so that insertion of a eukaryotic promoter into the *Hin*d III site is needed to produce transcription of the *cat* gene after transfection into eukaryotic cells. Similar vectors with a greater number of possible cloning sites are also available [170].

between prokaryotic and eukaryotic hosts). A variety of such SV40 vectors is available for different purposes. If one merely wishes to demonstrate that a DNA sequence encodes a particular protein, the vector can facilitate or optimize expression of this by providing promoter, enhancer and poly-adenylation signal (e.g. [168]). Alternatively one may wish to study the regulatory regions associated with a cloned gene, and require a vector lacking promoter or enhancer or both. In such a case it is advantageous for the regulatory sequences to be able to drive the transcription of a sequence coding for a protein that can be detected and quantified more easily than the authentic product of the gene, and it is for this purpose that the *CAT vectors* were developed. These allow the 5′-non-coding and untranslated regions of a gene to be linked to the coding region for the bacterial enzyme chloramphenicol acetyltransferase (CAT). *CAT assays*, as they are called, are extremely sensitive and are performed using [^{14}C]chloramphenicol as a substrate, and separating the acetylated chloramphenicol products from unreacted substrate by thin-layer chromatography [169]. An example of a CAT vector is presented in Fig. A.31.

(c) *Expression after integration into a cellular eukaryotic genome*

A small proportion of eukaryotic cells transfected with DNA actually become 'transformed': the DNA is integrated into the chromosome. If such transformed cells are selected from the vast majority of un-transformed cells, cell lines can be obtained in which to study the expression of stably integrated exogenous genes. It is not strictly necessary to incorporate the selectable marker into the same piece of DNA as the gene to be expressed, and *co-transfection* is often performed. However, there are advantages in cloning the DNA into a vector containing both the selectable marker and sequences that will enhance the expression of the integrated gene. Markers that allow selection in eukaryotic cells include *gpt*, the gene for *E. coli* xanthine–guanine phos-phoribosyltransferase (which confers resistance to mycophenolic acid), and *neo*, the gene for bacterial aminoglycoside phosphotransferase (which confers re-sistance to aminoglycoside antibiotics such as neomycin, kanamycin and geneticin). Sequences included to enhance expression include the SV40 enhancer and retrovirus LTRs. The vectors (e.g. [171]) generally include the origin of replication and am-picillin resistance gene of pUC18 for cloning and propagation in *E. coli*. In some cases the expression is under the control of a promoter inducible by glucocorticoids or heavy metals (e.g. [172]). The SV40 origin of replication may also be present in these vectors, which are often also used for transient expression; but it should be stressed that this origin has no role when the vector is used for stable integration. This latter point is underlined by the fact that with these vectors it is generally pre-ferable to linearize the DNA before trans-fection. This is not the case, however, for vectors based on defective retroviruses, which use the natural integration of retro-viral DNA into the chromosome of the host cell to achieve cleaner and more efficient integration (and perhaps better expression) [173].

A limitation of systems in which cloned genes are stably integrated into tissue culture cells arises when the major interest is in the tissue-specific expression of a particular gene. A technique in which this problem is overcome involves the micro-injection of cloned DNA into fertilized mouse eggs to produce so-called *transgenic mice* [174]. The foreign DNA is integrated at such an early stage that it is generally present in all daughter cells, and certainly in the progeny of those mice that pass it on through the germ line to the next generation. It is thus possible, using suitable constructs, to study whether a particular tissue-specific gene is expressed in a tissue-specific manner in its new location, and to investigate the regions of DNA which confer tissue-specificity of expression [175].

A.9.4 Mutagenesis *in vitro*

In studies of cloned genes of the type described in sections A.9.2 and A.9.3, conclusions regarding the importance of regulatory regions of DNA can be tested and extended if it is possible to produce mutations in the regions of interest. Such mutations are difficult to produce in higher eukaryotes *in vivo*, but powerful techniques are available to produce them in cloned DNA *in vitro*. The technique of mutagenesis *in vitro*, especially when directed to specific nucleotides (see below) can also be applied to the coding regions of genes, the products of which it is possible to express, purify and subject to structural or functional studies [176]. Some of the methods used to generate such mutations are described below.

Deletion mutagenesis may be performed using the enzyme *Bal* 31 nuclease (section 4.2.1). This enzyme has exonuclease activity against double-stranded DNA. Hence, if a plasmid containing the cloned gene is linearized by restriction endonuclease digestion at a point in the region of interest, *Bal* 31 nuclease will digest it in a way that produces a deletion emanating from this point in both directions. The size of the deletion is determined by the time allowed for digestion, after which the blunt ends (treatment with the Klenow fragment of *E. coli* DNA polymerase and dNTPs is

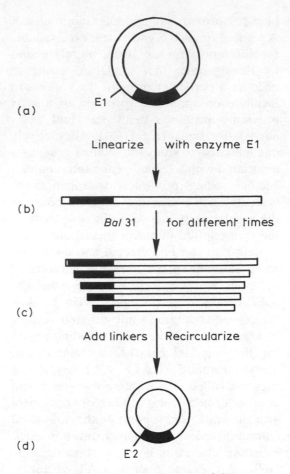

Fig. A.32 Use of exonuclease *Bal* 31 to generate bidirectional deletions. A hypothetical plasmid (a) is illustrated in a region of which (solid shading) it is desired to obtain deletions. A neighbouring restriction endonuclease site, E1, is used to linearize the plasmid (b) and this is digested for different times generating a spectrum of deleted molecules (c). These are recircularized using linkers for a restriction endonuclease, E2, giving plasmids such as (d).

employed to fill in any overhangs) are ligated to linkers (section A.8.3), cleaved, and ligated together with DNA ligase (Fig. A.32) [177]. One problem with such deletion mutants when used to assess the importance of deleted sequences is that they alter the relative proximity of the remaining regions of DNA. This disadvantage is overcome in the technique of *linker scanning mutagenesis* in which small regions of DNA may be replaced by a different sequence of the same length [178]. Descriptions of this method and a more convenient alternative strategy for achieving the same end [179] are, however, beyond the scope of this book.

It is most desirable to be able to produce mutations at specific sites (*site-directed mutagenesis*) and this can be achieved using synthetic oligonucleotides. The classic manner for doing this [180] involves sub-cloning the DNA into a bacteriophage M13 vector (Fig. A.14) that allows the generation of a single-stranded circular recombinant DNA. An oligonucleotide (e.g. 10–12 mer) is synthesized corresponding to the target of mutation and surrounding nucleotides, except that it contains the desired base-change. This is annealed to the single-stranded circular DNA, which is then converted into the replicative double-stranded form *in vitro* using the Klenow fragment of DNA polymerase I and DNA ligase. When *E. coli* are transfected with this DNA, phage containing both wild-type and mutant single-strands will result and give rise to 'plaques'. These can be distinguished by blotting onto nitrocellulose and hybridization to the ^{32}P-labelled oligonucleotide originally used, under conditions that enable a perfect match (the desired mutant) to be distinguished from a single mismatch (the wild type). This is shown diagrammatically in Fig. A.33. More sophisticated strategies (e.g. [181]) that optimize the yield and selection of mutants are now available. A modification of the procedure just described [182] may be used to generate short, specific deletions of a type hardly possible with *Bal* 31 nuclease. In this case the oligonucleotide synthesized is a hybrid of the sequences directly flanking the area to be deleted. This can hybridize to the

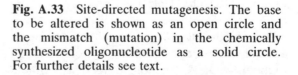

Fig. A.33 Site-directed mutagenesis. The base to be altered is shown as an open circle and the mismatch (mutation) in the chemically synthesized oligonucleotide as a solid circle. For further details see text.

single-stranded circular DNA if the area to be deleted loops out, and the desired mutant second strand can then be synthesized *in vitro*. More recently the polymerase

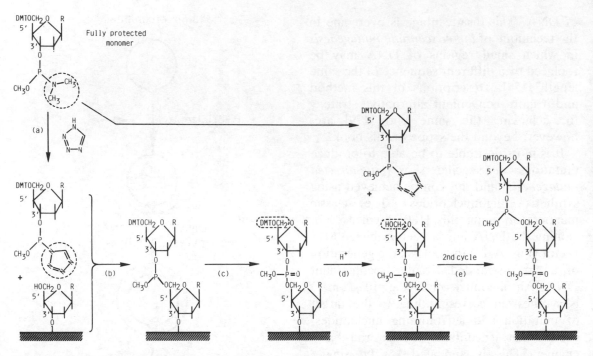

Fig. A.34 The phosphite triester method of oligonucleotide synthesis. (a) The protected phosphoramidite monomer is activated in the 3′-phosphate position by a suitable weak acid (e.g. tetrazole) allowing condensation (b) with 5′-activated monomer attached *via* its 3′-O to a solid support. The phosphite is then oxidized to a phosphate (c). Removal of the 5′-protecting DMT group of the dinucleotide with, for example, dichloroacetic acid (d) prepares this for the next cycle of condensation (e). R = benzoyl adenine, benzoyl cytosine, isobutyryl guanine or thymine.

chain reaction has been employed for the production of a variety of mutations by strategies that do not involve cloning into M13 [183].

A.10 CHEMICAL SYNTHESIS OF OLIGONUCLEOTIDES

The importance of chemically synthesized oligonucleotides for PCR (section A.7), in labelling DNA (section A.4.1), as primers for sequencing in M13 (section A.5.1(b)), for producing mutagenesis *in vitro* (section A.9.4) and in studying specific interactions between nucleic acids and proteins (section A.9.2) has already been mentioned. Custom

synthesis of oligonucleotides is now available commercially, as are machines for 'in house' synthesis. The chemistry involved derives from that developed by Khorana, the application of which culminated in the synthesis of a tRNA gene [184]. The method at present in greatest use, the *phosphite triester* method (Fig. A.34), will be outlined here. Further details can be found elsewhere [185].

Synthesis of oligonucleotides involves the formation of successive diester bonds between the 5′-OH group of one nucleotide derivative and 3′-OH of a second nucleotide derivative. In order to simplify the synthetic procedure the first nucleotide is directly linked to a solid support (e.g. silica gel) packed in a column. It is necessary to activate one of the components for the reaction, and this requires that other potentially reactive groups elsewhere in the molecules be 'protected' by reversible chemical modification. The reactive component is the free monomer in which a $3'$-OPO_4 has been substituted by dialkyl phosphoamidite (one of a variety of the possible dialkyl substitutions of the N is illustrated), and, after activation, this reacts with the free $5'$-OH group of the bound nucleotide or oligonucleotide to give a phosphite triester. This must be oxidized (e.g. with aqueous iodine) to a stable phosphotriester before the next synthetic step. The $3'$-OH of the immobilized reactant is protected by virtue of its attachment to the solid support, the $5'$-OH of the free monomer is protected by a dimethoxytrityl group, a methyl group is usually employed to protect the hydroxyl on the $3'$-OPO_4, and the individual bases are also protected. After reaction excess reagents are then washed off the column, any unreacted $5'$-OH groups are blocked or 'capped' using acetic anhydride, and the $5'$-O-dimethoxytrityl protecting group is removed by 80% acetic acid to allow the extended bound oligomer to react with another activated monomer in the next round of condensation. At the end of the synthesis the oligonucleotide must be chemically released from the column and deprotected. Removal of the oligonucleotide chain from the solid support depends, of course, on the nature of the linkage. Ester linkage via succinyl groups is quite common, in which case cleavage occurs at the same time as deprotection of the bases.

This phosphite triester method, which is used by most automatic synthesizers ('gene machines'), is relatively rapid (condensation is completed in about 2 min), and has a high enough efficiency to allow the synthesis of oligonucleotides containing more than 150 bases.

A.11 CELL-FREE SYSTEMS FOR TRANSCRIPTION AND TRANSLATION [8]

The transcription and translation of cloned genes in intact cells has already been discussed in section A.9.3. Here we consider methods for studying transcription and translation in cell-free systems, with particular reference to eukaryotes.

Systems for performing *transcription* of eukaryotic genes *in vitro* are somewhat limited. It is possible to use extracts of HeLa cell nuclei to study the transcription of genes by pol III [186], but for genes requiring pol II the method is more difficult to apply [187]. In the latter case it is more common to study the elongation of already initiated transcripts in isolated nuclei, using the so-called *nuclear run-off* assay [188] (also referred to as nuclear 'run-on').

As regards *translation*, the intact-cell system of microinjection into frog oocytes (or eggs) has the advantage that protein synthesis may proceed for several days [189]. Cell-free systems, however, are of greater utility to the non-specialized laboratory, even though these are not as simple to establish in eukaryotes as in *E. coli* (see below), and activity is easily lost during cell fractionation. Necessary components of the system may be lost with nuclei and mitochondria or adhere to intercellular structures. Furthermore, the disruption of the cytoskeleton may be a factor contributing to loss of activity. It is therefore perhaps not coincidental that the two most widely used systems for cell-free translation are derived from rather specialized cells. By far the most popular system is that derived from rabbit reticulocytes using hypotonic

Table A.3   Some inhibitors of DNA synthesis

Category	Examples	Reference
1. Reagents interacting with DNA*		
(a) Alkylating agents	Dimethyl sulphate	
	Mitomycin C	
	Nitrogen and sulphur mustards	section 7.2.2
	MNNG and NMS	
(b) Intercalating agents	Acridine dyes	section 7.2.3
	Actinomycins	section 7.2.3
	Adriamycin	[198]
	Anthracenes	section 7.2.2
	Benzpyrene	
	Ethidium bromide	section 7.2.3
	Propidium diiodide	
(c) Intertwining agents	Distamycin	[199]
	Netropsin	[199]
2. Analogues of bases, nucleosides etc.†	Acyclovir	section 3.7.7(c)
	Adenine β-1-D-arabinoside	section 7.2.1
	Amethopterin	sections 5.2 and 5.7
	Aminopterin	sections 5.2 and 5.7
	2-aminopurine	section 7.2.1
	Aphidicolin	section 6.4.7
	8-azaguanine	section 7.2.1
	Azaserine	section 5.2
	6-azauracil	section 7.2.1
	2′azido-2′-deoxynucleosides	[201]
	5-bromodeoxycytidine	[197]
	Cytosine β-1-D-arabinoside	section 7.2.1
	Diazooxynorleucine	section 5.2
	Dideoxynucleosides	section 6.4.7 and A.5.1(b)
	5-fluorodeoxycytidine	[197]
	5-fluorodeoxyuridine	section 5.2
	5-fluorouracil	section 7.2.1
	Hydroxyurea	section 5.2
	6-mercaptopurine	section 7.2.1
3. Inhibitors of topoisomerases	Coumermycin	section 6.5.4
	Nalidixic acid	
	Novobiocin	
	Oxolinic acid	
4. Inhibitors of cell division	Colcemid	[202]
	Colchicine	
	Vinblastine	
	Vincristine	

* These may be mutagens and/or agents that interfere with replication or transcription.
† These may interfere with nucleoside biosynthesis or, following incorporation into DNA, block further chain elongation or transcription.

Table A.4 Some inhibitors of RNA synthesis

Inhibitor	Remarks	Reference
Actinomycin D	Inhibits both prokaryotic and eukaryotic transcription by complexing with deoxyguanosine residues. Hence transcription of rRNA is more sensitive than that of mRNA	section 9.2.3
α-Amanitin and other fungal amatoxins	Inhibits eukaryotic RNA polymerase II specifically (also RNA polymerase III at very high concentration)	section 9.3
Cordycepin (3'-deoxyadenosine)	Inhibits poly(A) polymerase responsible for 3'-polyadenylation of eukaryotic mRNAs	section 11.5
Dichlororibofuranosyl benzimidazole	Inhibits appearance of eukaryotic mRNA in the cytoplasm, although only partially inhibiting synthesis of hnRNA	[203]
Rifampicin and streptovaricin	Bind to β subunit of *E. coli* RNA polymerase, inhibiting initiation	section 9.2.3
Streptolydigin	Bind to β subunit of *E. coli* RNA polymerase, inhibiting elongation	section 9.2.1

lysis followed by low-speed centrifugation to remove the cell membranes. This *rabbit reticulocyte lysate* has been further modified to allow more efficient utilization of exogenous mRNA. The endogenous globin mRNA is degraded by the Ca^{2+}-dependent micrococcal nuclease that can subsequently be inactivated by chelation with EGTA [137]. Such nuclease-treated lysates are commercially available. The other cell-free system still in general use is that derived from the post-microsomal supernatant of wheat germ [190], which has an intrinsically low content of endogenous mRNA. One limitation of the reticulocyte lysate and wheat-germ systems (but not of the frog oocyte) is the lack of endoplasmic reticulum for correct processing of translation products that contain signal peptides (section 12.8.1). This may be overcome by the addition of a microsomal membrane fraction, e.g. from dog pancreas [191].

Prokaryotic cell-free systems for translation were developed very early on; the ribosomes, tRNAs and soluble factors for translation of exogenous mRNA being present in the supernatant after centrifugation of disrupted cells at 30 000 g for *c*. 30 min. This system was originally developed by Nirenberg and Matthaei in their work on the genetic code and was subsequently developed for use in studying the translation of the stable mRNAs of the RNA coliphages [192]. The low endogenous mRNA activity of such systems is a reflection of the very short half-lives of bacterial mRNAs, which make it difficult to isolate bacterial mRNAs and study their translation directly. To overcome this problem a cell-free system for translation was devised in which the mRNA could be continually generated by simultaneous transcription of the gene. This coupled transcription–translation system [193] was originally

Table A.5 Some inhibitors of protein synthesis

Inhibitor	Prokaryotic/ eukaryotic	Remarks	Reference
Abrin	E	as for ricin (below)	[211]
Aurintricarboxylic acid	P, E	Inhibits initiation by preventing binding of mRNA to ribosome	[195, 196]
Chloramphenicol	P	Inhibits elongation at peptidyl transferase. Resistant mitochondria have altered large rRNA	section 12.6.2
Colicin E3	P	Specific nuclease for site on 16S rRNA	[197, 208]
Cycloheximide	E	Inhibits elongation and initiation, 'freezing' ribosomes on polysomes. Resistant yeast have altered 60S subunits	[195]
Diphtheria toxin	E	Inhibits elongation by inactivating EF-2 by ADP ribosylation	section 12.5.2
Edeine A	P, E	Inhibits initiation at either mRNA- or initiator tRNA-binding step	[195, 196]
Emetine	E	Inhibits elongation at translocation step. Resistant hamster cells have altered protein S14	[195, 207]
Erythromycin	P	Inhibits elongation at transpeptidation step	section 12.6.2
Ethionine	P, E	Causes synthesis of abnormal proteins with ethionine replacing methionine	[210]
Fluoride	(P?), E	Inhibits initiation in intact cells	[205]
5-fluorotryptophan	P, E	Prevents activation of tRNA^{Trp} by competitive inhibition of synthetase	[210]
Fusidic acid	P, E	Inhibits elongation, preventing release of EF-G . GDP complex from ribosome. Resistant bacteria have altered EF-G.	[195, 196]
Guanylyl methylene diphosphonate and Guanylyl imidodiphosphate	P, E	Non-hydrolysable analogues of GTP	section 12.5.1 and 12.4.2
Kanamycin	P, E	Inhibits elongation and causes misreading	[195, 196]
Kasugamycin	P	Inhibits initiation, $m_2^6Am_2^6A$ sequence of 16S rRNA involved	section 12.6.2
Kirromycin	P	Inhibits elongation, preventing release of EF-Tu . GDP complex from ribosome	section 12.6.2
O-methyl threonine	P, E	Causes synthesis of abnormal proteins with O-methyl threonine replacing threonine	[210]

Table A.5 *(continued)*

Inhibitor	Prokaryotic/ eukaryotic	Remarks	Reference
Modeccin	E	as for ricin (below)	[211]
Neomycin	P, E	Causes misreading and inhibits initiation and elongation. Used as a selectable DNA marker in eukaryotic cells	section A.9.3(c)
Norvaline	P, E	Prevents activation of tRNA^{Trp}	[210]
Pactamycin	P, E	Inhibits initiation, was used to determine gene order of proteins derived from picornaviral polyproteins	[204]
Paromomycin	P	Inhibits initiation. Resistant mitochondria have altered small rRNA	section 12.6.2
Puromycin	P, E	Inhibits elongation by binding to A-site and reacting with peptidyl-tRNA	section 12.4
Ricin	E	Inhibits elongation by specific *N*-glycosidase action on native 28S rRNA	[211]
α-sarcin	(P), E	Specific nuclease for 28S rRNA in ribosomes (non-specific with naked rRNA). Inactive against intact *E. coli*, but specifically cleaves 23S rRNA in ribosomes.	[209]
Shiga toxin	E	As for ricin (above)	[211]
Showdomycin	P, E	Inhibits initiation at the stage of ternary complex formation	[195]
Sparsomycin	P, E	Inhibits elongation at peptidyl transferase step	[195, 196]
Spectinomycin	P	Inhibits elongation at transpeptidation. Resistant ribosomes have altered protein S5	[195, 196]
Streptomycin	P	Binds to 30S subunit to cause misreading and (at higher concentrations) inhibition of initiation	section 12.6.3
Tetracycline	P	Inhibits elongation by blocking binding of aminoacyl-tRNA to the A-site on the 30S ribosomal subunit	[206]
Thiostrepton	P	Inhibits elongation, preventing binding of EF-G . GTP complex to ribosome	section 12.6.2
Trimethoprim	P	Prevents formation of fMet-tRNA by inhibition of synthesis of N^{10}-formyl-H$_4$-folate	[210]

Table A.6 Some *E. coli* genes relevant to nucleic acid metabolism

Gene symbol	Explanation	Reference
ada	Repair enzyme for removal of methyl groups	section 7.3.1
alaS	Alanyl-tRNA synthetase	section 12.3.2
alaT–W	Alanyl-tRNAs	section 12.3.1
apaH	AppppppA hydrolase	section 12.10.3
argS	Arginyl-tRNA synthetase	section 12.3.2
argT–X	Arginyl-tRNAs	section 12.3.1
asnS	Asparaginyl-tRNA synthetase	section 12.3.2
asnT–V	Asparaginyl-tRNAs	section 12.3.1
aspS	Aspartyl-tRNA synthetase	section 12.3.2
aspT–V	Aspartyl-tRNAs	section 12.3.1
*att*λ	Integration site for bacteriophage lambda (similarly *att*φ*80* etc.)	section 7.5.1
cca	tRNA nucleotidyl transferase	section 11.7.1
cmlA	Chloramphenicol acetyltransferase	section A.9.3(b)
cysS	Cysteinyl-tRNA synthetase	section 12.3.2
cysT	Cysteinyl-tRNA	section 12.3.1
dam	DNA adenine methylase	section 4.6.1
dcm	DNA cytosine methylase	section 4.6.1
dnaA (*B*, *C*, etc.)	Enzymes of DNA biosynthesis	Tables 6.3 and 6.5
dut	(*dnaS*) dUTPase	section 6.3.2
endA	DNA endonuclease I	Table 4.3
ffh	Component of signal recognition particle	section 12.8.1
ffs	4.5S RNA	section 12.8.1
ftsY	Receptor of signal recognition particle	section 12.8.1
fusA	Protein synthesis factor EF-G	section 12.4.2
glnS	Glutaminyl-tRNA synthetase	section 12.3.2
glnU–V	Glutaminyl-tRNAs	section 12.3.1
gltT–W	Glutamyl-tRNAs	section 12.3.1
gltX	Glutamyl-tRNA synthetase	section 12.3.2
glyS	Glycyl-tRNA synthetase	section 12.3.2
glyT–W	Glycyl-tRNAs	section 12.3.1
gpp	Guanosine pentaphosphatase	section 11.12
groEL, ES	Chaperone proteins	section 12.8
gyrA, B	(*cou*, *nalA*) DNA gyrase (two subunits)	section 6.5.4
hfl	High frequency of lysogenization by bacteriophage lambda	section A.8.2
hisR	Histidinyl-tRNA	section 12.3.1
hisS	Histidinyl-tRNA synthetase	section 12.3.2
hsdM, R, S	Host restriction/modification system (methylase, endonuclease and specificity components of *Eco*B and *Eco*K in B and K strains, respectively)	section 4.5.3
ileS	Isoleucyl-tRNA synthetase	section 12.3.2
ileT–V, X	Isoleucyl-tRNAs	section 12.3.1
infA, B, C	Protein synthesis factors IF-1, IF-2 and IF-3	section 12.4.1
lacA, I, Y, Z	Components of the *lac* operon	section 10.1.1
lepB	Signal peptidase I	section 12.8.1
leuS	Leucyl-tRNA synthetase	section 12.3.2
leuT–X, Z	Leucyl-tRNAs	section 12.3.1

Table A.6   *(continued)*

Gene symbol	Explanation	Reference
lig	(*dnaL*) DNA ligase	Table 6.5
lspA	Signal Peptidase II (specific for prolipoprotein)	section 12.8.1
lysS	Lysyl-tRNA synthetase	section 12.3.2
lysT, V	Lysyl-tRNAs	section 12.3.1
lysU	Lysyl-tRNA synthetase (inducible)	section 12.3.2
metG	Methionyl-tRNA synthetase	section 12.3.2
metT, Y, Z	Methionyl-tRNAs	section 12.3.1
mutD	(*dnaQ*) DNA polymerase III (ε)	Table 6.5
mutH, L, S	Uncharacterized proteins involved in DNA repair	section 7.3.3
nrdA, B	(*dnaF*) ribonuclease reductase (two subunits)	Table 6.5
nusA, B	Proteins involved in termination of transcription	section 9.2.4
nusC, D, E	= *rpoB*, *rho* and *rpsJ*, respectively	section 9.2
oriC	Origin of DNA replication	section 6.10.1
pheS, T	Phenylalanyl-tRNA synthetase (two subunits)	section 12.3.2
pheU, V	Phenylalanyl-tRNAs	section 12.3.1
phr	Deoxyribodipyrimidine photolyase	section 7.3.1
pnp	Polynucleotide phosphorylase	section 4.4
polA, B, C	DNA polymerases I, II and III (*dnaE*)	Tables 6.2 and 6.5
prfA, B	Protein synthesis factors RF-1 and RF-2	section 12.4.3
prlA, C, D, F, G	Proteins affecting protein export	section 12.8.1
prmA, B	Ribosomal protein methylases	section 12.6
proS	Prolyl-tRNA synthetase	section 12.3.2
proK–M	Prolyl-tRNAs	section 12.3.1
purA–N	Enzymes of purine biosynthesis	section 5.2
pyrA–I	Enzymes of pyrimidine biosynthesis	section 5.4
recA	Protein involved in recombination and repair	section 7.4.1
recB, C, D	DNA exonuclease V (three subunits)	Table 4.4 and section 7.4.2
recE	DNA exonuclease VIII	section 4.5.2
recF, J, N, O, Q	Proteins involved in DNA recombination and repair	section 7.4
relA	Stringent factor	section 12.10
rep	DNA helicase	section 6.5.3
rho	Transcriptional termination factor rho	section 9.2.4
rimB–L	Ribosomal protein modification enzymes (methylases, acetylases, etc.)	section 12.6.1
rna–rnh	Ribonucleases I–III, D, E and H	Table 4.1
rnpA–B	Ribonuclease P (protein and nucleic acid components)	section 11.3.4
rplA–Y, rpmA–J	50S subunit ribosomal proteins L1–L36	section 12.6.1
rpoA–D, H, N	RNA polymerase subunits α, β, β', σ^{70}, σ^{32} and σ^{60}	section 8.1.1
rpsA–U	30S subunit ribosomal proteins S1–S21	section 12.6.1
rrfA–E, G, H	5S rRNA genes of operons *rrnA–H*	section 8.4.4
rrlA–E, G, H	23S rRNA genes of operons *rrnA–H*	section 8.4.3

Table A.6   *(continued)*

Gene symbol	Explanation	Reference
rrnA – E, G, H	rRNA operons	section 8.4.3
rrsA – E, G, H	16S rRNA genes of operons *rrnA – H*	section 8.4.3
sbcB	DNA exonuclease I	Table 4.4
secA, B, D – F, Y	Proteins affecting protein export (*secA, E, Y = prlD, G, A*, respectively)	section 12.8.1
serS	Seryl-tRNA synthetase	section 12.3.2
serT – X	Seryl-tRNAs	section 12.3.1
spoT	Guanosine tetraphosphatase	section 12.10
ssb	DNA-binding protein	Table 6.5
tdk	Thymidine kinase	section 5.8
thrS	Threonyl-tRNA synthetase	section 12.3.2
thrT – W	Threonyl-tRNAs	section 12.3.1
thyA	Thymidylate synthetase	section 5.6
topA	Topoisomerase I	Table 6.5
traI	Helicase I (plasmid encoded)	section 6.5.3
trmA – F	tRNA methylases	section 11.7.2
trpS	Tryptophanyl-tRNA synthetase	section 12.3.2
trpT	Tryptophanyl-tRNA	section 12.3.2
tsf	Protein synthesis factor EF-Ts	section 12.4.2
tufA, B	Protein synthesis factor EF-Tu (two genes)	section 12.4.2
tyrS	Tyrosyl-tRNA synthetase	section 12.3.2
tyrT, U	Tyrosyl-tRNAs	section 12.3.1
umuC, D	Proteins involved in error-prone DNA repair	section 7.3.4
ung	Uracil-DNA-glycolase	section 6.3.2
uvrA – C	AP endonuclease (three components)	Table 4.3 and section 7.3.2
uvrD	Helicase II	section 6.5.3
valS	Valyl-tRNA synthetase	section 12.3.2
valT – W	Valyl-tRNAs	section 12.3.1
xthA	DNA exonuclease III	section 4.5.2
xseA, B	DNA exonuclease VII (two subunits)	Table 4.4

applied to genes cloned by more traditional genetic techniques using specialized transducing phages.

A.12 THE USE OF INHIBITORS IN THE STUDY OF GENE EXPRESSION

In the course of this book reference has been made to the use of antibiotics and other inhibitors of macromolecular synthesis. For the convenience of the reader the most common of these are summarized in Tables A.3 to A.5. Many inhibitors of transcription and translation are specific for eukaryotic or prokaryotic systems, and this specificity is indicated in Tables A.4 and A.5. Most inhibitors of DNA synthesis (Table A.3) do not discriminate between eukaryotes and prokaryotes; however, the dideoxynucleosides, on conversion into the triphosphates, do not inhibit eukaryotic DNA polymerase α. In contrast, a fairly specific inhibition of eukaryotic DNA polymerase α is obtained with aphidicolin. Further general [194–197] and specific [198–211] details may be found elsewhere.

A.13 *E. COLI* GENES RELEVANT TO NUCLEIC ACID METABOLISM

Much of the study of nucleic acid metabolism has been conducted in *E. coli*, and many of the components involved were identified by mutations in their genes. Because of this, and in view of the widespread use of specialized *E. coli* strains among non-geneticists undertaking recombinant DNA work, we felt it might be useful to provide a summary of the names given to some selected genetic loci of *E. coli* (Table A.6).

A few words of caution are necessary regarding Table A.6. Because genetic loci were often defined before the corresponding gene-product was known, there exist alternatives to some of the names listed. In some cases names have merely been rationalized to conform to a pattern such as *dna*, *rpo*, *rps*; and older nomenclature based on antibiotic resistance, for example, may still be found, especially in descriptions of genotypes. Thus, bacteria with a totally inactive gene for ribosomal protein S12 may be designated *rpsL*⁻, whereas bacteria with the mutation in the *rpsL* gene that conveys dependence on the antibiotic streptomycin (section 12.6.3) are generally designated Sm^D, the gene being originally called *strA*. Also worth mentioning are the *sup* genotypes, which define mutations in tRNA genes (e.g. *supF* is in *tyrT*) causing a particular termination codon to be read as an amino acid codon (suppression – section 12.9.6).

For fuller details of these and other gene symbols the reader is advised to consult the *E. coli* linkage map [212]. Finally, it is worth mentioning that sequences have been determined corresponding to many of these genetic loci [213].

REFERENCES

1 Cantoni, G. L. and Davies, D. R. (1971) *Procedures in Nucleic Acid Research*, vol. 2, Harper and Row, New York.
2 Birnie, G. D. and Rickwood, D. (1978) *Centrifugal Separations in Molecular and Cell Biology*, Butterworths, London.
3 Old, R. W. and Primrose, S. B. (1989) *Principles of Gene Manipulation* (4th edn), Blackwell, Oxford.
4 Walker, J. M. and Gaastra, W. (1983) *Techniques in Molecular Biology*, Croom Helm, London.
5 Kingsman, S. M. and Kingsman, A. J. (1988) *Genetic Engineering*, Blackwell, Oxford.
6 Habel, K. and Salzman, N. P. (1969) *Fundamental Techniques in Virology*, Academic Press, New York.
7 Sambrook, J., Fritsch, E. F. and Maniatis, T. (1989) *Molecular Cloning, a Laboratory Manual* (2nd edn), Cold Spring Harbor Laboratory, New York.
8 Hames, B. D. and Higgins, S. J. (1984) *Transcription and Translation: a Practical Approach*, IRL Press, Oxford.
9 Berger, S. L. and Kimmel, A. R. (1987) *Guide to Molecular Cloning Techniques*, Academic Press, New York.
10 Wu, R., Grossman, L. and Moldave, K. (1989) *Recombinant DNA Methodology*, Academic Press, New York.
11 Schneider, W. C. (1945) *J. Biol. Chem.*, **161**, 293.
12 Schmidt, G. and Thannhauser, S. J. (1945) *J. Biol. Chem.*, **161**, 83.
13 Fiske, C. and Subbarow, Y. (1929) *J. Biol. Chem.*, **81**, 629.
14 Berenblum, I. and Chain, E. (1958) *Biochem. J.*, **82**, 286.
15 Burton, K. (1956) *Biochem. J.*, **62**, 315.
16 Ceriotti, G. (1955) *J. Biol. Chem.*, **214**, 59.
17 Warburg, O. and Christian, W. (1942) *Biochem. Z.*, **310**, 384.

18 Lowry, O. H., Rosebrough, N. J., Farr, A. L. and Randall, R. J. (1951) *J. Biol. Chem.*, **193**, 265.

19 Bradford, M. M. (1976) *Anal. Biochem.*, **72**, 248.

20 Harspt, J. A., Krasna, A. and Zimm, B. H. (1965) *Biopolymers*, **6**, 595.

21 Krasna, A., Dawson, J. R. and Harspt, J. A. (1970) *Biopolymers*, **9**, 1017.

22 Krasna, A. (1970) *Biopolymers*, **9**, 1029.

23 Schmidt, V. A. and Hearst, J. E. (1969) *J. Mol. Biol.*, **44**, 143.

24 Kavenoff, R. and Zimm, B. H. (1963) *Chromosoma*, **41**, 1.

25 Crothers, D. M. and Zimm, B. H. (1965) *J. Mol. Biol.*, **12**, 525.

26 Lang, D. (1970) *J. Mol. Biol.*, **54**, 557.

27 Leighton, S. B. and Rubenstein, I. (1969) *J. Mol. Biol.*, **46**, 313.

28 Freifelder, D. (1970) *J. Mol. Biol.*, **54**, 567.

29 Eason, R. and Campbell, A. M. (1978) in *Centrifugal Separations in Molecular and Cell Biology* (eds G. D. Birnie and D. Rickwood), Butterworths, London, p. 251.

30 Bloomfield, V. A., Crothers, D. M. and Tinoco, I. (1974) *Physical Chemistry of Nucleic Acids*, Harper and Row, New York.

31 Cantor, C. R. and Schimell, P. R. (1980) *Biophysical Chemistry* Part II, W. H. Freeman, San Francisco.

32 Freifelder, D. (1982) *Physical Biochemistry* (2nd edn), W. H. Freeman, San Francisco.

33 Schmid, C. and Hearst, J. (1969) *J. Mol. Biol.*, **44**, 143.

34 Dublin, S. B., Benedek, G. B., Bancroft, F. C. and Freifelder, D. (1970) *J. Mol. Biol.*, **54**, 547.

35 Zimm, B. H. and Crothers, D. M. (1962) *Proc. Natl. Acad. Sci. USA*, **48**, 905.

36 Kavenoff. R., Klotz, L. C. and Zimm, B. H. (1973) *Cold Spring Harbor Symp. Quant. Biol.*, **38**, 1.

37 Kleinschmidt, A. (1968) *Methods Enzymol.*, **12B**, 361.

38 Inman, R. B. (1974) *Methods Enzymol.*, **12E**, 451.

39 Cairns, J. (1963) *Cold Spring Harbor Symp. Quant. Biol.*, **28**, 44.

40 Marmur, J. (1961) *J. Mol. Biol.*, **3**, 208.

41 Blin, N. and Stafford, D. W. (1976) *Nucleic Acids Res.*, **3**, 2303.

42 Birnboim, H. C. and Doly, J. (1979) *Nucleic Acids Res.*, **7**, 1513.

43 Holmes, D. S. and Quigley, M. (1981) *Anal. Biochem.*, **114**, 193.

44 Hirt, B. (1967) *J. Mol. Biol.*, **26**, 365.

45 Wallace, H. and Birnsteil, M. L. (1966) *Biochim. Biophys. Acta*, **114**, 296.

46 Aaji, C. and Borst, P. (1972) *Biochim. Biophys. Acta*, **269**, 192.

47 Maniatis, T., Jeffrey, A. and Kleid, D. G. (1975) *Proc. Natl. Acad. Sci. USA*, **72**, 1184.

48 Keller, W. (1975) *Proc. Natl. Acad. Sci. USA*, **72**, 4876.

49 Schwartz, D. C. and Cantor, C. R. (1984) *Cell*, **37**, 67.

50 Smith, C. L. and Cantor, C. R. (1987) *Trends Biochem. Sci.*, **12**, 284.

51 Maddox, J. (1990) *Nature (London)*, **345**, 381.

52 Birnie, G. D., Fox, S. M. and Harvey, D. R. (1972) in *Subcellular Components* (ed. G. D. Birnie), Butterworths, London, p. 235.

53 Roodyn, D. B. (1972) in *Subcellular Components* (ed. G. D. Birnie), Butterworths, London, p. 13.

54 Muramatsu, M., Hayashi, Y., Onishi, T., Sakai, M., Takai, K. and Kashiyama, T. (1974) *Exp. Cell Res.*, **88**, 345.

55 Estabrook, R. W. and Pullman, M. E. (1967) *Methods Enzymol.*, **10**.

56 Moldave, K. and Grossman, L. (1979) *Methods Enzymol.*, **59**.

57 Knowler, J. T., Moses, H. L. and Spelsberg, T. C. (1973) *J. Cell. Biol.*, **59**, 685.

58 McConkey, E. H. (1967) *Methods Enzymol.*, **12A**, 620.

59 Adelman, M. R., Blobel, G. and Sabatini, D. D. (1973) *J. Cell. Biol.*, **56**, 191.

60 Moldave, K. and Grossman, L. (1971) *Methods Enzymol.*, **20**.

61 Chomczynski, P. and Sacchi, N. (1987) *Anal. Biochem.*, **162**, 156.

62 Blackburn, P., Wilson, G. and Moore, S. (1987) *J. Biol. Chem.*, **252**, 5904.

63 Bergerm, S. L. and Bergermeister, C. S. (1979) *Biochemistry*, **18**, 5143.

64 Favoloro, J., Treisman, R. and Kamen, R. (1980) *Methods Enzymol.*, **65**, 718.

65 Aviv, H. and Leder, P. (1972) *Proc. Natl. Acad. Sci. USA*, **69**, 1408.

66 Adesnik, M., Solditt, N., Thomas, W. and Darnell, J. E. (1982) *J. Mol. Biol.*, **71**, 21.

67 DeWachter, R. and Fiers, W. (1971) *Methods Enzymol.*, **21**, 167.

68 Lehrach, H., Diamond, D., Wozney, J. M.

and Boedtker, H. (1977) *Biochemistry*, **16**, 4743.

69 Southern, E. (1980) *Methods Enzymol.*, **69**, 152.

70 Olszewska, E. and Jones, K. (1988) *Trends Genet.*, **4**, 92.

71 Botstein, D., White, R. L., Skolnick, M. and Davis, R. W. (1980) *Am. J. Hum. Genet.*, **32**, 314.

72 Marmur, J. and Doty, P. (1962) *J. Mol. Biol.*, **5**, 109.

73 Dove, W. F. and Davidson, N. (1962) *J. Mol. Biol.*, **5**, 467.

74 Bonner, T. I., Brenner, D. J., Neufeld, B. R. and Britten, R. J. (1973) *J. Mol. Biol.*, **81**, 123.

75 Laird, C. D., McConaughy, B. L. and McCarthy, B. T. (1969) *Nature (London)*, **224**, 149.

76 Alwine, J. C., Kemp, D. J. and Stark, G. R. (1977) *Proc. Natl. Acad. Sci. USA*, **74**, 5350.

77 Goldberg, D. A. (1980) *Proc. Natl. Acad. Sci. USA*, **77**, 5794.

78 Thomas, P. S. (1980) *Proc. Natl. Acad. Sci. USA*, **77**, 5201.

79 McMaster, G. K. and ·Carmichael, G. G. (1977) *Proc. Natl. Acad. Sci. USA*, **74**, 4835.

80 Langer, P. R., Waldrop, A. A. and Ward, D. C. (1981) *Proc. Natl. Acad. Sci. USA*, **78**, 6633.

81 Schaap, A. P. (1988) *Photochem. Photobiol.*, **47**, 505.

82 Rigby, P. W. J., Dieckmann, M., Knodes, C. and Berg, P. (1977) *J. Mol. Biol.*, **113**, 237.

83 Feinberg, A. P. and Vogelstein, B. (1983) *Anal. Biochem.*, **132**, 6.

84 Feinberg, A. P. and Vogelstein, B. (1984) *Anal. Biochem.*, **137**, 266.

85 Ricca, G. A., Taylor, J. M. and Kalinyak, J. E. (1982) *Proc. Natl. Acad. Sci. USA*, **79**, 724.

86 Green, M. R., Maniatis, T. and Melton, D. A. (1983) *Cell*, **32**, 681.

87 Craig, A. G., Nizetic, D. and Lehrach, H. (1989) *Nucleic Acids Res.*, **17**, 4605.

88 Rich, A. and RajBhandary, U. L. (1976) *Annu. Rev. Biochem.*, **45**, 805.

89 Randerath, K., Gupta, R. C. and Randerath, E. (1980) *Methods Enzymol.*, **65**, 638.

90 Chaconas, G. and van de Sande, J. H. (1980) *Methods. Enzymol.*, **65**, 75.

91 Challberg, M. D. and Englund, P. T. (1980) *Methods. Enzymol.*, **65**, 39.

92 Roychoudhury, R. and Wu, R. (1980) *Methods Enzymol.*, **65**, 43.

93 Gething, M. J., Bye, J., Skehel, J. and Waterfield, M. (1980) *Nature (London)*, **287**, 301.

94 Bishop, J. O., Rosbash, M. and Evans, D. (1974) *J. Mol. Biol.*, **85**, 75.

95 Maxam, A. M. and Gilbert, W. (1980) *Methods Enzymol.*, **65**, 499.

96 Sanger, F., Nicklen, S. and Coulson, A. R. (1977) *Proc. Natl. Acad. Sci. USA*, **74**, 5436.

97 Fu, C-P. D. and Wu, R. (1980) *Methods Enzymol.*, **65**, 620.

98 Sanger, F. and Coulson, A. R. (1978) *FEBS Lett.*, **87**, 107.

99 Biggin, M. D., Gibson, T. J. and Hong, G. F. (1983) *Proc. Natl. Acad. Sci. USA*, **80**, 3963.

100 Staden, R. (1977) *Nucleic Acids Res.*, **4**, 4037.

101 Devereux, J., Haeberli, P. and Smithies, O. (1984) *Nucleic Acids Res.*, **12**, 387.

102 Rubin, C. and Schmid, C. W. (1980) *Nucleic Acids Res.*, **8**, 4613.

103 Fritzsche, E., Hayatsu, H., Igloi, G. L., Iiada, S. and Koessel, H. (1987) *Nucleic Acids Res.*, **15**, 5517.

104 Ohmori, H., Tomizawa, J. and Maxam, A. M. (1978) *Nucleic Acids Res.*, **5**, 1479.

105 Church, G. M. and Gilbert, W. (1984) *Proc. Natl. Acad. Sci. USA*, **81**, 1991.

106 Saluz, H. P. and Jost, J. P. (1989) *Anal. Biochem.*, **176**, 201.

107 Saluz, H. P. and Jost, J. P. (1989) *Nature*, **338**, 277.

108 Gronenborn, B. and Messing, J. (1978) *Nature (London)*, **272**, 375.

109 Yanisch-Perron, C., Vieira, J. and Messing, J. (1985) *Gene*, **33**, 103.

110 Tabor, S. and Richardson, C. C. (1987) *Proc. Natl. Acad. Sci. USA*, **84**, 4767.

111 Henikoff, S. (1984) *Gene*, **28**, 351.

112 Sorge, J. A. and Blinderman, L. A. (1989) *Proc. Natl. Acad. Sci. USA*, **86**, 9208.

113 Brownlee, G. G. (1972) *Determination of Sequences in RNA*, North Holland, Amsterdam.

114 Donis-Keller, H., Maxam, A. M. and Gilbert, W. (1977) *Nucleic Acids Res.*, **4**, 2527.

115 Donis-Keller, H. (1980) *Nucleic Acids Res.*, **8**, 3133.

116 Danna, K. J., Sack, G. H. and Nathans, D. (1973) *J. Mol. Biol.*, **78**, 363.

117 Smith, H. O. and Birnstiel, M. L. (1976) *Nucleic Acids Res.*, **3**, 2387.

118 Rackwitz, H-R., Zehetner, G., Frischauf, A-M. and Lehrach, H. (1984) *Gene*, **30**, 195.

119 Saiki R. K., Scharf, S. J., Faloona, F., Mullis, K. B., Horn, G. T. *et al.* (1985) *Science*, **230**, 1350.

120 Saiki R. K., Gelfland, D. H., Stoffel, S., Scharf, S. J., Higuchi, R. *et al.* (1985) *Science*, **239**, 487.

121 White, T. J., Arnheim, N. and Erlich, H. A. (1989) *Trends Genet.*, **5**, 185.

122 Gelvin, S. B., Schilperoort, R. A. and Verma, D. P. S. (1989) *Plant Molecular Biology Manual* (2nd edn), Kluwer, Dordrecht.

123 Carter, B. L. A., Irani, M., MacKay, V. L., Seale, R. L., Sledziewski, A. V. *et al.* (1987) in *DNA Cloning, a Practical Approach*, vol. III (ed. D. M. Glover), IRL Press, Oxford.

124 Bolivar, F. (1978) *Gene*, **4**, 121.

125 Vieira, J. and Messing, J. (1982) *Gene*, **19**, 259.

126 Lederberg, E. M. and Cohen, S. N. (1974) *J. Bacteriol.*, **119**, 1072.

127 Messing, J. (1983) *Methods Enzymol.*, **101**, 20.

128 Hohn, B. (1979) *Methods Enzymol.*, **68**, 299.

129 Land, H., Grey, M., Hanser, H., Lindenmaier, W. and Shutz, G. (1981) *Nucleic Acids Res.*, **9**, 2251.

130 Okayama, H. and Berg, P. (1982) *Mol. Cell. Biol.*, **2**, 161.

131 Gubler, U. and Hoffman, B. J. (1983) *Gene*, **25**, 263.

132 Huynh, T. V., Young, R. A. and Davis, R. W. (1985) in *DNA cloning, a Practical Approach*, vol. I (ed. D. Glover), IRL Press, Oxford, p. 49.

133 Benton, W. D. and Davis, R. W. (1977) *Science*, **196**, 180.

134 Grunstein, M. and Hogness, D. (1975) *Proc. Natl. Acad. Sci. USA*, **72**, 3961.

135 Young, R. A. and Davis, R. W. (1983) *Proc. Natl. Acad. Sci. USA*, **80**, 1194.

136 Parnes, J. R., Velan, B., Felsenfeld, A., Ramanathan, U., Ferrini, U. *et al.* (1981) *Proc. Natl. Acad. Sci. USA*, **78**, 2253.

137 Pelham, M. R. B. and Jackson, R. J. (1976) *Eur. J. Biochem.*, **67**, 247.

138 Paterson, B. M., Roberts, B. E. and Kuff, E. L. (1977) *Proc. Natl. Acad. Sci. USA*, **74**, 4370.

139 Nagata, S., Taira, H., Hall, A., Johnsrud, L., Streuli, M., Ecsödi, J. *et al.* (1980) *Nature (London)*, **284**, 316.

140 Blattner, F. R., Williams, B. G., Blechl, A. E., Denniston-Thompson, K., Faber, H. E. *et al.* (1977) *Science*, **196**, 161.

141 Frischauf, A-M., Lehrach, H., Poustka, A. and Murray, N. (1983) *J. Mol. Biol.*, **170**, 827.

142 Hohn, B. and Collins, J. (1980) *Gene*, **11**, 291.

143 Sternberg, N., Ruether, J. and deRiel, K. (1990) *New Biologist*, **2**, 151.

144 Burke, D. T., Carle, G. F. and Olson, M. V. (1987) *Science*, **236**, 806.

145 Schlessinger, D. (1990) *Trends Biochem. Sci.*, **6**, 248.

146 Berk, A. J. and Sharp, P. A. (1977) *Cell*, **12**, 721.

147 Berk, A. J. and Sharp, P. A. (1978) *Cell*, **14**, 695.

148 Ghosh, P. K., Reddy, V. B., Swinscoe, J., Lebowitz, P. and Weissman, S. M. (1978) *J. Mol. Biol.*, **126**, 813.

149 Galas, D. J. and Schmitz, A. (1978) *Nucleic Acids Res.*, **5**, 3157.

150 Siebenlist, U., Simpson, R. B. and Gilbert, W. (1980) *Cell*, **20**, 269.

151 Nielsen, P. E. (1989) *BioEssays*, **11**, 152.

152 Becker, M. M. and Wang, J. C. (1984) *Nature (London)*, **309**, 682.

153 Nick, H. and Gilbert, W. (1985) *Nature (London)*, **313**, 795.

154 Church, G. M., Ephrussi, A., Gilbert, W. and Tonegawa, S. (1985) *Nature (London)*, **313**, 798.

155 Jordan, S. R. and Pabo, C. O. (1988) *Science*, **242**, 893.

156 Garner, M. M. and Revzin, A. (1986) *Trends Biochem. Sci.*, **11**, 396.

157 Amann, E. and Brosius, J. (1985) *Gene*, **40**, 183.

158 Studier, W., Rosenberg, A. H., Dunn, J. J. and Dubendorff, J. W. (1990) *Methods Enzymol.*, **185**, 60.

159 Bressan, G. M. and Stanley, K. K. (1987) *Nucleic Acids Res.*, **15**, 10056.

160 Ecker, D. J., Stadel, J. M., Butt, T. R., Marsh, J. A., Monia, B. P. *et al.* (1989) *J. Biol. Chem.*, **264**, 7715.

161 Smith, G. E., Summers, M. D. and Fraser, M. J. (1983) *Mol. Cell. Biol.*, **3**, 2156.

162 Moss, B., Elroy-Stein, O., Mizukami, T., Alexander, W. A. and Fuerst, T. R. (1990) *Nature (London)*, **348**, 91.

163 Mertz, J. E. and Gurdon, J. B. (1977) *Proc. Natl. Acad. Sci. USA*, **74**, 1502.

164 Melton, D. A., Kreig, P. A., Rebagliati, M. R., Zinn, K. and Green, M. R. (1984) *Nucleic Acids Res.*, **12**, 7035.

165 Graham, F. L. and van der Eb, A. J. (1973) *Virology*, **52**, 456.

166 Neumann, E., Schaeffer-Ridder, M., Wang, Y. and Hofschneider, P. H. (1982) *EMBO J.*, **1**, 841.

167 Gluzman, Y., (1981) *Cell*, **23**, 175.

168 Breatnach, R. and Harris, B. A. (1983) *Nucleic Acids Res.*, **11**, 7119.

169 Gorman, C. M., Moffat, L. F. and Howard, B. (1982) *Mol. Cell. Biol.*, **2**, 1044.

170 Luckow, B. and Schütz, G. (1987) *Nucleic Acids Res.*, **15**, 5490.

171 Spandidos, D. A. and Wilkie, N. M. (1984) *Nature*, **310**, 475.

172 Pouwels, P. H., Enger-Valk, B. E. and Brammar, W. J. (1987) *Cloning Vectors, Supplementary Update 1987*, Elsevier, Amsterdam.

173 Korman, A. J., Frantz, J. D., Strominger, J. L. and Mulligan, R. C. (1987) *Proc. Natl. Acad. Sci. USA*, **84**, 2150.

174 Gordon, J. W. and Ruddle, F. H. (1981) *Science*, **214**, 1244.

175 Palmiter, R. D. and Brinster, R. L. (1985) *Cell*, **41**, 343.

176 Winter, G., Fersht, A. R., Wilkinson, A. J., Zoller, M. and Smith, M. (1982) *Nature (London)*, **299**, 756.

177 Panayatos, N. and Truong, K. (1981) *Nucleic Acids Res.*, **9**, 5679.

178 McKnight, S. L. and Kingsbury, R. (1982) *Science*, **217**, 316.

179 Luckow, B., Renkawitz, R. and Schutz, G. (1987) *Nucleic Acids Res.*, **15**, 417.

180 Zoller, M. J. and Smith, M. (1983) *Methods Enzymol.*, **100**, 468.

181 Kunkel, T. A., Roberts, J. D. and Zakour, R. A. (1987) *Methods Enzymol.*, **154**, 367.

182 Gillam, S., Astell, C. R. and Smith, M. (1980) *Gene*, **12**, 129.

183 Jones, D. H., Sakamoto, K., Vorce, R. L. and Howard, B. H. (1990) *Nature (London)*, **344**, 793.

184 Khorana, H. G. (1979) *Science*, **203**, 614.

185 Gait, M. J. (1984) *Oligonucleotide Synthesis, a Practical Approach*, IRL Press, Oxford.

186 Shapiro, D. J., Sharp, P. A., Whali, W. W. and Keller, M. J. (1988) *DNA*, **7**, 47.

187 Dignam, J. D., Lebovitz, R. M. and Roeder, R. G. (1983) *Nucleic Acids Res.*, **11**, 1475.

188 Greenberg, M. E. and Ziff, E. B. (1984) *Nature (London)*, **311**, 433.

189 Gurdon, J. B., Lane, C. D., Woodland, H. R. and Marbaix, G. (1971) *Nature (London)*, **233**, 177.

190 Marcu, K. and Dudock, B. (1974) *Nucleic Acids Res.*, **1**, 1385.

191 Shields, D. and Blobel, G. (1978) *J. Biol. Chem.*, **253**, 3753.

192 Robertson, H. D. and Lodish, H. F. (1969) *J. Mol. Biol.*, **45**, 9.

193 Zubay, G., Chambers, D. A. and Cheong, L. C. (1970) in *The Lactose Operon* (eds J. R. Beckwith and D. Zipser), Cold Spring Harbor Monograph Series, p. 375.

194 Kersten, H. and Kersten, W. (1974) *Inhibitors of Nucleic Acid Synthesis, Biophysical and Biochemical Aspects*, Chapman and Hall, London.

195 Vázquez, D. (1979) *Antibiotic Inhibitors of Protein Biosynthesis*, Springer, Berlin.

196 Cundliffe, E. (1980) in *Ribosomes. Structure, Function and Genetics* (eds G. Chambliss, G. R. Craven, J. Davies, K. Davis, L. Kahan and M. Nomura), University Park Press, Baltimore, p. 555.

197 Cohen, P. and Van Heyningen, S. (1982) *Molecular Action of Toxins and Viruses*, Elsevier, Amsterdam.

198 Fritzsche, H., Triebel, H., Chaires, J. B., Dattagupta, N. and Crothers, D. M. (1982) *Biochemistry*, **21**, 3940.

199 Kopka M. L. Yoon, C., Goodsell, D., Pjura, P. and Dickerson R. E. (1985) *Proc. Natl. Acad. Sci. USA*, **82**, 1376.

200 Adams R. L. P. and Burdon, R. H. (1985) *Molecular Biology of DNA Methylation*, Springer-Verlag, New York.

201 Sjoberg, B-M., Graslund, A. and Eckstein, F. (1983) *J. Biol. Chem.*, **258**, 8090.

202 Adams, R. L. P. (1980) *Cell Culture for Biochemists*, Elsevier, Amsterdam.

203 Seghal, P. B., Darnell, J. E. and Tamm, I. (1976) *Cell*, **9**, 473.

204 Summers, D. F. and Maizel, J. V. (1971) *Proc. Natl. Acad. Sci. USA*, **68**, 2852.

205 Marks, P. A., Burke, E. R., Conconi, F. M., Perl, E. and Rifkind R. A. (1965) *Proc. Natl. Acad. Sci. USA*, **53**, 1437.

206 Bodley J. W. and Zieve, F. T. (1969) *Biochem. Biophys. Res. Commun.*, **36**, 463.

207 Madjar, J-J., Nielsen-Smith, K., Frahm, M. and Roufa, D. J. (1982) *Proc. Natl. Acad. Sci. USA*, **79**, 1003.

208 Boon, T. (1972) *Proc. Natl. Acad. Sci. USA*, **69**, 549.

209 Wool, I. G. (1984) *Trends Biochem. Sci.*, **9**, 14.

210 Nierhaus, K. H. and Wittmann, H. G. (1980) *Naturwiss.*, **67**, 234.

211 Endo, Y., Mitsui, K., Motizuki, M. and Tsurugi, K. (1987) *J. Biol. Chem.*, **262**, 5908.

212 Bachmann, B. J. (1990) *Microbiol. Rev.*, **54**, 130.

213 Rudd, K. E., Miller, W., Ostell, J. and Benson, D. A. (1990) *Nucleic Acids Res.*, **18**, 313.

214 Erlich, H. A., Gelfand, D. and Sninsky, J. (1991) *Science*, **252**, 1643.

Index

___